Lecture Notes in Control and Information Sciences

Edited by A. V. Balakrishnan and M. Thoma

For information about Vols. 1–21 please contact your bookseller or Springer-Verlag.

Lecture Notes in Control and Information Sciences
Edited by A.V. Balakrishnan and M. Thoma

67

Real Time Control of Large Scale Systems

Proceedings of the First European Workshop
University of Patras, Greece, July 9–12, 1984

Edited by
G. Schmidt · M. Singh · A. Titli · S. Tzafestas

Springer-Verlag Berlin Heidelberg GmbH

Editors
Günther Schmidt
Munich Technical University
FRG

Madan Singh
UMIST, Manchester
U.K.

André Titli
LAAS, Toulouse
France

Spyros Tzafestas
Patras University
Greece

ISBN 978-3-540-15033-6 ISBN 978-3-540-39219-4 (eBook)
DOI 10.1007/978-3-540-39219-4

Originally published by Springer-Verlag Berlin Heidelberg New York in 1985.

2061/3020-543210

PREFACE

This volume constitutes the proceedings of the *First European Workshop on the Real Time Control of Large Scale Systems*, University of Patras, Greece, July 9-12, 1984.

The Workshop was organized in the framework of a joint project (UMIST, LAAS, Munich Technical University, Patras University) sponsored by the Commission of the European Communities under the research part of the multiannual programme (1979-83) in the field of *Data Processing* aiming at promoting collaborative research work in the Community. It has really provided a unique opportunity to our colleagues from 24 countries for exchanging their experience, ideas and results in the field of large scale systems theory and applications.

Two reviews of the submitted 120 contributions were performed by the editors, one based on abstracts (Athens, 10 September 1983) and one based on full manuscripts (Toulouse, 20 April 1984). The final program of the Workshop contained 85 papers.

In this volume 67 papers are contained which are classified in the following parts:

I. Modelling of technological dynamics.
II. Hierarchical control systems.
III. Decentralized control systems.
IV. Miscellaneous large scale system techniques.
V. Computers in large scale systems.
VI. Applications: Power systems, Communication systems, Chemical and resource systems, and Management systems.

To comply with the publisher's general policy, some papers, not of acceptable typing quality, were not included here, but they can be found in the volume of Preprints.

We are grateful to the Commission of the European Communities (DGIII) for sponsoring the Workshop, and to the University of Patras for its hospitality and generous support.

Also we would like to deeply thank our invited speakers of the opening and plenary sessions for their coming at the Workshop, all authors for their high quality contributions, and the session chairmen for their excellent job.

We hope that the successful "First Workshop on RTC-LSS" will be followed by a second one very soon.

July 1984

Günther Schmidt
Madan Singh
André Titli
Spyros Tzafestas

OPENING SESSION OF THE
FIRST EUROPEAN WORKSHOP ON REAL TIME
CONTROL OF LARGE SCALE SYSTEMS

Welcome Address: Professor Spyros Tzafestas

Project Leader Address: Professor Madan Singh

Patras Univ. Address: Professor Andreas Philippou
Vice Rector

Athens Tech. Univ. Address: Professor Manolis Protonotarios
Vice Rector

CSEE Address: Professor Anastasios Venetsanopoulos
CSEE President

IMACS Address: Professor Robert Vichnevetsky
IMACS President

Opening Address: Professor Manfred Thoma
IFAC President

PLENARY PAPERS

M. ATHANS (MIT, USA)
Large scale system theoretic issues in command and control systems

G. SARIDIS (RPI, USA)
On the real-time control of intelligent robotic systems
(with K.P. Valavanis)

A. EPHREMIDES (Maryland Univ. USA)
Adaptive routing in radio communication networks

G. SCHMIDT (Munich Tech. Univ., FRG)
Real time hierarchical optimization and control:
An excerpt from the EEC Project

M. SINGH (UMIST, U.K.)
Real time control of large scale systems

A. TITLI (LAAS, FRANCE)
Decentralized control: A summary of the EEC Project

S. TZAFESTAS (Patras Univ. GREECE)
Reliability and fault detection techniques of large scale systems

A.P. SAGE (Virginia Univ., USA)
A learning approach for incorporation of imperfect knowledge in decision support system design (with A. Lagomasimo)

COUNTRIES OF PARTICIPANTS

Belgium	Egypt	Japan	Spain
Brazil	France	Kuwait	U.K.
Bulgaria	FRG	Netherlands	U.S.A.
Canada	Greece	Poland	U.S.S.R.
China P.R.	Iran	Portugal	Venezuella
Czechoslovakia	Italy	Romania	Yugoslavia

CONTENTS

I. INVITED PAPERS

Plenary Papers

Modelling of Technological Dynamics

II. HIERARCHICAL CONTROL SYSTEMS

III. DECENTRALIZED CONTROL SYSTEMS

Communication Systems

Chemical and Resource Systems

Management Systems

REAL TIME CONTROL OF LARGE SCALE SYSTEMS

by

Madan G. Singh
Professor of Control Engineering at
U.M.I.S.T., Manchester, U.K.

Abstract

In this note a brief description is given of the origins of the EEC real time control project and of this workshop. A summary is also given of two of the "state of the art" reports i.e. that on hierarchical optimisation and control and the one on Decentralised control. The final report on reliability and fault detection will be given by Professor Tzafestas at the workshop whilst the hierarchical and decentralised control will be covered respectively by Professor Schmidt and Professor Titli.

1. Introduction

The Commission of the European Communities (DG III) awarded a grant to myself at U.M.I.S.T. in Manchester in 1982 to collaborate with the groups of Professor André Titli in Toulouse, Professor Gunther Schmidt of the Technical University of Munich and Professor Spyros Tzafestas of the University of Patras in developing and examining the current state of the art of the important subject of real time control for large scale systems. It was agreed that the work of the four teams concentrates on the following three areas of real time control i.e. (1) Decentralised Control (2) Hierarchical Control and (3) Reliability and Fault Detection.

Since the award of the grant, major reports have already been provided to the EEC Commission describing the current state of the art of the real time control aspects of HIERARCHICAL CONTROL and DECENTRALISED CONTROL. My colleague Professor André Titli will present in our workshop a condensed version of our report on Decentralised Control whilst Professor Gunther Schmidt will do the same for our report on Hierarchical Control. Professor Tzafestas will be presenting here at this workshop our report on reliability and fault detection techniques for large scale systems. It was a condition of the award of the grant by the EEC Commission that a workshop be held at the conclusion of the grant to enable the four teams to present their "state of the art" reports to the international scientific community for comments and criticisms. It was agreed that the workshop be held in Patras and my colleague Professor Tzafestas very kindly agreed to organise it. In order to ensure that the results of the work on real time control are adequately disseminated within the EEC and at the same time, the work is seriously assessed by the wider international scientific community, it was agreed to open the workshop for participation from the United States and elsewhere. We are very pleased that the response from the international scientific community has been excellent and we welcome delegates from a large number of countries.

In the rest of this note, I will describe the basic problems of hierarchical and decentralised control which will be further elaborated by Professors Schmidt and Titli.

2. Decentralised Control

The decentralised control problem arises when a system is acted upon by a number of controllers which are unable to communicate amongst themselves on-line, even though they may have a set of *a priori* rules which allow them to have some *structural* knowledge about each other. In the last 15 years, a significant body of knowledge has been built up on the behaviour of such systems and ways of controlling them. In our "state of the art" report we provided a classification of this knowledge and touched upon some of the more interesting results which have been obtained.

It should be emphasised that, although decentralised controllers have been designed and used for controlling interconnected dynamical systems for over two decades, the design was based on *ad hoc* methods. For example, one usually assumed that the system comprised *weakly* interacting subsystems so that it was plausible to design the controllers independently for each subsystem. This is still the basis for most industrial control systems design. However, with the increased interest in optimisation techniques in the 1960s, there was an attempt to translate the notions of optimal centralised controller design to decentralised situations. It is only at this point that the intrinsic difficulty of decentralised control become apparent. For example, one of the best known and most useful results in the theory of control for centralised systems is the separation theorem. This, broadly speaking, states that for linear dynamical systems subject to Gaussian disturbances, it is possible to design a controller which minimises the expected value of a quadric cost function by designing separately an optimal state estimator and an optimal controller. Moreover, the state estimator is finite dimensional and the optimal controller is linear. These results fail in the case of decentralised control. A lot of the work on decentralised control could therefore be seen as an attempt at justifying the *ad hoc* design procedures used in current industrial practice and the new issues which emerged during these studies.

In the real world, decentralised decision making and control situations arise in many different fields. In Engineering, a typical example would be power systems control where different parts of the network are under the control of different authorities who have some *a priori* rules for acting based on previous knowledge but which do not have time or the physical possibility of exchanging information on-line. In management systems again certain *a priori* rules could define what each manager could do in a given situation, but there is very little scope for information exchange immediately before each decision is taken. There are also other interesting application areas where for reasons of cost it makes sense to take decisions in a decentralised way and it is interesting to examine the effect of the decentralisation constraint on the quality of the decision making. We did this in our state of the art report to some extent when we examined the imposition of this constraint on telephone routing.

2.1 The Classification Scheme

One convenient way of classifying the knowledge in this field is to consider first the split between stochastic decentralised control and deterministic decentralised control. In the former case, starting from team theory notions, we could consider the value of information and the optimality or otherwise of the control schemes which result when information constraints are imposed. Finally, we would need to consider fixed structure controllers as a practical way of designing the controls within a stochastic control framework.

In the case of deterministic decentralised control, the key notion in the analysis and controller design arises from the presence or absence of decentralised "fixed modes". We considered what they are and how we can get around their effects. We also considered certain design techniques which once again use fixed structure controllers.

Before decentralised control or decision making can be implemented, we often need to do decentralised observation/filtering. The state of the knowledge in this field was also examined.

Finally, as typical applications, a ship boiler system and a telephone routing system were studied.

An outline of our report is given in appendix 1.

3. Hierarchical Optimisation and Control

There are essentially two difficulties in solving problems of large scale systems using the techniques developed for multivariable systems. The first is a numerical one which arises due to the high dimensionality of the problem since the computations required increase by a high order (cubic or quartic) with a linear increase in the system dimension. The second problem is concerned with the possible lack of centrality in large scale systems i.e. all the available information about the system and all the calculations done based on this information are not brought to single central location. This is in marked contrast to the bulk of the modern computational and control procedures (e.g. standard static and dynamic optimisation techniques, tools for the design of multivariable controllers, algorithms for estimation and identification, etc.) which were developed with the explicit or implicit assumption of centrality and which were, in practice, meant only for solving relatively low order problems.

The lack of the centrality condition arises in many physical systems, e.g. various utility systems (gas, electricity, water), computer networks, traffic networks, etc. where due to the fact that the system is widely distributed in space, it becomes very expensive to centralise all information. This may also arise due to considerations of achieving real time control as in telephone networks where decentralised decision making provides virtually the only realistic solution.

As in any control design, there are four basic interacting stages and these also exist for large scale systems. These stages are: Modelling, Analysis, Design and implementation. In the case of large scale systems, the size of the problem imposes some modifications to this scheme. For example, in addition to the modelling, some model simplification may also be necessary. Thus the main research being done in the area of large scale systems can be roughly classified into the following types.

1) Procedures for simplifying the system model

2) Procedures for analysing and describing certain system properties (e.g. procedures for checking controllability or testing stability of large systems)

3) Procedures for solving complex computational problems (arising at the design stage - "off-line" problems - and arising when current "on-line" control decisions are made)

4) Attempts to work out appropriate controller structure for large systems.

It should be noted that the methodologies concerned with the specification of control structures ((4) above) will usually include applications of the results classified above in types (1) - (3). Similarly, a particular procedure of type (3) may make use of a simplified model or require a stability test for a complex system. Consider for example the case when we decide upon a decentralised regulatory control structure (type (4)), then use a hierarchical optimisation structure for computing the block diagonal matrix of the feedback gains and within the optimisation procedure, we need to test for the stability of the overall closed loop system. Thus all four types may well be required within a single design procedure.

In our state of the art report on hierarchical optimisation and control, we considered the first two types of results. As far as the remaining ones are concerned we concentrated on those computational procedures and control structures (strategies) which in our opinion constitute methods for optimisation and control. Although the field of hierarchical optimisation and control constitutes only a part of large scale systems theory it has itself received a lot of attention and many books as well as hundreds of papers have been devoted to it. In section II of our report we discussed the main concepts in this area; in section III we considered hierarchical optimisation methods; in section IV hierarchical control structures were described, while in section V we briefly examined some application. Finally, in section VI decoupling methods were introduced and in section VII distributed parameter systems were considered.

It should be noted that this report did not cover the whole variety of hierarchical structures and methods. In particular, management-type problems, in which the human factor is important and local goals are not in harmony with an overall system goal as understood by the center, were not considered in this report.

In appendix 2, we give an outline of our report which will be elaborated by Professor Schmidt.

The final area (on reliability and fault detection) will be reported upon by Professor Tzafestas.

4. Conclusions

The project was successful not only in the sense of it bringing together research workers from different EEC countries and cultures to work on a common problem of great importance and thus contributing in a small way to European Unity but also in a broader scientific sense. We now await the judgement of our peer group on the scientific aspects.

APPENDIX 1

CONTENTS OF THE DECENTRALISED CONTROL REPORT

APPENDIX 2

"STATE OF THE ART" REPORT ON HIERARCHICAL OPTIMISATION AND CONTROL

TABLE OF CONTENTS

DECENTRALIZED CONTROL:A SUMMARY OF THE EEC REPORT

A. TITLI
L.A.A.S.-C.N.R.S. and I.N.S.A. Toulouse

Introduction

This paper is a summary of the Status Report on Decentralised Control edited for EEC in the context of the grant : 003277 Real Time optimisation and control of large scale systems using distributed computing facilities.

The report contains four main parts (contribution of G. SCHMIDT, S. TZAFESTAS M.G. SINGH, A. TITLI and colleagues) :

1. Decentralised stochastic decision making and control.
2. Decentralised stabilisation (deterministic aspects).
3. Decentralised observers.
4. Applications of decentralised control.

The decentralised control problem arises when a system is acted upon by a number of controllers which are unable to communicate amongst themselves on-line. In the last 15 years, a significant body of knowledge has been built up on the behaviour of such systems and ways of controlling them. In the report we have provided a classification of this knowledge and touch upon some of the more interesting results which have been obtained.

In the real world, decentralised decision making and control situations arise in many different fields. In Engineering, a typical example would be power systems control where different parts of the network are under the control of different authorities who have some *a priori* rules for acting based on previous knowledge but which do not have time or the physical possibility of exchanging information on-line. In management systems again certain *a priori* rules could define what each manager could do in a given situation, but there is very little scope for information exchange immediately before each decision is taken. There are also other interesting applications areas where for reasons of cost it makes sense to take decisions in a decentralized way and it is interesting to examine the effect of the decentralisation constraint on the quality of the decision making (e.g. : telephone routing problems).

One convenient way of classifying the knowledge in this field is to consider first the split between stochastic decentralised control and deterministic decentralised control. In the former case, starting from team theory notions, we have considered the value of information and the optimality or otherwise of the control schemes which result when information constraints are imposed (non classical information pattern). Finally, we considered fixed structure controllers as a practical way of designing the controls within a stochastic control framework. This corresponds to the part I of the report.

Before decentralised control or decision making can be implemented, we often need to do decentralised observation/filtering. The state of the knowledge in this field is also examined in the report (Part. 3). Mainly four schemes for decentralised observation are presented and an example (rear-axle teststand) is given.

The last part of the report develops significant applications on a :
- ship boiler
- telephone routing system.

Here, more details will be given on the deterministic aspects of the decentralised control, mainly through the concept of "fixed modes".

II - The decentralised fixed modes

Consider a linear time-invariant multivariable system with local control stations described by :

$$\begin{aligned} \dot{x}(t) &= A\, x(t) + \sum_{i=1}^{\nu} B_i\, u_i(t) \\ y_i(t) &= C_i\, x(t),\ i = 1, \ldots, \nu \end{aligned} \tag{1}$$

where $x(t) \in R^n$ is the state, $u_i(t) \in R^{mi}$ and $y_i(t) \in R^{pi}$ are the input and output respectively of the ith local control station ($i = 1, \ldots, \nu$). The system is said to be a ν control agent system if we impose the following information flow constraint on the controller :

$$\begin{aligned} u_i &= K_i\, y_i + Q_i\, \xi_i \\ \dot{\xi}_i &= S_i\, \xi_i + R_i\, y_i \end{aligned} \qquad i = 1, \ldots, \nu \tag{2}$$

The following definition was made by Davison and Wang (1973).

- Definition (Davison and Wang, 1973) :

Given the system (1) where $A \in R^{mXn}$, $B_i \in R^{nXmi}$, $C_i \in R^{piXn}$ ($i = 1, \ldots, \nu$) let $\mathcal{K}$ be the set of block-diagonal matrices as follows :

$$\mathcal{K} = \{K / K = \text{block-diag}\, [K_1, \ldots, K_\nu],\ K_i \in R^{miXpi}\ (i = 1, \ldots, \nu)\}$$

Then, the set of fixed modes of (1) with respect to $\mathcal{K}$ is defined as follows ;

$$\Lambda(C, A, B, \mathcal{K}) = \bigcap_{k \in \mathcal{K}} \sigma(A+BKC)$$

where $\sigma(.)$ denotes the set of eigenvalues of (.) and where :

$$B \triangleq (B_1, B_2, \ldots, B_\nu) \qquad C = \begin{bmatrix} C_1 \\ C_2 \\ \vdots \\ C_\nu \end{bmatrix}$$

It is clear that the set of fixed modes includes any mode of the system which is not both controllable and observable (in a centralized sense), but in addition, generally includes other modes of the system also called decentralized fixed modes.

The following result obtained by Davison and Wang (1973) (also see Corfmat and Morse (1976, a, b) illustrates the importance of this concept in decentralized control systems design :

- Proposition 1 (Davison and Wang, 1973) :

The necessary and sufficient condition for the existence of a decentralized controller (2) for the system (1) such that the closed-loop system is asymptotically stable is that the decentralized fixed modes of (1) all lie in the open left half part of the complex plan.

Thus, the characterization and determination of the decentralized fixed modes has received attention in the litterature.

Davison and Wang (1973) give an algorithm to find the decentralized fixed modes of system (1) based directly on definition 1.

Anderson and Clements (1981) derive an interesting algebraic characterization of fixed modes :

- Proposition 2 (Anderson and Clements, 1981) :

Let be π the set $\{1, \ldots, \nu\}$ and define a partition of π into disjoint subsets $\beta = \{i_1, \ldots, i_k\}$ and $\pi-\beta = \{i_{k+1}, \ldots, i_\nu\}$. Define also the matrices :

$$B^{\beta} = [B_{i_1}, \ldots, B_{i_k}] \qquad B^{\pi-\beta} = [B_{i_{k+1}}, \ldots, B_{i_\nu}]$$

$$C^{\beta} = \begin{bmatrix} C_{i_1} \\ \vdots \\ C_{i_k} \end{bmatrix} \qquad C^{\pi-\beta} = \begin{bmatrix} C_{i_{k+1}} \\ \vdots \\ C_{i_\nu} \end{bmatrix}$$

Consider the system (1). Then a necessary and sufficient condition for $\lambda \in \sigma(A)$ to be a decentralized fixed mode of (1) is that :

$$\text{Rank} \begin{bmatrix} A-\lambda I & B^{\beta} \\ C^{\pi-\beta} & 0 \end{bmatrix} < n$$

Using this result, Davison and Ozguner (1983) give a recursive characterization of fixed modes which establishes that to study the characterization of the fixed modes of a ν - control agent system, it is really only necessary to examine the case $\nu = 2$.

Certain characterization of fixed modes exist also in the frequency domain. Anderson (1982) uses a result obtained by himself and Clements (1981) which gives a necessary and sufficient condition for the existence of fixed modes using matrix fraction description.

From this result, he derives conditions on the system transfert function matrix for the existence of fixed modes.

The transfert function matrix characterization proposed by Davison and Ozguner (1983) is given in the case where the system has distinct poles.

Widyasagar and Viswanadham (198) present another characterization of decentralized fixed polynomial, whose zeros are the fixed modes, in terms of the greatest common divisor of certain minors of the transfert function matrix and its characteristic polynomial.

From another point of view, Siljak and Sezer (1981) introduced recently the notion of structurally fixed modes. Decentralized fixed modes may have two origines : a perfect matching of system parameters or the special structure of the system. In the first case (which is very unlike from a physical point of view), a slight change of the parameters can eliminate the fixed modes. On the other hand, a structurally fixed mode can only be eliminated by changing the structure of the system or relaxing the constraints on the information flow between the local controllers, i.e. changing the structure of the feed-back matrix K.

Siljak and Sezer (1981) give a characterization of structurally fixed modes :

- Proposition 3 (Siljak and Sezer, 1981) :

The system (1) has structurally fixed modes with respect to the decentralized control of (3) if and only if either of the two following conditions holds :

(i) there exists a $\beta \subset \pi$ and a permutation matrix P such that :

$$P^T AP = \begin{bmatrix} A_{11} & \theta & 0 \\ A_{21} & A_{22} & 0 \\ A_{31} & A_{32} & A_{33} \end{bmatrix} \quad P^T B^{\pi-\beta} = \begin{bmatrix} 0 \\ 0 \\ B_3^{\pi-\beta} \end{bmatrix} \quad P^T B^{\beta} = \begin{bmatrix} B_1^{\beta} \\ B_2^{\beta} \\ B_3^{\beta} \end{bmatrix} \tag{3}$$

$$C^{\beta} P = [C_1^{\beta} \quad 0 \quad 0]$$

$$C^{\pi-\beta} P = [C_1^{\pi-\beta} C_2^{\pi-\beta} C_3^{\pi-\beta}]$$

(ii) there exists a $\beta \subset \pi$ such that :

$$\bar{\rho} \begin{bmatrix} A & B^{\beta} \\ C^{\pi-\beta} & 0 \end{bmatrix} < n$$

Notations (β, π, B^{β}, $B^{\pi-\beta}$, C^{β} $C^{\pi-\beta}$) are the same than in proposition 2. $\bar{\rho}(.)$ denotes the generic rank of (.)*.

This characterization presents two types of structurally fixed modes. In the case (i), it is clear that the eigenvalues of A_{22} remain fixed with respect to the decentralized control (2). In case (ii), the system (1) has a fixed mode at the origine.

* A structured matrix $\bar{M}$ is a matrix which has a number of fixed zeros at certain locations and arbitrary entries, say ν, elsewhere. With these entries, we associate a parameter space R^{ν} such that every data point $d \in R^{\nu}$ defines a matrix $M = \bar{M}$ (d). Conservely, to any matrix M, there corresponds a structured matrix $\bar{M}$ such that $M = \bar{M}$ (d) for some $d \in R^{\nu}$. If $\rho(M)$ denotes the rank of M, the generic rank $\bar{\rho}(M)$ is defined as :

$$\bar{\rho}\,(M) = \bar{\rho}\,(\bar{M}) = \max_{d \in R^{\nu}} \quad \{\rho[\bar{M}\,(d)]\}$$

- <u>Proposition 4</u> : (Tarras, Titli, 1984)

Using the notion of eigenvalue sensitivity with respect to the change of system parameters, a new algebraic characterization of decentralized fixed modes is provided.

Let $D = A + BKC$ the closed loop matrix of the system. A change $\delta D = B\ \delta KC$ in this closed loop matrix induces the following change in a distinct eigenvalue of D (Rosenbach, 1965) :

$$\delta\lambda_r = \frac{\text{trace}\left[\prod_{j(j\neq r)} (D-\lambda_j I)\delta D\right]}{\prod_{j(j\neq r)} (\lambda_r - \lambda_j)}$$

Then, for $K \in \mathcal{K}$, $\delta K \in \mathcal{K}$, if :

$$S_r = \text{trace} \prod_{j(j\neq r)} (D-\lambda_j I)\delta D = 0, \qquad \lambda_r \in \sigma(D)$$

then, λ_r is a fixed mode :

Ex : $A = \begin{bmatrix} 0 & 1 & 0 \\ 1 & 1 & 0 \\ 0 & 0 & 1 \end{bmatrix} \quad B = \begin{bmatrix} 1 & 0 \\ 0 & 0 \\ 0 & 1 \end{bmatrix} \quad C = \begin{bmatrix} 1 & 0 & 0 \\ 0 & 0 & 1 \end{bmatrix}$

$$K = \begin{bmatrix} k_1 & 0 \\ 0 & k_2 \end{bmatrix} = \begin{bmatrix} 1 & 0 \\ 0 & 1 \end{bmatrix} \qquad \sigma(A + BKC) = \{1, -1, 2\}$$

For $\delta k_1 = 1$, $\delta k_2 = 1$, we have :

$$S_r\ (\lambda = 1) = 0$$
$$S_r\ (\lambda = -1) = -4$$
$$S_r\ (\lambda = 2) = 2$$

Then the system under consideration has a fixed mode at $\lambda = 1$.

This approach has been extended to structurally fixed modes.

III - Synthesis of decentralised control in presence of instable fixed modes.

III.1 Decomposition of the system if it is not completely fixed

Let us assume that there exists some degree of freedom in choosing the subsystems and the resulting controller blocks. Then, using the different methods that we have of characterization of unstable fixed modes, we can test for different feasible decompositions and their associated controllers.

The absence of unstable fixed modes therefore becomes an essential criterion in this decomposition phase, although we can also use other secondary criteria (e.g. minimal interaction between the subsystems) to refine the chosen structure. Some work has been done in this direction by the use of information theory.

III.2 Structuring the feedback gain-matrix

III.2.1. Armentano and Singh (1982) approach, by adding certain non zero elements.

Armentano and Singh (1982) present a way of choosing a new structure for the feedback matrix such that the fixed modes are eliminated and such that the exchange of information between subsystems is reduced. They characterize fixed modes by means of block diagonally dominant matrices. Their procedure can be applied when we are in presence of a set of interconnected linear dynamical subsystems;

$$\dot{x}_i = A_{ii}\, x_i + B_i\, u_i + \sum_{\substack{j=1 \\ j\neq 1}}^{\nu} A_{ij}\, x_j \qquad i = 1, \ldots, \nu \tag{4}$$
$$y_i = C_i x_i$$

where $x_i \in R^{ni}$, $u_i \in R^{mi}$, $y_i \in R^{pi}$

Let $A = \{A_{ij},\ i = 1, \ldots, \nu,\ j = 1, \ldots, \nu\} \in R^{nXn}$

$B = \text{block diag}\ [B_1, \ldots, B_\nu] \in R^{nXm}$

$C = \text{block diag}\ [C_1, \ldots, C_\nu] \in R^{pXn}$

We can rewrite (4) as :

$$\dot{x} = Ax + Bu \qquad y = Cx \tag{5}$$

that is a particular case of system (1).
By applying the decentralised output feedback, $u_i = K_{ii}\, y_i \quad i , 1, \ldots, \nu$
or $y = Ky$ with $K = \text{block diag}\ [K_{11}, \ldots, K_{\nu\nu}]$
we obtain the following closed-loop matrix :

$$A + BKC = \begin{bmatrix} \hat{A}_{11} & A_{12} & \cdots & A_{1\nu} \\ A_{21} & \hat{A}_{22} & \ddots & \vdots \\ \vdots & & \ddots & \\ A_{\nu 1} & A_{\nu 2} & \cdots & \hat{A}_{\nu\nu} \end{bmatrix}$$

with $\hat{A}_{ii} = A_{ii} + B_i\, K_{ii}\, C_i \qquad i = 1, \ldots, \nu$

If the diagonal submatrices $\hat{A}_{ii}$ are non singular, and if :

$$\left\| \hat{A}_{ii}^{-1} \right\|^{-1} > \sum_{\substack{j=1 \\ j\neq 1}}^{\nu} \left\| A_{ij} \right\| \quad \text{for all } i = 1, \ldots, \nu$$

Then A + BKC is strictly block diagonally dominant.

THEOREM 1 : (Armento and Singh, 1982)
If the matrix A + BKC is strictly block diagonally dominant, then A + BKC is non singular.

COROLLARY : (Armento and Singh, 1982)
Let λ^*, a complex number, be a decentralized fixed mode.
Then : $\left\| \hat{A}_{ii} - \lambda^* I_i)^{-1} \right\|^{-1} < \sum_{\substack{j=1 \\ j \neq 1}}^{\nu} \left\| A_{ij} \right\| \ \forall K_{ii} \in R^{mi \times pi}$

for at least one i, i = 1, ..., ν. I_i is the ni X ni identity matrix.

Using this corollary, Armento and Singh derive a procedure to find the set of links between the subsystems (i.e. the blocks K_{ij} of K, i, j = 1,..., ν, i ≠ j) that can eliminate the fixed modes. Among this set, the choice can be made associating a cost to every link between the susbystem i and j. Note that the fact of considering the blocks Kij can result in a redundant structure of K in the sense that all the elements of the added block are not necessarily usefull. This fact is taken into account in the two following approaches.

III.2.2. Locatelli, Shiavoni and Tarantini (1977) approach :

Given the system :

$$\begin{aligned} \dot{x} &= Ax + Bu \\ y &= Cx \end{aligned} \qquad x \in R^n,\ u \in R^m,\ y \in R^p \tag{6}$$

the approach of Locatelli and co-workers (1977) is based on the definition of an associated graph $\Gamma s = (V_s, L_s)$ for (6). Define $M \triangleq \{1, \ldots, m\}$ and $p \triangleq \{(1,\ldots,p\}$. So, $S \subset P \times M$ is the set of the permitted feedback connections, i.e. if we consider the feedback matrix $K : R^p \rightarrow R^m$, $K = \{kij = 0(j, i) \notin S$.

For any set $S \subset P \times M$ define the sets : $V_s \triangleq V_{1_s} \cup V_{2_s}$

where : $V_{1_s} \triangleq \{i : (j,i) \in S \text{ for some } j\}$

$V_{2_s} \triangleq \{j : (j,i) \in S \text{ for some } i\}$

and $L_s \triangleq L_{1_s} \cup L_{2_s}$

where $L_{1_s} \triangleq \{(j,i) : (j,i) \in V_{1_s} \times V_{2_s}, G_{j,i}(s) \neq 0\}$

$L_{2_s} \triangleq \{(j,i) : (j,i) \in S\}$

$G_{j,i}$ is the transfert function between the ith input and the jth output. Thus, each node of Γ_s represents an input or output variable, while the arcs represent a non zero transfert function or a feedback connection.

Using this graph, Locatelli and co-workers (1977) give the following characterization of fixed modes with respect to the structure S of the feedback matrix :

THEOREM 2 : (Locatelli and co-workers, 1977)
An eigenvalue of system (6) is fixed with respect to S if and only if it is not a pole of any elementary cycle of Γs.

They treat the problem of finding the minimal set $S \subset \tilde{S}$ such that the set $\bar{\Lambda} \subset \sigma(A)$ is assignable, i.e. no element of $\bar{\Lambda}$ is a fixed mode of (6) with respect to S. The minimal set S is obtained considering the cost criterion R(S) = $\sum_{(i,j) \in S} r_{ij}$ (r_{ij} is a cost associated with every permitted feedback connection). This problem will have a solution if and only if the set of fixed modes of (6) with respect to $\tilde{S}$ is empty. It is solved by a boolean linear program (Locatelli co-workers, 1977, p. 118), using the characterization of fixed modes given by Theorem 2.

This boolean linear program is rather interesting for the fact it can be adapted, either as it stands or with slight modifications, to tackle a certain number of problems other than the one cited before. It can be used, for example, to determine the set of fixed modes of system (6) for any structure S of the feedback matrix.

Finally, this program can be used to solve the problem of eliminating the fixed modes in decentralized control systems.

III.2.3. Senning (1979) approach :

Given the system (1), Senning (1979) proposes to find a feasible decentralized control.

- Definition 1 : (Senning, 1979)

A control structure is feasibly decentralized if the system is stabilizable with this control structure and the cost of the exchange of information is minimal.

Senning treats simultaneously the classical optimisation problem based on the traditional quadratic optimisation criterion of linear systems and the search of an optimal control structure with respect to a criterion taking into account the system decomposition and the cost of the links between the control agents.

The solution to this problem will give a feasibly decentralized control :

$$u_i = K_{ii}\, y_i + \sum_{\substack{j=1\\ j\neq 1}}^{\nu} K_{ij}\, y_j \qquad i = 1, \ldots, \nu$$

For this propose, Senning defines an extended optimisation criterion (EOC) :

$$EOC = \int_0^{\infty} (x^T Q x + \sum^{\nu} u_i^T\, R_i\, u_i)dt + \sum_{i=1}^{\nu} m_i^2$$

with $Q \geqslant 0$, $R_i > 0$ $\qquad i = 1, \ldots, \nu$

The first part of the EOC is the traditional performance index (PI) and the second term takes into account the desired structure of the control by means of a weighted measure of the non-local information. This measure is taken as the vector function norm of the non-local part of the control weighted by certain factors γ_{ij} penalizing more or less the exchange of information between two control agents :

$$m_i = \left\| \sum_{\substack{j=1\\ j\neq 1}}^{\nu} \gamma_{ij}\; u_{ij} \right\| = \left\| \sum_{\substack{j=1\\ j\neq 1}}^{\nu} \gamma_{ij}\, K_{ij}\, y_j \right\| = \; K_i\, \Gamma_i\, y \,\|$$

With $K_i = [K_{i1}, \ldots, K_{i,i-1}, 0, K_{i,i+1} \ldots, K_{i\nu}]$

$$\Gamma_i = \begin{bmatrix} \gamma_{i1}\, I_{p1} & & & \\ & \ddots & & \\ & & 0 & \\ & & & \ddots \\ & & & & \gamma_{i\nu} I_{p\nu} \end{bmatrix}$$

I_{pi} denotes the matrix-unity of dimension $p_i x p_i$. The EOC becomes :

$$EOC = PI + \sum_{i=1}^{\nu} \| K_i \Gamma_i\, y\|^2$$

and we have the following optimisation task.
Find the optimal matrices $K^*_1, \ldots, K^*_\nu$ such that :

$$EOC(K^*_1, \ldots, K^*_\nu) \leqslant EOC(K_1, \ldots, K_\nu)$$

for all feasible matrice $K_1, \ldots, K_\nu$.

The solution of this optimisation problem is given by Senning (1979, p. 55) in terms of four equations to solve for each i, $i = 1, \ldots, \nu$.

He determines also the value of the optimal EOC. The procedure of Senning is rather interesting in decentralized control : it is the only one treating simultaneously the problem of structure and the classical optimisation problem. It permits to obtain an adequate output feedback matrix K with optimal gains and structure without needing to test before the existence of fixed modes. Of course, if the system has no fixed modes, the feasibly control resulting from this algorithm will be completely decentralized.

III.3 Use of vibrational control

Vibrational control theory was introduced by Meerkov (1973). It gives a solution for the cases where conventionnal control methods (based on feedback or feedforward principles) cannot be used because of lack of measurements. Thus, Meerkov showed that the introduction of certain vibrations on the dynamic system parameters can give a stabilizing effect. The decentralization constraints impose effectively the system to be controlled by controllers only using a reduced set of measures. Thus, vibrational control can be of help when the decentralized structure of the feedback matrix enables the stabilization of the system (i.e. there exists an unstable fixed mode).

Consider the time-invariant, linear system :

$$\dot{x} = Ax \qquad x \in R^n \tag{7}$$

and suppose it is not stable.

The principle of vibrational control consists in the introduction of vibrations on the parameters of A such that we obtain a time-varying system :

$$\mathring{x} = (A + D(t))x \tag{8}$$

with $D(t) = \|d_{ij}(t)\|^n_{i,j=1}$, $\quad d_{ij} = \alpha_{ij} \sin \omega_{ij} t$

If we determine now the "averaged description" of the system (8) (Meerkov, 1973, 1980) :

$$\mathring{x} = (A + \bar{D})x \tag{9}$$

and with $\bar{D}$ being a constant matrix, the behaviour of system (8) tends towards the behaviour of system (9) under certain conditions (Meerkov, 1980). The entries of $\bar{D}$ are functions of α_{ij} and ω_{ij}. So, the conditions of stability established for the time-invariant system (9) give the values of the vibration parameters that will permit to stabilise system (8).

The conditions on matrix A under which system (7) can be stabilized by vibrational control are given by Meerkov (1980).

Travé, Tarras and Titli (1983) show how vibrational control can be employed to cancel unstable fixed modes in decentralized control systems such that (1).

Vibrational control can be applied on the open-loop system or on the closed-loop system :

$$\dot{x} = (A + \sum_{i=1}^{\nu} B_i K_i C_i) x$$

The application of vibrational control results in a time-varying feedback matrix rejoining the result of Anderson and Moore (1981) presented in the following paragraph.

III.4 Time-varying feedback laws to eliminate no structurally fixed modes

Recently, it was pointed out (Anderson and Moore, 1981) that time-varying feedback laws can eliminate fixed modes given the satisfaction of certain connectivity conditions. Anderson (1982) give an interesting interpretation of this result for a 2-input, 2-output system.

Consider the system described by its rational transfert function matrix :

$$W(s) = \begin{bmatrix} W_{11}(s) & W_{21}(s) \\ W_{12}(s) & W_{22}(s) \end{bmatrix}$$

$$W_{12}(s) \neq 0 \qquad W_{21}(s) \neq 0$$

If we apply the feedback control law $u_2 = k_2\ y_2$, we can illustrate the resulting closed-loop system by the following sheme :

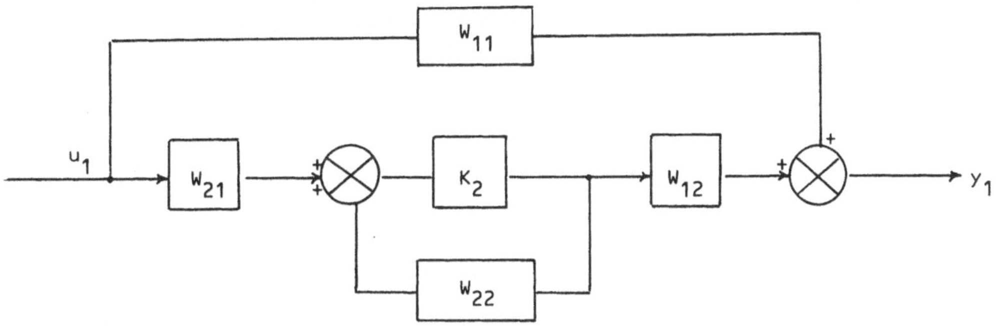

Figure 1

where $\frac{y_1}{u_1} = W_{11} + W_{12}(\frac{k_2\ W_{21}}{1-k_2W_{22}})$

- Proposition 5 : The 2-input, 2-output system with rational transfert function matrix W(s) and with a simple pole at s_o has a fixed mode at s_o if and only if W(s) or its transpose has the following form :

$$W(s) = \begin{bmatrix} \text{entry with no pole at } s_o & \text{entry with pole at } s_o \\ \text{entry with zero at } s_o & \text{entry with no pole at } s_o \end{bmatrix}$$

Fig. 1 shows that with k_2 time-varying, the pole-zero cancellation at s_o, the cause of the fixed mode, will no longer occur. (One cannot commute the time-varying block with an adjacent time-invariant block, and thereby juxtapose a cancelling pole-zero pair).

Fig. 1 also illustrates a result of Davison and Wang (1973) to the effect that if k_2 is replaced by a transfert function $k_2(s)$, the fixed mode is still present. Note that this result cannot be applied for a structurally fixed mode that is not provided by a cancellation.

Although these interpretations are less easy for a general m-input, p-output system, Anderson (1982) gives an extension of proposition 4 and Anderson and Moore (1981) show that for system (1), applying a control low :

$$u_i(t) = K_i(t)\ y_i(t), \qquad i = 1, \dots, \nu$$

with $K_i(t)$ periodic and piecewise constant, the system can be made uniformly controllable and observable from u_1 and y_1, even if there is a fixed mode.

For this purpose, $K_i(t)$ must have at least $\prod_{j=1}^{\nu} [\max(\dim u_j, \dim y_j)+1]$ values.

We note that the number of values becomes quickly huge. This may be a problem for large-scale systems.

On the same principle, but only for 2-input, 2-output systems, Purviance and Tylee (1982) show that a sinusoidal time-varying feedback is a better solution with respect to the degree of controllability and observability.

We want also to cite the proposition of Wang (1981) that suggests a design method using sample and hold.

The principle of time-varying controllers provides an interesting idea for the elimination of fixed modes but the design of such controllers seems to be difficult particularly when the system has a great number of variables.

IV - Design techniques for optimal decentralized and "quasi-decentralized" control in the absence of unstable fixed modes.

When the structure of the feedback matrix K has been determinated by one of the methods presented before, the problem consists in evaluating the gains with respect to some optimisation criterion generally quadratic.

Consider the linear, time-invariant system (1) writen as follows :

$$\dot{x} = Ax + Bu \qquad y = Cx \qquad x(o) = x_o \tag{10}$$

$$B \triangleq [B_1, \dots, B_\nu] \qquad C \triangleq \begin{bmatrix} C_1 \\ \vdots \\ C_\nu \end{bmatrix}$$

We seek a linear static feedback controller of the type :

$$u_i = -K_i y = -K_{ii} y_i - \sum_{\substack{j=1 \\ j \neq i}}^{\nu} K_{ij} y_j \qquad i = 1, \dots, \nu \tag{11}$$

The structure S of $K = \{k_{ij}\}_{\substack{i=1,\dots,m \\ j=1,\dots,p}} = [K_1, \dots, K_\nu]^T$ is fixed and the set of matrices having this structure is defined as $\Omega(S)$. The problem is to find a control of the form (11) so as to minimise a quadratic cost function :

$$\min_{K \in \Omega(S)} J(K) = \int_o^\infty [x^T Q x + u^T R u] \, dt$$

$$\text{s.t.} \quad \dot{x} = Ax + Bu \qquad x(o) = x_o$$

$$y = Cx$$

$$u = -Ky$$

$$\text{with} \quad Q \geq 0 \text{ and } R > 0$$

Geromel and Bernussou (1978) give an algorithm for the case $C = I_n$ (state feedback) and for a complete decentralized control, i.e. :

$$\Omega(S) = \{K \in R^{m \times p} / K = \text{block diag } [K_{11}, \dots, K_{\nu\nu}], \; K_{ii} \;\; R^{m_i \times p_i}\}$$

The initial matrix K_o is determined (Armentano and Singh, 1981) such that $A-BK_o$ is stable. An iteration of their algorithm consists in the following steps :

1. Calculation of the gradient of the criterion

$$\frac{\partial J(K)}{\partial K} = 2(RK - B^T P)L$$

$$\left\{ \begin{array}{l} (A-BK)^T P + P(A-BK) + Q + K^T RK = 0 \\ L(A-BK)^T + (A-BK)L + I = 0 \end{array} \right. \tag{12}$$

2. Projection of the gradient matrix on the set $\Omega(S)$:

$$D = \{k_{ij}\}_{\substack{i=1,\dots,m \\ j=1,\dots,p}} \quad \begin{cases} d_{ij} = \left(\frac{\partial J(K)}{\partial K}\right)_{ij} & \text{for } k_{ij} \neq 0 \\ d_{ij} = 0 & \text{for } k_{ij} = 0 \end{cases}$$

3. Progression in the direction of the gradient :

$$K^{\mu+1} = K^{\mu} - a\,D$$

where a is a step size such that $a > 0$ and $J(K - aD) < J(K)$ until $\frac{\partial J(K)}{\partial K} = 0$.

It is interesting to see that this algorithm can be extended for $C \neq I_n$ and used or adapted for a "quasi-decentralized" control (i.e. : only certain fixed k_{ij}, $i \neq j$ are nul).

We can also use a modified version of this algorithm proposed by Chen, Mahmoud and Singh (1980) which reaches the optimum with a considerably reduced CPU-time. The principle of this version is to begin with a full optimal matrix K and to annule successively the blocks that must be nul applying at each of these annulations the algorithm of Geromel and Bernussou for the considered matrix.

This approach has been generalized to the synthesis of decentralised control insuring some insensivity with respect to change of parameters (A. Tarras) or to structural disturbances (L. Travé).

VI - Conclusion

This paper gives a survey of problems connected to decentralized fixed modes. Several methods to avoid or to cancel them are presented. Then, the design of optimal decentralized control is treated.

It provides all the necessary elements that will permit to treat completely an optimal decentralized control problem. Although the existence of fixed modes is undesirable, we are having all the necessary tools to face this situation successfully with a minimum additional cost.

REFERENCES

1 - ANDERSON B.O.D. and MOORE J.B. (1981)
Time invariant feedback laws for decentralized control.
IEEE Trans. Auto. Control AC - 26 n° 5, pp. 1133-1139.

2 - ANDERSON B.O.D. and CLEMENTS D.J. (1981)
Algebraic characterization of fixed modes in decentralized control.
Automatica, vol. 17, n° 5, pp. 703-712.

3 - ANDERSON B.O.D. (1982)
Transfert function matrix description of decentralized fixed modes.
IEEE Trans. Auto. Control, vol. AC - 27, n° 26, pp. 1176-1182.

4 - ARMANTANO V.A. and SINGH M.G. (1981)
A new approach to the decentralized controller initialization problem.
Preprints IFAC/81 Congress, Kyoto (Japan).

5 - ARMANTANO V.A. and SINGH M.G. (1982)
A new decomposition approach to eliminating decentralized fixed modes.
IEEE Trans. Auto. Control AC - 27, n° 1, pp. 258-260.

6 - CORFMAT J.P. and MORSE A.S. (1976 a)
Decentralized control of linear multivariable systems. Automatica, vol. 12, n° 5, pp. 479-496.

7 - CORFMAT J.P. and MORSE A.S. (1976 b)
Decentralized control of linear systems through specified input channels. SIAM J. Contr. and optimisation, vol. 14, n° 1, pp. 163-175.

8 - CHEN Y., MAHMOUD M.S. and SINGH M.G. (1980)
An iterative block-diagonalization procedure for decentralized optimal control. CSC Report, n° 495, UMIST (Manchester, U.K.).

9 - DAVISON E.J. and WANG S.H. (1973)
On the stabilization of decentralized control systems. IEEE Trans. Auto. Control AC - 18, n° 5, pp. 473-478.

10 - DAVISON E.J. and OZGUNER U. (1983)
Characterization of decentralized fixed modes for interconnected systems. Automatica, vol. 19, n° 2, pp. 169-182.

11 - GEROMEL J.C. and BERNUSSOU J. (1974)
An algorithm for optimal decentralized regulation of linear quadratic interconnected systems. Automatica, vol. 15, pp. 489-491.

12 - LOCATELLI A., SCHIAVONI N. and TARANTINI A. (1977)
Pole placement : role and choice of the underlying information pattern. Ricerche di Automatica, Vol. 18, n° 1, pp. 107-126.

13 - MEERKOV S.M. (1973)
Vibrational control. Automation and Remote control. Vol. 34, pp. 201-209.

14 - MEERKOV S.M. (1980)
Principle of Vibrational control : theory and applications. IEEE Trans. Auto. Control, vol. AC - 25, n° 4, pp. 755-762.

15 - PURVIANCE J.E. and TYLEE J.L. (1982)
Scalar sinusoidal feedback laws in decentralized control. Procedings of the 21th IEEE Conference on Decision and Control, Florida (U.S.A.).

16 - SENNING M.F. (1979)
Feasibly decentralized control. Thesis, Zurich (Switzerland).

17 - SEZER M.E. and SILJAK D.D. (1981)
Structurally fixed modes. Systems and control letters. Vol. 1, n° 1, pp. 60-64.

18 - TRAVE L., TARRAS A.M. and TITLI A. (1983 a)
An application of vibrational control to cancel unstable decentralized fixed modes. Submitted for publication to IEEE Trans. Auto. Control.

19 - TARRAS A.M. and TITLI A.
An algebraic test for fixed modes. Internal report L.A.A.S., may 1984. Submitted for publication.

20 - VIDYASAGAR M. and VISWANADHAM N. (1982)
Algebraic characterization of decentralized fixed modes and pole assignement. Proc. 21th IEEE Conference on Decision and Control, pp. 501-505.

21 - WANG S.H. (1982)
Stabilization of decentralized control systems via time-varying controllers. IEEE Trans. Auto. Control, vol. AC - 27, n° 3, pp. 741-744.

REAL TIME HIERARCHICAL OPTIMIZATION AND CONTROL:

AN EXCERPT FROM THE EEC REPORT

Günther Schmidt
Lehrstuhl und Laboratorium für
Steuerungs- und Regelungstechnik
Technische Universität München
Arcisstraße 21, 8000 München 2
West-Germany

1. INTRODUCTION

Over the last two decades numerous papers and many excellent books, e.g. [1,2,3] have been written on advanced theoretical approaches to optimization and control of large scale as well as complex technical systems. It is however only more recently that inexpensive multiple microprocessor technology can provide an appropriate solution to the demand for computing power to meet new requirements and to support complex applications. Thus, one of the objectives of research in the joint EEC project was to work out computer-oriented approaches to real time optimization and control of large scale systems and to study their application to industrial problems. Problems considered by the different teams of the project were for example control of a sulphur production plant, supervision and control of gas transportation and distribution through pipeline networks [4], and hierarchical optimization and control of traffic flow along extended motorways. Because of lack of space we will constrain this paper to a presentation of selected results worked out for the latter application area. Thus, this paper is organized as follows. The mathematical traffic flow model that forms the base of various approaches to hierarchical optimization and control for motorway traffic is summarized in section 2. In section 3 basic objectives of a traffic flow control system are outlined. It is shown, how the control problems involved can be mathematically formulated as dynamic optimal control problems. With respect to real-time application, the large scale optimization problem is decomposed and implemented on a three-layer distributed computer control structure. Our main concern in this paper will be with the most challenging problems of optimization and control computations on the direct control layer. We discuss in section 4 central and hierarchical approaches to the solution of a nonlinear quadratic optimization problem and outline features of a multiple processor implementation in section 5. Simulation results and a comparison of the performance of various traffic flow control laws are presented in section 6.

2. A CONTROL-ORIENTED MOTORWAY TRAFFIC FLOW MODEL

It is only more recently that traffic flow on motorways is considered as a process in the control-theoretic sense, i.e. a more or less causal dynamic system with properties that can be described with sufficient accuracy by a set of mathematical equations. Motorway traffic flow is a highly nonlinear, large scale process with stable and unstable regimes of operation. From a control viewpoint there exist today various types of flow models with varying ranges of applicability. A detailed discussion of these models, their properties, relationships and possible applications can be found in [5,6].

For purposes of macroscopic mathematical modelling we assume an extended two-lane motorway to consist of N concatenated sections with at most one on- and one off-ramp, Fig. 1. Each section is in turn subdivided into segments. Assuming homogeneous traffic conditions in each segment the following aggregate traffic flow variables can be defined for discrete points of time $k \cdot T (k=0,1,2,\ldots)$

$c_i^j(k)$	traffic density (veh/km)
$v_i^j(k)$	space mean speed (km/h)
$q_i^j(k)$	traffic volume (veh/h) in the jth segment of the ith section
$r_i(k), s_i(k)$	on-ramp, off-ramp volumes in the ith section.

A deterministic, discrete time dynamical traffic flow model (originally presented in [7] and modified in [8] after validation for typical traffic conditions on the German Autobahn) can be based on the following set of difference equations

$$c_i^j(k+1) = c_i^j(k) + T/\Delta_i^j \cdot (q_i^{j-1} - q_i^j + \sigma_i^j)\Big|_{(k)} \tag{1}$$

with

$$\sigma_i^j(k) = \begin{cases} r_i(k) - s_i(k) & \text{if } j=1 \\ 0 & \text{else} \end{cases}$$

$$v_i^j(k+1) = v_i^j(k) + T/\tau \cdot [V(c_i^j) - v_i^j]\Big|_{(k)} + T/\Delta_i^j \cdot (v_i^{j-1} - v_i^j) v_i^j\Big|_{(k)} \tag{2}$$
$$- T/\Delta_i^j \cdot \nu/\tau \cdot (c_i^{j+1} - c_i^j)/(c_i^j + \kappa)\Big|_{(k)} .$$

These equations describe the dynamic evolution of traffic density and mean speed in the jth segment of ith section. Traffic volume q_i^j in (1) can be eliminated by use of a relation known from hydromechanics, i.e.

$$q_i^j(k) = v_i^j(k) \cdot c_i^j(k) . \tag{3}$$

V(c) denotes the steady state speed versus density characteristic, which can be approximated by the expression

$$V(c) = V_f \cdot [1 - (c/c_{max})^{l(3-2b_i)}]^m \cdot b_i , \tag{4}$$

where V_f means free speed, c_{max} is the jam density, l and m are positive real numbers, and $b_i(k)$ is a variable corresponding to a speed limitation indicated by a variable message sign in the ith section. The off-ramp volumes with known constant exit rates γ_i are given by

$$s_i(k) = \gamma_i \cdot q_{i-1}^{n(i-1)} . \tag{5}$$

n(i) is the number of segments in the ith section and the total number of segments is

$$n = \sum_{i=1}^{N} n(i) .$$

Typical values for major model parameters are presented in Table I. They were identified from real traffic data collected along a German Autobahn [8, 11].

In this model and in reality two sets of variables, $b_i(k)$ and $r_i(k)$, are available for purposes of control. We will constrain the following discussion to metering of on-ramp traffic volume $r_i(k)$ as the main control input while $b_i(k)$ will be kept constant and equal to 1. The admissible control region Ω can be defined by inequality constraints

$$0 < r_{,min} \leq r_i(k) \leq d_i + l_i(k)/T , \tag{6}$$

where $r_{i,min}$ are the minimal admissible values of the ramp volumes, d_i are the demands, and $l_i(k)$ means the length of the queue at the i-th on-ramp. The time evolution of the queues is described by an additional set of difference equations

$$l_i(k+1) = l_i(k) + T \cdot [d_i - r_i(k)] . \tag{7}$$

Summarizing, we end up with a nonlinear large-scale discrete-time state space model with state vector

$$\underline{x} = [l_1 \; c_1^1 \; n_1^1 \; c_1^2 \; v_1^2 \; \ldots \; c_N^{n(N)} \; v_N^{n(N)}]^T \tag{8a}$$

and control vector

$$\underline{u} = [r_1 \; \ldots \; r_N]^T . \tag{8b}$$

Typically, 100 states and 10 control inputs have to be considered for a 50 km long motorway.

3. BASIC OBJECTIVES OF MOTORWAY TRAFFIC CONTROL AND MULTILAYER CONTROL STRUCTURE

The main objective of a traffic flow control system must be to prevent the built-up of recurrent congestions caused by too high a demand at the on-ramps and to eliminate non-recurrent congestions occuring as a result of some local incident along the motorway. Prevention or elimination of congestions will usually increase the total throughput of vehicles and reduce the number of accidents on a motorway. In order to achieve these objectives, a control system must disperse the traffic entering a motorway in time and space by appropriate control of the input volumes r_i within their given limits.

Application of modern control methodology to the selection of reasonable or even optimal input volumes requires the traffic control problem to be formulated as a mathematical optimization problem. In our case minimization of the total travel time of all vehicles using the motorway (including total waiting time at the on-ramps) over a sufficiently long time-horizon is considered as one suitable objective of optimization. Thus we can describe the main goal of a traffic control system by the following optimization problem OP1

Select on-ramp volumes $r_i(k)$ from the admissible region (6) so that for the set of all vehicles on the motorway the total travel time

$$J_T = T \cdot \sum_{k=0}^{K-1} \sum_{i=1}^{N} \sum_{j=1}^{n(i)} [l_i(k) + c_i^j(k)\Delta_i^j] \tag{9}$$

over a specified time-horizon K is minimized subject to the model constraints (1) to (5) and (7).

OP1 could be solved by use of Pontrjagin's Maximum Principle leading to a TPBVP. Although such a solution can proof to be useful as a reference, a real-time traffic control system based on such a solution scheme would suffer from a couple of drawbacks, e.g.

(i) Extensive computer storage space and computer time is required for the solution of the resulting large-scale TPBVP. Occurence of singular control subarcs may lead to additional difficulties.
(ii) Accurate predictions of on-ramp demands and origin-destination rates must be provided.
(iii) Requirements for an adequate robustness of the control system will cause additional cost during implementation.

Most of the above mentioned drawbacks can be overcome by introducing a multilayer control structure for an approximate (suboptimal) solution of problem OP1 (see e.g. [9, 10, 11]). Multilayer control structures are developed so as to combine high efficiency of control (comparable to the one of the solution of problem OP1) with lower implementation cost. Often several less complicated, weakly coupled optimal control problems (e.g. for different time scales) are solved in the various layers of the traffic control system hierarchy.

In our case the solution for OP1 can be approximated by a three-layer control system [6], Fig. 3. A static or quasi-dynamic (time-of-day) control problem of the Linear Programming type is solved in an *optimization layer* giving nominal values $\bar{\underline{x}}$, $\bar{\underline{u}}$ for the input and state variables (8). An inferior *direct control layer* considers the fast process dynamics according to our model equations and drives the actual traffic state $\underline{x}(k)$ to the specified nominal values in spite of possible disturbances and model inaccuracies. Predicted values of system variables like demands, origin destination rates and exit rates are provided for the optimization layer by a supremal *adaption layer*. Fig. 2 indicates the overall multilayer control scheme including the modules for decentralized local data-processing and the Kalman-filter-like state-estimators [11], as well as the multiple computer system for its implementation.

Our further discussion will be directed to the most challenging part of the multilayer system, namely details of the implementation of the direct control layer.

4. OPTIMAL CONTROL PROBLEMS ON THE DIRECT CONTROL LAYER

The derivation of control actions on the direct control layer can be based on a quadratic performance index J_Q penalizing deviations of the actual state and control variables $\underline{x}(k)$, $\underline{u}(k)$ form its nominal values $\bar{\underline{x}}$, $\bar{\underline{u}}$

$$J_Q = \sum_{k=0}^{K=1} \underbrace{\frac{1}{2}(||\underline{x}(k) - \bar{\underline{x}}||_Q^2 + ||\underline{u}(k) - \bar{\underline{u}}||_R^2)}_{\phi} \tag{10}$$

with

$$Q = \mathrm{diag}(q_1\ q_c\ q_v\ q_c\ q_v\ \cdots\ q_c\ q_v),\quad R=\rho\cdot E$$
$$q_1=0,\ q_c=1,\ q_v=.3,\ \rho=.01,\ E = \text{unit matrix}\ .$$

Linear Feedback Control Law

After linearization of the model from section 2 around ($\bar{\underline{x}}$, $\bar{\underline{u}}$) the following standard LQ-problem OP2 can be formulated

Select on-ramp volumes $\Delta\underline{u}(k)$ so that (10) is minimized subject to the constraints

$$\Delta\underline{x}(k+1) = A\cdot\Delta\underline{x}(k) + B\cdot\Delta\underline{u}(k),\ \Delta\underline{x}(0) = \Delta\underline{x}_0\ ;\quad k = 0,1,\ldots K-1 \tag{11}$$

with $\Delta\underline{x} = \underline{x}-\bar{\underline{x}}$, $\Delta\underline{u} = \underline{u}-\bar{\underline{u}}$ and $\underline{x}$, $\underline{u}$ according to eqs. (8).

For sufficiently large K the solution for OP2 can be formulated as a time-invariant linear state-feedback control law

$$\Delta\underline{u}(k) = L\cdot\Delta\underline{x}(k) \tag{12}$$

with the gain matrix L being computed via backward Riccati recursion. A more detailed analysis of the traffic flow process shows that the centralized control law (12) can be decentralized with the result of N suboptimal local state-feedback laws for the Δu_is [12].

Nonlinear Control

If our interest is only in open-loop control $\underline{u}(k)$ for the direct control layer, a further optimization problem OP3 can be formulated

Select on-ramp volumes $\underline{u}(k)$ so that (10) is minimized subject to the constraints

$$\begin{aligned} &\underline{x}(k+1) = \underline{f}[\underline{x}(k), \underline{u}(k)],\quad \underline{x}(0) = \underline{x}^0\ . \\ &\underline{h}[\underline{x}(k), \underline{u}(k)] \geq 0;\quad k = 0,1,\ldots K-1 \end{aligned} \tag{13}$$

with $\underline{x}$, $\underline{u}$ and $\underline{h}$ [.] according to eqs. (8) and (7).

The numerical treatment of this problem will be approached by a central and a hierarchical solution scheme based on an appropriate decomposition of OP3.

Central Solution Scheme for OP3

If the Hamiltonian is defined as

$$H = \phi[\underline{x}(k), \underline{u}(k)] + \underline{\lambda}(k+1)^T \cdot \underline{f}[\underline{x}(k), \underline{u}(k)] \tag{14}$$

with $\underline{\lambda}(k)$ the costate vector, then the optimal solution of OP3 must satisfy the following necessary conditions [13]

$$\underline{x}(k+1) = \underline{f}[\underline{x}(k), \underline{u}(k)], \quad \underline{x}(0) = \underline{x}_o \tag{15}$$

$$\underline{\lambda}(k) - \partial\underline{h}^T/\partial\underline{x}(k) \cdot \underline{\mu}(k) = \partial\phi/\partial\underline{x}(k) + \partial\underline{f}^T/\partial\underline{x}(k) \cdot \underline{\lambda}(k+1); \quad \underline{\lambda}(K) = \underline{0} \tag{16}$$

$$-\partial\underline{h}^T/\partial\underline{u}(k) \cdot \underline{\mu}(k) = \partial\phi/\partial\underline{u}(k) + \partial\underline{f}^T/\partial\underline{u}(k) \cdot \underline{\lambda}(k+1) \tag{17}$$

$$\underline{h}[\underline{x}(k), \underline{u}(k)] \geq 0; \quad \underline{\mu}(k)^T \cdot h[\underline{x}(k), \underline{u}(k)] = 0; \quad \underline{\mu}(k) \leq 0, \quad k = 0,\dots,K-1 \ . \tag{18}$$

This equation set-up defines a large scale TPBVP which can be solved by use of iterative algorithms discussed in [14].

Decomposed Formulation of OP3

For decomposition of OP3 we introduce the following general notation

$$\underline{\pi}_i(k) = \sum_{j\neq i}^{N} \underline{g}_{ij}[\underline{x}_j(k), \underline{u}_j(k)] \tag{19a}$$

$$\underline{x} = [\underline{x}_1^T \dots \underline{x}_N^T]^T, \ \underline{u} = [\underline{u}_1^T \dots \underline{u}_N^T]^T \ . \tag{20a}$$

where $\underline{\pi}_i$ represents a vector of separable interconnection variables, N is the number of subsystems, $\underline{x}_j$ and $\underline{u}_j$ are local state and control variables. By means of these definitions the overall system of traffic flow state equations can be subdivided into N independent subsystems corresponding to the particular motorway sections. The problem-oriented interconnection variables and local state and control variables are given by

$$\underline{\pi}_i^T = [c_{i-1}^{n(i-1)} \ v_{i-1}^{n(i-1)} \ c_{i+1}^{1}] \tag{19b}$$

$$\underline{x}_i^T = [1_i \ c_i^1 \ v_i^1 \ c_i^2 \ v_i^2 \dots c_1^{n(i)} \ v_1^{n(i)}], \quad u_i = r_i \ . \tag{20b}$$

Next the overall problem OP3 can be reformulated in equivalent decomposed form as

$$\min_{\underline{u}} J_Q = \sum_{k=0}^{K-1} \sum_{i=1}^{N} \phi_i[\underline{x}_i(k), \underline{u}_i(k), \underline{\pi}_i(k)] \tag{21}$$

subject to eqs. (19) and

$$\underline{x}_i(k+1) = \underline{f}_i[\underline{x}_i(k), \underline{u}_i(k), \underline{\pi}_i(k)] \tag{22}$$

$$\underline{h}_i[\underline{x}_i(k), \underline{u}_i(k), \underline{\pi}_i(k)] \geq 0; \quad k=0,\dots,K-1; \quad i=1,\dots,N \ . \tag{23}$$

Hierarchical Solution Scheme Based on the Interaction Prediction Principle

The Hamiltonian of the overall problem in decomposed form is formulated by adjoining the interconnection constraints (19) with some Lagrange multiplier vector β to the original Hamiltonian (14). It is shown in [15] that the necessary conditions derived by use of the modified Hamiltonian are identical to the necessary conditions of the following subproblems:

$$\min_{\underline{x}_i,\underline{u}_i} J_i = \sum_{k=0}^{K-1} \{\phi_i[\underline{x}_i(k), \underline{u}_i(k), \underline{\pi}_i(k)] + \underline{\beta}_i(k)^T \underline{\pi}_i(k) - \sum_{j\neq i}^{N} \underline{\beta}_j(k)^T \underline{g}_{ji}[\underline{x}_i(k), \underline{u}_i(k)]\}$$

subject to (22), (23) and for given $\underline{\pi}_i, \underline{\beta}_j$, $j=1,\ldots,N$ (24)

with the coordination conditions

$$\underline{\pi}_i(k) = \sum_{j\neq i}^{N} \underline{g}_{ij}[\underline{x}_i(k), \underline{u}_i(k)] \tag{25}$$

$$\underline{\beta}_i(k) = -\partial\phi/\partial\underline{\pi}_i(k) - \partial\underline{f}_i^T/\partial\underline{\pi}_i(k)\cdot\underline{\lambda}_i(k+1) - \partial\underline{h}_i^T/\partial\underline{\pi}_i(k)\cdot\underline{\mu}_i(k) . \tag{26}$$

Let us assume that subproblems (24) have a solution which can be found by solving the corresponding TPBVPs. Then the overall problem can be solved by using the following two-level algorithm based on the interaction prediction principle [2]:

Step 1: Guess vector sequences $\underline{\pi}(k)$, $\underline{\beta}(k)$. Iteration index l := 1.

Step 2: Solve the independent subproblems (24) with given $\underline{\pi}(k)$, $\underline{\beta}(k)$ ("first level iteration") and specify the solution $\underline{x}^l(k)$, $\underline{u}^l(k)$, $\underline{\lambda}^l(k)$.

Step 3: Update $\underline{\pi}^{l+1}(k)$, $\underline{\beta}^{l+1}(k)$ by substituting the subproblems' solutions directly into (25), (26).

Step 4: If

$$||\underline{\pi}^{l+1}(k) - \underline{\pi}^l(k)|| > \epsilon_\pi,\ ||\underline{\beta}^{l+1}(k) - \underline{\beta}^l(k)|| > \epsilon_\beta , \tag{27}$$

$k = 0,\ldots,K-1$,

for some prescribed values ϵ_π, ϵ_β, go to *Step 2*,

else record $\underline{u}^l(k)$ as the optimal control trajectory and stop.

Fig. 3 indicates the basic structure of the hierarchical optimization procedure.

5. APPLICATION AND COMPUTER IMPLEMENTATION OF THE SOLUTION SCHEMES

For an evaluation of the efficiency of the central and hierarchical solution scheme a hypothetical two-lane motorway with five on-ramps and off-ramps, section length 4km and $\Delta_i^j = \Delta = 1$km is considered. Two sets of initial conditions are specified

(i) uncongested traffic, characterized by the initial values $c_i^j(0) = 20$, $v_i^j(0) = 100$, $j = 1(1)4$; $i = (1)5$.

(ii) congested traffic, characterized by the initial values

$$c_i^j(0) = \begin{cases} 120 \\ 80 \\ 50 \end{cases}, \quad v_i^j(0) = \begin{cases} 15 & \text{for } i = 3,\ j = 3 \\ 30 & \text{for } i = 3,\ j = 2,4 \\ 60 & \text{else.} \end{cases}$$

The nominal conditions in (10) are given by

$\bar{c}_1^j = 37;\ \bar{c}_2^j = 49;\ \bar{c}_i^j = 64,\ i = 3(1)5,\ j = 1(1)4$

$\bar{v}_1^j = 82;\ \bar{v}_2^j = 74;\ \bar{v}_i^j = 62,\ i = 3(1)5,\ j = 1(1)4$

$\bar{r}_1 = 3000;\ \bar{r}_2 = 700;\ \bar{r}_3 = 635;\ \bar{r}_4 = 217;\ \bar{r}_5 = 228$

and the corresponding demands and exit rates are set to

$d_1 = 3000;\ d_2 = d_3 = 700;\ d_4 = d_5 = 350.$

$\gamma_1 = 0;\ \gamma_2 = 0.05;\ \gamma_3 = 0.052;\ \gamma_4 = 0.054;\ \gamma_5 = 0.057.$

The optimization time horizon is 30 min (K = 120).

Mainframe Computer Implementation

To gain better insight into the features of the various solution schemes in connection with the example problem, both schemes are implemented on a mainframe computer (Cyber 175). The TPBVP in the central scheme was solved with a modified gradient algorithm [14]. For the hierarchical scheme both, a monoprocessor and a multiprocessor version (one processor for coordination and N processors for subsystem optimization, see Fig. 3), was studied.

In addition to the standard case specified above, the example was solved for a motorway with N = 2,3....7 sections with the order of the problems increasing proportionally with N. The major results are summarized in Fig. 4.

(i) The central solution scheme shows roughly a linear relationship between computation time and number of sections N. Computation time needed for a congested initial traffic state is generally higher than the one needed for uncongested initial conditions. This can be explained by the fact that the optimal input trajectories in the first case vary much stronger in time than in the latter. Thus more iterations are required for computation of the exact solutions.
(ii) The hierarchical solution scheme based on a multi-processor operation shows a reduction of computation time if N > 3 or 6 depending on the initial traffic state.
(iii) The value of computation time shown for the case of a hierarchical but monoprocessor solution indicates the inefficiency of this approach.

Further details and a more general theoretical treatment of the computation time, storage space, communication data requirements for the various solution schemes are reported in [15].
The following conclusions can be drawn from these experiments.

(i) A computation time reduction can be achieved with a multiprocessor system if the

order of the optimal control problem is sufficiently high.
(ii) The independent subproblems can be implemented on microcomputer systems.
(iii) The communication data rate and storage space requirements are comparatively modest.

These results provided a justification for the implementation of our example problem on a multiple microcomputer system.

Multiple Microcomputer Implementation

The multiprocessor used consists of three μcomputers (Z 80 + AMD 9511) and a 16 bit-minicomputer (Interdata M 70) connected in star configuration through a universal interface module (UIM) and DMA on the μcomputers' side, Fig. 3. Three of the N=5 subproblems of the traffic control problems are implemented (assembler) and solved in the three μcomputers. The minicomputer treats the remaining two subproblems and the coordination task. Since the minicomputer's execution time is much shorter than the μcomputers', the solution of the 5 independent subproblems can be considered as being parallel.

The computation time for one first-level iteration of the μcomputers has been found to be 9s. The data transfer at each second-level iteration for five subsystems amounts to 20 k Bytes corresponding to 2s of transmission time. Hence the data transfer times can be viewed as negligible for the prediction principle algorithms.

The total computation time required for the solution of the overall problem has been 12 min and 22 min for uncongested and the congested initial traffic respectively. These results equal those obtained on the main-frame computer.

22 min computation time is of course still too long compared to the 30 min long optimization horizon considered. However considerable reduction of computation time seems to be possible in view of the most recent developments of 16 and 32 bit microprocessor technology. With regard to the optimization horizon considered in our example a computation time in the order of 1 min will be sufficient for a real-time application of the optimal control scheme based on OP3.

6. SIMULATION RESULTS AND CONCLUDING REMARKS

To conclude we will mention some results of an evaluation of the performance of the various direct layer controls developed. Fig. 5 shows three graphs of time/space evolution of traffic density based on computer simulations with the nonlinear traffic flow model from section 2 and data as specified in section 4 for the congested initial state. Fig. 5a shows the case of the uncontrolled motorway and a (unstable) propagation of high densities in upstream direction, as caused by a severe incident in section 3. Fig. 5b shows traffic flow under control of a linear control law (12), i.e. the solution to OP2. Because of large initial deviations from the nominal conditions ($\bar{\underline{x}}$, $\bar{\underline{u}}$) and restrictions in the control variables, this control law cannot stabilize traffic and the motorway remains congested. A linear control law can only cope with light disturbances [16]. The results of an application of the nonlinear optimal control as developed for OP3 can be infered from Fig. 5c. In spite of the severe initial disturbance, traffic flow is brought back to its nominal conidtions rather smoothly and after about 10 min.

The selected results from a special application area reported in this paper demonstrate the general impact that nonlinear optimal control and hierarchical solution schemes together with multiple processor technology can have for the development of advanced control systems.

REFERENCES

[1] Singh, M.G.; Titli, A.: Systems Decomposition, Optimization and Control. Pergamon Press, 1978.
[2] Singh, M.G.: Dynamical Hierarchical Control, North Holland, 1980.
[3] Siljak, D.: Large Scale Dynamic Systems, North Holland, 1978.
[4] Schmidt, G.; Lappus, G.: Real Time Simulation, Network State Estimation and Predictive Control for Gas Transportation and Distribution Systems (in German). Messen, Steuern, Regeln (msr),pp.60-65, Febr. 1984.
[5] Papageorgiou, M,; Posch, B.; Schmidt, G.: Comparison of Macroscopic Models for Control of Freeway Traffic. Transportation Research B, vol. 17B, pp. 107-116.
[6] Papageorgiou, M.; Schmidt, G.: Freeway Traffic Modelling and Control. Preprints of 4th IFAC Conference on Control in Transportation Systems, April 1983, pp. 195-202.
[7] Payne, H.J.: Models of Freeway Traffic and Control, Simulation Council Proc., vol. 1, pp. 51-61, 1971.
[8] Cremer, M,; Papageorgiou, M.: Parameter Identification for a Traffic Flow Model. Automatica, vol. 17, pp. 837-843, 1981.
[9] Tabak, D.: Application of Modern Control and Optimization Techniques to Transportation Systems. Control and Dynamic Systems 10, Leondes, C.T., Ed., Academic Press, pp. 345-434, 1973.
[10] Athans, M.; e.a.: Stochastic Control of Freeway Corridor Systems. Proc. IEEE Conf. on Decision and Control, Dec. 10-12, 1975, pp. 676-685.
[11] Cremer, M.; Papageorgiou, M.; Schmidt, G.: Application of Control Equipment for the Improvement of Traffic Operations on Motorways (in German). Forschung, Straßenbau und Straßenverkehrstechnik, No. 307, 1980.
[12] Papageorgiou, M.: Applications of Automatic Control Concepts to Traffic Flow Modelling and Control. Springer Verlag, Berlin, 1983.
[13] Pearson, J.B.; Sridhar, R.: A Discrete Optimal Control Problem. IEEE Trans. Autom. Contr., vol. AD-11, pp. 171-174, 1966.
[14] Papageorgiou, M.; Schmidt, G.: On the Hierarchical Solution of Nonlinear Optimal Control Problems. Large Scale Systems, vol. 1, pp. 265-271, 1980.
[15] Papageorgiou, M.; Schmidt, G.: Implementation of a Hierarchical Optimization Algorithm on a Multimicrocomputer System. IEEE Transaction on Systems, Man, And Cybernetics, vol. 13, no. 1., pp. 11-18, 1983.
[16] Isaksen, L.; Payne, H.J.: Freeway Traffic Surveillance and Control. Proc. IEEE 61, pp. 526-536, 1973.

V_f	c_{max}	m	l	κ	ν	τ	T
123 km/h	200 veh/km	1.4	4	20 veh/km	21.6 km²/h	24 s	15 s

Table 1 Identified parameters of motorway model

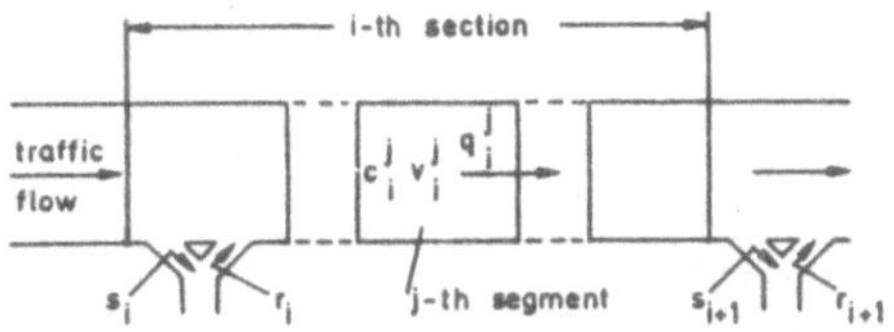

Fig.1 Motorway section and traffic flow variables

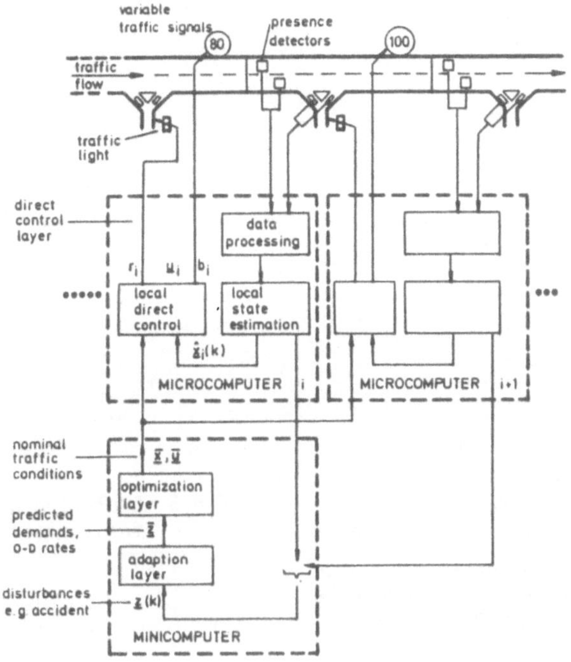

Fig.2 Three-layer traffic control system

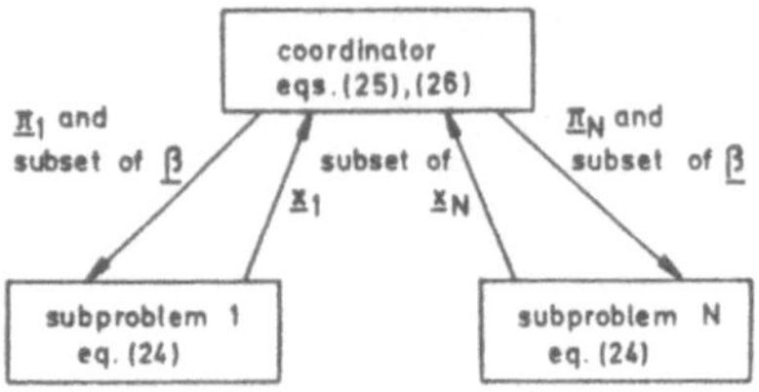

Fig.3 Hierarchical optimal control structure

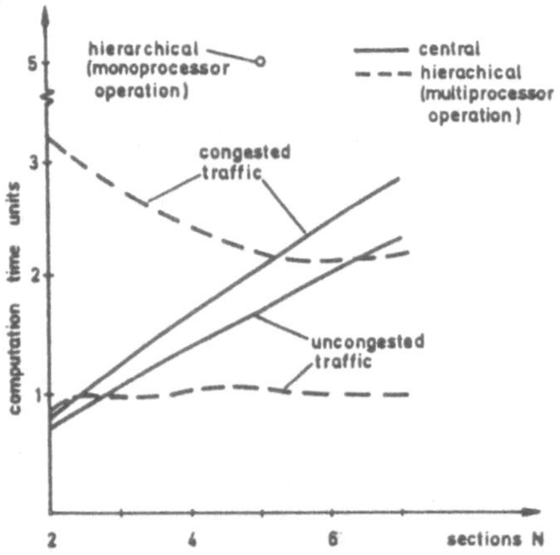

Fig.4 Computation time for central and hierarchical solution scheme

$$1 \text{ time unit} = \begin{cases} \text{10 sec mainframe computer} \\ \text{12 min multi-microcomputer system} \end{cases}$$

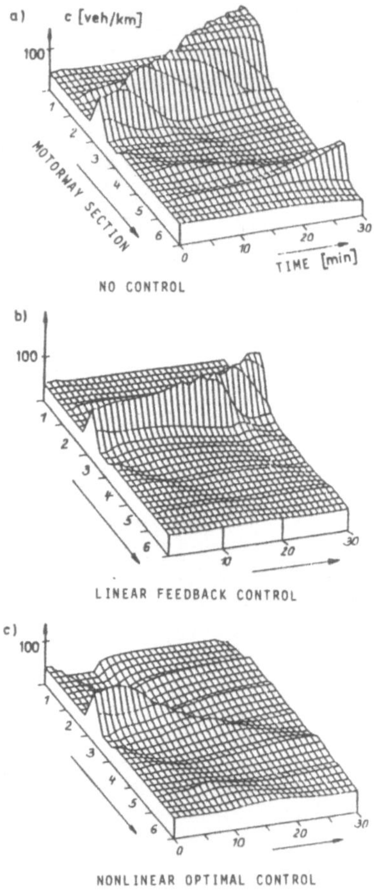

Fig.5 Evolution of traffic density for various controls

RELIABILITY AND FAULT DETECTION TECHNIQUES OF LARGE SCALE SYSTEMS
A Synthesis of the EEC Report

S.G. Tzafestas
Control Systems Laboratory
Electrical Engineering Department
University of Patras
Patras, Greece

ABSTRACT: This report provides a synthesis of the works on reliability, maintenance, fault detection and fault-tolerant controller design carried out by the members of the teams involved in the EEC real time control project. In the reliability area, a repairable standby system of assemblies with a hierarchical structure was studied and optimized, via the generalized minimum principle, by selecting appropriate maintenance control laws. A model of a repairable standby system with repair facilities subject to breakdowns was also studied. In the area of fault detection, an up-to-date survey of the available techniques was made, and a method was developed for detecting sensor failures in large scale systems using decentralized observers designed by the overlapping decomposition technique of Siljak. A similar observer-based failure detection scheme for distributed-parameter systems was also developed and applied to gas transportation and distribution systems. Finally, a fault-tolerant controller scheme was proposed, and tested both by simulation and by building an experimental set-up using Z-80 microprocessors. For completeness, a discussion of some related results concerning the stability and performance of hierarchical controllers, under structural perturbations, is also included.

1. INTRODUCTION

Among the primary problems in the planning, design and control of multicomponent systems, is the problem of using the available resources in the most effective way so as to maximize the overall system reliability/availability, or to minimize the consumption of resources subject to imposed reliability/availability constraints, or so as to achieve acceptable performance even if the system is subject to component or structural failures. The two standard ways for improving system reliability are (i) using redundancy and (ii) using repair/maintenance policies (preventive or not). Of particular importance in this framework are the techniques of failure detection and fault-tolerant system design.

During the recent years a considerable amount of work has been completed covering the whole repertory of reliability, repair and maintenance problems. The purpose of the present synthesis paper is to provide a short presentation of the work carried out by the four teams within the EEC project on real-time optimization and control of large scale systems, or closely related to it. An effort was made to make the presentation self-readable, but unavoidably many details of the derivations and results are left out. These can be found in the related publications. The report is organized as follows: Section 2 is devoted to the reliability and maintenance problems, section 3 presents the results on dynamic fault detection and fault-tolerant controller design, and finally section 4 provides some remarks on the problems of stability, stabilization and performance analysis of multilevel controllers under structural perturbations.

2. SYSTEM RELIABILITY AND MAINTENANCE OPTIMIZATION

Reliability and availability optimization problems have been solved by a variety of techniques such as the *Lagrange multiplier* method, *integer programming* technique, *unconstrained programming* method, etc. [1].

Here we shall give a brief presentation of the formulation and solution via standard optimal control theory [2,4].

A system is an assembly of components (or units) which for reliability/availability considerations are assumed to be connected in *series*, in *parallel* and in *series-parallel* combinations. In the *series* combination there are a number of stages, say n, statistically independent, and the total reliability is given by $R = \Pi_{k=1}^{n} R_k$, where $0 < R_k < 1$ is the reliability of the kth stage. In the *parallel* combination there are m statistically independent redundant units, and the total unreliability is equal to $Q = q^m$, where q is the unrealiability of each of the identical redundant units. In a *series-parallel* system there are m parallel subsystems, the kth one being a combination of n_k stages. The unrealibility Q of this system is given by

$$Q_k = \prod_{k=1}^{m} Q_k, \quad Q_k = 1 - \prod_{i=1}^{n_k} r_{ik}$$

where Q_k is the unreliability of the kth series subsystem, and r_{ik} is the reliability of the ith unit in the kth series subsystem. In a *parallel-series* system there are n stages in series, where the kth one has m_k parallel units. The total reliability in this case is

$$R = \prod_{k=1}^{n} R_k, \quad 1-R_k = q_k^{m_k}$$

where q_k is the unreliability of the identical parallel elements in the kth stage.

In a *general* (non-series/non-parallel)model the total reliability can be evaluated by two principal methods [6], namely: (a) *reliability network technique*, and (b) *Markovian (state transition) technique*.

In many cases not all the redundant (parallel) units are active at the same time, but are waiting for action (*standby* model). Very common in practice is the case of double failure systems, which have two types of failures, *open circuit*, and *short-circuit* failures. The reliability of a n-stage series model with m_k parallel units at stage k is equal to

$$R = \prod_{k=1}^{n} R_k = \prod_{k=1}^{n} \{(1-p_k)^{m_k} - q^{m_k}\}$$

where p_k and q_k are the "short-circuit" and "open-circuit" failure probabilities, respectively, of stage k. If $p_k = 0$ (no short circuit failures) R reduces exactly to the formula given above.

In more realistic models one includes the failure probabilities of the sensing and switching devices which put in operation the standby units whenever some operating units fail [17].

In general there are two kinds of optimization problems, namely:

(i) Reliability/availability maximization

(ii) Cost minimization

In the former, one wishes to choose the reliabilities r_k of the various units and their numbers m_k (k=1,2,...,n) so as to maximize the total reliability R of the system. In the simple series, parallel, series-parallel and parallel-series cases one uses the above expressions of R. In general more complex cases the reliability is given by a nonlinear expression of the type

$$R = (r_1(m_1),\ldots,r_n(m_n))$$

where $r_k(m_k)$ are the reliabilities of the subsystems involved.

In the latter problem (cost minimization) one uses some cost function representing a measure of the real cost of the system set-up. Thus, for example in a parallel-series system with n stages in series and m_k identical parallel units at stage k, the problem is to minimize the total (additive) cost

$$J = \sum_{k=1}^{n} L_k(m_k,u_k,d_k)$$

by choosing m_k, u_k(mean up-time of stage k), and d_k (mean down-time of stage k) within a permissible region, so as to meet a desired level of system availability A, where

$$A = \prod_{k=1}^{n} A_k, \qquad A_k = \frac{u_k}{u_k+d_k}$$

Here L_k is the cost, and A_k the availability of stage k. A typical design region for the kth stage is defined by $u_k \geq 4d_k$ and $d_k \leq 5$.

A full survey of the techniques used in solving reliability/cost optimization problems can be found in [1]. The solution of these problems through the *maximum principle* and the *dynamic* programming techniques can be found in [2].

Here we shall review the results of [3-5]. In [4] the reliability measures of the optimal configuration subject to budget constraints have been derived for a repairable standby system of assemblies with a hierarchical design. These measures, which depend on both the number of components in stock and the number of standby assemblies, were derived using the Markovian technique in which the system is described as a discrete-state continuous-time Markov process. A numerical procedure for solving the resulting set of linear equations was used, which does not require the matrix of the system to be stored. In [3] the problem of determining optimal maintenance policies for the above model was considered and solved by employing a generalized version of Pontryagin's minimum principle which is based on an integral Hamiltonian functional. The cost function used involves a maintenance cost term and a down-time cost term.

A short description of this model is as follows. Consider a repairable standby system of items (assemblies) with a hierarchical (or modular) structure in which the items have components that are also repairable. The system consists of a stock of m independent components and n assemblies, with one assembly operating and n-1 assemblies in standby (ready to replace failed assemblies). Items in standby are assumed failure-free. Each assembly failure is assumed to occur by a failure of at most a single component. A failed assembly is returned to operation by removing the defective component and replacing it by a functional one. The defective component is then repaired and put back to the standby stock of components. If no stock of operative components is available, the assembly repair cycle is lengthened until an operative component becomes available through the repair procedure.

Denote by S(k,i) the state where there are k failed components and i operative assemblies in the system. Then S(k,0), k=0,...,n+m are the states in which all n assemblies are failed. These failed states can be *absorbing* or *reflecting*. Now, under the assumption that at t=0 our system is in an operating state, the state probability equations of the system are

$$\frac{d}{dt}\begin{bmatrix} p_w(t) \\ \hline p_f(f) \end{bmatrix} = \left[\begin{array}{c|c} E & F \\ \hline G & H \end{array}\right] \begin{bmatrix} p_w(t) \\ \hline p_f(t) \end{bmatrix} \tag{1}$$

where $p_w(t)\{p_f(t)\}$ is the vector with components the probabilities that the system is in a working{failed state} and

$$E = \begin{bmatrix} A_0 B_1 & & & & \\ L_0 A_1 B_2 & & & 0 & \\ & \ddots & & & \\ & & L_{n+m-3} & A_{n+m-2} & B_{n+m-1} \\ 0 & & & L_{n+m-2} & A_{n+m-1} \end{bmatrix}, \quad F = \begin{bmatrix} F_0 \\ F_1 \\ \vdots \\ F_{n+m-1} \end{bmatrix}$$

$$G = [G_0 G_1 \cdots G_{n+m-1}]$$

$$H = \begin{bmatrix} (-n+m)\mu_c & & & & 0 \\ (n+m)\mu_c & \ddots & & & \\ & \ddots & -[k\mu_c + r(k)\mu_\alpha] & & \\ & & k\mu_c & \ddots & -\mu_c + n\mu_\alpha \\ 0 & & & & \mu_c - n\mu_\alpha \end{bmatrix} \tag{2}$$

with $\lambda(k,i)$, $\mu_\alpha(k,i)$, $\mu_c(k,i)$ being the failure rate, assembly service rate, and component service rate in state $S(k,i)$, respectively. Note that given the number k of failed components, the number i of operative assemblies assumes only the values $i=0,\ldots,r(k)$, where $r(k) = \min\{n,n+m-k\}$, $k=0,1,\ldots,n+m$. The matrices A_k, L_k, B_k, F_k, and G_k have the following form:

$$A_k = \begin{bmatrix} -(\lambda+k\mu_c) & \mu_\alpha & & & 0 \\ & -(\lambda+k\mu_c+\mu_\alpha) & 2\mu_\alpha & & \\ & & \ddots & & \\ & & -[\lambda+k\mu_c+(r(k)-2)\mu_\alpha] & [r(k)-1]\mu_\alpha & \\ 0 & & & -[\lambda+k\mu_c+(r(k)-1)\mu_\alpha] & \end{bmatrix}$$

$$L_k = \left[\begin{array}{c|c} 0 & 0 \\ \hline \Lambda_k & 0 \end{array}\right], \quad 0\le k\le m, \qquad L_k = [\Lambda_k|0], \quad m\le k<n+m-1 \tag{3}$$

$$\Lambda_k = \mathrm{diag}(\lambda), \quad \Lambda_k \in R^{(r(k)-1)\times(r(k)-1)},$$

$$B_k = M_k, \quad 0<k\le m, \quad B_k = \left[\frac{0}{M_k}\right], \quad m<k\le n+m-1, \quad M_k = \mathrm{diag}(k\mu_c), \quad M_k \in R^{r(k)\times r(k)}$$

$$G_k = \left[\frac{D_k}{0}\right], \quad D_k = \begin{bmatrix} 0 & \ldots & 0 \\ \vdots & & \vdots \\ 0 & \ldots & \lambda \\ \vdots & & \vdots \\ 0 & & 0 \end{bmatrix} \leftarrow (k+1), \quad 0\le k\le n+m-1, \quad D_k \in R^{(n+m)\times r(k)}$$

$$F_k = [0|V_k], \quad V_k = \begin{bmatrix} 0 & \ldots & 0 & \ldots & 0 \\ \vdots & & & & \vdots \\ 0 & \ldots & r(k)\mu_\alpha & \ldots & 0 \end{bmatrix}, \quad 0\le k\le n+m-1, \quad V_k \in R^{r(k)\times(n+m)}$$
$$\uparrow (k+1)$$

Denoting by p_{ki} the probability that the system is in state $S(k,i)$ at time $t=0$, where by assumption

$$\sum_{k=0}^{n+m-1} \sum_{i=1}^{r(k)} p_{ki} = 1, \quad p_{ki} = p(k,i,0) \tag{4}$$

the reliability measures of the system can be determined by solving (1) under the initial condition (4). One observes that E is a $q\times q$ block tridiagonal matrix ($q=n(m+1)+n(n-1)/2$ and that the column sums of the system state transition matrix $A = \begin{bmatrix} E & F \\ G & H \end{bmatrix}$ are equal to zero.

Here we assume unrestricted repair, i.e. $\lambda(k,i)=\lambda$, $\mu_\alpha(k,i)=[r(k)-i]\mu_\alpha$, $r(k)=\min\{n,n+m-k\}$, and $\mu_c(k,i) = k\mu_c$.

Let now $Q(t)$ be the probability that the system is in a failed state at time $t\ge 0$. Clearly

$$Q(t) = \sum_{k=0}^{n+m} p(k,0;t) \tag{5a}$$

Denoting by T the random variable representing the time interval from $t=0$ to system failure, the probability density function f(t) of T is given by

$$f(t) = \dot{Q}(t) = \sum_{k=0}^{n+m} \dot{p}(k,0;t) \tag{5b}$$

As is proved in [4] the mean value E(T) of T is given by the sum

$$E(T) = \sum_{k=0}^{n+m-1} \sum_{i=1}^{r(k)} z(k,i) \tag{5c}$$

where the q-dimensional vector z with components z(k,i) is the solution of the linear system $Ez = -p_w(0)$. Moreover, $MUT = u_q^T\pi_w/u_{n+m+1}^T G\pi_w$ and $MDT = u_{n+m+1}^T\pi_f/u_q^T F\pi_f$. In the above, π_w and π_f are the stationary values of $p_w(t)$ and $p_f(t)$, respectively.

In [4] a model was also developed for determining the optimal configuration of the system subject to some budget constraint. The model has the form max MTFF, subject to $k_\alpha n+k_c m\le B$, n,m integers, where n and m are the decision variables, k_α is the cost of each assembly, k_c is the cost of each component and B is the available budget. Given the values of n and m, the mean time to first failure is found from the relation

$$MTFF = \sum_{k=0}^{n+m-1} \sum_{i=1}^{r(k)} z(k,i)$$

where the vector z with components z(k,i) is the solution of the linear system $Ez = -(1,\dots,0)^T$, obtained using a numerical procedure which does not require a matrix to be stored.

The formulation of the optimal maintenance problem is as follows: Let $F\subset\Omega$ be the set of failed states (Ω is the system's finite state space), and $\tilde{F}\subset F$ be the set of reflecting failed states from which the system can be returned to a working state by a maintenance policy. Now if $P_\alpha(t)$ is the probability that the system is in state $S_\alpha e\Omega$, and $u_\alpha(t)$ is the maintenance (repair) rate vector $[\mu_\alpha,\mu_c]^T$, then the model (1) can be written in the compact form

$$\frac{dP(t)}{dt} = A(\tilde{u}_F,t)P(t), \quad P(0) = P_0 \tag{6}$$

where $P = \{P_\alpha,S_\alpha e\Omega\}$, $\tilde{u}_F = \{u_\alpha,S_\alpha e\tilde{F}\}$, and $\sum_{\Omega} P_\alpha(t) = 1, \quad t\ge 0$ (7)

The maintenance action $u_\alpha(t)$, which here plays the role of the control variable, is allowed to be fixed throughout the operation (control) interval [0,T] or variable depending on the failure rates of the individual components. The optimal maintenance control problem to be solved here is: For system (6) choose $u_\alpha(t)$ such that the following total cost is minimized

$$J = \frac{1}{T}\int_0^T \sum_{S_\alpha e\tilde{F}} [q_\alpha^T(t)P_\alpha(t) + r_\alpha^T(t)u_\alpha(t)]dt \tag{8}$$

subject to the constraints

$$0\le u_\alpha\le U_\alpha(t), \quad t\ge 0, \quad S_\alpha e\tilde{F} \tag{9}$$

Obviously, one can also treat integral constraints without special difficulty. The term involving $q_\alpha^T(t)P_\alpha(t)$ in (8) represents the average downtime cost, and the term involving $r_\alpha^T(t)u_\alpha(t)$ represents the average maintenance cost, over the desired operation period T. Care must be given in selecting the weighting cost functions $q_\alpha(t)$ and $r_\alpha(t)$.

Solution of the problem: The following two cases have been considered:

Case A: Maintenance action $u_\alpha(t)$, $t\epsilon[0,T]$ fixed.
Case B: Maintenance action $u_\alpha(t)$, $t\epsilon[0,T]$ variable.

To cover both the fixed and variable control cases we introduce the following generalized (integral) Hamiltonian functional (y is the costate vector):

$$\hat{H} = \frac{1}{T}\int_0^T H dt, \quad \text{where } H = y^T AP - \sum_{\alpha\epsilon \tilde{F}} (q_\alpha^T P_\alpha + r_\alpha^T u_\alpha) \tag{10}$$

which is to be maximized (since the minus sign is used) over all admissible maintenance control policies $u_{\tilde{F}}$ in $[0,T]$. Here H is the standard Hamiltonian functional.

The state and costate equations (in terms of H) are:

$$\frac{dP(t)}{dt} = \frac{\partial H}{\partial y(t)}, \quad P(0) = P_0 \tag{11a}$$

$$\frac{dy(t)}{dt} = -\frac{\partial H}{\partial P(t)} \quad \text{(Final state determined from transversality conditions)} \tag{11b}$$

In the variable maintenance case without integral constraints it is sufficient to maximize the standard Hamiltonian H, i.e. the optimal variable maintenance policy $u_{\alpha v}^0$ is given by

$$u_{\alpha v}^0 = \{u_\alpha : H \text{ is maximized}\} \tag{12a}$$

Correspondingly, the optimal fixed maintenance policy $u_{\alpha f}^0$ is given by

$$u_{\alpha f}^0 = \{u_\alpha : \hat{H} \text{ is maximized}\} \tag{12b}$$

Clearly, the integral Hamiltonian $\hat{H}$ is more general than H, but leads to weaker conditions. The maximum principle based on $\hat{H}$ is known as extended (or generalized) maximum principle [18]. The results have been applied to the following two cases:

(i) n=2 and m=1: System with two assemblies (one operating and one in failure-free standby) each one consisting of a single component.

(ii) n=2 and m=2: System with two assemblies each one consisting of m=2 independent components.

Now the results of [5] will be reviewed which concern a repairable standby system whose repair facilities are subject to breakdowns.

The repairable system consists of one operating unit and (n-1) units in standby ready for action. The time to failure of the operating unit is assumed to be exponentially distributed with mean $1/\lambda$, while the units which are in standby state are not subject to failure. The repair of failed units is done by n identical repair facilities, each one of which is able to repair one unit at a time. The repair times of successive failed units are independent identically distributed random variables with p.d.f. B(x) having finite mean b. Each repair facility can have random breakdowns occuring only when the facility is busy, for which the probability of breakdown at the next small time interval δt is of the order of $v\delta t+o(\delta t)$ and independent of previous breakdown occurrences. It turns out that breakdowns occuring at a facility possess a Poisson distribution (with rate v). During the repair period of a facility, this facility remains inactive and the unit whose repair is interrupted waits until the completion of the facility repair, so that no loss of service occurs. The repair times of successive breakdowns are independent identically distributed with p.d.f. C(x) having a finite mean c. According to Kendall's notation the above system is of the type M/G/n, where M indicates that failure and breakdown process is Poissonian, G indicates that the repair time has an arbitrary (general) distribution, and n stands for the number of servicing (repair) facilities.

The M/G/n and particularly the M/M/n repairable standby system without breakdowns of the repair facilities has been thoroughly studied in previous works [19,20]. For any $t(t\geq 0)$ define:

K(t): the number of busy facilities at time t;

M(t): the number of facilities under repair at time t;

$\underline{U}(t) = (U_1(t),\dots,U_{K(t)}(t))$: a random permutation of the past durations of the repairs of units which are in progress at time t;

$\underline{V}(t) = (V_1(t),\dots,V_{M(t)}(t))$: a random permutation of the past durations of the interrupted repairs of units at time t;

$\underline{W}(t) = (W_1(t),\dots,W_{M(t)}(t))$: a random permutation of the past durations of the repairs of the broken facilities at time t.

It is clear the process $\{(K(t),M(t),\underline{U}(t),\underline{V}(t),\underline{W}(t));\ t\geq 0\}$ is Markovian. Since all variables governing the system are finite, it follows that the system periodically empties and, therefore, statistical equilibrium is eventually established.

Let L denote the random variable describing the number of occupied (busy or under repair) facilities, when the system is in equilibrium state. The probability density function (P_ℓ), $\ell=0,1,2,\dots,n$ of L is identical with the equilibrium probability density of the number of busy servers in an M/G/n blocking system with arrival rate λ and mean service time $b(1+\nu c)$. Hence

$$P_\ell = P(L=\ell) = \frac{\gamma^\ell}{\ell!} \Big/ \sum_{i=0}^{n} \frac{\gamma^i}{i!} \quad (\ell=0,1,2,\dots,n) \tag{13a}$$

where γ is the generalized failure intensity, i.e. $\gamma = \lambda b(1+\nu c)$.

The joint probability density function (P_{km}), $k,m = 0,1,\dots,n;\ k+m\leq n$ of K and M is found to be [5]:

$$P_{km} = \frac{1}{k!m!}(\lambda b)^{k+m}(\nu c)^m \Big/ \sum_{i=0}^{n} \frac{\gamma^i}{i!} \qquad k,m = 0,1,2,\dots,n;\ k+m\leq n \tag{13b}$$

and the various reliability measures of the system are given by

$$A = \sum_{i=0}^{n-1} C_i^{n-1}\left(\frac{i!}{\gamma!}\right) \Big/ \left[\sum_{i=0}^{n-1} C_i^{n-1}\left(\frac{i!}{\gamma!}\right) + \frac{\gamma}{n}\right] \quad \text{(Availability)}$$

$$MDT = b(1+\nu c)/n \qquad \text{(Mean Down Time)}$$

$$MUT = \frac{1}{\lambda}\sum_{i=0}^{n-1} C_i^{n-1}\left(\frac{i!}{\gamma!}\right) \qquad \text{(Mean Up Time)}$$

where $C_k^m = m!/k!(m-k)!$

In [5] two models were also developed for determining the optimal configuration of the system subject to a budget constraint B. These are the following:

(i) $\max A(\lambda,\nu,n)$ subject to $n\{U(\lambda)+F(\nu)\} = B$ and $n,\lambda,\nu > 0$, n integer

(ii) For given n the model (1) becomes $\max A(\lambda,\nu)$ subject to $U(\lambda)+F(\nu) = \beta$, $\beta = \frac{B}{n}$ where $A(\lambda,\nu)$ is a concave function of λ,ν, and $U(\lambda)$, $F(\nu)$ are the purchase costs of each unit and repair facility as functions of λ and ν respectively.

The optimal solution of model (ii) for $U(\lambda)$ and $F(\nu)$ of the type $U(\lambda) = k_1/\lambda$, $F(\nu) = k_2/\nu$ was found to be

$$\lambda^o = m\left(1 - \frac{1}{1+\sqrt{1+m/ck}}\right), \quad \nu^o = \frac{k}{m}\left(1 + \sqrt{1+\frac{m}{ck}}\right)$$

where $m = \beta/k_1 = B/(nk_1)$ and $k = k_2/k_1$.

The optimal configuration for model (i) can be determined by computing λ^o, ν^o and $A(\lambda^o,\nu^o)$ of the model (ii) for $n = 2,3,\ldots$ until the system availability $A(\lambda,\nu,n)$ takes its maximum.

For a system with $B = 100$, $k_1 = 5$, $k_2 = 2$, $b=4$ and $c=2$ the maximum availability is achieved when $n^o=7$, $\lambda^o=0.514$ and $\nu^o=0.439$.

3. FAULT DETECTION AND FAULT-TOLERANT SYSTEMS

3.1 Dynamic Fault Detection

A comprehensive survey of the basic techniques developed over the recent years for the optimum supervision and detection/location of failures in dynamic technological systems was provided in [7]. Regarding the optimum supervision problem the following policies were considered: optimum selection of supervision frequency, optimum supervision policies, optimum selection of sequential tests and finally the effect of supervision capability upon the system availability. Regarding the optimum detection/location/identification of failures the following techniques have been investigated: majority technique, failure sensitive filter technique, generalized maximum likelihood ratio technique, output estimation error technique, dynamic redundancy technique with Luenberger observer, and finally jump processes technique (see also [9]).

Here we shall start with a brief presentation of the analytic redundancy concept through Luenberger observer.

Consider the closed loop system of Fig. 1 which contains the system under control, a duplex sensor device (AΔ), the feedback controller R, two Luenberger observers and the appropriate logic circuitry for the detection of the failures of the sensing devices. Each observer is fed with output measurements from the corresponding sensing device ($A\Delta_1$ or $A\Delta_2$). The two sensing devices are identical and so they contain exactly similar elements for the same variables. When no failure occurs the two measurement vector signals are exactly the same. But when there exist failures (represented by the disturbances V_1 and V_2) the values of Y_1 and Y_2 are different. The feedback controller (matrix) R is chosen such that the closed-loop system has desired dynamic performance and desired disturbance rejection policies. Each observer produces independently of the other, an estimate of the state vector ($\hat{x}_1$ or $\hat{x}_2$).

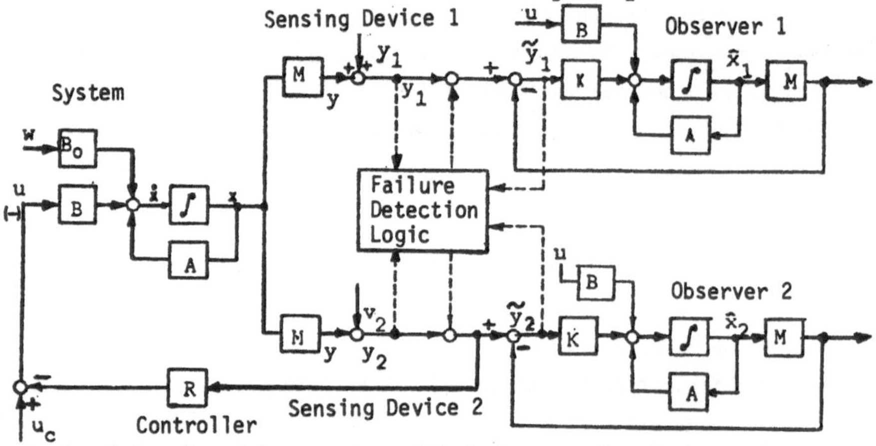

Fig. 1 Complete closed-loop system with duplex sensing device and two observers for analytic redundancy (Adapted from [21]).

Prior to the detection of a failure which is done by the "failure detection logic" the loop is closed exclusively from the output of the sensing device $A\Delta_2$. For the detection of failures of the devices $A\Delta_1$ and $A\Delta_2$ the residuals $\tilde{y}_i = y_i - M\hat{x}_i$ $(i=1,2)$ are fed into the "failure detection logic" as shown in Fig. 1. We remark, that here the observer outputs are used only for the failure detection and not for control purpose.

The system equations are:

Closed-loop equation:

$$\dot{x} = (A-BRM)x + B_0 w - BRv_2 + Bu_c, \quad y_1 = Mx+v_1, \; y_2 = Mx+v_2$$

Observer equations

$$\dot{\tilde{x}}_1 = (A-KM)\tilde{x}_1 + B_0 w - Kv_1, \; \tilde{y}_1 = M\tilde{x}_1 + v_1 \quad \text{(Observer 1)}$$

$$\dot{\tilde{x}}_2 = (A-KM)\tilde{x}_2 + B_0 w - Kv_2, \; \tilde{y}_2 = M\tilde{x}_2 + v_2 \quad \text{(Observer 2)}$$

We see that $\tilde{y}_1$, the output error of observer 1, depends only upon the failures of $A\Delta_1$, and similarly $\tilde{y}_2$ depends only upon the failures of $A\Delta_2$. Hence, assuming that the effect of the disturbances upon $\tilde{y}_1$ and $\tilde{y}_2$ is finite, we can define the thresholds on $\tilde{y}_1$ and $\tilde{y}_2$ as the maxima of their response to the disturbances. On the basis of the above the "failure detection logic" arrangement can be as shown in Fig. 2.

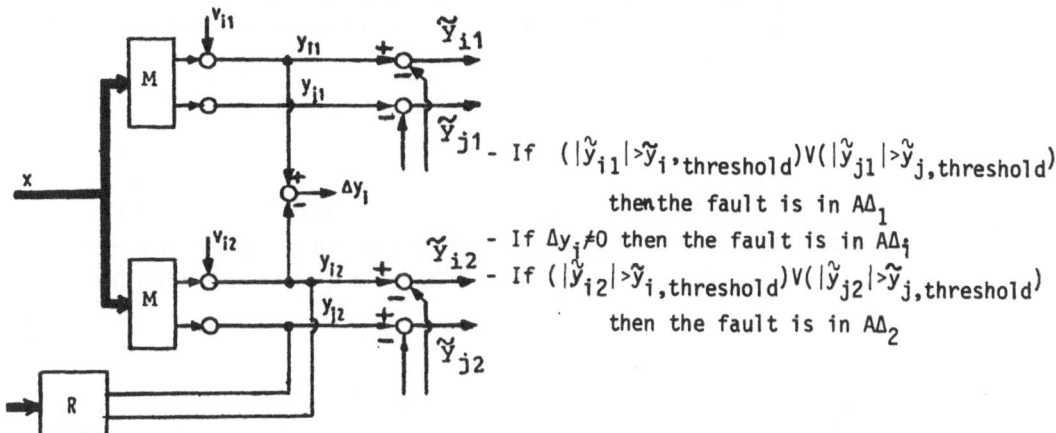

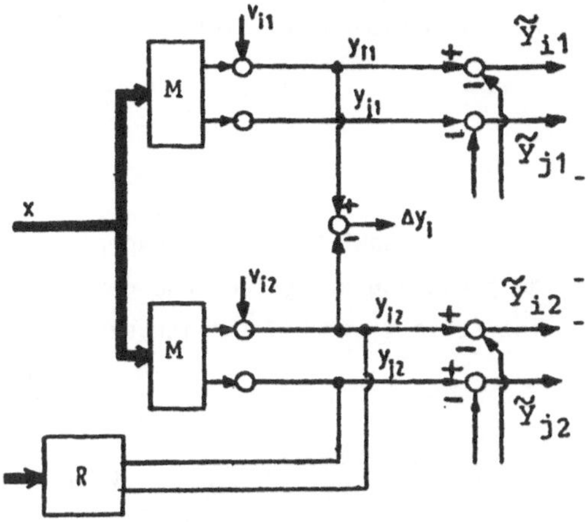

- If $(|\tilde{y}_{i1}|>\tilde{y}_{i,threshold}) \vee (|\tilde{y}_{j1}|>\tilde{y}_{j,threshold})$ then the fault is in $A\Delta_1$
- If $\Delta y_i \neq 0$ then the fault is in $A\Delta_i$
- If $(|\tilde{y}_{i2}|>\tilde{y}_{i,threshold}) \vee (|\tilde{y}_{j2}|>\tilde{y}_{j,threshold})$ then the fault is in $A\Delta_2$

Fig. 2 Failure detection logic of sensing devices (Adapted from [21]).

Clearly, there exists a failure when the difference $\Delta y_i = y_{i1}-y_{i2}$ is not zero. The failure occurs in the sensing device $A\Delta_1$ when at least one signal $\tilde{y}_{i1}$ (or $\tilde{y}_{j1}$) exceeds the threshold. The failure occurs in $A\Delta_2$ when $\tilde{y}_{i2}$ (or $\tilde{y}_{j2}$) exceeds the threshold.

When a failure is detected, the corresponding sensing element is replaced by the same sensing element of the other device. This principle has been applied for sensor failure detection in flight control systems [21].

Of particular importance is the multiple-model filter technique which has been used by Athans and his co-workers in the framework of the adaptive control of the aircraft F-8C (DFBW) [22] and by Willsky for detection and classification of cardiac arrythmia [23].

Next we describe the approach developed in [8] for sensor failure detection in large scale linear interconnected dynamic systems. The procedure is based on the use of the overlapping decomposition technique developed by Siljak and Ikeda [10], and employs decentralized observers for overlapping subsystems, which allow one to isolate failed sensors by comparing the discrepancies between estimates of the same state via different Luenberger observers (as in Fig. 1).

To start with, the concept of overlapping decomposition is first introduced. Consider a linear large-scale interconnected system S of the type

$$\dot{x} = Ax + Bu, \quad y = Cx \tag{14}$$

where $x \in R^n$ is the state vector, $u \in R^m$ is the control vector, $y \in R^q$ is the output measurement, A,B,C are the system matrices, with B and C being block diagonal matrices having N blocks, corresponding to the N subsystems.

The state and output equations of the ith subsystem, S_i, can be written as

$$\dot{x}_i = A_i x_i + B_i u_i + \sum_{\substack{j=1 \\ i\neq j}}^{N} A_{ij} x_j, \quad y_i = C_i x_i \tag{15}$$

where $x_i \in R^{n_i}$, $u_i \in R^{m_i}$, $y_i \in R^{q_i}$ are the state, control and output measurements of the ith subsystem, respectively, with $n=(n_1+\ldots+n_N)$, $m=(m_1+\ldots+m_N)$, $q=(q_1+\ldots+q_N)$. Here A_{ij} are the off-diagonal parts of the matrix A. By dividing the state vector of the ith subsystem, x_i, into two parts, i.e.

$$x_i = \begin{bmatrix} x_{i1} \\ \hline x_{i2} \end{bmatrix}$$

where $x_{i1} \in R^{n_{i1}}$, $x_{i2} \in R^{n_{i2}}$ and $n_i = n_{i1} + n_{i2}$, and expanding each subsystem, except the first one, to include the last state variable of the preceding subsystem, one gets an overlapping decomposition of the overall system. The state vector of the ith expanded subsystem is given by

$$\tilde{x}_i = \begin{bmatrix} x_{i-1,2} \\ \hline x_{i,1} \\ \hline x_{i,2} \end{bmatrix} \quad i=2,3,\ldots,N \tag{16}$$

As a result, the overall state vector of the new augmented or overlapped system $\tilde{S}$ is given by $\tilde{x}^T = [\tilde{x}_1^T \tilde{x}_2^T \ldots \tilde{x}_n^T]$ where $\tilde{x} \in R^{\tilde{n}}$, $\tilde{n} = n + \sum_{i=2}^{N} n_{i2}$. The vector $\tilde{x}$ is related to x as $\tilde{x} = Tx$, where T is the following $\tilde{n} \times n$ transformation matrix:

$$T = \begin{bmatrix} I_{11} & & & & & & \\ & I_{12} & & & & 0 & \\ & I_{12} & & & & & \\ & & I_{21} & & & & \\ & & & I_{22} & & & \\ & & & I_{22} & & & \\ & 0 & & & I_{N-1,2} & & \\ & & & & I_{N-1,2} & & \\ & & & & & I_{N,1} \end{bmatrix}$$

with

$$I_{i1} = \begin{cases} (n_i-n_{i,2}) \times (n_i-n_{i,2}) \\ \text{identity matrix for } i=1 \text{ and } N \\ (n_i-n_{i-1,2}-n_{i,2}) \times (n_i-n_{i-1,2}-n_{i,2}) \\ \text{identity matrix for } i=2,\ldots,N-1 \end{cases}$$

$I_{i2} = (n_{i,2}) \times (n_{i,2})$ identity matrix, $i=1,2,\ldots,N-1$.

The state and output equations of the expanded overall system $\tilde{S}$ are:

$$\dot{\tilde{x}} = \tilde{A}\tilde{x} + \tilde{B}u, \quad y = \tilde{C}\tilde{x} \tag{17}$$

where $\tilde{A} = TAT^+ + M$, $\tilde{B} = TB+N$, $\tilde{C} = CT^+ + L$, M,N,L are complementary matrices of appropriate dimension [10], T^+ is the generalised inverse of T.

The first step for detecting malfunctioning sensors is to design decentralized state observers (estimators) for the system (17), which consists by N interconnected subsystems of the type

$$\dot{\tilde{x}}_i = \tilde{A}_i\tilde{x}_i + \tilde{B}u_i + \sum_{\substack{j=1\\ i\neq j}}^{N} \tilde{A}_{ij}\tilde{x}_j, \quad y_i = \tilde{C}_i\tilde{x}_i \tag{18}$$

where $\tilde{x}_i \in R^{n_i}$, $u_i \in R^{m_i}$, $y_i \in R^{q_i}$ are the state, control and output vectors of the ith augmented subsystem. In order to estimate the state $\tilde{x}_i(t)$, one constructs subsystem observers of the form:

$$\dot{\hat{x}}_i = F_i\hat{x}_i + G_iy_i + \sum_{\substack{j=1\\ i\neq j}}^{N} F_{ij}\hat{x}_j + \tilde{B}_iu_i, \quad i=1,\dots,N \tag{19}$$

where the matrices F_i, F_{ij} are given by [8]

$$F_i = \tilde{A}_i - G_i\tilde{C}_i, \quad F_{ij} = \tilde{A}_{ij}, \; i=1,2,\dots,N \tag{20}$$

and G_i is chosen such that to minimize the functional

$$J = \sum_{i=1}^{N} \frac{1}{2}\int_0^\infty (||\bar{e}_i||^2_{\bar{Q}_i} + ||\bar{u}_i||^2_{\bar{R}_i})$$

under the constraint

$$\dot{\bar{e}} = \tilde{A}_i^T\bar{e}_i + \tilde{C}_i^T\bar{u}_i \tag{21}$$

where $\bar{Q}_i$ is an $\tilde{n}_i \times \tilde{n}_i$ symmetric positive semidefinite matrix and $\bar{R}_i$ is a $q_i \times q_i$ positive definite matrix. If the pair $(\tilde{A}_i^T, \tilde{C}_i^T)$ is completely controllable, one can choose $\bar{G}_i^T = \bar{R}_i^{-1}\tilde{C}_i\bar{P}_i$, in which case

$$\bar{u}_i = -\bar{G}_i^T\bar{e}_i \tag{22}$$

is the unique optimal control law, where $\bar{P}_i$ is an $\tilde{n}_i \times \tilde{n}_i$ symmetric positive definite solution of the matrix Riccati equation. Moreover, if the pair $(\tilde{A}_i^T, L_i)$ is completely observable, where $\bar{Q}_i = L_iL_i^T$, the closed-loop decentralized system given by $\dot{\bar{e}}_i = (\tilde{A}_i^T - \tilde{C}_i^T\bar{G}_i^T)\bar{e}_i$ is globally asymptotically stable.

In [8] it is also shown under what conditions the decentralized observer (19) leads to an error system, the equilibrium state e=0 ($e_i = \tilde{x}_i - \hat{x}_i$) of which is exponentially connectively stable in the large, according to [24], and thus $\hat{x} \to \tilde{x}$ as $t \to \infty$

From the definition of the overlapping decomposition according to (16) it follows that the state vectors $\tilde{x}_{i-1}$ and $\tilde{x}_i$ share the same state subvector $x_{i-1,2}$, namely

$$\tilde{x}_{i-1} = \begin{bmatrix} x_{i-2,2} \\ x_{i-1,1} \\ x_{i-1,2} \end{bmatrix}, \quad \tilde{x}_i = \begin{bmatrix} x_{i-1,2} \\ x_{i,1} \\ x_{i,2} \end{bmatrix}$$

This fact can be used for failure detection and location. Indeed if $L_1, L_2, \dots, L_N$ denote the N local observers, and $(\hat{x}_{i-1,2})_{L_{i-1}}$, $(\hat{x}_{i-1,2})_{L_i}$ denote the common substate $\hat{x}_{i-1,2}$ observed by the local observers L_{i-1} and L_i respectively, then during the normal operation of the overall system one has

$$(\hat{x}_{i-1,2})_{L_{i-1}} = (\hat{x}_{i-1,2})_{L_i} \tag{23}$$

whereas, when the ith subsystem sensor has a failure one has

$$(\hat{x}_{i-1,2})_{L_{i-1}} \neq (\hat{x}_{i-1,2})_{L_i} \quad \text{and} \quad (\hat{x}_{i,2})_{L_i} \neq (\hat{x}_{i,2})_{L_{i+1}} .$$

Thus, the validity or not of the relation (23) can be used not only to detect the sensor failure amongst the N subsystems, but also locate which one has failed.

The overlapping decomposition concept can also be used to design a set of overlapping state feedback controllers with a certain degree of redundancy, which results in reliability improvement.

In [8] a 6th-order system representing a ladder network has been worked out by the above technique. The matrices of this system are:

$$A = \left[\begin{array}{cc|cc|cc} -1.5 & 1.2 & 0 & 0 & 0 & 0 \\ 0.8 & -2.4 & 1.6 & 0 & 0 & 0 \\ \hline 0 & 1.6 & -2.4 & 0.8 & 0 & 0 \\ 0 & 0 & 0.6 & -1.2 & 0.6 & 0 \\ \hline 0 & 0 & 0 & 0.8 & -2.4 & 1.6 \\ 0 & 0 & 0 & 0 & 4.8 & -4.8 \end{array}\right] = \begin{bmatrix} A_1 & A_{12} & 0 \\ A_{21} & A_2 & A_{23} \\ 0 & A_{32} & A_3 \end{bmatrix}$$

$$C = \left[\begin{array}{cc|cc|cc} 0 & 1 & 0 & 0 & 0 & 0 \\ \hline 0 & 0 & 0 & 1 & 0 & 0 \\ \hline 0 & 0 & 0 & 0 & 0 & 1 \end{array}\right] = \begin{bmatrix} c_1' & 0 & 0 \\ 0 & c_2' & 0 \\ 0 & 0 & c_3' \end{bmatrix} \qquad B = \left[\begin{array}{c|c|c} 1 & 0 & 0 \\ 0 & 0 & 0 \\ \hline 0 & 1 & 0 \\ 0 & 0 & 0 \\ \hline 0 & 0 & 1 \\ 0 & 0 & 0 \end{array}\right] = \begin{bmatrix} b_1 & 0 & 0 \\ 0 & b_2 & 0 \\ 0 & 0 & b_3 \end{bmatrix}$$

The optimal control problem was to minimize an integral quadratic functional with Q a 6×6 identity matrix and R a 3×3 identity matrix. The increased system after performing the overlapping has the following matrices {see (17)}:

$$\tilde{A} = \left[\begin{array}{cc|ccc|ccc} -1.5 & 1.2 & 0 & 0 & 0 & 0 & 0 & 0 \\ 0.8 & -2.4 & 0 & 0 & 0 & 0 & 0 & 0 \\ \hline 0.8 & 0 & -2.4 & 1.6 & 0 & 0 & 0 & 0 \\ 0 & 0 & 1.6 & -2.4 & 0.8 & 0 & 0 & 0 \\ 0 & 0 & 0 & 0.6 & -1.2 & 0 & 0 & 0 \\ \hline 0 & 0 & 0 & 0.6 & 0 & -1.2 & 0.6 & 0 \\ 0 & 0 & 0 & 0 & 0 & 0.8 & -2.4 & 1.6 \\ 0 & 0 & 0 & 0 & 0 & 0 & 4.8 & -4.8 \end{array}\right] = \begin{bmatrix} \tilde{A}_1 & \tilde{A}_{12} & 0 \\ \tilde{A}_{21} & \tilde{A}_2 & \tilde{A}_{23} \\ 0 & \tilde{A}_{32} & \tilde{A}_3 \end{bmatrix}$$

$$\tilde{B} = \left[\begin{array}{c|c|c} 1 & 0 & 0 \\ 0 & 0 & 0 \\ \hline 0 & 0 & 0 \\ 0 & 1 & 0 \\ 0 & 0 & 0 \\ \hline 0 & 0 & 0 \\ 0 & 0 & 1 \\ 0 & 0 & 0 \end{array}\right] = \begin{bmatrix} \tilde{B}_1 & 0 & 0 \\ 0 & \tilde{B}_2 & 0 \\ 0 & 0 & \tilde{B}_3 \end{bmatrix}, \quad \tilde{x} = \left[\begin{array}{c} x_1 \\ x_2 \\ \hline x_2 \\ x_3 \\ x_4 \\ \hline x_4 \\ x_5 \\ x_6 \end{array}\right] = \left[\begin{array}{c} \tilde{x}_1 \\ \tilde{x}_2 \\ \hline \tilde{x}_3 \\ \tilde{x}_4 \\ \tilde{x}_5 \\ \hline \tilde{x}_6 \\ \tilde{x}_7 \\ \tilde{x}_8 \end{array}\right] = \begin{bmatrix} \tilde{x}_1 \\ \tilde{x}_2 \\ \tilde{x}_3 \end{bmatrix}$$

$$C = \left[\begin{array}{cc|ccc|ccc} 0 & 1 & 0 & 0 & 0 & 0 & 0 & 0 \\ \hline 0 & 0 & 0 & 0 & 1 & 0 & 0 & 0 \\ \hline 0 & 0 & 0 & 0 & 0 & 0 & 0 & 1 \end{array}\right] = \left[\begin{array}{c|c|c} \tilde{c}_1' & 0 & 0 \\ \hline 0 & c_2' & 0 \\ \hline 0 & 0 & \tilde{c}_3' \end{array}\right]$$

This system is separated into three obvious interconnected subsystems with state vectors $\underline{\tilde{x}}_1$, $\underline{\tilde{x}}_2$ and $\underline{\tilde{x}}_3$. Then three local observers are constructed with state vectors $\underline{\hat{x}}_1$, $\underline{\hat{x}}_2$, $\underline{\hat{x}}_3$ reproducing $\underline{\tilde{x}}_1$, $\underline{\tilde{x}}_2$ and $\underline{\tilde{x}}_3$, i.e.

Observer 1:

$$\dot{\underline{\hat{x}}}_1 = F_1\underline{\hat{x}}_1 + g_1 y_1 + F_{12}\underline{\hat{x}}_2 + b_1 u_1$$
$$F_1 = \tilde{A}_1 - g_1\tilde{c}_1', \; F_{12} = \tilde{A}_{12}, \; b_1 = \tilde{b}_1$$
$$g_1^T = [g_{11}, g_{12}]$$

Observer 2:

$$\dot{\underline{\hat{x}}}_2 = F_2\hat{x}_2 + g_2 y_2 + F_{21}\hat{x}_1 + F_{23}\hat{x}_3 + b_2 u_2$$
$$F_2 = \tilde{A}_2 - g_2\tilde{c}_2', F_{21} = \tilde{A}_{21}, \; F_{23} = \tilde{A}_{23}, \; b_2 = \tilde{b}_2$$
$$g_2^T = [g_{21}, g_{22}, g_{23}]$$

Observer 3:

$$\dot{\underline{\hat{x}}}_3 = F_3\underline{\hat{x}}_3 + g_3 y_3 + F_{32}\underline{\hat{x}}_2 + b_3 u_3$$
$$F_3 = \tilde{A}_3 - g_3\tilde{c}_3', \; F_{32} = \tilde{A}_{32}, \; b_3 = \tilde{b}_3$$
$$g_3^T = [g_{31}, g_{32}, g_{33}]$$

where the gain vectors g_1, g_2 and g_3 are chosen so as to guarantee the asymptotic stability of the observer with suitable α, $\bar{Q}_i$, $\bar{R}_i$, i=1,2,3 {see [8]}.
Since here the outputs $\underline{\hat{x}}_1$, $\underline{\hat{x}}_2$, $\underline{\hat{x}}_2$ of the local observers L_1, L_2 and L_3 are

$$\underline{\hat{x}}_1 = \begin{bmatrix}\hat{x}_1\\ \hat{x}_2\end{bmatrix}_{L_1}, \quad \underline{\hat{x}}_2 = \begin{bmatrix}\hat{x}_2\\ \hat{x}_3\\ \hat{x}_4\end{bmatrix}_{L_2}, \quad \underline{\hat{x}}_3 = \begin{bmatrix}\hat{x}_4\\ \hat{x}_5\\ \hat{x}_6\end{bmatrix}_{L_3}$$

we have the following

Normal Operation

$$|(\hat{x}_2)_{L_1} - (\hat{x}_2)_{L_2}| < \varepsilon_1$$
$$|(\hat{x}_4)_{L_2} - (\hat{x}_4)_{L_3}| < \varepsilon_2$$

Sensor of Subsystem 1 Failed

$$|(\hat{x}_2)_{L_1} - (\hat{x}_2)_{L_2}| > \varepsilon_1$$
$$|(\hat{x}_4)_{L_2} - (\hat{x}_4)_{L_3}| < \varepsilon_1$$

Sensor of Subsystem 3 Failed

$$|(\hat{x}_2)_{L_1} - (\hat{x}_2)_{L_2}| < \varepsilon_1$$
$$|(\hat{x}_4)_{L_2} - (\hat{x}_4)_{L_3}| > \varepsilon_2$$

Sensor of Subsystem 2 Failed

$$|(\hat{x}_2)_{L_1} - (\hat{x}_2)_{L_2}| > \varepsilon_1$$
$$|(\hat{x}_4)_{L_2} - (\hat{x}_4)_{L_3}| > \varepsilon_2$$

where the positive parameters ε_1 and ε_2 are selected by the designer.

The following figure 3 shows the observer trajectories.

Clearly if the situation of Fig. 3a occurs without any knowledge of the malfunctioning sensor,one can conclude that sensor of subsystem 1 has a failure. Similarly in the cases of Fig. 3b and Fig. 3c one can conclude that sensors of subsystems 2 and 3, respectively, are failed.

We close this section with a short account of the work on system supervision and failure detection described in [11-14]. These results have been derived for gas transportation and distribution systems, but they can also be used to other practical distributed-parameter systems. Before going into the supervision and failure detection aspects we give a review of the *control, simulation* and *state estimation* results.

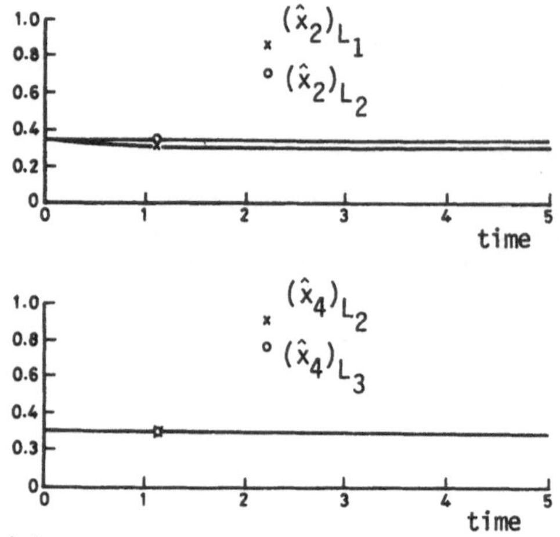

(a) C_{12} changed from 1 to 0.2 (sensor of subsystem 1 failed)

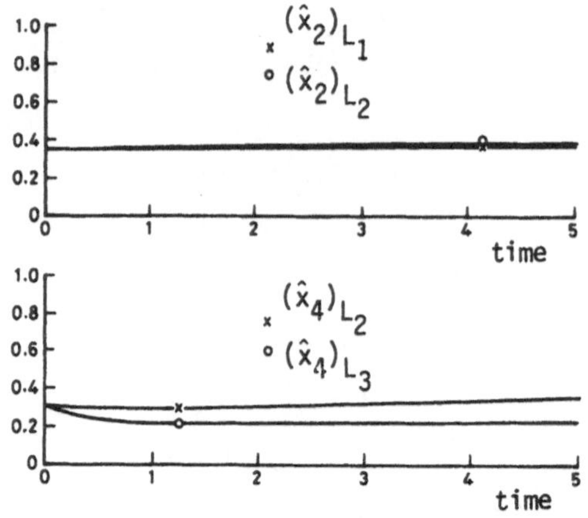

(b) C_{25} changed from 1 to 0.2 (sensor of subsystem 2 failed)

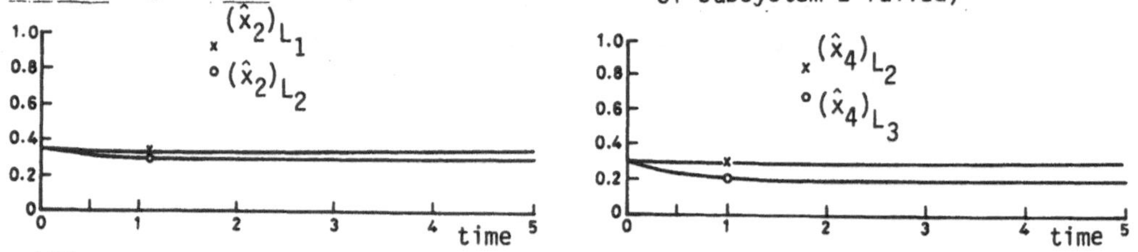

(c) C_{38} changed from 1 to 0.2 (sensor of subsystem 3 failed)

Fig. 3 Outputs of local observers under sensor failure conditions [8].

Gas transportation over hundreds or thousands of kilometers is done by one or more parallel pipelines with several compressor stations along the pipelines. Gas flow is unidirectional, loops do hardly occur, and there are only a few branches to other pipelines. Slightly transient flow conditions - at least during some time intervals - are due to time varying demand flows, e.g. ±5%...±20% of the daily meanvalue, small changes of the supply, and switching of compressor units. Control actions must be performed such that all pressure values will be bounded to given limits.

Regional gas distribution is done by complex pipeline networks with compressor stations, flow and pressure regulators, valves, and underground storages. There are only a few supply points, but a large number of offtakes. Gasflow may be bidirectional. A fail safe and economic control must keep all pressures within the given limits, i.e. maximal nominal pipeleg value(s) and contracted minimal values at the offtakes, and it should keep the supplies sufficiently constant in spite of all variations of the demand flows ("peak-shaving").

Due to the low values of gas velocity the dynamics of the transportation and distribution process are characterized by time constants of an order of many minutes up to several hours. The control scheme used is of the predictive type which needs reliable data on all actual network pressures and flows. The three basic elements of this type of control are predictive simulation, state estimation, and predictive control strategy optimization. The functional relationships of these three elements are obvious. At discrete locations of the network some pressures and flows are usually measured and telemetered to the control center in short time intervals, e.g. 5 min. Using these available data the actual state of the total network is first reconstructed and then permanently tracked by a state estimator. The estimated actual state can now be used as an initial state for a predictive simulation run. Thus, future flows and pressures (e.g. for a period of 4 up to 48 hrs) can be calculated taking into account forecasts of the offtakes and intended control actions. Such a predictive simulation run can be started automatically at fixed time intervals (e.g. 15 or 30 min) or manually on request by the dispatcher. The calculated future flows and pressures can be checked with respect to a fail safe and economic operation. If the results are not satisfactory, a new predictive simulation is started with the same initial conditions but with a different strategy. Thus, by an iterative scheme, a reasonable control strategy can be worked out, considering actual and future transient flow conditions as well as operational requirements. This search can be done manually by the dispatcher or automatically by some type of optimization procedure. The state estimator is also used for supervision and failure detection purposes. Unusual or unexpected large differences between measured and calculated data may be examined with respect to possible measurement errors or system damages, e.g. leakages.

Network simulation: The simulation of the gas network transient behaviour was carried out by a special program called GANESI [25]. For the simulation, the network under consideration is partitioned into subsystems of similar dynamics, such as pipelegs, regulators for pressure and flow, compressors (pressure or flow controlled), and valves. The dynamic behaviour of the total network model is primarily based on the mathematical model of gas flow through a long pipeleg, see Fig. 4. Under normal operational conditions one can assume a one-dimensional isothermal and compressible flow. It is modelled by a hyperbolic-type set of two quasi-linear partial differential equations for the state variables pressure $p(z,t)$ and mass flow $q(z,t)$, i.e.

$$\frac{\partial \underline{x}(z,t)}{\partial t} = \underline{f}_z(\underline{x}(z,t)), \quad z\epsilon(0,1), \; t\epsilon(t_0,\infty), \quad \underline{x}(z,t_0) = x_0(z) \tag{24}$$

where

$$\underline{x}(z,t) = \begin{bmatrix} p(z,t) \\ q(z,t) \end{bmatrix}, \quad \underline{f}_z(x,t) = \begin{bmatrix} a_1(p)\cdot \dfrac{\partial q(z,t)}{\partial z} \\ a_2(p)\cdot \dfrac{q\cdot|q|}{p} + a_3(p)\cdot p + a_4\cdot \dfrac{\partial p(z,t)}{\partial z} \end{bmatrix}$$

The boundary conditions at x=0,1 may be pressure defining functions $p_{0,1}(t)$, or flow defining functions $q_{0,1}(1)$, or coupling conditions for $p_{0,1}, q_{0,1}$.

In a transmission network single pipelegs are interconnected together with compressors, regulators and valves. The connection points are known as nodes. The coupling conditions at these nodes are derived from a mass and force balance. Hence, the

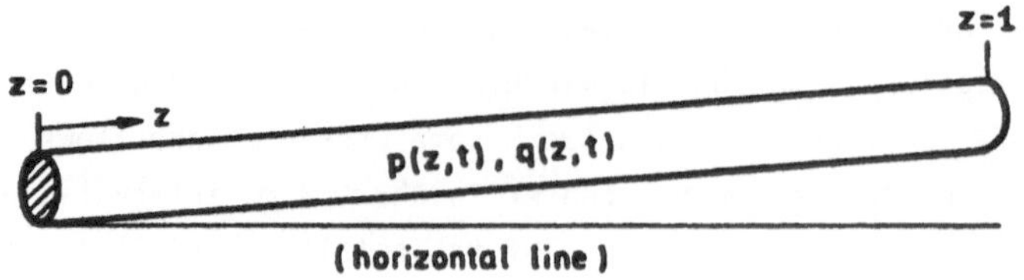

Fig. 4 Distributed-Parameter subsystem of gas network (pipeleg) [14].

total network model consists of a number of coupled subsystem models each describing one network element.

Numerical solution of the entire set of model equations was achieved by a space and time discretization using a modified Crank-Nicholson difference scheme. The resulting system of implicit, nonlinear equations is solved by a Newton-Raphson iteration method in connection with sparse matrix techniques.

The accuracy of the described network model and solution algorithm, realized in the simulation program GANESI, were successfully checked in several validation experiments. The necessary computation times are much faster than realtime even on standard process computers [25].

State Estimation: From an economic point of view it is advantageous to estimate the state using only a few measurement devices at fixed discrete locations of the network. This task can be solved by application of filter techniques, e.g. a *Kalman-type filter*, or by the usage of state reconstruction methods, e.g. a Luenberger-type observer, or by running a special sort of on-line simulation, which was called *Parallel Simulation*. From their experience the authors recommend to use the observer approach [14]. However, for simple networks with accurate measurements the Parallel Simulation method can also produce satisfactory estimates. As a result of a comparison test with field data the application of an extended Kalman-type filter proved not to be superior to the observer. On the contrary, at least up to now, the computational load involved seems to be uneconomically high.

Parallel simulation: Consider a gas transportation process as shown in Fig. 5a. Available measurements are the supply flow $q_1(t)$, the offtake flow $q_2(t)$, and the pressure $p_m(t)$ at an arbitrary location z_m along the pipeleg. Figure 5b shows a scheme for modeling the real process. This model (called model 1), is well suited for simulation purposes. The known variables supply $q_1(t)$ and offtake $q_2(t)$ are used as boundary conditions. The additionally available measurement $p_m(t)$ is *not* included into the simulation. (Such redundant measurements can be used for checking the simulation results.) From a systems-theoretic viewpoint, this model 1 shows an integrating behaviour which is due to the flow defining boundary conditions. Obviously, this model 1 cannot be used for state estimation purposes.

An alternative scheme for modelling the real process is shown in Fig. 5c. This model 2, which is actually used for the Parallel Simulation, is derived by dividing

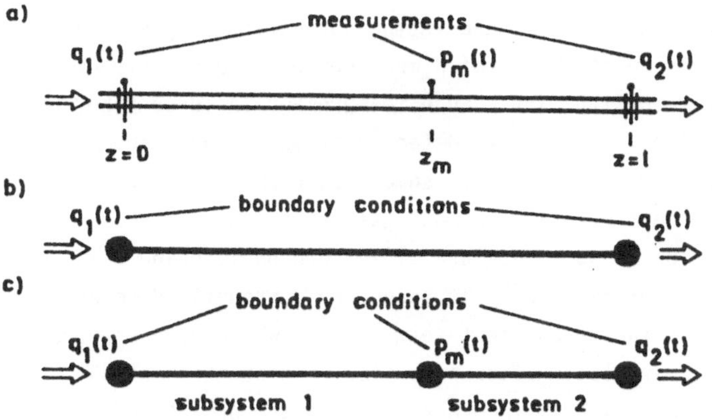

Fig. 5 State estimation by parallel simulation [14]
(a) Pipeline and available measurements, (b) Usual simulation,
(c) Parallel simulation.

the real pipeline into two subsystems being connected at the location z_m. Both subsystems are modelled according to Eqn. (1). The partitioning leads to an additional, artificial boundary point at z_m. Prescribing its state by the pressure measurement $p_m(t)$ results in a proportional behaviour of model 2. Consequently the influence of usually unknown initial conditions to the solution of the model equations decreases sufficiently fast with time. Thus, estimation and tracking of the real actual state can be achieved by the Parallel Simulation scheme. Some features of this technique and results of a field experiment are described by Lappus and Schmidt [11]. It is remarked that with the help of some theory one can show that Parallel Simulation is a special case of a Luenberger-type observer. Furthermore, it turns out that the state of model 2 is completely reconstructible whereas model 1 shows a lack of observability.

Observer design: State reconstruction by a state observer is a little more complicated than parallel simulation, but has more design freedom and several advantages with regard to modelling and measurement errors.

A nonlinear observer for the distributed-parameter model (24) with measured (known) boundary conditions, $q(o,t) = q_1(t)$, $q(1,t) = q_2(t)$, has the form

$$\underline{\hat{x}}_t(z,t) = \underline{f}_z(\underline{\hat{x}}(z,t)) + G(z)\cdot(\underline{y}(t) - \underline{\hat{y}}(t))$$
$$\underline{\hat{x}}(z,t_o) = \underline{\hat{x}}_o(z),\ \hat{q}(o,t) = \hat{q}_1(t);\ q(1,t) = q_2(t$$

The observer state $\underline{\hat{x}}(z,t)$ and the vector $\underline{\hat{y}}(t)$ are defined according to $\underline{x}(z,t)$ and $\underline{y}(t)$. The initial function $\underline{\hat{x}}_o(z)$ is assumed to be arbitrary, but physically reasonable. The elements of the $2 \times m$ matrix $\underline{G}(z)$ are space dependent weighting functions for the correction vector $(\underline{y}-\underline{\hat{y}})$. As usually, they should be chosen such that the reconstruction error shows asymptotic properties, i.e. $\underline{e}(z,t) = \underline{x}(z,t)-\underline{\hat{x}}(z,t) \to 0$ as $t \to \infty$.

It is noted that model parameter errors and faulty measurements in connection with the boundary conditions can partly be compensated by a proper weighting of available redundant pressure measurements at suitable locations. Also, having designed an observer with two well defined sets of measurements (set 1 is needed to define the boundary conditions and set 2 is used in the correction loop of the observer), every disconnection or gross disturbance of a measurement is obviously critical for permanent on-line state reconstruction. Such measurement failures (they use to occur rather frequently in practice!) are usually detected by modern data acquisition systems, but nevertheless the observer must include some "measurement failure control units" to deal with such transient errors. In most gas transmission networks the locations of some pressure and flow measurements are close together. Using - if possible - such "local redundant" measurements or - if necessary - predefined default values (functions of time) one reorders the measurement sets 1 and 2 whenever a measurement fails or a failed one becomes reavailable after some time. In such situations one must also adapt the weighting matrix G(z) in order to eliminate or, at least, reduce the effects of such failures on the reconstructed state.

The above observer can be expanded to complex gas transmission systems. The general structure of the corresponding large scale observer system has the form of Fig. 6, and has been realized through the simulation program GANESI. The resulting observer

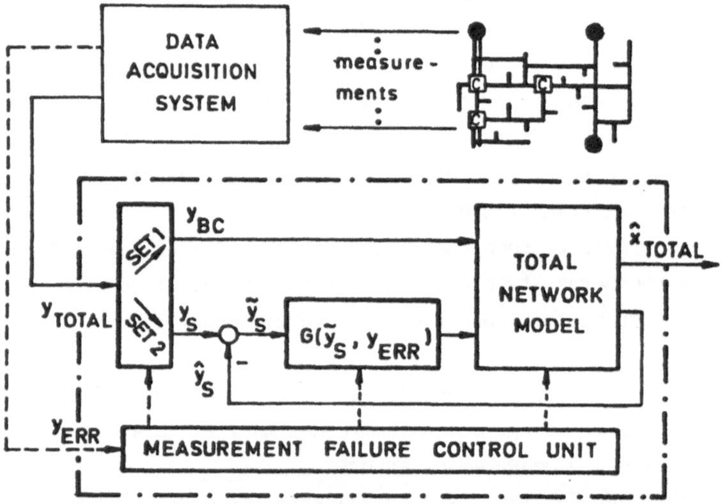

Fig. 6 Large scale observer for the gas transmission network

program was called GANBEO [13] and has been used in at least two industrial applications to medium-sized gas networks.

Failure detection: Failure detection can be achieved by using a special observer of the above type which is sensitive to modelling or measurement errors. Such an observer can be used as a failure detecting device if appropriate measurements are available. Failures under consideration are constant measurement errors, sudden changes in the network structure caused by mis-operated valves, and suddenly occuring leaks along a gas pipeline. Due to such failures the absolute values of parts of the reconstruction error vector e(z,t) become unusually large. Simulation studies and experience from industrial applications show that such failures can be detected by proper evaluation and processing of e(z,t). Thus, supervision and failure detection constitute a field for application of the discussed observer approach. The general structure of the resulting predictive supervision, failure detection and control scheme is shown in Fig. 7.

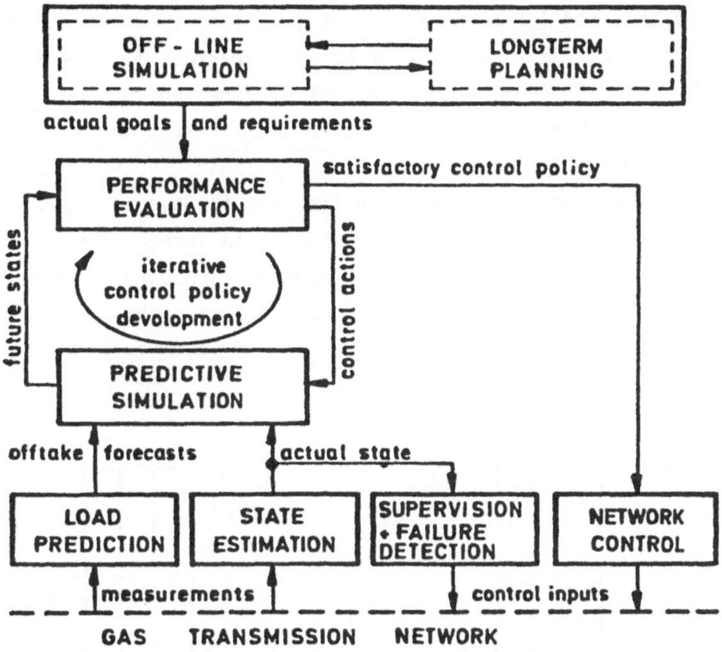

Fig. 7 Predictive supervision and control scheme.

3.2 Fault-Tolerant Controller Design

In [12] a fault-tolerant controller station was proposed which was structured as a decentralized modular system, composed of several microcomputer units performing control and supervisory functions, I/O-Modules, and a universal communication system (station bus) enabling each computer module to reach every process signal (see Fig. 8). Besides having other advantages, this structure allows to apply a certain redundancy concept called "global dynamic redundancy": the active computer units serve as back-up for each other according to a reconfiguration strategy; if necessary, spare computers can be added which may substitute any faulty computer unit. The reconfiguration strategy works such that in case of a computer failure, if no spare is available, the presently least important function is discarded and the function of the failing computer transferred to the corresponding computer.

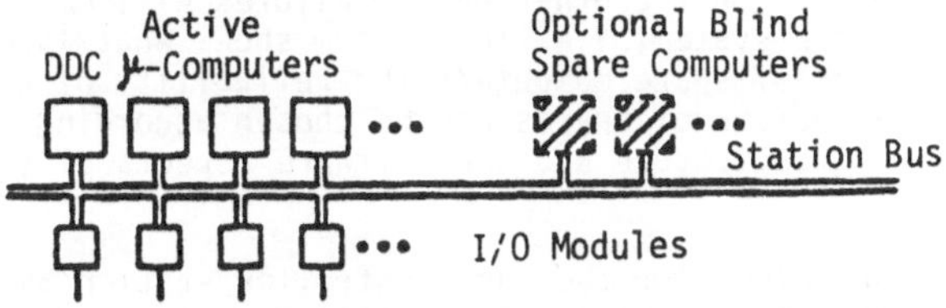

Fig. 8 Basic structure of the fault-tolerant controller station.

As an example a controller station consisting of 8 control computers and r=0 or 1 spare with a computer-MTBF of 10^5h and an MTTR of 100h has been considered. The 8 functions performed by the 8 computers may be numbered according to their importance with i=1 being the most important and i=8 being the least important function. Assuming ideal conditions like a failure-free "switching mechanism", perfect failure detection, and failure-free I/O-Modules, the MTBF of each control function in the station is given by curves (a) and (b) in Fig. 9a. For purposes of comparison, curve (c) shows the case where no mechanism for failure tolerance is available. Curves (d) and (e) are valid under the condition that only the spare computer is used as a back-up. Their difference results from the case of multiple computer failures: for curve (d) the spare replaces a failing computer until this computer is repaired, for curve(e)it is assumed that the spare substitutes the failing computer with the momentarily most important function[12].

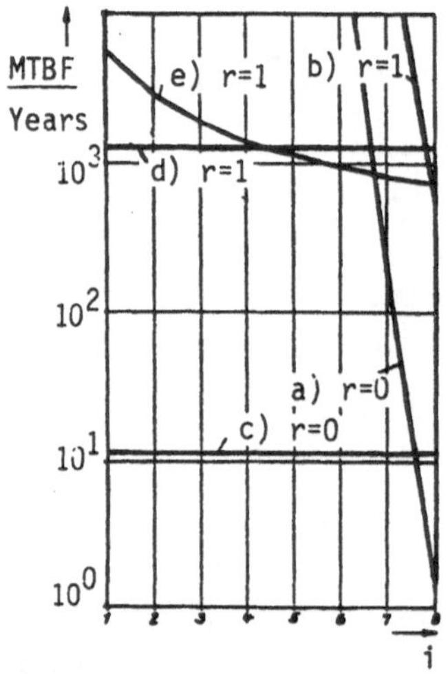

Fig. 9a MTBF values of the i-th DCC-function under simplified conditions. Influence of various redundancy concepts and reconfiguration strategies.

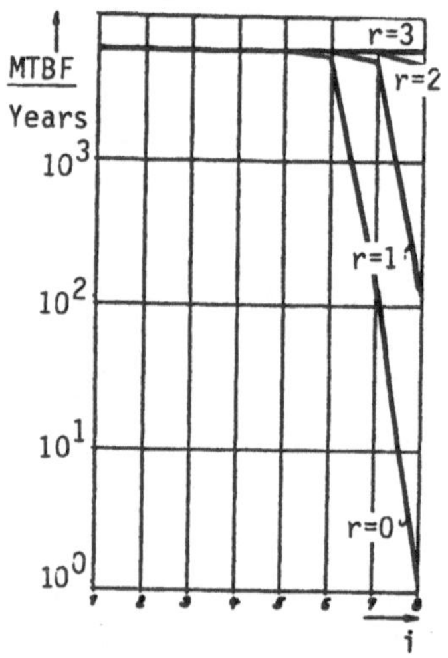

Fig. 9b MTBF values of the i-th DCC-function under more realistic conditions.

This theoretical example demonstrates some of the properties of "global dynamic redundancy": even without any spare at all (r=0) the reliability of most functions (i < 7) can be greatly enhanced. On the other hand, failures will mainly burden the least important function in the system, i.e. the system shows what is called a "defined failure effect". By addition of spare computers the reliability of all functions can be further improved, and the number of spares can be chosen according to the reliability demands. The hardware architecture and the software aspects of the controller station are described in [12].

Fig. 9b shows the MTBF values for the same controller station as Fig. 9a by including the reliability properties of the configuration computer and the bus arbiter. This makes the example more realistic. Due to the configuration computer the MTBF of the functions is now limited to a final value. It can be shown that this limiting value is identical to the MTBF of a duplex system, which means that with global dynamic redundancy and a few spares the same effect can be achieved as with the conventional duplication of each control computer. The reliability boundary existing in fig. 9b can be augmented by use of a spare configuration computer.

An experimental system of this type was built and tested with success at the Control Systems Laboratory of the Munich Technical University [12].

For a full survey of the design and properties of fault-tolerant computer systems the reader is referred to [26].

4. MULTILEVEL CONTROLLERS UNDER STRUCTURAL PERTURBATION

The present synthesis paper will be closed with a few words about some work which was recently carried out on the performance of multilevel controllers under structural perturbations [15,16]. The original work in this direction is due to Siljak and his collaborators [24,27]. In large scale systems the reliability could be improved by a completely decentralized control implementation, but there is considerable loss in optimality in "normal operation". Better results can be obtained using co-ordination in a multilevel implementation. In this case, the communication links which transmit information from the local controllers to the co-ordinator and vice-versa are often subject to structural perturbations, especially when the local controllers are widely distributed in space. This type of perturbations could lead to instability in the global system. Thus, the problem of deriving sufficient conditions which for the worst case ensure the global stability of the system is of primary importance.

Clearly, the performance of a system which is subject to structural perturbations shows a degradation. One way to treat this situation is to put an upper bound on this performance loss for the worst case, since if such a bound could be computed *a priori*, it may help in deciding whether it was worthwhile to use more reliable (and costly) communication channels.

In [15] the effect of information loss resulting from the breakdown of some of the links within the controller hierarchy is studied. Specifically, the stability of the resulting controller and the resulting loss of optimality in this system which has incomplete co-ordination is studied. In [16] these results are extended to the case when there is breakdown, not only in the link between co-ordinator and subsystems, but also in the physical interconnection between the subsystems. The results of [15] have been applied to a 11-machine interconnected system, where it has been verified the usefulness of the structural perturbation approach for testing the robustness of hierarchical control structures. Due to that the stability conditions and bounds are easily calculable, the designed controller can have a prespecified degree of stability. The results of [15,16] actually provide the link between the hierarchical optimization results and the robust controller results of Siljak and Sudereshan [24,27,28].

Acknowledgement: This paper was synthesized from contributions by G. Schmidt, M. Singh, A. Titli, S. Tzafestas and colleagues.

REFERENCES

1. S.G. Tzafestas, "Optimization of System Reliability: A Survey of Problems and Techniques", Int. J. Systems Sci., Vol. 11, No. 4, pp. 445-486, 1980.
2. S.G. Tzafestas, "Optimal Control Policies in System Reliability and Maintenance", in "Optimization and Control of Dynamic Operational Research Models" (S.G. Tzafestas, Editor), Ch. 12, pp. 395-434, North Holland, 1982.
3. S.G. Tzafestas and C.A. Botsaris, "Optimal Maintenance Policies for Modular Standby Systems", Proc. 11th IFIP Conf. on System Modelling and Optimization (P.Thoft-Christensen, Editor), LNCIS, Springer-Verlag, 1984.
4. C.A. Botsaris and S.G. Tzafestas, "Reliability Analysis and Optimization of a Repairable Standby System with Modularly Designed Items", Int. J. Systems Sci. 1984 (In Press).
5. C.A. Botsaris and S.G. Tzafestas, "Optimal Design of a Standby System with Repair Facilities subject to Breakdown", in "Methods and Applications of Measurement and Control" (S.G. Tzafestas, Editor), Vol. 2, pp. 632-634, ACTA Press, 1984.
6. S.G. Tzafestas, "System Reliability and Maintenance: Analysis, Optimization, and Control", In "Systems" (S.G. Tzafestas, Editor), IMACS-Greece, pp. 457-512, Athens, 1982 (in Greek).
7. S.G. Tzafestas, "Optimal Supervision and Failure Detection of Dynamic Technological Systems, DEMO No. 83/19G (in Greek), GAEC, Nuclear Research Center "Demokritos", Aghia Paraskevi, Attiki, Dec. 1983.
8. M.G. Singh, M.F. Hassan, Y.L. Chen and Q.R. Pan, "New Approach to Failure Detection in Large-Scale Systems", IEE Proceedings, Vol. 130, Pt. D, No. 5, pp. 243-249, 1983.
9. A. Wilsky, "A Survey of Design Methods for Failure Detection in Dynamic Systems", Automatica, Vol. 12, pp. 601-611, 1976.
10. M. Ikeda and D.D. Siljak, "Overlapping Decomposition, Expansion and Contraction of Dynamic Systems", J. Large Scale Syst., Vol. 1, pp. 29-38, 1980.
11. G. Lappus and G. Schmidt, "Supervision and Control of Gas Transportation and distribution Systems", in "Digital Computer Applications to Process Control", Preprints of IFAC/IFIP Conf. (R. Isermann and H. Kaltenecker, Editors) pp.251-258, Oct. 1980, Pergamon Press.
12. G. Schmidt and W. Sendler, "An Efficient Redundancy Concept for a Microcomputer Based Failure Tolerant Controller Station", ibid. pp. 545-547.
13. G. Lappus, "Computer Program GANBEO-Description and Validation of a Program for the Observation of Unsteady Flows in Gas Distribution Networks", KFK-PDV-215 (in German), Kernforschungszentrum, Karlsruhe.
14. G. Lappus and G. Schmidt, "Supervision and Control of Gas Transmission Networks", Encyclop. of Systems and Control, (M.G. Singh, Editor), Pergamon Press, 1984.

15. M.F. Hassan, A. Titli, M.G. Singh and R. Hurteau, "Stability, Stabilization and Performance of Multilevel Controllers Under Structural Perturbations", Part I: Cutting the links between co-ordinator and subsystems", IEE Proceedings, Vol. 127, Pt.D, No. 5, pp. 207-213, 1980.
16. M.F. Hassan and M.G. Singh, "Stability, Stabilization and Performance of Multilevel Controllers Under Structural Perturbations: Part II. Stabilization under any structural perturbation", IEE Proceedings, Vol. 127, Pt.D, No. 5, pp. 214-219, 1980.
17. R. Osaki, "On a 2-Unit Standby-Redundant System with Imperfect Switchover", IEEE Trans. Reliab., Vol. R-21, pp. 20-24, 1972.
18. H.H. Yeh and R. Kuhler, "Additional Properties of an Extended Maximum Principle", Int. J. Control, Vol. 17, pp. 1281-1286, 1973.
19. J.A. Buzacott, "Markov Approach to Finding Failure Times of Repairable Systems", IEEE Trans. on Reliability, Vol. R-19, No. 4, pp. 128-134, 1970.
20. B.A. Kozlov and I.A. Ushakov, "Reliability Handbook", Holt, Rinehart and Winston, New York, 1970.
21. N. Stuckenberg, "An Observer system for Sensor Failure Detection and Isolation in Digital Flight Control Systems", AGARD Conference Proc. No. 27: Advances in Guidance and Control Systems Using Digital Techniques", pp. 6.1-6.11, Ottawa, Canada, 8-11 May, 1979.
22. M. Athans et al., "The Stochastic Control of F-8C Aircraft Using a Multiple-Model Adaptive Control (MMAC) Method, Part I: Equilibrium Flight", IEEE Trans. Auto Control, Vol. AC- 22, No. 5, pp. 768-780, 1977.
23. D.E. Gustafson, A.S. Willsky and J.Y. Wang, "Final Rept.: No. R-920-Cardiac Arrythmia Detection and Classification Through Signal Analysis", The Charles Stark Draper Lab., Cambridge, Mass., July 1975.
24. D.D. Siljak, "Overlapping Decentralized Control", in "Handbook of Large Scale Systems Engineering Applications" (M.G. Singh and A. Titli, Eds.) North-Holland 1979.
25. A.Weimann, G. Schmidt, "Transient Simulation of Natural Gas Distribution Networks by Means of a Medium-Sized Process Computer ", 5th IFAC/IFIP Int. Conf. on Digital Computer Applications to Process Control (Van Nanta Lemke H.R., Verbruggen, H.B., editors) pp. 315-320, North-Holland, Amsterdam.
26. P.G. Depledge, "Fault-Tolerant Computer Systems", IEE Proc., Vol. 128, Pt.A, No.4, pp. 257-271, 1981.
27. D.D. Siljak, "On Reliability of Control", Proc. of IEEE CDC Conference, pp. 687-694, 1979.
28. D.D. Siljak and M.K. Sundareshan, "Large-Scale Systems: Optimality was Reliability" in "Directions in Decentralized Control"(Y.C. Ho and S.K. Mitter, Editors), Plenum, N.Y., 1976.

ON THE REAL-TIME CONTROL OF AN INTELLIGENT ROBOTIC SYSTEM

G. N. Saridis and K. P. Valavanis
Department of Electrical, Computer and Systems Engineering
Rensselaer Polytechnic Institute, Troy, New York 12181

Abstract

The high level decision making of the digital computer and the advanced mathematical modeling and synthesis techniques of system theory, provide the tools towards the realization of an Intelligent Robotic System, capable for performing complex and various but specific tasks. A unified approach of the different steps needed for the utilization of such a system is presented and the real-time problems are discussed. As an application of the approach suggested, a Unimation PUMA-600 series robot arm is used.

1. Introduction

Robotic systems demonstrate nonlinear variations of their dynamics which are almost impossible to model. Based on the fact that the dynamic performance of such a system is directly associated with the dynamic models derived for them, it is apparent that the results are imprecise control laws or strategies to achieve the desired response. This desired response of an arm is to maintain a prescribed motion along a prespecified time-based trajectory by applying corrective compensation torques to the actuators for adjustment of possible deviations of the arm from the trajectory.

The industrial approach for the control of such a robot arm is inadequate, since each joint of the arm is treated as a simple servomechanism, neglecting the possible changes of the motion and configuration of the whole arm mechanism. In order to improve the performance of such a system, sophisticated controllers and utilization of the power of the digital computer is needed. In the following sections two different configurations with the digital computer on-line will be presented.

2. The Controller of the Unimation PUMA Robot Arm

The PUMA-600 has six revolute joints as shown in Figure 1 (shown on next page) along with the axes and degrees of joint rotation. The most important component of the whole robot system is the controller whose components are: DEC LSI-11 computer, digital and analog servoboards, interface board, clock/terminator board, power supplies, power amplifier assembly, high power discharge board and arm cable board.

The LSI-11 system contains the LSI-11/02 processor, memory and communication boards. System software resides in EPROM and user program information is stored in RAM. Communication between the processor and the other components of the PUMA system is accomplished via a four-part asynchronous serial I/0 board, the DLV 11-J, and via a DRV 11 that provides parallel-line communication to and from the digital servoboards and links the processor to the interface board. The interface board connects the LSI-11 computer system of the controller with the servoboards that control the six joints of the arm. The behavior of the arm is thus controlled by the digital and analog servoboards as dictated by the LSI-11 commands. [1], [2].

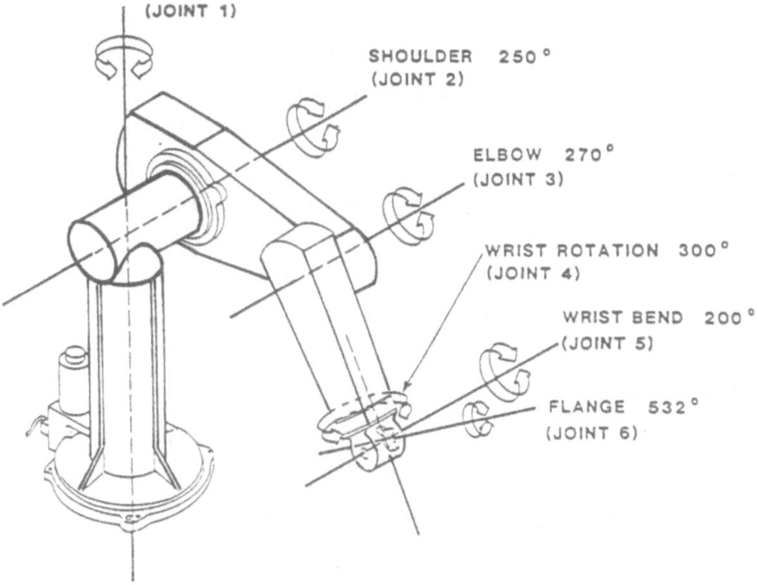

Figure 1 Robot Arm: Degrees of Joint Rotation

Given the structure of the controller, observe that it can be considered as "hierarchically arranged" and as the series connection through an interface board of a higher level - the computer level - and a lower level - the joint control level - whose performance is dominated by the six joint microprocessors, the 6503s, attached on the digital servoboards. [3], [4].

The purpose of the LSI-11/02 computer is to perform two major tasks:

i) on-line user interaction and subtask scheduling from the user's VAL (the language used by the PUMA robot arm) commands and

ii) subtask coordination with the six joint microprocessors to carry out the command.

The on-line interaction with the user includes parsing, interpreting and decoding the VAL commands and also report error messages to the user. Once a VAL command has been decoded, internal routines residing in the EPROM of the LSI-11/02 computer are activated to perform functions like coordinate systems transformations, joint-interpolated trajectory planning, that involves sending of incremental location updates every 28 ms to each joint, acknowledging from the joint microprocessors that each joint has completed its required incremental motion and two instructions look ahead to perform the continuous path interpolating if the robot is in the continuous path mode.

At the lower level, the joint control level, the components consist of the digital and analog servoboards and the power amplifiers. Each joint microprocessor directly controls each axis of motion of the arm joints and performs the following functions:

i) Every 28 ms, receive and acknowledge set-points from the LSI-11/02 computer and perform interpolation between the current joint value and the desired joint value.

ii) Every 0.875 ms, read the register value which stores the incremental values from the encoder mounted on each axis of rotation.

iii) Update the error actuating signals derived from the joint-interpolated set-points and the valves from the axis encoders.

iv) Convert the error actuating signal to voltage using the DAC's and send the voltage to the analog servoboard which move the joint.

Note that since the joint microprocessors expect new set-points every 28 ms. and read the value from the optical encoders every 0.875 ms., they divide the angle to be traveled into 32 equal increments and each time they servo the joints they add the increment and servo the joint to the updated angle value. The performance of the microprocessors is based on the set-points every 28 ms, meaning that any other rate different than that causes problems. If the set-points are acknowledged by the joint micros in time intervals less than 28 ms., then the joint microprocessors will not have time to complete their 32 servoings and a large step in position error will be the result when the new set-point is issued. If the set-points are acknowledged in time intervals greater than 28 ms., then the joint will reach the set point, stop and then start again when the new set-point is issued. The overall control scheme is shown in Figure 2 on the following page.

From Figure 2 see that there are two servo-loops for each joint control, the outer being the one providing the error information updated by the joint microprocessors every 0.875 ms. and the inner consisting of analog devices and a compensator with derivative feedback to put damping on the velocity variable. The gains of the servo-loops are constants and tuned to perform as a "critically-damped" joint system.

Concluding the description of the controller, it is repeated that position and velocity are interpolated linearly over 32 servo cycles, the commanded velocity is the difference between successive VAL set-points and that the acceleration is constant over a VAL cycle and is the difference between successive commanded velocities.

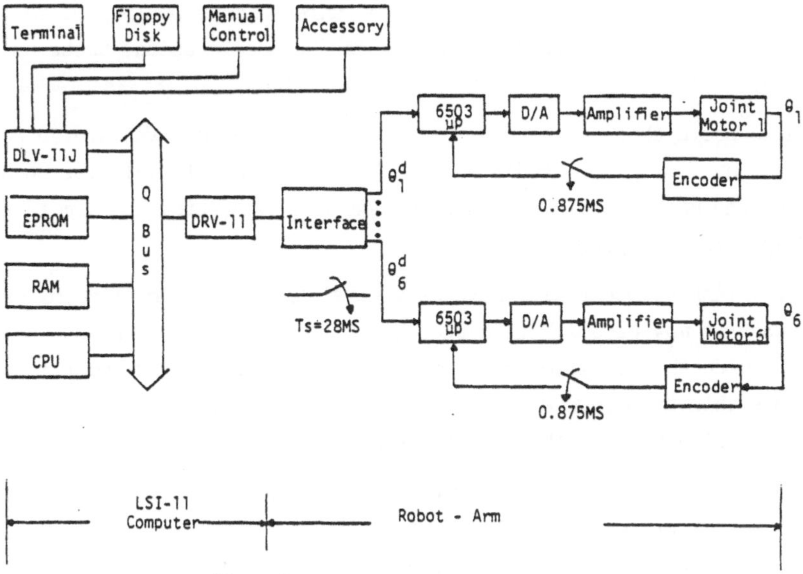

Figure 2 - Overall Control Scheme

3. Disadvantages of the PUMA Robot Arm Controller

The disadvantages of such a control scheme are summarized below:

i) All VAL commands are "open-loop" commands that send set-points to the joint microprocessors every 28 ms. and LSI-11/02 computer of the controller is not a part of the joint control servo-loops.

ii) VAL is almost exclusively Kinematic, while in order to apply modern control theories the dynamics of the arm must be derived and the modeling of the arm be based on them.

iii) The feedback gains are constant and prespecified and there is no flexibility of updating them under varying payloads.

iv) Position, velocity and/or acceleration data are not fed back to the computer level.

v) Because such a robot is a highly nonlinear system with much coupling between the joints, such a control scheme is inadequate even for simple tasks under varying speeds and payloads.

In order to improve the performance of the arm, first of all, an accurate mathematical model must be derived based on the dynamic equations of the arm and proper modifications in the controller be performed. The modifications include both modifications in the computer level and also in the joint control level.

4. Derivation of the New Controller

For the PUMA system, the problem to be solved in the smooth and anthropomorphic motion of the arm from an initial position in the three-dimensional space to a pre-specified one (end-point) with certain velocity and/or acceleration, based on modern control theories.

It has been shown above that neither the existing controller nor VAL are so powerful for such a complicated purpose. Additionally, VAL is not so flexible for such applications and the solution of disconnecting VAL and connecting the arm to a new controlling computer is adapted.

Such a new controlling computer will have many advantages over the existing one, summarized in that it may be quicker, especially if the host computer has a floating-point processor which the LSI-11/02 does not, more sophisticated and most important of all, it is possible to derive a new language, very flexible, to control the arm rather than restricting to the VAL operating system.

In order to control the PUMA robot system without VAL the LSI-11 computer that includes the LSI-11/02 processor the EPROM, the RAM, the DLV 11-J and the parallel board, the DRV 11, is disconnected from the joint control level. The new host computer is connected to the joint control level through the arm interface board via a new parallel board. **The first computer controlled scheme is presented in the sequel.**

4.1 A VAX-11/750 Computer-Controlled PUMA Robot Arm

The host computer is the DEC VAX-11/750, that represents the "higher-level" of the hierarchy, and communicates with the joint control level via a new parallel board, the DR 11-W. This means that the VAX computer can communicate directly with the joint microprocessors that control each joint of the arm.

The above is not an easy task because of the following two reasons:

i) Disconnecting the LSI-11 computer including the VAL operating system, we lose access to all the safety routines, initialization and calibration procedures that VAL runs in order to avoid situations that have to do with the hardware limits of the arm.

ii) The DRV 11 and the DR 11-W, though both are parallel boards, are not exactly compatible and a number of modifications must be done on the DR 11-W in order to interface the joint control level with the VAX computer.

Considering the safety routines, the initialization and the calibration procedures, the appropriate software must be written on the VAX and the same tests must be run every time the arm is initialized, exactly in the same way as with the old system.

Considering the interfacing problems, the situation is much more complicated.

The DRV 11 is a general purpose interface unit used for connecting parallel line TTL or DTL devices to the LSI-11 bus. It permits program-controlled data transfers at rates up to 40K words per second and provides LSI-11 bus interface and control logic for interrupt processing and vector generation. Data is handled by 16 diode-clamped input lines and 16 latched output lines. The unimation interface board transmits data from the joint microprocessors through the DRV 11 to the LSI-11 system and data from the LSI-11 system to the joint microprocessors. Of the 16 I/O lines of the DRV 11, two have to be mentioned for their importance. The Request A line that is used from the joint microprocessors to indicate "ready" situations so that data can be sent to or received from the joint microprocessors and the Request B line that is used to indicate error conditions. [5]

The DR 11-W is a general purpose, UNIBUS, direct memory access (DMA) device that provides the means to connect a user device to the UNIBUS of a VAX computer in either single or multiple DR 11-W system configurations. It can be operated in either a programmed I/O or DMA mode. In programmed I/O, data is moved to or from the user device under CPU program control. When operated in DMA mode, the DR 11-W becomes bus master via an NPR request and operates directly on the memory to satisfy requests originated at the user device. [6]

Replacing the DRV 11 with the DR 11-W, one needs a cross-connector interface box that matches the output lines of the unimation interface board to those of the DR 11-W and also modification for the Request B line that must be checked periodically for error conditions.

Since the timing and in general, the overall operation of the joint control level components is very difficult to change, it is obvious that the joint microprocessors expect new set points every 28 ms. This operation was controlled in the old configuration of the system by an event signal line that connected the clock/terminator board with the LSI-11/02 and generated an interrupt every 7 ms. Every fourth generated interrupt an error indication was sent from the joint microprocessors if a new set point was not received. With the new host computer, this problem can be solved by either using a software generated interrupt every 28 ms. or by using the ATTN H line of the DR 11-W that will connect the signal line from the clock/terminator board to the new host computer CPU. (See Figure 3)

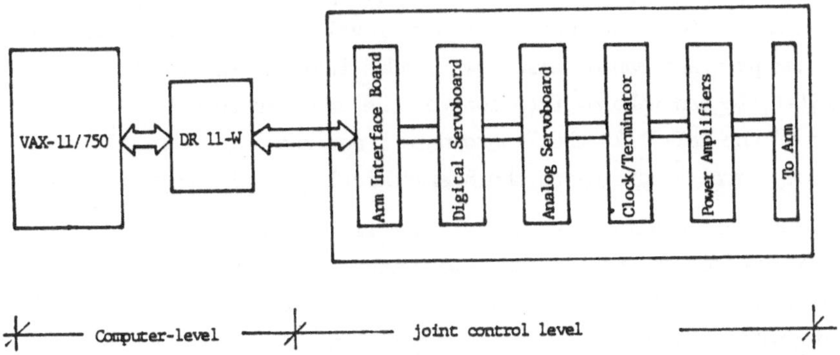

FIGURE 3

The advantage of such a configuration is that the host computer is a very powerful one and very flexible for any kind of complicated task. Despite the communication with the joint control level of the robot arm, the VAX computer can also communicate with other systems like the vision system that can provide "visual" information about the location of objects that are in the environment of the robot arm.

The disadvantage of such a configuration is that since the VAX computer is a time-shared machine every time a task has to be executed it is necessary to be performed on a "single user" basis. In addition to that, the VAX computer does all the computations and "grubby details" that need to be done before and during the execution of specific tasks, besides the interpretation of higher level commands and communication with other systems. For this reason an alternative approach is suggested.

4.2 A VAX-11/750, LSI-11/23 Computer Controlled PUMA Robot Arm

According to this configuration, the LSI-11/02 processor is replaced with an LSI-11/23 one that is faster and also has the option of floating point arithmetic that improves the accuracy in the calculations. The existing RAM card is replaced with a new 128K word memory card more than enough to store the user's programs. The EPROM card that had the VAL code is replaced with a new general purpose 64K card in which the new software resides, more flexible than VAL. The DRV 11 doesn't change. The LSI-11/23 automatically generates the 7ms interrupt signal exactly in the same way as in the old configuration with the LSI-11/02. This "update" of the LSI-11/02 to an LSI-11/23 system is only the first step. The second is the communication between the VAX-11/750 and the LSI-11/23 via a DR 11-W board that interfaces the UNIBUS to the LSI-11/23 Q-BUS. Such a scheme has two parallel interface boards. The DRV 11 that interfaces the LSI-11/23 with the joint control level and the Dr 11-W that interfaces the VAX with the LSI-11/23.

The advantage of this scheme is that all the software and all the calculations for the control of the arm are performed in the LSI-11/23 level and not on the VAX level. The VAX computer communicating with the LSI-11/23 computer only to report updating information from the vision system or other high level commands. The flexibility of the system is that programs can be also down loaded from the VAX to the LSI 11/23 and then the LSI-11/23 can take over to control the arm. In such a system the VAX is the "organizor" the LSI-11/23 the "coordinator" and the joint-control level the "hardware control" level according to the hierarchical control theory [7], [8].

The disadvantage of this scheme is that the connection between the VAX and the LSI-11/23 is very complicated and can be performed with either a DECNET or with the installation of a new operating system in the LSI-11/23, most likely the RT-11 one, and a number of protocol oriented communicating routines for the VAX LSI-11/23 connection. The configuration of the scheme is shown in Figure 4.

The above two schemes conclude the modification in the computer level.

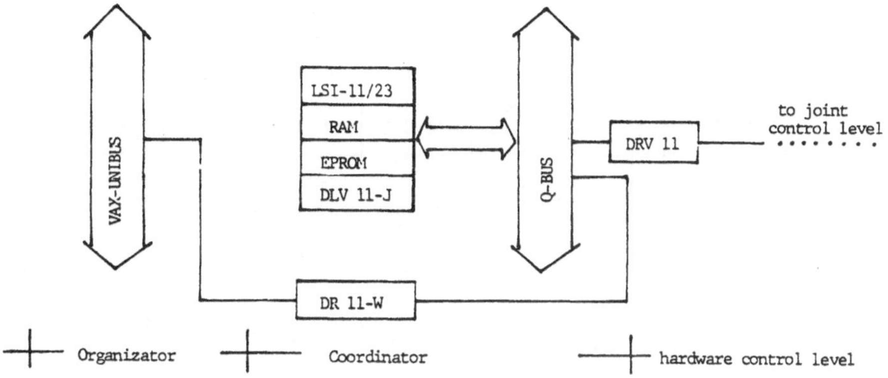

FIGURE 4

4.3 Modifications in the Joint-Control Level

Based on the fact that the control loops are closed only in the joint microprocessors level and no information is fed back to the host computer - either LSI-11/23 or VAX - and in order to apply modern control theories the host computer must be always on the line, modifications must be performed in the joint control level.

The modifications are summarized on a new card where using position information, differentiates twice it to obtain velocity and acceleration data that are fed back to the computer every .875 ms., each time the joint microprocessors servo the six joints of the arm.

Also due to the nonlinearities, the consideration of the member masses as point masses and the assumption that some system parameters are constant, though highly dependent on the relative position of the arm instead of trying to find a complicated mathematical model to improve the performance of the arm, a minor loop with a compensator built in to increase the robustness of the system is designed around each of the DC motors obtaining measurements through torque sensors. The torque servo is based on the joint torque sensor and is closed around the motor and gear reduction being independent of the configuration of the manipulator. The joint torque control loop has two servo-loops. The inner is a velocity servo whose purpose is to provide damping and the outer is a torque servo whose purpose is to provide a critically damped second order system such that the sensor output will track the desired torque.

A similar minor compensating loop based on measurement of joint accelerations rather than joint torques can be also designed compensating all nonlinear torques and ending up on simpler controller which can be very easily realized with analog circuits [9]. The overall controller configuration is shown in figure 5.

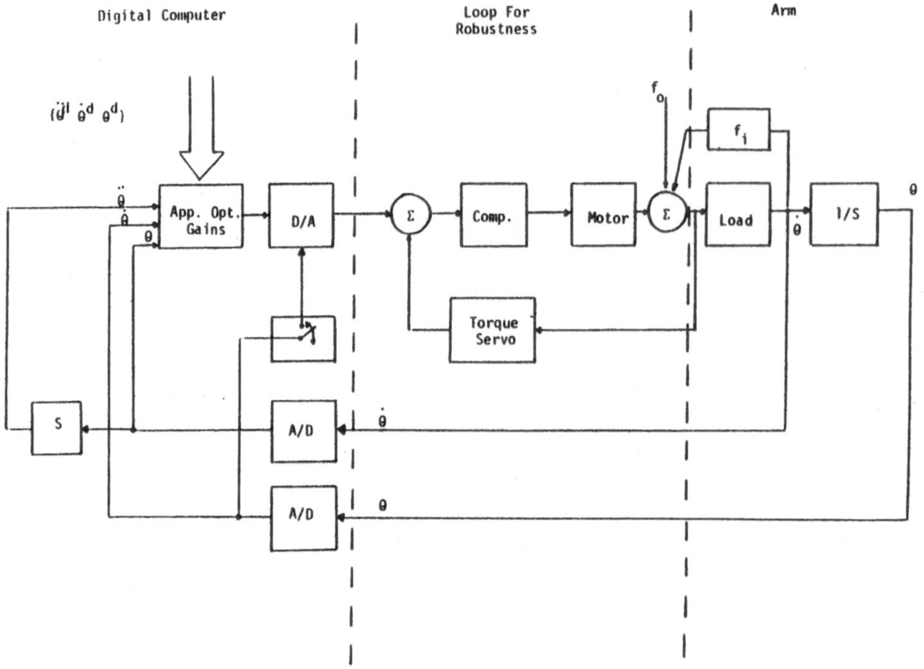

Figure 5. Suboptimal Arm Control with Minor Torque Compensating Loops

5. Advantages of the Derived Controller

i) The superiority of the VAX-750 and/or LSI-11/23 computer is obvious over the LSI-11/02 used by VAL. Utilization of the high-level decision making of the digital computer is applicable and flexibility in the language to be used for the applications of the various control methods available.

ii) The new software is based on the dynamics of the arm, resulting in a more accurate and more advanced mathematical model.

iii) The control loops are closed back to the computer level giving the capability to control the arm movements from the host-computer and the minor derivative loops guarantee the robustness of the arm.

iv) The gains are not constant and prespecified but adjustable according to the movements to be done.

v) Optimal and/or suboptimal control theories are applicable and comparison of the results obtained capable.

The above advantages prove the superiority of the derived computer-controlled system over the existing controller for the PUMA.

6. Disadvantages and Advantages of the Real-Control

There are certain time constraints that have to be met in every real-time based system. In this system the time limitations are due to the fact that the joint microprocessors expect new set points every 28 ms. Since their timing is very critical and no modification is encouraged, all real-time control methods must update the joint angle values within the interval of the 28 ms. This means that control laws and strategies based on the complete dynamic model of the arm with all the off-diagonal acceleration related and the nonlinear Cariolis and Centrifugal terms cannot be applied. In return, control laws based on simplified efficient models that update their data within 28 ms. must be derived and applied.

There is not a certain and clear method to overcome the above problem. It is the authors' intention that research in the area of parallel concurrent programming techniques will decrease the computation time to such a point so that one would be able to test almost all of the control methods based on the complete models.

The big advantages of the real-time control are that it can give definite answers for the potential of the various methods that have been tested so far only by simulation, and also about the overall performance of the arm.

7. Summary and Conclusions

The control scheme of the PUMA robot arm has been described and explained why it is inadequate for application of modern control theory. The new proposed controller is shown to be superior over the old one and its advantages are presented. It is believed that such a system will greatly improve the performance of the arm.

REFERENCES

[1] Kimon Valavanis "Unimate PUMA Manipulator Manual" Technical Report, RAL 3, March 1982, RPI, Troy, New York.
[2] UNIMATE PUMA™ Robot Manual 398H.
[3] Short Publications from Unimation.
[4] Kimon Valavanis "Controlling the PUMA with and/or without Using VAL" Technical Report, RAL 16, 5/83, RPI, Troy, New York.
[5] PDP-11 Microcomputer Interfaces Handbook by DEC.
[6] DR 11-W Direct Memory Interface Module by DEC.
[7] G. N. Saridis "Intelligent Robotic Control" Technical Report RAL 2, February 1982, RPI, Troy, New York.
[8] G. N. Saridis "Intelligent Robotic Control" IEEE Transactions on Automatic Control, Vol. AC-28, No. 5, May 1983.
[9] G. N. Saridis and K. P. Valavanis "Application of Optimal Control Theory" to the Unimation PUMA arm. SMC, Seattle, Washington, October 1982.

A LEARNING APPROACH FOR INCORPORATION OF IMPERFECT KNOWLEDGE IN DECISION SUPPORT SYSTEM DESIGN

ADOLFO LAGOMASINO ANDREW P. SAGE
Department of Systems Engineering
University of Virginia
Charlottesville, VA 22901

ABSTRACT

This paper discusses the decision support system design problem with a finite set of multiattributed alternatives and imperfect knowledge about: the decision situation structural model, about the impacts of alternative courses of action, and about the value perspectives of the decisionmakers. "Imperfect" knowledge refers to available information that may be imprecise relative to the degree of refinement with which the assessment is made, ambiguous in the sense of giving rise to inconsistencies with principles and laws of an assumed decision situation model, and incomplete in that otherwise needed elements are missing. Principles for analyzing decision problems with imprecise and incomplete information are presented for the case of outcome certainty. Extensions of this, for cases involving risk, are proposed. A procedure based on these results is developed to rank the set of alternatives. It can be formulated as a set of linear programming problems. Knowledge of the functional form of the utility function such as whether it is multilinear or additive, or equivalent independence conditions on the set of attributes such as preferential independence or mutual preferential independence, are not necessary for implementation of the resulting algorithms. The amount of information needed is reduced when some knowledge of these conditions is available. These results are

---

ACKNOWLEDGEMENT - The research on which this paper is based was supported by the Army Research Institute for the Behavioral and Social Sciences under Contract MDA-903-82-C-0124.

analogous to well known results in multiattribute utility theory in which precise and complete information is required. A decision support systems design methodology, that has implications for the knowledge base management subsystem and the model base management subsystem, results. The decision support design problem is formulated as a learning process in which the decisionmaker is encouraged to successively improve understanding of the decision situation, and adopt value judgements accordingly. Inconsistencies in the knowledge base are, for the most, the result of biases and inadequate heuristics that are used in the acquisition and representation of information. Extensions of existing methods and new approaches to identify, avoid, and resolve inconsistencies are discussed.

Special attention is paid to the complicated roles of the analyst in using the interactive screening procedure proposed. Due to the flexibility, and hence lack of structure, in assessing the required information, facilitation skills are required that go beyond those of more conventional and structured approaches. Identification of a minimal set of information required to induce a linear order on the set of alternatives is dicussed. A conceptual design of a proposed system to aid in the information processing activities of identification, acquisition, aggregation, evaluation, and interpretation of the required information for decision support is presented.

INTRODUCTION

Existing problems associated with the use of formal models for decision support are reviewed. The core of the problems studied here lie at the interface between the human and the formal model used to represent a particular decision situation. Difficulties associated with the design of a suitable dialog generation and management system provide the motivation for the adaptive learning approach for decision support discussed here. They concern the human information processing activities required for the use of these data based management and model based management subsystem models and the inability of people to organize them, in an unaided fashion, for efficient and effective decisionmaking. Our work is motivated by formal reasoning concerns, as contrasted with skill based and rule based reasoning that is appropriate in situations in which there exists considerable experiential familiarity [Sage84a,84b]. Formal models for decision support depend, for their use, on a very organized process to thoroughly pursue all the information surrounding an issue. They require that this information be precise relative to the degree of refinement with which the assessment is made, consistent in agreement with the principles and laws of the model, and complete in the sense of containing all that is necessary to obtain a linear preference ordering of alternatives.

It is often the case that various parameters of the model are unknown. Immeasurability, or perhaps partial knowledge of the parameters of the model, often impedes use of formal models for decision support, even in situations where this use is otherwise warranted. Information that is obtained is often discovered to be self contradictory after it is interpreted, or after recommendations of a preference order among alternatives and associated decision recommendation, based on use of a formal model, do not reflect some aspects of the decisionmaker's intuitive feelings of preferences among the alternatives. In this case, some aspects of the semantic character of the structure and information contained in the model have been improperly assessed.

Real life decision problems are so complex, unstructured, and poorly understood initially, that a formal process for decision support must allow and encourage learning so that the decisionmaker gradually improve understanding of the decision situation and associated value perspectives, and on the basis of these, to be able to make a more informed and hence better decision. Our research attempts to cope with the behavioral difficulties present in the formulation and resolution of complex decision problems by means of formal models for decisionmaking. Concepts and principles of rational decisionmaking, believed to make decision support models based on multiattribute utility theory (MAUT) more behaviorally meaningful, are presented. After a brief review of the MAUT based decision analysis paradigm, extensions are presented and discussed.

THE MAUT FRAMEWORK FOR DECISION ANALYSIS

In the decision analysis paradigm [Keen76] , it is assumed that a set of feasible alternatives $A=(a_1,...,a_m)$ and a set $(X_1,...,X_n)$ of attributes or evaluators of the alternatives can be identified. Associated with each alternative a in A, there is a corresponding consequence $X_1(a),X_2(a),...,X_n(a)$ in the n-dimensional consequence space $X=X_1 \times X_2 \times \cdots \times X_n$.

The decision maker's problem is to choose an alternative a in A so that the maximum pleasure with the payoff or consequence, $(X_1(a),\cdots,X_n(a))$, results. It is always possible to compare the values of each $X_i(a)$ for different alternatives, but in most situations, the magnitudes of $X_i(a)$ and $X_j(a)$ for i not equal j can not be meaningfully compared since they may be measured in totally different units. Thus, a scalar valued function defined on the attributes $(X_1,\cdots,X_n)$ is sought that will allow comparison of the alternatives across the attributes. The existence of the value function as a mechanism for representation and selection of alternatives in a utility space is based on the fundamental representation theorem of simple preferences which states that under certain conditions of rational behavior there exist a real-valued utility function U such that alternative a_1 is preferred to alternative a_2 if and only if the utility of a_1, denoted by $U(a_1)$, is greater than $U(a_2)$, the utility of a_2.

A primary interest in multiattribute utility theory (MAUT) is to structure and assess a utility function of the form

$$U[X_1(a),\cdots,X_n(a)]=f\{U_1[X_1(a)],\cdots,U_n[X_n(a)]\}$$

where U_i is a utility function over the single attribute X_i and f aggregates the values of the single attribute utility functions such as to enable one to compute the scalar utility of the alternatives. The utility functions U and U_i are assumed to be continuous, monotonic, and bounded. Usually, they are scaled by $U(x^*)=1$, $U(x^o)=0$, $U_i(x^*_i)=1$, and $U_i(x^o_i)=0$ for all i. Here $x^*=(x^*_1,x^*_2,\cdots,x^*_n)$ designates the most desirable consequence and $x^o=(x^o_1,x^o_2,\cdots,x^o_n)$ the least desirable. The symbols x^*_i and x^o_i refer to the best and worst consequence, respectively for each attribute X_i, i.e., $x^*_i=X_i(a^*)$ where a^* is the best alternative for attribute i, and $x^o_i=X_i(a^o)$ where a^o is the worst alternative for attribute i.

We have just described very briefly the case of certainty in a decision making framework. Associated with each alternative there is a known consequence that follows with certainty from implementation of the alternative. The foundations for decision making under risk are provided by the classical work of von Neumann and Morgenstern [Neum64] . The implications of this work are that probabilities and utilities can be used to calculate the expected utility of each alternative and that alternatives with higher expected utilities should be preferred over alternatives with lower ones.

Multiattribute utility theory provides representation theorems, based on some forms of independence across the attributes, that describe the functional form of the multiattribute utility U as an additive, multiplicative, or multilinear function of the conditional single attribute utility functions U_i. The books by Keeney and

Raiffa [Keen76] Krantz et al. [Kran71] and Fishburn [Fish64,70] are perhaps the most comprehensive works that deal with these representations. Recently, Tamura and Nakamura [Tamu83] have introduced a more general concept of convex dependence and associated convex decompositions of multiattribute utility functions which includes as special cases the additive and multiplicative representations.

The methodology of decision analysis, using multiattribute utility theory, is generally decomposed into four major steps [Keen82] as shown in Figure 1:

Step 1. Identification of the decision problem - This includes the generation of alternatives and the specification of objectives and hence attributes to be used in the evaluation of alternatives.

Step 2. Assessment of the possible consequences for each alternative - In the case of certainty, this consists only in specifying the unique known consequence that follows for sure from implementation of each alternative. When various possible consequences may occur, a probability distribution function over the set of attributes for each alternative must be determined.

Step 3. Determination of preferential information - The structure of the model is determined and quantification of its parameters is made. This step requires relevant, precise, and consistent information about value assessment, value tradeoffs, and risk attitude.

Step 4. Evaluation of alternative and sensitivity analysis - The information gathered is aggregated by use of the expected utility criterion. The alternative with the highest expected utility is the most desired. Finally, the sensitivity of the decision to a variety of changes is explored in order to gain some confidence concerning the recommended decision.

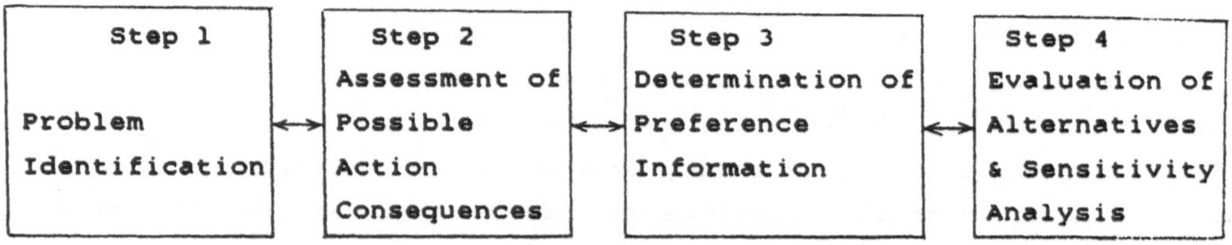

Figure 1. Decision Analysis Paradigm

Even though the theory and procedures of decision analysis are conceptually straightforward, there are other circumstances that make its implementation very complex. Putting the methodology into practice is much more involved than one might believe. Each of the foregoing four steps requires substantial interaction between the analyst and the decisionmaker. A very stressful thinking process is demanded of the decisionmaker while the analyst is in charge of coordinating a series of activities in order to facilitate this process. The analyst must obtain the minimum amount of relevant information about the decision problem to determine the various utility functions. Often redundant information should also be obtained in order to check for consistency. The analyst must be sensitive to biases and flawed heuristics that the decision maker may utilize; the analyst must be able to structure the decision problem regardless of the degree of complexity; and must, above all, retain the confidence of the decisionmaker with respect to the belief that a formal analysis of the problem will result is a more intelligent and informed decision. In a recent article, Howard [Howa80] described very vividly the roles of the analyst: "...The decision analyst is an elicitor of information and preferences, an engineer of logical models, and an evaluator of alternatives... He is skilled in constructing the decision set using his imaginary and colorful friends, the clairvoyant (who knows all and who helps with defining variables and events unambigously) and the wizard (who can do all and who help with value assignment)." The analyst is confronted with "his own" decision problem. It may involve a large number of objectives that may be initially unknown to him and uncertainties will always be present because he does not have a way to know what the responses or attitude of the decision maker be to his queries. With this view, decision makers and analysts are engaged in identical cognitive activities. The psychological research supporting the view that humans are not excellent processors of information [Kahn82, Slov77, Sage81] could be applied as well to the roles of the analyst facilitating the decision making process.

SCREENING MODELS FOR DECISION SUPPORT

The large amount and complexity of information required for complete specification of multiattribute utility functions and probabilities leads, especially in practice, to the use of simplified heuristics. Often, these are flawed [Sage81]. Even in prescriptive situations, screening procedures will often be needed to reduce the time, stress, and effort demanded from decisionmakers. Screening methods are intended to identify and reduce the nondominated set of alternatives, or Pareto optimal set, through use of behaviorally relevant and easily available information. One of the first to develop a screening method for decision making with incomplete knowledge of probabilities was Fishburn [Fish65] . He was concerned with the use of incomplete information on probabilities in comparing alternative strategies in a typical formulation of decisionmaking under risk. The criterion of choice or strategy that should normatively be used is the principle of maximum expected utility, so that the decisionmaker seeks a strategy a^* which is the maximum over i of the expected utility of alternatives a_i. The utility function $U_j(a_i)$ is a precisely assessed multiattribute utility function defined on the set of possible strategies $\{a_i\}$, and $p_j(\cdot)$ is a measure of the likelihood of the possible state of nature j given that a particular strategy was selected. The imprecise forms of the measure of probability that Fishburn considered are:

(1) No information about $p_j(\cdot)$

(2) Ordinal measure: an ordering of p_j, (eg. $p_1 \geq p_2 > p_3$)

(3) Linear inequalities: an ordering of sums of p_j, (eg. $p_1 + p_2 \geq p_3$)

(4) Bounded interval measure: (eg. $c_i \leq p_i \leq c_i + d_i$)

The search for the best alternative is, in this approach, performed by pairwise comparisons of expected utility among candidate strategies. Because of the restricted form of the available information on the p_j, the search for the best alternative can be put, in general, into a straightforward linear programming problem. Further work along this line of research resulted in the use of the concepts of statistical dominance [Fish64] and stochastic dominance [Whit78] .

Sarin [Sari77] proposed a screening procedure similar to that of Fishburn with the additional assumption of additive independence of the set of attributes. This assumption simplifies the search for dominance structures and results in a procedure which can be formulated as a mathematical programming problem. The parameters of the mathematical programming formulation include probabilities, importance weights, and single attribute utility functions. A simple procedure is then developed for the case when the probabilities and the importance weights are precisely known and utilities are stated in the form of linear inequalities, thereby resulting in simple linear programming formulations. Extensions to this research have been reported by White, Sage, et al. [Whit80,82,83a, Sage84a] who develop an Alternative Ranking Interactive Aid based on Dominance structural Elicitation (ARIADNE) concept. The general mathematical programming formulation of the search for dominance results in interactive solution of a large number of relatively simple linear programs. Several cases have been considered:

1. The probabilities (p_j) are known precisely, and the importance weights (w_i) and utilities (U_i) are described by linear inequalities.
2. The importance weights (w_i) are known precisely, and the utilities (u_i) and probabilities (p_j) are described by linear inequalities.
3. The importance weights (w_i) and utilities (u_i) are known precisely, and the probabilities (p_j) are described by linear inequalities.

In cases 1 and 2 the solution results in a set of hierarchically organized linear programming problems. The simplest formulation is that of case 3, which is equivalent to the problems solved by Sarin and Fishburn, resulting in a set of simple linear programming problems. Other approaches [Char78,79] have been developed, but they all rely on the existence of additive independence conditions to facilitate the computations.

Decision making problems in which there are several conflicting objectives have been formulated in terms of multiobjective programming problems. The mathematical programming formulation of the multiple objective decision problem is,

$$\max_{a} U[u_1(a),u_2(a),\cdots,u_n(a)]$$

where the u_i are the real-valued utility functions of the n objectives. The single attribute utility functions u_i are assumed known; the overall utility function U is unknown. Generally, the solution to this problem is not a unique alternative but a set of nondominated alternatives. Multiobjective programming techniques operate under the notion of dominance for generating the set of nondominated alternatives. Interactive algorithms to gradually gain knowledge about U and solve the above problem uniquely have been proposed by many researchers. The books by Cohon [Coho78], Chankong and Haimes [Chan84], and Zeleny [Zele82] present a number of these.

Screening procedures can, in general, be made to be interactive. Interactive approaches of this type assume that the decisionmaker can provide preference information on simple, often hypothetical, alternatives. Initially, a reduction of the nondominated set is made with the available information. If the decision maker can select an alternative from the nondominated set, then the process is stopped. Otherwise, further information is requested. Often, but not always, very little guidance is provided by these procedures about the information needed. The decisionmaker is asked to provide further information when a single nondominated alternative is not present. This information may be redundant, in that the decision model could have inferred it from previous information. In this case it could serve only as a check for consistency. The new information may be inconsistent with the existing information. Alternately, the decision maker may never recognize and provide information needed to reduce the nondominated set. All this may make decision support processes that are based on interactive multiple objective procedures ineffective and inefficient. The flexibility that the interactive screening, or scanning, procedures provide potentially result in a more effective support process. However, this flexibility complicates, to a considerable extent, the task of the decision analyst as a facilitator. In order to make efficient use of these new screening procedures, it seems very neccessary to provide the analyst with a suitable dialog generation and management system. Such a system must be designed with full knowledge of the parrticular data base and model based management system used to allow for the interactive dominance based scanning. At the very least, it is necessary to provide assistance to the analyst in determining what information is most needed, in terms of relevancy to the task at hand and with due consideration being given to information that is both important and cognitively easy to assess. The analyst should be aided in the evaluation of acquired information for consistency and in ways to avoid and resolve inconsistencies that do result. Finally, the information that is acquired must be represented and used within a model based management system that is valid and appropriate for decisionmaking.

BEHAVIORAL INTERPRETATION OF EXISTING THEORIES OF DECISIONMAKING

There exists a variety of literature describing the axiomatic bases for conventional multiattribute utility theory and decision analysis. Although interactive methods to screen candidate alternatives under realistic assumptions of imprecise and incomplete knowledge of preference and risk attitude now exist, there is no axiomatic development, indicating the circumstances under which this is a rationally correct and valid approach. The foundations of multiattribute expected utility theory provides useful models of normative behavior for decision making under risk. The assessment of precise utility functions in these models is mathematically justified by the existence of a real-valued utility function. Whether it is behaviorally justifies is yet another question. It is here desired to provide a similar development in terms of scanning and screening methods of representing imprecise utility functions by means of inequalities or bounded intervals.

Two of the most common real-valued utility representations for preferences are [Fish70] :

(1) $a \succ b \iff u(a)>u(b)$

(2) $ab \succ^* cd \iff u(a)-u(b)>u(c)-u(d)$

Utility representation (1) maps simple preference relations on the set of alternatives A into ordinal relations in the set of real numbers, while representation (2) reflects strength of preference relations among pairs of alternatives in A×A into difference orders in the real line. $ab\succ^*cd$ is taken to mean that the strength of preference of a over b is greater than the strength of preference of c over d. The common connection between the two representations is $a\succ b \iff ab\succ^* aa$, with aa interpreted as no difference between these two particular alternatives. This says that a is preferred to b if and only if the preference difference ab is "positive." Moreover u in (1) and (2) are strategically equivalent, i.e., they induce the same order on A. The utility function u in (1) is unique up to strictly monotonic transformations, and u′ in (2) is unique up to positive linear transformations. It is critically important to note that the preference relation $\succ$ is an ordinal relation. The relation $\succ^*$ is a cardinal relation.

Most studies of simple preference define an indifference relation ~ as the absence of strict preference. That is to say, you are indifferent between two alternatives a and b, if you do not prefer a to b and do not prefer b to a. Symbolically, this simply says that $a \sim b \iff$ not($a\succ b$) and not($b\succ a$). With this definition, indifference may arise in three different ways [Fish70] :

1. There may be no real difference between the two alternatives and an individual then may not state any definite preference among them.
2. An individual may find the comparison difficult and decline to commit themselves to a strict preference judgement.
3. Two alternatives may be incomparable on a preference basis.

While the first case is a generally accepted condition for indifference, the other two cases are not or at least should not be so accepted. They imply that ignorance and immeasurability of preferential judgements are equivalent to indifference in judgement.

An induced relation $\succsim$ on A is used to indicate preference or indifference relations ("is at least as preferred as"), where the or is given an exclusive interpretation. That is, a $\succsim$ b means that a is preferred or indifferent to b , but not both. The above definitions imply that if you prefer a to b then b is not preferred nor indifferent to a; the contrary being mathematically true also in that $a \succ b \iff \text{not}(b \succsim a)$. Although the first statement may be behaviorally meaningful, the contrary may not necessarily be accepted in general because of the same argument of ignorance and immeasurability not being equivalent to indifference, as was discussed in our earlier treatment of the indifference relation.

In addition to the several practical difficulties encountered in assessing precise utilities, we see that there are a number of semantic issues involved with the precise representations of preference judgements. One of the aims of our current research is to seek representation theorems that incorporate incomplete measurements of preference and risk attitude and to provide a behaviorally meaningful as well as rationally correct approach for decision support. Instead of assessing a real-valued utility function, we allow a "fuzzy" kind of imprecise representation of utilities, namely an interval number. Interval analysis is a well developed field in applied mathematics and is just, in a sense, a new language for inequalities [Moor79] .

EXTENSIONS TO INTERACTIVE DOMINANCE BASED SCANNING

The following interval-valued representations are needed for the development of our interactive approach for screening multiattribute alternatives. The change in notation here, in addition to serving the purpose of separating the cases of perfect and imperfect information, is made because of a different formulation of the decision problem. For example, the set of alternatives F_X contains the set of feasible alternatives A and a set of infeasible (hypothetical) alternatives. The existence of a single interval-valued function $Q: F_X \Rightarrow I$, where I is the set of all closed intervals in the real line, provides the mechanism for the imprecise representation of utility functions. When the set of alternatives F_X is finite, the relation $A^i \succsim A^j \iff Q(A^i) \geq Q(A^j)$ holds if and only if the relational system $(F_X, \succsim)$ is a simple order in that it is transitive, antisymmetric, and strongly complete. Also we conjecture that a different interval-valued function Q^* exists satisfying

$$A^i A^j \succsim^* A^k A^m \iff Q^*(A^i) - Q^*(A^j) \geq Q^*(A^k) - Q^*(A^m).$$

When the values of these interval-valued functions become degenerate intervals of the form [a,a], which are given through precise value assessment, then the representation results achieved here correspond to the standard representations of multiattribute theory.

The primary practical result from this is that the resulting decision analysis methodology could be formulated as a learning process, thereby allowing the decisionmaker to incorporate new alternatives, attributes, and outcomes at any time during the analysis and interpretation steps. As new alternatives, attributes, or outcomes are added to this problem, previously stated value judgements should not change. If a new set of attributes is added and they are preferentially independent, then several preferential inferences, on the new problem, could be made from previous judgements.

We can consider preferential comparisons among consequences in F_X that, although infeasible or hypothetical, may be cognitively easy to assess and at the same time could provide relevant information to infer some preferential ordering among the set of feasible alternatives. Whether this is justified or not depends strongly on the complexity of the problem. If during any structured analysis of a decision problem it becomes difficult to obtain the preferences of the decisionmaker because of the complexity of the problem, hypothetical questions, in simpler contexts, will generally have to be asked. The degree of hypothetical questioning required to avoid biases and possibly erroneous information processing heuristics is of particular concern in our research.

Assymetry and transitivity are usually interpreted as indication of rational behavior in a prescriptive sense. Another interpretation is the descriptive view which states that in order for measurement of preference to be possible, the decision maker's judgement must be assymetric and transitive. The assymetry and strongly complete conditions are assumptions that pertain to the structure of the decision situation, but also have an interpretation relevant to the behavioral attitude of the decisionmaker toward preferential assessments. The assymetry condition says that no two alternatives in F_X can have the same characteristics without being considered identical. The strong completeness condition indicates that for any two alternatives A^i and A^j in F_X, we must have either that A^i is at least as good as A^j or A^j is at least as good as A^i or both. This does not imply that the decisionmaker has to express a preference for all pairs of alternative, but that such preference exist though they may be immeasurable or just simply unknown. For instance, it is possible for the decisionmaker to state that A^i is not as good as A^j and that A^j is not as good as A^i. This statement infers that the decisionmaker is uncertain as to preference between A^i and A^j, and does not want to commit to either one at the present time. Alternatively, it might be the case that A^i and A^j are incomparable due to absence of some form of independence condition on the attributes, or for some other reason of immeasurability. So, it is assumed that a weak preference relation exists for all pairs of alternatives though this relationship may not be totally known.

An indifference relation ~ on F_X can be defined as follows. For any two alternatives A^i, A^j we say that the decisionmaker is equally happy with A^i or A^j, written $A^i \sim A^j$, if and only if $(A^i \succsim A^j) \cap (A^j \succsim A^i)$. The indifference relation is reflexive, symmetric, and transitive, so that it is an equivalence relation. Table 1 presents a summary of the above discussion, including the structural assumptions made about preference information and the interpretations for each combination of assessment that the decisionmaker may state.

For any two alternatives A^i, A^j in F_X

$(A^i \succsim A^j)$	$(A^j \succsim A^i)$	Assume	Interpretation
FALSE	FALSE	FALSE	Indecision, Ignorance, Immeasurability
FALSE	TRUE	TRUE	A^j at least as good as A^i
TRUE	FALSE	TRUE	A^i at least as good as A^j
TRUE	TRUE	TRUE	A^i & A^j equally preferred

Table 2. Behavioral interpretation of preference assessments.

The only logical connection between the relation $\succ$ and $\succsim$ is $A^i \succ A^j \Rightarrow \sim(A^j \succsim A^i)$. The contrary is not true. Stating that $\sim(A^i \succsim A^j)$ yields no implication about the validity of $(A^j \succsim A^i)$. In general, the negation of any form of preference relation does not have any implication on the validity of the other form of preference relation. This concept is not standard in utility theory which presumes, due to the existence of a real-valued utility function, that the logical implication holds in both directions in that the existence or absence of one form of preference negates or supports the existence of the other. Under such conditions, indifference between alternatives A^i and A^j may arise due to immeasurability. It is common to find the definition of indifference as $A^i \sim A^j \Leftrightarrow \sim(A^i \succ A^j) \cap \sim(A^j \succ A^i)$. The condition in the right hand side of the implication may be caused by immeasurability of preference among the alternatives. The results presented here avoid this problem by allowing imprecision and incompleteness in the assessments, and therefore are more behaviorally meaningfull in a situation in which partial information only is available.

The set of consequences F_X is a subset of the power set of X. F_X may contain, depending on the decision situation, consequences that are infeasible. An interesting proposition that is useful both for generating the efficient frontier of the range set of A, R(A), and for directing and controlling the assessment protocol consists of defining an extended efficient frontier as the set of consequences of F_X that are not dominated. Observe that R(A) is a subset of F_X and since we have identified the efficient frontier of F_X, we consequently may identify the efficient frontier of R as the set of feasible alternatives in the extended efficient frontier. The problem of generating the extended efficient frontier using the concept of dominance should be then the simplest case of the problem of investigating preferences among the alternatives for, as it implies, only the ordinal preference of the alternatives for each consequence need to be considered.

The logical connection between the strict preference relation $\succ$ and the "is at least as preferred as" relation $\succsim$ thus indicates a somewhat more meaningful interpretation in the judgement of preferences as compared with that previously stated. The negation of one relation does not have any implication on the validity of the other relation. That is, $A^i \succ A^j \Rightarrow \text{not}(A^j \succsim A^i)$, and the contrary is not true. Likewise, indifference between two alternatives can not arise by the absence of strict preference but only by stating that each alternative is as good as the other. Thus we have $A^i \sim A^j \Leftrightarrow (A^i \succsim A^j)$ and $(A^j \succsim A^i)$.

A result concerning dominance and preferential independence is of major use here. In a multiattribute decision problem, dominance as an optimal decision rule is equivalent to preferential independence of the attributes . Three interpretations immediately follow from this result:

1. Multiobjective programming formulations of complex decision problems depend on the concept of dominance to generate the set of nondominated alternatives. In their development it is often stated that the form of the overall utility function $U(u_1,\cdots,u_n)$ is unknown and, on occasions,

it is also claimed that no independence assumption on the attributes need be made. Our results show that these methods provide optimal results if and only if the attributes are preferentially independent, so that U is a multilinear function of the single attribute utility functions (u_i). By assuming a multilinear form of U and assessing the parameters of that function by means of regression techniques, as suggested in Hammond's work on social judgement theory [Hamm80] , results identical to those obtained by multiple objective optimization theory should be obtained as in the work of Dewispelare and Sage [Dewi81].

2. It provides a very natural test for preferential independence of the attributes. If the decision maker can state preferences among alternatives for each attribute, and clearly state that an alternative would be selected as the best if it is superior to all others for each attribute, we have good reasons to believe that the attributes, for that specific problem, are preferentially independent.

3. The proof of this statement, together with results stated earlier, provides a very simple procedure to check for preference patterns. Its formulation is

   $\min \; Q(A^i) - Q(A^j) = W$

   subject to: Explicit assessments

   Inferred assessments from independence conditions

   Decision rule: if $W \geq 0 \Rightarrow A^i \succeq A^j$.

The explicit and inferred assessments are in the form of linear inequalities, making this a simple linear programming problem. Obviously, the procedure does not rely on independence assumptions since inferred assessments will not be included in the program, but the amount of information required for reducing the nondominated set of alternatives is greatly diminished when these conditions hold. This procedure is believed to be more powerful than those studied earlier because it requires no assumption whatsoever, typical of these other methods, concerning preferential independence and at the same time retaining computational simplicity.

Interpretation 1 will not be pursued any further here. Interpretations 2 and 3 form the basis of the interactive approach for screening candidate alternatives. Together with the methodology for directing and controlling assessment of the required information, this comprises the heart of the dialog generation and management subsystem that is our adaptive inquiry decision support system.

In summary, we have proved the following result: Given that the decision maker is consistent in providing preferential assessments, the alternative that is superior in each attribute is clearly the most preferred alternative if and only if the attributes are preferentially independent. This result has significant implications concerning the validity of the notion of dominance as an optimal decision rule. For instance, suppose attribute X_1 is coffee and X_2 is sugar and alternatives $a_i=(x_i^1,x_i^2)$ are different mixes of coffee and sugar. You might prefer $a_1=$(6oz,1tspoon) over $a_2=$(8oz,4tspoon); yet by the dominance concept $a_2 \succ a_1$. This is really an argument against the structural modeling of the problem that lead to attributes that are not preferentially independent. Consequently, dominance can not be used as a decision rule unless we have preferential independence. Instead, if we let attribute X_1 be the amount of coffee and X_2 be a differential proportion of coffee and sugar, we may have preferential independence of the attributes.

The result presented here relates the behavioral attitude of the decision maker toward expressing preferences and the existence of some independence conditions on the attributes. Thus, it gives us a very natural heuristic test for preferential independence - if the decision maker can state preferences among alternatives for each attribute, and clearly states preference for an alternative as the best if it is superior to all others for each attribute, we have good reasons to belief that the attributes, for that specific decision problem, are preferentially independent.

INCONSISTENCIES

When assessing and representing human judgements about the value or utility of certain actions and outcomes, concerns relative to veridicality usually arise. We posit the existence of a mapping Q that corresponds to the assessments of outcomes in the extended consequence space into a utility space. A method to resolve the decision situation has been developed for the case where the assessments are imprecise and incomplete, but consistent. A necessary and sufficient condition for the assessements to be consistent is that the mapping Q be single valued and monotonic.

A case of inconsistency could be identified when obtained assessments result in Q being a multiple valued mapping. It should be possible, at least conceptually, to examine the judgement process to determine and explain to the decisionmaker which judgements are the results of errors in measurement, response, processing, or framing effect, and to suggest corrective procedures that will encourage single and double loop learning [Agr83] with respect to both decisionmaker values and perceptions relative to the decision situation.

CORRECTING INCONSISTENT JUDGEMENTS

Methods to reconcile inconsistencies have been developed primarily for assessments concerning probabilities; relatively less research is available with respect to utility assessments. While methods to combine probability judgements from disparate sources is closely related to our interest here, we will restrict attention in this paper to resolution of incoherent judgements coming from a single decision maker, but obtained by different response mechanisms or with instability over time due to learning effects.

The simplest way to deal with inconsistencies is to avoid them, for example by obtaining a minimally specified set of information sufficient to enable conclusions about an issue [Brow82] . This can, of course, lead to selective perception of a problem. It eliminates the possibility of inconsistent judgements, but at the same time prevents the decision maker from looking at a problem from a variety of different perspectives such as to enhance confidence in the resulting decision. A dual to this approach consists of partitioning a large set of inconsistent information into maximally specified subsets that are internally consistent [Resc70]. A maximally consistent subset (MCS) S_i of a larger set S of inconsistent propositions is defined as follows:

1. S_i is a nonempty subset of S
2. S_i is consistent
3. No element of S that is not a member of S_i can be added to it without generating an inconsistency.

The basic idea here is to identify the MCS's and select, on the basis of some indexing criteria, that subset which represents best the decision situation extant. Dialectic approaches [Maso81], for example, rely on promoting a dialogue by systematically generating a constructive dissagreement with the existing propositions. The multiple representation, multiple perspective view of this approach has been proposed by various authors as a descriptive as well as a prescriptive approach to study issues that are complex and not well understood initially [Lins81] , and as a means to deal with errors in utility and probability judgements [Krzy80], [Brow82]. A way of selecting a representation is suggested by Brown and Lindley [Brow82].

They encourage "digging in the subjects psychological

field." The claim is that people frequently make judgements without accounting for all the important facts that they already know. People, by looking at a problem from different perspectives and resolving any inconsistencies by analyzing their own thougths should presumably provide sounder judgements. As Brown and Lindley point out, this approach is inadequate when people have problems in resolving inconsistencies in an unaided manner. It could be a viable means for improving the quality of judgements in aided situations.

There are approaches intended to resolve inconsistent judgements whenever people have no apparent way of resolving them. They are based primarily on estimation techniques that force the parameters assessed to be changed sufficiently, such as to reconcile any inconsistencies present. Approaches such as these has been applied in situations involving assessments of probabilities and utility values [Free81], [Lind79], [Rosi81].

The method proposed here consists in transforming inconsistencies back into imprecise judgements. Since the approach previously discussed can deal with imprecise judgements it will consequently be able to handle inconsistencies if we can perform a suitable transformation. A simple and appropriate transformation is that of defining a new interval that contains the elements of the disjoint set of intervals causing Q to be multiple valued. In our notation, if Q is in two disjoint subsets I_i and I_j in I, then we can define a new interval. This is a very conservative transformation because it does not involve any adjustment of previously stated judgements, as is typical of estimation approaches. Therefore it may be implemented without the specific consent of the decision maker if it does not change any previously accepted conclusions.

The steps in the approach envisioned here consist of activities which result in:

1. a minimally specified set of information,
2. multiple representations,
3. estimation, and
4. transformation to imprecision.

The use of these approaches and the information we could obtain from each of them provides significant insights concerning the nature of the error in judgement causing the inconsistency, and may provide feedback information to the decision maker as an aid to help revise previous judgements or to discover new information. For instance, if an inconsistency is transformed to imprecision and this does not change any previously obtained conclusion, then the inconsistency may be irrelevant such that there is no immediate need to resolve it. The sum of squared errors in a reconciliation by estimation methods indicate the magnitude of the inconsistency and may suggest if the method, in a specific instance, is succeeding or not. Presentation of various sets of minimally and maximally consistent propositions to the decision maker should serve as an aid to determination of the most appropriate knowledge representation.

ASSESSING THE REQUIRED INFORMATION

Like most interactive procedures, the one described here lacks a structured process for the determination and acquisition of relevant information. As we have noted, lack of structure in the assessment step usually results in inefficiency and ineffectiveness. Effectiveness implies the identification of a minimal information set for a data base, such that when the date is aggregated by the model base management system, there results an "optimal" recommendation to the decisionmaker. Efficiency relates to performing the above task in accordance with some predefined performance criteria which could include time, cognitive effort, understandability, and many others that could be dependent of the contingency task structure of the decision situation.

The goal of the proposed inquiring system is to identify a set of minimal information necessary to induce a linear order in the set of candidate alternatives. This minimal set will constitute the inquiry pattern, or set of prompts that the system will present to the decision maker in order to learn about value perspectives and relevant factual information perceived as relevant to the decision situation. At any time, the decisionmaker may decide not to respond to any prompt. The decision support system will then search for another inquiry pattern leading to another, possibly the same, linear order. This approach is aimed at achievement of effectiveness of the overall process. Effectiveness, in this case, clearly requires adaptation and learning.

This goal seeking behavior can be formulated as a set of simple deterministic dynamic programming problems that correspond to the knowledge base diagram of Figure 2. The states S_k represent the possible orders of the alternatives. These are represented in Figure 2 by rectangles in which the initial state S_o represents no order on the set of alternatives, the final state S_N is a set of possible linear orders, and intermediate states S_k are partial orders. The policies m_k represent a set of admissible inquiries. There are a

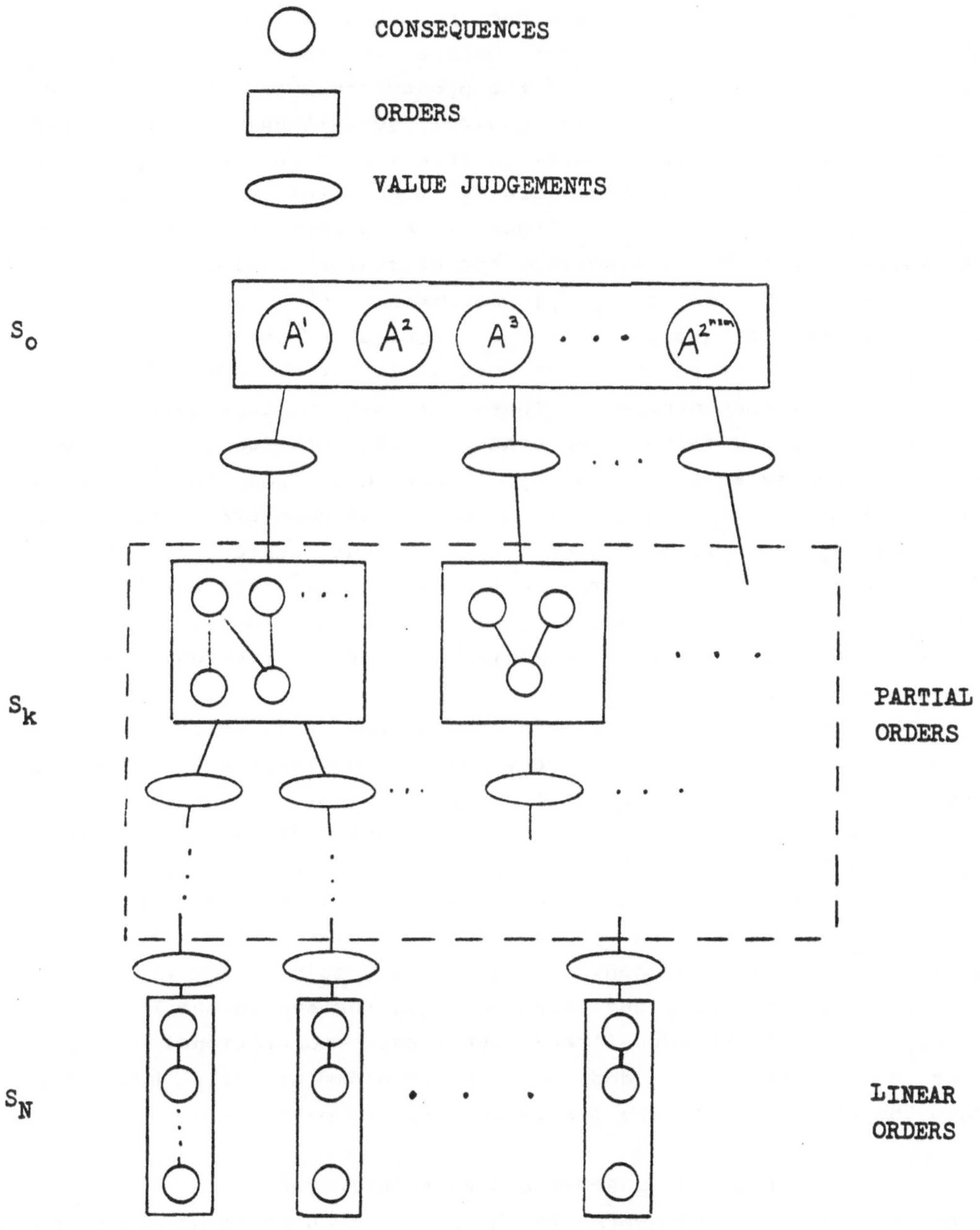

Figure 2. Knowledge base

large number of possible preference comparisons in F_X, some of which may be more susceptible than others to the use of biases and flawed heuristics. Through use of the procedures suggested here, we are potentially able to obtain general guidelines, based on the existing psychological literature, to select from the set of possible inquiries those that are behaviorally meaningful to the decision maker. These are represented in Figure 2, a conceptual representation of the dynamic programming algorithm for search, as ellipses.

The cost function $g_k(S_k, m_k)$ represents the cost associated with responding to query m_k at a particular state. There will generally exist queries that are more difficult to understand, and hence respond to, than others. There will also be some queries that lead to a final state faster than others. The cost function reflects this. The system equation $f(S_k, m_k)$ is obtained from the decision model for interactive aiding. It computes the next state (order) of the system $S_{k+1} = f(S_k, m_k)$ given the current state (order) and the respond to query m_k. The objective of all of this is to minimize the cost of the inquiring process through identification of the optimal inquiry patterns, or minimal set of information, required to provide a recommendation.

The dynamic programming formulation for the search in the knowledge base may result in a very complex and large set of dynamic programming problems. The set of final states grows very rapidly with the number of alternatives (m) and the number of attributes (n). There are $2^{nm}!$ possible linear orders. This may render this approach inefficient due to the computational complexity required each time a response to a query is made. Complexity is reduced when some information is learned. Thus, it is possible to start the process with heuristically generated queries and search the knowledge base by means of breadth-first/depth-first operations. This combination of heuristic and dynamic programming will hopefully improve the efficiency of the search for inquiry patterns.

DECISION SUPPORT SYSTEM ARCHITECTURE

An appropriate framework in which knowledge could be organized and utilized efficiently and effectively is desired. An inquiring system based on the concepts of representation, acquisition, and aggregation of knowledge presented here will potentially accomplish this. A conceptual organization of the resulting decision support system is presented in Figure 3. The knowledge base, described by Figure 2, represents and organizes knowledge about value perspectives and factual information. The dialogue generation and management

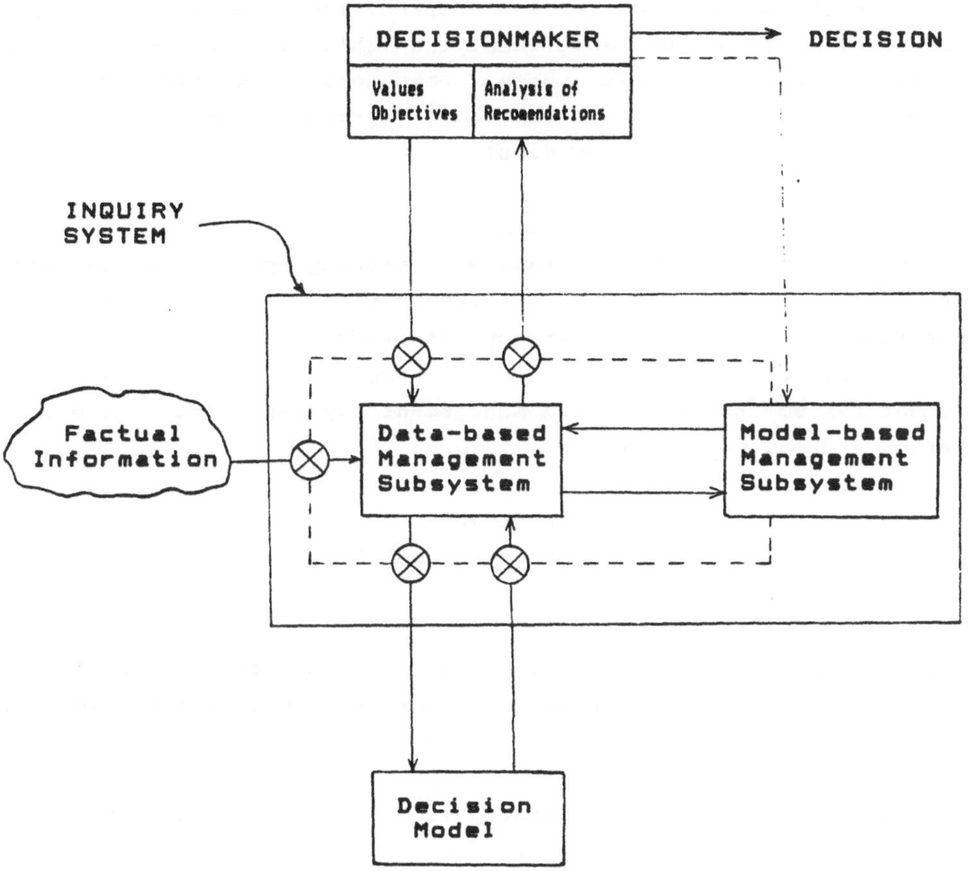

Figure 3. Proposed inquiry system for decision support.

system is in charge of directing and controlling the assessment of the required information and supplying this to the model base management system. These three components constitute decision support system, including the inquiry system or interface between decision maker and the decision model.

SUMMARY

We have described an interactive screening approach to decision support that allows for the incorporation of imprecision, incompleteness and inconsistency of information in the three components that constitute a decision support system: the data base management system, the model base management system, and the dialog generation and management system.

REFERENCES

[Agry82] Agyris, C. Reasoning, Learning and Action, Jossey-Bass, San Francisco, 1982.

[Brow82] Brown, R.V. and Lindley, D.V. "Improving Judgement by Reconciling Incoherence," Theory and Decision , Vol.14, 1982, pp.113-132.

[Chan84] Chankong, V., and Haimes, Y. Y., Multiobjective Decision Making: Theory and Methodology, North Holland, 1984.

[Char78] Charnetski, J.R. and Soland, R.M., "Multiattribute Decision Making with Partial Information," Naval Research Logistic Quarterly , Vol 25, No. 2, June 1978, pp. 279-288.

[Char79] Charnetski, J.R. and Soland, R.M., "Multiattribute Decision Making with Partial Information: The Expected Value Criterion," Naval Research Logistic Quarterly , Vol 26, No. 2, 1979, pp. 249-256.

[Coho78] Cohon, J.L., Multiobjective Programming and Planning , Academic Press, New York, 1978.

[DeWi81] DeWispelare, A., and Sage, A. P., "On Combined Multiple Objective Optimization Theory and Multiple Attribute Utility Theory for Evaluation and Choice Making," Large Scale Systems, Vol. 2, No. 1, February 1981, pp. 1-19.

[Einh81] Einhorn, H.J. and Hogarth, R.M., "Behavioral Decision Theory: Processes of Judgement and Choice," Annual Review of Psychology , Vol. 32, 1981, pp. 53-88.

[Fish64] Fishburn, P.C., Decision and Value Theory , Wiley, New York, 1964.

[Fish65] Fishburn, P.C., "Analysis of Decisions with Incomplete Knowledge of Probabilities," Operations Research , Vol 13, No. 2, March-April 1965, pp. 217-237.

[Fish70] Fishburn, P.C., Utility Theory for Decision Making , Wiley, New York, 1970.

[Free81] Freeling, A.N.S., "Reconciliation of Multiple Probability Assessments," Organizational Behavior and Human Performance , Vol. 28, 1981, pp. 395-414.

[Hamm80] Hammond, K.R., McClelland, G.H., and Mumpower, J., Human Judgement and Decision Making: Theories Methods and Procedures , Hemisphere/Praeger, New York, 1980.

[Howa80] Howard, R.A., "An Assessment of Decision Analysis," Operations Research , Vol 28, No. 1, Jan-Feb 1980, pp. 4-27.

[Kahn82] Kahneman, D., Slovic, P., and Tversky, A. (Eds) Judgement Under Uncertainty: Heiristics and Biases , Cambridge Univ. Press, New York, 1982.

[Keen76] Keeney, R.L. and Raiffa, H., Decisions with Multiple Objectives: Preferences and Value Tradeoffs , Wiley, New York, 1976.

[Keen82] Keeney, R.L., "Decision Analysis: An Overview," Operations Research , Vol. 30, No. 5, Sept-Oct 1982, pp. 803-838.

[Kmie81] Kmietowicz, Z.W. and Pearman, A.D., Decision Theory and Incomplete Knowledge , Gower Publishing Co., England 1981.

[Kran71] Krantz, D.H., Luce, R.D., Suppes, P. and Tversky, A., Foundations of Measurement , Vol. I, Academic Press, new York, 1971.

[Krzy80] Krzysztofowicz, R. and Duckstein, L. "Assessment Errors in Multiattribute Utility Functions," Organizational Behavior and Human Performance , Vol. 26, 1980, pp. 326-348.

[Lind79] Lindley, D.V., Tversky, A. and Brown, R.V., "On the Reconciliation of Probability Assessments," Journal of the Royal Statistical Society Series A, Vol. 142, part 2, 1979, pp. 146-180.

[Lind83] Lindley, R.V., Review of Kmietowicz, Z.W. and Pearman, A.D., Decision Theory and Incomplete Knowledge , European Journal of Operational Research , Vol. 13, No. 3, July 1983, pp. 334-335.

[Lins81] Linstone, H.A. "The Multiple Perspective Concept," Technological Forecasting and Social Change , Vol. 20, 1981, pp. 275-325.

[Mask79] Maskin, E., "Decision Making under Ignorance with Implications for Social Choice," Theory and Decision , Vol 11, 1979, pp. 319-337.

[Maso81] Mason, R.O. and Mitroff, I.I., Challenging Strategic Planning Assumptions , J. Wiley, 1981.

[Moor79] Moore, R.E., Methods and Applications of Interval Analysis, SIAM, Philadelphia, 1979.

[Nadl83] Nadler, G. "Human Purposeful Activities for Classifying Management Problems," OMEGA , Vol. 11, No. 1, 1983, pp. 15-26.

[Neum64] von Neumann, J. and Morgenstern, O., Theory of Games and Economic Behavior , 3rd Edition, Wiley, New York, 1964.

[Resc70] Rescher, N. and Manor, R. "On Inference from Inconsistent Premisses," Theory and Decision , Vol. 1, 1970, pp. 179-217.

[Robe79] Roberts, F.S., Measurement Theory , Addison-Wesley Publishing Co., 1979.

[Rosi81] Rosinger, E.E., "Interactive Algorithm for Multiobjective Optimization," Journal of Optimization Theory and Applications , Vol. 35, No. 3, 1981.

[Sage81] Sage, A. P., "Behavioral and Organizational Considerations in the Design of Information Systems and Processes for Planning and Decision Support," IEEE Transactions on Systems, Man and Cybernetics, Vol. 11, No. 9, September 1981, pp. 640-678.

[Sage84a] Sage, A. P., and White, C. C., "ARIADNE: A Knowledge Based Interactive System for Planning and Decision Support," IEEE Transactions on Systems, Man, and Cybernetics, Vol. 14, No. 1, January 1984.

[Sage84b] Sage, A. P., Galing, B., and Lagomasino, A., "Methodologies for Determination of Information Requirements for Decision Support Systems," Large Scale Systems, Vol. 5, No. 3, 1984.

[Sari77] Sarin, R.K., "Screening of Multiattribute Alternatives," OMEGA , Vol 5, No. 4, 1977, pp. 481-489.

[Schw83] Schwenk, C. and Thomas, H., "Formulating the Mess: The Roles of Decision Aids in Problem Formulation," Omega , Vol 11, No. 3, 1983, pp. 239-252.

[Slov77] Slovic, P., Fischhoff, B., and Lichtenstein, S., "Behavioral Decision Theory," Annual Review of Psychology, Vol. 28, 1977, pp. 1-39.

[Tamu83] Tamura, H. and Nakamura, Y., "Decompositions of Multiattribute Utility Functions Based on Convex Dependence," Operations Research , Vol 31, No. 3, May-June 1983, pp. 488-506.

[Whit80] White, C.C. and Sage, A.P., "A Multiple Objective Optimization Optimization Based Approach to Choice Making," IEEE Trans. on Systems Man and Cybernetics , SMC-10, No. 6, June 1980, pp. 315-326.

[Whit82] White, C.C., Sage, A.P., and Scherer, W.T., "Decision Support with Partially Identified Parameters," Large Scale Systems , Vol 3, 1982, pp. 177-189.

[Whit83a] White, C.C., Dozono, S., and Scherer, W.T., "An Interactive Procedure for Aiding Multiattribute Alternative Selection," OMEGA , Vol 11, No. 2, 1983, pp. 212-214.

[Whit83b] White, C.C., Sage, A.P., Dozono, S., and Scherer, W.T., "Performance Evaluation of a Decision Support System," Large Scale System , Vol. 5, No. 2, 1983.

[Whit78] Whitmore, G.A. and Findlay, N.C. (Eds) Stochastic Dominance , Lexington Books, Mass. 1978.

[Zele82] Zeleny, M. Multiple Criteria Decision Making , McGraw Hill, New York, 1982.

ADAPTIVE ROUTING IN RADIO COMMUNICATION NETWORKS

Anthony Ephremides
Department of Electrical Engineering
University of Maryland
College Park, Maryland 20742

Abstract

After a brief review of the field of routing algorithms for communication networks this paper focuses on adaptive routing for radio networks of mobile nodes. Two algorithms are proposed and their performance is evaluated and compared to that of flooding.

I. Introduction

Few problems in the field of computer communication networks have received as early and extensive attention as the routing problem. The reason for this is twofold. First, there have been numerous studies of flow problems in network theory that provide a basis for modeling the routing problem in a mathematically tractable way and, second, routing has proved to be a necessary and fundamental design choice for the operation of message or packet switched networks.

In its simplest form the statement of the routing problem is as follows. Consider a node that receives messages from an incoming link and has a choice of forwarding them via one of two outgoing links. Let λ be the average rate of the arrival message stream. By implementing a specific routing decision the node produces two outgoing streams the average rates of which are λ_1 and λ_2. Of course in order to maintain equilibrium and finite delays it is necessary that

$$\lambda_1 + \lambda_2 = \lambda$$

and thus λ_1 and λ_2 can be represented as

$$\lambda_1 = \phi\lambda \quad , \quad \lambda_2 = (1-\phi)\lambda$$

where $0<\phi<1$. The routing problem consists of choosing ϕ to maximize a given performance measure which is usually related to the average delay per message in the network. A variety of constraints usually accompany this optimization problem. A more detailed formulation will follow later. Now we provide an early view of the classification of routing algorithms according to various solution philosophies. These philosophies emerge according to the answers chosen for the following questions:

1) How is ϕ chosen (i.e. by whom and on the basis of what information)?

2) How often is ϕ updated?

3) Are these any limitations in the range of values ϕ can take?

The first question leads to a distinction between centralized and distributed routing algorithms.

The second question distinguishes static from dynamic algorithms.

Finally the third question separates fixed from alternate routing algorithms.

II. Early Analytical Approaches

Unlike other communication network design problems, such as topological design or error and flow control, the routing problem is amenable to analytical formulation, if certain assumptions are made. We will describe here a few of the most notable analytical approaches.

In [1] the first mathematical model was proposed to accomplish centralized, static, alternate routing. To appreciate the degree of difficulty associated with the analytical treatment of the routing problem, we must emphasize that even though the centralized, static, alternate problem is the least complex of the routing problems, the model proposed in [1] falls short of obtaining a true solution even in the presence of strong, idealized assumptions.

Consider a network of N nodes and M links. Let $r_i(j)$ represent the average rate of exogenous message traffic entering the network at node i and destined for node j for $i,j = 1,\ldots,N$. These quantitites represent the traffic load matrix. Each entry represents the value of one commodity; that is each source destination pair corresponds to a separate commodity. Let C_i represent the capacity of link i, $i=1,\ldots,M$ in bits/s.

Assume that the average length of each message is $\frac{1}{\mu}$ bits. Suppose that the routing variables are chosen so as to induce traffic of average rate λ_i on link i, $i=1,\ldots,M$. Obviously the value of λ_i is determined by the $r_i(j)$'s and by the routing variables. If the assumption is made that link i behaves like an $M|M|1$ queueing system independently of all other links and with customer arrival rate λ_i and service rate μC_i, then it is possible to express the average delay per message for traversing that link as

$$T_i = \frac{1}{\mu C_i - \lambda_i}$$

for $\lambda_i < \mu C_i$. The preceding assumption is a very strong one and requires some discussion. It was observed early by Kleinrock [2] that packet-switched or, more generally, store-and-forward communication networks can be modeled as queueing networks. Each link can be thought of as an autonomous service system with service time composed of three components: propagation time which is independent of message length and is negligible for short distances), transmission time (which depends on link capacity and message length), and processing time (often quite small and partly dependent on message length). Messages arriving at a node are stored in a buffer of sufficiently large capacity, so that overflows are rate. Thus each link has all the attributes of a queueing system. Unfortunately only a very specialized class of interconnected queueing systems is amenable to analysis. This is the class of Jacksonian networks, so termed because of the pioneering study of them by Jackson [3]. Recently a great deal of attention has been paid to non-Jacksonian networks for the purpose of extending to them the analytical treatment techniques that were successful in the case of Jacksonian networks. A Jacksonian network consists of <u>independent</u> exponential servers that are interconnected so that a customer may visit several of them in succession. The exogenous customer arrival streams are assumed to be Poisson processes. In order for a communication network to be usefully modeled by a Jacksonian network it is therefore necessary to assume the following:

a) the exogenous message arrival streams are Poisson processes.

b) the service time for each message on each link is exponentially distributed with rate μC_i.

c) the service times at different links for the <u>same</u> message are independent.

The last assumption is the truly problematic one since each message retains its length as it traverses successive links. Thus the "service" times it encounters on different links are strongly correlated. Making the crucial <u>independence</u> assumption permits the use of Jacksonian models for communication networks. For certain types of, so called, "large-and-balanced" networks experimental results have shown that the independence assumption can be tolerable if used with care and applied judiciously [4].

In any event, with the independence assumption the routing

problem becomes

$$\min_{\{\lambda_i\}} T$$

where $T = \sum_{i=1}^{M} \lambda_i \frac{1}{\mu C_i - \lambda_i}$; that is, minimize the weighted total average delay subject to the constants $0 \leq \lambda_i < \mu C_i$. The optimization can be carried out by a variety of classical methods since T is a convex function of the λ_i's. The flow deviation method (a form of steepest descent algorithm) was used in [1] to produce a set of optimizing link flow rate values. For this reason the approach was called the flow assignment procedure. Note that just by obtaining a set of optimal flows we have not yet solved the routing problem. The next step is to find the values of the routing variables (the ϕ's) which when applied to the given input variables (the $r_i(j)$'s) will produce the desired link flows (the λ_i's). It is clear that for certain topologies of the network and for certain regions of values of the input variables it is not possible to meet the "λ" values obtained by solving the minimization problem. In those cases the already complicated mathematical programming problem of matching the inputs to the induced flows by means of the appropriate splitting fractions "ϕ" becomes an approximation problem in which the objective becomes not to find exactly, but simply to come "close" to the desired values.

The next major breakthrough in the routing problem was Gallager's algorithm [5] which appeared in the literature in 1977. In that algorithm a solution was obtained for a distributed, alternate, quasi-static routing problem that minimized weighted average delay, that guaranteed the absence of loops, and that did not explicitly require the independence assumption.

Using the notation introduced above and defining the following additional quantities we can briefly state the nature of the algorithm's result. Let

$r_i(j) \triangleq$ average rate of exogenous traffic entering at i for j as before),

$t_i(j) \triangleq$ average rate of total traffic entering i for j,

$\lambda_{ik} \triangleq$ average rate of combined traffic of all commodities on link (i,k), i.e. on the link that connects nodes i and k (replacing what was called λ_i before); this quantity is zero if there is no link (i,k), and

$\phi_{ik}(j) \triangleq$ the routing variable of node i for link (i,k) and for commodity (destination) j; equivalently, the percentage of the commodity j that node i routes to neighbor k (if there is no link (i,k), then $\phi_{ik}(j) = 0$).

Clearly the following are true:

$$t_i(j) = r_i(j) + \sum_{k=1}^{N} \phi_{ki}(j) t_k(j), \quad \forall\, i,j$$

$$\lambda_{ik} = \sum_{j=1}^{N} t_i(j) \phi_{ik}(j) \quad , \forall\, i,k$$

$$\phi_{ik}(j) \geq 0 \quad , \forall\, i,k,j$$

$$\sum_{k=1}^{N} \phi_{ik}(j) = 1 \quad , \forall\, i,j$$

Instead of assuming that the average delay T_{ik} per message on link (i,k) is given by the M|M|1 formula used before, which would be tantamount to making the independence assumption, it is simply assumed that the delay is an increasing, convex function of the link flow f_{ik} only*. Then the routing problem becomes

$$\min_{\{\phi\text{'s}\}} T$$

where

---

* it can be argued that this assumption is equivalent to assuming independence because only for a network of interconnected <u>exponential</u> servers is it true that the average delay in each stage depends on the average flow in that stage only and not on the average flows in the other stages. However, Gallager's algorithm has been generalized to allow arbitrary convex increasing dependence of link delay on all of the flows [6].

$T \Delta \sum\sum_{i,k} f_{ik} T_{ik}$; that is minimizing the average weighted total delay as before. So the problem looks like the earlier one except that it attempts minimization directly with respect to the routing variables rather than indirectly by means of the link flows, and that it does not make the independence assumption; that is, it still looks like a static, centralized, alternate routing problem, but in a less restricted framework. It is during the attempt to solve this problem that it became clear that a distributed and quasi-static implementation of the solution was feasible, thus transforming the algorithm into a distributed, quasi-static, alternate routing scheme. Convergence of that implementation was shown and a scheme to avoid loops at all stages of the algorithm was incorporated in the solution. A major difficulty that had to be resolved was the non-convex dependence of the objective function on the routing variables. Another non-trivial complication associated with this algorithm is the need to find an initial feasible and loop-free set of flows by other means. There exist algorithms and methods to achieve this goal, however, and thus it is possible to establish the initial conditions the algorithm needs. The nature of the algorithm is briefly the following. Each node, between periodic updates, measures a local "distance measure" to his neighbors. It then passes to, and receives from, his neighbors information about those measurements. This exchange of information must follow specific rules of order in order to avoid loops and to guarantee convergence. Then each node alters the previous values of the ϕ_{ik}'s by appropriate step sizes so that more traffic is sent towards those neighbors that promise less delay, and less traffic is routed via the neighbors that exhibit tardiness. If the rate of change in the input variables is less than the convergence rate of the algorithm, it becomes possible to "track" the shifting optimum set of values of the routing variables.

An important drawback of this algorithm is that it cannot be implemented in its pure form because of the difficulty in obtaining well-behaved estimates of the quantities that each node must measure and because of the problems associated with its convergence. Its chief value lies in the fact that it showed how analytical treatment can "drive" the choices of practical routing algorithms, thus making total reliance on heuristics unnecessary. It also inspired subsequent work on routing that eventually led to very practical and robust schemes that can be, and have been, implemented.

It was later shown by Bertsekas [7,8] that Gallager's algorithm can be viewed as a special case of a much more general class of optimization techniques known as projected Newton methods and Goldstein-Levitin-Poljak gradient projection methods. These methods have been adequately described in the literature but have failed so far to be translated into useful routing algorithms. It is noteworthy that the end-to-end flow control problem which was previously unanalyzed and unmodeled, was reduced to a routing problem in the context of the Gallager-Bertsekas class of models by means of a simple ingeneous observation explained in [9].

III. The Radio-Network Case

The discussion so far has implicitly assumed that the networks considered are of the store-and-forward type with land-line links. An equally important class of networks consists of the so called radio-networks. The links in these network are radio-based. The key difference in this case is that a radio link is a multiple access and broadcast channel; that is a node's transmission is subject to interference from the transmissions of other nodes and it can be also received by nodes other than the destination node. These properties of the radio links give new dimensions to the routing problem. Furthermore in a radio network a node is not necessarily stationary. Mobility of the nodes may induce frequent topological changes that make the routing problem much more difficult and challenging.

In this section we will describe two routing algorithms that are designed for a special type of a radio network and we will compare several attributes of their performance to those of their chief competitor, namely flooding [10].

Consider the case of N mobile nodes that wish to remain interconnected via radio links. Suppose that the communication range is variable and cannot cover the entire geographical area of node dispersion. Many factors may account for such circumstances: limited power, interference, variable antenna orientation, node motion, physical objects, etc. In such an environment the first question is how these nodes can organize themselves into a reliable network by means of a distributed algorithm. This question corresponds to the physical layer design according to the OSI architecture model. The second question is how the discovered links can be activated by each node, again in a distributed way, and without interference. This question addresses the multiple access aspect of the radio environment and corresponds to the link layer of the OSI model. These questions can be handled in a variety of ways. In references [11-14] some solutions and algorithms are described that can take care of these fundamental issues.

Next comes the questions of routing that resides naturally in the third (or network) layer of the OSI architecture. How should a node send a message to a remote destination? Clearly due to the assumed volatility of the connectivities in the network it is not possible to employ any of the existing methods of routing. In other words we cannot assume that a node may know the location of the desination node or his distance from that node. The only exception is flooding which can still be used. It must be noted, however, that in a radio environment flooding has additional disadvantages since it generates unnecessary interfering and bandwidth consuming traffic. Nevertheless flooding will ensure the delivery of the message to its destination. The question is at what cost relative to that of other candidate algorithms. Let us consider one alternative.

Suppose that node i wishes to send a message to node j. There is no prior knowledge about the location of j nor of his distance from i in terms of any specific distance measure. A natural thing to do is to send a short query to all neighbors of i. Instead of uncontrollably flooding this query message further out, i's neighbors perform the steps of a structured process. Each neighbor of i passes on the received query to every neighbor of his except to those from which he has already received the query. Furthermore he doesn't pass on that query to anyone if he had originated a query about the same destination node himself (and to which query he hasn't received a response yet). In this manner the query will start propagating in all directions away from the originating node. The first node who, on receipt of the query, finds that he has knowledge of the whereabouts of the destination node currently under search, either because that node is a direct neighbor or because he has already found out by means of a response to an earlier query, immediately generates a response which he transmits to all nodes from which he has received a query about that destination node. Each node who receives a response to a query passes it on to all those neighbors from which he has received queries. In this manner the response propagates back toward the inquiring node in a controlled fashion without unnecessary flooding. Clearly care must be taken in specifying the precise rules of order in such a process of propagation in order to avoid cycles and deadlocks under any possible relative ordering of the steps of this basically asynchronous process of transmissions and retransmissions of the same query. The procedure described above succeeds in avoiding cycles and deadlocks and terminates with the response to the query received by the originating node. It is possible however that there will be redundant responses since the response message may propagate along different parallel paths back to the origin.

The reason that a query, rather than the main message itself, is forwarded in this manner to the destination is twofold. First, it is a much shorter message since it consists merely of a flag and of the source destination identity. Secondly, and most importantly, all nodes that participate in this search process gain knowledge of the whereabouts of that particular destination node, eliminating thus the need for future queries of their own should they later desire to send a message to that node themselves. The path actually selected for routing the message after receipt of the response to the query is not important here. It could be simply the one via the neighbor through whom the response was received first, or a selection process can be implemented that is based on some distance measure. It is assumed that an acknowledgement message mechanism exists so that a node can detect a change of connectivities that destroys the previously sought and established path and can reinitiate a query. It is clear that the number of steps involved in a second search will be in general less than in the original one, since nodes with valid existing paths to the destination are likely to be encountered in the close vicinity of the inquiring node.

A simple modification to the algorithm just described may generate additional knowledge about the connectivities in the network at a rather negligible additional overhead cost that may eventually save in total overhead. It was assumed that the query message consisted only of a flag and of the identity of the destination node. Suppose it is enlarged to include the identity of the inquiring node as well. Furthermore, as an additional option, it may include a field for updating a "distance" value that could be simply the number of "hops". Each intermediate node augments the distance field entry before passing it on (if the distance measure is simply the number of hops, the entry is augmented by one). In this manner, as the query propagates outward, several nodes gain knowledge of the whereabouts of the <u>source</u> node, plus of its distance from them. This information can be stored at each intermediate node for future use. Thus many future queries become unnecessary. As far as the response message is concerned, it, too, may be enlarged to include an entry for distance. Thus, as it propagates back toward the source node, the distance entry is appropriately augmented to provide knowledge to all nodes that can be used to choose amongst alternate paths. In all other respects the algorithm remains unchanged.

In order to evaluate the performance of these algorithms it was necessary to utilize simulation techniques. It is impossible to analyze mathematically the performance, no matter what idealized assumptions are made, simply because there is strong dependence on topological layouts and on mobility scenarios neither of which can be analytically modeled. These simulation results were reported in [10].

REFERENCES

1. L. Fratta, M. Gerla, L. Kleinrock, "The Flow Deviation Method - An Approach to Store-and-Forward Communication Network Design", Networks, Vol. 3, pp. 97-133, 1973.
2. L. Kleinrock, Communication Nets - Stochastic Message Flow and Delay, McGraw-Hill, New York, 1964.
3. J.R. Jackson, "Networks of Waiting Lines", Operations Research, Vol. 5, pp. 518-521, 1957.
4. L. Kleinrock, Queueing Systems, Vol. II, Wiley-Interscience, New York, 1976.
5. R.G. Gallager, "A Minimum Delay Routing Algorithm Using Distributed Computation", IEEE Trans. on Communications, Vol. COM-25, pp. 73-85, 1978.
6. A. Ephremides, "Extension of an Adaptive Distributed Routing Algorithm to Mixed Media Networks", IEEE Trans. on Communications, Vol. 26, No. 8, August 1978.
7. D.P. Bertsekas, "A Class of Optimal Routing Algorithms for Communication Networks", Proc. 5th International Conference on Computer Communications, Atlanta, GA, October 1980, pp. 71-76.
8. D.P. Bertsekas, "Projected Newton Methods for Optimization Problems with Simple Constraints", SIAM Journal on Control and Optimization, Vol. 20, pp. 221-246, 1982.
9. R.G. Gallager, S.J. Golestaani, "Flow Control and Routing Algorithms for Data Networks", Proc. 5th International Conference on Computer Communications, Atlanta, GA, October 1980, pp. 779-784.

10. M. Weber, A. Ephremides, "A Simulated Performance Study of Some Distributed Routing Algorithms for Mobile Radio Networks," Proc. of Johns Hopkins Conference, Baltimore, MD, March, 1983.

11. J.E. Wieselthier, D.J. Baker, A. Ephremides, D.N. McGregor, "Preliminary System Concept for an HF Intra-Taska Force Communication Network", NRL Report 8637, August, 1983.

12. D.J. Baker, A. Ephremides, "The Architectural Organization of a Mobile Radio Network via a Distributed Algorithm", IEEE Trans. on Communications, Vol. COM-29, pp. 1694-1701, November, 1981.

13. D.J. Baker, A. Ephremides, J.E. Wieselthier, "A Distributed Algorithm for Scheduling the Activation of Links in a Self-Organizing, Mobile, Radio Network", Proc. of the ICC, Philadelphia, PA, June 1982.

14. A. Ephremides, "Distributed Protocols for Mobile Radio Networks", Proc. of NATO Advanced Study Institute on the Impact of Processing Techniques to Communications, Chateau Bonas, France, July, 1983.

MODELLING ALTERNATIVE STRUCTURES FOR TIME CRITICAL CORPORATE ADAPTATION

Bernard P. Zeigler and Robert G. Reynolds
Computer Technology Modelling Project
Department of Computer Science
Wayne State University
Detroit, MI 48202

Abstract

Our goal is to relate the organizational structure of a firm to the structure of the problem environment in which it operates. To this end a formal model of a prototypical organization adapting to changes in its problem environment is presented. Based upon distributed computer system concepts, the organization is modelled as a hierarchically co-ordinated modular system. The model firm's environment is time critical, i.e., it generates a sequence of problems which must be solved within a limited time after arrival. Interpreted for technologically-based firms, problems represent technology related product development opportunities, and the problem deadlines bound response windows within which opportunities must be capitalized upon. We study the ability of the organization to adjust its structure to accomodate changes in the parameters of its problem environment. In this context, organizational flexibility is characterized and shown to be in a trade-off relation to speed of response. Implications are suggested for the controversy surrounding the relative innovation potential of small firms versus large corporations.

Introduction

Although it is not often cited in the technological innovation literature, studies of the evolution of corporations suggest that information processing is an important factor in determining system structure. Based on extensive case studies, Chandler (1962) suggested that as a firm grows and diversifies, its administrative structure eventually suffers an information overload. Sherman (1983) presents a simulation model in which such overload is explicitly represented as noise in decision maker intercommunication. Such overload leads to a search for new organizational forms which limit the amount of information that individual managers receive and consequently, the number of decisions they must make. Moss (1981) suggests that discovery of such a form reestablishes managerial efficiency so that managers can devote attention to further growth and diversification. The innovative structure is most likely to be discovered by the largest and most successful firms, to be imitated and adopted by others, and thus to lead to a general increase in the level of output in the economy.

Although the environment is not an explicit actor in Chandler's theories, it is not a large step to reinterpret them in a manner consistent with ours: to grow, viz., to capitalize on the environmental opportunities available to it, a system must appropriately transform its information processing structure. Moreover, once the role of the environment is recognized it becomes possible to decompose it along dimensions that may enable us to get a better handle on its relation to information processing structure.

Another avenue that information processing might have entered into the conceptual framework of technological innovation is that of hierarchical control theory (Mesorovic, Macko and Takahara, 1970) where information flows co-ordinate goal seeking at the various levels. However, in his review of the literature, Gijsbrechts (1983) reports that a large gap persists between the ideal structures of the theory and the actual structures that seem to characterize real corporate hierarchies. Moreover, the relation of the information structure to the time critical nature of its environment has been little studied.

Our approach derives from perspectives in computer science that concern designs for distributed processing and flexible computer architectures (Carver and Conway, 1908; Kung, 1980; Kartashev and Kartashev, 1982; Hockney and Jesshope, 1981). Such designs suggest possible information processing structures such as pipelines. The performance (e.g. throughput) of such structures has been studied in relation to imposed workloads (Kogge, 1981). However, our characterization of the time critical environment and the relation to processor organization that this concept facilitates, while rather elementary, seems to be novel to the queueing theory-based performance modelling literature. First suggested by Reynolds (1984) for modelling human cultural evolution, the characterization is elaborated upon here to apply to corporate evolution as well.

This paper is a first attempt to model the evolution of technologically based firms from a point of view in which information processing is of the essence. All other considerations that might impact such evolution, such as economic viability or productivity, play no explicit role in the model, although they may be included implicitly in assumptions supporting the model interpretation. The questions that we wish to address with this approach concern structural transformations that firms may undergo in interacting with their environments. Actually, our characterization of these environments is a crucial feature of the approach. An environment is modelled as a succession of problems, each of which must be solved by the system within a given time span. The *opportunity rate* (frequency of problem arrival), the *response window* (time span) and the *problem variety* are taken to be characteristic parameters. Such problems represent opportunities that must be exploited by the firm within the response window in order to succeed. We suppose (and to a certain extent can show) that an environment has an optimum structural organization that matches it and that the system will evolve to that optimum structure over time. Such evolution must necessarily be constrained by structural transformations that are feasible (in which we include such considerations as capacity to effect the change, stability of the resulting system, etc.). Although the task of characterizing transformations is initiated here, the task of justifying their feasibility is left for further work.

Studying the evolution of problem solving systems in time-critical environments is intended to throw some light on issues that have been raised in the literature on technological innovation, for example: are small firms more innovative than large ones? can mature corporations re-organize themselves to survive, and even spearhead, major technological shifts? In view of the abstract character of our model, the light cast is suggestive, rather than definitive. But to our knowledge, it is the first attempt to establish a formal mechanism with which to interpret the many empirical studies heretofore characteristic of the literature of technological innovation and to critically examine the conclusions derived therefrom. First suggested by Reynolds (1984) for modelling human cultural evolution, the characterization is elaborated upon here to apply to corporate evolution as well.

Hierarchical Framework for Structure Transformation

The basic system formalism is illustrated in Figure 1. Problems, interpreted as opportunities, arrive every t time units and must be solved within a span of time T of arrival in order to be capitalized upon. We shall say that a system is *responding adequately* to its environment if each problem deadline is met. Unless otherwise specified, all systems mentioned in the sequel will be assumed to be responding adequately. Environmental parameters: interarrival time t, the response window T, and the problem variety will either be constant, or at most piecewise constant, over time. We shall call the environment invariant in the first case and subject to step changes in the second.

Our emphasis will be on relating the structure of the system inside the box of Figure 1 to its environmental characteristics. To this end, we shall employ the *canonical structure* specification shown in Figure 1b). The black box is realized as a coupled system, i.e., as a set of component systems (black boxes) that are interconnected by means of a coupling scheme (Zeigler, 1984). To implement the coupling, a co-ordinator, shown as an oval, is added with the task of 1) directing the information flow, specified by the coupling, between environment and the system components, and among themselves, and 2) synchronizing the activities of the components, so that the proper sequencing is enforced (the model is inherently concurrent in operation). The information flows linking environment, co-ordinator and components required to carry out tasks 1) and 2), are referred to as communication and synchronization, flows, respectively. In our interpretation, the components are considering to be functional in nature, while the co-ordinator is managerial or supervisory. The original function has now been decomposed into subfunctions whose activity must be co-ordinated by a supervisor.

In Figure 2a), the canonical structure is recursively applied to itself. Note that components which were previously considered functional are now given a realization as co-ordinated coupled systems(b) This results in a two level hierarchy of co-ordinators and a lowest level of functional components. Applying this recursion a finite number of times, ℓ, results in a hierarchical tree of depth $\ell+1$ in which the root and interior nodes represent managers while the leaves represent undecomposed functional elements. Note however, that each node, whether it is a leaf or not, also represents a functional component, viz., the resultant, shown in dotted lines in a), of coupling the system components represented by the children of the node. A more complete formal description of these concepts is given in Zeigler (1984).

It should be apparent that the above approach generates a tree structure for a system, which can be considered to be a hierarchy either of control (Mesarovic et. al., 1970) or of description (boxes within boxes) (Simon, 1969) depending on nodes are associated with their co-ordinators or with the resultant of the subtree they dominate.

Fundamental Performance Relations

We shall now derive some simple, but significant, relations between environmental parameters and the structure of any system that responds adequately to it. Throughout, we shall assume that each problem is solved in the same length of time. We take this length to be the response window, T, in effect assuming that the system works no faster than is necessary. In this case, the number of problems in process within the system at any time after initial start up is T/t.

Let there be n components in the coupled system in Figure 1b) . In this context we shall refer to the components as processors. We shall assume that a processor can handle at most one problem at a time. Under these circumstances, for the system to respond adequately, $n >= T/t$.

The *granularity* g refers to the fraction of problem handled by a processor. (This is a function-based definition as opposed to the structure-based definition in (Kung, 1980)). Thus each problem visits 1/g processors. Sojourning a total of T units in the system, it spends T/(1/g) at each. Thus, each processor takes p = T*g time units to process a problem.

There are many cases in which the lower bound on n is realized. As illustrated in Figure 3, there are at least two types of configurations that can respond adequately. In the first, called a *dedicated* configuration, there are T/t processors, each with a processing time of T. Upon its arrival, each problem is assigned a processor and remains with it until T has elapsed after which it is considered solved and leaves the system. As this happens, a new problem arrives and is assigned to this the freed up processor. Alternatively, in the *pipeline* configuration, each problem visits all T/t processors, spending t units at each.

Properties of Structures

It is evident that there may be many internal structures that meet the demands of a given environment. This remains true even if we require that the number of processors be the minimum possible. Such structures differ in the granularity and speed of their components, and in the complexity involved in co-ordination and synchronization.

Table 1 compares the above mentioned structures in this regard.

Table 1. Comparison of Dedicated and Pipeline Structures

	Granularity	Processing time	Complexity
Dedicated	1	T	medium
Pipeline	1:n	t	low

Note: n is the number of processors, T/t.

Judgments concerning complexity in the table are informal and based on the fact that in the pipeline architecture, the co-ordination and synchronization can be easily "pushed down" (see next section) so that a physical embodiment of the co-ordinator is not necessary. In other words, global control can be fully dispensed with in this case.

Basic Structural Transformations

Properties such as the above and others to be introduced may have import for the course of evolution of a firm as it adapts to an environment subject to step changes in its characteristic parameters. For example, in Figure 4, the opportunity rate is double that of Figure 1. Since t is halved, the number of processors required for adequate response is doubled.

The figure illustrates two basic forms of structural transformation. The first, depicted in a), replicates without change the existing components and puts them under control of the co-ordinator.

The second form is that of change by stepwise refinement. This employs the recursive transformation of Figure 2, with the additional property that processing speeds are considered as well. In b) each of the first level processors is given a coupled system realization in which the two second level processors operate at twice the speed of the first level predecessor. This realization is based upon a decomposition of the original function into two subfunctions, the combined performance of which is equivalent to the original. Such decomposition is called "divide and conquer" (Ziegler, 1983). It is reasonable to expect, although not necessary, that the time required for performing a subtask is substantially less than that required for the original. The co-ordination required for the equivalence is assumed to be sequential (first one, then the other) leading to the pipeline structure at level 2. Co-ordination would be more extensive, if the case where subtasks must be iterated before converging to a solution (c.f. Mesarovic, et. al., 1970). In b) note the order in which the highest level coordinator must direct incoming problems to each processor. A second example of the refinement transformation is shown in c) where each of the pipeline processors is transformed into a second level pipeline with faster components.

The transformations so far discussed, call them *replicating* and *splitting* respectively, are "up-building", i.e., increase the number of components. However, organizational adjustments may also be in the "down sizing" direction. For example, consider the inverses of the just-mentioned transformations. *Dropping* undoes replicating by eliminating components. *Consolidating* replaces the coupled systems realization of a function by an undifferentiated equivalent.

Flattening is a transformation that replaces part of the co-ordinating interior of a hierarchy by a single co-ordinator which implements the equivalent co-ordination and synchronization tasks. Interior co-ordinators are eliminated by absorbing their roles within those of superiors and inferiors. In a special case of such absorption, roles are "pushed down", i.e. absorbed by inferiors. In the extreme case, called "control distribution", the root level co-ordinator is eliminated and its synchronization and communication functions pushed down to the individual components.

Response-time/Flexibility Tradeoff

We represent the *flexibility* of a hierarchy as the number of alternative coupled system structures it can assume under given constraints on the elementary modifications permitted at its nodes. Consider Figure 1b), where we suppose that the number of possible coupling schemes that the co-ordinator is capable of assuming is c. This number may be much smaller than the number of possible schemes definable with respect to the components. For n components, this latter number is at least as large as the number of digraphs on n nodes which is superexponential in n (2^{n^2}). The ratio of c to the total number of possible coupling schemes can be taken as a measure of the "softness" of the co-ordinator. A small value of c represents a rigidity of the co-ordination task usually associated in the computational context with hardware, as opposed to software, implementation.

In Figure 1b), we fix one of the c possible coupling schemes and suppose that each of the components has a flexibility of m, i.e., it is possible to change each one to assume m mutually exclusive alternatives. Then the number of possible system configurations with the coupling fixed is m^n, assuming independence of component modifications from each other. Also, assuming independence of coupling scheme modifications from component modifications, the number of structural alternatives becomes m^n*c. The "independence of modification" assumption is consistent with the concept of "modularity" (Zeigler, 1984) and is justified by assuming that the I/O interfaces of the components are not altered in any modification.

Consider a hierarchy of depth ℓ, with branching factor taken as 2 for convenience. Let each of the leaf components have a flexibility factor of m and let each non-leaf node be capable of assuming c alternative coupling structures. Assuming independence of modification as above, the flexibility of the hierarchy, the number of alternative structures it is capable of assuming, is:

$$N(\ell) = m^{2^{\ell}} c^{2^{\ell-1}}$$

Proof of this relation may be obtained by induction on ℓ. The case $\ell=1$ is just that of the coupled system discussed above with n=2. For the inductive step, recall that a hierarchy of depth $\ell+1$, can be considered to be a coupled system co-ordinated by the root node with components realized by the substructures of depth ℓ. By induction, each substructure has a flexibility factor $N(\ell)$, so that the flexibility of the hierarchy is $N(\ell)^2*c$. Substitution of the expression for $N(\ell)$ verifies that for $N(\ell+1)$.

Note that the flexibility so defined is the number of alternative structures, not the number of alternative functions realizable by the hierarchy which may be smaller due to functional equivalence of structures. However, we shall assume that such equivalence is unlikely to change the superexponential dependence of the flexibility on depth.

As the depth of a hierarchy grows, its size (the number of nodes its contains) grows exponentially and the flexibility as defined above increases superexponentially. Moreover, let us assume (realistically) that the time taken by the co-ordinator at each level is finite. Then, we should expect that the overall co-ordination time grows linearly with depth. Co-ordination time here includes the time taken by the co-ordinator to decide on its next action, whether synchronization or communication related, and the time for this decision to reach subordinates. Linearity of co-ordination time with depth is consistent with the assumption that information flows a fixed number of times between root and leaf nodes, with each level being traversed in parallel.

Thus as a system grows in depth, its size and flexibility grow much faster than its internal communication speed degrades. Still, at some point this speed may become significant relative to the processing speed of the leaf elements and the system may seek to reduce its depth by flattening its hierarchy. In this case, the flexibility available to the system is greatly reduced for each unit gain in co-ordination speed.

Recall that consolidation was described as a transformation that is the inverse of splitting, i.e., the substructure at a node in the hierarchy is replaced by a functionally equivalent (undecomposed) component. Such a transformation can be viewed as the flattening of the substructure, and therefore, likely to increase speed.

Summary of Environmental-Evolutionary Relationships

Let us summarize the correlations between environmental characteristics and the options for adaptive re-organization suggested by our model. A system subjected to high environmental parallelism (large response window relative to opportunity interarrival time), is likely to develop a deep hierarchy with fast acting functional components. Such a system will likely flatten its hierarchy to prevent co-ordination time from becoming a bottleneck in its response. Alternatively, a system may opt for the lower overhead and efficiency afforded by flattening if its environment is perceived to be invariant. Such efficiency is bought at the price of greatly reduced flexibility which may be needed if the environmental problem variety increases in the future (see below).

A second form of environmental speed-up is to decrease the response window, T. In this case, environmental parallelism, T/t decreases and fewer components are required for adequate response. If processing time p, can not be reduced in proportion to the window reduction, the system must increase processor granularity, g in order to maintain adequate response ($p = T*g$). It can both reduce the number of components and increase granularity by consolidation. Once more flexibility is sacrificed in this transformation.

Change in the nature of the opportunities (problems) presented to the system may also be studied. One such change is an increase in problem size. This reduces to previous cases if it manifests itself in a change in one or both of the environmental time related parameters. For example, as an organization grows, its total resource usage may increase proportionately, while the time available for replenishment remains constant, c.f, an organization which pays the same average salary as it grows. This might be viewed as an increase in problem frequency since more resource captures of constant value must be made in the same time span. Previous discussion indicates that the organization must have enough processors to retain adequate response. Depending on phasing, this may trigger an exponentially exploding cycle of autocatalytic growth.

Environmental changes have so far been limited to those involving opportunity rate or response window. On the basis of our flexibility concepts, we may also consider changes in problem variety. Suppose that an organization is adequately responding to a sequence of problems by realizing a function f. Consider a step change in the environment in which opportunity rate and window remain fixed but the problem class now requires realization of a function f' in addition to f. Such an increase in variety may for example, be associated with the emergence of a new technological opportunity.

Intuitively, the "distance" between f and f' will determine the extent of the re-organization required by the system to accommodate f'. Let us take this "distance" as infinite if f' is not realizable by any of the structures within the flexibility class of the current structure, i.e., we consider only the structures accessible by modification of the coupling schemes (at the root and interior nodes) and functional leaf components as feasible candidates. We can obtain a crude metric for this class by comparing the current structure with one that realizes f' (to make matters simpler, assume that only one such structure exists in the accessible class). Starting at the root we recursively compare the two structures asking at each node, whether the difference lies 1) in the coupling only, 2) in the subfunctions only or 3) in both coupling and subfunctions. It seems natural to take the last case as one requiring major transformation at the node. Let d be the reorganization depth, i.e., the minimum depth (distance from the root) at which a case 3) node occurs. The larger this depth, the less fundamental is the major transformation required. In the extreme, only changes in the leaf functional components are required. On the other hand, a depth of zero indicates that a major transformation is required at the root.

Now consider the current hierarchy modified so that all case 3) nodes have been consolidated. By definition to realize f' fundamental modifications are required, at most, in the leaf nodes in this structure. A possible adaptive strategy is to replicate this hierarchal structure, possibly flattening it for efficiency, and giving it "quasi-autonomous" status to deal with the problem requiring function f'. Such global structural transformations seem a promising basis for modelling those strategies characteristic of "neotonous" corporations as discussed by Baba, et. al. (1984). Further study of adaptation to change in problem variety is planned.

Summary and Conclusions

Studying the re-organizations that are required for technologically based firms to remain competitive, we formalized the environments typical of those that such firms face and the information processing structures that evolve to match them. The environment was characterized in terms of the opportunity rate, the response window, and the problem variety. The product of opportunity rate and response window, or equivalently the response window divided by the opportunity inter-arrival time, was called the environmental parallelism. An adequately responding system must have at least a size equal to this parallelism. Two manifestations of environmental speed up, opportunity rate increase and response window narrowing do not have symmetrical effects on optimal system structure. Response to increase in opportunity rate was characterized as deapening of this hierarchy, while window narrowing response was characterized by the inverse process called consolidation.

We showed that the flexibility of the system (the variety of reconfigurations it can assume) is superexponentially related to the depth of the hierarchy. Thus optimal adaptation to increased opportunity rate has the concomitant effect of greatly increasing system flexibility. On the other hand, such adaptation to response window narrowing has the opposite effect: flexibility rapidly declines with consolidation. However, we showed that an organization subject to an invariant environment which has developed a deep hierarchy may opt for a "flattening" of its decision structure to reduce overhead and communication delay. Such flattening has an effect on flexibility which is similar to that of consolidation. Thus an organization that has optimally adapted to increased opportunity rate would seem to be in a better long term position than one which had adapted to response window narrowing or has exploited the invariance of its environment to achieve efficiency. Since, all things being equal, the former organization type would be larger than the latter, and therefore much more flexible, our results seem at variance with folkloric correlations claiming that large firms are more rigid and less able to capitalize on technology opportunities than small ones. However, the deeper analysis of empirical results that we shall discuss in a moment tends to support the model.

Our model requires us to make a distinction between flexibility, a structural property, and capacity to match the environmental problem variety, a behavioral property. This latter capacity may be smaller due to fact that while a problem may be solvable with an reconfiguration within the flexibility class of the current structure, this reconfiguration is too difficult for the system to discover or put into effect. We provided a natural metric for the difficulty of reconfigurations that helps characterize the conditions under which a large organization will be motivated to replicate and externalize a smaller version of itself, rather than attempt a reconfiguration, in order to capitalize on new opportunities potentially within its problem solving capacity. Such structural transformations correspond to "spin-off" and similar reorganizations elaborated upon in Baba, Zeigler, and Reynolds (1984).

As indicatesd above, analysis of our model indicate that small firms are not necessarily more innovative than large ones as might be commonly accepted. Empirical evidence gathered by Rothwell and Zegveld(1983) throws light on this situation. Having examined the innovation rates in a variety of industries in different epochs, these authors conclude that any innovation advantage of small firms is limited to the early phase of development of an industry where change is rapid and development is fluid. As consolidation to an "dominant design" (Utterback, 1974) takes hold and competitive entry becomes more difficult for small firms, innovation is largely the province of the surviving large firms and may, or may not, decline.

We believe that further analysis of our model will find that it successfully illuminates these findings. We intend to characterize the growth and development of an industry by two unimodal curves of opposite tendency: opportunity rate being related to the derivative of growth, having a maximum, and response window width, related to competition, having a minimum. Thus, a typical industry may exhibit all combinations of these two parameters in a, more or less, predictable order of succession: (low opportunity rate, large window), (moderate opportunity rate, moderate window), and so on. We shall attempt to corelate the structural reorganizations suggested by our model with with the above mentioned empirical findings.

References

Baba, M.L., B. P. Zeigler, and R. G. Reynolds (1984), "Managing Technological Turbulence: Strategies for Neoteny in Technology-based Corporations", In: Proc. IX Triennial World Congress of the International Federation of Automatic Control, Pergamon Press, Oxford.

Hockney, R.W. and C.R. Jessape (1981). Parallel Computers, Adam Hilger Press

Gijsbrechts, E. (1983), Hierarchical Approaches to Large Scale Organizational Problems: an Overview, Department Economische Wetenschappen, Unversiteit Antwerpen.

Kartashev, S.P. and S.I. Kartashev, (1982) eds, Designing and Programming Modern Computers and Systems, Vol. 1: LSI Modular Computer Systems, Prentice Hall, N.J.

Kogge, P.(1981), The Architecture of Pipeline Computers, McGraw Hill, NY.

Kung, H.T. (1980), "The Structure of Parallel Algorithms", In: Advances in Computers (ed: M. Yovitz), pp. 65-112, Academic Press, NY.

Mead C. and L. Conway (1980), Introduction to VLSI Systems, Addison Wesley Press, MA.

Mesarovic, M.D., D. Macko, and Y. Takahara (1970), Theory of Hierarchical Multilevel Systems, Academic Press, NY.

Moss, S.J. (1981), An Economic Theory of Business Strategy, Wiley, NY.

Reynolds, R.G. (1984), "A Computational Model of Hierarchical Decision Systems", J. Theo. Archaelology.

Rothwell, R. and W. Zegveld (1983), Innovation and the Small and Medium Sized Firm, Kluwer-Nijhoff Publishing Co., Hingham, MA.

Sherman, M.D. (1983), "MANAGE- An Interactive Simulation Model for Evaluating the Efficiency of Management Structures on Organizational Performance", Simulation, August.

Simon, H.A. (1969), The Sciences of the Artificial, MIT Press, Cambridge, MA.

Utterback, J.M. (1974), Innovation in Industry and the Diffusion of Technology", Science, 183: 620-626.

Zeigler, B.P. (1984), Multifacetted Modelling and Discrete Event Simulation, Academic Press, London.

Ziegler, C.A. (1983) Programming Systems Methodology, Prentice Hall,NJ.

a)

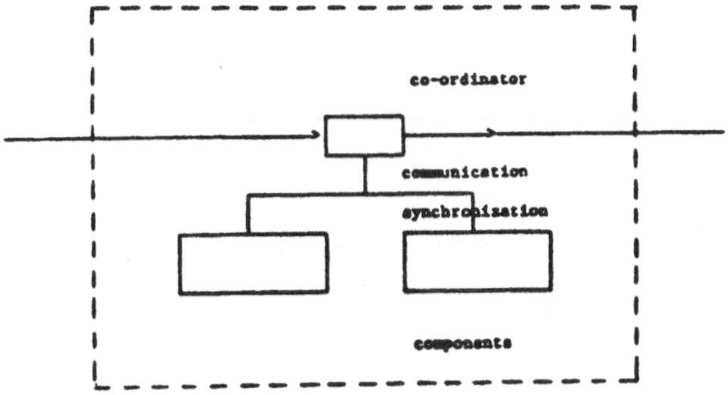

b)

Figure 1. a) Basic framework: problems denoted by numerals arrive every t units and must be solved by system in box within T units. b) Canonical Structure (co-ordinated coupled system) realization of box.

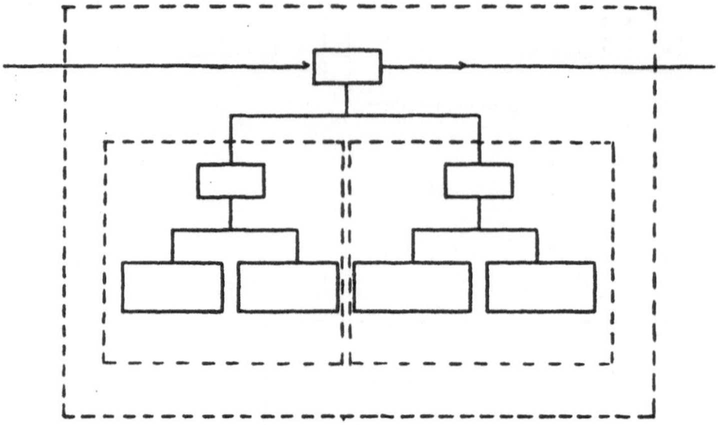

a)

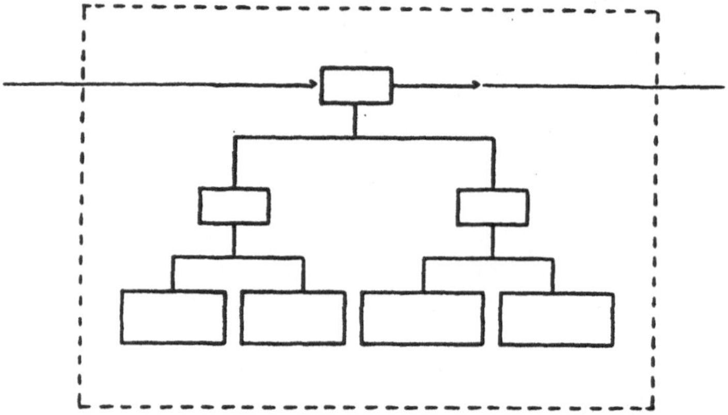

b)

Figure 2. a) Recursive application of canonical structure to components of Figure 1b). b) Interpretation of result of refinement as a hierarchically co-ordinated system.

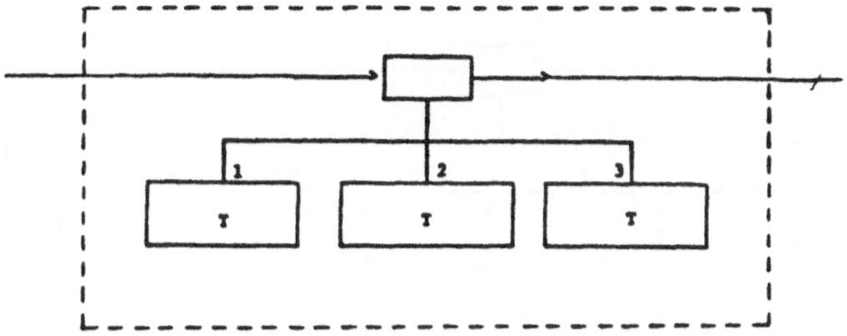

a)

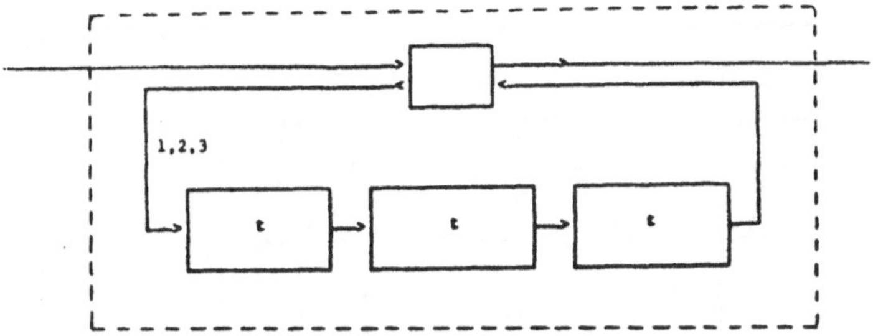

b)

Figure 3. Coupled system structures that adequately respond to the problem environment of Figure 1. a) each processor is dedicated entirely to a problem. b) classical pipeline configuration.

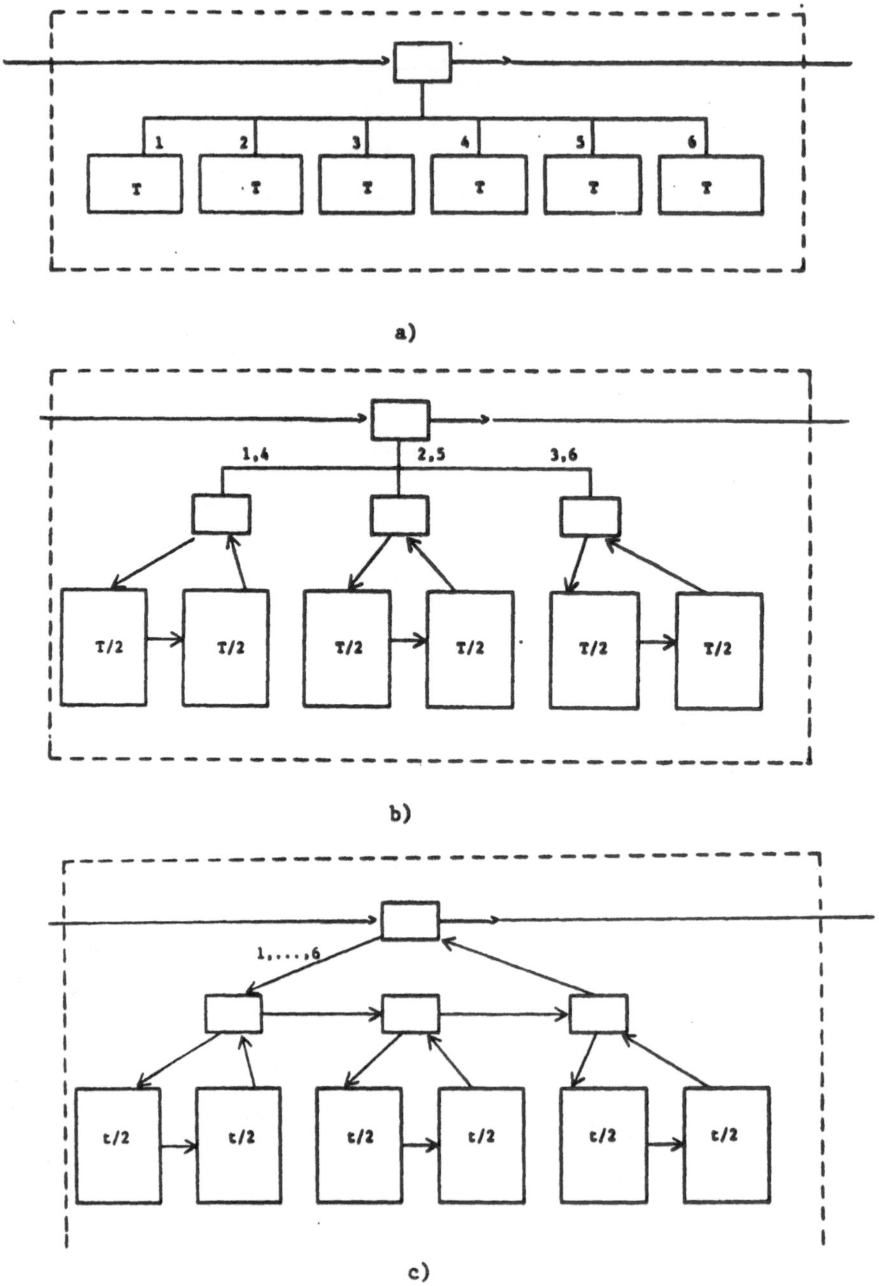

Figure 4. Hierarchical coupled system structures that adequately respond to an environment with interarrival time t/2 and response window T. a) dedicated. b) dedicated at level 1, pipeline at level 2. c) pipeline at both levels.

EVOLUTION OF CORPORATE STRUCTURE IN TURBULENT TECHNOLOGICAL FIELDS

M. L. Baba
Department of Anthropology
Wayne State University
Detroit, Michigan 48202

This paper presents a model of co-evolution among technology-based firms and their technological fields. Evolutionary trends in the technological environment, generated in part by aggregate organizational forces, place strong selective pressures on the structure and behavior of technology-based corporations. Corporate structures appear to be adapting to these pressures through the process of dynamic restructuring and creative recombination. The basic units of corporate adaptation (quasi-autonomous subassemblies) may, in turn, generate new forces of environmental change by enabling mature corporate hierarchies to survive and exploit technological discontinuities.

I. INTRODUCTION

Significant advance in organization theory has been achieved through application of explanatory models drawn from the biological sciences. In particular, the central concepts of Darwinian selection and population ecology have proven to be powerful analytic models in the study of organizational change. Like living organisms in the natural world, organizations may be viewed as open systems whose structures and behaviors evolve in response to changing environmental pressures, particularly those related to resource availability and competitive threats (Aldrich, 1979).

The dynamic interplay of forces generated by organizational systems and their environmental fields may underlie the co-evolution of organizational-environmental complexes. In a classic paper on the causal texture of organizational environments, Emery and Trist (1965) describe a progressive series of environmental types which display varying patterns of resource distribution, rates of change and structural complexity. The most complex form of environment is the "turbulent field", in which dynamic properties arise not only from the interaction of organizations within the environment, but from the field itself. Turbulent fields display increasing rates of interconnection among environmental components and increasingly steep gradients of change, leading to a gross increase in the degree of relevant environmental uncertainty. Emery and Trist predicted that organizational stability in turbulent fields would be highly problematic due to the fact that no single organization, however large, could successfully adapt to the constant force of unpredictable change. They postulated that turbulence would "demand some form of organization that is essentially different from the hierarchically structured forms to which we are accustomed" (Emery and Trist, 1965, p. 29).

This paper presents a preliminary model of co-evolution among technology-based corporations and turbulent technological environments. The accelerating pace of technological change is found to create new selective pressures for large, hierarchically ordered corporations whose process and output structures originally evolved under more stable conditions. Recent corporate adaptations to technological dynamics appear to involve the creation of new organizational structures which enhance market flexibility and decrease the time needed to respond to environmental demands.

II. TURBULENT TECHNOLOGICAL FIELDS

Technological change is a powerful socio-economic force which now marks the landscape of all modern organizational environments. Two interrelated aspects of technological change are known to create severe environmental turbulence for technology-based corporations (defined here as firms whose competitive position rests upon the development and utilization of advanced technological products and processes). These aspects include: (1.) the dynamics of the technological life cycle for a given product-process configuration; and (2.) the periodic emergence of qualitatively new forms of technology (i.e. technological discontinuities).

First, the continuous birth, maturation and decline of products and their embedded technologies creates marked environmental instability, a key feature of turbulent fields. The "technological life cycle" can be described as beginning with a lengthy period of high-risk R&D, followed by a phase of rapid product development, market growth and rationalized production, and finally terminates with product stagnation near the inherent developmental limits of the technology in question (Foster, 1983). The inevitable trajectory of the product life cycle is thought to follow a sigmoidal curve during a "lifetime" which averages 10 to 60 years in most industries, but may telescope to 3 or 5 years in the electronics industry (Girifalco, 1983; Merrifield, 1984).

Turbulence is also generated by technological discontinuity, a phenomenon that greatly increases the degree of relevant environmental uncertainty. Discontinuities occur when radically different or new generations of technology become commercially viable, and by virtue of increased capabilities, reduced cost, or both, render the previously dominant technology obsolete. Radical new technologies tend to emerge in clusters, and to produce severe shock waves throughout the economic structure. Although major clusters of innovation appear to emerge on the average every 50-60 years, the pace of technological change seems to be increasing. In the past 100 years, both the frequency of major discontinuities, and the rate at which new technologies reach the marketplace, have doubled (Girifalco, 1983). Because new generations of technology typically do not evolve incrementally from the dominant technological field, but appear to emerge suddenly and often unexpectedly from marginal areas of R&D, their arrival is difficult to predict and even more troublesome to accomodate.

III. SELECTIVE PRESSURE AND CORPORATE FITNESS IN TURBULENT TECHNOLOGICAL FIELDS

A long-term trend toward acceleration in the tempo of technological change may be expected to generate continuously increasing turbulence in the technological field (i.e. greater instability, uncertainty and loss of control), and to place unprecedented selective pressures on technology-based corporations, as predicted by Emery and Trist in 1965. Accelerated rates of technological change are translated in the marketplace to produce two distinctive forms of selective market pressure: (1.) an increase in the rate of new market opportunities; and (2.) a constriction in the amount of time available for responding to new market opportunities. The "opportunity rate" is defined here as the number of new market applications for a given technology that arise during a designated period of time. An increase in this rate results from the telescoping of both product and technological life cycles which, in turn, can generate a very rapidly emerging sequence of new market possibilities for a new or existing technology. The increased "opportunity rate" will be limited and complicated, however, by a growing constriction in the period of time during which a company may respond to a new opportunity. "Available response time" is defined as the optimal period during which a new technological concept or application must be introduced to the market if an emerging opportunity is to be successfully exploited. The available response time has been narrowing steadily during recent years, (particularly in new industries such as electronics), largely as a result of rapid technological turnover, and the increasing competition and escalating entry costs that follow technological discontinuity (Wise et al, 1980).

According to the evolutionary paradigm, we would expect firms to display differential "fitness values" (i.e. long-term adaptive potential) with respect to the mounting selective pressures described above. Within the total range of organizational variability, certain corporate structures and behaviors should permit a superior response to the increased "opportunity rate", while other forms should enable more rapid response to available opportunities. Finally, some organizational forms might confer a dual advantage, allowing corporations to seize larger numbers of opportunities within ever more constricted periods of time. Companies possessing this dual advantage should be most "fit" in turbulent technological fields, i.e. most likely to demonstrate sustained growth in a rapidly changing technological domain, and to survive technological discontinuities, which typically require rapid, anticipatory responses.

It must be noted, however, that a dual capacity to exploit rapidly emerging technological opportunities, and to maneuver quickly through narrowing windows of time, may require very different, even contradictory, organizational structures and behaviors. Superior technological performance in environments characterized by an increasing "opportunity rate" are most likely to be displayed by firms that can flexibly integrate the innovation chain. As the product cycle unfolds more quickly, the identification of new market applications also quickens, as does the need to move products rapidly through the innovation system, from R&D, to pilot production, and finally to large-scale marketing and distribution. Firms that can integrate the entire innovation system, and can effectively couple the diverse segments of this system, will be in the best position to speed the movement of many new products from the research lab to the marketplace, and thereby to exploit an expanding array of new technological possibilities. Flexible, vertical integration of the innovation chain within a single company implies a highly differentiated, hierarchical structure with multiple linkages and feedback loops between various functional components (see Tornatzky et al, 1983). The features of functional differentiation and complex, hierarchically ordered structures are, of course, most commonly found in large, mature corporations (Tosi, 1984; also see Baba et al, 1984).

On the other hand, high fitness values in environments shaped by constricted response windows may demand a simplified management structure, capable of making decisions, reorganizing priorities, and changing strategies more rapidly than is possible in large, hierarchical organizations (Abernathy and Utterback, 1978). The structural and dynamic prerequisites of rapid environmental response time have most frequently been displayed by mission-oriented groups of entrepreneurs and "organically" structured, youthful start-up firms (Burns and Stalker, 1961; Tornatzky et al, 1983; Baba et al, 1984). Furthermore, a readiness for technological discontinuity often derives from a "corporate culture" that is oriented toward radical, high-risk R&D ventures, another important feature of small, start-up companies (Jacobs, 1982).

IV. THE EVOLUTION OF CORPORATE STRUCTURES IN TURBULENT TECHNOLOGICAL ENVIRONMENTS

A review of recent organizational experiments undertaken by large corporations in the U.S. suggests that market forces in turbulent technological environments may be selecting for new corporate structures that combine key innovation features of both mature and youthful organizations. Specifically, it appears that mature hierarchical corporations are developing special structures that enhance organizational flexibility and permit improved response times. The central hypothesis of this paper holds that the development of such new structures constitutes an evolutionary adaptation to turbulent field forces which may ultimately increase the stability of large technology-based corporations.

The Quasi-Autonomous Subassembly (QAS)

The emergence of new corporate structures that accomodate accelerated technological change appears to occur through a process of dynamic organizational restructuring. This process utilizes the mechanism of partial externalization to form quasi-autonomous subassemblies (QAS) which serve as smaller, youthful extensions of the parent corporation. For the purposes of this paper, subassemblies may be defined either as structural-functional units, or as functional aspects of structural units, which constitute a hierarchically ordered system. The subassemblies which together comprise a mature organization have been described as "loosely coupled"; that is, they maintain some partially independent identity (see Aldrich, 1979). Subassemblies within a hierarchy may be partially externalized, that is "broken out" of the hierarchy and charged by the parental body with a separate and distinctive mission.

Externalized subassemblies maintain information/resource linkages with the parent, but perform mission-oriented rather than simply function-specific tasks. Partial externalization may be acheived by assigning a specific mission to a physically isolated working group, to a legally separate affiliate, and/or to a collaborative body formed in conjunction with several external organizations. The externalization process should leave the basic hierarchic operation intact (i.e. the core functional elements of an externalized unit will also be retained within the hierarchy), and should further empower the mature organization with certain features normally found in smaller, more youthful companies.

Some of the youthful properties typically displayed by QAS units are: (1.) a flexible network structure; (2.) an independent external mission, together with some decision making authority; (3.) unique personnel composition, (e.g. high concentration of proven entrepreneurs); and (4.) small and/or high-risk market niche orientation (Burns and Stalker, 1961; Baba et al, 1984). Furthermore, QAS units appear to possess a unique structural capacity for "recombination" with other external entities.

Adaptive Potential of QAS

The adoption of QAS structures may confer multiple advantages to mature hierarchical firms. First, such structures could strengthen the basic flexibility of an hierarchical organization, and thereby expand the number of market opportunities that a given firm can exploit successfully. For example, QAS may be used to create a "technology portfolio" that can enhance a firm's ability to capture the full potential of all product cycle phases. Corporations can delegate to a QAS portfolio the responsibility for particularly high risk or troublesome projects (e.g. radical R&D ventures, phase-out of stagnating product lines), with the explicit understanding that parental support will be terminated under specific conditions (Jacobs, 1982; Stumpe, 1983). Furthermore, because QAS frequently display the flexible network structure of small firms, it is also possible that such units may out-perform the parental body during the very early stages of product development.

Secondly, QAS units may improve corporate response times and provide for advance positioning in frontier market niches. Because external subassemblies do not appear to be critical to the basic maintenance operations of the parent body, the formation and dissolution of such units should be relatively less difficult than the creation and destruction of integral structural-functional units. Thus, QAS units can be rapidly deployed to explore fast-breaking opportunities, and can also be quickly dismantled when no longer needed. In addition, the flexible network structure and unique personnel composition of QAS units permits rapid decision making and aggressive pursuit of new market possibilities, particularly in the critical high-risk period of a technological discontinuity.

Finally, the ability of QAS to "recombine" creatively could provide a significant organizational advantage in turbulent technological fields. Recombination is defined here as the structural-functional coupling of QAS units formed by two or more parental organizations. Recombined subassemblies can establish vital information and resource linkages between different types of organizations, thereby reducing environmental uncertainty and loss of local control, while simultaneously permitting several organizations to pool capital and human resources to solve complex technological problems. Emery and Trist (1965) predicted that recombined (or "matrix") organizations would emerge as a fundamental form of adaptation to increasing environmental turbulence.

Basic Types of QAS Formations

A review of organizational experiments now underway in several U.S. corporations has revealed three basic types of QAS formations. These types vary by nature and complexity of mission, degree of independence from the parent corporation, and nature of relations with external entities. All of the subassembly types, however, appear to strengthen organizational flexibility and to improve potential response time. General characteristics of each subassembly type are described below.

TYPE I. The Entrepreneurial Portfolio

Type I neotenous subassemblies are formed when a single parental corporation creates QAS offspring by externalizing a combination of different functional components necessary to achieve a highly specific objective. The mission usually involves exploitation of some phase of the product development cycle (e.g. creating a new product, establishing a new product market, phasing-out a stagnant product line, etc.), and frequently represents a departure from past corporate practice. The best known examples of Type I QAS formation are wholly-owned "spin-off" firms or internal mission-oriented project teams which harness the talents of entrepreneurial individuals by enabling them to "run their own businesses" (Jacobs, 1982; Stumpe, 1983). Of all QAS types, these off-spring units most nearly resemble youthful companies in structure and behavior, and are most capable of rapidly exploiting fast-breaking market opportunities at the periphery of the dominant product field. A constellation of such offspring units may serve as a technological portfolio, careful management of which can secure advance standing in a wide array of high-risk/high-potential areas, thereby improving the "discontinuity readiness" of mature corporations.

TYPE II. The R&D Consortium

Type II QAS are formed when a group of parental organizations jointly externalize and recombine a set of similar functional components in order to achieve a high-risk, technology-specific mission that no single organization could achieve alone. The most common Type II subassembly formation in the U.S. is the R&D consortium (See Cromie, 1982; Colton, 1982). In this form of Type II assemblage, a variety of different organizations (e.g. several large corporations and research universities) pool a similar set of resources (financial, scientific, technological) in order to generate the critical mass necessary to solve major technological problems that are of interest to all parties. While Type II subassemblies do not appear to directly enhance market activities, they do establish superior information linkages with the rapidly evolving "knowledge construct" (e.g. through close ties to leading universities), thereby extending the basic research scope of a given company and increasing its chances of surviving a discontinuity. Participation in a Type II QAS formation may also enable a wider range of individual corporations to maintain an advance position at the forefront of technological change, since many organizations bear the financial risk should R&D prove unsuccessful.

TYPE III. The Technological QUANGO (T-QUANGO)

The most complex type of QAS formation is the Technological QUANGO (or T-QUANGO), following the British acronym for "Quasi-Autonomous Non-Governmental Organ". Type III subassembly formations represent the epitome of organizational flexibility -- they are highly mutable in composition, structure and mission. T-QUANGOs are created when a variable set of public and/or private sector organizations (large and small companies, government agencies, venture capital investors, universities) form "partnerships", with each "partner" externalizing a different set of functional components in order to accomplish one or more technological objectives from a diverse assortment of possible benefits (see Baba et al, 1984). Prime examples of T-QUANGOs that are now common in the U.S. include the university-based "high technology" research park, and the research and development limited partnership (RDLP) (Merrifield, 1982; Thompson, 1983). T-QUANGOs appear to be an excellent means of extending the market range of participating organizations, since the information links created in such assemblies bring to light new market opportunities and may also facilitate the exploitation of such opportunities. In addition, T-QUANGOs may be used to sustain radical, high-risk research projects during long incubation periods. In general terms, the Type III QAS may represent the most powerful form of organizational adaptation to turbulent field forces. Because such recombined assemblies link the operations of many diverse organizational actors, they may serve to reduce the total amount of unpredictability and loss of control experienced within a local environment.

V. CONCLUSIONS

The evolution of corporate structures in turbulent fields has significant implications for organization theory. QAS units, whether in their singular portfolio form, or in recombined assemblages, may reflect an evolutionary trend toward greater organizational complexity in turbulent fields. QAS units increase organizational complexity by expanding the number of structural-functional organs which are part of the total organizational system, and by forcing the development of new integrating mechanisms which can coordinate (if not control) the action of these new structural-functional organs. In addition, QAS units appear to increase the complexity of developmental trajectories (i.e. growth curve and "life cycle" phases) displayed by large corporations, since they serve both to strengthen and extend the basic characteristics of organizational maturity, while simultaneously infusing the mature phase with some of the structural and dynamic properties of youth. The emergence of increasingly complex structures in large technology-based firms demonstrates the fundamental plasticity of mature hierarchic organizations, and further confirms the role of environmental forces in shaping the evolution of organizational systems.

REFERENCES

Abernathy, W. J. and J. M. Utterback. 1978. Patterns of industrial innovation. Technology Review. 80:40-47.

Aldrich, H. E. 1979. Organizations and Environments. Englewood Cliffs, NJ: Prentic Hall, Inc.

Baba, M. L., B. P. Zeigler and R. G. Reynolds. 1984. Managing technological turbulence: strategies for neoteny in technology-based corporations. Proceedings of the IX Triennial World Congress of the International Federation of Automated Control, Pergamon Press.

Burns, T. and G. M. Stalker. 1961. The Management of Innovation. London: Tavistock Publications.

Colton, R. M. (Ed.) 1982. Analyses of Five National Science Foundation Experiments to Stimulate Increased Technological Innovation in the Private Sector. Washington, DC: National Science Foundation.

Cromie, W. 1982. University-industry-state government consortia in microelectronics research. In: University-Industry Research Relationships: Selected Studies. (National Science Board, ed.) Washington, DC: National Science Foundation.

Emery, F. E. and E. L. Trist. 1965. The causal texture of organizational environments. Human Relation. 18:21-32.

Foster, R. N. 1983. Boost the R & D payoff. Chem Tech. 13(12):720.

Girifalco, L. A. 1983. Technological Dynamics for Corporations. University of Pennsylvania. Unpublished Manuscript.

Jacobs, M. 1982. The physicist as entrepreneur. Physics Today. January, 1982. pp. 34-40.

Merrifield, D. B. 1982. Information and Steps Necessary to Form Research and Development Limited Partnerships. Washington, DC: United States Department of Commerce. Unpublished Manuscript.

Price, W. J. and L. W. Bass. 1969. Science. 164:802.

Stumpe, W. R. 1983. Entrepreneurship in R & D. Research Management. 25(1):13-17.

Thompson, P. 1983. A chip on an old block: Metropolitan Center for High Technology. The Detroiter. September, 1983. pp. 61-64.

Tornatzky, L. G., J. D. Eveland, M. G. Boylan, W. A. Hetzner, E. C. Johnson, D. Roithman, and J. Schneider. 1982. The Process of Innovation: Analyzing the Literature. Washington, DC: National Science Foundation.

Tosi, H. 1984. Theories of Organization. Second Edition. New York: John Wiley & Sons, Inc.

Wise, K. D., K. Chen, and R. E. Yokely. 1980. Microcomputers: A Technology Forecast and Assessment to the Year 2000. New York: John Wiley and Sons, Inc.

HIERARCHICAL APPROACHES TO MATHEMATICS OF TECHNOLOGICAL CHANGE

Louis A. Girifalco
University of Pennsylvania
Philadelphia, PA 19104, U.S.A.

I. Introduction and Definitions

Most quantitative analyses of technological change have been relatively simple, at least from a mathematical viewpoint. These include analyses of the rate of adoption of a new industrial innovation(1), the rate of substitution of a new product for an old one(2,3), the lifetime of a given technology(4), the frequency of inventions and innovations(4), the probabilities of success of R & D projects(1) and the rate of increase in technical performance or figures of merit(5,6). Only one or two dependent variables occur in such studies and they are generally connected to time as the independent variable through a simple differential equation.

Large scale computer models of technological change, comparable to those of econometrics, have not yet been constructed. The nearest thing to such models are energy utilization and energy optimization models(7,8). The function of these, however, is not to model technological change but to analyze economic and use factors for scenarios defined by such inputs as fuel supply at a given cost, inventories, interest rates and energy demand. Nevertheless, these models are of use in pointing the way to possible analogous methods of studying technological change.

Technological change is so broad, and covers so many activities, that it is essential to carefully identify the objects of study and their important parameters. In this paper, the required definitions will be based on specification of three categories: technical devices, technologies and technological enterprises.

The definition of a technical device is obvious and its important parameters are those that describe how well it performs its function. An audio amplifier is an example and its important parameters are gain, power, peak current, bandwidth, frequency response, distortion, signal-to-noise ratio, etc. We will also include such objects as numerical control programs and computer software in the definition of technical devices.

The term "a technology" is often used loosely, without precise definition and in a variety of ways in the literature of technological change. To avoid confusion it is necessary to define precisely what is meant by "a technology" for the specific case at the beginning of each study of technological change. We start by defining a technology to be a set of devices, all designed to perform the same general function and all based on the same general scientific principles. Actually, this defines hierarchies of technological systems, each of which can be defined as a technology. The set of solid state stereo amplifiers from the 1960's to the present day can all be defined as a technology based on solid state electronics, in distinction to the amplifier technology based on radio tubes. Taken together, all amplifiers can be defined as an amplifier technology based on the laws of electricity and magnetism. There is thus a longitudinal hierarchy of technologies, each defined by the level of definition of the scientific and engineering principles on which they are based. There is also a vertical hierarchy. Combining the amplifier with other components creates a high fidelity stereo system which represents an audio technology, composed of sets of other technologies. The amplifier consists of components whose performance depends on metallurgy, machining, ceramic science, vapor deposition, chemistry, etc., all of which define sets of other technology. Thus there is a linkage hierarchy of technologies that relates one technology to another, and whenever a technology is studied a decision must be made as to what is included.

A distinction is often made between process and product technologies. Our definitions focus on product technology, but can be extended to include processes by considering a process technology to be the set of all devices required to make the product.

Technologies are connected systems. The longitudinal hierarchy connects them in time, the vertical hierarchy connects them in function and the linkage hierarchy connects them to their environments. Clearly, technologies are open systems because they exchange information and material with each other.

A technological enterprise is defined as an institution or social organization that performs the functions of creating, improving or producing technological systems. Again, the precise definition must be constructed at the start of a specific study. Examples are R & D teams, R & D laboratories, corporations and entire nations. Technological enterprises exist in longitudinal, vertical and linkage hierarchies in a manner completely analogous to technologies. However, a technological enterprise can, and often does, have non-technological functions, as in the case of corporations. Also, technological enterprises have varying degrees of organization and specificity of purpose. A manufacturing plant is highly organized to a limited specific end, but the same is not true for a nation. This is important because non-technological factors can have large effects on technological change.

The important parameters in technological enterprises are those that specify their functions. These are not necessarily technical parameters but might include productivity, market share of a product, frequency of innovation, return on investment in innovation, numbers of patents and even national statistics such as life expectancy or GNP.

In terms of technological systems (defined as technologies or technological enterprises) the study of technological change is the analysis of the time evolution of the important parameters of the system. The question arises as to the degree of regularity of these parameters. If they vary smoothly with time*, there is some hope of finding mathematical representations that reflect underlying causal factors; but if they are completely random, only broad statistical statements can be made. This is an empirical question. A priori, we would expect both regular (causal) and random variables to exist. This is born out by historical data which exhibit many instances of smooth technological change in technological performance (i.e. efficiency of electric lighting, speed of aircraft, etc.) as well as individual events such as the accidental discovery of penicillin or the Second World War which had important effects on technological change. There are also variables that exhibit mixed characteristics such as the time lag between inventions and innovations.** This time lag can vary from a few to a hundred years. However, in the past century both the mean time lag and the difference between maximum and minimum time lags have shown a regular decrease by a factor of two[9]. As a working hypothesis we assume that both kinds of variables describe technological systems: regular (or causal) variables that vary smoothly with time and random variables.

Any analysis of technological change must start with a specification of the object of that analysis. The hierarchical definitions given above should provide a sound basis for studying many aspects of technological change. However, no one set of definitions can capture

---

*Of course, on a fine enough scale, all technological change is discrete. But with a sufficient degree of regularity in these changes, they can be approximated by continuous functions.

**Inventions are defined by the time they are first made public. Innovations are defined by the time they are first commercialized.

all aspects of interest in technological dynamics. A new chemical reaction, for example, can cause far reaching change and its essence does not reside in a device. Also, process technologies are not necessarily best described in terms of devices. Nevertheless, the definitions given here are useful in a number of ways. Also, the concept of three dimensional hierarchies should be valid for alternative or broader definitions of technologies.

In this paper, we will focus on the description of vertical and longitudinal hierarchies since this is the simplest illustration of the definitions. The basic mathematics is trivial. The major difficulties in applications are defining the technologies and acquiring the data needed to estimate the relevant parameters.

II. Limits and Life Cycles

Many technological activities exhibit organic life cycles that can be approximated by S-shaped curves. The classical example of this is the substitution of a new technological product for an old one in the marketplace[(2)]. Once the new product has captured a few percent of the market, its market share grows rapidly up to a point where the growth starts to slow down and approaches zero as the market becomes saturated. Similar patterns of initial slow growth, rapid acceleration, slowing down and finally approaching a limit are characteristic of many technological parameters. The imitation of innovations, the efficiency and performance of devices, the productivity of R & D and even the overall advance of industrial societies seem to follow these patterns. We therefore recognize that the parameters describing technological systems have intrinsic limits that arise either from the capacity of the system (such as market share or the number of firms that exist to adopt an industrial innovation) or from scientific and engineering limits (such as the laws of thermodynamics or the theoretical strength of materials). Many technological parameters therefore have the general form of an s-shaped curve and their rates have a bell-shape. The simplest representation of this is the logistic equation whose differential equation is

$$\dot{P} \equiv \frac{dp}{dt} = aP(P_L - P) \qquad (1)$$

with the solution

$$P = \frac{1}{e^{a(t_{\frac{1}{2}}-t)}+1} \qquad (2)$$

where P is the parameter in question, P_L is its limiting value, α is the initial growth rate and $t_{1/2}$ is the half-life, i.e. when $P = P_L/2$.

The importance of this concept can hardly be overestimated since it provides a basis for a mathematical description of technological systems. In any analysis, the theoretical limits of the parameters need to be identified if absurdities are to be avoided. Of course, the precise form of the S-shaped curve is an empirical question, but equation (2) is a surprisingly good representation in many cases.

It is interesting to note that the logistic curve also represents biological growth, the demographic transition, cultural diffusion and the spread of epidemics to a reasonable approximation.

In a longitudinal hierarchy of technologies, we expect that each member has its own performance limits. Thus, if we define a hierarchy of technologies for a given general function (such as lighting or air travel), the overall time evolution will be a series of S-curves, each with a higher limit than the others, the envelope of which represents the hierarchy as a whole and approaches a limit of its own. In a vertical hierarchy, each component technology exhibits a logistic type of evolution which determines the S-shaped growth of the next higher technology in the hierarchy. In a linkage hierarchy, the growth curves of the technologies outside the defined technology describe the environmental changes that affect that technology.

III. Technological Performance and Figures of Merit

If we are interested in a single performance parameter (such as the maximum efficiency of incandescent lamps) as a function of time, then the analysis is straightforward. But this gives limited information. Mean time between failures, size and frequency distribution are also important parameters in incandescent lamp technology. In general, a set of parameters are required to adequately describe a technology and the progress of that technology is related to all of the parameters. We can, of course, treat the parameters individually but it is often desirable to have a single figure of merit that combines the performance parameters and that can be used as a measure of the overall technological performance. A simple approach to constructing a figure of merit is to assume it is composed of a linear combination of some of the performance parameters and a multiplicative combination of the others[5,6]. The figure of merit would then have the form

$$P_{FM} = P_1^{a_1} P_2^{a_2} \cdots P_R^{a_R} \left[b_1 P_{R+1} + b_2 P_{R+2} + \cdots \right] \quad (3)$$

The parameters in the product are those for which a zero value indicates that the technology is useless. Efficiency and mean time between failures are of this type since if either of these are zero the technology does not function. The parameters in the sum are those that are desirable but not essential. The inverse of size might be such a parameter. While a small device may be desirable, large size alone (within certain limits) does not destroy its function. The a's and b's are weights that describe the relative importance of the parameters. The a's must enter as powers because the relative scale defined by equation (3) is not changed if it is multiplied by a numerical factor. This is equivalent to a linear weighting of the logarithms of the multiplicative parameters.

The construction of such a figure of merit is arbitrary and contains a strong subjective element. It is, for example, a matter of judgement as to how to weight the relative advantages of extended bass response and harmonic distortion of a loudspeaker. However, the purpose of the technology is to satisfy human desires. The best that can be done is to estimate how users of the technology value the effects of the various parameters.

Even the description of a single member of the three dimensional hierarchy of technologies is seen to have a degree of complexity. The relationship between overall technological performance, which requires some subjectivity, and the objective individual performance parameters is not trivial.

IV. Vertical Hierarchies of Technology

The definition of vertical hierarchies recognizes the truism that a device consists of components and that a technology consists of subtechnologies. The vertical hierarchy is a fundamental statement of the structure of a technology. A straightforward way to determine this structure is to construct a kind of input-output matrix in which the performance parameters constitute the output and the components which determine those performance parameters are the input. An example is shown in Table 1 which gives a simplified technology input-output matrix for the single-lens reflex camera. Each plus sign entered into this matrix indicates that there is a functional relationship between the performance parameter of the camera and the performance parameters of the component parts. Of course, each component can be defined as a sub-technology and analyzed into its parts. Thus, for example, an input-output table can be constructed for the meter system and its performance can be related to the performance of its components.

In general, if P_j^A are the performance parameters of a technology labeled A and if $(P_{n_1}^{A,1}, P_{n_2}^{A,2}, \ldots, P_{n_c}^{A,C}, \ldots)$ are the performance parameters of the component technology labeled (A,1), (A,2),...,(A,C) then the jth performance parameter of A is a function of some subset of all the performance variables of all the component technologies, i.e.

$$P_j^A = F_j^A \{P_{n_c}^{A,C}\} \tag{4}$$

where $\{P_{n_c}^{A,C}\}$ is a subset of the array of component performance parameters:

$$\{P_{n_c}^{A,C}\} = \begin{Bmatrix} P_1^{A,1}, & P_2^{A,1}, & P_3^{A,1} \cdots \\ P_1^{A,2}, & P_2^{A,2}, & P_3^{A,2} \cdots \\ P_1^{A,3}, & P_2^{A,3}, & P_3^{A,3} \cdots \\ \vdots & \vdots & \vdots \end{Bmatrix} \tag{5}$$

Obviously, each component technology can be represented in an analogous fashion thereby giving a hierarchy of functional relationships in which each dependent variable at one level is determined by dependent variables at a lower level.

To apply these ideas, it is clearly necessary that a series of systems analyses be performed to construct the input-output tables and that the functional relationship between the performance parameters of a technology and the performance of its components be specified. This is admittedly complex and a great deal of detailed work must be done based on a sound engineering knowledge of the technology. But this just reflects the actual complexities of technological systems. The vertical hierarchy has the advantage of explicitly displaying the technical origins of technological performance and shows how technological change in a technology is dependent on technological change in its components. Also, study of a vertical hierarchy can identify the relative importance and state of development of the component technologies thereby identifying those limiting factors in technological evolution.

V. Longitudinal Hierarchies and Generalized Fisher-Pry Theory

The Fisher-Pry model regards technological progress as the substitution of a product or process by a new one which is technically superior[2]. Their model is embodied in three assumptions:

1. Technological advance is the competitive substitution of a new way of satisfying a need for an old.
2. If a substitution has gone as far as a few percent, it will go to completion.
3. The fractional rate of fractional substitution is proportional to the fraction remaining to be substituted.

The assumptions require some comment. Clearly, new "needs" arise as technology advances. There was no need for larger computer memory before the age of computers. This objection can be mitigated to some extent by defining needs broadly, or by simply accepting the fact that new needs arise and then applying the model to them as new technologies develop to fulfill these needs. The second assumption really consists of two parts. The first is that after a few percent success, a new product or process will continue to capture market share until it levels off at some maximum value. The second part is that the maximum value is 100 percent substitution. The first part of the second assumption has not been tested empirically. Such a test would require identifying innovations that had some small success but then died out. The second part is known to be incorrect for some cases. Fluorescent lighting, for example, never fully replaced incandescent lighting. This second assumption is not serious since it can be met by segmenting the market or by assigning a value to the maximum substitution that is less than 100 percent.

The third assumption just states that the differential equation for technological substitution is

$$\frac{df}{dt} = \beta f(1-f) \qquad (6)$$

where f is the market share of the new product at time t and β is a rate constant. The solution of (4) is

$$f=\left[e^{\beta(t_{1/2}-t)}+1\right]^{-1} \tag{7}$$

where $t_{1/2}$ is the time at which $f = 1/2$. Equation (4) and (5) are identical in form to equations (1) and (2). Other substitution equations have been proposed but they all have the sigmoid form of equation (5) and the general theory presented here can readily accomodate these modifications. Note that equation (5) is proportional to the number of units sold only if the market is static. If the market is growing or decreasing (because of population growth, for example) further information on the time dependence of market size is needed.

In spite of its shortcomings, the Fisher-Pry model is useful and has been successfully applied to a large number of technologial substitutions.

The Fisher-Pry model is binary in that it applies only to a pair of products competing in the marketplace. Sharif and Kabir[9] have extended the substitution model to the multi-level case in which there is a series of new products competing with each other and applied it to a sequence of three products. The assumption here is that a given product competes with all earlier products and is displaced by all later products. An alternative assumption is that all products are in pair-wise competition with each other. Sharif and Kabir tried to find the pair-wise substitution rates although they adopted the assumption given above. But the two assumptions are inconsistent and the pair-wise assumption leads to severe difficulties in determining rate constants and market shares. We therefore develop a longitudinal hierarchy of technological change as follows:

Let a series of innovations 1, 2, 3,..N to satisfy a particular need be introduced sequentially at times t_1, t_2,...t_N with $t_N > t_{N-1} > \ldots > t_1$. Since the last innovation substitutes for all the others, the (N-1) substitutes for all except the Nth, etc., the quantities that satisfy the substitution equation are

$$f_2=\frac{n_2}{n_1+n_2}$$

$$f_3=\frac{n_3}{n_1+n_2+n_3}$$

$$\vdots \qquad \vdots \tag{8}$$

$$f_j = \frac{n_j}{n_1+n_2+\cdots+n_j}$$
$$\vdots \qquad \vdots$$
$$f_N = \frac{n_N}{n_1+n_2+\cdots+n_j} \tag{8}$$

$$n_1+n_2+n_3+\cdots+n_N = 1 \tag{9}$$

where n_j is the market share of the jth product.

If we adopt the Fisher-Pry equation, then the time evolutions of the f_j are given by

$$\dot{f}_j = \beta_j f_j(1-f_j) \tag{10}$$

for the substitution of the jth product for all that went before. The rate constant β_j can be determined from initial substitution data, provided only that the next product (j+1) has no significant market share until a trend for the jth product is established.

The quantities of interest are the market shares n_j so it is necessary to solve equations (8) for the n_j in terms of the f_j. This is readily done by rewriting (8) in the form

$$f_N = n_N$$
$$f_{N-1} = \frac{n_{N-1}}{1-n_N}$$
$$f_{N-2} = \frac{n_{N-2}}{1-n_N-n_{N-1}} \tag{11}$$
$$\vdots \qquad \vdots$$

From (11), it easily follows that

$$\left.\begin{aligned}
n_N &= f_N \\
n_{N-1} &= f_{N-1}(1-f_N) \\
n_{N-2} &= f_{N-2}(1-f_{N-1})(1-f_N) \\
&\;\vdots \\
n_{N-j} &= f_{N-j}\prod_{i=0}^{j-1}(1-f_{N-i}) \\
&\;\vdots \\
n_2 &= f_2(1-f_3)(1-f_4)\ldots(1-f_N)
\end{aligned}\right\} \qquad (12)$$

Note that f_1 is not defined, so that n_1 is obtained from the normalization condition of equation (9).

Equations (12) answer the question of relative market share of competing technological products. To get the actual number of such products, the n's must be multiplied by the total size of the market captured at time t. Thus if Q_j is the number of units of type j, and Q_t is the total number, then

$$Q = Q_T(t)\, n_j \qquad (13)$$

A knowledge of the time dependence of the total market is needed. Four simple cases for this are: a static, saturated market, for which Q_t is a constant; a static unsaturated market, for which Q_t is growing but ultimately approaches a limit; a saturated, growing market, for which Q_t keeps pace with population growth; an unsaturated, growing market in which Q_t is growing more rapidly than population and tends to a growth rate proportional to the population. Representative functions that describe these cases are:

case 1)

$$Q_T = \text{CONSTANT} \tag{14}$$

case 2)

$$Q_T = \frac{Q_L}{e^{r(t_{1/2}-t)} + 1} \tag{15}$$

case 3)

$$Q_T = K\,e^{at} \tag{16}$$

case 4)

$$Q_T = \frac{Q_L\,K\,e^{at}}{e^{r(t_{1/2}-t)}+1} \tag{17}$$

Another question of interest in a longitudinal hierarchy is that of the level of the best technology in service. That is, for a given hierarchy of technologies, what are the best performance parameters commercially available at any given time? To simplify the discussion, we assume that the performance can be described by a single figure of merit. The generalization to multiple performance parameters is straightforward. The answer is simple. Of the technologies in a hierarchy at a given time, just determine which has the best performance parameter.

A slightly more complex question is that of the technological level of society as a whole with respect to the hierarchy. Several measures can be chosen for this. The simplest is the mean performance parameter for all users. Then, if $P_j(t)$ is the performance parameter (figure of merit) for the jth member of the hierarchy, we define the mean

$$\bar{P}(t) = \sum_j n_j\,P_j(t) \tag{18}$$

as a measure of the social technological level. A more detailed measure can be described by a distribution function which gives the number of units in service with performance greater than some specified value. To get this, consider the jth technology in the hierarchy. After some time t_1, its performance parameter exceeds the specified value $\tilde{P}$ and the number of units in service at time t, with $P_j > \tilde{P}$ is

$$\int_{t_1}^{t} Q_j(t)dt \tag{19}$$

so the total number of units with performance parameter greater than $\tilde{p}$ at time t is

$$\tilde{Q}(\tilde{p},t) = \sum_{j(p_j > \tilde{p})} \int_{t_1}^{t} Q_j(t')dt' \tag{20}$$

Application of the Fisher-Pry approach to the idea of the longitudinal hierarchy results in a mathematical description of the evolution of maximum technological performance, the market shares of competing technologies and the level of technological performance in place for the technology users.

Conclusions and Summary

Technological change is an enormously complex and diverse phenomenon. To deal with it quantitatively, a classification scheme is needed that reflects this complexity and yet permits separate aspects of technological change to be described. In this paper, a hierarchical approach is presented that is suitable for describing many kinds of technological change. The hierarchy is three dimensional and connects technologies in time, in function and to each other. The virtues of this classification are that it clearly defines the objects of study, it can limit the scale of the analysis by treating particular members of the hierarchies and it can progressively extend the analysis by including more members of the hierarchies. The study of technological change should then start with a systems analysis to determine the structure of the hierarchies of interest for a particular technology or set of technologies. Once this is done, a member or members of the hierarchy are selected for analysis and the time evolution can be described in terms of systems of equations for the parameters defining the members of the hierarchy. The structure of the hierarchy determines the relation of the equations among various technological systems. For technologies considered as sets of devices, this procedure leads to nested sets of coupled equations. The evolution of performance parameters and the capture of market share by a new product are known to follow logistic type equations. Thus, while the overall complexity of the mathematics reflects the complexity of technological change, the individual parts of the mathematical structure are straightforward.

References

1. See, for example, Mansfield, E. "Industrial Research and Technological Innovation"; W.W. Norton & Co. Inc., New York(1968).

2. Fisher, J.C. and Pry, R.H., "A Simple Substitution Model of Technological Change" in "Industrial Applications of Technological Forecasting", ed. by M. Cetron and C.A. Ralph, Wiley-Interscience, New York(1971).

3. Linstone, H.A. and Sahal, D. eds, "Technological Substitution", Elsevier, New York(1976).

4. Mensch, G., "Stalemate in Technology", Ballinger Publishing Co., Cambridge, Mass.(1979), translated from the German, first published in 1975.

5. Gordon, T.J., Munson, T.R., Technological Forecasting and Social Change, 20, p. 1-26(1981).

6. Martins, J.P., "Technological Forecasting for Decision Making", North-Holland, New York, 2nd edition(1983), Ch. 6.

7. Kydes, A.S. and Rabinowitz, J., Resources and Energy, 3, p. 65 (1981).

8. Kydes, A.S., Minesi, M.J. and Hudson, E.A.; Energy Modeling and Simulation, North-Holland(1983), p. 49.

9. Sharif, M.N. and Kabir, C., "System Dynamics Modeling for Forecasting Multilevel Technological Substitution", p. 21 in reference 3.

Table I-Technology Input-Output Table for Single-Lens Reflex Camera

Components	Performance Parameters					
	P_1	P_2	P_3	P_4	P_5	P_6
C_1	+	+	0	0	+	+
C_2	0	0	0	0	0	+
C_3	0	0	+	0	0	0
C_4	0	0	+	+	+	+
C_5	0	0	0	0	+	+
C_6	0	0	0	+	0	0

P_1 = lens speed

P_2 = lens resolution

P_3 = shutter speeds

P_4 = meter resolution

P_5 = prism image brightness

P_6 = focusing accuracy

. .

. .

. .

C_1 = lens system

C_2 = lens mount

C_3 = shutter system

C_4 = mirror

C_5 = prism

C_6 = meter system

. .

. .

. .

ORGANIZATIONAL STRUCTURES FOR FACILITATING PROCESS INNOVATION.

Maurice S. Elzas, chairman,
Department of Computer Science
Agricultural University
Hollandseweg 1, 6706 KN Wageningen
The Netherlands

When one reflects on the nature of existing organizations, especially those organizations which are large (encompassing many people) and have been in existence for a long time, an obvious - though somewhat iconoclastic - conclusion one may reach is that the concept of organization is in itself a model that is made to fit - willy-nilly - to an already existing conglomerate of persons, functions and processes that have been carrying on together for some time for some (set of) purpose(s) under a common name.

In such a conglomerate, hierarchies have - more often than not - developed more or less autonomously, based on individual abilities and personalities of the human being's concerned, the common goal being only somewhat defined as an afterthought at a later date e.g. whenever the organization ran into trouble and had to be adjusted because of internal disputes or external pressures.

Indeed, one might - cynically - comment that the boom in management- and organization-consulting firms, is mainly due to this state of affairs.
Our "older" organizations have grown like if they were ecological systems, which - when they are not interfered with - are apparently self-organizing. In this sense the "ecological" system is *"in itself the vast computer that gets the answers right (or roughly so, give or take a few plagues, famines and so forth). But it has no programme, no planning department, no licences to breed, no bureaucracy. It just works. We, the intelligent humans, interfere with this system, unbalancing it for our own needs"*, (Beer, 1981).

This, clearly, is in flagrant contrast with the definitions of "organization" that can be found e.g. in Webster's Third Dictionary:"the administrative and functional structure of a business (political party, military unit, etc.) including the established relationships of personnel through lines of authority and responsibility with delegated and assigned duties" and "a group of people that has a more or less constant membership, a body of officers, a purpose and usualy a set of regulations".
Thus, two basic hypotheses can lie at the base of an organization:

a. the autopoietive*) hypothesis: which states that the organization is a homeostat**) in which the critical condition held steady is the systems' own organization. (Maturana, 1980);

b. the teleologic***) hypothesis: which is based on the assumption that an organization is an adaptable system which is governed by the prerequisite to achieve a goal or purpose and where the structure is uniquely geared toward allowing optimal control of the (most efficient) course to reach the goal of the system.

It is deemed evident that there is no organization containing human beings which is either fully autopoietic or fully teleologic.
Let us suffice by remarking that older (and especially large) institutions tend to behave in the autopoietic way, while newer and and (smaller) units can come close to being teleologic.
In everyday practice, a systems approach (including modelling) can be used with a reasonable chance of success for designing or changing an organization in the teleologic sense. Which working hypothesis is a must for the survival of any innovation impetus, autopoiesis being the main obstacle for all change of this kind.

To date humankind has developed no skills whatsoever to "design" organizations that can survive by autopoiesis, however desirable this might socially and psychologically be for the individuals that are affected.
The term "affected" is used in the previous sentence to focus attention on the fact that any practical social system has its homeostatic tendencies, and that therefore any change in organization is perceived by the individual in the system as a change of the system itself, especially as it affects internal structure, thus the relation between the individual and the other components of the system (Beer, 1981).

In any teleological system consisting of partly autonomous components (what human beings are par excellence) "control to a purpose" implies channeling of information (in general aggregated "bottom-up") and broadcasting of commands (in general "top-down") in such a way that the whole stays manageable.

----

*) Autopoiesis: the faculty of producing something (especially creatively) oneself autonomously.

**) Homeostat: a (social) system which tends to maintain (relatively) stable conditions with respect to (external) disturbing factors and competing tendencies and -powers within itself.

***) Teleology: the fact or character of being directed toward an end or shaped by a purpose.

This statement of fact leads by necessity to a *layered structure of the system, including at least a hierarchy in the nature- and flow of information and in the kinds and levels of command.*

This prerequisite is a direct consequence of Ashby's law of requisite variety, which states that control can be achieved if, and only if, *"the variety of the controller is at least as great as the variety of the situation to be controlled"* (Ashby, 1956). (For a clear explanation of the concept of variety the reader is referred to Beer, 1979, chapter 2).
To illustrate this point let us look at fig. 1, a rough model of the classical organization of an average manufacturing firm, like it is still "operational" in many places today.
Let us assume that an extremely crude information system exists in the company under study: the only information that every level in the organization provides to its higher echelon is OK or NOT OK (meaning all is within the goal set a priori for the total operation or this is not the case) similarly controls are chosen to be binary: GO ON or STOP to reconsider.

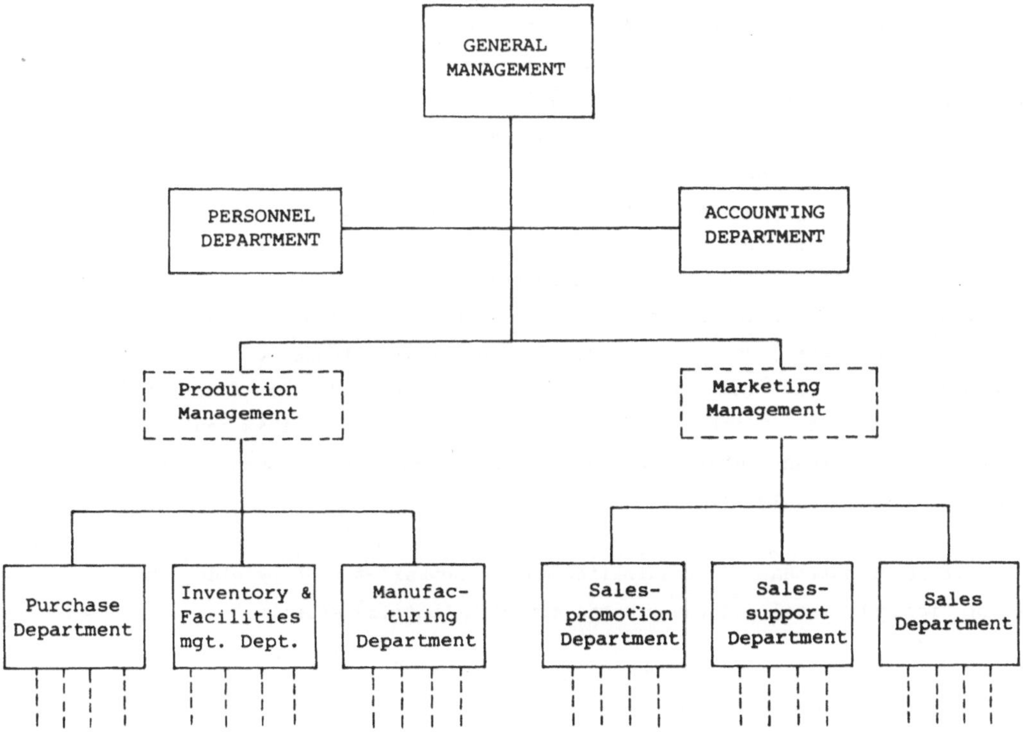

fig. 1

The reason for creating production and marketing management positions on the organizational graph are then evident from a variety-control view-point. If these - sectorial management positions - were not available, management would have to be able to control (and foresee) $2^8 = 256$ different possible "input" situations and provide associated responses (outputs), which brings the number of possible (not necessarily adequate) alternatives to be chosen from to: $2 \times 2^8 = 512$. Inserting (central) production management and marketing management reduces the selection of control measures problem to 1 in 32 (4 inputs, 2 outputs), if one leaves central personnel and accounting departments in place as indicated in fig. 1.
If personnel and accounting are converted into two separate subdepartments of production and marketing, the variety to be controled is reduced to 1 in 8*).
So variety control tells us that management needs variety attenuators/control amplifiers (and thus an adequate organizational structure) to stay sane.
Figure 1 shows us more than only the basic reason (variety matching) for modelling (and possibly changing) organizations as they have emerged from history: the personnel and accounting departments "float on thin air" in our model. What is worse is that accounting provides additional ("consolidated") information to management which might very well be extreme in its lack of variety (thus often giving a biased view) and conflicting with the information received through direct channels. (Possibly a reason for traditional managers to overrate the accounting information and disregard direct signals?)
The presence of a personnel department shows another typical trend: functional specialisation.
Apparently it is felt that the function of personnel-selection/salary-harmonization cannot be trusted to the individual departments. Moreover classical organizations often develop the problems indicated in figure 2.
This has been the reason in the last decade for the development of a new form of organisational model, which - considered objectively - tends to concentrate on integration of processes that take place in the firm instead of on functions and hierarchies (the phrase "matrix management" is often coined in this context).
(For detailed background-information refer to a.o. Kampfraath & Marcelis 1981 (in Dutch), Kampfraath 1971).

So, two basic mainstreams of organization models now exist side by side: the function-hierarchy model and the process-oriented (matrix) model.

----

*) In fact, stopping the whole outfit to reconsider operations is a far to rough type of control. If the response is taken to be binary PER DEPARTMENT, the ratios would be $1 : (256)^2 = 1 : 65536$ (instead of 1 : 512) without middle level, respectively 1 : 1024 (instead of 1 : 32) even with the first type of middle management level, a far more dramatic improvement!

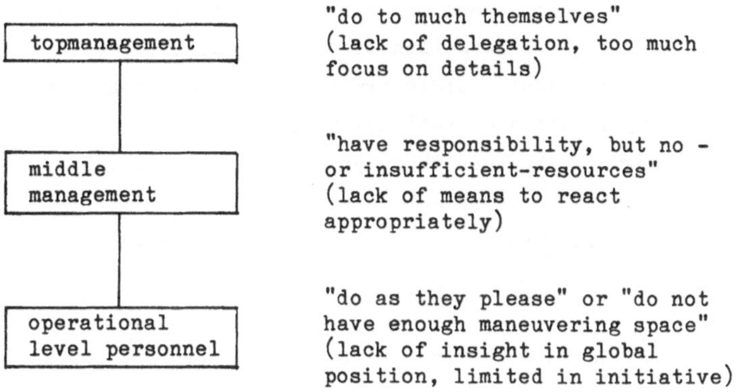
topmanagement
"do to much themselves"
(lack of delegation, too much
focus on details)
middle
management
"have responsibility, but no -
or insufficient-resources"
(lack of means to react
appropriately)
operational
level personnel
"do as they please" or "do not
have enough maneuvering space"
(lack of insight in global
position, limited in initiative)

fig. 2

Before elaborating these models, and, discussing their influence on (re)design of organizations, it is appropriate to mention that - in practical situations - neither of these models are actually amenable to implementation in their "pure" form. The main reason for this being the prevailing homeostatic tendency in social groups, which leave such strong autpoietic seeds in any organization that teleologic changes (especially those which are *completely* based on rationale) have a strong tendency to be stifled.
Returning now to the properties of the two contending organization models, let us try to describe their basic construction and -elements from a modelling methodology viewpoint.

1. The FUNCTION-HIERARCHY MODEL.

As pointed out earlier on: this still is the prevailing organizational model nowadays.
The basic element in the model is the "function" which represents a person (or group of persons) with specific, predefined, capabilities for fulfilling a *specific* task in *a fixed* place in the hierarchy of command and control.
The model does not only require the system (i.e. the institution or firm) to be decomposable into functions (which in practice is erroneously considered to be a "trivial" condition) but also - to be able to design, assess and adapt the teleological organization - requires the goal or purpose of the system to be decomposable in an appropriate set of subgoals befitting each and every function. The basis concept being that function aggregation in the hierarchy - "automatically" - leads to goal-aggregation along the same lines and thus if "every man is in the right place" the organization pursues its global purpose "all by itself".

One directly notices the strong homeostatic characteristics of this model, but it took the advent of (modern) information processing tools to achieve general awareness of the inadequacy of the model, for information-aggregation purposes and consequently for the succesfull application of rational tools, for global planning and policy analysis purposes. Mainly, because these activities require decomposition into operations - rather than functions - in the system.
So, the main problem with this type of organization might well be the "function" in the model.

Most of the functions which occur in practice are historical products of schooling, uniform (industry-wide) remuneration and the need for understandable job-descriptions in recruiting operations, rather than decomposites of the goal of the organization. The most difficult problem in any set-up using human labour is establishing adequate payment for individual work in an objective way.
Reasoning in a strict teleological way renumeration should be proportional to the contribution to the overall goal of any individual employee.
It is clear that this - in itself - is a problem that is difficult to solve for two main reasons:

- Often decomposition of roles in the organization with respect to goal-contribution is impossible, or an extremely subjective undertaking at best. While in organizations subject to innovative impulses, these decompositions vary widely in time.

- Although human labour can sometimes be evaluated - as to its commercial value - in an absolute sense by comparing it to the full cost of replacing this labour by other means (e.g. machines), this principle is only applicable for a very small part of the labour-force.

Hence renumeration of - the greater majority of - the employees on basis of qualifications and experience for a - more or less widely accepted - function is the way out that is generaly choosen. In fact this is the only way in which career-planning and mobility can be achieved and understood.
An interesting operations-research oriented basis for these activities, if the hierarchy can accomodate the notions of "rank" and "seniority" on top of the available functions, can be found in Fraguio (1981), who shows that in such a case career-planning can be proven to be a a strictly deterministic process.
It is deemed to be clear that - where innovation takes place - the problem of personal rewards is extremely difficult to solve.

2. The PROCESS-ORIENTED ORGANIZATION MODEL.

Since the process of arriving at decisions in organizations has been under study, especially since the advent of powerfull information processing devices, it has become clear that the information needed follows rather from the aggregation of operational processes than from the hierarchic aggregation of functions and their associated information filters.
This problem has best been approached in a more-or-less "cybernetic" sense by Beer and Ackoff (Beer 1981, Ackoff 1967, Ackoff & Emery 1972).
The advent of frequent - and far reaching - mergers in the past years, has increased the experience with restructuring organizations to operate in a new environment. These mergers are - more often than not - undertaken with financial/marketing objectives in mind and often result in an organization that has to cater for a large amount of loosely connected products and activities.
Therefore, when the time comes that the two companies have to join their workforces into a joint operation, considerable human difficulties can arise in the elaboration of a new organization. In fact the problem is that a new system is created without establishing a well-founded purpose beforehand, which reflects back to a situation in which - sometimes divergent - goals have to be merged into a common purpose which - because of diverging initial interests - is extremely difficult to assess.
In the case two process-oriented organizations are merged, the task is much easier (a.o. for the personnel concerned) because it is easier to merge and concentrate processes than to eliminate functions or change hierarchies.
The same situation is present when - in an existing organization - innovation causes new activities to emerge as new shoots on the old stem.

Interestingly enough, at the recent International Working Conference on Model Realism (April 20-23, 1982, Bad Honnef, FRG)(Wedde, 1982) most of the lectures addressing organizational problems where oriented towards process-models of organizations rather than function-hierarchy models.
The power of the process-model of organizations is that the organization is considered to be an aggregation of basic cells which are uniform in nature (fig. 3).

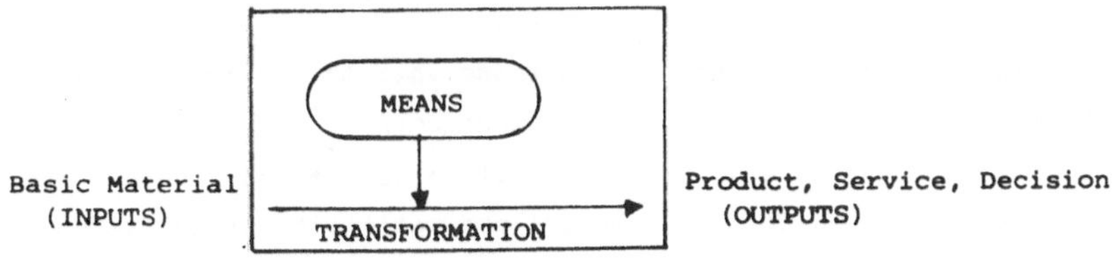

fig. 3

The cell consists of (existing) means which are put to the disposition of the process by the firm or institution, and a transformation process that uses the means in order to "convert" basic material into products, services or decisions.
The term "means" is used to denote any resource (tools machines, finances, labour, know-how, local inventory, etc.) which is available to be applied in the process of converting basic material (raw materials, semi-finished goods, data, etc.) into goods (products, services, information, decisions, etc.) in the quantity and of the quality required by the <u>local</u> goal (which is the localised decomposition of the overall goal).
Accepting for a moment that all the (local) processes are sufficiently provided of means, than proceses are uniquely linked together through their inputs (basic materials) and outputs (products, services, decisions) without an a priori necessity for hierarchy other than the one indicated by the laws of aggregation. Thus this model allows "chaining" of operations in such a sense that producing certain outputs for the whole of the organization can be achieved by production in stages, where every cell provides the partial transformation which is best geared to the <u>means</u> at its disposal.
Thus every cell can "work" on "parts" of several "products" at any one time. Putting this pattern into a symbolical (possibly graphical) representation gives rise to a network-like pattern, hence - possibly - the term "matrix-organization".

To clarify these concepts, let us briefly envisage a few example organizations, of which the first two can easily be found in practice and the third one endeavours to implement the process-oriented organizational model in its extreme form.
In the first place a process-type of organization for a firm, whose object it is to sell several (technical) products made elsewhere, will be considered.
The process-element comes in by implementing a so called "productline" set-up in the day-to-day activities of the firm, without totally removing the remnants of the classical functional subdivision. This concept is illustrated in fig. 4.
In this example (which is a simplified reflection of a number actually existing organizations), the general management is also responsible for the personnel and financial management aspects of the whole firm, while the internal allocation of resources is arrived at through periodic discussions between the heads of the departments and the product managers.

Now, proceeding along the lines of assessing variety control, as sketched earlier in this chapter, and assuming that general management gets more detailed information this time, let us try to evaluate the addition of the "horizontal" elements to the corporate structure. To do this more or less realistically let us consider the case where general management hinges its GO/NO GO decisions on OK/NOT OK assesment of the personnel, capital resources, turnover, cost of turnover and the allocation of resources to products. The first two and the last one of these information items

pertain to the vertical departments (marketing and sales, accounting and administration, customer support).
The turnover aspects per product clearly belong to the horizontal productlines. The resulting problem of choice of relevant control measures, calculated on the basis of a response per department/productline as before - and including moreover the possibility to STOP to reconsider the whole operation, shows a variety of approximately 1 : 1.34 * 10^8 for the matrix organization shown in fig. 4.

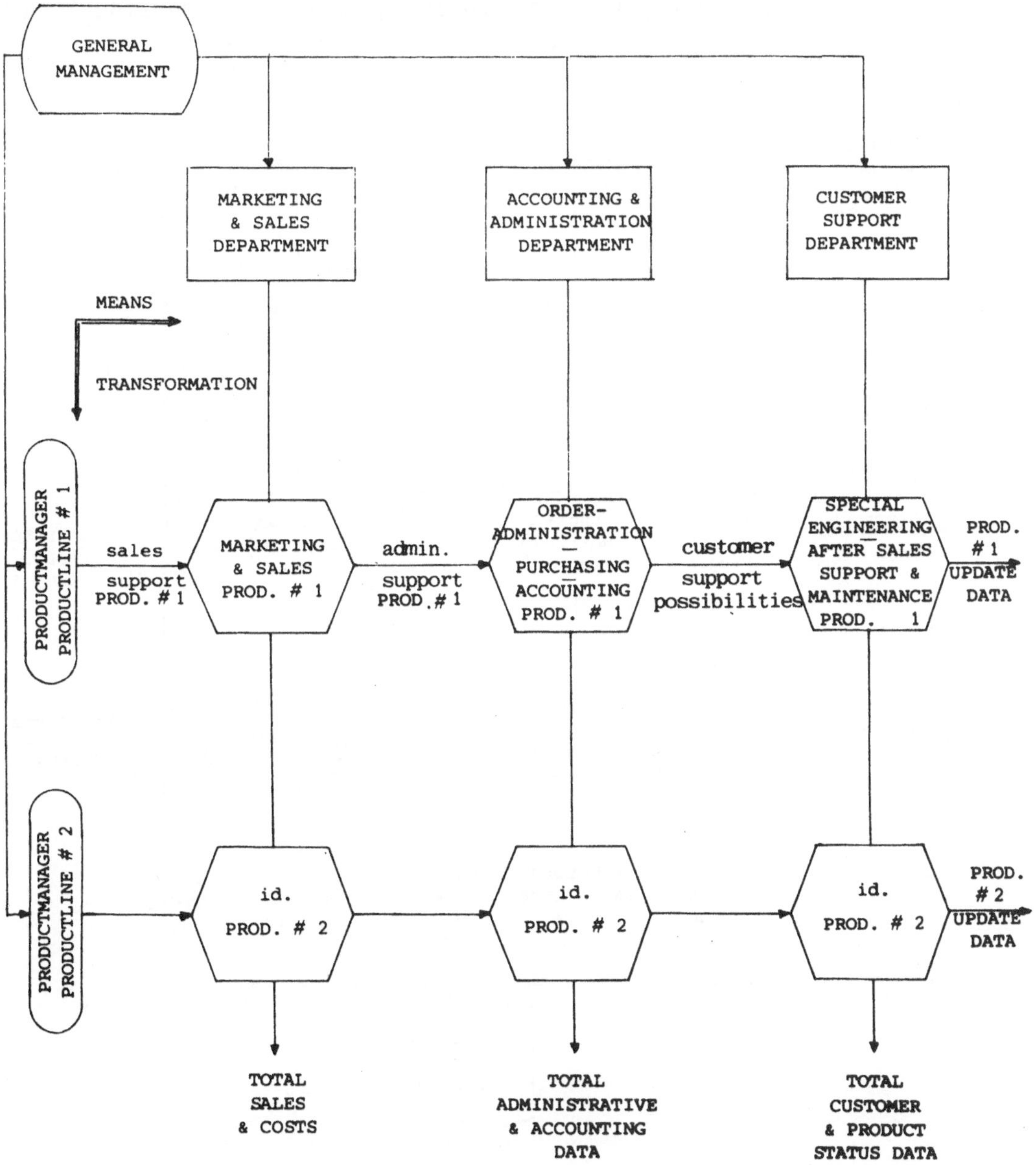

fig. 4

Eliminating the product management raises this variety of possible general management decision choices to 1 : $2.15 * 10^9$. Thus the productline-type of organization shows a beneficial - though not extremely dramatic - influence.

Another example of a similar matrix set-up is shown in fig. 5 for a contracting firm that has an organization "matricised" by contracted project.
The variety control advantages can be calculated in the same way as before, and are larger because of the greater number of "vertical" departments, thus the increase in variety of about 16 in the previous case grows to about 4000 in this case.

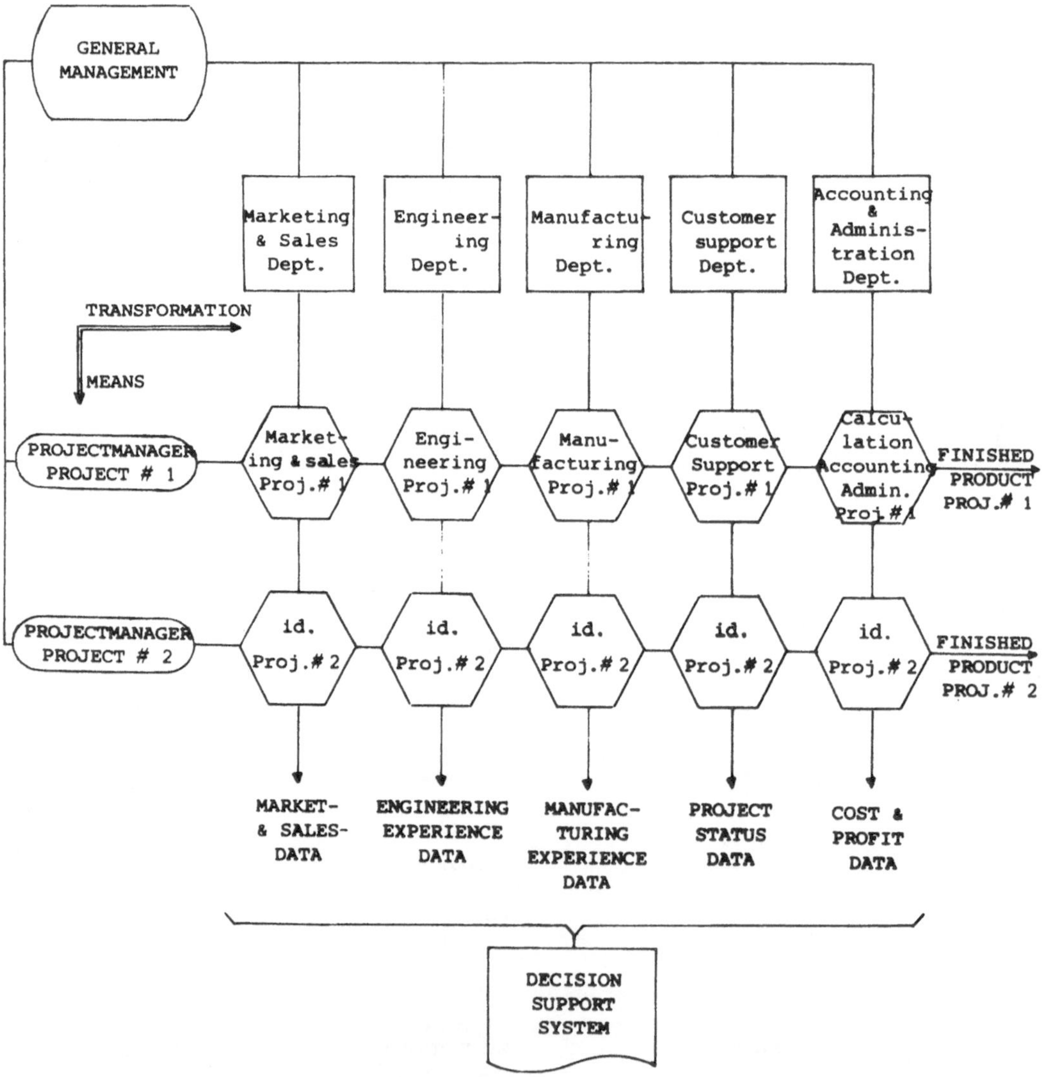

fig. 5

Apparently the larger the number of departments in the original (classical) organization, the more advantage is to be had by superimposing the matrix-type of crosscurrent-control.
As, roughly, sketched in fig. 5, the data flows which are inherent in this type of organizational model, facilitate the introduction of decision-support information systems. As depicted the system can supply relevant information as to the utilisation of the company's means in the different sectors for the different projects, and - in this way - provide adequate historical data for future projects. So, the setup is useable in situations where several innovations are in a "try-out" stage.

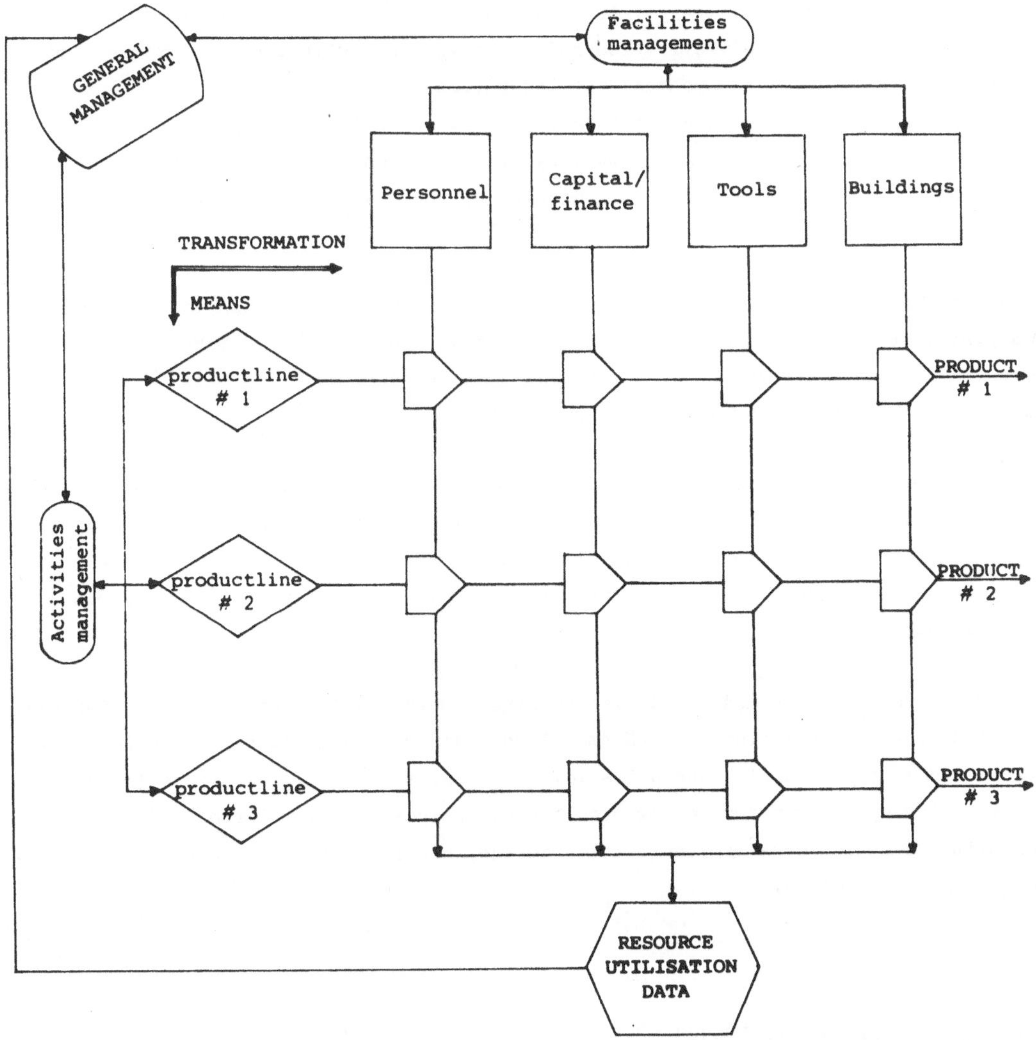

fig. 6

These two examples are typical of the "state-of-the-art" of process-oriented organizational models: they yield the present image of the matrix organization because they are implanted a posteriori on a classical, functional, organization. This "hybridization" may be one of the reasons for the relatively low degree of success of matrix-organization-concepts. The "pure" process-organization-model would certainly yield a different structure in the organizational set-up, as depicted in fig. 6.
In this illustration the functional characteristics have completely "disappeared", because every "cell" associated with specific product-transformation activities uses the mix of facilities best suited to its requirements. General management is left to its main purpose in life: the strategic aspects of the organization, that is: choosing the right mix of activities and providing them with the necessary mix of facilities. An "ideal" environment for innovation cells!
The tactical considerations and the daily operation of the organization are delegated to the level where they belong. Naturally, periodic consultation between the different levels of management (as indicated by the bidirectional connections) remains a prerequisite to keep the total in line.

Most of the actual organizations of this type will (gradualy change to) become losely coupled distributed-systems, where special precautions have to be taken to ensure corporate goals to be achieved. These requirements can readily be evaluated by modelling. Schiffers (1982), has shown that the concept of *interaction systems* can be usefull for this purpose.

The study of the process-type of organization, which still - at this moment because its relative novelty - requires modelling to evaluate, can benefit from certain network-oriented modelling formalisms because of the very nature of the organizational concept.

The Petri-Net approach and its derivates like Superposed Automata Nets (de Cindio et al. 1982), Decision-Action Nets (Kramer & Schmidt 1982) and Place Transaction Nets (Fuss 1982) has especially shown its value in addressing - at least on a formal basis - the specific coordinating problems of the adaptively interrelated distributed-action organizational units as they are found in process-based organizational models.

3. BIBLIOGRAPHY and REFERENCES.

Ackoff, R.L. (1967), "Management Misinformation Systems", Management Science, Vol.14, no.4, pp 147-156.

Ackoff, R.L. and Emery, F.E. (1972), On Purposeful Systems, Aldin-Atherton, N.Y.

Alter, S.L. (1980), Decision Support Systems, current practice and continuing challenges, Addison Wesley, Philippines.

Ashby, W.R. (1956), An Introduction to Cybernetics, Chapman and Hall.

Beer, S. (1975), Platform for Change, John Wiley & Sons, London.

Beer, S. (1979), The heart of the enterprise, John Wiley & Sons, Chichester.

Beer, S. (1981), Brain of the Firm, Second Edition, John Wiley & Sons, Chichester.

de Cindio, F. et al. (1982), "Real System Modelling: a Formal but Realistic Approach to Organization Design", in H.Wedde (ed.), Model Adequacy: Proceedings of the International Working Conference on Model Realism, Springer-Verlag, Heidelberg, F.R.G.

Elzas, M.S. (1980), "Simulation and the Processes of Change", in Oren, Shub, Roth (Eds.), Simulation with Discrete Models: A State-of-the-Art View, IEEE, N.Y., pp. 3-18.

Elzas, M.S. (1982a), "The Use of Structured Design Methodology to Improve Realism in National Economic Planning", in H.Wedde (ed.), Model Adequacy, Proceedings of the International Working Conference on Model Realism, Springer-Verlag, Heidelberg, F.R.G.

Fraguio, C.J. (1981), "A mathematical model of hierarchies", in J.P.Brans (ed.), Operations Research '81, North-Holland, Amsterdam.

Fuss,H. (1982), "Reversal Simulation with Place-Transactor-Nets", in H.Wedde (ed.), Model Adequacy: Proceedings of the International Working Conference on Model Realism, Springer-Verlag, Heidelberg, F.R.G.

Kampfraath, A.A. (1971), "The provision of organizational conditions for project management", in Proc. 10th Congress of the European Federation of Productivity Services, The Hague (Netherlands).

Kampfraath, A.A. and Marcelis, W.J. (1981), Besturen en Organiseren, Kluwer, Deventer (Netherlands).

Kramer, B. and Schmidt, H.W. (1982), "A High level Net language for Stepwise Modelling of Organizational Systems", in H.Wedde (ed.), Model Adequacy: Proc. of the Intern. Working Conf. on Model Realism, Springer-Verlag,Heidelberg,F.R.G.

Maturana, H. (1980), Autopoietic Systems, Harvard Monographs.

Schiffers, M. (1982), "An Exercise in Achieving Goals in Distributed Systems", in H.Wedde (ed.), Model Adequacy: Proc. of Intern. Working Conf. on Model Realism, Springer-Verlag, Heidelberg, F.R.G.

Umpleby, S. (1982), "A Group Process Approach to Organizational change", in H.Wedde (ed.), Model Adequacy: Proc. of Intern. Working Conf. on Model Realism, Springer-Verlag, Heidelberg, F.R.G.

Wedde, H. (1982), (ed.), Model Adequacy: Proc. of the Intern. Conf. on Model Realism, Springer-Verlag, Heidelberg, F.R.G.

Wiener, A. (1978), Magnificent Myth - patterns of control in post-industrial society. Pergamon Press, Oxford.

THE JOINT COORDINATION METHOD WITH APPLICATION TO THE ON-LINE STEADY STATE CONTROL OF NON-STANDARD PROBLEMS

J.E. Ellis, H. Michalska and P.D. Roberts
Control Engineering Centre
School of Electrical Engineering and Applied Physics
The City University,
London ECIV OHB

1. INTRODUCTION

It has been shown how, for the situation where the global objective is the sum of local objectives, the interaction balance method (IBM) for the optimizing control of large scale systems (Findeisen and co-workers, 1980), can be modified to give a joint coordination method (JCM) (Michalska and co-workers, 1984). By utilizing ideas from integrated system optimisation and parameter estimation for single processes (Roberts, 1978, 1979), JCM introduces model parameter estimation at the local decision problem level and produces optimal solutions for a wide class of problems. The JCM also acts on the given local objectives directly, elminating the requirement for local problems to be separable.

However, when the local objective functions appear in the global objective as non-additive terms, it is again not possible to obtain separable local problems and in these circumstances the JCM cannot be applied in its original form. In this paper, an extension to JCM is presented which enables such non-additive global problems to be dealt with. This is achieved by introducing an appropriate substitution global objective function. The extension of JCM produces optimal solutions for a wider class of large scale optimizing control problems.

2. EXTENSION OF JCM

2.1. The Control Problem

Consider the steady-state optimizing control problem for a large scale process, composed of N interconnected sub-processes. The global objective function, Q is related to the local objectives, Q_i, $i \in \overline{1, N}$ by,

$$Q(\underset{\sim}{c}, \underset{\sim}{u}, \underset{\sim}{y}) = \Phi\left(Q_1(\underset{\sim}{c}_1, \underset{\sim}{u}_1, \underset{\sim}{y}_1), \ldots, Q_N(\underset{\sim}{c}_N, \underset{\sim}{u}_N, \underset{\sim}{y}_N)\right) \tag{1}$$

where $\underset{\sim}{c}_i$, $\underset{\sim}{u}_i$, $\underset{\sim}{y}_i$ are the sub-system controls, interconnection inputs and outputs respectively, defined in the appropriate spaces. For the global system,

$$\left.\begin{aligned} \underset{\sim}{c} &\triangleq (\underset{\sim}{c}_1, \ldots\ldots, \underset{\sim}{c}_N) \\ \underset{\sim}{u} &\triangleq (\underset{\sim}{u}_1, \ldots\ldots, \underset{\sim}{u}_N) \\ \underset{\sim}{y} &\triangleq (\underset{\sim}{y}_1, \ldots\ldots, \underset{\sim}{y}_N) \end{aligned}\right\} \tag{2}$$

The problem is considered where the mapping of the local objective functions, Φ in (1), appear as non-additive terms. The behaviour of each subsystem is described by the mapping,

$$\underset{\sim}{y}_i = \underset{\sim}{F}_{*i} (\underset{\sim}{c}_i, \underset{\sim}{u}_i) , i \in \overline{1, N} \quad (3)$$

The structure of the interconnections between the subsystems is assumed to be known exactly and is given as,

$$\underset{\sim}{u}_i = \sum_{j=1}^{N} H_{ij} \underset{\sim}{y}_j , \quad i \in \overline{1, N} \quad (4)$$

where H_{ij} are interconnection matrices whose elements are either zero or unity. For the system as a whole,

$$\underset{\sim}{y} = F_* (\underset{\sim}{c}, \underset{\sim}{u}) \quad (5)$$

$$\underset{\sim}{u} = H \underset{\sim}{y} \quad (6)$$

It will be further assumed that (5) and (6) are uniquely solvable with respect to the controls, so that the system input-output relationship can be represented by the single mapping,

$$\underset{\sim}{y} = K_* (\underset{\sim}{c}) \quad (7)$$

Because of the lack of perfect knowledge of the real system mapping, or the desire to simplify the input-output relations, a model is used of the form,

$$\underset{\sim}{y}_i = \underset{\sim}{F}_i (\underset{\sim}{c}_i, \underset{\sim}{u}_i, \underset{\sim}{\alpha}_i) \quad , \quad i \in \overline{1, N} \quad (8)$$

where $\underset{\sim}{\alpha}_i$ is the i'th sub-system model parameter, which is available for estimation purposes within the proposed scheme. In a similar manner to the system, a global model mapping can be formed,

$$\underset{\sim}{y} = \underset{\sim}{F} (\underset{\sim}{c}, \underset{\sim}{u}, \underset{\sim}{\alpha}) \quad (9)$$

where, $\underset{\sim}{\alpha} \triangleq (\underset{\sim}{\alpha}_1, \ldots\ldots, \underset{\sim}{\alpha}_N)$ (10)

Each subsystem is also subject to constraints,

$$\underset{\sim}{g}_i (\underset{\sim}{c}_i, \underset{\sim}{u}_i, \underset{\sim}{y}_i) \leqslant \underset{\sim}{o} , \quad i \in \overline{1, N} \quad (11)$$

where the mappings, $\underset{\sim}{g}_i$ are assumed to be known exactly. Again, globally these can be written,

$$\underset{\sim}{g} (\underset{\sim}{c}, \underset{\sim}{u}, \underset{\sim}{y}) \leqslant \underset{\sim}{o} \quad (12)$$

The problem is thus to minimise the global index, (1) subject to model (9) and constraints, (12)

2.2 Structure for Non-Additive Global Problems.

Using the coupling equations, (4), (6), and the models, (8), (9) the model based optimisation problem, from (1), becomes,

$$\min_{\underset{\sim}{c}} Q(\underset{\sim}{c}, \underset{\sim}{u}, \underset{\sim}{\alpha}) = \min_{\underset{\sim}{c}} \Phi\, (Q_1(\underset{\sim}{c}_1, \underset{\sim}{u}_1, \underset{\sim}{\alpha}_1), \ldots\ldots, Q_N(\underset{\sim}{c}_N, \underset{\sim}{u}_N, \underset{\sim}{\alpha}_N)) \tag{13}$$

subject to constraints (11) and (12) which using (8) become,

$$\underset{\sim}{g}_i\, (\underset{\sim}{c}_i, \underset{\sim}{u}_i, \underset{\sim}{\alpha}_i) \leqslant \underset{\sim}{o}, \; i \in \overline{1, N} \tag{14}$$

$$\underset{\sim}{g}\, (\underset{\sim}{c}, \underset{\sim}{u}, \underset{\sim}{\alpha}) \leqslant \underset{\sim}{o} \tag{15}$$

If measurements of all outputs are available and if the model, (9) is of the required structure, (7) and (9) can be used in the simple parameter identification procedure to yield estimates of the parameter, $\underset{\sim}{\alpha}$:

$$\underset{\sim}{y}(\underset{\sim}{c}, \underset{\sim}{u}, \underset{\sim}{\alpha}) = \underset{\sim}{y}_*\, (\underset{\sim}{c}) \;\rightarrow \underset{\sim}{\alpha} \tag{16}$$

where $\underset{\sim}{y}_*\, (\underset{\sim}{c})$ are the system outputs. It can be seen that the optimisation problem, (13) and the estimation problem (16) are coupled by controls, $\underset{\sim}{c}$, interactions, $\underset{\sim}{u}$ and parameters, $\underset{\sim}{\alpha}$. To decouple these problems, additional variables are introduced into the relevant problems,

$$\left.\begin{array}{l} \underset{\sim}{v} = \underset{\sim}{c} \\ \underset{\sim}{w} = \underset{\sim}{u} \\ \underset{\sim}{\sigma} = \underset{\sim}{\alpha} \end{array}\right) \tag{17}$$

with,

$$\left.\begin{array}{l} \underset{\sim}{v} \triangleq (\underset{\sim}{v}_1, \ldots\ldots, \underset{\sim}{v}_N) \\ \underset{\sim}{w} \triangleq (\underset{\sim}{w}_1, \ldots\ldots, \underset{\sim}{w}_N) \\ \underset{\sim}{\sigma} \triangleq (\underset{\sim}{\sigma}_1, \ldots\ldots, \underset{\sim}{\sigma}_N) \end{array}\right) \tag{18}$$

From the global optimisation problem, (13) and (6), (11), (16) and (19), a global Lagrangian can be formed to given an integrated optimisation and identification problem,

$$L(.) = Q(\underset{\sim}{c}, \underset{\sim}{u}, \underset{\sim}{\sigma}) + \underset{\sim}{\lambda}'(\underset{\sim}{u} - H\underset{\sim}{y}\, (\underset{\sim}{c}, \underset{\sim}{u}, \underset{\sim}{\sigma}\,)) + \sum_{i=1}^{N} \underset{\sim}{\gamma}_i'\, \underset{\sim}{g}_i\, (\underset{\sim}{c}_i, \underset{\sim}{u}_i, \underset{\sim}{\sigma}_i)$$

$$+ \underset{\sim}{\zeta}'\, (\underset{\sim}{v} - \underset{\sim}{c}) + \underset{\sim}{\xi}'\, (\underset{\sim}{w} - \underset{\sim}{u}) + \underset{\sim}{\mu}'\, (\underset{\sim}{\sigma} - \underset{\sim}{\alpha}) + \underset{\sim}{\eta}'\, (\underset{\sim}{y}(\underset{\sim}{v}, \underset{\sim}{w}, \underset{\sim}{\alpha}) - \underset{\sim}{y}_*\, (\underset{\sim}{v})) \tag{19}$$

where $\underset{\sim}{\lambda}$, $\underset{\sim}{\gamma}_i$, $\underset{\sim}{\zeta}$, $\underset{\sim}{\xi}$, $\underset{\sim}{\mu}$ and $\underset{\sim}{\eta}$ are Lagrangian multipliers. Due to Q being non-additive, it is not possible to resort to the first order necessary optimality conditions of (19) and obtain separable sub-problems. However, equivalent stationarity conditions of (19), omitting dependencies where appropraite, with respect to $\underset{\sim}{c}$ and $\underset{\sim}{u}$ can be written:

$$\frac{\partial \tilde{Q}}{\partial \underset{\sim}{c}} + \left(\frac{\partial Q}{\partial \underset{\sim}{c}} - \frac{\partial \tilde{Q}}{\partial \underset{\sim}{c}}\right) - \frac{\partial \underset{\sim}{y}'}{\partial \underset{\sim}{c}} H' \underset{\sim}{\lambda} + \sum_{i=1}^{N} \frac{\partial \underset{\sim}{g}_i'}{\partial \underset{\sim}{c}} \underset{\sim}{\gamma}_i - \underset{\sim}{\zeta} = \underset{\sim}{o} \tag{20}$$

$$\frac{\partial \tilde{Q}}{\partial \underset{\sim}{u}} + \left(\frac{\partial Q}{\partial \underset{\sim}{u}} - \frac{\partial \tilde{Q}}{\partial \underset{\sim}{u}}\right) + \underset{\sim}{\lambda} - \frac{\partial \underset{\sim}{y}'}{\partial \underset{\sim}{u}} H' \underset{\sim}{\lambda} + \sum_{i=1}^{N} \frac{\partial \underset{\sim}{g}_i'}{\partial \underset{\sim}{u}} \underset{\sim}{\gamma}_i - \underset{\sim}{\xi} = \underset{\sim}{o} \tag{21}$$

where $\tilde{Q}$ represents an appropriate additive performance objective. In a similar manner to JCM in its original form (Michalska and co-workers, 1984), for given $\underset{\sim}{v}$, $\underset{\sim}{w}$, $\underset{\sim}{\sigma}$ and $\underset{\sim}{\lambda}$, the conditions (20) and (21) together with the other stationarity conditions of (19) provide a modified optimisation problem:

$$\min_{\underset{\sim}{c},\, \underset{\sim}{u}} \{\tilde{Q}(\underset{\sim}{c}, \underset{\sim}{u}, \underset{\sim}{\sigma}) + (\underset{\sim}{\phi}_1' - \underset{\sim}{\zeta}_*') \underset{\sim}{c} + (\underset{\sim}{\phi}_2' - \underset{\sim}{\xi}_*') \underset{\sim}{u} + \underset{\sim}{\lambda}' (\underset{\sim}{u} - H\underset{\sim}{y} (\underset{\sim}{c}, \underset{\sim}{u}, \underset{\sim}{\sigma}))\} \tag{22}$$

solved subject to the modified constraints:

$$\underset{\sim}{g}_i(\underset{\sim}{c}_i, \underset{\sim}{u}_i, \underset{\sim}{\alpha}_i) - \beta_{ic}' (\underset{\sim}{c} - \underset{\sim}{v}) - \beta_{iu}' (\underset{\sim}{u}, \underset{\sim}{w}) \leqslant \underset{\sim}{o}, \; i \in \overline{1, N} \tag{23}$$

The new variables appearing in (22) and (23) are also obtained from the stationarity conditions of (19) and are given by,

$$\underset{\sim}{\zeta}_* = \left[\frac{\partial \underset{\sim}{y}'}{\partial \underset{\sim}{v}} - \frac{\partial \underset{\sim}{y}_*'}{\partial \underset{\sim}{v}}\right] \left[\frac{\partial \underset{\sim}{y}'}{\partial \underset{\sim}{\alpha}}\right]^{-1} \left[\frac{\partial Q}{\partial \underset{\sim}{\sigma}} - \frac{\partial \underset{\sim}{y}'}{\partial \underset{\sim}{\sigma}} H' \underset{\sim}{\lambda}\right] \tag{24}$$

$$\underset{\sim}{\xi}_* = \frac{\partial \underset{\sim}{y}'}{\partial \underset{\sim}{w}} \left[\frac{\partial \underset{\sim}{y}'}{\partial \underset{\sim}{\alpha}}\right]^{-1} \left[\frac{\partial Q}{\partial \underset{\sim}{\sigma}} - \frac{\partial \underset{\sim}{y}'}{\partial \underset{\sim}{\sigma}} H' \underset{\sim}{\lambda}\right] \tag{25}$$

$$\beta_{ic}' = \left[\frac{\partial \underset{\sim}{y}'}{\partial \underset{\sim}{v}} - \frac{\partial \underset{\sim}{y}_*'}{\partial \underset{\sim}{v}}\right] \left[\frac{\partial \underset{\sim}{y}'}{\partial \underset{\sim}{\alpha}}\right]^{-1} \left[\frac{\partial \underset{\sim}{g}_i'}{\partial \underset{\sim}{\sigma}}\right] \tag{26}$$

$$\beta_{iu}' = \frac{\partial \underset{\sim}{y}'}{\partial \underset{\sim}{w}} \left[\frac{\partial \underset{\sim}{y}'}{\partial \underset{\sim}{\alpha}}\right]^{-1} \left[\frac{\partial \underset{\sim}{g}'}{\partial \underset{\sim}{\sigma}}\right] \tag{27}$$

The vectors, $\underset{\sim}{\phi}_1$ and $\underset{\sim}{\phi}_2$ are correction terms to be treated as constant at each step, of what will be seen to be, and iterative procedure. Also, if measurements, $\underset{\sim}{u}_*$ of interactions are available, these can be used to evaluate these vectors. This will then permit subsequent decomposition of the global problem.

$$\underset{\sim}{\phi}_1 = \left[\frac{\partial Q}{\partial \underset{\sim}{c}} - \frac{\partial \tilde{Q}}{\partial \underset{\sim}{c}}\right]_{\underset{\sim}{c} = \underset{\sim}{v}} \tag{28}$$

$$\underset{\sim}{\phi}_2 = \left[\frac{\partial Q}{\partial \underset{\sim}{u}} \quad \frac{\partial \tilde{Q}}{\partial \underset{\sim}{u}}\right]_{\underset{\sim}{u} = \underset{\sim}{u}_*} \tag{29}$$

If the modified optimisation problem, (22) and (23) are continuously differentiable, the multiplier, $\underset{\sim}{\lambda}$ can be found by applying a steepest accent approach so that the interation balance equation (6) is satisfied at the k'th iteration of the procedure (Findeisen and co-workers, 1980).

$$\underset{\sim}{\lambda}^{(k)} = \underset{\sim}{\lambda}^{(k-1)} + K_{\lambda}^{(k)} \left[\underset{\sim}{u}^{(k)} - H \underset{\sim}{y}^{(k)}\right] \tag{30}$$

where $K_{\lambda}^{(k)}$ is a positive definite matrix of gain parameters.

In the usual on-line application, it is found convenient to replace the first of (17) by the difference equation,

$$\underset{\sim}{v}^{(k)} = \underset{\sim}{v}^{(k-1)} + K_{v}^{(k)} \left[\underset{\sim}{c}^{(k)} - \underset{\sim}{v}^{(k-1)}\right] \tag{31}$$

where $\underset{\sim}{v}^{(k)}$ represents the applied contols and $K_{v}^{(k)}$ is a matrix, usually diagonal, which can be generally chosen to ensure stability of the procedure.

So far, the global problem has been considered. The modified constraints, (23) prevent the modified optimisation problem, (22) from being decomposed into independent sub-problems due to the coupling present in β_{ic} and β_{iu}. However, by setting $\beta_{ic} = \beta_{iu} = o$, (23) can be replaced by,

$$\underset{\sim}{g}_i(\underset{\sim}{c}_i, \underset{\sim}{u}_i, \underset{\sim}{\sigma}_i) \leqslant \underset{\sim}{o} \ , \ i \in \overline{1, N} \tag{32}$$

and independent local modified optimisation problems are obtained

$$\min_{\underset{\sim}{c}_i, \underset{\sim}{u}_i} \{\tilde{Q}_i + (\underset{\sim}{\phi}'_{1i} - \underset{\sim}{\zeta}'_{*i})\underset{\sim}{c}_i + (\underset{\sim}{\phi}'_{2i} - \underset{\sim}{\xi}'_{*i})\underset{\sim}{u}_i + \underset{\sim}{\lambda}'_i \underset{\sim}{u}_i - \sum_{j=1}^{N} \underset{\sim}{\lambda}'_j H_{ji} y_i (c_i, u_i, \sigma_i)\} \tag{33}$$

because

$$\left.\begin{aligned}
\underset{\sim}{\phi}_1 &\triangleq (\underset{\sim}{\phi}_{11}, \ldots, \underset{\sim}{\phi}_{1N}) \\
\underset{\sim}{\phi}_2 &\triangleq (\underset{\sim}{\phi}_{21}, \ldots, \underset{\sim}{\phi}_{2N}) \\
\underset{\sim}{\zeta}_* &\triangleq (\underset{\sim}{\zeta}_1, \ldots, \underset{\sim}{\zeta}_{*N}) \\
\underset{\sim}{\xi}_* &\triangleq (\underset{\sim}{\xi}_{*1}, \ldots, \underset{\sim}{\xi}_{*N})
\end{aligned}\right) \tag{34}$$

Clearly by neglecting β_{ic} and β_{iu} some approximation is made to the structure when general constraints are present and this in turn will generally lead to some approximation in the final solution. If, however, the constraints are independent of outputs, it can be seen from (26) and (27) that β_{ic} and β_{iu} disappear and no approximation is involved. Also, by neglecting the terms in β_{ic} and β_{iu} in the modified constraints,

(23), there is less danger of violating real constraints.

Also, as measurements, $\underset{\sim}{u}_*$, of interactions are available, the second of (17) can be replaced by,

$$\underset{\sim}{w} = \underset{\sim}{u}_* \tag{35}$$

Finally, if $K_v^{(k)}$ is diagonal, (31) and (35), the last of (17) and the identification problem (16) can be easily decomposed:

$$\left.\begin{aligned}
\underset{\sim}{v}_i^{(k)} &= \underset{\sim}{v}_i^{(k-1)} + K_{vi}^{(k)} (\underset{\sim}{c}_i^{(k)} - \underset{\sim}{v}_i^{(k-1)}) \\
\underset{\sim}{w}_i &= \underset{\sim}{u}_{*i} \\
\underset{\sim}{\sigma}_i &= \underset{\sim}{\alpha}_i \\
\underset{\sim}{y}_i &= \underset{\sim}{y}_{*i} + \underset{\sim}{\alpha}_i
\end{aligned}\right) \tag{36}$$

The scheme is shown in Fig. 1, with θ_i the required information transfer.

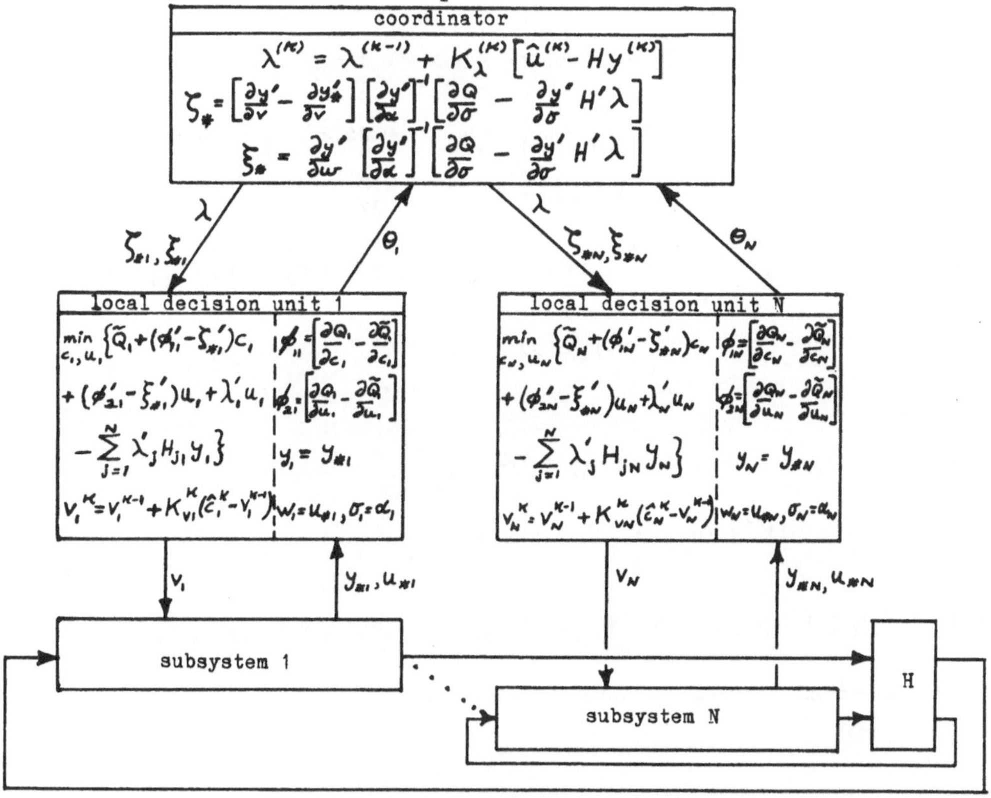

Fig. 1. Information Structure of JCM for non-standard Problems.

It can be seen from Fig. 1. that coordination is achieved jointly by the prices, $\underset{\sim}{\lambda}$, which seek to achieve interaction balance and the variables, or modifiers, $\underset{\sim}{\zeta}_*$ and $\underset{\sim}{\zeta}_*$ which exert additional influence on the local performance objectives. Evaluation of the prices and modifiers takes place at the coordinator level while the parameter estimation and calculation of correction terms is performed at the local level.

3. EXAMPLE

The performance of the technique is now examined when applied to a three sub-process example. The system used is that given by Findeisen and co-workers (1980).

Subsystem 1.

$$y_{11} = F_{*11}(\underset{\sim}{c}_1, \underset{\sim}{u}_1) = 1.3\, c_{11} - c_{12} + 2\, u_{11} + 0.15\, u_{11} c_{11}$$

$$Q_1(\underset{\sim}{c}_1, \underset{\sim}{u}_1) = (u_{11} - 1)^4 + 5\,(c_{11} + c_{12} - 2)^2$$

$$\text{subject to, } CU_1 = \{(\underset{\sim}{c}_1, \underset{\sim}{u}_1) \in \mathcal{R}^3 : c_{11}^2 + c_{12}^2 \leqslant 1 \wedge o \leqslant u_{11} \leqslant 0.5\}$$

Subsystem 2.

$$y_{21} = F_{*21}(\underset{\sim}{c}_2, \underset{\sim}{u}_2) = c_{21} - c_{22} + 1.2\, u_{21} - 3\, u_{22} + 0.1\, c_{22}^2$$

$$y_{22} = F_{*22}(\underset{\sim}{c}_2, \underset{\sim}{u}_2) = 2c_{22} - 1.25\, c_{23} - u_{21} + u_{22} + 0.25\, c_{22}\, c_{23} + 0.1$$

$$Q_2(\underset{\sim}{c}_2, \underset{\sim}{u}_2) = 2\,(c_{21} - 2)^2 + c_{22}^2 + 3c_{23}^2 + 4\, u_{21}^2 + u_{22}^2$$

$$\text{subject to, } CU_2 = \{(c_2, u_2) \in \mathcal{R}^5 : 4c_{21}^2 + 2c_{21}\, u_{21} + 0.4\, u_{21} + c_{21}\, c_{23} + 0.5\, c_{23}^2 + u_{21}^2 \leqslant 4 \wedge 0.5\, c_{21} + c_{22} + 2c_{23}^2 \leqslant 1\}$$

Subsystem 3

$$y_{21} = F_{*31}(\underset{\sim}{c}_3, \underset{\sim}{u}_3) = 0.8c_{31} + 2.5\, c_{32} - 4.2\, u_{31}$$

$$Q_3(\underset{\sim}{c}_3, \underset{\sim}{u}_3) = (c_{31} - 1)^2 + (u_{31} - 1)^2 + 2.5.\, c_{32}^2$$

$$\text{subject to } CU_3 = \{(\underset{\sim}{c}_3, \underset{\sim}{u}_3) \in \mathcal{R}^3 : c_{31} + u_{31} + 0.5 \geqslant o \wedge o \leqslant c_{32} \leqslant 1\}$$

The interconnections between the subsystems are given by,

$$\underset{\sim}{u} = H\ \underset{\sim}{y} = \begin{bmatrix} 0 & 1 & 0 & 0 \\ 1 & 0 & 0 & 0 \\ 0 & 0 & 0 & 1 \\ 0 & 0 & 1 & 0 \end{bmatrix} \underset{\sim}{y}$$

with the global performance objective given as,

$$Q(\underset{\sim}{c}, \underset{\sim}{u}) = Q_1 \cdot Q_2 + Q_3$$

which has a reality optimum, $\hat{Q}_*$ of 7.291.
The additive approximation used is

$$\tilde{Q}\ (\underset{\sim}{c}, \underset{\sim}{u}) = Q_1 + Q_2 + Q_3$$

which does, in fact, give a reality optimum, $\hat{\tilde{Q}}_*$ of 6.314. Following the suggestions in Ellis and Roberts (1982), simple shift models are used for each of the subsystems,

$$y_{11} = F_{11}(\underset{\sim}{c}_1, \underset{\sim}{\alpha}_1) = c_{11} + \alpha_{11}$$

$$y_{21} = F_{21}(\underset{\sim}{c}_2, \underset{\sim}{\alpha}_2) = c_{21} + \alpha_{21}$$

$$y_{22} = F_{22}(\underset{\sim}{c}_2, \underset{\sim}{\alpha}_2) = c_{22} + \alpha_{22}$$

$$y_{31} = F_{31}(\underset{\sim}{c}_3, \underset{\sim}{\alpha}_3) = c_{31} + \alpha_{31}$$

In the example, all required process derivatives were estimated by perturbing the controls by 0.005 and applying finite differences. The gain matrices, $K_\lambda^{(k)}$ and $K_v^{(k)}$ were set:

$$K_\lambda^{(k)} = K_\lambda = 0.1 I_4$$

$$K_v^{(k)} = K_v = 0.05 I_7$$

The convergence behaviour is shown in Fig. 2. where, even though the additive $\tilde{Q}$ is used instead of the original non-additive, Q, the method produces the reality optimum solution, $\hat{Q}_*$. It is also found that, for this example, the optimum solution, $\hat{Q}_*$ lies on the constraint boundary as each of the first written members of the sets of subsystem constraints is active.

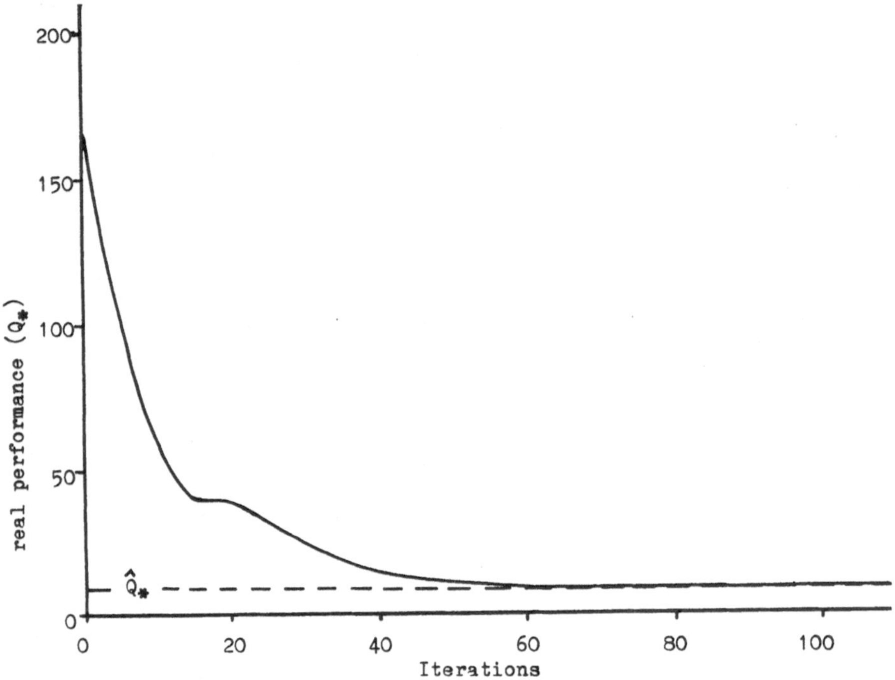

Fig. 2. Three Subsystem Example Performance.

4. CONCLUDING COMMENTS

The technique is a direct extension of JCM for additive global objectives. Clearly, the original form of JCM can be obtained, when additive global objectives have to be dealt with, by setting $Q = \tilde{Q}$, to give $\underset{\sim}{\phi}_1 = \underset{\sim}{\phi}_2 = o$ in (28) and (29).

In (23) an approximation has been made to the modified constraints so that the modified local optimisation problems become separable. However, if the original constraints are independent of outputs, no approximation is involved as β_{ic} and β_{iu}, given by (26) and (27), vanish naturally.

If the matrices K_λ and K_v, in (30) and (31), are able to be chosen to ensure convergence then, by inspection of the converged form of the modified optimisation problem, defined by (22) - (29), it can be seen that the optimal reality solution is obtained when all active constraints are independent of outputs. If any constraint containing an output is active, then, because β_{ic} and β_{iu} are both set to zero in the modified constraints, (25), some sub-optimality must be accepted in the solution.

A drawback of the method is the requirement to calculate process output derivatives with respect to controls for the calculation of modifier, $\underset{\sim}{\zeta}_*$. This is most easily achieved by applying perturbations to the controls and using finite differences. However, in practice, the effects of process noise may influence the derivative values and degrade the algorithm performance. To avoid measuring real process output derivatives, a similar procedure to that recently proposed by Brdyś and Roberts (1984) can be adopted and a third level in a hierarchical structure introduced to evaluate the required modifier.

Finally, it should be noted that it is possible to deal with such non-additive problems by, for example, introducing pseudo-variables to replace the sub-problem coupling variables (Titli and co-workers, 1973). However, the technique presented avoids the difficulty of having to determine the minimum set of coupling variables and also does not increase the dimensionality of the problem.

ACKNOWLEDGEMENT

The authors wish to thank the S.E.R.C., U.K., for supporting this work.

REFERENCES

Brdyś, M, and Roberts, P.D. (1984). Optimal structures for steady state adaptive optimizing control of large scale industrial processes. The City University Control Engineering Centre, research memorandum CEC/MB-PDR/1

Ellis, J.E., and Roberts, P.D. (1982). Measurement and modelling trade-offs for integrated system optimization and parameter estimation. Large Scale Systems 3, 191-204.

Findeisen, W. Bailey, F.N., Brdyś, M., Malinowski, K. Tatjewski, P., and Wozniak, A., (1980). Control and Coordination in Hierarchical Systems, John Wiley.

Michalska, H., Ellis, J.E., and Roberts, P.D. (1984). Joint coordination method for the optimizing control of large scale systems. The City University, Control Engineering Centre, research memorandum, CEC/HM-JEE-PDR/2

Roberts, P.D. (1978). Algorithms for integrated system optimisation and parameter estimation. Electronics Lett., 14, 196-197.

Roberts, P.D., (1979). An algorithm for steady-state optimisation and parameter estimation. Int. J. of Syst Sci., 10, 719-724

Titli, A., Lefevre, T., and Richetin, M., (1973). Multilevel optimization methods for non-separable problems and application Int. J. of Syst. Sci., 4., 865-880.

Multiple criteria optimization with adaptive partition

Y. DELALIEUX
Université Libre de Bruxelles
Service d'Automatique -CP 165
Avenue F.D. Roosevelt,50
B-1050 Bruxelles-Belgium

1. Introduction

Some large scale systems are hardly optimized : their optimal operating point cannot be directly found, because the system models are unknown or difficult to be approximated.
In that case, real-time optimization can be achieved by choosing successive operating points in such a way that the system operates nearer its optimal operating point.
These various trial points inform implicitly on the relationships between the inputs and the outputs variables.
However, real-time choice prevents too large variations, which move the system far from its optimal operating point.
In order to fulfil such purposes, hierarchical optimization algorithms are presented in this paper, which allow to optimize simultaneously several criteria without needing analytical expression of these ones.

2. Hierarchical algorithms

The algorithms are splitted into two levels : the upper level concerns how to choose the input variables used to optimize each criterion and the lower level concerns the values taken by these inputs.
As shown schematically in figure 1, at the upper level, a partition of inputs e_i is defined so that for each criterion c_j, corresponds an inputs subset v_j including the variables e_i which will be modified in the lower level of the algorithm.
It is necessary to make this partition because there are too many input variables to consider at the same time and because only few of these variables really act upon every criterion.
Moreover, this partition allows to take account of the alterations which appear in the structure of the system.
The partition is constructed in order that each partition subset includes, according to a causality index, the inputs, which will be controlled for optimizing each criterion.

For example, this index is calculated from the crosscorrelations between the outputs c_j and the inputs e_i.
But correlations are functions requiring long computing time, often prohibitive time.
Therefore, the causality index can be estimated roughly by keeping in each partition subset, the input variables with which,each criterion has the best value :

$e_i \in v_j$ if j corresponds to the maximum of

$\frac{c_{jT}-c_{j0}}{T_{ij}}$ where T_{ij} : time for which e_i was varying in order to maximize c_j

c_{jT} and c_{j0} : final and initial values of the criterion c_j.

Evidently, because no dynamic interaction is wanted between the two levels, the frequency of the upper level which modifies the subsets v_j, must be much smaller then the execution frequency of the optimization algorithms.

3. Optimization with learning theory

Thanks to the N subsets v_j defined above, the multiple criteria optimization is reduced to N single criterion optimization problems, for which no mathematical model is known.
The optimization is then performed by seeking the optimal operating point with successive trial points, to which corresponds the measured value of the criterion.
That informs how the next trial points must be chosen. In fact, at each step of the optimization algorithm, the point chosen is a decision made from what is learned with the previous points, about the criterion model.
Consequently, the trial points choice in the optimization algorithm can be considered as a learning process, by which we attempt to find the best value of an a priori unknown function.
The used learning principle consists of rewards and penalties : reward means that the probability of a decision making is increased because its result is favourable to a fixed goal, and reversely, penalty means that the probability of a decision making is decreased because unfavourable result is observed.
With a mathematical formulation, this principle is described by:

Define,

$D = \{d_1, \ldots, d_n\}$ the set of feasible decisions d_i

P = the set of all transposed vectors $p^t = (p_1, \ldots, p_n)$ where p_i is the probability of a decision d_i chosing, thence satisfying, of course, the two relations :

$$0 \leq p_i \leq 1 \quad (1)$$

$$\sum_{i=1}^{n} p_i = 1 \quad (2)$$

R = {success, failure} the set of the two possible outcomes over the performance criterion, due to the chosen decision.

The learning algorithm $P \times D \times R \rightarrow P$ progresses by this way :

Assume that at the step k, the decision d_j is chosen as a random sample from the probabilities vector $p(k)$.

Either, the result is a success and the probability of d_j choosing is increased :

$$p_j(k+1) = p_j(k) + f_{jj} \{p(k)\}$$

$$p_i(k+1) = p_i(k) - f_{ij} \{p(k)\} \quad \text{for all } i \neq j$$

or the result is a failure and the probability of d_j is decreased :

$$p_j(k+1) = p_j(k) - g_{jj} \{p(k)\} \quad \text{and}$$

$$p_i(k+1) = p_i(k) + g_{ij}\{p(k)\} \quad \text{for all } i \neq j$$

with f_{jj}, f_{ij}, g_{jj}, g_{ij} functions of p such that the relations (1) and (2) remain verified at the following step k+1.

In optimization, success means that a better criterion value is found and failure that a worse one is found.

More details about the decisions d_i and the functions f_{ij} and g_{ij} to be used will now be given.

4. Single variable criterion

If there is only one optimizing variable e, the learning algorithm takes, for instance, this particular form.

The set D is the pair of decisions d_1 and d_2, where d_1 is increasing the variable e by an increment E and d_2 is decreasing e by the same value E, and, in the case where the criterion is maximized, the probabilities variations are so formulated :

at the step k,
if $c(k) > c(k-1)$ results from the decision d_1
then $p_1(k+1) = p_1(k) + a\{1-p_1(k)\}$
else $p_1(k+1) = p_1(k) - a\, p_1(k)$

the other probability p_2 deriving simply from

$p_2(k+1) = 1 - p_1(k+1)$

The coefficient a which has to be chosen between 0 and 1, is a parameter filtering the probabilities variations :
if $a = 0$ the probabilities remain unchanged and if $a = 1$ the decisions choice is determined only by the difference of $c(k)$ and $c(k-1)$.
Thanks to the value of the coefficient a the optimization algorithm is adapted to the magnitude of the measured criterion noise.

5. Multiple variables criterion

When the subset v_j includes more then one variable, the criterion c_j is optimized by means of a vector u, the components of which are the optimizing c_j variables.
In this case, the next trial point $u(k+1)$ depends upon the point $u(k)$, and upon the magnitude and direction which are chosen for the vector

$$t(k) = u(k+1) - u(k) \qquad (3)$$

(see figure 2)
Because there is an infinite number of possible directions for seeking the optimal point, the decision set will be replaced by the calculation of the vector t direction u_t from the previous vector t direction u_t and from a random direction u_r

$$u_t(k) = \frac{z(k)}{|z(k)|} \qquad (4)$$

with $|z|$ magnitude of the vector z.

$z(k) = w(k)\, u_t(k-1) + (1-w(k))\, u_r(k)$
where w is a weighting coefficient at step k.
Instead of computing the probabilities p_i associated to the decisions d_i, the learning process of the direction to which the optimal point will be found, is progressing by w varying.
For example,
if $c(k) > c(k-1)$ i.e. result = success

then w(k) = 1

else w(k) = b w(k-1)

where b is a filtering parameter ($0<b<1$)

In other words, the vector t direction is unchanged when the criterion is ameliorated and it is the previous direction mixed with a random one otherwise.

The magnitude chosen for the vector t is a random number A lying between the bounds A_{min} and A_{max} which fix the domain where the next point can be chosen.

Finally, the next trial point is obtained by

$u(k+1) = u(k) + t(k)$

with $t(k) = A(k)\ u_t(k)$

and the relations (4), (5) and (6).

6. Concluding remarks

Because no a priori criterion model is needed, the presented optimization method is specially suitable when the optimized system is very complex.

Two main features of this method are the robustness in regard with the disturbances, by adjusting the filtering parameters, and the easeness for real-time implementation, more particularly when the partitionning level has very reduced the number of variables to consider for each optimized criterion.

7. References

1. Applied non-linear programming
David HIMMELBAU
Mac Graw Hill - 1972.
2. Learning algorithms theory and applications
S. LAKSHMIVARAHAN
Springer-Verlag - 1981
3. Commande optimale d'un aérocondenseur
Y. DELALIEUX
Travail de spécialisation en automatique
Université Libre de Bruxelles - 1982.

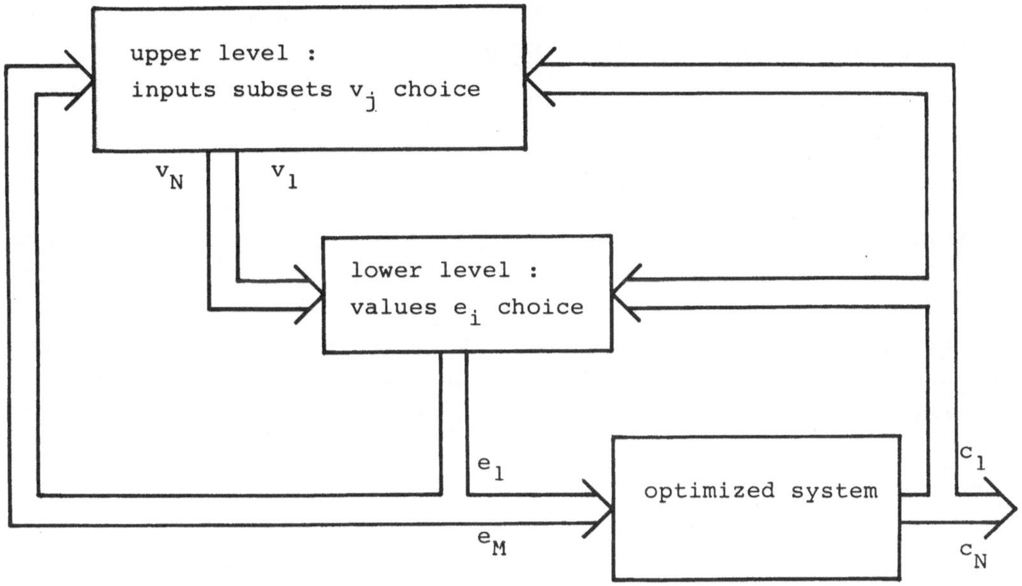

Figure 1.

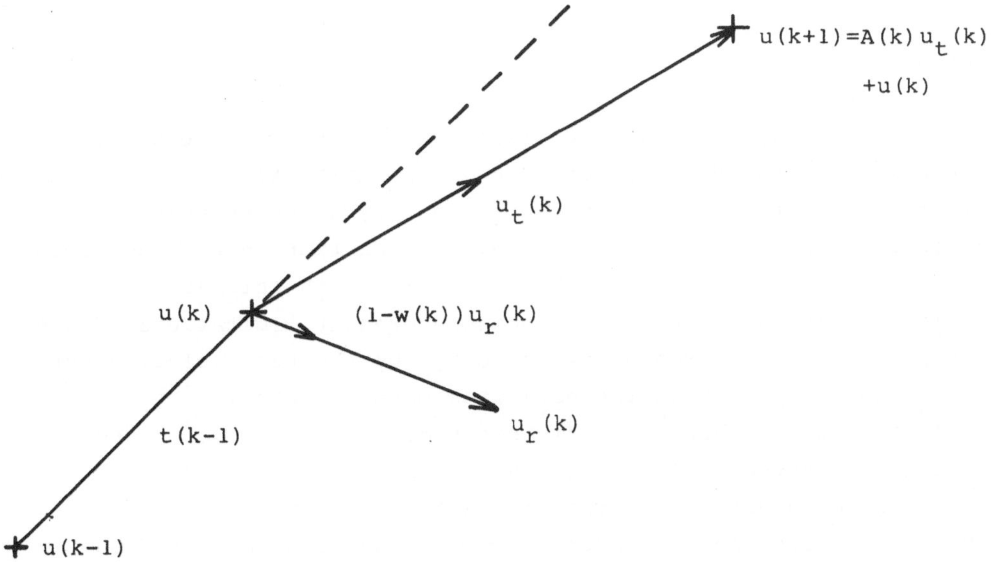

Figure 2.

MULTICRITERION ANALYSIS OF DYNAMIC PROPERTIES OF HIERARHICAL DISTRIBUTED PROCESS CONTROL SYSTEM STRUCTURE

Lj.Vlačić and B.Matić
Energoinvest - Institute for
Automatic control and computer science
Sarajevo, Yugoslavia

Abstract: The objective of the subject paper is to present the results of performed multicriterion analysis of dynamic properties of a hierarhical distributed process control system structure.

1. Introduction

The application of multicriterion approaches in the analysis of many complex situations has been remarkably intensified in the course of last ten years /1/. Apart from their dispersiveness, the following steps can be observed with any approch:

- defining the set of decision-making criteria;
- defining the set of decision-making variables;
- procedure of finding appropriate solution.

All multicriterion approaches can be broadly classified into two categories, namely (1) multi-attribute decision-making approach (MADM) and (2) multi-objectives decision-making approach (MODM). It should be noted that there exists no differentiation between MADM and MODM in many papers /2/.

The goal of MADM approach is the selection of one from the set of possible solutions on the basis of inter- and intra-attribute comparison wherein the attributes needn't necessarily be measurable.

MODM approach is not aimed at defining the solution being predominant; its goal is to find out the best solution with obligatory consideration of various interactions between present limitations. Hence MODM approach is an approach toward designing the best solution but not its selection as it is the case with MADM approach.

The selection of hierarchical, distributed, microprocessor-based process control system topology and the defining of correspondence between system components is a typical example of a multicriterion situation /3/, considering that such correspondence can vary from purely decentralized to purely centralized case.

The objective of the subject paper is to present the results of performed multicriterion analysis of dynamic properties of a hierarhical distributed process control system structure. The need for hierarhical structure /4/, will not be considered, because the hierarhical structure of process control systems has already been detaily explained, for the case of single-objective approach in /4/, /5/, /6/, and in /1/, /2/, /3/, /7/ for the case of multicriterion approach respectively.

2. Basic Assumptions

The distribution of information is an immanent property of industrial processes. Functional and topological distribution are specifically pronounced in processing and power generating systems. With the development of effective data transmission systems, the picture of control systems has drastically changed. Modern technology permits data processing within the system to be run almost parallelly with natural flow the industrial process.

The specifications of any process control system cover the aspects of topology, functions and system performances. System topology results out of the plant topology and means the property of the system which defines its organization in the sense of defining the type of system components interconnection. System associated functions comprise all functional tasks expected to be fulfilled by the system toward the process and the operator. System performance is the property of the system defined as the measure of effectiveness in performing the associated functions. Hence, it becomes quite esential to extablish the correlation between system performances, topology and associated function in control system analysis, and is of vital interest in both the stage of system development and in the stage of reconfiguration of an existing system.

Consequently, any topology is defined by the attributes the performance of which determine its suitabillity for specific application.

2.1. The Attributes

The control system attributes under observation are those related with dynamic properties of the process and for which tha analytical models have been derived.

These are:

. system utilisation factor (ϱ) . in the sense of definition and analytical model presented in /4/;
. waiting time in queue (W) - in the sense of definition and analytical model presented in /8/;
. probability that arrival must wait longer than T_o time units $G(t \geq T_o)$ - in the sense of definition and analytical model presented in /8/;
. system response time (T_r) - in the sense of definition and analytical model presented in /8/;
. extent of system centralization (δ) - in the of definition presented in /9/.

An analysis and quantification of heuristic attributes and their performances is beyond the scope of subject paper.

2.2. The Goal

The goal of multicriterion analysis was aimed at finding the hierarhical structural form which is effective *) in the sense of optimizing the vector criterion of effectiveness

$$E(\varrho, W, T_r, G, \delta) = \text{extr}.$$

3. The Results

In /4/ there has been demonstrated that the hierarchical structure from fig. 1 can be represented by equivalent queueing model from fig.2 covering the aspects of change of both hierarchical levels and number of servers within one and the same hierarchical level. The results calculated in /4/ are presented in graphical form in figs. 3 to 6.

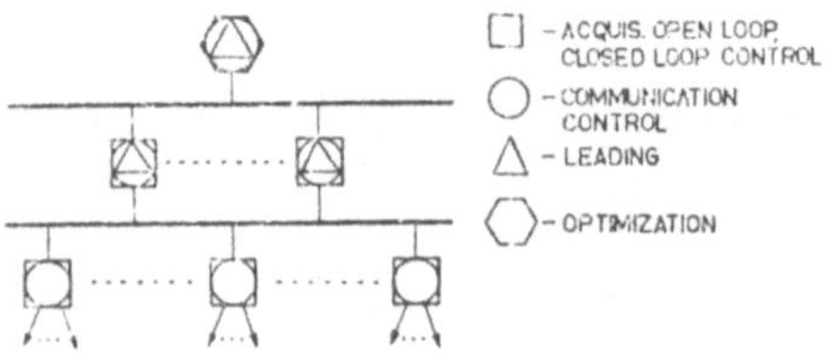

Fig.1

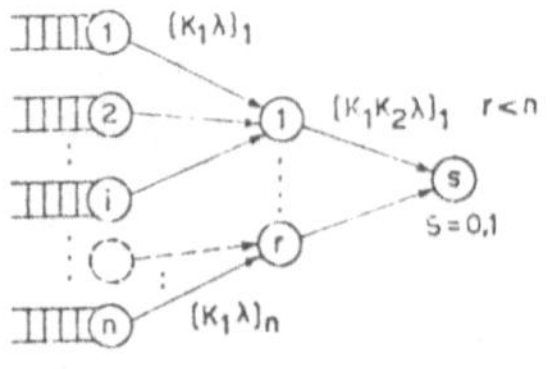

Fig.2

*) The solution $x^o \in X_o$ is effective, if there exist no such $x \in X$ for which is $E(x) \geq E(x^o)$, where X means the set of possible solutions.

n - number of first level servers;

r - number of second level servers;

s - number of third level servers;

In all graphs the parameter of change is presented by the extent of complexness of hierarchical structure (the variables n, r and s). Fig.3 illustrates the change of system utilisation factor (the variable ρ). The change of probability of losing information (W) is illustrated in fig. 4, and fig. 5 indicates the change in system response time (T_r) with increasing the number of control loops.

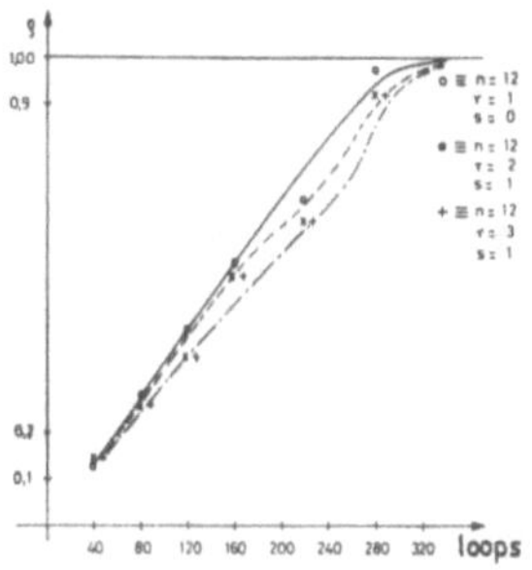

Fig.3

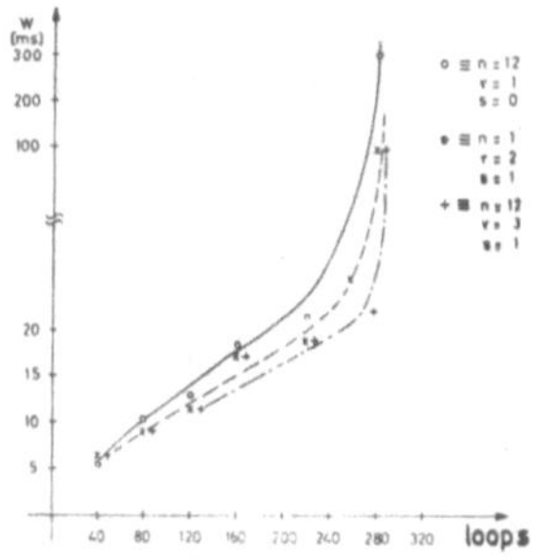

Fig.4

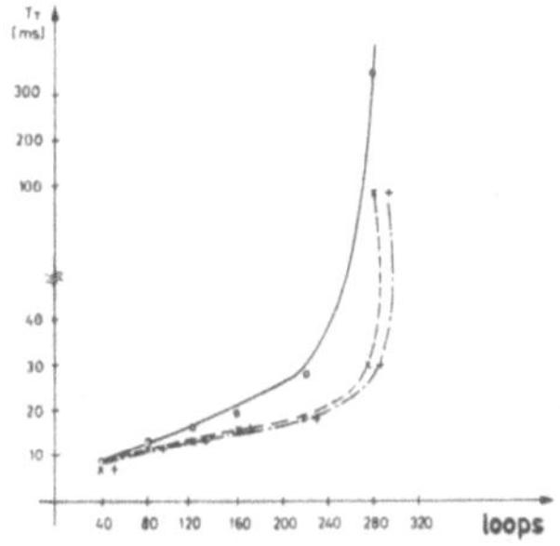

Fig.5

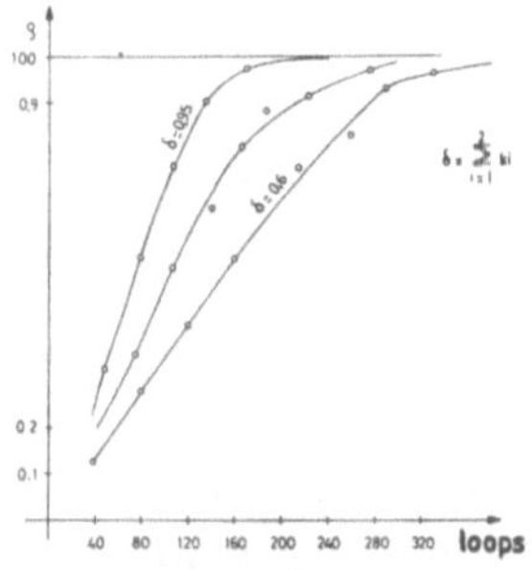

Fig.6

4. <u>The Conclusion</u>

An analysis of results obtained in /4/ indicates that:

- no remarkable increase can be expected in the number of control loops covered by the system when the system topology is rendered more complex;
- the probability of losing information on process status is decreasing with increasing the complexity of system hierarchy;
- with increasing the complexity of system hierarchy practically on improvement of its applicability from the aspect of response time is achieved.

By generalizing the approach described in /4/ (investigations covered the case $K_1 = K_2 = 0,33$) to a wide range of change of K_1/K_2 ratio, it has been estimated that the influence of change in the extent of system centralization upon the complexity of multi-level structure is very important and has the form as in fig.6.

Thus, the obtained results complement the conclusion that an increase in the extent of centralization, though resulting to an increase of system controllability, reduces the number of control loops which can be covered by the system.

5. References

1. M.Sakawa; F.Seo
"Interactive MODM in environmental systems using sequential proxy optimization techniques"
Automatica, vol.18, N°2, pp 155-165, 1982.

2. C.L.Hwang; A.S.Md Masud
"Multiple objective decision making methods and applications"
Springer - Verlag, 1979.

3. W.Findeisen
"Decentralized and hierarchical control under consistency or disagreement of interests"
Automatica, vol.18, N°6 pp 647-664, 1982.

4. Lj.Vlačić; B.Matić
"Performance evaluation of distributed process control systems based on Foster - Perros model"
IEEE-Melecon '83 conference, athens 1983.

5. M.F.Mesarovic; D.Macko; Y.Takahara
"Theory of hierarchical multilevel systems"
Academic Press, N.Y. 1980.

6. Y.Y.Haimes
"Hierarchical analyses of water resources systems modeling and optimization ot Large - scale systems"
McGraw-Hill, N.Y. 1977.

7. K:Tarvainen, Y.Y.Haimes, I.Lefkowitz
"Decomposition methods in multiobjective discrete-time dynamic problems"
Automatica, vol.19, N°1 pp.15-28, 1983.

8. H.Kobayashi
"Computer performance evaluation an introduction"
Prentice-Hall, Inc. 1980.

9. A.D.Cvirkun
"Osnovi sinteza strukturi složnih sistem"
Nauka, 1982.

HIERARCHICAL NON-ITERATIVE CONTROL OF LARGE-SCALE MECHANICAL SYSTEMS

L.K. Mikhailov, J.D. Zaprianov
Central Laboratory of Control Systems, Bulgarian Academy of Sciences
Acad. G. Bonchev str., bl. 107, Sofia 1113, Bulgaria

N.D. Naplatanov
Derartment of Automatics, Higher Institute of Mechanical and Electrical Engineering, Sofia 1156, Bulgaria

INTRODUCTION

In the past few years there has been an increased interest in dynamical control methods of mechanisms, in which articulated systems of levers are individually powered by actuators, operating at each joint, to perform different tasks in locomotion and manipulation. This paper deals with the control of industrial robots and manipulators, which are typical examples of such systems.

Some authors (Vucobratovic and others, 1977; Mahil, 1979; Patarinski and Danev, 1980) reduce the control synthesis of such mechanical systems to the standart linear-quadratic problem. Although this approach meets the process quality requirements, in practice the optimization methods are not used for robot control system synthesis. The main reason consists in the high dimension of the linear model, which leads to considerable computation expences. In (Naplatanov and others, 1979) an hierarchical method is proposed to decrease these difficulties, by means of an appropriate decomposition of the mechanical system into subsystems, corresponding to each joint. A two-level hierarchical structure is used to generate an optimal stabilising control, based on the interaction prediction principle (Singh and Titli, 1978). This control method however requires an iterative coordination procedure and may suffer from convergence difficulties.

In this paper a non-iterative sheme for near-optimum hierarchical control for large-scale mechanical systems is proposed, based on the stuctural decomposition of the system. On the first level optimal decentralized subcontrollers are synthesized for each isolated subsystem. By means of Liapunov functions, which give an exact estimation of the exponential stability of the subsystems, aggregate vectors of the interactions between the subsystems are formed, following the proposed in (Mahalanabis and Singh, 1980) approach. These vectors are used to obtain an additional control from the upper level, which ensures asympthotic stability of the overall system and satisfactory performance.

CONTROL PROBLEM

Let us consider a dynamical non-linear model of the robot arm and the corresponding actuators

$$\dot{X}(t) = G(X) + H(X)U(t) \tag{1}$$

where $X=(X_m^T\ X_d^T)^T \varepsilon\ R^n$ is the state vector, the first component of which represents the generalized coordinates and their velocities, $X_m=(q^T\ \dot{q}^T)^T$; $X_m \varepsilon\ R^{2N}$, $q=(q_1\ q_2 \ldots q_N)^T$, N- number of degrees of freedom. The second component of the state vector $X_d \varepsilon\ R^p$ gives us the actuator system states. The vector G and the matrix H are with dimensions n, n x N accordingly; $U \varepsilon\ R^N$ is the control vector. Obviously, the dimension of the non-linear system is $n=2N+p$.

The problem of the robot arm control is to find such control U, which will ensure the motion of the arm in the δ-neighbourhood of the desired joint trajectories q_i, $i=1,2,\ldots,N$, for $t \varepsilon T$, where $T=(t_o\ ,\ t_1)$ denotes the control time interval. When the technological considerations require sufficiently small radius δ and at the same time motions with a high speed are to be performed, then the dynamics of the robot arm, described by Eq.(1) has to be taken into consideration.

To solve this problem we have to obtain the vector of the desired states $X^*(t)$ and the corresponding nominal controls $U^*(t)$, using a non-linear dynamical model of the robot arm and taking into consideration the planned joint trajectories $q^*(t)$, their velocities and accelerations. Because of the unadequate model and the presence of external disturbances it is necessary the motion to be stabilized with an additional feedback control, synthesized as a function of the error between the actual and the desired motions. If these errors are small enough, linear methods can be used to obtain this control law.

The linearized model of the robot arm along the desired states $X^*(t)$ and the computed nominal controls $U^*(t)$ can be obtained from Eq.(1):

$$\dot{\tilde{x}}(t)=\tilde{A}(t)\tilde{x}(t)+\tilde{B}(t)u(t) \tag{2}$$

where $\tilde{x}(t)=X(t)-X^*(t) \in R^n$ is the state vector of the linear system, $u(t)=U(t)-U^*(t) \in R^N$ is the stabilizing control vector. The matrices $\tilde{A}$ and $\tilde{B}$ are with appropriate dimensions and are defined as follows:

$$\begin{aligned}\tilde{A}(t) &= \left.\frac{dG(X)}{dX}\right]_{X(t)=X^*(t)} + \left.\frac{dH(X)}{dX}\right]_{X(t)=X^*(t)} U^*(t)\\ \tilde{B}(t) &= H(X^*)\end{aligned} \tag{3}$$

Futher we leave the argument t for brevity, excluding the cases, when misunderstanding may occure.

HIERARCHICAL CONTROL SYSTEM SYNTHESIS

Let us introduce the nonsingular linear transformation $x=T_p\tilde{x}$, where the new state vector is $x=(x_1^T\ x_2^T \dots x_N^T)^T$, $x_i \in R^{n_i}$, $\sum_{i=1}^{N} n_i = n$ and T_p is permutation matrix, which arranges the rows and the columns of the linear system (2) in such way, that each state vector x_i includes variables, corresponding only to the ith joint. Then, the linear system (2) can be represented as

$$\dot{x} = Ax + Bu \tag{4}$$

where $B=T_p\tilde{B}$ = block diag (B_i), $B_i \in R^{n_i}$,

$$A=T_p\tilde{A}T_p^{-1}=\begin{bmatrix} A_{11} & A_{12} & \cdots & A_{1N} \\ & & \cdots & \\ A_{N1} & A_{N2} & \cdots & A_{NN}\end{bmatrix}, \quad A_{ii} \in R^{n_i x n_i}, \quad A_{ij} \in R^{n_i x n_j}.$$

Then the linear system can be decomposed into N subsystems, corresponding to each joint of the robot arm:

$$\begin{aligned}\dot{x}_i &= A_{ii}x_i + B_iu_i + z_i\\ z_i &= \sum_{j=i}^{N} A_{ij}x_j \qquad , \quad i=1,2,\dots,N.\end{aligned} \tag{5}$$

where $z_i \in R^{n_i}$, $i=1,2,\dots,N$ are interconnection vectors.

Let us chose local quadratic performance index for each subsystem

$$J_i = \frac{1}{2}\int_{t_o}^{t_1} (x_i^T Q_i x_i + u_i^T R_i u_i)dt \qquad (6)$$

where $Q_i = Q_i^T \geq 0$, $R_i > 0$, $i=1,2,\dots,N$.

If we suppose, that the interconnections between the subsystems are neglectable, optimal local controls can be found

$$u_i^L = -R_i^{-1} B_i^T P_i x_i \quad , \; i=1,2,\dots,N, \qquad (7)$$

since the pairs (A_{ii}, B_i) are completely controllable, where $P_i = P_i^T > 0$ is solution of the matrix Riccati equation

$$-\dot{P}_i = P_i A_{ii} + A_{ii}^T P_i - P_i B_i R_i^{-1} B_i^T P_i + Q_i \quad , \; P(t_1)=0, \; i=1,2,\dots,N. \qquad (8)$$

If the pairs $(Q_i^{1/2}, A_{ii})$ are completely observable, each closed-loop isolated subsystem

$$\dot{x}_i = (A_{ii} - B_i R_i^{-1} B_i^T P_i) x_i \quad , \; i=1,2,\dots,N \qquad (9)$$

is globally exponentially stable. But due to the interconnections the control (7) is not optimal for the initial system (5). Moreover, when the strength of the interconnections is large enough, the closed-loop system

$$\dot{x}_i = A_{oi} x_i + \sum_{j=i}^{N} A_{ij} x_j \quad , \; i=1,2,\dots,N \qquad (10)$$

where $A_{oi} = A_{ii} - B_i R_i^{-1} B_i^T P_i$, may become unstable. It is necessary therefore to introduce an additional global controller, which will ensure stability of (10):

$$u^G = -Mx \qquad (11)$$

In (Siliak and Sundareshan, 1976), where this idea was proposed first, the global controller synthesis is carried out, solving algebraic minimization problem. For the considered mechanical system however, the rank conditions which are given there are not met.

Let us choose Liapunov functions for each subsystem

$$V_i(x_i) = x_i^T \bar{P}_i x_i \quad , \; i=1,2,\dots,N, \qquad (12)$$

where $\bar{P}_i = \lim_{t\to\infty} P_i(t)$. Due to the observability of the pair $(Q_i^{1/2}, A_{ii})$,

$\overline{P}_i$ is symmetric, positive definite matrix. Obviously, $\overline{P}_i$ satisfies the following Liapunov equation

$$(A_{ii}-B_iR_i^{-1}B_i^T\overline{P}_i)^T\overline{P}_i + \overline{P}_i(A_{ii}-B_iR_i^{-1}B_i^T\overline{P}_i) = -\overline{Q}_i \tag{13}$$

where $\sigma(A_{ii} - B_iR_i^{-1}B_i^T\overline{P}_i) \in C^-$ and $\overline{Q}_i = Q_i + \overline{P}_iB_iR_i^{-1}B_i^T\overline{P}_i$ is positive definite matrix.

As it is shown in (Mahalanabis and Singh, 1980), the asymptotic stability of the overall system (10) can be evaluated by means of the following comparison system

$$\dot{x}=\overline{A}_o x + (H-BM)x \tag{14}$$

where $\overline{A}_o$=block diag$(\overline{A}_{o1},\overline{A}_{o2},\ldots,\overline{A}_{oN})$, $\overline{A}_{oi}=A_{ii} - B_iR_i^{-1}B_i^T\overline{P}_i$, H = block diag$(H_1,H_2,\ldots,H_N)$, $H_i=h_iI$. The effective interconnection coefficients h_i can be obtained via the aggregation procedure, proposed in (Mahalanabis and Singh, 1980). These coefficients depend on the interconnection matrices A_{ij}, i=1,2,...,N, j=1,2,...,N, j≠i and $\overline{P}_i$, i=1,2,...,N.

The matrix M has to ensure eigenvalues with negative real parts of the matrix $A_c = \overline{A}_o + H - BM$, so that $\sigma(A_c) = \sigma(A_{o1}) \cup \sigma(A_{o2}) \cup\ldots\cup \sigma(A_{oN})$, where $\sigma(.)$ denotes the spectrum of the corresponding matrix. This problem can easily be solved, using standart pole assignment techique. Then the overall initial system is stable too, as follows from the aggregation procedure.

EXPERIMENTAL RESULTS

To demonstrate the applicability of the proposed hierarchical approach to robot control synthesis, digital simulation was carried out using IBM 370 computer. Robot arm with four degrees of freedom was chosen and non-linear dynamic equation (1) was derived via Lagrange's principle.

A computer program, which simulates robot arm behaviour was written, based on the continuous system modelling package S/360 CSMP. This program carries out the following functions:

- generates the desired joint trajectoris, their velocities and accelerations;
- computes the nominal state vector X^* and the corresponding nominal control U^*, using the complete or a simplified non-linear models;
- computes the stabilizing control components u_i, i=1,2,3,4, by means

of the proposed hierarchical approach;

- forms the r.h.s. of the nonlinear differential equation (1);
- integrates the nonlinear system equation, describing the actual robot dynamics, when the computed controls are applied. The 4th order Runge-Kutta method is used with an appropriate integration step.

A 16th order linearized model of the robot arm (4) is decomposed into four subsystems of 4th order, which values for the final point of the desired trajectory are

$$\dot{x}_1 = \begin{bmatrix} 0 & 1 & 0 & 0 \\ 0 & 0 & .013 & 0 \\ 0 & 0 & -5 & 5 \\ 0 & 0 & 0 & -8 \end{bmatrix} x_1 + \begin{bmatrix} 0 & 0 & 0 & 0 \\ -.002 & .003 & .005 & 0 \\ 0 & 0 & 0 & 0 \\ 0 & 0 & 0 & 0 \end{bmatrix} x_2 + \begin{bmatrix} 0 & 0 & 0 & 0 \\ -.03 & .1 & 0 & 0 \\ 0 & 0 & 0 & 0 \\ 0 & 0 & 0 & 0 \end{bmatrix} x_4 + \begin{bmatrix} 0 \\ 0 \\ 0 \\ 2 \end{bmatrix} u_1$$

$$\dot{x}_2 = \begin{bmatrix} 0 & 1 & 0 & 0 \\ -.067 & -.03 & .4 & 0 \\ 0 & 0 & -5 & 5 \\ 0 & 0 & 0 & 0 \end{bmatrix} x_2 + \begin{bmatrix} 0 & 0 & 0 & 0 \\ 0 & 0 & .049 & 0 \\ 0 & 0 & 0 & 0 \\ 0 & 0 & 0 & 0 \end{bmatrix} x_1 + \begin{bmatrix} 0 & 0 & 0 & 0 \\ -.78 & -.12 & -.001 & 0 \\ 0 & 0 & 0 & 0 \\ 0 & 0 & 0 & 0 \end{bmatrix} x_4 + \begin{bmatrix} 0 \\ 0 \\ 0 \\ 2 \end{bmatrix} u_2$$

$$\dot{x}_3 = \begin{bmatrix} 0 & 1 & 0 & 0 \\ 0 & 0 & .05 & 0 \\ 0 & 0 & -5 & 5 \\ 0 & 0 & 0 & -8 \end{bmatrix} x_3 + \begin{bmatrix} 0 \\ 0 \\ 0 \\ 2 \end{bmatrix} u_3$$

$$\dot{x}_4 = \begin{bmatrix} 0 & 1 & 0 & 0 \\ .049 & -.02 & .05 & 0 \\ 0 & 0 & -5 & 5 \\ 0 & 0 & 0 & -8 \end{bmatrix} x_4 + \begin{bmatrix} 0 & 0 & 0 & 0 \\ 0 & 0 & -.003 & 0 \\ 0 & 0 & 0 & 0 \\ 0 & 0 & 0 & 0 \end{bmatrix} x_1 + \begin{bmatrix} 0 & 0 & 0 & 0 \\ -.06 & .1 & -.011 & 0 \\ 0 & 0 & 0 & 0 \\ 0 & 0 & 0 & 0 \end{bmatrix} x_2 + \begin{bmatrix} 0 \\ 0 \\ 0 \\ 2 \end{bmatrix} u_4$$

The following weighting matrices were chosen for each subsystem

$$Q_1 = \begin{bmatrix} 10.37E{+}5 & 0 & 0 & 0 \\ 0 & 64.8E{+}3 & 0 & 0 \\ 0 & 0 & .27 & -.04 \\ 0 & 0 & -.04 & .04 \end{bmatrix}, \quad Q_2 = \begin{bmatrix} 688 & 0 & 0 & 0 \\ 0 & 43 & 0 & 0 \\ 0 & 0 & .27 & -.04 \\ 0 & 0 & -.04 & .04 \end{bmatrix},$$

$$Q_3 = \begin{bmatrix} 67.1E{+}3 & 0 & 0 & 0 \\ 0 & 41.9E{+}2 & 0 & 0 \\ 0 & 0 & .27 & -.04 \\ 0 & 0 & -.04 & .04 \end{bmatrix}, \quad Q_4 = \begin{bmatrix} 65.3E{+}3 & 0 & 0 & 0 \\ 0 & 40E{+}2 & 0 & 0 \\ 0 & 0 & .27 & -.04 \\ 0 & 0 & -.04 & .04 \end{bmatrix},$$

R_1=0.1E-2, R_2=0.1, R_3=0.1E-2 and R_4=0.1E-3.

After solving the linear-quadratic problem, the following local controls (7) were found

$$u_1^L = -(32.2E{+}3 \;\; 16.1E{+}3 \;\; 24.9 \;\; 9.4)x_1 \;, \quad u_2^L = -(82.8 \;\; 63.7 \;\; 4.7 \;\; 2.3)x_2 \;,$$

$u_3^L = -(8.2E+3\ \ 4.1E+3\ \ 24.9\ \ 9.4)x_3$, $u_4^L = -(25.6E+3\ \ 11.7E+3\ \ 68.2\ \ 23.5)x_4$.

The eigenvalues of the isolated subsystems (9) are

$$s^1_{1,2} = -9.49 \pm j2.42,\ s^1_{3,4} = -6.44 \pm j1.1,\ s^2_{1,2} = -2.29 \pm j2.19,$$
$$s^2_3 = -7.9,\ s^2_4 = -5.1,\ s^3_{1,2} = -9.5 \pm j2.43,\ s^3_{3,4} = -6.03 \pm j1.125,$$
$$s^4_{1,2} = -6.03 \pm j1.26,\ s^4_3 = -8.718,\ s^4_4 = -29.285.$$

The values of the effective interconnection coefficients h_i, i=1,2,3,4 are

$$h_1 = 0.599E-2,\ h_2 = 5.28,\ h_3 = 0,\ h_4 = 0.23.$$

The comparison system (14) eigenvalues, when M=0 are

$$s^1_{1,2} = -9.48 \pm j2.42,\ s^1_{3,4} = -6.43 \pm j1.1,\ s^2_{1,2} = 3.28 \pm j2.19,$$
$$s^2_3 = -2.414,\ s^2_4 = 0.489,\ s^3_{12} = -9.5 \pm j2.43,\ s^3_{3,4} = -6.03 + j1.125,$$
$$s^4_{1,2} = -5.8 \pm j1.2,\ s^4_3 = -8.48,\ s^4_4 = -29.05.$$

It can be seen, that the second comparison subsystem is unstable and it is necessary to apply global control. The following global control is obtained

$$M = \text{block diag } (M_1, M_2, M_3, M_4), \text{ where}$$
$$M_1 = (96.63\ \ 36.23\ \ 24.9\ \ 13.4),\ M_2 = (1872.3\ \ 721.2\ \ 41.8\ \ 14.9),$$
$$M_3 = (0\ \ 0\ \ 0\ \ 0),\ M_4 = (2805.2\ \ 901.4\ \ 71.97\ \ 27.8).$$

The desired trajectory q_2^* for the second joint of the robot arm and the actual trajectory q, when hierarchical stabilizing control is applied, are shown on the next figure. As it can be seen, this control ensures precise tracking of the desired trajectory.

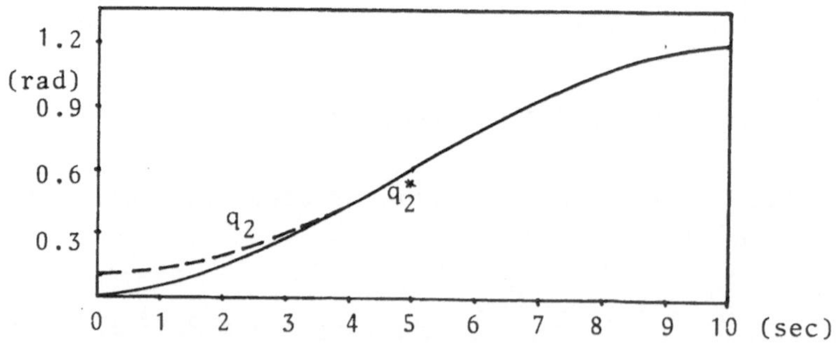

CONCLUSION

In this paper a two-level non-iterative sheme for control of large-scale multibody mechanical systems is presented, based on the structural decomposition of the system. On the first level optimal decentralized subcontrollers are synthesized for each isolated subsystem. An additional control is obtained on the upper level, which ensures asymptotic stability of the overall system and satisfactory near-optimum performance.

The proposed method decreases the computational expences, in comparison with the iterative hierarchical methods and allows multi-microprocessor control structure to be realized.

REFERENCES

Vukobratovic M., D. Stokic, D. Hristic and N. Gruhalic (1977). Dynamical manipulator control. Automatika, Zagreb, vol.1-2,43-51, in Serbian.

Mahil S.S. (1979). Modelling techniques in the control of anthropomorphic industrial manipulators. Ph.D. Thesis, Coventry Polytechnic, England.

Patarinski S.P. and Danev B.V. (1980). Program suite for computer aided design of industrial manipulator control systems. In L.Dekker (Ed.), Simulation of Systems'79, North- Holland Publ. Co., 307-310.

Naplatanov N.D., Zaprianov J.D. and Mikhailov L.M. (1979). Hierarchical two-level control of industrial robots. In Proc. of First Conf. on Robotics, St. Zagora, June 1979, Bulgaria.

Singh M.G. and A. Titli (1978). Systems:decomposition, control and optimization. Pergamon Press, Oxford.

Mahalanabis A.K. and R. Singh (1980). On decentralized feedback stabilization of large-scale interconnected systems. Int.J. Control, vol. 32, No 1, 115-126.

Siliak D.D. and M.K. Sundareshan (1976). A multilevel optimization of large-scale dynamic systems. IEEE Trans. Autom. Control, vol. AC-21, 79-84.

OPTIMAL REAL-TIME CONTROL OF SOME LINEAR LARGE SCALE SYSTEMS

Henryk GÓRECKI
ACADEMY OF MINING AND METALLURGY
Institute of Automatic System Engineering and Telecommunication
CRACOW, POLAND

Henryk POŁCIK
FOUNDRY RESEARCH INSTITUTE
CRACOW, POLAND

Summary

The optimal real-time control for the system which can be expressed in the form $\dot{x} = Ax + Bu$ is considered in this paper. The minimum-time problem for heat equation can be transferred to this form. The control problem of such systems takes place in a three-level structure of the control of the foundry production process.

The discussed problems are illustrated by numerical examples.

INTRODUCTION

In the theory of large scale systems and lately also in the industrial practice the multi-level control structures are applied. This type of structure can also be used to control a foundry plant. The three-level structure of the control of a given foundry process is considered in the work [2]. The highest, i.e. the third level, contains global optimization and planning of the production with a long time horizon. The results of optimization in the form of partial tasks are directed to the particular production sections. The co-ordination of these segments belongs to the tasks of the second level, whereas the control of the individual operations comes within the tasks of the first level. Some of the objects of the lowest level can be described by linear dynamic systems with distributed parameters. To the objects which are the most important and at the same time very difficult to control belong such operations as: metal melting, pouring into moulds and heat treatment of castings. Criteria of the optimization of the controlled operations can be various. In the systems which control the charge heating /metal melting, heat treatment/ it is important to minimize the time in order to obtain the required temperature. The optimal control of the above mentioned systems described by parabolic equations may prove unreal because of the length of time needed for its calculation.

In this paper we present a methodology of constructing rapid algorithms for control systems with concentrated parameters. At the same time we discuss - taking as an example parabolic equations - the calculations of controls for systems with distributed parameters.

OPTIMAL CONTROL OF THE LINEAR DYNAMIC SYSTEMS IN REAL-TIME

The classical time optimal control problem we can now describe as follows:
Find a control $u_{op}(t)$ which transfers the system

$$\dot{x}(t) = A(t)x(t) + B(t)u(t) \tag{1}$$

from its initial point x_o to the origin $x_k = 0$ in the minimum time.

If we assume that $|u(t)| \leqslant 1$, then - as it is well known - the components of the time optimal control are of the bang-bang type.

To solve this problem some numerical procedures were designed using Neustad's method and special program in FORTRAN was elaborated.

This program being called PMT1 can be utilized as:

- an optimal regulator for a wider class of linear nonstationary dynamic systems,
- a simulating method for the analysis of dynamic systems.

A more detailed description of this program can be found in [1].

Now we shall deal with the, so-called, rapid algorithm which can be applied to the control system (1) in real-time. We shall discuss it taking as an example an object of the second order. The work of the proposed algorithm consists in three parts:

1° - computing the optimal control /by using PMT1/ for the initial points x_{oj} which belong to R_o region and which are chosen in a definite manner,

2° - approximation of the relations $t_{op} = t_{op}(x_o)$ and $t^1 = t^1(x_o)$, basing on the results from 1° /where: t^1 - the first switching time/,

3° - computing the approximate optimal control /according to 2°/ for the current control of the object, or - if it is needed - making the net denser, using the PMT1 program.

Let us take for the particular case of the discussed system (1)

$$A = \begin{bmatrix} 0, & 1 \\ -1, & 0 \end{bmatrix}, \quad B = [0, 1]^T, \quad |u(t)| \leqslant 1 \quad \text{for} \quad t \geqslant 0 \tag{2}$$

The region of the initial conditions is defined by:

$$R_o: \{ x_1 \in [-8, 8] , \quad x_2 \in [-8, 8] \} \tag{3}$$

In this region we have chosen 74 points and the optimal controls have been computed using the PMT1 program. Some of the results gave the following approximate relations:

$$\begin{aligned} t_j^1 &= 1.772 - 0.272x_1 + 0.002x_2 + 0.009x_1^2 + 0.011x_1x_2 - 0.001x_2^2 \\ t_{opj} &= 19.80 - 3.755x_1 + 0.433x_2 + 0.249x_1^2 + 0.204x_1x_2 - 0.060x_2^2 \end{aligned} \tag{4}$$

According to the results of the simulation, the sequent time switching can be obtained from the relation:

$$t_j^i = t_j^{i-1} + \pi ; \qquad t_j^i \leq t_{opj} \tag{5}$$

where:

$$i = 2, \dots, \quad \text{ENTIER} \left[\frac{t_{opj} - t_j^i}{\pi} \right]$$

and

t_j^i - the ith time switching for the x_{oj} point.

The controls in the intervals $(t^{i-1}, t^i]$ are equal:

$$u^i(t) = -1 \cdot u^{i-1}(t) \tag{6}$$

Calculation of the control $u^1(t)$ can be done in several ways. We have assumed the following procedure:

1/ Let $u^1(t) = 1$ for $t \in (0, t_1]$

2/ Let us calculate $x(t_{opj})$ taking the sequence +1, -1, +1,... as the controls.

3/ Let us test the relation:

$$\| x(t_{opj}) \| \leq \varepsilon \tag{7}$$

where: ε - the given precision of transferring (1) to the terminal point (0),

- if (7) is satisfied we assume $u^1(t) = +1$
- if not $u^1(t) = -1$

The procedure 1/ 2/ 3/ realizes part 3° of our algorithm.

As a result of the numerical simulation of the system (1), some of its properties can be computed. They are helpful - as it has already been shown above - to construct rapid algorithms. The PMT1 program can be utilized to calculate the control for the systems of heat flow. This problem will be discussed in the next part of the paper.

MINIMUM-TIME PROBLEM FOR HEAT EQUATION

One-dimensional problem

In the work [3] problems of constructing simplified mathematical models of processes described by partial differential equations are discussed. For simplified models the adequate optimal controls can be calculated in an easier way. As a result of discretization, for instance of space variables of the partial equations, we can obtain a simplified model - a linear system of differential equations.

Let us take the parabolic equation:

$$\frac{\partial V}{\partial t} = \frac{\partial^2 V}{\partial x^2} \tag{8}$$

where $V(x,t)$ is determined and continuous in the closed region $\bar{\Omega}$: $0 \leqslant x \leqslant 1$, $0 \leqslant t \leqslant t_1$. The function $V(x,t)$ satisfies (8) in the open region $0 < x < 1$, $0 < t < t_1$, as well as the initial and boundary conditions:

$$V(x,0) = \varphi(x), \quad x \in \bar{\Omega} \tag{9}$$

$$V(0,t) = V_0(t), \quad V(1,t) = V_{M+1}(t), \quad 0 \leqslant t < t_1 \tag{10}$$

where $\varphi(x)$, $V_0(t)$, $V_{M+1}(t)$ are continuous functions which satisfy conditions:

$$\varphi(0) = V_0(0), \quad \varphi(1) = V_{M+1}(0) \tag{11}$$

Assuming for the variable x the discretization step h we can expand the functions $V(x+h,t)$ and $V(x-h,t)$ in a Taylor series, by adding the both sides of which we can obtain:

$$\frac{\partial^2 V}{\partial x^2} = \frac{1}{h^2}\left[V(x-h,t) - 2V(x,t) + V(x+h,t)\right] + \frac{0(h^4)}{h^2} \tag{12}$$

Considering (9), (10), (11) and (12) we shall receive a system of linear differential equations:

$$\begin{bmatrix} \dot{V}_1 \\ \dot{V}_2 \\ \cdot \\ \cdot \\ \cdot \\ \dot{V}_M \end{bmatrix} = \frac{1}{h^2} \begin{bmatrix} -2 & 1 & 0 & . & . & . & 0 \\ 1 & -2 & 1 & & & & . \\ . & & . & . & . & & . \\ . & & & . & . & . & . \\ . & & & & . & . & . \\ 0 & . & . & . & . & 1 & -2 \end{bmatrix} \begin{bmatrix} V_1 \\ V_2 \\ \cdot \\ \cdot \\ \cdot \\ V_M \end{bmatrix} + \frac{1}{h^2} \begin{bmatrix} V_0 \\ 0 \\ \cdot \\ \cdot \\ \cdot \\ V_{M+1} \end{bmatrix} \tag{13}$$

with the initial condition:

$$V_n(0) = \varphi(nh) \tag{14}$$

we shall treat the functions $V_n(t)$ as an approximation of the function value $V(x,t)$ in the points of the space $x_n = n \cdot h$, where: $n = 1, 2, \dots, M$, $h = \frac{1}{M+1}$.

The equation (8) presents the heat flow in the one-dimensional space in the normalized form.
We can consider the minimum-time control problem for this system assuming the initial condition:

$$V(x,0) = -1, \qquad 0 \leqslant x \leqslant 1 \tag{15}$$

and the constraint

$$|u(t)| \leqslant 1 \tag{16}$$

According to [3], [6] the time-optimal control belongs to the boundary of the set of controls, what means it is of the bang--bang type.
Let us consider the case of the simultaneous control of both boundaries. Hence

$$V(0,t) = V(1,t) = u \tag{17}$$

The aim of the control will be the transfer of the system (8) to the zero point.
For such a problem the equation (8) may be approximated by the model (13), the solution of which can be done by numerical methods with the utilization of the PMT1 program, which has been discussed earlier.
As it is shown in [3] the eigenvalues of the system are in the form:

$$\lambda_s = -\frac{4}{h^2}\sin^2\frac{\varphi_s}{2}, \qquad \varphi_s = \frac{s \cdot \pi}{M+1}, \quad s = 1, 2, \dots, M \tag{18}$$

Using (18) it was easy to apply the PMT1 program to calculate the optimal control.

The calculation of the controls takes usually a long time. There are also some difficulties in finding a proper integration step and other parameters which are necessary for the PMT1 program. For these reasons we present numerical results for only $M = 4$, $\varepsilon = 0.0005$, $H = 0.001$. The time of the first switching $t_1 = 0.0802$, whereas the time of the optimal control equals: $t_{op} = 0.094$. As a result of the control the final state after the time t_{op} will take the value:

$$v_1 = -0.098, \quad v_2 = -0.039, \quad v_3 = -0.039, \quad v_4 = -0.098$$

These values are adequate to the results presented in [6] where other numerical methods were applied.

Two-dimensional problem

The solution of the problem (8), (9) and (10) may be approximated by the solution of the system (13) with the given initial conditions. It can be done with such an arbitrary precision as we require. The proof of the following theorem can be found in [5] ; for the given ε there exists $h(\varepsilon)$, that $|V(x,t) - V_n(t)| < \varepsilon$ for all n, when $h < h(\varepsilon)$ in the system (13).

In the similar way the problem which is described by parabolic equations can be solved. It is obvious that some assumptions must be satisfied. The presented method of solving the minimum-time problem for the heat equation can be applied for the n-dimensional case.

Let us consider the two-dimensional problem:

$$\frac{\partial u}{\partial t} = k\left(\frac{\partial^2 u}{\partial x^2} + \frac{\partial^2 u}{\partial y^2}\right) \tag{19}$$

for the region $0 \leqslant x \leqslant a$, $0 \leqslant y \leqslant b$ where u is known initially at all points of this region and the boundary conditions are given.

The optimization problem for this object can also be approximated by equation $\dot{x} = Ax + V$. The matrix A and the vector V are of a more complicated form.

For the assumed discretization net /Fig. 1/ and for the denotations

$$x = ih, \quad y = jh, \quad t = n\delta t, \quad u(ih, jh, n\delta t) = V_{i,j,n}$$

the difference representation of equation (19) given in [4] is in the form:

$$\frac{V_{i,j,n+1} - V_{i,j,n}}{\delta t} = \tag{20}$$

$$= \frac{k}{h^2}\left[V_{i-1,j,n} - 4V_{i,j,n} + V_{i+1,j,n} + V_{i,j-1,n} + V_{i,j+1,n}\right]$$

For the scheme of the net shown in Fig. 1

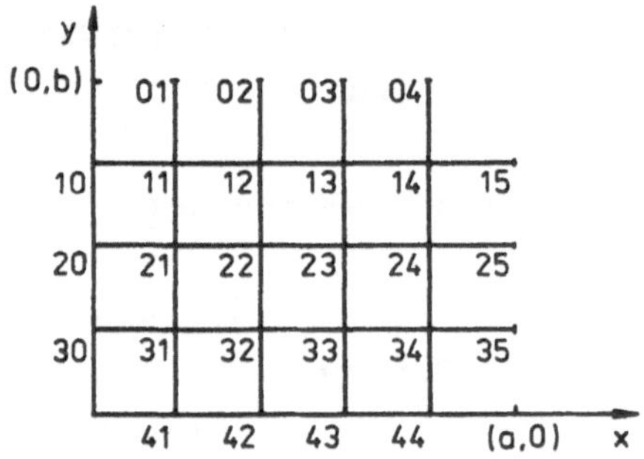

Fig. 1

considering (20), for k = 1 the elements of equation $\dot{x} = Ax + V$ are given

$$A = \frac{1}{h^2}\left[\begin{array}{c|c|c} B & I & \\ \hline I & B & I \\ \hline & I & B \end{array}\right] \qquad B = \begin{bmatrix} -4 & 1 & 0 & 0 \\ 1 & -4 & 1 & 0 \\ 0 & 1 & -4 & 1 \\ 0 & 0 & 1 & -4 \end{bmatrix}$$

where: I - unit matrix, while the elements of the matrix A which are not written equal zero.

$$x = \left[V_{11}, V_{12}, V_{13}, V_{14}, V_{21}, V_{22}, V_{23}, V_{24}, V_{31}, V_{32}, V_{33}, V_{34}\right]^T$$

$$V = \frac{1}{h^2}\left[V_{10} + V_{01}, V_{02}, V_{03}, V_{15} + V_{04}, V_{20}, 0, 0, V_{25}, V_{30} + \right.$$

$$\left. + V_{41}, V_{42}, V_{43}, V_{35} + V_{44}\right]^T$$

In the work [3] algorithms which can be used for the numerical solution of these problems with the application of the methods already discussed in this paper are given in detail.

Considering the approximation of the optimal problems which are described by parabolic equations, it is necessary to take into account such problems as: the existence and the univocality of the optimal solution. In many cases these problems can be solved in the experimental way.

FINAL COMMENTS

The presented idea of constructing control algorithms in real-time can be applied to control of technological operations which are described by linear dynamic systems with concentrated parameters. Parabolic equations which describe some operations in the foundry production process can be transferred to this form. The construction of control algorithms with the application of the PMT1 program requires a lot of experimental calculations. The above mentioned problems are connected with the time-optimal control. It will be interesting to apply this methodology to other problems, e.g. to the minimum effort problems.

REFERENCES

1. Górecki H., Połcik H.: Numeryczne wyznaczanie sterowania czaso-optymalnego liniowych układów dynamicznych i koncepcja sterowania w czasie rzeczywistym. Z.N. Automatyka, nr 24, Akademia Górniczo-Hutnicza, Kraków 1982.
2. Marcinkowski J., Połcik H., Wójcik T.: Hierarchiczna struktura sterowania produkcją odlewów. W: Podstawy komputerowych systemów sterowania odlewniczymi procesami technologicznymi. Seminarium II. Skrypt. Instytut Odlewnictwa, Kraków 1983.
3. Mitkowski W.: Sterowanie obiektów o parametrach rozłożonych w oparciu o aproksymację modelem dyskretnym. Rozprawa doktorska. Akademia Górniczo-Hutnicza, Kraków 1973.
4. Smith G.D.: Numerical solution of partial differential equations. Oxford University Press, London 1965.
5. Tichonov A.N., Samarski A.A.: Equations of mathematical physics. PWN, Warszawa 1963 /in Polish/.
6. Turnau A.: Sterowanie suboptymalne ze względu na czas dla układów opisywanych równaniami różniczkowymi cząstkowymi. Praca doktorska. Akademia Górniczo-Hutnicza, Kraków 1974.

APPLICATION OF HIERARCHICAL APPROACH FOR SHORT TERM ECONOMIC DISPATCHING OF A LARGE SCALE HYDROELECTRIC SYSTEM

T. Lefevre, M. Aveledo
Sección Planificación Energética. Departamento de
Procesos y Sistemas, Universidad Simón Bolivar
Valle de Sartenejas - Baruta - Caracas - Venezuela

A. Titli, J.L. Calvet
Laboratoire d'Automatique et d'Analyse des Systèmes du CNRS
7, avenue du Colonel Roche - 31077 Toulouse Cédex - France

Abstract

The economical short term dispatching of hydroelectric system is a very actual problem and its resolution consists in solving a large scale optimization problem; the classic methods fail when the complexity and the dimension of the problem are very large. The multilevel control techniques used in this work distribute between two control levels the calculation of optimum set points of the system by decomposition-coordination of the overall optimization problem.

For the non-separable problem examined in this paper, it is necessary to use a formulation of the decomposition-coordination principle different from that used for separable problems, and it needs to introduce pseudo-variables. Due to the special structure (serie-parallel) of the global optimization problem, it is possible to define algorithms both centralized and decentralized which accelerate the global convergence.

The decomposition in time and space of the overall optimization problem is presented and the coordination algorithms of the two control levels are described and applied to the case of Bajo-Caroni Hydroelectric System of Venezuela.

I. INTRODUCTION

This paper is devoted to the numerical solution of optimization problem on large scale discrete dynamical systems. Mainly concerned with the problem of dimensionality, it deals with :

- iterative (indirect) methods for solving large scale equations
- hierarchical approach of large scale systems

and is directed toward distributed computing. One of the purposes is to exploit some interconnection structure to provide usefull communication schemes in order to coordinate the decisions locally elaborated by subsystems, in a decentralized fashion. In the reference (1), we have discussed the application of such a methodology to the quadratic optimization on linear continuous (and distributed-lag discrete) dynamical systems, interconnected in a serial-parallel pattern. In the present paper, we will limit ourselves to the discrete case for which, we will mainly emphasize partitioning and interative coordination as numerical application.

Iterative methods of decomposition : The studied decomposition methods apply to the discrete case the double concept of approximation and iteration and are usually called : iterative methods, successive approximations method, indirect method,... The approximation is carried out by means of splitting decomposition in both time and space, as by additional techniques improving search directions.
This leads to the construction of an approximating sequence $x^{(k+1)}$:

$$x^{(k+1)} = T \left| x^{(k)} \right|$$

where $k \geqslant 0$ is the iteration index and $x^{(o)}$ is the initial guess of x, which converges toward a fixed point of the mapping T, i.e. toward a solution x of :

$$x = T \left| x \right|$$

Semi iterative methods : This concept, introduced by Varga (2) defines non stationary methods :

$$x^{(k+1)} = T_{k+1} \left| x^{(k)} \right|$$

with a view to improving the conditions of applicability, or convergence.
Among the most popular, let us mention :

- The alternating directions methods

$$x^{(k+1/2)} = T_1 \left| x^{(k)} \right|$$
$$x^{(k+1)} = T_2 \left| x^{(k+1/2)} \right|$$

- The generalized Richardson methods

$$x^{(k+1)} = x^{(k)} + \varepsilon_{k+1} D_{k+1} r^{(k)}$$

where : $\varepsilon_{k+1} \in \mathbb{R}$
D_{k+1} is a non singular matrix
r^k expresses the residual vector at step k

Different decomposition algorithms can thereby be constructed after partitioning, to iteratively approximate a solution x^*, taking into account such miscellaneous features as improvement of convergence, adaptation to the computing tool (parallel, vector computers), use of distributed computers,...

Let us mention the basic schemes illustrated by Figure 1 - 3.

Parallel decomposition (Jacobi type algorithms)

$$x_i^{(k+1)} = T_i \mid x_j(K) \mid \quad j \neq i;\ i,j=1,\dots,N$$

Sequential decomposition (Gauss-Seidel type algorithms)

$$x_i^{(k+1)} = T_i \mid x_j^{(k+1)},\ x_l^{(k)} \mid \quad j < i\ ; \quad l > i; \qquad i=1 \rightarrow N$$

Sequential decomposition with symetric alternated directions

$$x_i^{(k+1/2)} = T_i \mid x_j^{(k=1/2)},\ x_l^{(k)} \mid \qquad i= 1 \rightarrow N \quad j < i; \quad l > i$$

$$x_i^{(k+1)} = T_i \mid x_j^{(k+1/2)},\ x_l^{(k+1)} \mid \qquad i= N \rightarrow 1$$

II OPTIMIZATION OF AN INTERCONNECTED RESERVOIR NETWORK

II.1. Network description (3), (4)

Let the hydroelectric network described in the Figure 4, composed by a set of dams serially interconnected along the same river. The dams provide the necessary volume of water for the production of the electric energy generated, in order to respond to the demand for the whole hydroelectric system.

Let us define the following variables :

$h_{1i}(t)$: level of the ith dam at the time instant t(m)

$h_{2i}(t)$: level of the backwater in the ith dike at the instant t(m)

$X_i(t)$: volume in the ith dam at the instant t(m^3)

$H_i(t)$: net waterfall usable in the ith central (powerhouse) at the instant t(m).

$Y_i(t)$: inflow of water in the ith dam at the instant t(m^3/S)

$qi^{(t)}$: flow of the turbinated water at the ith central at the instant t(m^3/S).

$V_i(t)$: flow of the spilled water in the ith central at the instant t.

$e_i(t)$: flow of the evaporated water in the ith dam at the instant t(m^3/S)

$Pi^{(t)}$: electric power generated by the ith central at the instant t(W)

$\eta Ti^{(t)}$: global effectiveness of the set turbine-generator at the ith central at the instant t.

$\delta_i(t)$: losses due to the effect of the load in the stress turbines of the ith central at the instant t(m).

$d_i(t)$: flow of the natural contribution in the dam at the instant $t(m^3/S)$

$P_D(t)$: global demand of the electric energy at the instant t(W).

$t\in[t_o,t_f]$: time variable into the horizon t_o,t_f

$E_p(t)$: potential energy of the dam system at the instant t(Ws)

$\xi_i(t)$: energy generated in the ith central by unit of turbinated water flow at the instant $t(W/m^3)$.

$\Delta E_p(t)$: potential energy losses at instant t(Ws)

$\tau_{i-1,1}(t)$: travelling time of the water between the (i-1)th dam and the ith one.

Since the measure of the reservoir system performance to be done over regular time periods is relatively long (in the range of hours approximately), then it is possible to discretize the time horizon in N periods of duration Δt.

$K\in[1,\dots,N]$ with K : discretized time variable

The equations of the ith dam may be written as follows :

$$h_{1i}(K) = ai\ (bi - e^{-ciXi(K)})$$
$$X_i(K) = X_i(K-1) + [Y_i(K) - V_i(K) - q_i(K) - ei(K)].\Delta t$$
$$X_i(1) = X_i^o$$
$$X_i(N) \geqslant X_i^f$$
$$h_{2i}(K) = l_i\ (q_i(K) + \alpha_i(K)\ V_i(K)) + oi$$
$$H_i(K) = h_{1i}(K) - h_{2i}(K) - \delta_i(K)$$
$$\eta T_i(K) = \rho .g. {}_i(K) \qquad (O_i,\ \delta_i : \text{residual terms})$$

In addition the following inequality constraints have to be added :

$$0 \leqslant \underline{Xi} \leqslant Xi(K) \leqslant \overline{Xi}$$
$$0 \leqslant \underline{qi} \leqslant qi(K) \leqslant \overline{qi} = pi\sqrt{\frac{H_i(K)}{r_i}} \qquad (2)$$
$$0 \leqslant \underline{Vi}(K) \leqslant Vi(K)$$
$$Vi(K) \leqslant \eta i.h_{1i}(K) + oi$$
$$0 \leqslant \underline{Pi} \leqslant Pi(K) \leqslant Pi$$

The equations of the hydraulic coupling between the (ith) reservoir and the (i-1)th one are :

$$Yi(K) = q_{i-1}1(K - \delta_{i-1,i}) + V_{i-1}(K - \tau_{i-1,i}) + d_i(K)$$

with :

$$q_o(K - \tau_{o,1}) = 0$$
$$V_o(K - \tau_{o,1}) = 0 \qquad (3)$$
$$d_1(K) \neq 0$$
$$i\in\{1,\dots,n\}\ ;\ K\in\{1,\dots,N\}$$

and the equation of the balancing power is :

$$P_D(K) = \sum_{i=1}^{n} P_i(K) \qquad (4)$$

with :

$$K \in \{1,\dots,N\}.$$

II.2 Time performance of the network

The objective of the optimization problem may be summarized as : to respond, in the best way, to the demand P_D over a given horizon while minimizing the potential energy losses of the system. The set of conditions, defined at a planification level in a long term basis, for each reservoir, will be taken into account. Mathematically the problem may be stated as :

$$\min_{q,V,H,P,X,y} \left[\sum_{K=1}^{N} \Delta E_p(K) \right]$$

s.t. the equations (1), (3), (4) and the restrictions (2)

with $$\Delta E_p = \sum_{i=1}^{n} \left[\sum_{d=i}^{n} \zeta d(K) + \frac{\partial \zeta i(K)}{\partial Xi(K)} \sum_{u=1}^{i-1} X_u(K) \right] \Delta X_i(K)$$

$$\Delta X_i(K) = X_i(K) - X_i(K-1)$$

and

$$\zeta_i(K) = H_i(K)\, \eta_i(K)$$

The Lagrange function corresponding to the global problem is shown in the table n° 1.

The problem stated in this way is a large scale one which is a function of : 6nN main variables :

$q_i(K)$, $V_i(K)$, $Pi(K)$, $Xi(K)$, $Hi(K)$, $Yi(K)$ $\quad K \in [1,\dots N] \quad i \in [1,\dots.n]$

To solve the above problem we need to decompose it and to use multilevel techniques (5) (6). The lagrangian of the general problem is a highly non-separable function :

- in time (index K) because of the presence of terms with indexes

  $K-1$, $K-\tau_{i-1,i}$: $X_i(K-1)$, $q_{i-1}(K-\tau_{i-1,i})$, $V_{i-1}(K-\tau_{i-1,i})$

- in space (index i) due to the presence of indexes

  $(i-1)$: $q_{i-1}(K-\tau_{i-1,i})$, $V_{i-1}(K-\tau_{i-1,i})$

Thus, we are going to introduce a time domain decomposition in order to obtain an additive separable form of the lagrangian for the general problem : $L = \sum_{K=1}^{N} L_K$ in which the Lagrange functions L_K will depend only upon the index K for given coordination variables. Due to the non-separability of the reservoir equations and the reservoir interconnection, it is necessary to introduce pseudo-variables in order to provide for the decomposition of the Lagrangian function L an additive separable form (7). Thus these pseudo-variables are :

$$X''_i(K-1) = X_i(K-1) \quad , \quad i\in\{1,n\}, \quad K\in\{2,N\}$$

$$\left.\begin{array}{l} q''_{i-1}(K-\tau_{i-1,i}) = q_{i-1}(K-\tau_{i-1,i}) \\ V''_{i-1}(K-\tau_{i-1,i}) = V_{i-1}(K-\tau_{i-1,i}) \end{array}\right\} \quad i\in\{2,n\} \quad K\in\{\tau_{i-1,i}+1,N\}; \; N>\tau_{i-1,i}$$

and we have to add the following equations to the optimization problem previously defined :

$$X_i(K) - X''_i(K) = 0 \quad i\in\{1,n\}; \quad K\in\{2,N\}$$

$$\left.\begin{array}{l} q_i(K) - q''_i(K) = 0 \\ V_i(K) - V''_i(K) = 0 \end{array}\right\} \quad i\in\{1, n-1\}; \quad K\in\{1, \quad N-\tau_{i,i+1}\} \; ; \; N> \tau_{i,i+1}$$

The decomposition can be done by handling at the coordination level (coordination in time) the following variables and Lagrange parameters.

$$X''_i(K), \; \lambda_{6i}(K) \; : \quad K<N, \quad N>1$$

$$\left.\begin{array}{l} q''_i(K), \; \lambda_{7i}(K) \\ V''_i(K), \; \lambda_{8i}(K) \end{array}\right\} \quad \begin{array}{l} K\leq -\tau_{i,i+1} \\ N>\tau_{i,i+1} \end{array}$$

The introduction of the pseudovariables and their associated Lagrange parameters as coordination variables allows the use of direct iteration (8).

This decomposition is shown in the table N° 3 and it corresponds to a two-level optimization structure (Fig. 5) in which :

- at the lower level, we solve N optimization problems, each one corresponding to the following Lagrange functions :

  $L_K, \; K\in\{1,\ldots,N\}$

- at the top level we use a direct iteration algorithm on :

  $X''_i, \; \lambda_{6i}, \; q''_i, \; \lambda_{7i}, \; V''_i, \; \lambda_{8i}$

The Lagrange function L_K, corresponds to a large scale optimization problem, which can be difficult to handle by using standard non linear optimization algorithms. This is why a new decomposition, now in the space domain, may be proposed: With this second decomposition, an optmization sub-problem may be associated to each reservoir "i". Then we will try to obtain for L_K the following separable additive form :

$$L_K = \sum_{i=1}^{n} L_{Ki}$$

where L_{Ki} will correspond to the Lagrange function of the optimization sub-problem associated with the reservoir N°i.

However in the case of small systems (3 or 4 cascade reservoirs), space decomposition is not compulsory since there are some nonlinear programming algorithms that allow to deal with problems of such size.
Again, as it was done for the time-decomposition in order to decompose L_K into separable sub-problems, it is necessary to define the following pseudo-variables :

$$X'_u(K) = X_u(K);\ u \in \{1,n-1\},\quad n > 1$$
$$H'_d(K) = H_d(K);\ d \in \{2,n\}\quad n > 1$$

and consequently we have to add to the optimization problem, (defined table n° 4) the following equations :

$$X_i(K) - X'_i(K) = 0\ ;\ i \in \{1,n-1\};\quad n > 1$$
$$H_i(K) = H'_i(K) = 0\ ;\ i \in \{2,n\};\quad n > 1$$

Then the decomposition of L_K will be obtained by managing, at the coordination level, the following variables and Lagrange parameters :

$$X'_i(K), \lambda_{9i}(K)\ ;\quad i \in \{1,n-1\};\quad n > 1$$
$$H'_i(K), \lambda_{10i}(K);\quad i \in \{2,n\};\quad n > 1$$
$$P_i(K),\ \lambda_1(K);\quad i \in \{1,n\}$$

This decomposition is given in table n° 5, and it corresponds to two level optimization structure (figure n° 6) in which

- at the lower level, we solve "n" optimization problems corresponding each one to the computation of the min-max of the Lagrange functions L_{Ki}.
- at the top level, the problem consists to :

$$\min_{\lambda_1,\lambda_9,\lambda_{10}}\ \max_{X',H',P}\ [L_K]$$

In figure 7 we can see the three level structure corresponding to the resolution of the global problem, where :

- level n° 1 : optimization of the sub-problems i, $i \in \{1,n\}$
- level n° 2 : spatial coordination
- level n° 3 : temporal coordination

II.3 Implementation of the optimization structure

The implementation of this optimization structure is done in the following way :

i) Second coordination level (temporal coordination)

At this level we will use a direct iteration algorithm for the variables : (X'', λ_6), (q'', λ_7), (V'', λ_5) (Table n° 7a)
However in order to accelerate the convergence, it is possible to use a structure similar to the one shown in figure 3. In this case (table n° 7b), for each hour K, the values of the variables corresponding to

this coordination level will be analyzed using the results obtained :
- in the hour K if K=1 → N for X", q", V"
- in the hours K, K+1, $K+\tau_{i,i+1}$ if K=N → 1 for λ_6, λ_7, λ_8

Figure 8 shows the implementation of the decentralized structure at the second level of coordination.

ii) First coordination level (spatial coordination)

At this level we will solve the problem

$$\min_{\lambda_1\ \lambda_9\ \lambda_{10}}\ \max_{X',H',P}\ [L_K]$$

using a classic non-linear programming algorithm. However it is possible to handle the variables X', λ_9 and H', λ_{10} with a direct iteration algorithm, remaining unsolved the problem of :

$$\min_{\lambda_1}\ \max_{P}\ [L_K]$$

Because of the sequential structure of the stated problem it is possible, in order to accelerate the convergence at this coordination level, to use a structure as the one shown in figure 3. In this case, every time a sub-problem (say the ith one) is solved, the new values of the variables at this coordination level will be calculated, using the results obtained :
- in the sub-problem 1, 2,...i-1 if i=1 → n for H" d, λ_{9i}
- in the sub-problem i+1,...,n if i=n → 1 for X' u, λ_{10i}

Figure 9 shows the implementation of this decentralized structure at the first coordination level.

iii) Lowest level

At the lowest level we are going to use classic non-linear algorithms since the optimization sub-problems are relatively simple. In this case we will use the GRG package implemented by LASDON [9]. The problem to be solved is very simple (table 6) and corresponds to an optimization problem of :

6 variables
2 equality constraints
8 inequality constraints

III APPLICATION EXAMPLE AND ANALYSIS OF RESULTS

The optimization model was applied to the "Bajo Caroni" hydroelectric system. This system is composed of 4 cascade reservoirs :

GURI - TOCOMA - GARUACHI and MACAGUA

III.1 General characteristics of the system

GURI reservoir
Maximum elevation : 270,0 mts

TOCOMA reservoir
Maximum elevation : 125 mts.

Minimum elevation : 236,5 mts
N° of house power : 2
N° of units : Home 1 : 10
Home 2 : 10

Minimum elevation : 120 mts.
N° of house power : 1
N° of units : 12

CARUACHI reservoir
Maximum elevation : 91 mts
Minimum elevation : 86 mts
N° of house power : 2
N° of units : 14

MACAGUA reservoir
Maximum elevation : 54.5 mts
Minimum elevation : 50.5 mts
N° of house power : 2
N° of units : Home 1 : 6
Home 2 : 12

III.2 Initial conditions of the system for the studied cases

Starting elevations :

GURY	269.60 mts
TOCOMA	125.00 mts
CARUACHI	91.00 mts
MACAGUA	54.50 mts

Natural inflows :

GURI	4771 m^3/s in 24 h
TOCOMA	0 m^3/s in 24 h
CARUACHI	0 m^3/s in 24 h
MACAGUA	0 m^3/s in 24 h

Available units : All the units were available for the four cases of study.

Travelling time between the reservoirs : the travelling time was considered constant :

- Between Guri and Tocoma .... + 2 hours
- Between Tocoma and Caruachi ...... + 10 hours
- Between Caruachi and Macagua ..... + 5 hours

System global demand :

Hours : 1, 2, 3, 4, 5, 6, 7, 8, 9, 10, 11, 12, 13, 14, 15, 16, 17 18, 19, 20, 21, 22, 23, 24

Demand (MW) 8640, 8340, 8100, 8100, 8280, 8900, 9800, 11720, 12720, 12200, 13560, 13000, 10700, 13400, 13600, 13400, 13040, 13020, 13500, 13560, 12760, 11680, 10140, 9300.

Maximum demand : 13660 MV in the hour 15.

Total energy : 271.34 GWH

Travelling flows before the optimization

- Guri's discharges in Tocoma (2 hours before) : 5000 m^3/s were discharged constantly
- Tocoma's discharges in Caruachi (10 hours before) : 5000 m^3/s were discharged constantly
- Caruachi's discharges in Macagua (5 hours before) : 5000 m^3/s were discharged constantly

Minimum obligatory spill :

Guri : 0 m^3/s, Tocoma : 0 m^3/s, Caruachi : 0 m^3/s, Macagua : 1000 m^3/s.

Tests (Optimization)

The following results were performed :

1) Optimization of the cascade reservoir system, having as the minimum elevation constraint the minimum design elevation of the reservoir.
2) Optimization of the reservoir system, keeping Tocoma at its maximum elevation during the 24 hours, and others reservoirs as in case n° 1.
3) Optimization of the reservoir system, keeping Tocoma and Caruachi at their maximum elevations and others reservoirs as in case n° 21.
4) Optimization of the reservoir system maintaining all the reservoir after Guri : Tocoma, Caruachi and Macagua at their maximum elevation during the 24 hours.

These tests were done for the same initial conditions, demand curve and natural inflows.

III.3 Experimental results

- CASE # 1 (figures 4, 5, 6 and 7)

The reservoir system losses 1025 GWH of potential hydroelectric energy in order to satisfy the global demand.
It can be observed that at Macagua and Caruachi, the optimization algorithm assigns a practically constant generation ratio among the reservoir, during the whole time of study. The reservoirs "below" Guri are kept at high plant factor. It is noticed also that there is no spill for any reservoirs except Macagua, where the spill is kept at a constant value equal to 1000 m^3/s. This value is just the obligatory minimum value assigned to the spill in all the study cases.

- CASE # 2 (Figures 14, 15, 16 and 17)

In this second case, the system losses 1642 GWH in potential energy. The fact that Tocoma was forced to stay at its maximum elevation during the 24 hours entails also that the turbine works only with the incomming water. As in the first case, the reservoirs "below" Guri are kept at high plant factor. The spill is zero in all the reservoirs except at Tacagua where it is equal to the minimum obligatory value.

- CASE # 3 (Figures 18, 19, 20 and 21)

In this case, to maintain the same energy production level, it is necessary to tolerate a total loss in potential energy equal to 4344 GWH. Notice that Tocoma and Caruachi are maintained at their maximum elevation all the time. The plant factors of the reservoirs "below" Guri are high as in the previous cases.

- CASE # 4 (Figures 22, 23, 24 and 25)

Now the losses in potential energy reach the value of 7762 GWH.

Tocoma, Caruachi and Macagua are kept at their maximum elevation. Notice that is this case, the optimal solution in the distribution of the generation implies that none of the reservoirs is to work with a block of constant energy, and therefore at a constant net fall. However, again the reservoirs "below" Guri work at a high plant factor. Finally it is pertinent to remark that the optimal solution tolerates certain amount of spill in Tocoma and Caruachi, meanwhile at Macagua, it is necessary to spill only 100 m^3/s above the minimum obligatory value.

BIBLIOGRAPHIE

1. CALVET, J.L., Décomposition sur la base de la structure Série-parallele de systèmes dynamiques - Congrès AFCET Automatique, Nantes, France, 27-28 Octobre 1981.
2. VARGA, R.S., Matrix iterative analysis. Printice Hall (1962).
3. BECKER, L., YEH, W., Optimization of real time operation of a multiple - Reservoir System. Water Resources Research. December 1974.
4. GARCIA, J.M., TURGEON, A., Gestion optimale d'un ensemble hydroélectrique : aspects long terme et court terme. RAIRO Automatique - System Analysis and Control, 1981.
5. TITLI, A., Contribution à l'étude des structures de commande hiérarchisée en vue de l'optimisation des processus complexes. Thèse de Doctorat ès-Sciences Physiques. Université Paul Sabatier, Toulouse, Juin 1972.
6. LEFEVRE, T., Etude et mise en oeuvre des algorithmes de coordination dans les structures de commande hiérarchisée. Thèse de Docteur-Ingénieur, Université Paul Sabatier, Toulouse, Décembre 1972.
7. TITLI, A., LEFEVRE, T., RICHETIN, M., Multilevel optimization methods for non separable problems and applications. International Journal of Systems Sciences, vol. 4, n° 6, 1973.
8. LEFEVRE, T., Etude d'un algorithme mixte de coordination dans les structures hiérarchisées à deux niveaux. Revue RAIRO, 13, pp 145-150, Novembre 1973.
9. LASDON, L., GRG2-A Fortran - IV non linear programming system description and summary of capabilities.

Remark Due to space limitation some figures are missing: a complete set of illustrations can be obtained on request, or during the workshop.

TWO LEVEL HIERARCHICAL CONTROL OF COMPLEX DYNAMIC SYSTEM AND ITS APPLICATION

J. D. Zaprjanov, S. B. Boeva
Central Laboratary of Control Systems
Bulgarien Academy of Sciences, Sofia 1113
Acad. G. Bonchev str. bl.107, BULGARIA

Introduction

The use of clasical methods for optimal control of complex dynamical systems described by high order mathematical models often results in regulators of complicated structure which cannot be easily realized. To overcome these difficulties one often applies hierarchicaly decentralized synthesis.

Such structures have a number of advantages I3I concerning the executation times for the control algorithm, the flexibility of the configuration, the operational reliability of the system and the reduction of its cost.

In view of the broad spectrum applications of digital techniques and microprocessors in practice a discrete description in the state space is made use of.

The reliability of the algorithm proposed here is illustrated by the synthesis of a two level hierarchicaly-decentralized system for the stabilization of the motion of a flying vehiclealong a prescribed trajectory.

2. The synthesis of a two-level hierarchical regulator for direct digital control

Assume that the system is described by the following system of difference equations in the state space:

$$x_i(t+1)=A_{ii}x_i(t)+B_{ii}u_i(t)+\xi_i(t) \tag{1}$$

$$\xi_i(t)=\sum_{j\neq i}^{M}(A_{ij}x_j(t)+B_{ij}u_j(t))$$

$i=1,\dots,M$

where - $x_i(t) \in R^{n_i}$ are the state vectors of the individual subsystems
$u_i(t) \in R^{m_i}$ are the control vectors of the individual subsystems
$\xi_i(t) \in R^{n_i}$ are the vectors of the interconnections
$x_i(0)$-are the initial states of the subsystems
A_{ij}, B_{ij} are constant matrices of compatible dimensions

Thi description has been obtained through the space decomposition of the global description I5I.

We require the controllability of both the global system and the subsystems. $(A_{ii}, B_{ii}), i=1,\ldots,M$.

The ptoblem is to synthesize the controls $u_i(t)$ that hold the the system at the origin of the coordinate system and minimize a quadratic performance index, The decompozed form of which can be written as:

$$J=\frac{1}{2}\sum_{i\neq j}^{M}(x_i'(N)Q_i x_i(N) + \sum_{t=0}^{N-1}(x_i'(t)Q_i x_i(t)+u_i'(t)R_i u_i(t))) \tag{2}$$

The Hamiltonian for equation (2) subject to the constraints (1), is:

$$H(t)=\sum_{i=1}^{M}(1/2(x_i'(t)Q_i x_i(t)+u_i'(t)R_i u_i(t)+\lambda_i(t)(A_{ii}x_i(t)+$$
$$+B_{ii}u_i(t)+\xi_i(t))+\pi_i(t)(\xi_i(t)-\sum_{i\neq j}^{M}(A_{ij}x_j(t)+B_{ij}u_j(t))$$

where $\lambda_i(t) \in R^{n_i}$ are the vectors of the Lagrange multipliers;
$\pi_i(t) \in R^{n_i}$ are the covectors;

As far as H(t) is an additive-separable function, it can be decomposed as follows:

$$H(t)=\sum_{i=1}^{M}H_i(t) \tag{3}$$

and
$$\min H(t)=\sum_{i=1}^{M}\min H_i(t)$$

respectively, where:

$$H_i(t)=x_i'(t)Q_i x_i(t)+u_i'R_i u_i(t)+\lambda_i(t)(A_{ii}x_i(t)+B_{ii}u_i(t)+\xi_i(t))+$$
$$\pi_i(t)(\xi_i(t)-\sum_{j\neq i}^{M}(A_{ij}x_j(t)+B_{ij}u_j(t))$$

The necessary conditions quaranteeing optimality of both the global system and the individual subsystems have the form I3,5I:

$$-\lambda_i(t-1)=\frac{\partial H_i(t)}{\partial x_i(t)} \qquad \lambda_i(t-1)=-Q_i x_i(t)+\lambda_i'(t)A_{ii}+\sum_{j\neq i}^{M}\pi_j'(t)A_{ji} \tag{4}$$

$$x_i(t+1)=\frac{\partial H_i(t)}{\partial \lambda_j(t)} \tag{5}$$

$$\frac{\partial H_i(t)}{\partial u_i(t)}=0 \qquad u_i(t)=R_i^{-1}\Big(-\lambda_i(t)B_{ii}+\sum_{J\neq i}^{M}\pi_j'(t)B_{ji}\Big) \tag{6}$$

$$\frac{\partial H(t)}{\partial \xi(t)}=0 \qquad \pi_i^*(t+1)=-\lambda_i(t) \tag{7}$$

$$\frac{\partial H(t)}{\partial \pi(t)}=0 \qquad \xi_i^*(t+1)=\sum_{j\neq i}^{M}\big(A_{ij}x_j(t)+B_{ij}u_j(t)\big) \tag{8}$$

In the course of minimization of the performance index one assumes that the quantities $_i(t)$ and $_i(t)$ are known. The prediction is performed by means of the previous step of the coordinator, following the relations (7) and (8).

Assume, that the Lagrangian multipliers are of the following structure:

$$\lambda_i(t)=P_{ii}(t)x_i(t)+q_i(t) \tag{9}$$

where - $P_{ii}(t)\in R^{n_i x n_i}$

$q_i(t)\in R^{n_i}$

From the above relations, one has the controls:

$$u_i(t)=-R_i^{-1}B_{ii}'P_{ii}(t)x_i(t)-R_i^{-1}q_i(t)+R_i^{-1}\sum_{j\neq i}^{M}\pi_j'(t)B_{ji} \tag{10}$$

$i=1,\ldots,M$

From equation (10) it can be seen that the control has two components-the state feedback and the compensation one. To find the matrices $P_{ii}(t)$ and $q_i(t)$ the folloming stages are involved: substitute equation (9) into (6); in the resulting equation $x_i(t)$ is expressed through the system's equations (1), while the control $u_i(t)$ is expressed through equation (10). Finally, the terms containing $x_i(t-1)$ are grouped and the corresponding coefficients from the right and left hand sides of the equation are equated to obtain the necessary relations for $P_{ii}(t)$.

$$P_{ii}(t-1)=(I-(A'_{ii}P_{ii}(t)-Q_i)B_{ii}R_i^{-1}B'_{ii})^{-1}(A'_{ii}P_{ii}(t)-Q_i)A_{ii} \quad (11)$$

The boundary condition for this recurrence equation can be found from the requirement that at the (N-1)-st moment the control should be such that I4I $x_i(N)=0$,thet is:

$$P_{ii}(N-1)=B_{ii}(B'_{ii}B_{ii})^{-1}R_i(B'_{ii}B_{ii})^{-1}B'_{ii}A_{ii} \quad (12)$$

It is obvious that equation (11) can be solved in advance because their solution is independent of the system and the control and from practical point of view it determines the feedback in the corresponding control laws.

To find the compensation component one has the folloving recurrence relations:

$$q_i(t-1)=(I-(A'_{ii}P_{ii}(t)-Q_i)B_{ii}R_i^{-1}B'_{ii})^{-1}q_i(t)(A'_{ii}P_{ii}(t)-Q_i)(B_{ii}R_i^{-1}\sum_{j\neq i}^{M}B_{ij}\pi_j(t)$$

$$+\xi_i(t-1)) \quad (13)$$

Obviously its values depend on the current states of the subsystems, and they are not known. A number of approximate methods for the determination of the compensation component, such as to solve the problem for n linearly independent initial conditions I3I or to solve the global optimization problem I3I.

For on-line applications one has a convinient algorithm, based on the requirement that at the next moment the control renders a zero value of state . The following recurence relation is obtained for $q_i(t)$:

$$q_i(t)= -B^+_{ii}(\sum_{j\neq i}^{M}B'_{ji}\pi^*_j(t)+R_iB^+_{ii}\xi^*_i(t)) \quad (14)$$

with the boundary conditions $q_i(0)=0$.

It is obvious that through this relation the equation is solved for positive times and the future unknown controls and states of the system are absent in it.

In briev, the algorithem for direct diggtal control synthesized here, involves the following steps:

The value ofP$P_{ii}(t)$ can be found in advance from equation (11). with the boundary condition equation (12). Next, at each step one finds:

1.$q_i(t)$ from equations (14),with boundary condition $q_i(0)=0$.

2.$\lambda_i^1(t)$ from equation (9).

3. ^{1} $u_i(t)$ from equation (5).

The coordinator operats the folloving relations, equation (7,8).

The number of steps necessary to take the largest of the subsy-systems into the zero state is prescribed in advance. The process is repeated until the required result is obtained.
3.

3. The synthesis of a two-level hierarchicaly decentralized regulator for the stabilization of the motion of a flying vehicle

The motion of the fl²ing vechicle is described by non-linear differential equations under the corresponding assumptions for the for the coordinate s²stems I1,2I. The motion can be decomposed into a lateral and a longitudinal component if the inertial aerodynamical and giroscope interrelations are neglected and this is not possible for more complicated motions.

The linearized vector equation of motion of flying vehicle is time-dependent and in the vicinity of a given trajectory one obtains the folloving system of equations:

Longitudinal motion:

$$\begin{aligned}
\dot{\vartheta} &= 0.5\omega_z \\
\dot{\alpha} &= -2.47\alpha + 3\omega_z - 7.68v \\
\dot{\omega}_z &= -2.43\alpha - 1.47\omega_z + 1.26v - 32.62\delta_b \\
\dot{h} &= 0.016\vartheta - 0.00134\alpha \\
\dot{v} &= -0.0833\vartheta - 0.0071\alpha - 0.13v + 0.0388\delta_d \\
\dot{\delta}_b &= -3.333\delta_b + u_b \\
\dot{\delta}_d &= -3.333\delta_d + u_d
\end{aligned} \tag{15}$$

Lateral motion:

$$\begin{aligned}
\dot{\psi} &= 0.0833\omega_y \\
\dot{\gamma} &= 0.5\omega_x \\
\dot{\beta} &= 2.948\gamma - 0.103\beta + 0.048\omega_x - 3\omega_y \\
\dot{\omega}_x &= -14.6\beta - 125\omega_x - 1.18\omega_y - 57.33\delta_h - 90.69\delta_e \\
\dot{\delta}_e &= -3.333\delta_e + u_e \\
\dot{\delta}_n &= -3.333\delta_n + u_n
\end{aligned}$$

After discretization with an appropriate period of both motions, considered here as two subsystems, we obtain the folloving parameters:

$$A_{11}=\begin{bmatrix} 0.92 & -0.72 & 0.11 & 0.0 & -0.38 & -1.2 & -0.006 \\ 0.005 & 0.15 & -0.13 & 0.0 & 0.02 & 0.17 & -0.006 \\ -0.065 & 0.92 & 0.039 & 0.0 & 0.59 & 0.21 & 0.007 \\ -0.27 & 0.29 & -0.1 & 1.0 & 0.38 & 0.87 & 0.005 \\ 0.083 & -0.08 & 0.02 & 0.0 & 0.98 & -0.18 & -0.08 \\ 0.0 & 0.0 & 0.0 & 0.0 & 0.0 & 0.19 & 0.0 \\ 0.0 & 0.0 & 0.0 & 0.0 & 0.0 & 0.0 & 0.19 \end{bmatrix} \qquad B_{11}=\begin{bmatrix} -3.0 & -0.15 \\ -0.73 & -0.003 \\ -10.0 & 0.01 \\ -0.11 & 0.0006 \\ 0.78 & 0.067 \\ 2.7 & 0.0 \\ 0.0 & 5.8 \end{bmatrix}$$

$$A_{22}=\begin{bmatrix} 1.0 & -0.11 & -1.8 & -0.005 & 0.062 & 0.038 & -0.33 \\ 0.0 & 0.96 & -0.37 & 0.07 & -0.22 & -1.3 & -0.23 \\ 0.0 & 0.01 & -0.87 & 0.003 & 0.046 & -0.05 & -0.34 \\ 0.0 & -0.09 & 4.3 & -0.017 & -0.048 & 1.2 & 1.3 \\ 0.0 & -0.45 & -2.3 & -0.035 & -0.83 & 0.25 & 0.38 \\ 0.0 & 0.0 & 0.0 & 0.0 & 0.0 & 0.19 & 0.0 \\ 0.0 & 0.0 & 0.0 & 0.0 & 0.0 & 0.0 & 0.19 \end{bmatrix} \qquad B_{22}=\begin{bmatrix} 0.012 & -1.1 \\ -2.4 & -0.98 \\ -0.08 & -1.1 \\ -8.6 & -1.4 \\ 0.24 & -2.8 \\ 1.6 & 0.0 \\ 0.0 & 1.6 \end{bmatrix}$$

$$A_{12}=\begin{bmatrix} 0.0 & -0.002 & 0.0 & 0.0 & 0.0 & 0.0 & 0.002 \\ 0.0 & 0.01 & 0.006 & 0.005 & 0.04 & 0.0 & 0.005 \\ 0.0 & 0.001 & 0.0 & 0.002 & 0.0 & 0.0 & 0.0 \\ 0.0 & 0.002 & 0.0 & 0.02 & 0.006 & 0.0 & 0.04 \\ 0.0 & 0.005 & 0.0 & 0.001 & 0.0 & 0.0 & 0.0 \\ 0.0 & 0.0 & 0.0 & 0.002 & 0.0 & 0.02 & 0.005 \\ 0.005 & 0.04 & 0.0 & 0.005 & 0.0 & 0.001 & 0.0 \end{bmatrix}$$

$$A_{21}=\begin{bmatrix} 0.0 & 0.0 & 0.0 & 0.0 & 0.0 & 0.0 & 0.0 \\ 0.0 & 0.0 & 0.0 & 0.0 & 0.04 & 0.0 & 0.005 \\ 0.0 & 0.0 & 0.0 & 0.0 & 0.0 & 0.04 & 0.0 \\ 0.01 & 0.0 & 0.0 & 0.02 & 0.006 & 0.0 & 0.0 \\ 0.02 & 0.01 & 0.0 & 0.001 & 0.01 & 0.006 & 0.0 \\ 0.0 & 0.02 & 0.0 & 0.002 & 0.001 & 0.0 & 0.0 \\ 0.0 & 0.0 & 0.0 & 0.005 & 0.002 & 0.0 & 0.0 \end{bmatrix}$$

$$B_{12}=B'_{21}=\begin{bmatrix} 0.02 & 0.006 \\ 0.001 & 0.0 \\ 0.002 & 0.0 \\ 0.005 & 0.0 \\ 0.0 & 0.0 \\ 0.04 & 0.0 \\ 0.0 & 0.0 \end{bmatrix}$$

For the subsystems we chose the following weight matrices:

Q_1=diag(0.1 0.1 0.1 10.0 100.0 0.1 0.1)

Q_2=diag(10.0 0.1 0.1 0.1 0.1 0.1 10.0)

The transient responce for the initial states:

x_{10}=(0.2 0.18 0.0 0.0 -0.2 0.2 0.0)

x_{20}=(0.5 -0.2 0.0 -0.5 0.2 0.0 0.0)

shown on fig 1.

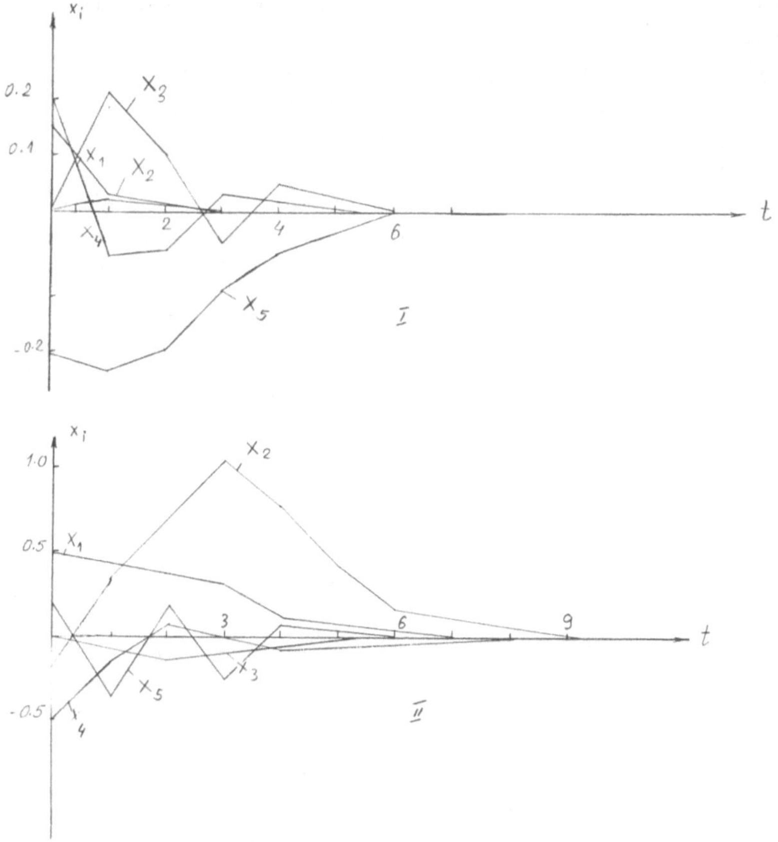

fig.1

Conclusions

A two level hierarchical algorithm for on-line control of complex dynamical systems has been proposed.It is chosen the conection-prediction coordination principle of Takahara, which is characterized by its rapid convergence and simplicity. The reliability of the algorithm is illustrated by the synthesis of a regulator for the stabilization of the motion of a flying vehicle along a prescribed trajectory.

REFERENCES

1 Bodner V. Control systems for flying vehicles,Mashinostroenie, Moskau,1973.

2 Ostoslavskii,I.,I.Strazeva,Dynamika of flying,Mashinostroenie, Moskau, 1969.

3 Singh M. Dynamical Hierarchical Control,North Holland publ. Comp.

4 Zaprjanov J.,Sv. Boeva,Hierarchical Decentralized Control of Industrial Robots,3th Trienal World Congress IFAC,August 24-28, 1981,Kyoto,Japan

5 Titli A.,Partitioning and Time Decomposition for The Control of Interconnected Systems,IFAC Symphosium,Oct.15-19,1979,Varna,Bulg.

REAL-TIME CONTROL ON THE COMPLEX SYSTEMS LOCALLY CONTROLLED

K.B.Czechowicz and J.K.Hunek
Institute of Control and Systems Engineering and
Institute of Organization and Management
Technical University of Wrocław, Wyb.Wyspiańskiego 27
50-370 Wrocław, Poland.

Summary. The paper discusses the problems of optimal control in a complex of operations. The case under consideration concerns the situation in which operations, i.e. certain operations running in time is additionally controlled by respective control quantities /tasks, resources/ from a upper level.

1. Introduction

In problems connected with the control of production, research or organizational processes there occurs often the problem of distribution /allocation/ of limited resources, such as, f.e.: energy, financial resources, manpower, material reserves etc. to particular elements of these processes. Describing them as certain operations running in time, i.e. as operations with determined initial and final states, and then analyzing and describing definite time relations among these operations we obtain in result a complex system of the complex of operations type. Problems connected with the control of such a system were solved for a number of cases, in which operation models were deterministic or probabilistic [1,2,3] , the operations were realized independently or dependently or the structure of the complex was determined or random [1] .The multilevel /two-level/ complex of operations was eventually determined by its structure or by the decomposition of the global task of determining the control of complex of operations. Then, depending on the form of mathematical model of an operation either a deterministic or a probabilistic optimization [1,3] is obtained on a given level.
The case considered in this paper concerns the situation, in which particular operations of a complex are additionally controlled by local control devices, whereas the tasks or resources assigned to these operations can influence the local parameters of an object, its initial and final state or the elements of the set of limitations imposed on the control. Such case for probabilistic models of the complex of independent operations was discussed in [2,3] , while the present considerations deal with the

operations on deterministic models in parallel structure. A general scheme of such a system can be presented as in Fig.1.

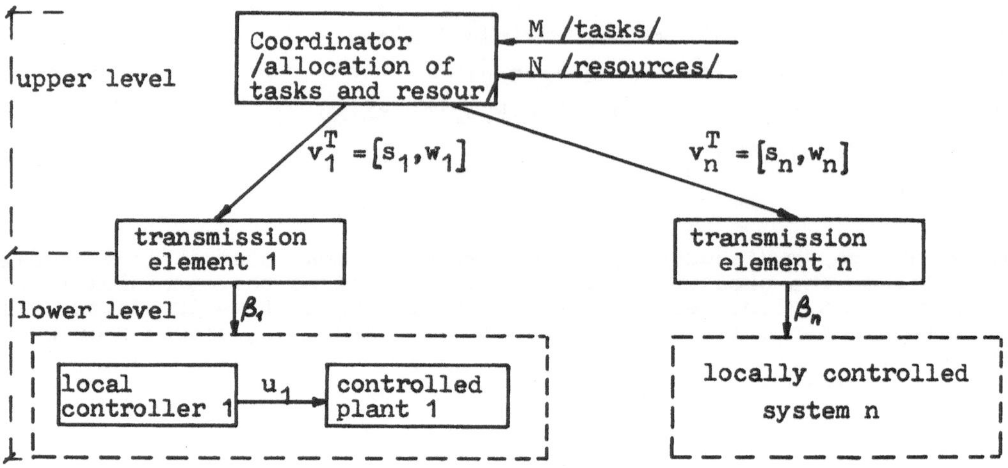

Fig.1. General scheme of a system.

2. Formulation of a problem

Mathematical descriptions of particular elements of a system presented in Fig.1. are following:

- description of the i-th locally controlled object:

$$\dot{x}_i(t) = f_i[x_i(t), u_i(t)], \quad x_{i0}, \; x_i^*, \quad i=1,2,\ldots,n \qquad /1/$$

where: $x_i(t)$ - state of the i-th object at the moment t,
$u_i(t)$ - local control of the i-th object at the moment t,
x_{i0} - initial state of the i-th object,
x_i^* - final state of the i-th object,
n - number of locally controlled objects.

- description of the i-th intermediary element /transmission unit/:

$$\beta_i = h_i(v_i) \;, \quad v_i^T = [s_i, w_i] \;, \quad i=1,2,\ldots,n \qquad /2/$$

and $\beta_i^T = [\alpha_i, x_{i0}, x_i^*, \gamma_i]$,

where: α_i - parameters of i-th locally controlled object,
γ_i - elements of the set of limitations imposed on local control,
v_i - control quantities /tasks, resources influencing the i-th local system.

In particular, the control quantity v_i can influence only a part of a vector β_i, f.e.:

$\beta_i = \alpha_i = h_i'(v_i)$ influence on the parameters of the i-th object, or

$\beta_i = x_{i0} = h_i''(v_i)$ - influence on the initial state of the i-th object

While controlling the i-th operation locally at the lower level, i.e.: transferring the i-th object from the inital state x_{i0} to final state x_i^* /assuming full controllability of objects/, we obtain the realization time of the i-th operation /the moment in which the i-th object reaches the state $x_i(t)|_{t=T_i} = x_i^*$ / in form:

$$T_i = F_i'[u_i(\cdot)_{0,T_i}, \beta_i] = F_i''[u_i(\cdot)_{0,T_i}, h_i(v_i)] = F_i[u_i(\cdot)_{0,T_i}, v_i] \qquad /3/$$

where $u_i(\cdot)_{0,T_i}$ - is the local control of the i-th object in time $0 \leq t \leq T_i$

Dependence /3/ is a model of the i-th operation for further considerations. The problem of optimal control at this level is following: taking v_i as a parameter, determine such optimal control $u_i^*(\cdot)_{0,T_i}$ for which T_i is minimal i.e. $T_i = T_{i,min}$.

Assuming that v_i is constant during the course of the i-th operation, i.e. v_i = const., for $0 \leq t \leq T_i$ we obtain a classical problem of determining the time-optimal control at the lower level. Let us now formulate the problem of controlling the whole system, composed of the operations described above. The criterion of the system control is defined as the time in which the realization of all operations of a complex is completed, i.e. the time after which all operations reach their final states x_i^*, $i=1,2,\ldots,n$. In general case it is a functions

$T = G(T_1, T_2, \ldots, T_n)$.

For the case of a complex of independent operations, considered in this paper, this function has the form $T = \max(T_1, T_2, \ldots, T_n) = \max_{1 \leq i \leq n} T_i$ /4/

The problem of time-optimal control in a complex system of the type complex of operations can be formulated as follows:

determine such control functions $[u_i(\cdot)_{0,T_i}, v_i]$, $i=1,2,\ldots,n$ for which T is minimal, with following limitations:

Constrain 1: $u_i(\cdot)_{0,T_i} \in \Gamma_{iu} \subset U_{if}$, $i=1,2,\ldots,n$ /5/

where: Γ_{iu} - space of allowable controls, e.g. in form:

$$\Gamma_{iu} = \{u_i(\cdot)_{0,T_i} : |u_i^k(\cdot)_{0,T_i}| \leq M_i^k, \quad k=1,2,\ldots,l\}$$

U_{if} - space of the function $u_i(\cdot)_{0,T_i}$

Constrain 2: $v_i \in V$, $V = \{v = [v_1, v_2, \ldots, v_n]^T, \sum_{i=1}^{n} v_i \leq N, \ v_i \geq 0, \ i=1,..,n\}$ /6/

Using /3/-/6/, the problem of control optimization of the system can be formulated in form:

$$\min_{\substack{[u_i(\cdot)_{0,T_i}, v_i] \\ 1\le i\le n, Con.1,2}} T = \min_{\substack{[u_i(\cdot)_{0,T_i}, v_i] \\ 1\le i\le n, Con.1,2}} \left\{ \max_{1\le i\le n} F_i[u_i(\cdot)_{0,T_i}, v_i] \right\} \qquad /7/$$

The above optimization problem is a mixed problem of dynamic-statical optimization, in which dynamic optimization occurs at the lower level /in a local control system/ and statical optimization at the upper level. This problem was solved with the application of a functional decomposition method and the identity of the direct problem /7/ and decomposed problem solutions was proved.

3. Algorithm for the solution of a problem

Functional decomposition of the optimization problem in direct form /7/ is based on the determination of optimal controls on two levels $[u_i^*(\cdot)_{0,T_i}, v_i^*]$ for all i=1,2,...,n, which minimize:

- on the lower level $\min\limits_{\substack{u_i(\cdot)_{0,T_i} \\ 1\le i\le n, Con.1}} T_i$
- and on the upper level $\min\limits_{\substack{v_i \\ 1\le i\le n, Con.2}} T.$

The identity of the direct problem /7/ and decomposed problem solutions can be demonstrated on the example of the following lemmas [2,3].

Lemma 1 Let F(u) be an arbitrary function of the class C^0 and α=const. a random parameter. Then $\min\limits_u \{ \max[F(u), \alpha]\} = \max\{\min\limits_u F(u), \alpha\}$.

The proof of this lemma is immediate:

$$\min_u[\max\{F(u),\alpha\}] = \begin{cases} \min\limits_u F(u) & \text{for } u \in \{u: \min\limits_u F(u) > \alpha\} \\ \alpha & \text{for } u \in \{u: \min\limits_u F(u) < \alpha\} \end{cases} = \max\{\min_u F(u), \alpha\}$$

Lemma 2 Let $\bigwedge\limits_{1\le i\le n} F_i(u_i)$ be arbitrary functions of the class C^0. Then

$$\min_{\substack{u_i \\ 1\le i\le n}} \left\{ \max_{1\le i\le n} [F_i(u_i)] \right\} = \max \left\{ \min_{\substack{u_i \\ 1\le i\le n}} F_i(u_i) \right\}$$

Prof: The thesis of the lemma 2 can be expressed in the form:

$$\min_{u_1} \min_{u_2} \ldots \min_{u_n} \left\{ \max[F_1(u_1), F_2(u_2), \ldots, F_n(u_n)] \right\} =$$

$$= \max\left\{ \min_{u_1} F_1(u_1),\ \min_{u_2} F_2(u_2), \ldots, \min_{u_n} F_n(u_n) \right\}.$$

Using, the induction principle we obtain:

- for one variable u_1: $\min_{u_1}\left\{\max_{1\le i\le n}\left[F_i(u_i)\right]\right\} = \min_{u_1}\left\{\max\left[F_1(u_1), \max_{2\le i\le n} F_i(u_i)\right]\right\}$

Denoting $\max_{2\le i\le n} F_i(u_i) = \alpha$ and using Lemma 1 we have:

$$\min_{u_1}\left\{\max\left[F_1(u_1), \alpha\right]\right\} = \max\left\{\min_{u_1} F_1(u), \alpha\right\}.$$

- let the thesis of the Lemma 2 be true for j variables $u_1, u_2, \dots, u_j$:

$$\min_{u_1}\min_{u_2},\dots,\min_{u_j}\left\{\max_{1\le i\le n} F_i(u_i)\right\} = \max\left\{\min_{u_1} F_1(u_1),\dots,\min_{u_j} F_j(u_j), \max_{j+1\le i\le n} F_i(u_i)\right\}$$

- we will prove the truth of the above thesis for j+1 variables :

$$\min_{u_1}\min_{u_2},\dots,\min_{u_{j+1}}\left\{\max_{1\le i\le n} F_i(u_i)\right\} = \min_{u_{j+1}}\left\{\max\left[\min_{\substack{u_i\\1\le i\le j}} F_i(u_i), \max_{j+1\le i\le n} F_i(u_i)\right]\right\} =$$

$$= \min_{u_{j+1}}\left\{\max\left[F_{j+1}(u_{j+1}), \bar{\alpha}\right]\right\}, \text{ where } \bar{\alpha} = \max\left\{\min_{\substack{u_i\\1\le i\le j}} F_i(u_i), \max_{j+2\le i\le n} F_i(u_i)\right\}$$

Basing on the Lemma 1 we have:

$$\min_{u_{j+1}}\left\{\max\left[F_{j+1}(u_{j+1}), \bar{\alpha}\right]\right\} = \max\left\{\min_{u_{j+1}} F_{j+1}(u_{j+1}), \bar{\alpha}\right\} = \max\left\{\min_{\substack{u_i\\1\le i\le j+1}} F_i(u_i), \max_{j+2\le i\le n} F_i(u_i)\right\}$$

Thus, basing on the induction principle, we have proved that the Lemma 2 is true. On the basis of the above lemmas it is possible to prove the following theorem, enabling the functional decomposition of the problem of direct optimization /7/.

<u>Theorem 1</u>: Let $\bigwedge_{1\le i\le n}\left\{F_i\left[u_i(\cdot)_{0,T_i}, v_i\right] \subset C^o\right\}$. Then

$$\min_{\substack{[u_i(\cdot)_{0,T_i} v_i]\\1\le i\le n,\text{Con.}1,2}}\left\{\max_{1\le i\le n} F_i\left[u_i(\cdot)_{0,T_i}, v_i\right]\right\} = \min_{\substack{v_i\\1\le i\le n,\text{Con.}2}}\left\{\max_{1\le i\le n}\left[\min_{\substack{u_i(\cdot)_{0,T_i}\\1\le i\le n,\text{Con.}1}} F_i\left[u_i(\cdot)_{0,T_i}, v_i\right]\right]\right\}$$

This theorem can be proved immediately by the application of the Lemmas 1 and 2, namely:

$$\min_{\substack{[u_i(\cdot)_{0,T_i}, v_i]\\1\le i\le n,\text{Con.}1,2}}\left\{\max_{1\le i\le n} F_i\left[u_i(\cdot)_{0,T_i}, v_i\right]\right\} = \min_{\substack{v_i,\text{Con.}2\\1\le i\le n}}\left[\min_{\substack{u_i(\cdot)_{0,T_i}\\1\le i\le n,\text{Con.}1}}\left\{\max_{1\le i\le n} F_i\left[u_i(\cdot)_{0,T_i}, v_i\right]\right\}\right] =$$

$$= \min_{\substack{[v_i,\text{Con.}2]\\1\le i\le n}}\left\{\max_{1\le i\le n}\left[\min_{\substack{u_i(\cdot)_{0,T_i}\\1\le i\le n,\text{Con.}1}}\left[F_i\left[u_i(\cdot)_{0,T_i}, v_i\right]\right]\right]\right\}.$$

The algorithm for determination of optimal controls on two levels, based on the Theorem 1, will have the following steps:

Comment: There are given descriptions of elements /1/,/2/ and sets of constrains /5/, /6/.

Step 1. Determination of time-optimal controls $u_i^*(\cdot)_{0,T_i}$, $i=1,2,\ldots,n$ and time of realization for operations T_i on the lower level. Then we obtain the operation models in the form:

$$T_{i,min} = T_i' = \min_{\substack{u_i(\cdot)_{0,T_i} \\ \text{Con.1}}} F_i[u_i(\cdot)_{0,T_i}, v_i] = \varphi_i(v_i), \quad i=1,2,\ldots,n. \qquad /8/$$

Step.2. Determination of time-optimal controls v_i^*, $i=1,2,\ldots,n$ on the upper level, i.e.

$$T = \min_{\substack{v_i \\ 1\ i\ n,\text{Con.1}}} \left\{ \max_{1\leq i\leq n} \varphi_i(v_i) \right\} \qquad /9/$$

Depending on the form of descriptions /1/,/2/,/5/,/6/ the above algorithm can be more or less complicated. In the step 1 the maximum principle was used for determination of the controls $u_i^*(\cdot)_{0,T_i}$, $i=1,2,\ldots,n$ whereas in the step 2 the algorithms of the control of complex of independent operations [2] was applied.

4. Conclusion

The paper deals with relatively narrow class of time-optimal problems of resources allocation in system of complex of operations type, with operations locally controlled. The problem becomes wore complicated when some inter-operation relations occur, i.e. in case of the complex of dopendent operations or when the operation models or the description of relations structure have a probabilistic character. Essential difficulties occur in the phase of elaboration of satisfactory effective calculation algorithms. The above presented ideas have many practical applications. Such problems are encountered for instance in the control of a line of drying chambers in an industrial process of wood drying, in the control of fuel /petroleum, coal/ distribution into parallelly operating turbo-generators in a power plant or at last in the control - through respective catalysts - of parallelly running chemical processes.

References

[1] Bubnicki Z., Two-level optimization and control of the complex of operations, Proc. 6th IFAC Congress, Pergamon Press,Helsinki, 1978.

[2] Czechowicz K., Time-optimal algorithms of control of the complex of independent operations /in Polish/, Podstawy Sterowania, t.4, 1974.

[3] Czechowicz K., Malave N., A complex of operations approach to the optimal control of the set of plants,Systems Science,Vol.5,No4,1979.

INVESTIGATING A TWO-LEVEL DYNAMIC ESTIMATION ALGORITHM FOR LARGE-SCALE ELECTRIC POWER SYSTEMS

P. Rousseaux, T. Van Cutsem, M. Ribbens-Pavella
Department of Electrical Engineering
University of Liège, Sart-Tilman B28, B - 4000 Liège, Belgium

ABSTRACT. A dynamic hierarchical state estimator designed to meet the on-line requirements of electric power systems is considered and its performances are explored and assessed. The estimator derives from the conjunction of an extended Kalman filter, embedded in a hierarchical scheme. The investigations concern three distinct series of simulations. This dynamic hierarchical estimator is shown to perform perfectly well with respect to accuracy and reliability, while complying with on-line computing needs which had never been met till now by other dynamic estimators.

NOTATION

N.B. All symbols are fully defined at the place they are first introduced. As a convenience to the reader we have collected below the more frequently used symbols in several places.

With some obvious exceptions lower case italic letters indicate vectors and capital italic letters denote matrices.

k denotes a time sample.

Δt = time interval.

z : m-dimensional measurement vector.

v : measurement noise vector.

R : covariance matrix of the measurement noise vector.

$\sigma(j)$: standard deviation of the j -th measurement noise.

x : n-dimensional state vector.

$h(\cdot)$: nonlinear vector function relating measurements z to x ; $h(\cdot) : \mathbb{R}_n \to \mathbb{R}_m$

w : system noise vector.

Q : covariance matrix of the system noise vector.

q(j) : maximum rate of change of the j -th state variable.

P : covariance matrix of the estimated state vector.

K : Kalman gain.

$b \sim N[m,C]$ indicates that vector b is a gaussian (normal) random vector with mean m and covariance matrix C .

ISE : "Integrated" State Estimation, i.e. estimation of the system taken as a whole.

HSE : "Hierarchical" State Estimation, i.e. estimation provided by a hierarchical scheme, through system decomposition.

DISE : Dynamic ISE.

DHSE : Dynamic HSE

EKF : Extended Kalman Filter.

$\hat{x}$: estimated value of x provided through an ISE.

$\tilde{x}$: estimated value of x provided at the end of the first level of the HSE.

$\approx{x}$: estimated value of x provided at the end of the second level of the HSE.

δ_{ij} : Kronecker's delta : $\delta_{ij} = 1$ (0) if $i = j$ ($i \neq j$) .

E = expectation.

diag = diagonal matrix.

$A^T = A$ transposed.

1. INTRODUCTION

Dynamic state estimation of electric power systems has long been considered to be quite problematic, if at all feasible. Indeed, this dynamic estimation involves quite heavy algorithms and therefore important computational burden which moreover amplifies considerably with the system dimensions. In electric power systems, which by essence are large-scale systems, the computational burden becomes quickly prohibitive; the difficulties increase even more rapidly because the state estimator is called to run on-line and in parallel with the other computer programs implemented in the power system control center.

To remove the above difficulties, we have recently proposed a hierarchical state estimator [1]. It results from the conjunction of an extended Kalman filter and of a hierarchically-organized estimator. The latter may in principle comprise more than two levels. In this paper the simplest, two-level version, has been considered. Indeed, the essential objective sought here is to show that the above dynamic hierarchical state estimator behaves totally satisfactorily with respect to reliability and accuracy, while complying with on-line computing requirements. The investigations are based on three sets of simulations, and their results are assessed and discussed. The sets differ by the type of operating conditions considered. The system used throughout these simulations is the IEEE 30-bus test-system. This is obviously not a typical large-scale system; it allows however making easy comparisons and illustrating potentialities of the dynamic hierarchical estimator.

2. DYNAMIC STATE ESTIMATION (DSE) IN ELECTRIC POWER SYSTEMS

2.1. The "Dynamic Integrated State Estimation" (DISE) problem

Consider the "integrated" electric power system comprising N nodes or buses.

For this system, taken as a whole, the DISE consists in finding an estimate $\hat{x}(k)$ of its state vector x at the sample interval k, from the observation vector $z(k)$ possibly corrupted by noise, where

* z is the m-dimensional measurement vector, generally consisting of active/reactive line power flows and nodal power injections, as well as some voltage magnitudes;
* x is the n-dimensional state vector, comprising N nodal voltage magnitudes and N-1 nodal phase angles referred to a reference node or slack bus.

The above general filtering problem can be solved by means of a recursive filter with state transition equation of the general form

$$x(k+1) = f[x(k), w(k)] \tag{1}$$

and with the measurement equation

$$z(k+1) = h[x(k+1)] + v(k+1) \tag{2}$$

Within the quasi steady-state system approximation, the state transition eq. (1) takes on the simplified form [1,2]

$$x(k+1) = x(k) + w(k) \tag{3}$$

where

$$w(k) \sim N[0, \delta_{k\ell}\, Q(k)] \tag{4}$$

$$Q(k) = \alpha^2 (\Delta t)^2 \text{ diag } [q^2(j)] \tag{5}$$

In the latter expression, parameter $q(j)$ is the maximum rate of change of the j-th state variable, to be determined by past variational records of the node voltages; parameter α will also be determined off-line, by simulating a ramp variation according to eq. (3) with $w(k) = q\Delta t$, and by minimizing the cost quadratic function [1,2]:

$$I = \lim_{k\to\infty} E\{[\hat{x}(k) - x(k)]^T [\hat{x}(k) - x(k)]\} \tag{6}$$

Thus, within the quasi steady-state system approximation, the system is described by the state transition eq. (3) and the measurement eq. (2) where the "noise" vectors are modelled according to eqs. (4) and (7) :

$$v(k) \sim N(0, \delta_{k\ell} R) \quad \text{with } R = \text{diag } [\sigma^2(j)] \tag{7}$$

2.2. DISE Algorithm

One seeks to minimize the performance index

$$J(k) = \{z(k) - h[x(k)]\}^T R^{-1} \{z(k) - h[x(k)]\} + [x(k) - \bar{x}(0)]^T P(0)^{-1} [x(k) - \bar{x}(0)] \tag{8}$$

where $x(0)$ denotes the state at the time origin $t = 0$ and where

$$\bar{x}(0) = E[x(0)] \quad , \quad P(0) = E[x(0)\, x(0)^T] \tag{9}$$

The DISE algorithm is an Extended Kalman Filter (EKF) described by the following equations [1] :

$$\hat{x}(k+1/k) = \hat{x}(k) \tag{10}$$

$$P(k+1/k) = P(k) + Q(k) \tag{11}$$

$$K(k+1) = P(k+1/k)\ H(k+1)^T\ [R + H(k+1)\ P(k+1/k)\ H(k+1)^T]^{-1} \tag{12}$$

$$\hat{x}(k+1) = \hat{x}(k+1/k) + K(k+1)\ [z(k+1) - h[\hat{x}(k+1/k)]] \tag{13}$$

$$P(k+1) = [I - K(k+1)\ H(k+1)]\ P(k+1/k) \tag{14}$$

$$H(k+1) = \left.\frac{\partial h}{\partial x}\right|_{x=\hat{x}(k+1/k)} \tag{15}$$

$H(k+1)$, the Jacobian matrix of h , is evaluated for $x = \hat{x}(k+1/k)$, i.e. for the predicted value of vector x at time $(k+1)$ through measurements provided from $t=0$ till $t = k\Delta t$: $[z(0),\ldots,z(k)]$.

$P(k+1)$ is the covariance matrix of the estimated state vector $\hat{x}(k+1)$.

$K(k+1)$ is the gain matrix of the Kalman filter.

The estimation procedure is assumed to start at $t(0) = 0$ with initial condition $x(0)$ as defined above.

2.3. The "Dynamic Hierarchical State Estimation" (DHSE) problem

Rather than estimating the overall system as a whole, we intend to decompose it into smaller, easier to handle, estimation subproblems. Following [3], we consider in this paragraph the simplest case of a two-level structure and we define convenient decomposition rules. Next § 2.4 is devoted to the derivation of an appropriate DHSE algorithm.

Let the system be decomposed into s nonoverlapping areas (or regions, or subsystems) $A^1,A^2,\ldots,A^s$. These areas are linked by tielines which in the power system are electrical lines or transformers. The two ends of each tieline are buses (nodes) belonging to two distinct areas, called *boundary nodes*. The set of all boundary nodes defines an additional (pseudo-) area, the *interconnection area* labelled $(s+1)$.

In the i-th subsystem $(i = 1,2,\ldots,s)$, we define :

* x_c^i vector consisting of the voltage magnitudes and phase angles at the boundary nodes;
* x_r^i vector consisting of the voltage magnitudes and phase angles at internal nodes;
* x^i , the composite state vector : $x^i = [(x_c^i)^T, (x_r^i)^T]^T$ (16)
* x_c , the state vector relative to the boundary nodes :

$$x_c = [(x_c^1)^T, (x_c^2)^T, \ldots, (x_c^s)^T]^T \tag{17}$$

In each area, we choose a local reference : hence, to determine completely the state of the system, we introduce the vector of angular differences :

$$u = (u^1,u^2,\ldots,u^{s-1})^T \tag{18}$$

where u^i represents the phase of the reference node of the i-th area with respect to that of the s-th area, arbitrarily chosen as a general reference for the entire system $(u^s = 0)$.

Further, we decompose the measurement vector accordingly into $s+1$ subvectors :

$$z = [(z^1)^T, (z^2)^T, \ldots, (z^s)^T, (z_u)^T]^T \tag{19}$$

where z^i and z_u denote the measurement vectors of the i-th subsystem ($i = 1,\ldots,s$) and of the interconnection area respectively. z_u consists of active and/or reactive power flows in tielines.

2.4. The DHSE algorithm

Generally speaking, the number of algorithmic levels equals the number of structural decomposition levels of the power system. The estimation algorithms used in the various levels are of two types : First Level Estimation (FLE) type and second or Coordination Level Estimation (CLE) type. In what follows, we describe these two algorithms in the particular case of § 2.3 where the power system has been (simply) decomposed in a two-level structure.

2.4.1. The FLE algorithm

It derives from the EKF algorithm described by eqs. (10) to (15), straightforwardly adapted to the s subsystems whose independence is ensured thanks to the use of a local slack bus for each one of them. Therefore, the EKF algorithm is run independently, and either in parallel or sequentially for each subsystem $i = 1,2,\ldots,s$.

At the end of this FLE one gets the estimates $\tilde{x}^i$ of x^i of form (16), for $i = 1,2,\ldots,s$. As it will be seen below, the part $\tilde{x}^i_c$ of $\tilde{x}^i$ relative to the boundary nodes is sent and used at the CL, along with matrix $P^i_c(k)$. The latter is the part of the error covariance matrix of the estimated state vector x^i :

$$P^i(k) = \begin{bmatrix} P^i_r(k) & P^i_{rc}(k) \\ P^i_{cr}(k) & P^i_c(k) \end{bmatrix} \tag{20}$$

relative to the state subvector x^i_c .

2.4.1. The CLE algorithm [4]

The essential task of this algorithm is the evaluation of the vector u components, defined by (18). Indeed, due to the decomposition and to the use of a slack bus for each subsystem, use of the observations contained in z_u and estimation of the tieline power flows impose estimating u in addition to x_c . The CLE algorithm uses (for justification and details, see [1]) :

* the CL state vector $x_s(k) = [(u(k))^T, (x_c(k))^T]^T$ (21)
* the CL "measurement" vector $z_s(k) = [(z_u(k))^T, (\tilde{x}_c(k))^T]^T$ (22)

It has two steps, consisting respectively of an EKF algorithm and of a static estimation one, and performed sequentially [1].

First step. Upon fixing $x_c = \tilde{x}_c$ solve the EKF where :

* the measurement equation is $z_u(k+1) = h_u[u(k+1), \tilde{x}_c(k+1)] + v_u(k+1)$ (23)
* the state transition equation is $u(k+1) = u(k) + w_u(k)$ (24)

h_u denotes the part of $h(x)$ relative to the measurements z_u , and

$$w_u(k) \sim N[0, \delta_{k\ell} Q_u(k)] \quad \text{with} \quad Q_u(k) = \alpha_u^2 (\Delta t)^2 \text{ diag } [q_u^2(j)] \tag{25}$$

In the above expressions, parameter α_u is determined on an off-line basis so as to minimize the cost function

$$I = \lim_{k\to\infty} \{ [\tilde{x}_s(k) - x_s(k)]^T [\tilde{x}_s(k) - x_s(k)] \} \tag{26}$$

Estimation of this first step yields $\tilde{u}$.

Second step. Upon fixing $u = \tilde{u}$ solve the linear static estimation algorithm

$$z_u(k+1) = h_u[\tilde{u}(k+1), x_c(k+1)] + v_u(k+1) \tag{27}$$

$$\tilde{x}_c(k+1) = x_c(k+1) + r_c(k+1) \tag{28}$$

At the end of this second step calculation, one gets $\tilde{\tilde{x}}_c$.

2.5. Hierarchical vs. integrated DSE

There are two main aspects of concern : estimation accuracy and computer core storage and time requirements.

2.5.1. Estimation accuracy

Compared to the ISE scheme, HSE is suboptimal with regard to accuracy. This lies essentially in the loss of information implied by the impossibility of using the observations contained in z_u at the FLE. This causes a decrease in the mean redundancy of FLE as compared to the ISE, especially in the neighbourhoods of the boundary nodes. Generally, however, this loss of information is rather minor; moreover, the EKF algorithm is by essence quite insensitive to (small) redundancy variations. Apart from this loss of information and its rather minor effects, the FLE algorithm behaves as well as the ISE one. Simulations confirm this conclusion.

2.5.2. Computer requirements

As is well known, EKF techniques are extremely demanding with respect to both computer storage and time requirements. This is mainly due to the full character of the implied matrices; of the most demanding among the various computations is certainly their inversion. And quite obviously, decomposing a system into subsystems and hence partitioning the EKF matrices allows substantial computer savings. In what follows we attempt at giving only a first approximative appraisal of the expected gains of the DHSE structure with respect to the DISE.

To begin, we shall concentrate on the FLE, assuming that the entire network is decomposed into s equal-sized subnetworks. Of course, the comparisons depend upon whether the various subsystems are processed *sequentially* by the same computer or *in parallel* by distinct, distributed ones. In what follows we discuss the former - and less advantageous - case.

Computer storage : roughly speaking, this is proportional to the number n^2 of state variables, i.e. to $(2N-1)^2 \simeq (2N)^2$. Decomposing the entire system into s sub-systems leads to dividing the required core storage by the factor

$$\frac{(2N)^2}{s\,(2N/s)^2} = s \tag{29}$$

Computer time : roughly speaking, it is proportional to n^3 . Decomposing the entire system into s subsystems allows dividing the computing time by the factor

$$\frac{(2N)^3}{s\,(2N/s)^3} = s^2 \tag{30}$$

Overall computer savings : their appraisal should account for the additional computational efforts required by the CL scheme. This however is only a few percentage of those required by the FLE of a subsystem; in a first rough approximation we can neglect it and stick to the above approximate evaluations.

3. SIMULATION INVESTIGATIONS AND RESULTS

The investigations reported in this paragraph have essentially sought three objectives and have accordingly involved three distinct series of simulations. The first one has attempted to compare DHSE with DISE and with a static WLS approach under normal operating conditions whereas the two others have concentrated on the exploration of DHSE performances under two different types of stringent operating conditions. All these simulations have been carried out on the same system, viz. the well-known IEEE 30-bus test system. This choice is motivated by the fact that its size is small enough to avoid prohibitive computing burden of the DISE scheme, used here as a benchmark, and yet large enough to allow realistic comparisons.

3.1. System and data description

Fig. 1 and Table I provide the main characteristics of the system along with the adopted measurement configuration. This latter has been randomly chosen upon fixing an overall redundancy $\eta \simeq 2.1$; the number of state variables being 59, this redundancy corresponds to 122 measurements for the ISE scheme. Moreover, a reasonable (realistic) proportion of voltage magnitudes/active-reactive power measurements has been chosen, then the measurement topology has further been readjusted so as to avoid unobservable islands and to obtain an even measurement distribution. As concerning the values of the selected measurements, they have been determined by using a standard load flow program and by adding to the values thus computed, gaussian noises with zero means and standard deviations fixed as follows :

* for power measurements : $\sigma = 1.5$ (0.8) MW/MVAr at 132 kV (33 kV) ;
* for voltage measurements : $\sigma = 0.005$ p.u. ;
* for injection pseudo-measurements : $\sigma = 0.2$ MW/MVAr .

For DHSE purposes, the 30-bus system has been decomposed into two areas. Note that this extreme decomposition case has been chosen to get realistic subsystems' size; this does not distort the coordination level's features and computations. Note also that the decomposition caused a slight decrease of the subsystems' redundancy.

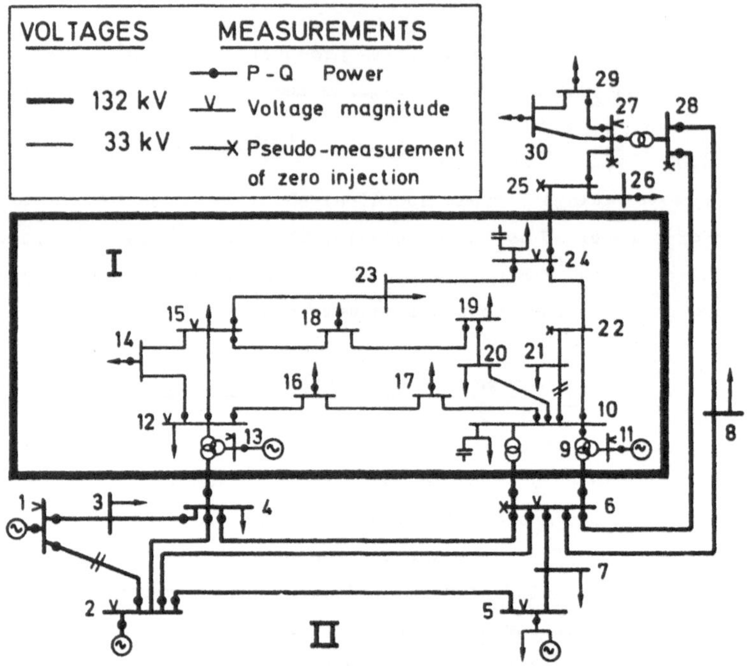

FIGURE 1 - IEEE 30-bus test system

TABLE I. GENERAL CHARACTERISTICS OF THE TEST SYSTEM AND OF THE SUBSYSTEMS

	Number of nodes	Number of lines	Volt. level	Number of bound. nodes
Entire system	30	41	132 kV 33 kV	
Area I	16	19	33 kV	4
Area II	14	18	132 kV 33 kV	3
Inter. area	7	4	132 kV 33 kV	7

	Number of measurements			η
	Voltage magnitudes	Power flows	Power injections	
Entire system	10	76	36	2.1
Area I	5	28	14	1.5
Area II	5	40	18	2.3
Inter. area		8		$\eta_c = 1.6$ $\eta_u = 8.0$

3.2. Preliminary computations. Initial EKF conditions

According to the developments of Section 2, use of an EKF algorithm implies the preliminary (off-line) determination of parameter α for the DISE, and α^1, α^2 and α_u for the DHSE. These have been computed via Monte Carlo simulations [1]. In all these simulations, Δt is set equal to 1 min. Ramp state variations have been considered with slopes $q(j) = 0.015$ p.u./min (voltage magnitudes) and $q(j) = 1.5°$/min (voltage phases). Therefrom, the following values of α^i have been computed [1] :
$\alpha = 6$ for the entire system ; $\alpha^1 = 8$ (for A^1) ; $\alpha^2 = 5$ (for A^2) ; $\alpha_u = 3$ (for the CL) .

On the other hand, the initial conditions adopted when running the EKF algorithm, i.e. conditions at $k=0$ are $x(0) = 1$ p.u./0° and $P(0) = 10^{-2}\,U$ where U is the unit matrix; this particular covariance matrix corresponds to assume that the state variables are uncorrelated and that $\sigma = 10^{-1}$ p.u./rad , i.e. $\sigma = 10^{-1}$ p.u. , 5.73° .

Finally, note that with the above initial conditions and under normal operating conditions, the EKF algorithm reaches a satisfactory accuracy after $k=2$; the results presented in Tables II and III below correspond to the time sample $k=3$.

3.3. Performance criteria

The estimation accuracy is assessed by means of Monte Carlo simulations. The estimated state variables and line power flows (denoted hereafter by $h_j(\hat{x})$) are accordingly assessed on a statistical basis using 50 simulations and involving the following quantities :

* the mean bias (which theoretically must be zero) :

$$\frac{\sum_{p=1}^{50} h_{jp}(\hat{x})}{50} - h_j(x)$$

* the standard deviation with respect to the exact value :

$$\sigma_e(h_j) = \left\{ \frac{\sum_{p=1}^{50} [\, h_{jp}(\hat{x}) - h_j(x)\,]^2}{50} \right\}^{1/2}$$

* the mean standard deviation $\bar{\sigma}_e$ of all quantities of the same type (voltage magnitudes, voltage phases, active/reactive power flows/injections);
* the filtering ratio (mean filtering ratio) for each measured quantity (respectively for all measured quantities of the same type) :

$$f = \frac{\sigma_e}{\sigma} \quad .$$

3.4. First set of simulations [1]

This set attempts to compare estimation results provided under normal operating conditions by the DHSE algorithm, with those provided by the DISE algorithm and also by a static WLS algorithm, also of the "integrated type" (SISE). Table II summarizes the main results of this comparison. One can see that despite the loss of information imposed by the decomposition, the DHSE (FLE) mean standard deviations and filtering ratios are very satisfactory as such, and in a very good agreement with these of DISE and of SISE.

More detailed results relative to the state variables of the interconnection area are reported in Table III. Comparison of the accuracy (DHSE vs. DISE) of the power flow estimates on the tielines is also given. Although these are the less accurately estimated quantities of the DHSE, one can see that their corresponding DHSE standard deviations and filtering ratios are still satisfactory. Observe also that the less accurate among them are related to nodes having the poorest local measurement configuration.

TABLE II : MEAN STANDARD DEVIATION (MEAN FILTERING RATIO IN %)

Area	Voltage magni. (p.u.)			Voltage phase (deg.)			Active power flow (MW)			React.power fl.(MVAr)			Active injection (MW)			React.inject.(MVAr)		
	SISE	DISE	DHSE	SISE	DISE	DHSE	SISE	DISE	DHSE	SISE	DISE	DHSE	SISE	DISE	DHSE	SISE	DISE	DHSE
I	0.0021 (43)	0.0021 (41)	0.0022 (45)	0.146	0.124	0.094	0.507 (72)	0.505 (71)	0.527 (75)	0.587 (76)	0.560 (69)	0.613 (78)	0.819 (46)	0.797 (42)	0.820 (46)	0.917 (65)	0.899 (64)	0.956 (75)
II	0.0024 (38)	0.0022 (38)	0.0030 (48)	0.115	0.114	0.142	0.605 (43)	0.665 (48)	0.712 (51)	0.634 (42)	0.631 (45)	0.713 (51)	0.824 (49)	0.930 (55)	1.075 (66)	1.048 (52)	0.919 (52)	1.098 (63)
CL	0.0020 (40)	0.0019 (40)	0.0023 (44)	0.144	0.108	0.141	0.592 (46)	0.495 (40)	0.864 (68)	0.481 (38)	0.542 (42)	0.805 (62)	0.946	0.988	1.750	0.964	0.992	1.746

TABLE III : ESTIMATING THE INTERCONNECTION AREA.
STANDARD DEVIATIONS UNDER NORMAL OPERATING CONDITIONS. k = 3.

	Voltage magnitude (p.u.) at node # (x 10^{-4})							Voltage phase (degree) at node # (x 10^{-3})							u (degree)
	4	6	9	10	12	24	25	4	6	9	10	12	24	25	
FLE	24	23	22	21	22	25	43	83	84	109	97	121	117	289	
CLE	21	23	20	21	20	23	30	112	136	108	172	130	111	215	0.201

	Active power flow (MW) on line # (filtering ratio in %)				Reactive power flow (MVAr) on line # (filtering ratio in %)			
	4-12	6-9	6-10	24-25	4-12	6-9	6-10	24-25
DISE	0.795 (53)	0.435 (29)	0.247 (16)	0.499 (62)	0.771 (51)	0.579 (39)	0.300 (20)	0.453 (57)
DHSE (CLE)	1.049 (70)	1.155 (77)	0.554 (37)	0.696 (87)	0.935 (62)	1.047 (70)	0.471 (31)	0.689 (86)

3.5. Second set of simulations

This series of simulations intends to explore the DHSE accuracy under stringent operating conditions, namely sudden and significant variations simultaneously imposed on the state variables (voltage magnitudes and phases) of four nodes (two in each area). The severity is increased by choosing two of them to be boundary nodes. Figs. 2 to 5 sketch the shape of the time variations of the state variables of these four nodes : in solid lines are the actual variations and in dotted lines the estimated ones . The actual variations at all four nodes have been taken as follows :

* for voltage magnitudes : 0.015 p.u./min ;
* for phase angles : 0.5°/min in A^1 , 1°/min in A^2 .

Note that the variations of voltage magnitudes are the largest admissible ones with regard to the q(j) values (see § 3.2). Despite the severity of these conditions, figs. 2 to 5 show that there is a very good agreement between actual and estimated time variations, apart from the first two time samples (k = 1,2).

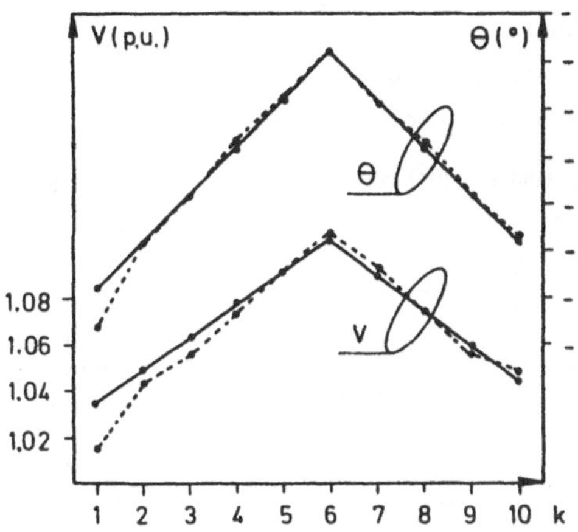

FIGURE 2

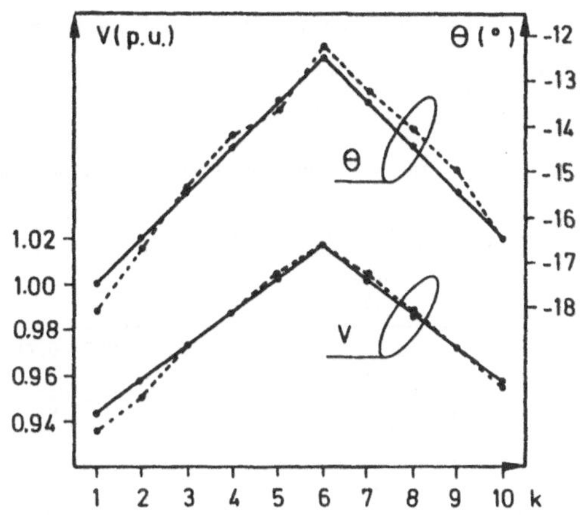

FIGURE 3

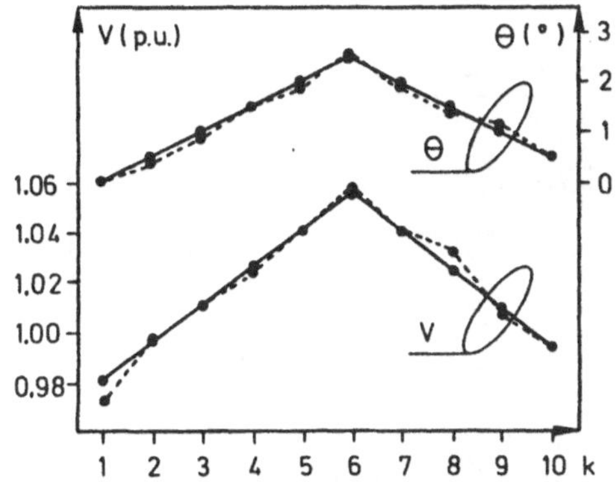

FIGURE 4

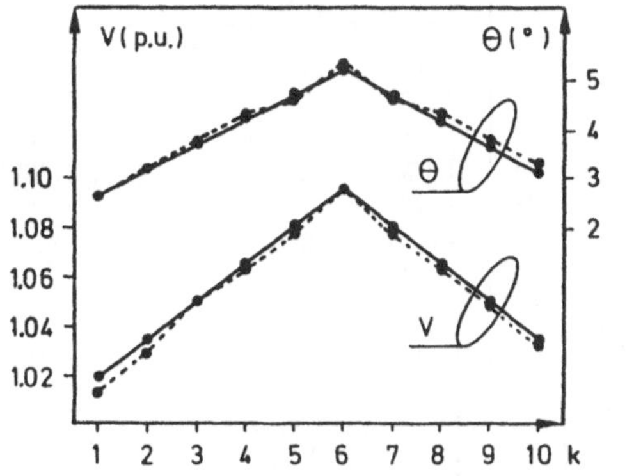

FIGURE 5

Table IV reproduces some important estimation results provided by DHSE and allows comparisons with their corresponding DISE, for three different time samples. As under normal operating conditions, the DHSE mean standard deviations (filtering ratios) are still in a very good agreement with the DISE ones. Table V provides additional results relative to the interconnection area for the same time samples. Again, comparisons are made with the DISE corresponding ones.

3.6. Third set of simulations

This set aims at showing the behaviour of the dynamic hierarchical state estimator under variations of active and reactive power injections at different nodes of the system. The nodes of concern and the variations imposed are reported in Table VI. Note that the disturbances of the state variables induced by these loads changes with time lie within the maximum rate of change allowed by the simulation conditions : 0.015 p.u./min for voltage magnitudes and 1.5° for voltage phases. Table VI gives the estimated and the actual values of the power injections at the three nodes. Their close agreement is obvious.

Table VII shows the mean accuracy realized in each area in terms of mean standard deviations and mean filtering ratios. Results relative to the interconnection area of the FLE and CLE algorithms are given in Table VIII.

TABLE IV : MEAN STANDARD DEVIATION (MEAN FILTERING RATIO IN %). DISTURBED STATE of § 3.5

Area	k	Voltage magn. (p.u.)		Voltage phase (de)		Active power flow (MW)		React. power flow (MVAr)		Act.inject. (MW)		React.inject. (MVAr)	
		DISE	DHSE	DISE	DHSE	DISE	DHSE	DISE	DHSE	DISE	DHSE	DISE	DHSE
I	3	0.0016 (32)	0.0022 (45)	0.135	0.093	0.481 (68)	0.534 (76)	0.554 (70)	0.615 (77)	0.799 (58)	0.828 (52)	0.875 (88)	0.960 (75)
	6	0.0022 (43)	0.0025 (50)	0.101	0.090	0.470 (68)	0.542 (78)	0.607 (75)	0.597 (72)	0.796 (53)	0.853 (57)	0.976 (93)	0.916 (71)
	7	0.0021 (42)	0.0023 (46)	0.171	0.088	0.502 (70)	0.536 (75)	0.573 (71)	0.620 (75)	0.839 (66)	0.808 (50)	0.847 (87)	0.961 (79)
II	3	0.0024 (42)	0.0030 (48)	0.115	0.142	0.603 (44)	0.757 (56)	0.619 (44)	0.728 (52)	0.945 (100)	1.176 (71)	0.902 (84)	1.130 (66)
	6	0.0019 (31)	0.0029 (48)	0.088	0.134	0.616 (46)	0.774 (54)	0.592 (42)	0.687 (49)	0.879 (79)	1.254 (69)	0.883 (71)	1.073 (62)
	7	0.0030 (51)	0.0031 (51)	0.122	0.136	0.542 (37)	0.705 (51)	0.598 (45)	0.708 (50)	0.910 (97)	1.127 (64)	0.791 (79)	1.115 (67)
INTERCONNECTION	3	0.0016 (29)	0.0023 (43)	0.116	0.143	0.665 (54)	0.908 (70)	0.478 (40)	0.798 (62)	1.323	1.931	0.901	1.769
	6	0.0017 (37)	0.0023 (45)	0.079	0.138	0.494 (39)	0.998 (77)	0.428 (33)	0.779 (61)	1.058	2.028	1.186	1.881
	7	0.0022 (41)	0.0022 (41)	0.134	0.131	0.703 (54)	0.950 (72)	0.508 (42)	0.826 (63)	1.132	1.918	0.913	1.814

TABLE V : ESTIMATING THE INTERCONNECTION AREA. STANDARD DEVIATIONS DURING DISTURBANCES

k	LEVEL	Voltage magnitude (p.u.) at node # (x 10^{-4})							Voltage phase (degree) at node # (x 10^{-3})							u (degree)
		4	6	9	10	12	24	25	4	6	9	10	12	24	25	
3	FLE	24	23	22	21	22	25	43	83	83	108	98	121	116	290	
	CLE	22	22	20	21	20	23	30	105	135	108	187	144	109	214	0.198
6	FLE	20	22	22	24	26	23	46	50	67	94	87	115	119	315	
	CLE	21	21	22	22	22	24	30	98	125	90	220	113	121	196	0.137
7	FLE	22	23	21	22	22	25	41	61	74	95	88	107	128	302	
	CLE	21	20	19	24	18	23	26	97	135	93	170	105	124	191	0.155

k		Active power flow (MW) on line # (filtering ratio in %)				Reactive power flow (MVAr) on line # (filtering ratio in %)			
		4-12	6-9	6-10	24-25	4-12	6-9	6-10	24-25
3	DISE	0.966 (64)	0.673 (45)	0.365 (24)	0.654 (82)	0.858 (57)	0.498 (33)	0.334 (22)	0.386 (48)
	DHSE	1.102 (73)	1.196 (80)	0.635 (42)	0.695 (87)	0.950 (63)	1.009 (67)	0.446 (30)	0.692 (87)
6	DISE	0.884 (59)	0.459 (31)	0.254 (17)	0.377 (47)	0.775 (52)	0.338 (23)	0.251 (17)	0.321 (40)
	DHSE	1.138 (76)	1.264 (84)	0.868 (58)	0.719 (90)	0.834 (56)	1.040 (69)	0.397 (26)	0.726 (91)
7	DISE	1.301 (87)	0.684 (46)	0.343 (23)	0.483 (60)	0.745 (50)	0.313 (21)	0.227 (15)	0.647 (81)
	DHSE	1.262 (84)	1.216 (81)	0.696 (46)	0.626 (78)	1.125 (75)	0.970 (65)	0.545 (36)	0.598 (75)

TABLE VI : POWER INJECTIONS VARIATIONS

k	Active injection (MW) at node #			Reactive injection (MVAr) at node #			Estimated active injection (MW) at node #			Estimated reactive injection (MVAr) at node #		
	8	1	2	8	1	2	8	1	2	8	1	2
1	0	243.7	3	0	-24.0	32.8	1.3	270.9	-1.9	-1.8	-20.6	53.0
2	0	243.7	3	0	-24.0	32.8	-2.4	245.2	1.5	-0.5	-22.9	34.3
3	0	243.7	3	0	-24.0	32.8	-0.4	245.1	4.2	1.0	-24.3	32.4
4	-20	255.6	13	-10	-23.2	40.5	-20.7	254.5	14.3	-8.1	-24.3	41.0
5	-40	267.8	23	-20	-22.3	49.0	-41.8	266.8	23.2	-19.7	-23.4	50.3
6	-60	280.5	33	-30	-21.2	58.0	-59.8	279.7	36.4	-28.7	-21.4	57.9
7	-80	293.7	43	-40	-19.8	68.2	-80.4	293.2	45.5	-39.7	-18.9	69.1
8	-60	280.5	33	-30	-21.2	58.0	-59.4	280.7	32.2	-30.9	-19.1	59.6
9	-40	267.8	23	-20	-22.3	49.0	-40.2	266.1	23.6	-20.1	-22.3	49.4
10	-20	255.6	13	-10	-23.2	40.5	-20.4	254.7	14.0	-11.6	-21.9	39.3
11	0	243.7	3	0	-24.0	32.8	1.3	242.8	2.5	0.6	-24.3	33.9

TABLE VII : MEAN STANDARD DEVIATION (MEAN FILTERING RATIO IN %). DISTURBED STATE of § 3.6

k	Voltage magnitude (p.u.)			Voltage phase (degree)			Active power flow (MW)			Reactive power flow (MVAr)			Active injection (MW)			Reactive injection (MVAr)		
	I	II	INT.	I	II	INT.	I	II	INT.	I	II	INT.	I	II	INT.	I	II	INT.
3	0.0026 (50)	0.0030 (48)	0.0023 (45)	0.102	0.108	0.161	0.518 (74)	0.656 (47)	0.826 (65)	0.604 (73)	0.666 (49)	0.850 (66)	0.770 (45)	1.076 (63)	1.719	0.913 (80)	1.051 (65)	1.682
4	0.0027 (51)	0.0030 (52)	0.0020 (46)	0.088	0.141	0.128	0.530 (76)	0.684 (50)	0.724 (57)	0.617 (78)	0.656 (47)	0.868 (67)	0.800 (44)	1.075 (66)	1.608	0.950 (73)	1.059 (66)	1.880
5	0.0024 (47)	0.0031 (55)	0.0022 (41)	0.102	0.140	0.194	0.526 (75)	0.664 (49)	0.835 (64)	0.590 (75)	0.692 (50)	0.733 (56)	0.795 (47)	1.038 (65)	1.612	0.930 (70)	1.113 (65)	1.740
6	0.0024 (46)	0.0028 (47)	0.0023 (42)	0.092	0.143	0.163	0.508 (73)	0.723 (53)	0.747 (58)	0.583 (74)	0.645 (48)	0.816 (60)	0.804 (46)	1.136 (70)	1.618	0.931 (68)	1.020 (60)	1.818
7	0.0026 (49)	0.0030 (50)	0.0022 (42)	0.103	0.119	0.159	0.528 (76)	0.643 (46)	0.796 (62)	0.583 (74)	0.669 (49)	0.761 (60)	0.812 (46)	1.061 (65)	1.725	0.909 (74)	1.055 (63)	1.631
8	0.0026 (51)	0.0028 (44)	0.0023 (42)	0.096	0.125	0.146	0.519 (75)	0.681 (49)	0.789 (61)	0.607 (76)	0.709 (51)	0.831 (63)	0.818 (45)	1.078 (65)	1.658	0.897 (74)	1.117 (66)	1.677
9	0.0026 (50)	0.0029 (48)	0.0022 (40)	0.098	0.131	0.164	0.541 (77)	0.662 (48)	0.703 (57)	0.633 (80)	0.665 (48)	0.775 (60)	0.812 (44)	1.033 (61)	1.590	0.981 (78)	1.116 (69)	1.730
10	0.0026 (47)	0.0029 (48)	0.0022 (41)	0.098	0.131	0.154	0.527 (77)	0.704 (52)	0.748 (59)	0.588 (74)	0.681 (48)	0.806 (61)	0.830 (51)	1.104 (65)	1.628	0.920 (70)	1.109 (64)	1.798
11	0.0027 (51)	0.0028 (45)	0.0022 (42)	0.087	0.125	0.151	0.507 (73)	0.670 (48)	0.829 (67)	0.599 (75)	0.671 (47)	0.920 (71)	0.805 (46)	1.041 (64)	1.638	0.896 (73)	1.127 (67)	1.863

TABLE VIII : ESTIMATING THE INTERCONNECTION AREA. STANDARD DEVIATIONS DURING DISTURBANCES.

k		Voltage magnitude (p.u.) at node # ($\times 10^{-4}$)							Voltage phase (degree) at node # ($\times 10^{-3}$)							u (deg.)
		4	6	9	10	12	24	25	4	6	9	10	12	24	25	
3	FLE	25	24	25	25	24	30	38	62	60	104	107	136	128	195	
	CLE	22	21	23	24	20	26	30	62	61	96	108	116	127	182	0.172
4	FLE	26	26	29	27	22	31	37	75	75	111	94	95	141	262	
	CLE	22	23	23	27	20	26	24	74	81	126	88	100	129	209	0.162
5	FLE	26	26	23	22	24	25	38	92	97	116	110	116	138	223	
	CLE	23	22	19	20	18	22	30	90	96	113	116	115	130	209	0.210
6	FLE	23	23	24	24	21	25	40	61	72	90	97	116	121	260	
	CLE	21	21	21	24	22	21	31	62	79	97	100	111	123	180	0.174
7	FLE	24	25	25	26	23	26	36	57	60	129	110	126	139	222	
	CLE	21	20	21	25	19	24	26	57	66	125	116	118	142	195	0.167
8	FLE	20	21	27	26	23	27	37	67	72	89	80	139	114	222	
	CLE	20	20	21	25	21	23	32	68	75	93	83	126	103	152	0.149
9	FLE	21	22	25	26	24	28	37	73	73	104	92	124	137	224	
	CLE	21	18	21	25	18	25	26	76	75	100	92	112	133	178	0.172
10	FLE	22	21	26	27	20	28	40	68	72	112	97	114	126	230	
	CLE	19	21	20	25	18	23	30	68	79	121	93	118	125	178	0.150
11	FLE	22	23	28	28	24	27	33	64	72	89	81	109	120	245	
	CLE	18	18	24	28	21	25	25	66	74	87	81	107	114	201	0.149

k		Active power flow (MW) on line # (filtering ratio in %)				Reactive power flow (MVAr) on line # (filtering ratio in %)			
		4-12	6-9	6-10	24-25	4-12	6-9	6-10	24-25
3		1.076 (72)	1.122 (75)	0.416 (28)	0.689 (86)	1.008 (67)	1.155 (77)	0.484 (32)	0.710 (89)
4		0.948 (63)	0.960 (64)	0.355 (24)	0.632 (79)	0.920 (61)	1.181 (79)	0.565 (38)	0.745 (93)
5		0.981 (65)	1.302 (87)	0.450 (30)	0.608 (76)	0.820 (55)	0.991 (66)	0.382 (25)	0.643 (80)
6		0.960 (64)	1.021 (68)	0.395 (26)	0.612 (76)	1.026 (68)	1.124 (75)	0.491 (33)	0.521 (65)
7		0.949 (63)	1.190 (79)	0.401 (27)	0.644 (80)	0.868 (58)	1.008 (67)	0.496 (33)	0.638 (80)
8		1.038 (69)	1.063 (71)	0.427 (28)	0.630 (79)	1.088 (73)	1.071 (71)	0.424 (28)	0.648 (81)
9		0.849 (57)	0.898 (60)	0.386 (26)	0.677 (85)	0.897 (60)	0.990 (66)	0.490 (33)	0.655 (82)
10		0.816 (54)	1.111 (74)	0.395 (26)	0.670 (84)	0.889 (59)	1.186 (79)	0.504 (34)	0.599 (75)
11		0.916 (61)	1.104 (74)	0.445 (30)	0.690 (86)	1.041 (69)	1.336 (89)	0.581 (39)	0.706 (88)

4. CONCLUSIONS

A dynamic hierarchical state estimator has been investigated and its reliability assessed under various operating conditions. It was found quite accurate and appropriate for on-line purposes, even in the case of extremely stringent situations. Admittedly, many questions remain still open and further research appears necessary to provide it with the interesting properties and possibilities generally expected from a dynamic estimator. Nevertheless, the results obtained by now are very encouraging and promising.

5. REFERENCES

[1] P. ROUSSEAUX, T. VAN CUTSEM, M. RIBBENS-PAVELLA, "Dynamic State Estimator for large electric power systems", 9th IEEE World Congress, Budapest, Hungary, July 2-6, 1984.

[2] A. DEBS, R. LARSON, "A dynamic estimator for tracking the state of a power system", *IEEE PAS-89*, No.7, September/October 1970.

[3] T. VAN CUTSEM, J.L. HORWARD, M. RIBBENS-PAVELLA, "A two-level static state estimator for electric power systems", *IEEE PAS-100*, No.8, August 1981.

[4] P. ROUSSEAUX, T. VAN CUTSEM, M. RIBBENS-PAVELLA, "Multi-level dynamic state estimation for electric power systems", 8th PSCC, Helsinki, Finland, 19-24 August, 1984.

A METHOD FOR REAL TIME HIERARCHICAL STATE ESTIMATION OF LINEAR DYNAMIC SYSTEMS

G.R. Sotirov
Department of Automation of Production
Higher Institute of Mechanical
and Electrical Engineering
Sofia, 1156/Bulgaria

INTRODUCTION

Kalman's filter is a basic method for data processing in estimating the dynamic systems state. When the systems are of large dimension, in practical elaboration of the well known estimation algorithms, based on Kalman's filter and its non-linear modifications, some computational difficulties arise (Aoki, 1967).

The paper proposes an estimation method, applied in case, when the dimension of the linear dynamic system state vector is large. For this purpose, decentralized estimations of the states of the linear dynamic systems, forming interconected subsystems, are defined.

PROBLEM DEFINITION

We consider a linear dynamic system, which consists of interconected subsystems, where each of them is described by the equations

$$x_i(t+1)=\Phi_i(t)x_i(t) + \sum_{j\neq i}^{N} \Phi_{ij}(t)x_j(t) + v_i(t); \tag{1}$$

$$y_i(t)=C_i(t)x_i(t) + \sum_{j\neq i}^{N} C_{ij}(t)x_j(t) + e_i(t), \tag{2}$$

where: $x_i(t)$ is the state vector of the ith-subsystem of n_i-dimension; $\sum_{i=1}^{N} n_i = n$; $y_i(t)$ - m_i-vector; $\sum_{i=1}^{N} m_i = m$; $\Phi_i(t)$, $\Phi_{ij}(t)$, $C_i(t)$, $C_{ij}(t)$ - matrixes; $\{v_i(t)\}$, $\{e_i(t)\}$ - noise in the object and in the observation. We suppose, that the noise $\{v_i(t)\}$ and $\{e_i(t)\}$ are white Gaussian random sequences, with :

$$E\{v_i(t)v_i^T(t)\}=Q_i \quad , \quad E\{e_i(t)e_i^T(t)\}=R_i \quad ;$$

$$Q_i(t)\geqslant 0; \; R_i(t)\geqslant 0, \; i=\overline{1,N}.$$

The initial vector $x_i(0)$, $i=\overline{1,N}$ is a Gaussian one with parameters :

$$E\{x_i(0)\}=0 \; ; \; E\{x_i(0)x_i^T(0)\}=P_i(0) \; .$$

The noise in the object and in the measurements, as well as the initial states, are not correlated for the different subsystems and besides

$$E\{v_i(t)e_i^T(t)\} = E\{v_i(t)x_i^T(0)\} = E\{e_i(t)x_i^T(0)\} = 0 ,$$

for all $i=\overline{1,N}$, $t=0,1,2,\ldots$

Having in mind these conditions, it is necessary to determine an optimal estimation for the system state $x(t)=[x_1(t),x_2(t),\ldots,x_N(t)]^T$ at the t moment, using the sequence of measurements $y_i(1),y_i(2),\ldots,y_i(t)$, $i=\overline{1,N}$ of the subsystems states.

Unlike (Leondes, 1976) we propose to determine decentralized estimations of the subsystems states, using measurements $y_i(1),\ldots,y_i(t)$ of the subsystem under estimation. Estimating the state of a given subsystem no statistic information for the remaining subsystems is used.

FILTER STRUCTURE

The structure of the filter for the ith subsystem is determined as follows :

$$\hat{x}_i(t+1) = F_i(t+1)\hat{x}_i(t) + K_i(t+1)y_i(t+1) + a_i(t) , \qquad (3)$$

where : $F_i(t),K_i(t)$ are matrixes, that have to be determined ; $a_i(t)$ - a vector, counting the shift of the decentralized estimation (3) due to the systems interconected. It comes from (1),(2) and (3), that estimation (3) won't be shifted when

$$F_i(t+1) = [I - K_i(t+1)C_i(t+1)]\Phi_i(t) , \quad E\{x_i(0)\}= E\{\hat{x}_i(0)\},$$

$$a_i(t) = [I - K_i(t+1)C_i(t+1)] \sum_{j\neq i}^{N} \Phi_{ij}(t)\hat{x}_j(t)$$

$$- K_i(t+1)\sum_{j\neq i}^{N} C_{ij}(t+1)\hat{x}_j(t+1/t) ,$$

where :

$$\hat{x}_j(t+1/t) = \Phi_j(t)\hat{x}_j(t) + \sum_{r\neq j}^{N} \Phi_{jr}(t)\hat{x}_r(t) . \qquad (4)$$

In that case, the nonshifted filter will be

$$\hat{x}_i(t+1) = \hat{x}_i(t+1) + K_i(t+1)[y_i(t+1) - C_i(t+1)\hat{x}_i(t+1/t)$$

$$- \sum_{j\neq i}^{N} C_{ij}(t+1)\hat{x}_j(t+1/t)] , \qquad (5)$$

and the equation for the error $\varepsilon_i(t) = x_i(t) - x_i(t)$ of the nonshifted estimation of the ith subsystem state is noted down as

$$\varepsilon_i(t+1) = [I - K_i(t+1)C_i(t+1)]\Phi_i(t)\ \varepsilon_i(t) + [I - K_i(t+1)X$$

$$XC_i(t+1)]\,d_i(t) - K_i(t+1)d_i(t+1/t) + [I - K_i(t+1)X$$

$$XC_i(t+1)\big]v_i(t) - K_i(t+1)e_i(t) ,$$

where :

$$d_i(t) = \sum_{j\neq i} \Phi_{ij}(t)\varepsilon_j(t) ;$$

$$d_i(t+1/t) = \sum_{j\neq i} C_{ij}(t+1)\varepsilon_j(t+1/t) ;$$

$$\varepsilon_j(t+1/t) = x_j(t+1) - \hat{x}_j(t+1/t) .$$

DETERMINATION OF THE GAIN

The gain matrixes of the K_i, $i=\overline{1,N}$ local filters are determined according to the condition for the minimum of the decentralized estimation (4) summary dispersion

$$D = \sum_{i=1}^{N} D_i = \sum_{i=1}^{N} tr\,\{P_i(T)\} , \tag{6}$$

where $P_i(T)$ is the covariance matrix of the estimation error of the ith subsystem.
The covariances P_i, $i=\overline{1,N}$ are determined from the equation :

$$\begin{aligned} P_i(t+1) = {} & \big[I - K_i(t+1)C_i(t+1)\big]P_i(t+1/t)\big[I - K_i(t+1)C_i(t+1)\big]^T \\ & - \big[I - K_i(t+1)C_i(t+1)\big]\sum_{j\neq i} P_{ij}(t+1/t)C_{ij}^T(t+1)K_i(t+1) \\ & - K_i(t+1)\sum_{j\neq i} C_{ij}(t+1)P_{ji}(t+1/t)\big[I - K_i(t+1)C_i(t+1)\big]^T \\ & + K_i(t+1)\sum_{j\neq i}\sum_{r\neq i} C_{ij}(t+1)P_{jr}(t+1/t)C_{ir}^T(t+1)K_i^T(t+1) \\ & + K_i(t+1)R_i(t+1)K_i^T(t+1) , \end{aligned} \tag{7}$$

where :

$$P_{ij}(t) = E\{\varepsilon_i(t)\,\varepsilon_i^T(t)\} ,\ P_{ii}=P_i ;$$

$$P_{ij}(t+1/t) = \Phi_i(t)P_{ij}(t)\Phi_j^T(t) + \overline{P}_{ij}(t) + Q_{ij}(t) , \tag{8}$$

$$Q_{ij} = 0 , \text{ when } i \neq j,\ i,j=\overline{1,N} ;$$

$$\begin{aligned} \overline{P}_{ij}(t) = {} & \sum_{r\neq j} \Phi_i(t)P_{ir}(t)\,\Phi_{jr}^T(t) + \sum_{r\neq i} \Phi_{ir}(t)X \\ & XP_{rj}(t)\,\Phi_j^T(t) + \sum_{s\neq i}\sum_{r\neq j} \Phi_{is}(t)P_{sr}(t)\,\Phi_{jr}^T(t),\ i\neq j. \end{aligned} \tag{9}$$

The conditions for necessity and sufficiency, on which the estimation (4), minimizing the dispersion (6), coincides with Kalman's estimation of the state vector of the system as a whole, are determined from the following

<u>Theorem</u>: For the system, described by the equations (1) and (2), let the conditions $C_{ij}=0$, $P_{ij}(0)=Q(t)=0$, $R_{ij}(t)=0$ when $i \neq j$, $i,j=\overline{1,N}$ be fulfilled and K_i be determined from the condition for the dispersion minimum (6). Then the linear decentralized estimation (4) will be a Kalman one, when $P_{ij}(t) = 0$, on $i \neq j$, $i,j=\overline{1,N}$, or

$$\sum_{s=1}^{N} \Phi_{is}(t)\,\Phi_{js}^{T}(t) = 0,$$

where : $\Phi_{ii} = \Phi_i$, $i,j=\overline{1,N}$, $t=0,1,2,\ldots$

Using, in equation (7), $P_{ij}(t)=0$ and $P_{ij}(t+1/t)=0$, for $i \neq j$, $i,j=\overline{1,N}$ we get :

$$P_i(t+1) = \left[I - K_i(t+1)C_i(t+1)\right]P_i(t+1/t)\left[I - K_i(t+1)\times\right.$$
$$\left.\times C_i(t+1)\right]^{T} + K_i(t+1)W_i(t+1/t)K_i^{T}(t+1)$$
$$+ K_i(t+1)R_i(t+1)K_i^{T}(t+1)\,, \tag{10}$$

where :

$$W_i(t+1/t) = \sum_{j\neq i}^{N} C_{ij}(t+1)P_j(t+1/t)C_{ij}^{T}(t+1)\,. \tag{11}$$

The matrixes $P_i(t+1/t)$ are determined from equation (9) for $i=j$, where :

$$P_i(t) = \sum_{j\neq i}^{N} \Phi_{ij}(t)P_j(t)\,\Phi_{ij}^{T}(t)\,. \tag{12}$$

So, the determination of the optimal coefficients K_i requires the solution of N interconnected optimizing problems, minimizing for K_i the local functional $tr\{P_i(t)\}$, when dynamic constraints (10) and interconnection constraints (11) are available. In general, the obtained double-level hierarchical optimization problem is solved, using one of the known coordination principles. It may be proved, however, that in our case, the coordinated solution of the optimization problems coincides with the uncoordinated one and the optimal gain of the local filter is determined from the equation :

$$K_i(t+1) = P_i(t+1/t)C_i^{T}(t+1)\left[C_i(t+1)P_i(t+1/t)C_i(t+1)\right.$$
$$\left.+ W_i(t+1/t) + R_i(t+1)\right]^{-1}\,. \tag{13}$$

The equations (4), (5), (8), (10), (11) and (12) with initial conditions $E\{x_i(0)\} = \hat{x}_i(0)$, $E\{\varepsilon_i(0)\,\varepsilon_i^{T}(0)\} = P_i(0)$, totally determine

the local decentralized filter.
We want to mention, that filter (5) with gain (13) will be an optimal Kalman's filter for subsystem (1), (2) model expressed as :

$$\tilde{x}_i(t+1) = \Phi_i(t)\tilde{x}_i(t) + \hat{d}_i(t) + \tilde{v}_i(t) , \tag{14}$$

$$\tilde{y}_i(t+1) = C_i(t+1)\tilde{x}_i(t+1) + \hat{d}_i(t+1/t) + \tilde{e}_i(t+1) , \tag{15}$$

where $\tilde{v}_i(t)$ and $\tilde{e}_i(t)$ are white Gaussian noise with covariances

$$E\{\tilde{v}_i(t)\tilde{v}_i^T(t)\} = Q_i(t) + \bar{P}_i(t) = \tilde{Q}_i(t) ,$$

$$E\{\tilde{e}_i(t)\tilde{e}_i^T(t)\} = R_i(t) + W_i(t/t-1) = \tilde{W}_i(t/t-1) .$$

$\hat{d}_i(t)$, $\hat{d}_i(t+1/t)$ - interaction estimations :

$$\hat{d}_i(t) = \sum_{j\neq i}^{N} \Phi_{ij}(t)\hat{\tilde{x}}_i(t) ,$$

$$\hat{d}_i(t+1/t) = \sum_{j\neq i}^{N} C_{ij}(t+1)\hat{\tilde{x}}_j(t) .$$

$\hat{\tilde{x}}_i(t)$ and $\hat{\tilde{x}}_i(t+1/t)$ are the estimations of the vectors, obtained from equations (4) and (5) with coefficient (13).

ESTIMATION OF THE REAL COVARIANCES

We write equation (10) as follows

$$P_i(t+1) = [C_i^T(t+1)W_i^{-1}(t+1/t)C_i(t+1) + P_i^{-1}(t+1/t)]^{-1} .$$

Then (5) becomes the following

$$\begin{aligned}\hat{x}_i(t+1) = {} & P_i(t+1)\{C_i^T(t+1)\tilde{W}_i^{-1}(t+1/t)y_i(t+1) \\ & - C_i^T(t+1)\tilde{W}_i^{-1}(t+1/t)\hat{d}_i(t+1/t) + P_i^{-1}(t+1/t)X \\ & X[\Phi_i(t)\hat{\tilde{x}}_i(t) + \hat{d}_i(t)]\} .\end{aligned}$$

Therefore, the error is determined from equation

$$\begin{aligned}x_i(t+1) - \hat{\tilde{x}}_i(t+1) = {} & P_i(t+1)P_i^{-1}(t+1/t)v_i(t) + P_i(t+1)X \\ & XP_i^{-1}(t+1/t)\,\Phi_i(t)[x_i(t) - \hat{\tilde{x}}_i(t)] + P_i(t+1)X \\ & XP_i^{-1}(t+1/t) \sum_{j\neq i} \Phi_{ij}(t)\,\varepsilon_j(t) - P_i(t+1)C_i^T(t+1)X \\ & X\tilde{W}_i^{-1}(t+1/t) \sum_{j\neq i} C_{ij}(t+1)\varepsilon_j(t+1/t) \\ & - P_i(t+1)C_i^T(t+1)\tilde{W}_i^{-1}(t+1/t)e_i(t+1) .\end{aligned} \tag{16}$$

From (16) we define the real covariance matrix

$$P_i^*(t+1) = P_i(t+1)P_i^{-1}(t+1/t)P_i^*(t+1/t)P_j^{-1}(t+1/t)P_i(t+1) -$$
$$- P_i(t+1)P_i^{-1}(t+1/t)\sum_{j\neq i} P_{ij}^*(t+1/t)C_{ij}^T(t+1)X$$
$$X\tilde{W}_i^{-1}(t+1/t)C_i(t+1)P_i(t+1) - P_i(t+1)C_i^T(t+1)X$$
$$X\tilde{W}_i^{-1}(t+1/t)\sum_{j\neq i} C_{ij}(t+1)P_{ji}^*(t+1/t)P_i^{-1}(t+1/t)X$$
$$XP_i(t+1) + P_i(t+1)C_i^T(t+1)\tilde{W}_i^{-1}(t+1/t)\ X$$
$$X\sum_{j\neq i}\sum_{r\neq i} C_{ij}(t+1)P_{jr}^*(t+1/t)C_{ir}^T(t+1)\tilde{W}_i^{-1}(t+1/t)X$$
$$XC_i(t+1)P_i(t+1) + P_i(t+1)C_i^T(t+1)\tilde{W}_i^{-1}(t+1/t)X$$
$$XR_i(t+1)\tilde{W}_i^{-1}(t+1/t)C_i(t+1)P_i(t+1)\ , \qquad (17)$$

where :

$$P_{ij}^*(t+1/t) = \Phi_i(t)P_{ij}^*(t)\ \Phi_j^T(t) + V_{ij}^*(t) + Q_{ij}(t)\ ; \qquad (18)$$

$$P_i^*(t+1/t) = P_{ij}^*(t+1/t);\ Q_{ij}(t) = 0,\ \text{when } i\neq j;$$

$$V_{ij}^*(t) = \sum_{r\neq j} \Phi_i(t)P_{ir}^*(t)\Phi_{jr}^T(t) + \sum_{r\neq i} \Phi_{ir}(t)X$$
$$XP_{rj}^*(t)\ \Phi_j^T(t) + \sum_{s=i}\sum_{r=j} \Phi_{is}(t)P_{sr}^*(t)\ \Phi_{jr}^T(t), \qquad (19)$$

$$V_i^* = V_{ii}^*,\ i,j=\overline{1,N},\ P_{ij}^*(0) = P_{ij}(0).$$

Theorem 2: If the model (14), (15) is stochastically controlled and observed and $P_{ij}(0)$ are limitted in the sense of

$$\|P_{ij}(0)\| = \max_{i,j} |P_{ij}^{(ij)}(0)| < \beta,\ \beta \geq 0,\ i,j=\overline{1,N}$$

where β is a limit constant, and $P_{ij}^{(ij)}(0)$ is an element of the $P_{ij}(0)$ matrix, then the estimation (5) with a gain (13) is uniformly asymptotically stable in a large sense and $P_i(t)$ remains limitted for $t\to\infty$

CONCLUSIONS

Equations (17), (18) and (19) may be used for comparing the decentralized filter with the optimal filter of Kalman. If several variants of decomposition are allowed for the system, then it follows, that the variant, with the smallest covariances P_{ij}^* for $i\neq j$, should be chosen. These covariances are the quality measure of system decomposing. We will remind, that using the decentralized filter, we cut down the computations and the needed memory.

EXAMPLE

We consider a system, consisting of two-dimensional subsystems :

$$\begin{bmatrix} x_1(t+1) \\ x_2(t+1) \end{bmatrix} = \begin{bmatrix} 0,951 & 0,624 \\ 0,743 & 0,301 \end{bmatrix} \begin{bmatrix} x_1(t) \\ x_2(t) \end{bmatrix} + \begin{bmatrix} 0,052 & 0,203 \\ -0,451 & -0,1 \end{bmatrix} \begin{bmatrix} x_3(t) \\ x_4(t) \end{bmatrix} + \begin{bmatrix} v_1(t) \\ v_2(t) \end{bmatrix}$$

$$y_1(t) = \begin{bmatrix} 0 & 1 \end{bmatrix} \begin{bmatrix} x_1(t) \\ x_2(t) \end{bmatrix} + e_1(t)$$

$$\begin{bmatrix} x_3(t+1) \\ x_4(t+1) \end{bmatrix} = \begin{bmatrix} 0,882 & 0,342 \\ 0,891 & 0,904 \end{bmatrix} \begin{bmatrix} x_3(t) \\ x_4(t) \end{bmatrix} + \begin{bmatrix} 0,534 & -0,062 \\ -0,04 & 0,095 \end{bmatrix} \begin{bmatrix} x_1(t) \\ x_2(t) \end{bmatrix} + \begin{bmatrix} v_3(t) \\ v_4(t) \end{bmatrix}$$

$$y_2(t) = \begin{bmatrix} 1 & 0 \end{bmatrix} \begin{bmatrix} x_3(t) \\ x_4(t) \end{bmatrix} + e_2(t)$$

$$Q_1 = Q_2 = I; \; R_1 = R_2 = 0,5; \; P_1(0) = P_2(0) = I.$$

For this system, we compare the estimation, obtained by means of the decentralized filter with that by the Kalman filter. We compare the values of the real dispersions of the estimation error. The results, obtained, are shown in the table.

est. \ $\hat{x}_i$	$\hat{x}_1$	$\hat{x}_2$	$\hat{x}_3$	$\hat{x}_4$
subopt.est.	2,351	0,412	0,426	3,413
opt.est.	2,121	0,404	0,420	3,095

REFERENCES

Aoki, M.(1967). Optimization of Stochastic Systems, Academic Press, New York.

Leondes, C.T.(Ed.)(1976). Control and Dynamic Systems. Academic Press, New York.

DECENTRALIZED CONTROLS BASED ON ENERGETIC DECOMPOSITION

Z. Jacyno
Department of Physics
University of Quebec at Montreal
Montreal, Canada, H3C 3P8

INTRODUCTION

Decomposition of complex control systems into simpler subsystems can be done in many ways. Physical structure of a system, formed of distinctive machinery, may by itself indicate on possible divisions. Flow of materials and processing stages have also to be considered. Convenience of mathematical treatment of certain types of subsystems may be an another factor. One or another approach usually offer some advantages. They all carry though a degree of arbitration. Each choice bears subjectivity marks and design becomes an art, strongly relying on acquired experience. Furthermore, there is no assurance whether particular decouplings were best available. All this, because techniques used do not rely consciously on internal mechanisms governing dynamical systems.

Changes of state, no matter what actual physical form they take, require the use of force, power or energy of some kind, supplied by system, by outer controlling action or both. State of system and resulting observable output signals, are of the outmost importance in design and operation, but they are merely display of occuring "power game". The energetic approach to system dynamics stresses fact, that energetic relations are causes and state of system is an effect, translating these relationships into signals, through a particular configuration and a chosen set of structural parameters. Ability of system to accumulate energy will be translated into a tendency towards oscillations. Systems with grater capacity for energy exchange will be characterized by sharp variations of state. Energy absorption from outside during dynamical process, produces instability and dissipation of energy stabilizes the system. In the same way, control action may tend toward increase or decrease of inner energy, affecting correspondingly state of the system.

Such energetic approach seems attractive for more than one reason. It evedently enhances our understanding of systems dynamics. Aside philosophical aspect, it provides a simple tool to evaluate transients and steady state, using a scalar entity and its time derivatives. This global characterization presents simultaneously advantage and hindrance: simplicity of energetic representation shifts focus from individual state components to a total state. To recover lost informations on components, energetic decomposition according to subsystems of interest must be introduced. Thus, instead of studying system as a whole, energetic relations within subsystems are analyzed.

In spite of possible advantages, the energetic method is hampered by its main foe: energy notion, seemingly so obvious in a concept, is very illusive in practice, especially for high order, more complex systems. Some effort has been made for general linear systems, given in state space form, but nonlinear systems remain still almost untouched.

ENERGETIC DECOMPOSITION

Linear, constant-coefficient plant in state space

$$\dot{X} = AX + MU, \quad X \in R^n, U \in R^m, m \leq n, \tag{1}$$

may be transfered into general phase space by applying an appropriate linear operator P, to yield a new state vector

$$Y = PX \tag{2}$$

and a new state equation

$$\dot{Y} = BY + PMU. \tag{3}$$

Choice of P is made to assure the decomposition of Y into two subvectors, [1],

$$Y = [Y_1^T \; Y_2^T]^T, \tag{4}$$

with properties of phase space coordinates

$$\dot{Y}_1 = CY_2. \tag{5}$$

As a result, initial system (3) can be replaced by a second-order representation with respect of either Y_1 or Y_2, or it may be put into a parametric form, [2],

$$(CY_2)^T C dY_2 + Y_1^T (CG)^T dY_1 = -(CHY_2)^T dY_1 + (CV_2)^T dY_1 - V_1^T C dY_2, \tag{6}$$

where

$$PMU = [V_1^T \; V_2^T]^T. \tag{7}$$

The latter provides energetic decomposition of the plant into internal energy (left-hand side) and energy exchanged with the environment through dissipation/absorption and controls acting on static and velocity modes of the plant, Fig. 1.

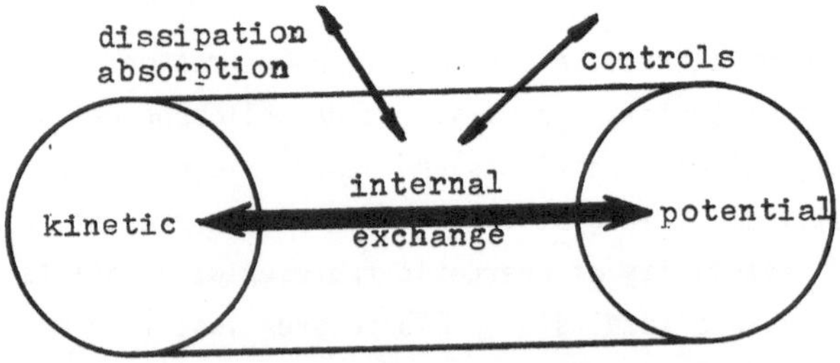

Fig. 1. Global energetic decomposition of linear plant.

In adiabatic state, with all outside energetic ties severed

$$(CY_2)^T C dY_2 + Y_1^T (CG)^T dY_1 = 0, \qquad (8)$$

parametric equation becomes integrable. Indeed, after the diagonalization of CG, (8) clearly represents the law of energy conservation. If, in addition, CG > 0, i.e. is positive definite, the quadratic form (8) describes an oscillating system with a trajectory placed on a closed and convex equienergetic surface

$$E(Y_2, Y_1) = \tfrac{1}{2} Y^T \Phi Y = \text{const.} \qquad (9)$$

The increment of state vector, (3),

$$dY = BYdt \qquad (10)$$

and the gradient to an equienergetic surface

$$DE = F, \qquad (11)$$

where

$$D = \frac{d}{dY},$$

F - generalized force,

form two vector fields, [3]. Green's theorem requires vanishing of the line integral along a closed, piecewise smooth path in such field space

$$\oint F^T dY = 0. \qquad (12)$$

Condition (12) is equivalent to

$$D(DE) = \Phi, \qquad (13)$$

with

$$\Phi = \Phi^T \qquad (14)$$

and is obviously fulfilled for (8) under the assumption CG > 0. Consequently, phase space Y may be seen as a conservative field and quadratic form (9), as its potential. Taking action integral in space Y

$$Q = \int_{t_1}^{t_2} L(Y)dt \qquad (15)$$

and the Lagrangian

$$L = E_k - E_p, \qquad (16)$$

with E_k and E_p as they appear in (8), principle of the least effort applies to Q. The minimization of Q is achieved, if its first variation vanishes

$$\frac{d}{dt}\frac{\partial Q}{\partial Y_2} - \frac{\partial Q}{\partial Y_1} = 0 \qquad (17)$$

and its second variation remains positive definite

$$\frac{\partial^2 Q}{\partial Y_1^2} > 0, \qquad (18)$$

that is, if

$$\det \Phi \neq 0 \qquad (19)$$

and

$$\Phi > 0, \qquad (20)$$

both conditions beeing satisfied for (8) under assumptions previously made. This leads to the following conclusions: 1) relation (9) represents the Hamiltonian for system (3) in adiabatic state, 2) relations (8-9) define kinetic, potential and total energy of such system. Furthermore, total energy (9) generates the equation for the dynamic system, by means of (17) and (18) and the trajectory of the system is a projection of a given energetic level onto the phase space, Fig. 2.

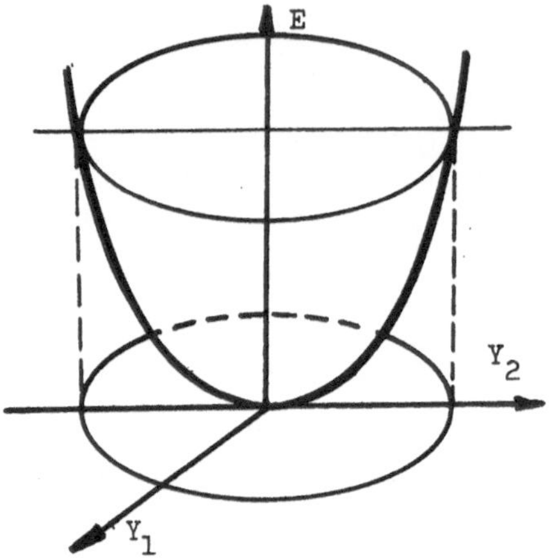

Fig. 2. Trajectory of the system as a projection of its energetic level.

Systems subjected to control actions and dissipating energy to the surroundings will continuously change their energy levels, but still their trajectories may be interpreted as projection of those varying energetic levels.

ELEMENTARY BLOCS

Instead of projecting total energy on the whole phase space as in Fig. 2, it will be more productive for complex plants to consider several projections on some chosen subspaces. In this way set of equations (17-18) will generate a few subsystems of dimensions lower than initial high order system. Such energetic decomposition may be done arbitrarily, taking subsets of Y in a sufficient number to cover whole phase space. However our approach will be to seek the simplest possible elementary blocks, which are still representative of all energetic phenomena occuring within the system. Those are second-order subsystems, eventually complemented with some first-oder ones, needed in case of odd order plants.

Dynamic equation (3) in phase space can be brought to an equivalent second-order representation with respect to one of subvectors of Y, (4), [4],

$$\ddot{Y}_2 + H\dot{Y}_2 + GCY_2 = \dot{V}_2 - GV_1. \qquad (21)$$

In adiabatic state of plant, this equation produces energetic levels (8), serving as a reference for decomposition. Under control action and with energy exchange taking place, energetic relations are given by (6). Consequently (21) provides an another description of plant as generated by projected energy levels, but this time projection is made with respect to phase space subvectors (4). Subvector Y_2 has the dimension

$$p = \dim Y_2 = \begin{cases} \frac{1}{2}n, & \text{if } n \text{ is even,} \\ \frac{1}{2}(n-1), & \text{if } n \text{ is odd,} \end{cases} \tag{22}$$

and (21) is equal to a set of p equations, given in terms of elementary components of Y_2. They represent elementary second-order blocs used for energetic decomposition of complex plant. Each bloc is described by

$$\ddot{y}_i + h_i\dot{y}_i + g_i c_i y_i = v_i + w_i, \tag{23}$$

where

$y_i \in Y_2$, $i = 1, \ldots, p$,

v_i - control action applied to y_i,

w_i - interactions from other subsystems.

As a result, initial high order plant (3) is replaced by a set of elementary blocs as shown in Fig. 3. Parameters h_i and $g_i c_i$ determine energy dissipated and stored within an elementary bloc.

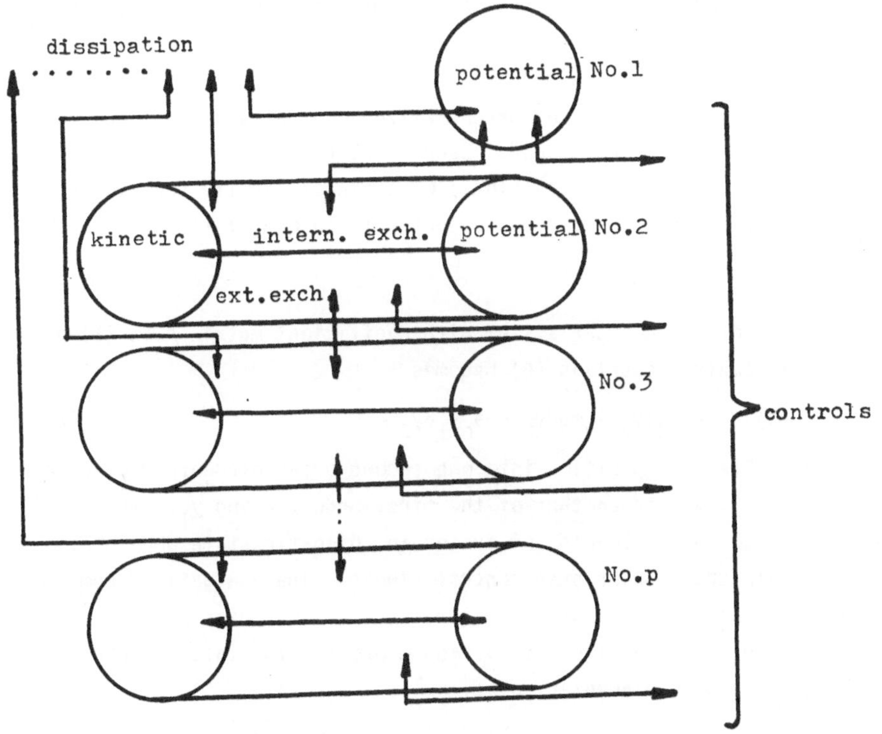

Fig. 3. Decomposition of complex plant into elementary energetic blocs.

DECENTRALIZED CONTROLS

Typical unconstrained control problem for plant (1) in state space consists in minimization of quadratic performance index

$$J = \int_0^{\infty}(X^TSX + U^TRU)dt, \tag{24}$$

with

$$S \geq 0,\ R > 0,$$
$$S = S^T,\ R = R^T.$$

In phase space this problem becames

$$J = \int_0^{\infty}(Y^TP^{-T}SP^{-1}Y + U^TRU)dt. \tag{25}$$

When tracking problem is considered, second term in J is dropped.

All conditions for matrix S in (24) are met, if it is chosen to yield in phase space a corresponding quadratic form given by

$$Y^TP^{-T}SP^{-1}Y = \begin{bmatrix} Y_1 \\ Y_2 \end{bmatrix}^T \begin{bmatrix} (CG)^T & 0 \\ 0 & C^TC \end{bmatrix} \begin{bmatrix} Y_1 \\ Y_2 \end{bmatrix}. \tag{26}$$

Comparison of (26) with (6) reveals, that minimization of performance index (25) requires minimization of internal energy stored in the plant. This conclusion is even more evident for optimal tracking problem. Taking into account the additivity characteristics of performance index, the same requirement holds for any elementary bloc forming plant.

Based on these remarks, a strategy for decentralized controls, as applied to decomposed system in Fig. 3, can be developed. Each local control action must tend to minimize internal energy of elementary bloc. On a hierarchically superior level, the allocation of local controls must be subjected to an overal limits imposed on control vector U.

Let take one example of third-order system with coefficient matrix A in (1) of companion type. Parametric equation (6) becomes

$$\tfrac{1}{2}d(y_3^2 + a_1y_2^2) = -a_2y_3dy_2 + mudy_2 - a_0y_1dy_2\,. \tag{27}$$

It indicates a possible decomposition into one second-order elementary bloc, defined by phase coordinates y_2-y_3 and another of the first-order, along y_1. In Fig. 4 are shown reference equienergetic levels for system in adiabatic state and in Fig. 5 - trajectories in phase space under step control signal. The energetic decomposition (27) and Fig. 5 point out, that parameter a_2 represents dissipating factor in subspace y_2-y_3 and, that coefficients a_0 and m provide for intersubspace and control couplings with this elementary bloc.

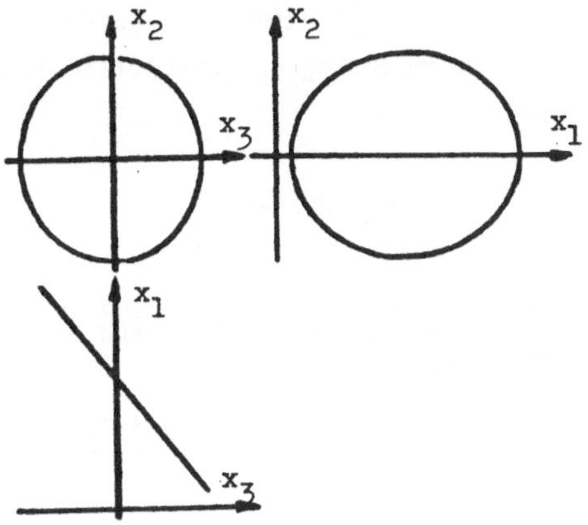

Fig. 4. Equienergetic levels for third-order system in adiabatic state, as simulated on analog computer.

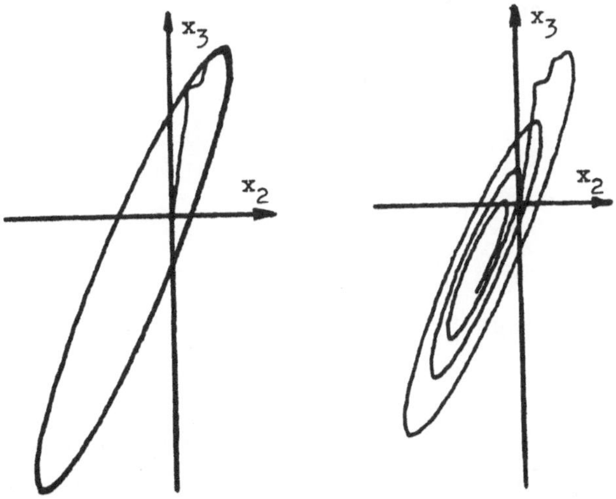

Fig. 5. Trajectories of third-order system subjected to step control signal (on the left - system without dissipation).

CONCLUDING COMMENTS

Approach to decentralized controls, based on phase space representation of plants and resulting from that energetic decomposition, as outlined in this paper, offer some conceptual advantages. Usual intuitive feelings about the role of energy and its relation to control performance index are put in evidence. Resulting decentralized control strategies become clearly defined. In addition, plant decomposition process itself is put under straightforward guidelines. On the other hand though, practical problems in computing an appropriate transformation to phase space need to be overcome.

REFERENCES

[1] Z. Jacyno, "Energetic Relations in General Dynamic Systems", Proceedings of 1982 IEEE International Symposium on Circuits and Systems, vol. 2, p. 389-392; May 10-12, 1982, Rome, Italy.

[2] Z. Jacyno, "Automatic Controls: A Hyperphase Space Approach", Proceedings of IV Congress de Informatica y Automatica, vol. I, p. 14-22; October 16-19, 1979, Madrid, Spain.

[3] Z. Jacyno, "An Energetic Decomposition of Dynamic Systems", Proceedings of 1979 IEEE International Symposium on Circuits and Systems, p. 154-157; July 17-19, 1979, Tokyo, Japan.

[4] Z. Jacyno, B. Lemaire, "Automatic Control of Interconnected Dynamical Systems", Proceedings of V Congreso de Informatica y Automatica, vol. I, p. 17-20; May 4-6, 1982, Madrid, Spain.

NUMERICAL SYNTHESIS OF OPTIMUM REDUCED-ORDER DYNAMIC REGULATORS

G. Capitani, M.E. Penati, M. Tibaldi
Dept. of Electronics, Systems and Computer Science
University of Bologna, 40136 Bologna (Italy)

Abstract

This work presents a recursive algorithm which gives a suitable starting reduced-order regulator which stabilizes the over-all system. This procedure applies a model-reduction approach to a full-order regulator built-up by the Kalman optimum observer and the optimum algebraic regulator. A periodical check of the suitability of the chosen structure is introduced.

1. INTRODUCTION

As already pointed out by many authors (see e.g.: Penati, Tibaldi, Bertoni 1983 [11], Franceschi, Penati, Tibaldi 1983 [10]), the need of introducing dynamic regulators for high-order dynamic systems is getting stronger in many fields of engineering and design.

Owing to the high order of the plant, this kind of problem cannot be solved by the classic approach which implies a Kalman filter and a Kalman algebraic regulator (*full-order regulator*).

A first approach to the solution is to implement the regulator as a full-order regulator using a reduced-order model of the plant, i.e. a model more simplified than the one used to represent completely the true plant. This kind of approach gives regulators which do not work properly owing to the well known *observation spillover* and *control spillover* effects.

A straightforward approach to the solution is to implement a reduced-order regulator which takes into account the full-order model of the plant.

Penati, Tibaldi, Bertoni 1983 [11] show a procedure to numerically optimize a reduced-order regulator with respect to a quadratic performance index. This procedure exhibits two major shortcomings: i) the starting configuration must be a reduced-order regulator which stabilizes the over-all system, ii) the choice of the structure of the reduced-order regulator is quite critical, in fact a poor one can prevent to achieve optimality.

This paper presents a recursive procedure which gives the starting configuration by properly reducing the above mentioned full-order regulator. Furthermore the authors suggest a way to choose a suitable structure for the reduced-order regulator.

Section 2 states the problem. Section 3 presents the model reduction approach. Section 4 uses this approach to implement the starting configuration of the reduced-order regulator. Section 5 deals with the choice of a suitable structure for the regulator and then summarizes the whole procedure.

2. THE DYNAMIC REGULATOR PROBLEM

Consider the following linear time-invariant system (*plant*):

$$\begin{aligned} \dot{\underline{x}}_p(t) &= \underline{A}_p\underline{x}_p(t) + \underline{B}_p\underline{u}_p(t) + \underline{D}_p\underline{w}_p(t), \qquad \underline{x}_p(t_o) = \underline{x}_{po}, \\ \underline{y}_p(t) &= \underline{C}_p\underline{x}_p(t) + \underline{v}_p(t), \end{aligned} \tag{1}$$

where $\underline{x}_p(t)$ is the state vector (n x 1), $\underline{u}_p(t)$ is the input vector (r x 1), $\underline{y}_p(t)$ is the output vector (m x 1), $\underline{w}_p(t)$ and $\underline{v}_p(t)$ are zero-mean white gaussian time-invariant stochastic processes with $\underline{R}_{wp}$ and $\underline{R}_{vp}$ as covariance matrices and, finally, $\underline{x}_{po}$ is a zero-mean stochastic vector with covariance matrix $\underline{P}_o$. $\underline{A}_p$, $\underline{B}_p$, $\underline{C}_p$ and $\underline{D}_p$ are matrices of suitable dimensions.

Let the system $(\underline{A}_p, \underline{B}_p)$ be completely controllable and the system $(\underline{A}_p, \underline{C}_p)$ completely reconstructable.

Given the performance index:

$$J = \lim_{t_f \to \infty} \frac{1}{t_f - t_o} E\{\int_{t_o}^{t_f} [\underline{x}_p^T(t)\underline{Q}_p\underline{x}_p(t) + \underline{u}_p^T(t)\underline{R}_p\underline{u}_p(t)]dt\}, \tag{2}$$

where $\underline{Q}_p$ is a (n x n) symmetric nonnegative-definite matrix, $\underline{R}_p$ is a (r x r) symmetric positive-definite matrix and E is the expected-value operator, find a dynamic regulator (*controller*), i.e. a linear time-invariant system of the type:

$$\begin{aligned} \dot{\underline{x}}_c(t) &= \underline{A}_c\underline{x}_c(t) + \underline{B}_c\underline{y}_p(t), \qquad \underline{x}_c(t_o) = \underline{x}_{co}, \\ \underline{u}_p(t) &= \underline{C}_c\underline{x}_c(t), \end{aligned} \tag{3}$$

which minimizes J. The state vector $\underline{x}_c(t)$ has p ($\leq$ n) components.

When p = n, the well known solution is the steady-state Kalman filter followed by the Kalman algebraic regulator:

Kalman filter:

$$\begin{aligned} \underline{A}_c &= \underline{A}_p - \underline{P}\,\underline{C}_p^T\,\underline{R}_{vp}^{-1}\,\underline{C}_p + \underline{B}_p\,\underline{C}_c, \\ \underline{B}_c &= \underline{P}\,\underline{C}_p^T\,\underline{R}_{vp}^{-1}, \\ \underline{x}_{co} &= \underline{0}, \end{aligned} \tag{4}$$

where $\underline{P}$ is the symmetric nonnegative-definite matrix which satisfies the steady-state Riccati-type equation:

$$\underline{A}_p\,\underline{P} + \underline{P}\,\underline{A}_p^T - \underline{P}\,\underline{C}_p^T\,\underline{R}_{vp}^{-1}\,\underline{C}_p\,\underline{P} + \underline{D}_p\,\underline{R}_{wp}\,\underline{D}_p^T = \underline{0}. \tag{5}$$

Kalman algebraic regulator:

$$\underline{C}_c = -\,\underline{R}_p^{-1}\,\underline{B}_p^T\,\underline{S}, \tag{6}$$

where $\underline{S}$ is the symmetric nonnegative-definite matrix which satisfies the steady-state Riccati-type equation:

$$\underline{S}\,\underline{A}_p + \underline{A}_p^T\,\underline{S} - \underline{S}\,\underline{B}_p\,\underline{R}_p^{-1}\,\underline{B}_p^T\,\underline{S} + \underline{Q}_p = \underline{0}. \tag{7}$$

When p < n, many authors (Martin, Bryson 1980 [4], Mukhopadhyay, Newsom, Abel 1981 [7], 1982 [8], Franceschi, Penati, Tibaldi 1983 [10]) suggest a numerical solution, which consists on minimizing the performance index J with respect to the p(m+r) independent elements of the matrices $\underline{A}_c$, $\underline{B}_c$ and $\underline{C}_c$.

To this aim, the gradient of J with respect to the elements of the matrices $\underline{A}_c$, $\underline{B}_c$ and $\underline{C}_c$ can be easily computed, as shown by Kwakernaak, Sivan 1972 [1].

All suggested procedures start from a controller which stabilizes the plant and then optimize it. Unfortunately the synthesis of such a stabilizing controller is not a trivial task.

Mukhopadhyay et al. 1981 [7], 1982 [8], build up the reduced-order regulator treating it as a partial estimator of p *key* states of the plant, but they do not give any criterion for a selection of these key states. Following a similar phylosophy in Section 4 we will show that the reduced-order regulator can be obtained just *truncating* the full-order optimal controller (Kalman + Kalman). We will show that the selection of the states to be truncated can be done following the *model reduction* procedure shown in the next section.

3. THE MODEL REDUCTION

As it is well known, the model reduction problem can be defined in many ways (Skelton 1980 [5]). For systems without deterministic inputs, the model reduction problem can be stated as follows.

Given the linear time-invariant n-order system:

$$\underline{\dot{x}}(t) = \underline{A}\ \underline{x}(t) + \underline{D}\ \underline{w}(t), \qquad \underline{x}(t_o) = \underline{x}_o,$$

$$\underline{y}(t) = \underline{C}\ \underline{x}(t),$$

where $\underline{w}(t)$ is a zero-mean white gaussian time-invariant stochastic process with covariance matrix $\underline{V}$ and $\underline{x}_o$ is a zero-mean stochastic vector with covariance matrix $\underline{P}_o$, find a truncated model of the given system in order to minimize the variation of the performance index:

$$H = \lim_{t_f \to \infty} \frac{1}{t_f - t_o} E \{\int_{t_o}^{t_f} \underline{y}^T(t)\ \underline{y}(t)\ dt\}.$$

The most obvious way consists on splitting the performance index value into n contributions, each one related to a single state x_i, i=1,..., n, and then truncating those states whose contributions are less significant.

As it is well known (see e.g.: Bertoni, Beghelli, Capitani, Tibaldi 1973 [2]), if the given system is asymptotically stable, the steady-state covariance matrix $\underline{P}$ of the state vector $\underline{x}(t)$ satisfies the Lyapunov-type equation:

$$\underline{A}\ \underline{P} + \underline{P}\ \underline{A}^T + \underline{D}\ \underline{V}\ \underline{D}^T = \underline{0}.$$

Since the performance index H can be rewritten as follows:

$$H = \lim_{t_o \to -\infty} E\{\underline{y}^T(t)\underline{y}(t)\}$$

$$= \lim_{t_o \to -\infty} tr\{\underline{C}^T \underline{C} E[\underline{x}(t)\underline{x}^T(t)]\}$$

$$= tr\{\underline{C}^T \underline{C} \underline{P}\}$$

$$= \sum_{i=1}^{n} \{ \sum_{j=1}^{n} (p_{ij} \underline{c}_i^T \underline{c}_j)\},$$

where $tr(\underline{A})$ means the trace of the matrix $\underline{A}$ and $\underline{c}_i$ is the i-th column of the matrix $\underline{C}$, we can treat the performance index H as the sum of n costs:

$$H = \sum_{i=1}^{n} h_i,$$

where:

$$h_i = \sum_{j=1}^{n} h_{ij}, \qquad h_{ij} = p_{ij} \underline{c}_i^T \underline{c}_j.$$

Note that the term $\underline{c}_i^T \underline{c}_j$ weights the term p_{ij}, so that we can give the term h_{ii} the meaning of *direct cost* of the state x_i, and the general term h_{ij}, with $i \neq j$, the meaning of *induced cost* of the state x_i due to the interaction with the state x_j; we can easily see that $h_{ij} = h_{ji}$ for all $i=1,\dots,n$, $j=1,\dots,n$. Therefore we can conclude that each cost component h_i, $i=1,\dots,n$, represents the contribution of the corresponding state x_i to the performance index value.

Moreover note that, unless the matrix $\underline{C}$ contains only orthogonal column vectors (in this case $\underline{C}^T\underline{C}$ is a diagonal matrix and $\underline{C}^T\underline{C}\ \underline{P}$ is nonnegative-definite, but of course $rank(\underline{C})=n$ must hold), some cost components h_i may be negative.

Reordering the states according to their contribution to the performance index:

$$|h_1| \geq |h_2| \geq \dots \geq |h_{n-1}| \geq |h_n|,$$

and truncating the least significant states, we can get a ***reduced-order model*** of the original system, following the performance index criterion given by H.

Obviously, the new value of H for the reduced-order model is *not* equivalent to the previous value minus the contribution of the truncated states. In general the new value of H may be less than, equal to or greater than the previous one.

4. THE CONTROLLER REDUCTION

The over-all closed-loop system using the full-order optimal regulator (i.e.: plant + Kalman filter + Kalman algebraic regulator) is the following:

$$\dot{\underline{x}}(t) = \underline{A}\,\underline{x}(t) + \underline{D}\,\underline{w}(t), \qquad \underline{x}(t_o) = \underline{x}_o,$$

where:

$$\underline{x}(t) = \begin{bmatrix} \underline{x}_p(t) \\ \underline{x}_c(t) \end{bmatrix}, \quad \underline{w}(t) = \begin{bmatrix} \underline{w}_p(t) \\ \underline{v}_p(t) \end{bmatrix}, \quad \underline{x}_o = \begin{bmatrix} \underline{x}_{po} \\ \underline{x}_{co} \end{bmatrix},$$

$$\underline{A} = \begin{bmatrix} \underline{A}_p & \underline{B}_p\underline{C}_c \\ \underline{B}_c\underline{C}_p & \underline{A}_c \end{bmatrix}, \quad \underline{D} = \begin{bmatrix} \underline{D}_p & \underline{0} \\ \underline{0} & \underline{B}_c \end{bmatrix}.$$

The performance index (2) can be rewritten as follows:

$$J = \lim_{t_f \to \infty} \frac{1}{t_f - t_o} E\{\int_{t_o}^{t_f} \underline{y}^T(t)\underline{y}(t)dt\},$$

where:

$$\underline{y}(t) = \underline{C}\,\underline{x}(t),$$

$$\underline{C} = \begin{bmatrix} \underline{Q} & \underline{0} \\ \underline{0} & \underline{R}\,\underline{C}_c \end{bmatrix}, \quad \underline{Q}^T\,\underline{Q} = \underline{Q}_p, \quad \underline{R}^T\,\underline{R} = \underline{R}_p.$$

Applying the model reduction approach to this system, we can get a reduced-order regulator by truncating the least significant states *belonging to the controller*, i.e. among $\underline{x}_c(t)$ components (obviously *not* among $\underline{x}_p(t)$ components).

In this case we can argue that the new value of the performance index J cannot be less than the previous one, because the Kalman + Kalman controller *is* the optimal solution.

Truncation stops either when asymptotic stability cannot be achieved or when the predefined reduced order is obtained.

This procedure seems to be somehow related to that one proposed by

Skelton 1980 [5].

The outlined procedure for the controller reduction can be summarized as follows:

ALGORITHM I

1) Set p=n (the order of the plant).
2) Compute the n+p cost components h_i.
3) Truncate the least significant state belonging to the controller, according to the cost components ordering; set p=p-1.
4) Check the asymptotic stability of the over-all system (plant + controller).
5) If the over-all system is not asymptotically stable, the controller to be used is the (p+1)-order controller previously obtained; stop.
6) If p is sufficiently small, stop; otherwise go to step 3).

A more complex version consists on:

6) ...; otherwise go to step 2).

This latter version is heavier from the computational point of view but, in some cases, it allows to achieve a lower-order stabilizing controller.

5. THE ILL-CONDITIONING PROBLEM

Since the main property of the reduced-order regulator obtained in the previous section is the capability of stabilizing the closed- loop system, it can be used as a starting configuration for the minimizing procedure of the performance index J.

As already mentioned in Section 2, a minimum of J can be sought using the gradient of J with respect to the elements of the controller matrices $\underline{A}_c$, $\underline{B}_c$ and $\underline{C}_c$. To this aim Franceschi, Penati, Tibaldi 1983 [10] transform these matrices into a suitable canonical form in order to emphasize their p(m+r) independent elements and to simplify the numerical procedure.

Unfortunately the controller representation can be ill-conditioned and the canonical transformation can even make this ill-conditioning worse, so preventing to achieve the optimum.

To avoid these shortcomings, multistructural models, introduced by Guidorzi 1982 [9], can be used. The main results of Guidorzi's paper are:

i) the definition of a procedure to obtain one of the possible transformations giving a multistructural model.

ii) the definition of a conditioning index in order to compare different multistructural models of the same system.

iii) an algorithm to transform a given multistructural model into an *adjacent* one.

These results can be used in the following step-by-step procedure:

ALGORITHM II

1) Given the plant (1) and the performance index (2), determine the full-order regulator (3) applying (4)÷(7).
2) Determine a reduced-order stabilizing regulator using Algorithm I.
3) Determine a multistructural model of the dynamic controller.
4) Starting from the multistructural model previously obtained and using algorithm (iii), search the *best*-conditioned model among the adjacent ones.
5) Search the optimum of the performance index (2), periodically testing the conditioning index and, in case, go to step 4).

6. EXAMPLES

In this section we shall apply the steps 1) and 2) of the algorithm II to some text-book examples.

6.1 *3rd-order system*

Given the plant:

$n = 3,\ m = 1,\ r = 1,$

$$\underline{A}_p = \begin{bmatrix} 0.1 & 1 & 1 \\ 0 & -1 & -0.5 \\ 0 & 0 & -2 \end{bmatrix}, \quad \underline{B}_p = \begin{bmatrix} 0 \\ 0 \\ 1 \end{bmatrix}, \quad \underline{D}_p = \begin{bmatrix} 0 \\ 0 \\ 1 \end{bmatrix},$$

$$\underline{C}_p = [\ 1 \quad 0 \quad 0\ ], \quad \underline{R}_{wp} = [\ 1\ ], \quad \underline{R}_{vp} = [\ 1\ ],$$

$$\underline{Q}_p = \begin{bmatrix} 1 & 0 & 0 \\ 0 & 1 & 0 \\ 0 & 0 & 1 \end{bmatrix}, \quad \underline{R}_p = [\ 1\ ],$$

the Kalman + Kalman regulator is:

$$\underline{A}_c = \begin{bmatrix} -0.298 & 1 & 1 \\ 0.0518 & -1 & -0.5 \\ -1.16 & -1.15 & -2.62 \end{bmatrix}, \quad \underline{B}_c = \begin{bmatrix} 0.398 \\ -0.0512 \\ 0.0906 \end{bmatrix},$$

$$\underline{C}_c = \begin{bmatrix} -1.51 & -1.15 & -0.620 \end{bmatrix},$$

with associated performance index value:

$$J = 1.58.$$

The component costs are:

$$h_4 = 0.570, \quad h_5 = 0.0672, \quad h_6 = -0.124.$$

Truncating the state 5 and applying the more complex version of the algorithm I, we obtain a 5th-order asymptotically stable over-all system with performance index value:

$$J = 2.04$$

and component costs:

$$h_4 = 0.641, \quad h_5 = -0.118.$$

Truncating the (new) state 5 we obtain a 4th-order asymptotically stable over-all system and a minimum-order regulator with performance index value:

$$J = 2.75.$$

The poles of the over-all system are:

$$s_1 = -2.20, \quad s_2 = -0.708, \quad s_3, s_4 = -0.147 \pm 0.365j.$$

6.2 *4th-order system*

Given the plant:

$$n = 4, \quad m = 1, \quad r = 1,$$

$$\underline{A}_p = \begin{bmatrix} 0.1 & 1 & 2 & 3 \\ 0 & -1 & -1 & 1 \\ 0 & 0 & -2 & -2 \\ 0 & 0 & 0 & -3 \end{bmatrix}, \quad \underline{B}_p = \begin{bmatrix} 0 \\ 0 \\ 0 \\ 1 \end{bmatrix}, \quad \underline{D}_p = \begin{bmatrix} 0 \\ 0 \\ 0 \\ 1 \end{bmatrix},$$

$$\underline{C}_p = \begin{bmatrix} 1 & 0 & 0 & 0 \end{bmatrix}, \quad \underline{R}_{wp} = [\ 1\ ], \quad \underline{R}_{vp} = [\ 1\ ],$$

$$\underline{Q}_p = \begin{bmatrix} 1 & 0 & 0 & 0 \\ 0 & 1 & 0 & 0 \\ 0 & 0 & 1 & 0 \\ 0 & 0 & 0 & 1 \end{bmatrix}, \quad \underline{R}_p = [\ 1\ ],$$

the Kalman + Kalman regulator is:

$$\underline{A}_c = \begin{bmatrix} -0.753 & 1 & 2 & 3 \\ -0.206 & -1 & -1 & 1 \\ 0.131 & 0 & -2 & -2 \\ -1.30 & -0.882 & -0.568 & -4.07 \end{bmatrix}, \quad \underline{B}_c = \begin{bmatrix} 0.853 \\ 0.206 \\ -0.131 \\ 0.112 \end{bmatrix},$$

$$\underline{C}_c = \begin{bmatrix} -1.18 & -0.882 & -0.568 & -1.07 \end{bmatrix},$$

with associated performance index value:

$J = 3.05.$

The component costs are:

$h_5 = 0.950, \quad h_6 = 0.0117, \quad h_7 = 0.0350, \quad h_8 = -0.184.$

Truncating the state 6, we obtain a 7th-order asymptotically stable over-all system with performance index value:

$J = 3.91,$

and component costs:

$h_5 = 1.22, \quad h_6 = 0.0901, \quad h_7 = -0.283.$

Truncating the (new) state 6, we obtain a 6th-order asymptotically stable over-all system with performance index value:

$J = 3.10,$

and component costs:

$h_5 = 0.877, \quad h_6 = -0.163.$

Truncating the (new) state 6, we obtain a 5th-order asymptotically stable over-all system and a minimum-order regulator with performance index value:

$J = 8.56$.

The poles of the over-all system are:

$$s_1 = -3.56, \quad s_2,s_3 = -1.46\pm0.794j, \quad s_4,s_5 = -0.0838\pm0.749j.$$

Note that in both the examples shown the choice of the states to be truncated would be the same in both the versions (step 6) of the algorithm I.

7. REFERENCES

[1] Kwakernaak, Sivan: *Linear Optimal Control Systems*. Wiley-Interscience, New York, 1972.

[2] Bertoni, Beghelli, Capitani, Tibaldi: *Teoria e tecnica della regolazione automatica*. Pitagora, Bologna, 1973.

[3] Skelton, Gregory: *Measurement Feedback and Model Reduction by Modal Cost Analysis*. Proc. Automatic Control Conf., Denver, 1979.

[4] Martin, Bryson: *Attitude Control of a Flexible Spacecraft*. AIAA J. of Guidance and Control, vol. 3, no. 1, Jan.-Feb. 1980.

[5] Skelton: *Cost Decomposition of Linear System with Application to Model Rdeuction*. Int. J. of Control, vol. 32, no. 6, 1980.

[6] Skelton, Hughes: *Modal Cost Analysis for Linear Matrix-Second-Order Systems*. J. of Dynamic Systems, Measurement and Control, ASME, vol. 102, Sept. 1980.

[7] Mukhopadhyay, Newsom, Abel: *A Method for Obtaining Reduced-Order Control Laws for High-Order Systems Using Optimization Techniques*. NASA Technical Paper 1876, Aug. 1981.

[8] Mukhopadhyay, Newsom, Abel: *Reduced-Order Optimal Feedback Control Law Synthesis for Flutter Suppression*. AIAA J. of Guidance and Control, vol. 5, no. 4, July-Aug. 1982.

[9] Guidorzi: *Multistructural Model Selection*. Cybernetics and Systems Research, R. Trappl ed., North-Holland Publishing Co., 1982.

[10] Franceschi, Penati, Tibaldi: *I regolatori dinamici di ordine ridotto: analisi e sintesi ottima per sistemi deterministici e stocastici*. Rapp. int. Istituto di Automatica, Università di Bologna, Genn. 1983.

[11] Penati, Tibaldi, Bertoni: *Reduced-Order Regulators for Large Scale Systems: A Numerical Implementation for Stochastic and Deterministic Systems*. IFAC/IFORS Symp., Warsaw, Poland, July 1983.

DECENTRALIZED EIGENVALUE ASSIGNMENT FOR DELAYED DYNAMICALLY INTERCONNECTED SYSTEMS

A.El-Kashlan and M.El-Geneidy
Department of Electrical Engineering,Faculty of Engineering,
Alexandria University,Alexandria,EGYPT

1.ABSTRACT

A characterization for the class of interactions among dynamically interconnected large-scale system,that are necessary for eigenvalue assignment is given. The system is transformed to an equivalent,non-delayed system. Subsystems-interconnections share global eigenvalue assignment by solving repeatedly appropriate eigenvalue assignment problem at their level. A procedure based on these results is given for global eigenvalue assignment and improving the prescribed eigenvalues sensitivity.

2. INTRODUCTION

In recent past,there has been an increasing interest in large-scale systems which are composed of several interconnected subsystems that may be physically distinct. Ramakrishna and Viswan [1], derived sufficient conditions for decentralized stabilization for a class of dynamically interconnected systems. In [2] the present authors presented a procedure for eigenvalue assignment for such class of systems. Guangquan and Lee [3], reformulated the control design problem as an optimization problem. However such systems may be geographically separated and/or dispersed in space. Several related new conceptual difficulties arise,among these is the associated delayed interconnections. Anderson [4] , Ikeda and Siljak [5],looked at this important problem from the stabilization point of view.

The purpose of this paper is to characterize in algebraic setting the interactions among the subsystems that lead to achieve a prespecified spectrum to the global system containing delayed interconnections,and to improve the sensitivity of the prescribed eigenvalues.

The key role to by-pass the delay difficulties is to expand the delayed system into an extended state space,where delay disappears. The equivalent,non-delayed system controllability originates from subsystems-interconnections controllability. Increased dimensionality is no longer a problem since the global problem is decomposed at the subsystems-interconnections level.

The procedure utilizes the **kernel space formulation** for the subsystems-interconnections to solve a series of appropriate eigenvalue assignment problems. Individual solutions are then coordinated to yield the global solution. The important role played by the eigenvectors to share achieving the desired spectrum together with minimizing its sensitivity is indicated in the procedure.

3.EIGENVALUE ASSIGNMENT PROCEDURE

Let the large-scale system under consideration be composed of S subsystem,each is described as

$$x_r(k+1) = A_r\, x_r(k) + B_r\, u_r(k)$$
$$y_r(k) = C_r\, x_r(k) \qquad (r=1,\dots,S) \qquad (1)$$

where $x_r(k)\in R^{n_r}$, $u_r(k)\in R^{m_r}$ and $y_r(k)\in R^{p_r}$ are the state vector,input vector,and output vector of subsystem r ,respectively. The matrices A_r , B_r and C_r are of appropriate dimensions. Let the associated interconnection be described as

$$z_r(k+1) = M_r\, z_r(k) + \sum_{q=1}^{S} L_{rq} y_q(k)$$
$$w_r(k) = N_r\, z_r(k) + \sum_{q=1}^{S} R_{rq}\, x_q(k) \qquad (2)$$

where $z_r(k)\in R^{a_r}$, $w_r(k)\in R^{m_r}$ are the state vector and output vector of the r-th interaction subsystem. The matrices M_r , N_r ,L_{rq} and R_{rq} are of appropriate dimensions. Such representation models many practical large-scale systems.

Assume that

$$(A_r, B_r) \text{ is a controllable pair } \forall\, r=1,\dots,S$$
and
$$(M_r, L_{rr}) \text{ is a controllable pair } \forall\, r=1,\dots,S \qquad (3)$$

Since subsystems are widely separated,let the subsystems be interconnected according to

$$u_r(k) = w_r(k-1) \tag{4}$$

Therefore the r-th subsystem becomes

$$x_r(k+1) = A_r\, x_r(k) + B_r \sum_{q=1}^{S} R_{rq}\, x_q(k-1) + B_r\, N_r\, z_r(k-1) \tag{5}$$

The augmented delayed interconnected system may be expressed as

$$\begin{bmatrix} X(k+1) \\ \hline Z(k+1) \end{bmatrix} = \left[\begin{array}{c|c} A & 0 \\ \hline L\,C & M \end{array}\right] \begin{bmatrix} X(k) \\ \hline Z(k) \end{bmatrix} + \left[\begin{array}{c|c} B\,R & B\,N \\ \hline 0 & 0 \end{array}\right] \begin{bmatrix} X(k-1) \\ \hline Z(k-1) \end{bmatrix} \tag{6}$$

where $X^T(k) = \left[x_1^T(k)\,, \ldots, x_S^T(k)\right]$, $Z^T(k) = \left[z_1^T(k), \ldots, z_S^T(k)\right]$

A , B , C , M and N are block diagonal matrices;

$$L = \left[L_{rq}\right] \quad , \quad R = \left[R_{rq}\right] \qquad r,q = 1,\ldots,S$$

Let

$$n = \sum_{r=1}^{S} n_r \quad , \quad m = \sum_{r=1}^{S} m_r \quad , \quad a = \sum_{r=1}^{S} a_r$$

To expand (6) in a larger state space where delay disappears,define the state vectors

$$S_1(k) = X(k-1) \quad \text{and} \quad S_2(k) = Z(k-1)$$

Also define the extended state vector $X_e(k)$ as

$$X_e^T(k) = \left[S_1^T(k) \quad S_2^T(k) \quad X^T(k) \quad Z^T(k)\right]$$

Therefore the equivalent non-delayed system is given by

$$X_e(k+1) = \Gamma \; X_e(k) \tag{7}$$

where

$$\Gamma = \left[\begin{array}{cc|cc} & 0 & I_n & \\ & & & I_a \\ \hline B\,R & B\,N & A & 0 \\ 0 & 0 & L\,C & M \end{array}\right] \tag{8}$$

Equation(7) is a typical representation for distributed lag system, Conlisk [6] . It is remarked that Γ in(8) has ($\xi = 2(n+a)$) eigenvalue. Here only distinct eigenvalues will be considered. For some eigenvalue λ_i of Γ ,the following eigenequation must be satisfied

$$(\Gamma - \lambda_i\, I)\; V^i = 0 \tag{9}$$

Where V^i is the corresponding eigenvector, $i=1,\ldots,\xi$. Hereafter we will show how the eigenvectors interact to share achieving the desired spectrum and how they improve its sensitivity against perturbations. Let the eigenvector V^i ,which corresponds the eigenvalue λ_i, be partitioned as

$$V^i = \left[V_{i1}^T \; V_{i2}^T \;\; V_{i3}^T \;\; V_{i4}^T\right]^T \tag{10}$$

where $V_{i1}, V_{i3} \in R^n$ and $V_{i2}, V_{i4} \in R^a$

Using (8),(10) in (9) one gets

$$\text{and} \quad V_{i3} = \lambda_i V_{i1} \quad , \quad V_{i4} = \lambda_i V_{i2} \tag{11a}$$

$$(A - \lambda_i I)\, V_{i3} + B\, R\, V_{i1} + B\, N\, V_{i2} = 0 \tag{11b}$$

$$(M - \lambda_i I)\, V_{i4} + L\, C\, V_{i3} = 0 \tag{11c}$$

Let us repartition both V_{i1} and V_{i2} to S subeigenvectors to conform with the subsystems and the interconnections,respectively. Thus we may write

$$V_{i1}^T = \left[v_{i1}^{1T},\; v_{i1}^{2T},\ldots,\; v_{i1}^{rT},\ldots,\; v_{i1}^{ST}\right] \quad ; \quad v_{i1}^r \in R^{n_r} \tag{12a}$$

$$V_{i2}^T = \left[v_{i2}^{1T},\; v_{i2}^{2T},\ldots,\; v_{i2}^{rT},\ldots,\; v_{i2}^{ST}\right] \quad ; \quad v_{i2}^r \in R^{a_r} \tag{12b}$$

After some matrix manipulations,(11b) may be decomposed to S matrix equation at the subsystem level. Each of these equations has the form

$$\left[(A_r - \lambda_i I) \mid B_r\right] \begin{bmatrix} \lambda_i \, v_{i1}^r \\ \hline w_{ir} \end{bmatrix} = 0 \tag{13}$$

where

$$w_{ir} = N_r \, v_{i2}^r + \sum_{q=1}^{S} R_{rq} \, v_{i1}^q \tag{14}$$

Similarly (11c) may be decomposed to S matrix equation at the interaction level. Each has the form

$$\left[(M_r - \lambda_i I) \mid L_{r1} C_1 \ldots L_{rr} C_r \ldots \mid L_{rS} C_S\right] \left[v_{i2}^{rT} \quad V_{i1}^T\right]^T = 0 \tag{15}$$

Inspecting (13),(15) reveals that the global eigenvalue assignment problem is decomposed to several lower order subproblems which subsystems- interactions are asked to solve.

Vectors associated with these equations lies within the kernel spaces of the indicated matrices.

Parameterizing the n_r+m_r vector in(13),to have a solution,one may freely choose m_r element(cf. Sebakhy and Sorial, [7]).

By repeated use of (13) for r=1,...,S ,the vector V_{i1}in (12a) contains m free parameter.

Similarly equation(15) represents a_r linear equations in(a_r+n) parameter, of which n element; those contained in V_{i1} are already expressed in terms of the m free elements. Therefore the choice that solves (13), also leads to solving (15). Such an interesting feature will be exploited to solve the problem.

Now , construct the following matrices;

$$W_r = \left[w_{1r} , w_{2r} , \dots , w_{\xi r}\right] \qquad R^{m_r \times \xi} \tag{16}$$

$$V_r = \left[v_{11}^r , \dots , v_{\xi 1}^r\right] \qquad R^{n_r \times \xi} \tag{17}$$

$$\bar{V}_r = \left[v_{12}^r , \dots , v_{\xi 2}^r\right] \qquad R^{a_r \times \xi} \tag{18}$$

Using these definitions together with (14), we get

$$W_r = N_r \bar{V}_r + \sum_{q=1}^{S} R_{rq} v_{i1}^q \tag{19}$$

Next define

$$W = \begin{bmatrix} W_1 \\ \cdot \\ \cdot \\ \cdot \\ W_S \end{bmatrix} , \quad V = \begin{bmatrix} V_1 \\ \vdots \\ V_S \\ \bar{V}_1 \\ \cdot \\ \cdot \\ \bar{V}_S \end{bmatrix} = \begin{bmatrix} V_{Su} \\ \hline V_{In} \end{bmatrix} \tag{20}$$

Therefore, using(16) and (19) we can write

$$W = N V_{In} + R V_{Su} \tag{21}$$

while(11c) gives

$$M V_{In} + L C V_{Su} = V_{In} Q \tag{22}$$

Where the subscripts Su, In stands for subsystems and interconnections, respectively. The matrix $Q = \text{diag}(\lambda_1 , \dots\dots , \lambda_\xi)$ is the desired spectrum.

Using (21),(22) one gets a solution in terms of the composite eigenvectors as;

$$\begin{bmatrix} W \\ \hline V_{In} Q \end{bmatrix} = \left[\begin{array}{cc|cc} R & N & 0 & 0 \\ \hline L C & M & 0 & 0 \end{array}\right] \begin{bmatrix} V_{Su} \\ V_{In} \\ \hline V_{Su} Q \\ V_{In} Q \end{bmatrix} \tag{23}$$

Since distinct eigenvalues are considered,their corresponding eigenvectors are linearly independent,hence

$$\begin{bmatrix} W & \\ V_{In} & Q \end{bmatrix} \begin{bmatrix} V_{Su} & \\ V_{In} & \\ V_{Su} & Q \\ V_{In} & Q \end{bmatrix}^{-1} = \begin{bmatrix} R & N & 0 & 0 \\ L\,C & M & 0 & 0 \end{bmatrix} \tag{24}$$

To improve eigenvalues sensitivity,the eigenvectors are choosen to be maximally orthogonal,Bhattacharyya and de Souza[8].

Define a sensitivity measure J as

$$J = \text{trace}\ (I - V^T V)^2 \tag{25}$$

Finally,judicious selection for the parameters contained in the composite eigenvectors that achieves matching of(24) together with minimizing(25),gives the desired and necessary amount of information R and N .

4. CONCLUSION

The eigenvalue assignment for dynamically delayed interconnected system is presented and solved. Characterization for the class of necessary interconnections is given. The system is transformed to an equivalent,non-delayed system. A practical eigenvalue assignment procedure is then applied through exploiting subsystems-interconnections eigenspace formulation. An interesting feature is that the proposed procedure plays an important role in both theoritical and practical applications of large-scale systems.

5. REFERENCES

[1] A.Ramakrishna and N.Viswanadham,Decentralized Control Of Interconnected Dynamical Systems,IEEE Transactions on Automatic Control, Vol. AC-27,pp.159-164,February 1982.

[2] A.El-Kashlan and M.El-Geneidy,Use Of Subsystems Eigenstructure in Dynamically Interconnected Systems For Eigenvalue Assignment,9-th IFAC Congress,Budapest,Hungary,July 2-6,1984.

[3] L.Guangquan and G.Lee,Decentralized Control Of Large-Scale Dynamic Interconnected Systems, Int.J.Control,Vol.37,pp.775-786,1983.

[4] B.D.O.Anderson,Time Delays In Large-Scale Systems,Proceedings 18-th IEEE Conf.on Decision and Control,Fort Lauderdale,FL.1979,655-660.

[5] M.Ikeda and D.D.Siljak,Decentralized Stabilization Of Large-Scale Systems With Time Delay,Large-Scale Systems,Vol.1,pp. 273-279,1980.
[6] J.Conlisk,Quick Stability Checks and Matrix Norms,Economica, Vol.40,pp.402-409,November 1973.
[7] O.A.Sebakhy and N.N.Sorial,Optimization Of Linear Systems With Prescribed Closed Loop Eigenvalues,IEEE Transactions on Automatic Control,Vol.AC-24,pp.355-357,1979.
[8] S.P.Bhattacharyya and E.de Souza,Pole Assignment Via Sylvester's Equation,Systems and Control Letters,Vol.1,pp.261-263,1982.

DECENTRALIZED CONTROL OF INPUT-OUTPUT STOCHASTIC MODELS

E.P. Melo
COPPE/UFRJ C.P. 68504
21.944 Rio de Janeiro - RJ, BRASIL

J.C. Geromel
FEC/UNICAMP C.P. 6166
13.100 Campinas - SP, BRASIL

ABSTRACT : A new approach for decentralized control design of input-output stochastic models is discussed. The control law is obtained from the minimization of the output plus control signals variance of each subsystem. For the decentralized control implementation, it becomes necessary to predicte the interconnection variables among subsystems. For that, it is used an ARMA model updated though on-line local identification. Within computational limits, the procedure presents great facility for practical implementation since only local informations are handled at each subsystem level. A simple interconnected system is solved and discussed in details.

1. Introduction

In this paper, we present a new approach to design decentralized minimum variance control for input-output stochastic systems. This study is motivated mainly for two reasons. First, it can be found in the literature several results concerning decentralized control of dynamic systems with state variables representation: stochastic [1, 2] and deterministic [3] models; however, almost nothing can be found for input-output models. Second, the classical minimum variance control [4-6] is largely used in practice due mainly its solution and implementation simplicity. Thus, it appears to be important to study this technique subjected to additional information constraints (for instance decentralized control).

The proposed scheme consists to find a decentralized control law based on local adaptive prediction of the interconnection variables among subsystems. Each subsystem is supposed to be represented by an input-output stochastic model and the decentralized control minimizes its output plus control variances.

The paper is divided as follows. In second and third sections, the decentralized minimum variance control is obtained. However, it depends on future occurences of the interconnection variables. Therefore, we suggest to predict it from an ARMA model updated though on-line identification. The model and the predictor are analysed in the fourth section. In section five an example of an overall system composed by two interconnected subsystems is solved and simulated by action of the proposed control. Finally in section six we conclude the paper.

2. Statement of the Problem

Let us consider an input-output interconnected system, composed by N subsystems, each of one having the following model

$$A_i y_i(k) = B_i u_i(k - p_i) + C_i z_i(k) + D_i w_i(k) \quad ; \quad i = 1 \ldots N \tag{1}$$

where A_i, B_i, C_i and D_i, $i = 1 \ldots N$, are known polinomials in q^{-1}, q being the one step ahead operator. By assumption, these polinomials are of order n_i and satisfy

$$\begin{aligned} A_i(0) &= D_i(0) = 1 \\ B_i(0) &= b_{io} \neq 0 \end{aligned} \tag{2}$$

At instant $k = 1, 2, \dots$; $y_i(k)$, $u_i(k)$ and $w_i(k)$ are respectively the *output* , the *control* and *white noise input* of subsystem i.

The variables $z_i(k)$, $i = 1 \dots N$, are called *interconnection variables* and it depends on the output variables of other subsystems, that is:

$$z_i(k) \triangleq \sum_{j \neq i}^{N} L_{ij} y_j(k) \tag{3}$$

where L_{ij} $i \neq j = 1 \dots N$ are known polinomials in q^{-1}.

The control delay $p_i \geq 1$ is a known integer parameter. From (1 - 3) it is clear that the overall dynamic system is composed by N interconnected monovariable subsystems and it is important to note that the assumption of that the polinomials in (1) have the same degree n_i does not imply any loss of generality.

At each instant k, we suppose that all available informations related to the ith subsystem are

$$I_i(k) = \{y_i(t), z_i(t), u_i(t-1) \; ; \; t = k, k-1, \dots\}, \quad i = 1 \dots N \tag{4}$$

Of course, the task of each subsystem controller is to regulate its output to some steady state value (for instance the origin). In this sense they are found by solving N local minimum variance control problems:

$$\underset{u_i(k)}{\text{Min}} \quad E\{y_i(k + p_i)^2 + \bar{\lambda}_i u_i(k)^2 / I_i(k)\} \; , \; i = 1 \dots N \tag{5}$$

It is important to note that the set of available informations of each subsystem controller is naturally constrained only to *local* measurements, for this reason we call it decentralized minimum variance controller. Then

$$u_i(k) = \gamma_i(I_i(k)) \qquad i = 1 \dots N \tag{6}$$

In the next two sections, it is proved that at any instant k, the knowledgement of the information set $I_i(k)$ does not suffice to determine *exactly* the control law (6), since it depends on future occurences of the interconnection variables $z_i(t)$, $t = k+1, k+2, \dots, k+p_i$. To eliminate this difficulty, we propose to estimate the future occurences of $z_i(\cdot)$ using local informations only

$$\hat{z}_i(t) \triangleq E\{z_i(t) / I_i(k)\} \qquad t = k+1, \dots, k+p_i \tag{7}$$

and, in this case, the decentralized control law, turns to be

$$u_i(k) = \hat{\gamma}_i(\hat{z}_i(t), I_i(k)) \tag{8}$$

The structure of each local controller is given in fig. 1 bellow.

Remark : Since $z_i(\cdot)$ is an element of the set $I_i(k)$, we suppose that it is directly measurable at each subsystem level. Of course this is not the more general case, however it occurs in many practical problem. It is possible to eliminate $z_i(\cdot)$ of $I_i(k)$, in this case (7 - 8) remain true, however the determination of $\hat{\gamma}_i(\cdot, \cdot)$ becomes much more difficult to be found. In this paper we do not consider this situation.

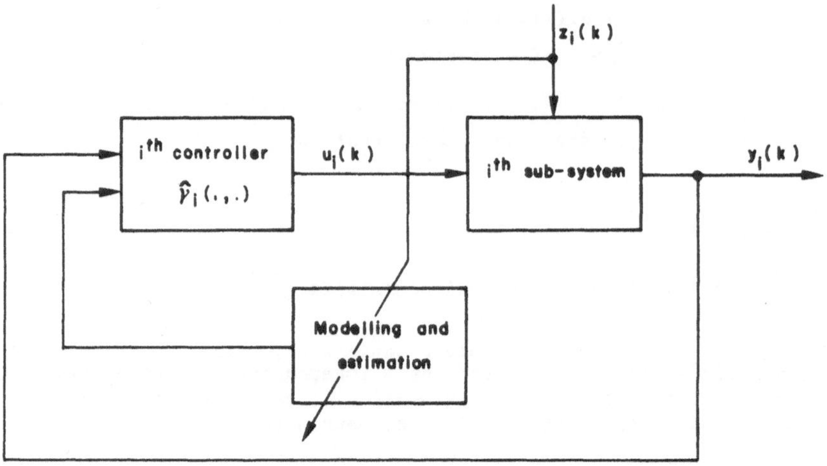

Fig. 1 - Local minimum variance control.

3. Decentralized Minimum Variance Control

In order to solve the problem stated before, we consider first that the interconnection variables $z_i(t)$ are exactly known at each instant k, even for $t > k$. With this assumption, problem (5) subjected to (1) is a classical one [4-6].

Let us consider the following polinomial identity

$$D_i = A_i R_i + q^{-p_i} S_i \qquad i = 1 \dots N \tag{9}$$

where R_i and S_i are polinomials in q^{-1} with degrees $(p_i - 1)$ and $(n_i - 1)$ respectively and $R_i(0) = 1$. With (9) we can write

$$y_i(k + p_i) = \psi_i(k + p_i / k) + R_i w_i(k + p_i) \tag{10}$$

where

$$\psi_i(k + p_i/k) \triangleq \frac{B_i R_i}{D_i} u_i(k) + \frac{C_i R_i}{D_i} z_i(k + p_i) + \frac{S_i}{D_i} y_i(k) \tag{11}$$

Since the two terms of the right hand side of (10) are uncorrelated, the local minimum variance problem (5) turns to be equivalent to

$$\min_{u_i(k)} \{\psi_i(k + p_i / k)^2 + \bar{\lambda}_i u_i(k)^2 + cte\} \qquad i = 1 \dots N \tag{12}$$

and the optimality condition (necessary and sufficient) for it can be readly write as

$$\psi_i(k + p_i / k) \frac{\partial \psi_i}{\partial u_i(k)} + \bar{\lambda}_i u_i(k) = 0 \qquad i = 1 \dots N \tag{13}$$

Now, defining $\lambda_i = \bar{\lambda}_i / b_{io}$ we obtain the optimal control law related to each subsystem:

$$M_i u_i(k) + N_i y_i(k) + P_i z_i(k + p_i) = 0 \qquad i = 1 \ldots N \tag{14}$$

where M_i, N_i and P_i are polinomials in q^{-1} given by

$$\begin{aligned} M_i &\triangleq B_i R_i + \lambda_i D_i \\ N_i &\triangleq S_i \\ P_i &\triangleq C_i R_i \end{aligned} \tag{15}$$

It is important to note that the control law given by (14) is not the optimal solution of (5) because at instant k, the future occurences of $z_i(\cdot)$ are not actually known. We claim however that if at the place of $z_i(\cdot)$ we put $\hat{z}_i(\cdot)$ given by (7) then, (14) provides $\hat{\gamma}_i(\cdot, \cdot)$. For this, it is necessary to estimate on-line an adequate model for the interconnection variables which will be analysed in the next section.

It remains to define the parameter $\bar{\lambda}_i > 0$ for each subsystem. First, let us consider each isolated closed loop subsystem $z_i(\cdot) = 0$ with the control law given by (14)

$$y_i(k) = \frac{M_i}{B_i + \lambda_i A_i} w_i(k) \qquad i = 1 \ldots N \tag{16}$$

The parameters $\bar{\lambda}_i$ is determined such that the roots of the characteristic equations $B_i + \lambda_i A_i = 0$ lie inside the unit circle. In some cases this choice does not suffice to guarantee the overall system stability with the interconnections given by (3). However it can be expected that if $\hat{z}_i(\cdot)$ is a "good" estimator then this choice is adequate for the stability of the overall multivariable closed-loop system (see example at section 5).

A numerical method to determine the parameters $\bar{\lambda}_i$, if one exists, though the realization of the characteristic equation as an output feedback problem is presented in [7] by using an algorithm given in [8].

4. Interconnection Modelling

As pointed-out before, the decentralized control (8) is given by

$$M_i u_i(k) + N_i y_i(k) + P_i \hat{z}_i(k + p_i) = 0 \tag{17}$$

where $\hat{z}_i(\cdot)$ can be estimated from the available informations about the overall system, that is

$$I(k) \triangleq I_1(k) \oplus \ldots. \oplus I_N(k) \tag{18}$$

In the case of completely decentralized control we must actually constraint the knowledgement of each subsystem as being $I_i(k)$ (completely decoupled).

To construct the interconnection predictor associated to each subsystem, let us suppose that the local interconnection can be modelled, in a general form, by an ARMA model

$$F_i z_i(k) = G_i e_i(k) \qquad i = 1 \ldots N \tag{19}$$

where F_i and G_i are polinomials in q^{-1} with degrees m_i and such that $F_i(0) = G_i(0) = 1$. At each instant of time $k = 1, 2, \ldots$ (19) provides us the local interconnection variable $z_i(k)$ and $e_i(k)$ represents a white noise sequence.

Supposing that the degree m_i is known (through off-line experiments), the parameters of polinomials $F_i(\cdot)$ and $G_i(\cdot)$ can be estimated by using the extended least square method [9, 10]. Based on the estimation of those parameters on previous steps and on the set of available informations $I_i(k)$, a minimum quadratic error predictor [11] is used to define $\hat{z}_i(\cdot)$.

For that, let us consider the following polinomial identity

$$G_i = F_i V_i + q^{-j} T_i \qquad i = 1 \ldots N, \quad j = 1 \ldots p_i \tag{20}$$

where V_i and T_i are polinomials in q^{-1} such that $V_i(0) = 1$ with degrees $(j - 1)$ and $(m_i - 1)$ respectively. Note that, at each step j, of prediction, we have different polinomials V_i and T_i given by (20).

Making use of (20), the model (19) can be rewrite as

$$z_i(k + j) = V_i e_i(k + j) + \frac{T_i}{G_i} z_i(k) \tag{21}$$

Provide $E\{e_i(k + j) / I_i(k)\} = 0$, $j = 1 \ldots p_i$, the solution of (7) is obtained immediately from the calculation of the conditional expectation of (21) and we get to:

$$G_i \hat{z}_i(k + j) \overset{\Delta}{=} T_i z_i(k) \qquad i = 1 \ldots N, \quad j = 1 \ldots p_i \tag{22}$$

clearly, (22) provides us the estimation of each local interconnection $j = 1 \ldots p_i$ steps ahead.

Remark : With regards to the identification procedure (extended least square) it can be commented that its convergence to the actual values of the model parameters (19) is not always sure. However in practice this procedure has been used with great success since the closed loop performance is frequently acceptable [12].

From these results, at each instant of time k, the decentralized control related to any subsystem can be calculated as follows

* Local measurements of $y_i(k)$ and $z_i(k)$ which enable us to construct the information set $I_i(k)$
* Identification of the local interconnection model parameters
* Determination of V_i and T_i polinomials for $j = 1 \ldots p_i$
* Prediction of the local interconnection variables $\hat{z}_i(k + j)$
* Determination of R_i and S_i polinomials and calculation of the decentralized control $u_i(k)$.

As it can be seen, the procedure is very simple to be implemented. We should note that decentralized control using adaptive prediction as showed here, has been also used recently [13] but with another control methodology.

5. Numerical Example

Let us consider an interconnected system composed by two identical, *non-minimum fase* subsystems defined by

$$
\begin{aligned}
A_i(q^{-1}) &= 1 + 2q^{-1} + 1.5q^{-2} + 0.5q^{-3} + 0.06q^{-4} \\
B_i(q^{-1}) &= 0.01 - 0.01q^{-1} \quad ; \quad p_i = 3 \qquad\qquad i = 1,\ 2 \\
C_i(q^{-1}) &= 0.05q^{-2} - 0.08q^{-3} + 0.03q^{-4} \\
D_i(q^{-1}) &= 1 + 2.2q^{-1} + 1.81q^{-2} + 0.66q^{-3} + 0.09q^{-4}
\end{aligned}
\tag{23}
$$

the interconnection variables are $z_i(k) = y_j(k)$, $i \neq j = 1, 2$. With the numerical procedure given in [7] we get $\lambda_1 = \lambda_2 = 0.024$. The closed-loop, isolated subsystems poles are $-0.34 \pm j\ 0.76$ (module 0.84) and $-0.22 \pm j\ 0.10$ (module 0.25). With these values of λ's the polinomials (15) turn to be

$$
\begin{aligned}
M_i(q^{-1}) &= 0.034 + 0.046q^{-1} + 0.041q^{-2} + 0.017q^{-3} + 0.002q^{-4} \\
N_i(q^{-1}) &= 0.040 + 0.065q^{-1} + 0.033q^{-2} + 0.005q^{-3} \qquad\qquad i = 1,\ 2 \\
P_i(q^{-1}) &= 0.050 - 0.070q^{-1} + 0.010q^{-2} + 0.013q^{-3} - 0.003q^{-4}
\end{aligned}
\tag{24}
$$

Though off-line experiments, the interconnection model of each subsystem was choosen as a second order autoregressive (AR)

$$F_i z_i(k) = e_i(k) \tag{25}$$

where $F_i(q^{-1}) = 1 + f_{i1}q^{-1} + f_{i2}q^{-2}$. For simulation purpose it has been considered the initial conditions $y_i(-1) = 1$; $y_i(k) = 0$ for $k = -2, -3, -3$ for $i = 1$ and 2. Also, both subsystems stochastic inputs has been choosen with zero mean and variance equal to 0.1. Fig. 2 shows the time evolution of the parameters of each interconnection model and its output when the decentralized control (17) is applied.

This figure shows also that the output of both subsystems are satisfactorily regulated around zero and the parameters estimators converge to constant values which enables us to conclude about the good performance of the proposed decentralized scheme.

6. Conclusions

In this paper, we presented a procedure to determine *on-line* decentralized control for input-output stochastic models based on the classical minimum variance control problem.

It has been shown by means a simulation of a interconnected system composed by two subsystems that the proposed control law gives good results since the outputs of each subsystem is satisfactorily regulated around the set point.

We must note however, that many additional properties of the proposed scheme need to be exactly determined which of course deserves future research efforts in this area. For instance, stabilizability and suboptimality index of the proposed decentralized stochastic control law.

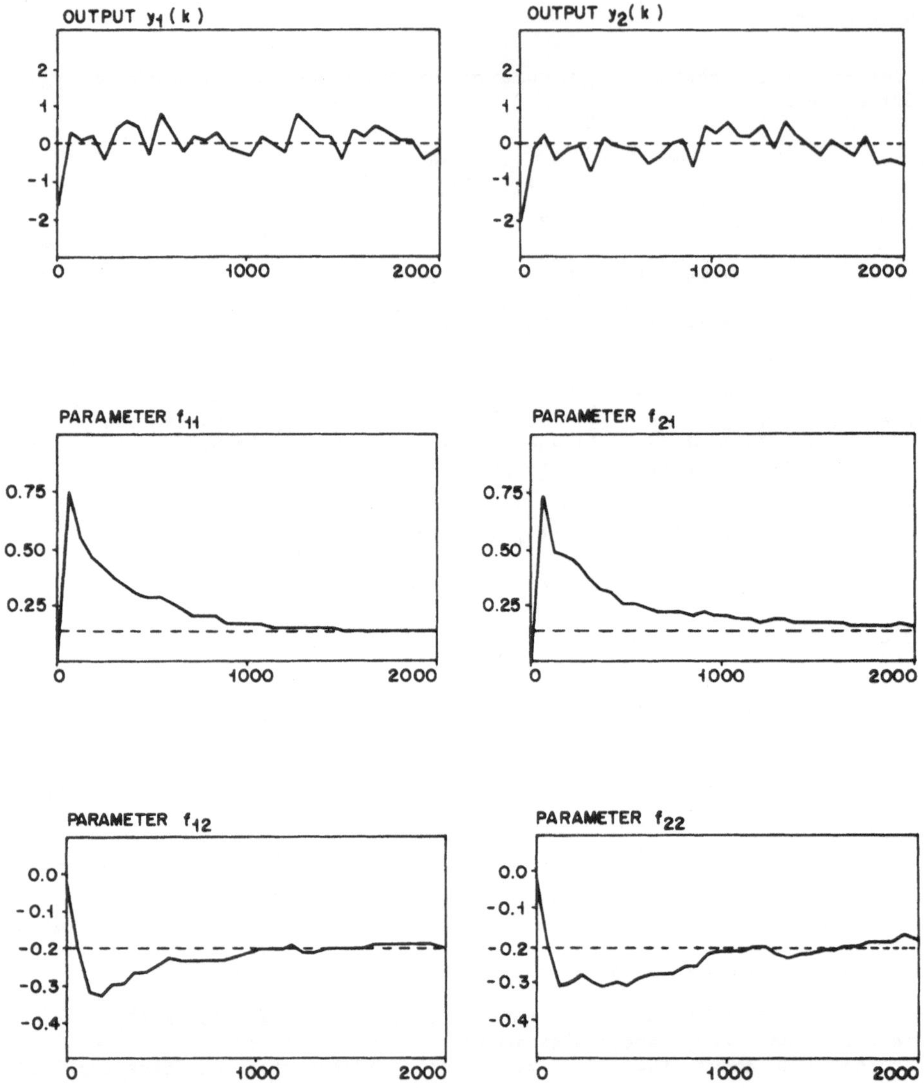

Fig. 2 - Simulation of the overall interconnected system.

Acknowledgement - This research was developed in part with the financial support of the Conselho Nacional de Desenvolvimento Científico e Tecnológico (CNPq-Brazil) under grant no. 301373/80.

7. References

[1] Aoki, M. - On Decentralized Linear Stochastic Control Problems with Quadratic Cost, IEEE Trans. on Automatic Control, vol. 18, nº 3, pp 243-249, June 1973.

[2] Chong, C.Y. and Athans, M. - On the Periodic Coordination of Linear Stochastic Systems, Automatica, vol. 12, nº 4, pp 321-335, July 1976.

[3] Geromel, J.C. and Bernussou, J. - Optimal Decentralized Control of Dynamic Systems, Automatica, vol. 18, nº 5, pp 545-557, September 1982.

[4] Astrom, K.J. and Wittenmark, B. - On Self-Tuning Regulators, Automatica, vol. 9, nº 2, pp 185-199, March 1973.

[5] Astrom, K.J., Borisson, U. and Ljung, L. - Theory and Applications of Self-Tuning Regulators, Automatica, vol. 13, nº 5, pp 457-476, September 1977.

[6] Astrom, K.J. - Self-Tuning Regulators Design, Principles and Applications, Narendra and Monopoli (eds.), Applications of Adaptive Control, Academic Press, N.Y., 1980.

[7] Melo, E.P. and Geromel, J.C. - Decentralized Minimum Variance Control, Note Interne nº 61/83, FEC/UNICAMP, Campinas, October 1983 (in portuguese).

[8] Geromel, J.C. and Melo, E.P. - Structural Constrained Controllers for Linear Discrete Dynamic Systems, Proceedings of the 9th IFAC World Congress, Hungary, July 1984.

[9] Panuska, V. - An Adaptive Recursive Least Squares Identification Algorithm, Proceedings of JACC, 1968.

[10] Young, P.C. - Comments on On-Line Identification with Applications to Kalman Filtering, IEEE Trans. on Automatic Control, vol. 17, pp 269-271, April 1972.

[11] Wittenmark, B. - A Self-Tuning Predictor, IEEE Trans. on Automatic Control, vol. 19, nº 6, pp 848-851, December 1974.

[12] Ljung, L., Saderstrom, T. and Gustavsson - Counterexamples to General Convergence of a Commonly Used Recursive Identification Method, IEEE Trans. on Automatic Control, vol. 20, pp 643-652, 1975.

[13] Funabashi, M., Masui, S. and Ohnari, M. - A Reliable Scheme for Controlling Large Scale Plants and its Application to Multiple Reservoir Operation, Proceedings of 8th IFAC World Congress, Japan, 1981.

DECENTRALIZED CONTROL OF WATER QUALITY IN THE RIVER NILE

Magdi S. Mahmoud
Electrical Engineering Department, Kuwait University
P.O. Box 5969, KUWAIT

Mohamed F. Hassan
Electronics and Communications Engineering Department
Cairo University, Giza, EGYPT

ABSTRACT

A linearized version of a recently developed water quality dynamic model of the River Nile is considered. Two control strategies are designed in a decentralized way to optimally adjusting the concentration of water quality constituents. Simulation results of the decentralized control structure are presented.

1. INTRODUCTION

With the extension of civilization and development, major environmental problems, which interact with nature and life have been created. One of these problems is conserned with preserving water quality standards in streams. This problem is of vital importance to maintaining the balance of the ecological system; otherwise living species will continually diminish. Industrial wastes and sewage are often the main sources of pollution in streams. In turn this affects the level of one or more of water quality constituents like biochemical oxygen demand (BOD), dissolved oxygen (DO), ammonia, nitrite, phosphate, temperature, radioactive materials, toxic metals, ..etc [1-3].

In Egypt, the industrial expansion has increased the level of pollution in the River Nile [4]. The construction of the High Dam has added a new important source of pollution in the River Nile; namely algae. The increase of algae deaths produces toxic constituents, through a series of complex biological processes, which has a damaging effect on all aspects of life. It therefore seems mandatory to seek appropriate water quality policies with the purpose of maintaining pollution levels within reasonable bounds. The practical implementation of such policies, in the form of control strategies, would seem to constitute a basic task to the decision-makers.

A quantitative approach to study pollution problems in streams is usually based on modeling and system analysis. Mathematical models of water quality problems are numerous and, in general, nonlinear expressing variations in the concentration of water quality constituents along the stream. A brief discussion of some of these models is presented in [4]. The most appealing of such models is a modified version of the simulation model QUAL II [5], which has been reported in [6,7]. In this paper, a linearized version of the revised QUAL II model is developed, by considering two control actions. The first is through the stream velocity which can be implemented by regulating the water discharge into the basin. The second is through the adjustment of the concentration of water quality constituents discharged from polluter stations. A decentralized control structure is then developed and the unknown gains are determined by a three-level algorithm. Simulation results on typical data are presented.

2. AN OVERVIEW OF STREAM QUALITY MODELS

Water streams can be used either as a source for potable supply or for recreation purposes. In both cases, it is of prime importance to maintain pollution levels within reasonable bounds. This should meet the community needs and should yield satisfactory ecological balance. Such bounds delineate water quality standards.

Various mathematical models have been applied to simulate water quality conditions in rivers [1]. These models vary considerably in their degree of sophistication. Low resolution models represent general trends for a few coupled, dependent variables over a limited set of boundary conditions. A typical example is the use of Streeter-Phillips equation to exhibit average monthly DO concentration in the San Antonio river [2]. On the other hand, high resolution models may be used to describe the response of a large number of linked dependent variables over a wider set of boundary conditions. A typical example is the application of HEC model [3] for simulating 20 variables in the Boise river system.

Modifications of the classical model of Streeter and Phillips have been reported in [8,9]. The above models share the following properties:

(a) They are dynamic: that is, accepting time-varying input functions and operating upon them to produce time-varying response

(b) They provide a reasonable mathematical approximation of the physico-chemical changes occuring in river systems.

(c) They characterize a quantity of fluid mixing in stream as a convenient parameter which is time-varying and spatially distributed.

In view of property (c), one can distinguish between the presently available physico-chemical models:

(1) Partial Mixing

This is represented by a second-order partial differential equation of the form

$$\partial \underline{c}(z,t)/\partial t = \underline{f}_1\left[\partial^2 \underline{c}(\underline{z},t)/\partial z^2,\ \partial \underline{c}(\underline{z},t)/\partial z,\ \underline{c}(z,t)\right] \qquad (1)$$

$$z_0 \leq z \leq z_1 \quad ; \quad t_0 \leq t \leq t_f$$

(2) No Mixing

In this case the model is represented by a first-order partial differential equation of the type:

$$\partial \underline{c}(z,t)/\partial t = \underline{f}_2\left[\partial \underline{c}(z,t)/\partial z,\ \underline{c}(z,t)\right] \tag{2}$$

$$z_0 \leq z \leq z_1 \quad ; \quad t_0 \leq t \leq t_f$$

(3) Perfect Mixing

Here the model is given by a first-order ordinary differential equation with associated transportation delay T_d to allow for distributed effects between spatial ordinates z_0 and z_1:

$$d\underline{c}(z_1,t)/dt = \underline{f}_3\left[\underline{c}(z,t),\ \underline{c}(z_0,t - T_d)\right] \tag{3}$$

$$t_0 \leq t \leq t_f$$

In (1)-(3), t is the independent variable of time, z is the independent variable of distance, $\underline{c}$ is the vector of dependent concentration variables, $(\underline{f}_1,\underline{f}_2,\underline{f}_3)$ are vector functions, (z_0,z_1) representing the spatial boundaries of the system and (t_0,t_f) representing the initial and final times, respectively.

From a theoretical view point, model (1) would appear to represent the system dynamics most accurately. For control purposes however, model (3) seems to be the most attractive one [4]. In the light of this remark, we recall that the water quality models developed in [10-12] are based on model (3). The simulation model QUAL II [5] is of the type (1) described above. Further refinements of the model and some case studies are given in [13,14].

Next we examine the case of the River Nile.

3. WATER QUALITY MODEL OF THE RIVER NILE

It has been reported in [5,13,14] that QUAL II model simulates the in-stream interaction of the chemical and biochemcial constituents in steady flow rivers. The fundamental one-dimensional transport (mass balance) equation

$$(A_x\ dx)\partial \underline{c}/\partial t = \partial/\partial x\left[A_x D_L \partial \underline{c}/\partial \underline{x}\right]dx - \partial/\partial x\left[A_x\ u_m \underline{c}\right] dx + (A_x dx)d\underline{c}/dt \pm S \tag{4}$$

is of the type (1) and used to represent the temporal change in the concentration of the constituents within the river. In (4), $\underline{c}$ is the concentration, x is the distance, t is the time, A_x is the cross-sectional area at distance x, D_L is the dispersion coefficient, S is a source or sink and u_m is the mean stream velocity. We note that the first term in (4) represents the transport of material by turbulence effects, and the second term represents the advection of the constituents by the flowing water. The third term describes the individual constituents changes due to biochemical transformation whereas the fourth term designates external input sources.

It has been pointed out [4] that the application of the QUAL II model in some streams indicated some limitations; mainly because it is a distributed parameter system. Recently, a revised form of QUAL II has been developed [6,7] which, to a great extent, overcomes the major difficulties encountered in [5].

3.1 Basic Dynamics

The revised QUAL II model [6,7] traces a drop of water along the basin from the most upper stream part till the most down stream part. It is based on the use of (4) under steady state conditions ($\partial \underline{c}/\partial t=0$) and for turbulence-free streams ($\partial/\partial x\ [A_x D_L \partial \underline{c}/\partial x]=0$). Both assumptions are justified by extensive computer simulation studies [4,6,7]. It includes however, the effect of input junctions, waste load points and output canals. Under these assumptions, the model (4) simplifies into the nonlinear dynamic model:

$$
\begin{aligned}
&d(v\ a)/dx=(\mu-\rho-\sigma_1/D)a+S_a/A_x\ dx\\
&d(v\ n_1)/dx=\alpha_1\rho a-\beta_1 n_1+\sigma_3/A_x+S_{n_1}/A_x d_x\\
&d(v\ n_2)/dx=\beta_1 n_1-\beta_2 n_2+S_{n_2}/A_x dx\\
&d(v\ n_3)/dx=\beta_2 n_2-\alpha_1\mu a+S_{n_3}/A_x dx\\
&d(v\ p)/dx=\alpha_2(\rho-\mu)a-\sigma_2/A_x\ +S_p/A_x dx\\
&d(v\ h)/dx=\ -(k_1+k_3)h\ +S_h/A_x dx\\
&d(v\ q)/dx\ =\ -k_2 q+(\alpha_3\mu-\alpha_4\rho)a-k_1 h+\alpha_5\beta_1 n_1-\alpha_6\beta_2 n_2+k_2 q^s-k_4/A_x+S_q/A_x dx
\end{aligned}
\qquad (5)
$$

where

(1) a is the algae concentration with growth rate μ, respiration rate ρ, settling rate σ_1, D is the average stream depth and v is the stream velocity.

(2) n_1 is the ammonia nitrogen concentration with biological oxidation rate β_1. α_1 is the fraction of respired algae biomass which is resolubilized as ammonia by bacterial action,σ_3 is the corresponding benthos source rate.

(3) n_2 is the nitrite nitrogen concentration with oxidation rate β_2.

(4) n_3 is the nitrate nitrogen concentration

(5) p is the phosphate phosphorous concentration. α_2 is the fraction of algae biomass which is phosphorous and σ_2 is the benthos source rate for phosphorous.

(6) h is the BOD concentration with rate of decay k_1 and k_3 is the loss rate due to settling.

(7) q is the DO concentration with saturation level q^s. α_3 is the rate of oxygen production per unit of algae (photosynthesis),α_4 is the rate of oxygen uptake per unit of algae respired,α_5 is the rate of oxygen uptake per unit of ammonia oxidation,α_6 is the rate of oxygen uptake per unit of nitrite nitrogen oxidation and k_2 is the areation rate.In(5),S_m is the source or sink of constituent m. The growth rate of algae μ has the form [4]:

$$\mu=\Phi\hat{\mu}\ (2\Pi/24\lambda D)(n_3/n_3+k_n)(p/p+k_p) \qquad (6)$$

where $\hat{\mu}$ is the maximum specific growth rate,λ is the light half-extension coefficient, k_n is the nitrogen half-saturation constant and k_p is the phosphorous half-saturation constant.Φ is a compound factor depending upon the photo period, amount of daily radiation and light saturation level.

To manipulate the model (5), we have to consider the stream velocity v. If in a given reach v is constant then it implies that $x=v\tau$ and $dx=vd\tau$ with τ being the transportation time. It is then easy to see that $d/dx[v\ m]=dm/d\tau$ for any concentration m. Using this result in (5) and (6), it yields:

$$d\ \underline{X}/d\tau = \underline{f}\ [\underline{X}] + B\underline{Z} + \underline{g} \tag{7}$$

where $\underline{X} = [\ a\ \ n_1\ n_2\ n_3\ p\ h\ q\ ]^t$; the vector of concentration levels, Z is the input vector = $[z_a\ z_{n1}\ z_{n2}\ z_{n3}\ z_p\ z_h\ z_q]^t$; $z_m=S_m/(A_x\ dx)$, $B= I_7$ (unit matrix of order 7) and $\underline{g}$ is an external input= $[0\ \sigma_3/A_x\ 0\ 0\ -\sigma_2/A_x\ \ 0\ \ (k_2q^s - k_4/A_x)]^t$.

Variable stream velocity in reaches correspond to $(\partial Q/\partial x=0)$ where Q is the discharge into the river. It turns out that the substitution of this condition in (4) using (5) leads to a state-variable model of the type (7), but $\underline{Z}$ in this case is a function of distance. The nonlinear function $\underline{f}[.]$ is given in [4].

The model (7) exhibits the variation of concentration of water quality constituents in an isolated reach of the river Nile which is, loosely speaking, an idealistic situation. To reflect the practical issues, let us consider a reach which receives one major controlled effluent discharge from a sewage or industrial waste treatment facility. We assume for simplicity that perfect mixing has taken place in this reach. Define Q_e as the flow rate of the effluent in the reach, Q as the stream flow rate, $\underline{r}$ as the vector of concentrations of the water quality constituents in the effluent discharging in the reach, V_0 as the volume of water in the reach, and $\underline{Y}$ as the vector of concentrations of the water quality constituents in the preceding reach.

Then the model (7) can be modified into [7]:

$$\dot{\underline{X}}=\underline{f}[X]+[Q/V_0]\underline{Y} - [Q+Q_e/V_0]\underline{X} +[Q_e/V_0]B\ \underline{r} + B\underline{Z}+\underline{g} \tag{8}$$

which describes the water quality dynamics in an arbitrary reach of the River Nile.

3.2 Linearization

For control purposes, it would seem desirable to linearize the model (8) about an equilibrium (steady-state) point. In doing this we consider that

(a) the effluents are pretreated to a desired level $\underline{r}^*$ which maintains water quality requirements $\underline{X}^*$, before their discharge into the river. We simply take the point $(\underline{X}^*,\underline{r}^*)$ as the equilibrium point, information about which can be obtained from the records of polluter stations.

(b) the concentration level Y has reached a steady-state level. Thus it can be considered as a constant disturbance acting upon the system. This is justified based on the geographical layout of polluter stations.

(c) the water flow into the reach is made variable by appropriately changing the stream velocity.

Now, we define $\underline{x}=\underline{X} - \underline{X}^*$, $\underline{w} = \underline{r} - \underline{r}^*$ as perturbations from the equilibrium and perform first-order expansions of the model (8) about $(\underline{X}^*,\underline{r}^*)$ to yield the linearized model:

$$\underline{\dot{x}} = F\,\underline{x} + E\,\underline{w} + \underline{b}\,s + \underline{d} \tag{9}$$

where

$$F = \left\{ \partial \underline{f}[\underline{X}]/\partial \underline{X} - (Q+Q_e)/V_0 \right\}_{\underline{X}=\underline{X}^*} \tag{10a}$$

$$E = [Q_e/V_o]B \tag{10b}$$

$$\underline{b} = (\underline{Y}^* - \underline{X}^*)/V_0 \tag{10c}$$

$$\underline{d} = B\underline{Z} + \underline{g} + \underline{f}\,[\underline{X}^*] \tag{10d}$$

and s is a command signal; proportional to the change in stream velocity. It has been observed [4] that only algae is directly affected by stream velocity, and that the variables (ammonia, nitrate, phosphate, phosphorous, BOD) are only controllable through waste water treatment. In the light of this important observation and using measured data (see [4], Ch.10), the model (9) reduces to:

$$\underline{\dot{x}} = F\,\underline{x} + G\,\underline{u} \tag{11}$$

where

$$F = \begin{bmatrix} -.89825 & 0 & 0 & .00298 & .02587 & 0 & 0 \\ .005 & -1.0457 & 0 & 0 & 0 & 0 & 0 \\ 0 & .3 & -1.2457 & 0 & 0 & 0 & 0 \\ -.0031 & 0 & .5 & -.746 & -.00259 & 0 & 0 \\ -.00019 & 0 & 0 & -.00003 & -.74596 & 0 & 0 \\ 0 & 0 & 0 & 0 & 0 & -.9757 & 0 \\ -.00681 & -.966 & -.56 & .00387 & .03363 & -.23 & -1.2457 \end{bmatrix} \tag{12a}$$

$$G = \begin{bmatrix} .01 & 0 & 0 & 0 & 0 \\ 0 & .1 & 0 & 0 & 0 \\ 0 & 0 & 0 & 0 & 0 \\ 0 & 0 & .1 & 0 & 0 \\ 0 & 0 & 0 & .1 & 0 \\ 0 & 0 & 0 & 0 & .1 \\ 0 & 0 & 0 & 0 & 0 \end{bmatrix} \tag{12b}$$

$$\underline{u} = \begin{bmatrix} s & w_1 & w_2 & w_3 & w_5 \end{bmatrix}^t \tag{12c}$$

We should emphasize that the entries of the constant vector $\underline{d}$ are of negligeable magnitudes, hence it is set to zero in (11).

4. DECENTRALIZED CONTROL

Our purpose is to derive the control signal $\underline{u}^0(t)$ which derives the linearized model (11) while minimizing

$$J = \tfrac{1}{2} \int_0^\infty \left(\|\underline{x}\|^2_{R_1} + \|\underline{u}\|^2_{R_2} \right) dt; \quad R_1 \geq 0 \ , \ R_2 > 0 \tag{13}$$

where $||\underline{a}||^2_V = \underline{a}^t V \underline{a}$. In view of the geographic separation of the control actions (adjustment of the stream velocity s and variation of level of water treatment $\underline{W}$), it is quite natural to seek a decentralized control structure. To achieve this, we decompose the seventh-order model (12) into two submodels: one of first-order representing the dynamics of algae and the second of sixth-order representing the dynamics of other constituents. The optimal control is of the form

$$\underline{u}^0(t) = - G_b \underline{x}(t) - T \dot{\underline{x}}(t) \tag{14}$$

where G_b is the block diagonal matrix obtained by solving a Riccati equation independently for each submodel and T is a full matrix obtained by a hierarchical technique [13]. One way of obtaining a decentralized control is to constrain the matrix T to be a diagonal matrix T_d. Thus we need to find the T_d which minimizes

$$J = \tfrac{1}{2} \int_0^\infty (||\underline{x}||^2_{R_1} + ||\underline{x}||^2_R)\, dt \tag{15a}$$

where

$$R = G_b^t R_2 G_b + G_b^t R_2 T_d + T_d^t R_2 G_b + T_d^t R_2 T_d \tag{15b}$$

subject to

$$\dot{\underline{x}} = (F - G G_b) \underline{x} - G T_d \underline{x} \tag{16}$$

The problem at hand can be solved by a three-level computation structure [13]. With t_f=4 days, it was found [4] that steady-state trajectories can be obtained. Using $R_1=I_7$ and $R_2=I_5$, the simulation results are

$$G_b = \begin{bmatrix} .05549 & 0 & 0 & 0 & 0 & 0 & 0 \\ 0 & .06867 & 0 & 0 & 0 & 0 & 0 \\ 0 & 0 & 0 & .06673 & 0 & 0 & 0 \\ 0 & 0 & 0 & 0 & .06676 & 0 & 0 \\ 0 & 0 & 0 & 0 & 0 & .05208 & 0 \end{bmatrix}$$

$$T_d = \begin{bmatrix} .02456 & 0 & 0 & 0 & 0 & 0 & 0 \\ 0 & -.02076 & 0 & 0 & 0 & 0 & 0 \\ 0 & 0 & 0 & .00058 & 0 & 0 & 0 \\ 0 & 0 & 0 & 0 & .00161 & 0 & 0 \\ 0 & 0 & 0 & 0 & 0 & -.00079 & 0 \end{bmatrix}$$

The corresponding optimal state and control trajectories are displayed in Figs.(1)-(5). We observe that the implementation of the decentralized control $\underline{u}^0(t) = -(G_b + T_d)\, \underline{x}(t)$ avoids state informations transfer. Thus, it is quite suitable for utilization in preserving water quality standards in the River Nile.

5. CONCLUSIONS

We have presented a linearized version of a recently developed water quality model for the River Nile. The linearized model has two control actions: one is concerned with the adjustment of stream velocity and the other is related to the variation of level of water treatment. A decentralized control structure is designed using a three-level algorithm. Simulation results for a typical data are presented.

6. REFERENCES

[1] Grenney, W.J., et al.
"Water Quality Relationships to Flow Stream and Estuaries", in (Methodologies for the Determination of Stream Resource Flow Requirement: An Assessment), edited by C.B. Stalnaker, U.S. Fish and Wildlife Service, Office of Biological Service, Western Water Allocation, Washington DC, 1973.

[2] Texas Water Development Board.
"Simulation of Water Quality in Streams and Canals", Program Documentation and Users Manual, EPA-OWP-TEX-DOSAGI, NTIS, 1970.

[3] Chen, C.W. and J. Wells
"Boise River Water Quality Ecologic Model for Urban Planning Study", Tetra Tech. Reports No. TC-368&TC-605, Tetra Tech.Inc., Lafayette, CA, 1966.

[4] Mahmoud, M.S., M.F. Hassan and M.G. Darwish
"Large Scale Control Systems: Theories and Techniques", Marcel Dekker, N.Y.,1984.

[5] E.P.A.
"Computer Program Documentation for the Stream Quality Model QUAL II", Systems Development Branch, Washington DC, 1972.

[6] Hassan, M.F., M.I. Younis and K.H. Mancy.
"A Developed Stream Water Quality Model: A Case Study on the River Nile", Proc. IFAC Systems Approach for Development, Rabat, MOROCO, 1980.

[7] Hassan, M.F., M.I. Younis and K.H. Mancy.
"Model Development and Optimization of the River Nile Water Quality", Proc. Water Resource Management, Cairo, EGYPT, 1981.

[8] Dobbis, W.E.
"B.O.D. and Oxygen Relationships in Streams", ASCE J. Sanitary Eng. Div., Vol. 90, 1966.

[9] Camp, T.R.
"Field Estimates of Oxygen Balance Parameters", ASCE J. Sanitary Eng. Div., Vol. 91, 1967.

[10] Young, P., B. Beck and M.G. Singh
"The Modelling and Control of Pollution in a River System", Report CUED/8-Control/TR32, University of Cambridge, 1972.

[11] Tamura, H.
"A Discrete Dynamic Model with Distributed Transport Delays and its Hierarchical Optimization for Preserving Stream Quality", IEEE Trans. Systems, Man and Cybernetics, Vol. SMC-4, 1974, pp. 424-431.

[12] Singh,M.G. and M.F. Hassan
"Closed-Loop Hierarchical Control for River Pollution", Automatica, Vol.12, 1976, pp. 261-266.

[13] Singh, M.G. and A. Titli
"Systems: Decomposition, Optimization and Control", Pergamon Press,Oxford, 1978.

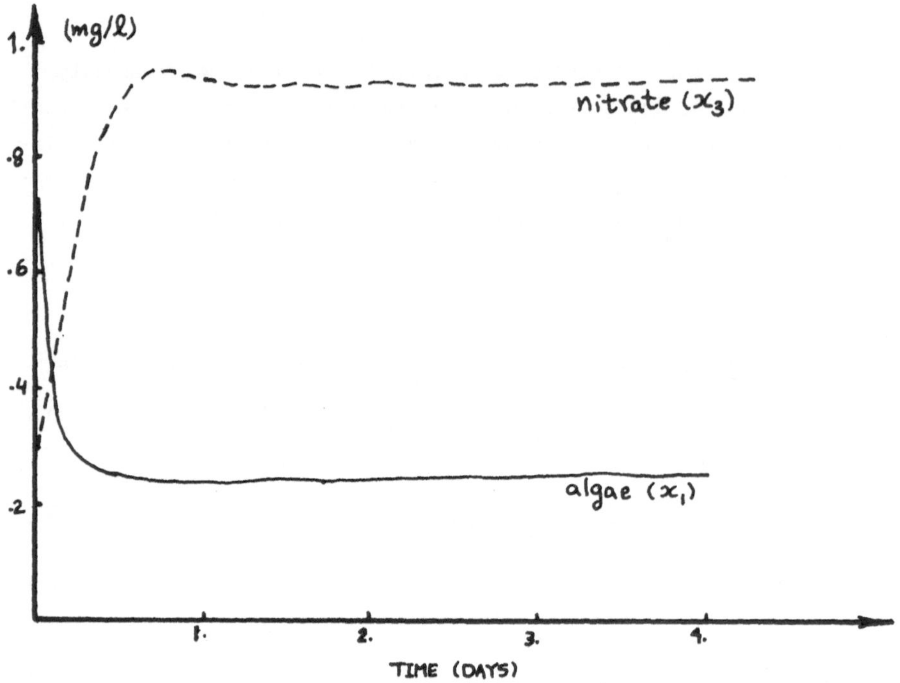

Fig.(1): Optimal variations of algae and nitrate

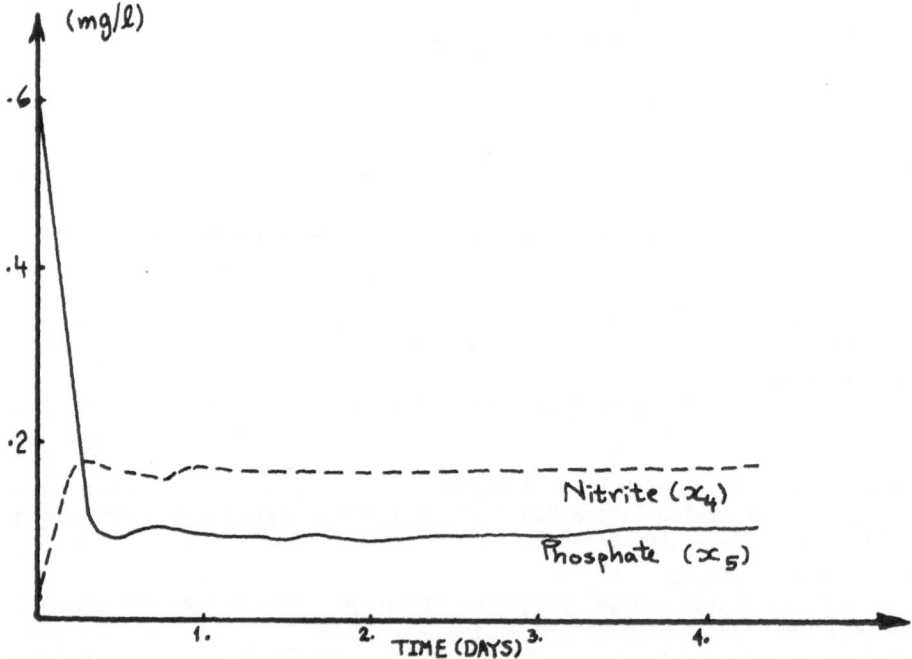

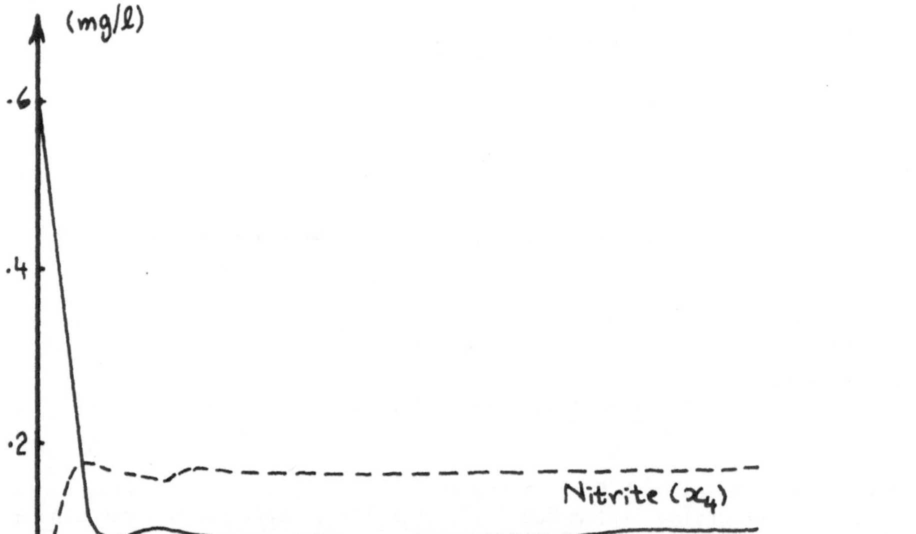

Fig.(2): Optimal variations of nitrite and phosphate phosphorous

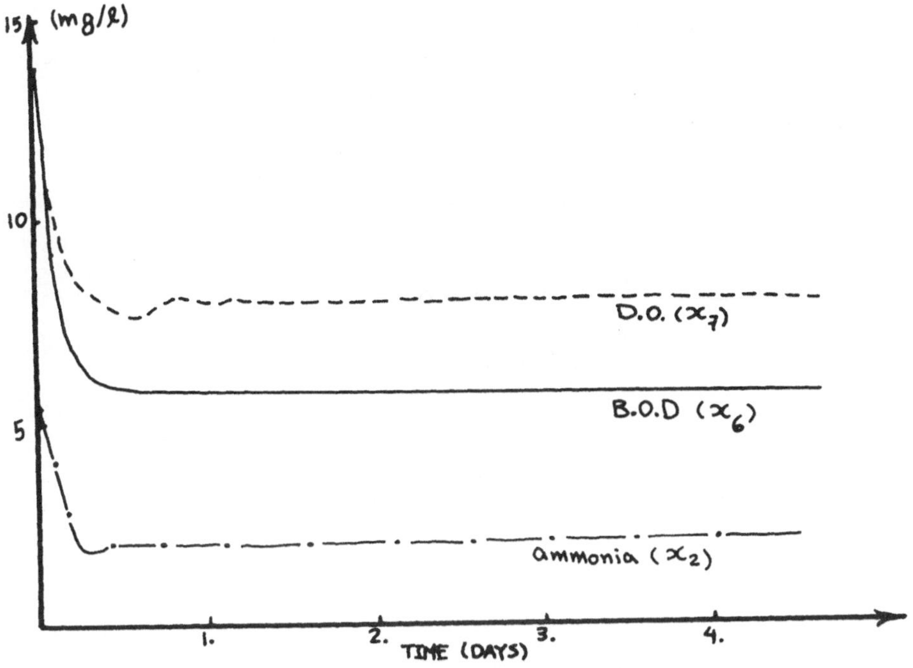

Fig.(3): Optimal variations of ammonia, B.O.D. and D.O.

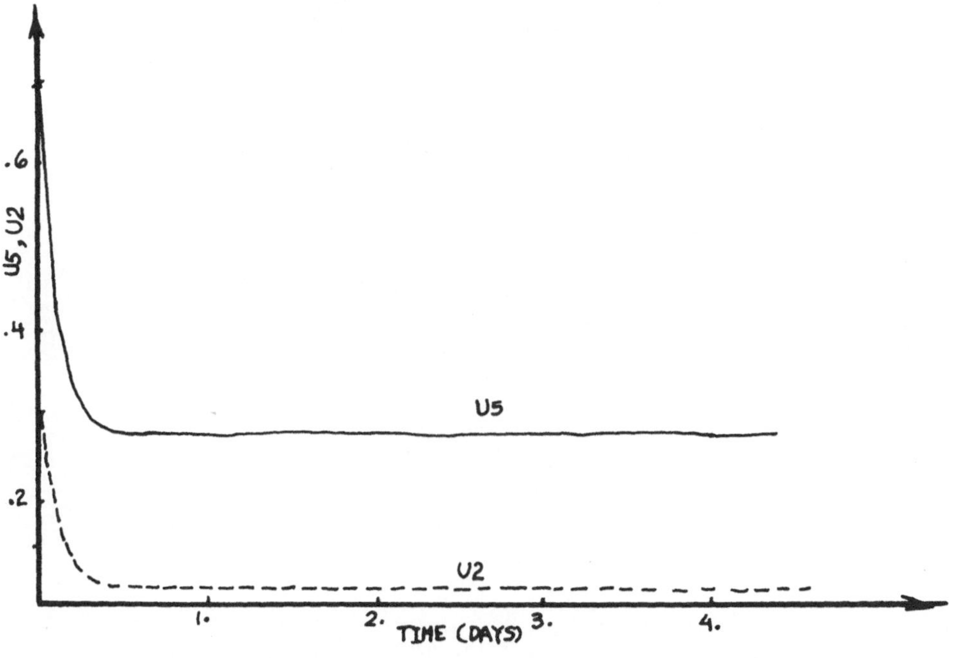

Fig.(4): Optimal variations of control signals u_2 and u_5

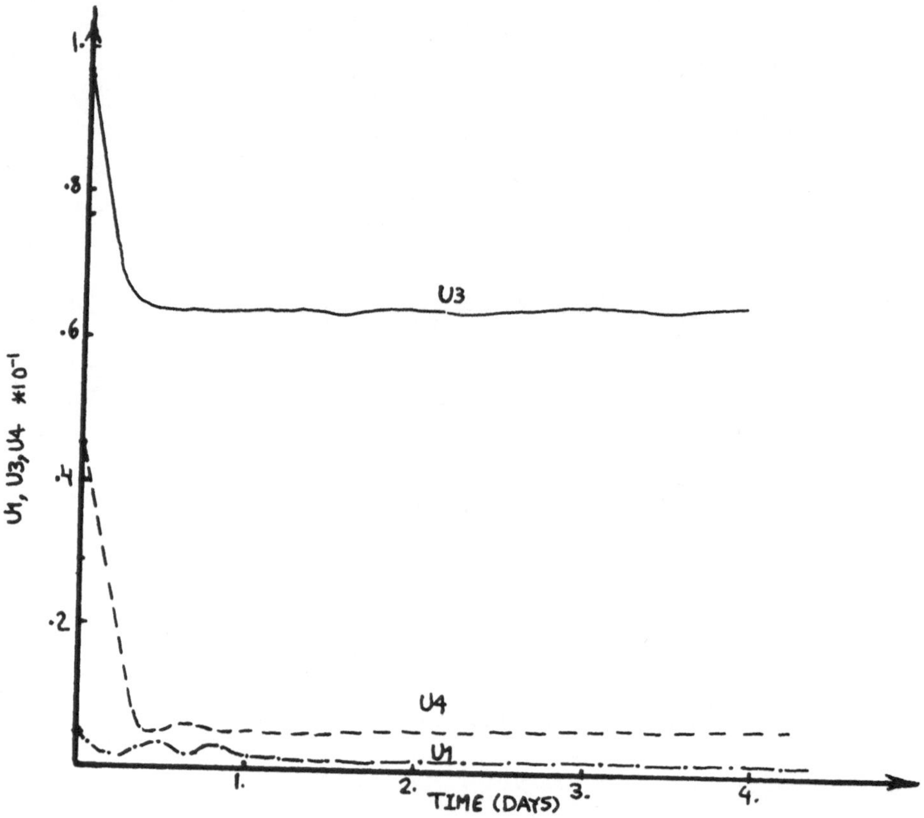

Fig.(5): Optimal variations of control signals u_1 , u_4 and u_5

ADAPTIVE CONTROL OF LARGE SCALE SYSTEMS

WITH REDUCED MODELS*

by

Petros Ioannou
University of Southern California
Dept. of Electrical Engineering-Systems
Los Angeles, CA 90089-0781

ABSTRACT

This paper considers the robust redesign of decentralized adaptive controllers for a class of large scale systems formed of an arbitrary interconnection of subsystems with unknown parameters. In the design, a lower order model is assumed for each decoupled subsystem. However, the overall adaptive control scheme is analyzed in the presence of bounded external disturbances and unmodeled plant uncertainties. It is shown that, by properly choosing some design parameters in the adaptive laws, the decentralized adaptive control scheme is globally stable in the sense that all the signals in the closed loop are bounded for any bounded initial conditions, and the tracking/parameter errors converge to a small residual set.

1. Problem Statement

Consider a system which is described as an interconnection of N subsystems, i.e.,

$$\dot{x}_{si} = A_{si}x_{si} + b_{si}u_i + \varepsilon_{si}E_{si}z_i + d_{si} + \sum_{j\neq i}^{N} A_{sij}x_{sj} \tag{2.1}$$

$$\dot{z}_i = A_{sii}z_i + b_{sii}u_i + \sum_{j=1}^{N} E_{sij}x_{sj} + d_{sii} \tag{2.2}$$

$$y_{si} = h_{si}^T x_{si} + \mu_{si} g_i^T z_i + d_{oi}, \quad i = 1,2,\ldots,N \tag{2.3}$$

where for the i-th subsystem: $x_i \in R^{n_i}$, $z_i \in R^{m_i}$ are the state vectors, $u_i \in R^1$ is the control variable, $y_i \in R^1$ is the output and $d_{si} \in R^{n_i}$, $d_{sii} \in R^{m_i}$ and $d_{oi} \in R^1$ are bounded external disturbances. The parameters of (2.1) to (2.3) are constant but unknown and ε_{si}, μ_{si} are small unknown constant scalars.

Without changing the input-output characteristics, we can also represent (2.1) to (2.3) as

$$\dot{x}_i = A_i x_i + b_i u_i + d_i + \sum_{j\neq i}^{N} A_{ij}x_j + \sum_{j=1}^{N} \varepsilon_j H_{ij} z_j \tag{2.4}$$

---

*This work was supported in part by the U.S.C. Faculty Research and Innovation Fund, University of Southern California, Los Angeles, CA 90089 and in part by the U.S. Department of Energy, Electric Energy Systems Division under Contract DE-AC01-81RA-50658, with Dynamic Systems, Inc., P.O. Box 423, Urbana, IL 61801.

$$\dot{z}_i = A_{ii}z_i + b_{ii}u_i + d_{ii} + \sum_{j=1}^{N} E_{ij}x_j + \sum_{j\neq i}^{N} \varepsilon_j M_{ij} z_j \qquad (2.5)$$

$$y_i = [1\ 0\ .\ .\ 0]x_i = h_i^T x_i \qquad (2.6)$$

where $\varepsilon_j = \max[\varepsilon_{sj}, \mu_{sj}]$, by assuming that $\dot{d}_{oi}$ is bounded. A similar representation (2.4) to (2.6) can be obtained by filtering the local outputs y_{si} using first order filters [1,2] and without having to assume boundedness for $\dot{d}_{oi}$.

Assuming that

$$\text{Re}\lambda(A_{ii}) < 0\ , \quad i = 1,2,\ldots,N \qquad (2.7)$$

we can simplify (2.4) to (2.6) by neglecting (i) the effects of the dynamics of (2.5) (i.e., $\varepsilon_j = 0$), (ii) the interconnections between subsystems ($A_{ij} = 0$), and (iii) the effects of the external disturbances ($d_i = 0$, $d_{ii} = 0$). Under the assumptions (i), (ii), and (iii), (2.4), (2.6) become

$$\dot{\bar{x}}_i = A_i\bar{x}_i + b_i u_i \qquad (2.8)$$

$$\bar{y}_i = h_i^T \bar{x}_i\ , \quad i = 1,2,\ldots,N \qquad (2.9)$$

The simplification of (2.4) to (2.6) to the lower order decoupled subsystems is quite common in practice and has several advantages: First the subsystems (2.5) are difficult to control due to the weak observability of the states z_i in the plant output. Furthermore, the order m_i of z_i might be too large or unknown. Second, the disturbances are unknown and therefore they cannot be accurately modeled. In most cases, these disturbances are of low-level and do not affect the performance of the system very much. Third, the lack of communication between subsystems or computational constraints [3,4] demand a decentralized solution rather than a centralized one. In the decentralized case, each subsystem is assumed to be isolated (i.e., $A_{ij} = 0$) and is not allowed to exchange information with the other subsystems.

For the reduced-order system (2.8), (2.9) we make the following assumptions:

A1 : The order n_i is known

A2 : The triples (A_i, b_i, h_i) are completely controllable and observable

A3 : In the transfer function

$$W_i(s) = h_i^T(sI - A_i)^{-1}b_i = k_i\,\frac{Z_i(s)}{R_i(s)} \qquad (2.10)$$

$Z_i(s)$ is a monic Hurwitz polynomial, $R_i(s)$ is a monic polynomial of degree n_i, the sign of k_i and the relative degree n_i^* of $W_i(s)$ are known. Without loss of generality, we assume that k_i is positive.

No assumptions, however, are made about the degree m_i of the neglected subsystems and the relative degree of the overall system. The parameters of the reduced-order as well as the full order system are assumed to be constant but unknown.

Our objective is to design local adaptive controllers using only local information so that the output y_i of each subsystem is regulated to zero or it tracks the output y_{mi} of a corresponding local reference model

$$\dot{x}_{mi} = A_{mi}x_{mi} + b_{mi}r_i \tag{2.10}$$

$$y_{mi} = c_{mi}^T x_{mi} \tag{2.11}$$

whose transfer function

$$W_{mi}(s) = c_{mi}^T(sI - A_{mi})^{-1}b_{mi} = k_{mi}\frac{Z_{mi}(s)}{R_{mi}(s)} \tag{2.12}$$

is chosen to be strictly positive real, r_i is a uniformly bounded reference input signal, and (2.10) have the same order as the corresponding reduced-order decoupled subsystem (2.8), (2.9).

In the absence of disturbances, interconnections and plant uncertainties, i.e., when the plant is (2.8), (2.9) instead of (2.4) to (2.6) the solution follows directly from the single-input single-output (SISO) adaptive control problem. However, as it was shown in [1], [5-8] when disturbances, interconnections or unmodeled dynamics are present, some of the present adaptive control schemes can become unstable.

In the following section, we will introduce a modified adaptive controller whose design is based on the reduced-order decoupled subsystem (2.8), (2.9) and is robust when applied to the real system (2.4) to (2.6) with disturbances, plant uncertainties and interconnections.

2. Decentralized Adaptive Controller

Each control station assumes that the simplified system (2.8), (2.9) is a valid model for the actual system (2.4) to (2.6) and therefore designs its own adaptive controller based on (2.8), (2.9) rather than (2.4) to (2.6). We consider the simple adaptive control case where $n_i^* = 1$. The output y_i and input u_i are used to generate the $(2n_i - 2)$ vector $v_i = [v_i^{(1)T}, v_i^{(2)T}]^T$, i.e.,

$$\dot{v}_i^{(1)} = \Lambda_i v_i^{(1)} + g_i u_i \tag{3.1}$$

$$w_i^{(1)} = c_i^T(t)v_i^{(1)} \tag{3.2}$$

$$\dot{v}_i^{(2)} = \Lambda_i v_i^{(2)} + g_i y_i \tag{3.3}$$

$$w_i^{(2)} = d_{oi}(t)y_i + d_i^T(t)v_i^{(2)} \tag{3.4}$$

where Λ_i is an $(n_i - 1) \times (n_i - 1)$ stable matrix, $g_i = [0,0\ldots,1]^T$, (Λ_i, g_i) is a contrallable pair and Λ_i is chosen such that $\det(sI - \Lambda_i) =$

$Z_{mi}(s)$. The vectors $C_i(t)$, $d_i(t)$ and scalar $d_{oi}(t)$ are time varying and adjustable.

The local input is chosen as

$$u_i = \theta_i^T w_i + \dot{\theta}_i^T \xi_i \tag{3.5}$$

$$\dot{\xi}_i = -a_i \xi_i + w_i, \quad \xi_i(0) = 0 \tag{3.6}$$

$$\dot{\theta}_i = -\sigma_i \theta_i - \Gamma e_{oi} \xi_i \tag{3.7}$$

where

$$\theta_i = [k_{oi}(t), C_i^T(t), d_{oi}(t) d_i^T(t)]^T$$

$$w_i = [r_i, v_i^{(1)T}, y_i, v_i^{(2)T}] \text{ and}$$

$$e_{oi} = y_i - y_{mi} \tag{3.8}$$

and a_i, σ_i are positive constant scalars to be designed. It can be shown [9,10] that a constant vector $\theta_i^* = [k_{oi}^*, C_i^{*T}, d_{oi}^*, d_i^{*T}]^T$ exists such that for $\theta_i = \theta_i^*$ and $\dot{\theta}_i = 0$ the transfer function $\bar{y}_i(s)/r_i(s)$ of the reduced order decoupled i-th subsystem (2.8), (2.9) with controller (3.1) to (3.4) matches that of the i-th reference model given by (2.12).

Let us now apply (3.1) to (3.5) to the actual system (2.4) to (2.6) with disturbances, interconnections, and unmodeled dynamics. Introducing

$$Y_i^T = [x_i^T, v_i^{(1)T}, v_i^{(2)T}] \text{ and}$$

$$\phi_i = \theta_i - \theta_i^* \tag{3.9}$$

we can write the closed loop system as

$$\dot{Y}_i = A_{ci} Y_i + b_{ci}(\phi_i^T w_i + k_{oi}^* r + \dot{\phi}_i^T \xi_i) + D_{ci} + \sum_{j \neq i}^{N} \bar{A}_{ij} Y_j + \sum_{i=1}^{N} \varepsilon_j \bar{H}_{ij} z_j \tag{3.10}$$

$$\dot{z}_i = A_{ii} z_i + b_{ii}(\phi_i^T w_i + \dot{\phi}^T \xi_i) + b_{ii} \theta_i^{*T} w_i + d_{ii} + \sum_{j=1}^{N} \bar{E}_{ij} Y_j + \sum_{j \neq i}^{N} \varepsilon_j M_{ij} z_j \tag{3.11}$$

$$y_i = [1\ 0\ .\ .\ .\ 0\ ] Y_i = h_{ci}^T Y_i \tag{3.12}$$

where

$$A_{ci} = \left[\begin{array}{c|c|c} A_i + d_{oi}^* b_i h_i^T & b_i c_i^{*T} & b_i d_i^{*T} \\ \hline g_i d_{oi}^* h_i^T & \Lambda_i + g_i c_i^{*T} & g_i d_i^{*T} \\ \hline g_i h_i^T & 0 & \Lambda_i \end{array}\right], \quad b_{ci} = \left[\begin{array}{c} b_i \\ \hline g_i \\ \hline 0 \end{array}\right]$$

$$D_{ci} = [d_i \mid 0 \mid 0\ ] \tag{3.13}$$

and $\bar{A}_{ij}$, $\bar{H}_{ij}$, $\bar{E}_{ij}$ are appropriately defined.

3. Error System and Stability Analysis

For $\theta_i = \theta_i^*$, $D_{ci} = 0$, $\bar{A}_{ij} = 0$, $\bar{H}_{ij} = 0$, (3.10) is a non-minimal representation of the i-th reference model

$$\dot{x}_{ci} = A_{ci}x_{ci} + b_{ci}k_{oi}^* r_i \tag{4.1}$$

where

$$x_{ci} = [x_{mi}^T, v_{mi}^{(1)T}, v_{mi}^{(2)T}] \tag{4.2}$$

Similarly, for $\theta_i = \theta_i^*$ and $Y_i = x_{ci}$ (3.11) is a representation of

$$\dot{z}_{mi} = A_{ii}z_{mi} + b_{ii}\theta_i^{*T}w_{mi} + d_{ii} + \sum_{j=1}^{N} \bar{E}_{ij}x_{cj} + \sum_{j\neq i}^{N} \varepsilon_j M_{ij} z_{mj}, \quad z_{mi}(0) = z_i(0) \tag{4.3}$$

and (3.6) is a representation of

$$\dot{\xi}_{mi} = -a_i\xi_{mi} + w_{mi}, \quad \xi_{mi}(0) = 0 \tag{4.4}$$

where

$$w_{mi} = [r_i, v_{mi}^{(1)T}, y_{mi}, v_{mi}^{(2)T}]^T. \tag{4.5}$$

We should note that systems (4.1) to (4.5) are not available and they cannot be used for implementation. However, they can be used for analysis since $z_{mi}(t)$ and $\xi_{mi}(t)$ are bounded and can also be considered as the desired equilibrium trajectories of $z_i(t)$ and $\xi_i(t)$ respectively.

Defining

$$\begin{aligned} e_i &= Y_i - x_{ci} \\ e_{mi} &= z_i - z_{mi} \\ e_{fi} &= \xi_i - \xi_{mi} \end{aligned} \tag{4.6}$$

we can obtain from (3.6), (3.10) to (3.12) and (4.1) to (4.5)

$$\dot{e}_i = A_{ci}e_i + b_{ci}(p + a_i)(\phi_i^T\xi_i) + D_{ci} + \sum_{j\neq i}^{N} \bar{A}_{ij}(e_j + x_{cj}) + \sum_{j=1}^{N} \varepsilon_j \bar{H}_{ij}(e_{mj} + z_{mj}) \tag{4.7}$$

$$\dot{e}_{mi} = A_{ii}e_{mi} + b_{ii}(p + a_i)(\phi_i^T\xi_i) + b_{ii}\theta_i^{*T}\bar{e}_i + \sum_{j=1}^{N} \bar{E}_{ij}e_j + \sum_{j\neq i}^{N} \varepsilon_j M_{ij}e_{mj} \tag{4.8}$$

$$\dot{e}_{fi} = -a_i e_{fi} + \bar{e}_i \tag{4.9}$$

$$e_{oi} = h_{ci}^T e_i \tag{4.10}$$

where p is the differential operator $\frac{d}{dt}$ and

$$\bar{e}_i = [0, e_i^T]^T \tag{4.11}$$

Equations (4.7) to (4.10) together with

$$\dot{\phi}_i = -\sigma_i(\phi_i + \theta_i^*) - \Gamma e_{oi}\xi_i \tag{4.12}$$

for $i = 1,2,\ldots,N$ describe the stability properties of the overall decentralized control scheme. A property of the triples (A_{ci}, b_{ci}, h_{ci}) which is crucial for stability is the strict positive realness of the transfer function

$$h_{ci}^T(sI - A_{ci})^{-1}b_{ci} = \frac{k_i}{k_{mi}} W_{mi}(s) \tag{4.13}$$

Using the Kalman-Yacobovich lemma, it follows from (4.13) that a matrix $P_{ci} = P_{ci}^T > 0$ exists such that

$$P_{ci}A_{ci} + A_{ci}^T P_{ci} = -q_i q_i^T - \nu_i L_i \tag{4.14}$$

$$P_{ci}b_{ci} = h_{ci} \tag{4.15}$$

are satisfied for some vector q_i, matrix $L_i = L_i^T > 0$ and $\nu_i > 0$.

The stability of matrices A_{ii} imply the existence of matrices $P_i = P_i^T > 0$ and $Q_i = Q_i^T > 0$ such that

$$P_i A_i + A_i^T P_i = -Q_i \tag{4.16}$$

is satisfied. Using (4.14) to (4.16) the following theorem establishes the stability properties of (4.7) to (4.12).

Theorem 1: Let

$$\lambda_i = \frac{1}{2}\nu_i \min\lambda(L_i)$$

$$a_{ij} = \| P_{ci}\bar{A}_{ij} \| \tag{4.17}$$

If there exists an N-vector $\alpha = [\alpha_1, \alpha_2, \ldots, \alpha_N]^T$ with positive elements such that the $N \times N$ matrix S with elements

$$s_{ij} = \begin{cases} \alpha_i \lambda_i & i = j \\ -(\alpha_i a_{ij} + \alpha_j a_{ji}) & i \neq j \end{cases} \tag{4.18}$$

is positive definite, then there exists positive constants $\sigma_i > 0$, $a_i^* > 0$, $\rho_o > 0$ and $\varepsilon^* > 0$ such that for each $|\varepsilon_i| \in [0, \varepsilon^*]$ and a_i satisfying

$$a_i^* < a_i < O\left(\frac{1}{|\varepsilon_i|}\right) \tag{4.19}$$

the solution $e_i(t)$, $e_{mi}(t)$, $e_{fi}(t)$, $\phi_i(t)$, $i = 1,2,\ldots,N$ is uniformly bounded and converges with a rate at least as fast as $\exp(-\frac{1}{2}\rho_o t)$ to the residual set

$$D_r = \{e_i, e_{mi}, e_{fi}, \phi_i : V(e_i, e_{mi}, e_{fi}, \phi_i) \le \frac{k_o}{\rho_o}\} \tag{4.20}$$

where

$$V(e_i, e_{mi}, e_{fi}, \phi_i) = \frac{1}{2} \sum_{i=1}^{N} \Big\{ \alpha_i \Big[\left(e_i - b_{ci}\phi_i^T\xi_i\right)^T P_{ci} \left(e_i - b_{ci}\phi_i^T\xi_i\right) + a_i \phi_i^T \Gamma_i^{-1} \phi_i \Big] + |\varepsilon_i| \Big[\left(e_{mi} - b_{ii}\phi_i^T\xi_i\right)^T P_i \left(e_{mi} - b_{ii}\phi_i^T\xi_i\right) \Big] + a_i |\varepsilon_i| e_{fi}^T e_{fi} \Big\} , \tag{4.21}$$

$$K_o = \frac{\sigma_i a_i}{2\gamma_i} \|\Gamma_i^{-1}\theta_i^*\|^2 + \frac{(2+N)}{2\lambda_s} \alpha_i^2 m_{oi}^2 + \frac{3(N+1)}{2a_i g_i} \alpha_i^2 n_{oi}^2 , \tag{4.22}$$

γ_i, λ_s, g_i are positive constants, $\Gamma_i = \Gamma_i^T > 0$,

$$n_{oi} = \sup_t [\| D_{ci}(t) + \sum_{j \ne i}^{N} \bar{A}_{ij} x_{cij}(t) \|] , \tag{4.23}$$

$$m_{oi} = \|P_{ci}\| \left(n_{oi} + \sup_t [\sum_{j=1}^{N} |\varepsilon_j| \|\bar{H}_{ij} z_{mj}(t)\|]\right) \tag{4.24}$$

$$a_i^* = \frac{(2+N)}{g_i \lambda_s} \alpha_i^2 \left(\|A_{ci}\|^2 + \sum_{j \ne i}^{N} \|A_{ij}\|\right) \tag{4.25}$$

<u>Proof</u>: Consider the positive definite function V given by (4.21). The time derivative of V along the solution of (4.7) to (4.12) is

$$\dot{V} = \sum_{i=1}^{N} \alpha_i \Big[\left(e_i - b_{ci}\phi_i^T\xi_i\right)^T P_{ci} \Big(A_{ci}e_i + a_i b_{ci}\phi_i^T\xi_i + D_{ci} + \sum_{j \ne i}^{N} \bar{A}_{ij}(e_j + x_{cij}) + \sum_{j=1}^{N} \varepsilon_j \bar{H}_{ij}(e_{mj} + z_{mj}) \Big) - a_i \sigma_i \phi_i^T \Gamma_i^{-1}(\phi_i + \theta_i^*) - a_i e_{oi}\phi_i^T\xi_i \Big] + \sum_{i=1}^{N} |\varepsilon_i| \Big[\left(e_{mi} - b_{ii}\phi_i^T\xi_i\right)^T P_i \Big(A_{ii}e_{mi} + a_i b_{ii}\phi_i^T\xi_i + b_{ii}\theta_i^{*T}\bar{e}_i + \sum_{j=1}^{N} \bar{E}_{ij}e_j + \sum_{j \ne i}^{N} \varepsilon_j M_{ij} e_{mj} \Big) \Big] + \sum_{i=1}^{N} a_i |\varepsilon_i| \left(-a_i e_{fi}^T e_{fi} + e_{fi}^T \bar{e}_i\right) \tag{4.26}$$

Using (4.14) to (4.16), (4.26) can be rewritten as

$$\dot{V} = \sum_{i=1}^{N} \alpha_i \Big[-\frac{1}{2} e_i^T (q_i q_i^T + \nu_i L_i) e_i + e_i^T P_{ci} \sum_{j \ne i}^{N} \bar{A}_{ij} e_j + e_i^T P_{ci} \Big(D_{ci} + \sum_{j \ne i}^{N} A_{ij} x_{cij} + \sum_{j=1}^{N} \varepsilon_j H_{ij} z_{mj} \Big) + e_i^T P_{ci} \sum_{j=1}^{N} \varepsilon_j \bar{H}_{ij} e_{mj}$$

$$- h_{ci}^T A_{ci} e_i \phi_i^T \xi_i - a_i b_{ci}^T P_{ci} b_{ci} (\phi_i^T \xi_i)^2$$

$$- \phi_i^T \xi_i h_{ci}^T \left(D_{ci} + \sum_{j \neq i}^{N} \bar{A}_{ij} x_{cij} \right) - \phi_i^T \xi_i h_{ci}^T \sum_{j \neq i}^{N} \bar{A}_{ij} e_j$$

$$- \phi_i^T \xi_i h_{ci}^T \sum_{j=1}^{N} \varepsilon_j M_{ij} e_{mj} - \sigma_i a_i \phi_i^T \Gamma_i^{-1} \phi_i - \sigma_i a_i \phi_i^T \Gamma_i^{-1} \theta_i^* \Big]$$

$$+ \sum_{i=1}^{N} |\varepsilon_i| \Big[-\frac{1}{2} e_{mi}^T Q_i e_{mi} + a_i e_{mi}^T P_i b_{ii} \phi_i^T \xi_i + e_{mi}^T P_i b_{ii} \theta_i^{*T} \bar{e}_i$$

$$+ e_{mi}^T P_i \sum_{j=1}^{N} \bar{E}_{ij} e_j + e_{mi}^T P_i \sum_{j \neq i}^{N} \varepsilon_j M_{ij} e_{mj} - \phi_i^T \xi_i b_{ii}^T P_i A_{ii} e_{mi}$$

$$- a_i b_{ii}^T P_i b_{ii} (\phi_i^T \xi_i)^2 - \phi_i^T \xi_i b_{ii}^T P_i b_{ii} \theta_i^{*T} \bar{e}_i$$

$$- \phi_i^T \xi_i b_{ii}^T P_i \sum_{j=1}^{N} \bar{E}_{ij} e_j - \phi_i^T \xi_i b_{ii}^T P_i \sum_{j \neq i}^{N} \varepsilon_j M_{ij} e_{mj} \Big]$$

$$+ \sum_{i=1}^{N} \Big[-a_i^2 |\varepsilon_i| e_{fi}^T e_{fi} + a_i |\varepsilon_i| e_{fi}^T \bar{e}_i \Big] \tag{4.27}$$

Defining
$$\lambda_{qi} = \tfrac{1}{2} \min \lambda (Q_i),$$

$$q_{ij} = \|P_{ci} \bar{H}_{ij}\| \ , \quad p_i = \|P_i b_{ii} \theta_i^{*T}\|$$

$$s_i = \|P_i b_{ii}\| \ , \quad \tau_{ij} = \|P_i \bar{E}_{ij}\|$$

$$J_{ij} = \|P_i M_{ij}\| \ , \quad \pi_i = \|b_{ii}^T P_i A_{ii}\|$$

$$f_{ij} = \|b_{ii}^T P_i M_{ij}\| \ , \quad g_i = b_{ci}^T P_{ci} b_{ci}$$

and using (4.17), (4.23), (4.24), we can write (4.27) as

$$\dot{V} \leq \sum_{i=1}^{N} \alpha_i \Big[-\lambda_i \|e_i\|^2 + \|e_i\| \sum_{j \neq i}^{N} (\|P_{ci} \bar{A}_{ij}\| \|e_j\|) + \|e_i\| m_{oi}$$

$$+ \|e_i\| \sum_{j=1}^{N} (|\varepsilon_j| q_{ij} \|e_{mj}\|) + \|A_{ci}\| \|e_i\| \|\phi_i^T \xi_i\| - a_i g_i (\phi_i^T \xi_i)^2$$

$$+ |\phi_i^T \xi_i| n_{oi} + |\phi_i^T \xi_i| \sum_{j \neq i}^{N} (\|A_{ij}\| \|e_j\|) + |\phi_i^T \xi_i| \sum_{j=1}^{N} (|\varepsilon_j| \|M_{ij}\| \|e_{mj}\|)$$

$$- \sigma_i a_i \phi_i^T \Gamma_i^{-1} \phi_i - \sigma_i a_i \phi_i^T \Gamma_i^{-1} \theta_i^* \Big]$$

$$+ \sum_{i=1}^{N} |\varepsilon_i| \Big[-\lambda_{qi} \|e_{mi}\|^2 + a_i s_i |\phi_i^T \xi_i| \|e_{mi}\| + p_i \|e_{mi}\| \|e_i\|$$

$$+\|e_{mi}\|\sum_{j=1}^{N}\tau_{ij}\|e_j\|+\|e_{mi}\|\sum_{j\neq i}^{N}(|\varepsilon_j|J_{ij}\|e_{mj}\|)+\pi_i|\phi_i^T\xi_i|\|e_{mi}\|$$

$$+\ell_i|\phi_i^T\xi_i|\|e_i\|+|\phi_i^T\xi_i|\sum_{j=1}^{N}c_{ij}\|e_j\|+|\phi_i^T\xi_i|\sum_{j\neq i}^{N}|\varepsilon_j|f_{ij}\|e_{mj}\|\Big]$$

$$+\sum_{i=1}^{N}a_i|\varepsilon_i|\Big[-a_i\|e_{fi}\|^2+\|e_{fi}\|\|e_i\|\Big] \qquad (4.28)$$

Defining

$$\lambda_s = \min\lambda(S)$$

$$\gamma_i = \min\lambda(\Gamma_i^{-1})$$

and completing the squares, we have

$$\dot{V}\leq\sum_{i=1}^{N}-\Bigg\{\|e_i\|^2\Bigg[\frac{\lambda_s}{2}-\frac{3(N+1)}{2}\sum_{j=1}^{N}\left(|\varepsilon_j|^2\frac{c_{ji}^2}{a_jg_j}\right)-\frac{2(N+1)}{2}\sum_{j=1}^{N}\left(|\varepsilon_j|\frac{\tau_{ji}^2}{\lambda_{qj}}\right)$$

$$-|\varepsilon_i|\frac{(2N+1)}{2\lambda_{qi}}\;p_i^2\Bigg]+(\phi_i^T\xi_i)^2\Bigg[\frac{a_ig_i}{2}-\frac{(2+N)}{2\lambda_s}\Big(\alpha_i^2\|A_{ci}\|^2$$

$$+\alpha_i^2\sum_{j\neq i}^{N}\|A_{ij}\|\Big)+|\varepsilon_i|^2\ell_i^2)\Bigg]+|\varepsilon_i|\|e_{mi}\|^2\Bigg[\frac{\lambda_{qi}}{2}$$

$$-|\varepsilon_i|\left(\frac{(2+N)}{2\lambda_s}\sum_{j=1}^{N}(q_{ji}^2\alpha_j^2)+\frac{3}{2}\frac{(N+1)}{a_ig_i}\sum_{j=1}^{N}(\alpha_j^2\|M_{ji}\|^2)+\frac{3}{2}\frac{(N+1)}{g_i}a_is_i^2\right.$$

$$\left.+\frac{3}{2}\frac{(N+1)}{a_ig_i}\;f_i+\frac{(2N+1)}{2\lambda_{qi}}|\varepsilon_i|J_i\right)\Bigg]$$

$$+a_i^2|\varepsilon_i|\|e_{fi}\|^2\left[1-|\varepsilon_i|\frac{(2+N)}{2\lambda_s}\right]+\frac{\sigma_ia_i}{2\gamma_i}\|\phi_i\|^2\Bigg\}+K_o \qquad (4.29)$$

where

$$f_i=\pi_i^2+\sum_{j=1}^{N}f_{ji}^2 \text{ and}$$

$$J_i=\sum_{j=1}^{N}J_{ji}^2 .$$

In view of (4.29), it is clear that if a_i is chosen as in (4.19), then there exists positive constants μ_o to μ_3 and ε^* such that for each $|\varepsilon_i|\in[0,\varepsilon^*]$

$$\dot{V}\leq\sum_{i=1}^{N}-\Big(\|e_i\|^2\mu_o+(\phi_i^T\xi_i)^2\mu_1+|\varepsilon_i|\|e_{mi}\|^2\mu_2$$

$$+ |\varepsilon_i| \|e_{fi}\|^2 \mu_3 + \frac{\sigma_i a_i}{2\gamma_i} \|\phi_i\|^2 \Big) + K_o \qquad (4.30)$$

Defining

$$f_{ci} = \alpha_i \max\lambda(P_{ci})$$

$$f_{pi} = \max\lambda(P_i)$$

we can write (4.30) as

$$\dot{V} \le -\rho_o V - \sum_{i=1}^{N} \Big\{ \|e_i\|^2 \big[\mu_o - \rho_o f_{ci}\big] + |\phi_i^T \xi_i|^2$$

$$\big[\mu_1 - \ell_o (f_{ci}\|b_{ci}\|^2 + |\varepsilon_i| f_{ii}\|b_{ii}\|^2)\big] + |\varepsilon_i| \|e_{mi}\|^2 \big[\mu_2 - \rho_o f_{ii}\big]$$

$$+ |\varepsilon_i| \|e_{fi}\|^2 \Big[\mu_3 - \rho_o \frac{a_i}{2}\Big] + a_i \|\phi_i\|^2 \Big[\frac{\sigma_i}{2\gamma_i} - \rho_o \alpha_i \|\Gamma_i^{-1}\|\Big] \Big\} + K_o \qquad (4.31)$$

where

$$\rho_o = \min_i \left[\frac{\mu_o}{f_{ci}}, \frac{\mu_1}{(f_{ci}\|b_{ci}\|^2 + |\varepsilon_i| f_{ii}\|b_{ii}\|^2)}, \frac{\mu_2}{f_{ii}}, \frac{2\mu_3}{a_i}, \frac{\sigma_i}{(2\gamma_i \alpha_i \|\Gamma_i^{-1}\|)} \right] \qquad (4.32)$$

Hence

$$\dot{V} \le -\rho_o V + K_o \qquad (4.33)$$

and therefore

$$V(t) \le [V(o) - \frac{K_o}{\rho_o}]\exp(-\rho_o t) + \frac{K_o}{\rho_o} \qquad (4.34)$$

and the proof is complete.

Remark 1: The design parameters a_i in (3.5) have to be selected such that (4.19) is satisfied. However, a_i^* given by (4.25) is not known since g_i, λ_s, A_{ci}, A_{ij} are unknown. If an upper bound for a_i^* cannot be calculated, then large values for a_i will satisfy (4.25), provided $|\varepsilon_i|$ is small enough.

Remark 2: The design parameters $\sigma_i > 0$ in the local adaptive laws are also essential for stability. In the absence of uncertainties, i.e., $m_{oi} = n_{oi} = 0$, however the $\sigma_i > 0$ produces a non-zero tracking error as indicated by (4.22) and (4.20).

Remark 3: The size of the residual set D_r depends on the magnitude of the external disturbances, the strength of the interconnections, and the design parameters σ_i and a_i. When disturbances, interconnections or plant uncertainties are present, such a residual set cannot be avoided even if the parameters of the reduced-order decoupled subsystems are exactly known.

Remark 4: We should emphasize that boundedness of the solution $e_i(t)$, $e_{mi}(t)$, $e_{fi}(t)$, $\phi_i(t)$ implies boundedness for all the other signals in the overall closed loop system.

4. Conclusions

A decentralized robust adaptive control scheme for a class of large scale systems is presented and analyzed in the presence of external disturbances, unmodeled plant dynamics and interconnections. We show that by properly selecting some design parameters in the local adaptive controllers we can achieve boundedness for all the signals in the closed-loop system, given any bounded initial conditions and small plant uncertainties (i.e., small $|\varepsilon_i|$). Furthermore, we also guarantee that the tracking and parameter errors converge exponentially fast to a bounded residual set. The size of this residual set depends on the strength of the interconnections, the magnitude of the disturbances and on some design parameters.

An important point in our approach is the choice of the design parameters σ_i and a_i to be used for the implementation of the proposed scheme. The parameters σ_i have to be positive and small so that their contribution to the residual set D_r is small. In order to choose the parameters a_i, lower and upper bounds for a_i are needed. The lower bound depends on the interconnections and on some unknown parameters, whereas the upper bound is of $O(\frac{1}{|\varepsilon_i|})$. If an upper bound on the interconnections is known, then the lower bound for a_i can be calculated and therefore for $|\varepsilon_i|$ small a_i can be easily chosen. In general, if $|\varepsilon_i|$ is relatively small, a_i can be large and the conditions for stability will be satisfied.

A further investigation of the proposed decentralized adaptive control scheme is a topic for future research.

5. References

[1] P. A. Ioannou and P. V. Kokotovic, Adaptive Systems with Reduced Models, Springer-Verlag, 1983.

[2] P. A. Ioannou and P. V. Kokotovic, "Decentralized Adaptive Control in the Presence of Multiparameter Singular Perturbations and Bounded Disturbances," American Control Conference, San Francisco, CA, June 1983.

[3] A. N. Michel and R. K. Miller, Quantitative Analysis of Large Scale Dynamical Systems, Academic Press, 1977.

[4] D. D. Siljak, Large-Scale Dynamic Systems: Stability and Structure, Elsevier North-Holland, New York, 1977.

[5] P. A. Ioannou and P. V. Kokotovic, "Improvement of Robustness of Adaptive Schemes," Proceedings of the Third Yale Workshop on Applications of Adaptive Systems Theory, Yale University, June 1983.

[6] C. E. Rohrs, L. Valavani, M. Athans, and G. Stein, "Analyti cal Verification of Undesirable Properties of Direct Model Reference Adaptive Control Algorithms," Proc. 20th IEEE Conf. on Decision and Control, San Diego, CA, December 1981.

[7] B. Egardt, "Stability Analysis of Adaptive Control Systems with Disturbances," Proc. JACC, San Francisco, CA, 1980.

[8] P. A. Ioannou, "Design of Decentralized Adaptive Controllers," Proc. 22nd IEEE Conf. on Decision and Control, San Diego, CA, December 1983.

[9] P. A. Ioannou, "Decentralized Adaptive Control of Interconnected Systems," U.S.C., EE-Systems, Report 84-02-1, February 1984.

[10] K. S. Narendra and L. S. Valavani, "Stable Adaptive Controller Design-Direct Control," IEEE Trans. on Automatic Control, Vol. AC-23, No. 4, August 1984.

A DESIGN APPROACH FOR DECENTRALIZED OBSERVERS

Udo Kuhn and Günther Schmidt

Lehrstuhl und Laboratorium für
Steuerungs- und Regelungstechnik

Technische Universität München
Arcisstraße 21, 8000 München 2
West-Germany

1. INTRODUCTION

This paper is based on the well-known deterministic observer approach for estimating unmeasurable states of a linear system. We consider in particular the problem of observing a large scale system by decentralized observers. Detailed studies concerning appropriate structures of decentralized observers were reported in /1,2/. In this paper we discuss decentralized observer design with respect to application of the observers for purposes of surveillance. Decentralized observers as part of a state feedback control concept are not considered here.

Starting point of decentralized observation in the sense of this paper is a large scale LTI system consisting of several coupled subsystems. In section 2 a general form of a decentralized observer for estimating the corresponding subsystem's states is introduced. The optimization of the free matrices or parameters of this "subobserver" can be treated as some sort of optimal output-feedback-design problem for a general linear system. For this reason in section 3 a solution of this more general problem is given. Its application to the design of an optimal subobserver is shown in section 4. Concluding, the results of this new design approach are illustrated in section 5 by means of an example.

2. GENERAL FORM OF A DECENTRALIZED OBSERVER

We consider the observation of a large scale LTI system with an overall description given by

$$\dot{\underline{x}} = A\underline{x} + B\underline{u} \tag{1}$$

$$\underline{y} = C\underline{x} \tag{2}$$

with state vector $\underline{x} \in \mathbb{R}^n$, input vector $\underline{u} \in \mathbb{R}^r$ and output vector $y \in \mathbb{R}^m$. Furthermore, we assume that the system (1,2) is assymptotically stable and consists of s interconnected subsystems of the form

$$\dot{\underline{x}}_i = A_i\underline{x}_i + F_i\underline{z}_i + B_i\underline{u}_i \tag{3}$$

$$\underline{y}_i = C_i\underline{x}_i \qquad i=1,\ldots,s \ , \tag{4}$$

where $\underline{z}_i \in \mathbb{R}^{p_i}$ represents the interactions of the i-th subsystem with all other subsystems, i.e.

$$\underline{z}_i = H_i\underline{x} \ . \tag{5}$$

Based on this subsystem description we introduce the notion of a "sub(system)observer": This is an observer that estimates the state vector $\underline{x}_i$ of the i-th subsystem. Thus, decentralized observation in this paper means reconstruction of the complete state $\underline{x}$ or parts of it by means of a set of subobservers. It is obvious that the main difficulty with decentralized compared to centralized observation is the existence of the interaction vector $\underline{z}_i$ which can be considered an additional subsystem input. There are two basic approaches to deal with this problem

- making the subobserver independent of $\underline{z}_i$, thus leading to the so-called interaction-invariant subobserver /2,3/,
- providing information about the interaction to the subobserver: four schemes devoted to this idea are discussed in /1,2/.

As a main result of this paper we will develop an observation scheme based on the second approach. As a first step we introduce a general interaction (variable) model. This is a dynamical system with an output $\underline{z}_i^+(t)$ that at least approximates the actual interaction vector $\underline{z}_i(t)$. It is given by

$$\dot{\underline{x}}_{zi} = A_{zi}\underline{x}_{zi} + B_{ziu}\underline{u} + B_{ziy}\underline{y} + B_{zix}\underline{x}_i \tag{6}$$

$$\underline{z}_i \approx \underline{z}_i^+ = Z_i\underline{x}_{zi} + D_{ziu}\underline{u} + D_{ziy}\underline{y} \ , \tag{7}$$

with $\underline{x}_{zi} \in \mathbb{R}^{q_i}$ the state of the interaction model, which must be observable from $\underline{z}_i^+$. This rather general interaction model contains various observation schemes known from literature /1,2/:

(i) It is possible to model by (6,7) that part of the overall system (1,2) which is observable from $\underline{z}_i$. Thus the interaction model gives the exact value of $\underline{z}_i$, provided that the initial values of $\underline{x}_{zi}$ are chosen appropriately.
If this model is too complex to be included in a subobserver (as described

later) its order can be reduced by well-known reduction methods. In this case $\underline{z}_i^+$ will be only an approximation to $\underline{z}_i$.
It should be noted that in each of these schemes the global output $\underline{y}$ is not included in (6), i.e. $B_{ziy} = 0$, $D_{ziy} = 0$

(ii) Often some elements of the interaction vector $\underline{z}_i$ are measurable, which results in a reduced interaction model,since the measurable part of $\underline{z}_i$ can be computed by the relationship $D_{ziy} \cdot \underline{y}$.

(iii) In order to construct a strictly decentralized subobserver, utilization of the global variables $\underline{u}$ and $\underline{y}$ in the interaction model must be avoided. If in addition $\underline{x}_i$ is not included in (6), we end up with a homogeneous interaction model as presented in /1,2/. This model proves to be useful in those cases where the interaction variables are nearly constant with time, i.e. $\dot{\underline{z}}_i \approx \underline{0}$, and/or contain destinctive oscillatory modes with known frequencies. On the other hand, retaining the subsystem's state $\underline{x}_i$ in eq. (6) leads to the so-called closing-model ("Abschlußmodell") approach as discussed in /4/.

For our further discussion all non-zero matrices of the interaction model are assumed to be known. In this case we can form a subobserver by adding to the identity observer for the subsystem (3) the interaction model. Consequently, we get for the complete subobserver the following description

$$\dot{\hat{\underline{x}}}_i = A_i\hat{\underline{x}}_i + F_i\hat{\underline{z}}_i + B_i\underline{u}_i + L_{ix}(\underline{y}_i - C_i\hat{\underline{x}}_i) \tag{8}$$

$$\dot{\hat{\underline{x}}}_{zi} = A_{zi}\hat{\underline{x}}_{zi} + B_{ziu}\underline{u} + B_{ziy}\underline{y} + B_{zix}\hat{\underline{x}}_i + L_{iv}(\underline{y}_i - C_i\hat{\underline{x}}_i) \tag{9}$$

$$\hat{\underline{z}}_i = Z_i\hat{\underline{x}}_{zi} + D_{ziu}\underline{u} + D_{ziy}\underline{y} \quad , \tag{10}$$

where $\hat{\underline{x}}_i \in \mathbb{R}^{ni}$ is the estimate of the subsystem's state $\underline{x}_i$, and $\hat{\underline{x}}_{zi} \in \mathbb{R}^{qi}$ is an estimate of $\underline{x}_{zi}$. L_{ix} and L_{iv} are weighting matrices for the deviations between the subsystems' and subobservers' output. These matrices contain the free observer parameters that have to be chosen appropriately.

For an evaluation of the above described observer scheme let us compute the suboberver error

$$\tilde{\underline{x}}_i = \underline{x}_i - \hat{\underline{x}}_i \ . \tag{11}$$

Since the interaction model will not be exact in general, i.e. $\underline{z}_i^+$ is only an approximation of $\underline{z}_i$, and the subobserver error depends on the properties of the total system to be observed. Consequently, $\tilde{\underline{x}}_i$ is given by the following set of equations

$$\begin{bmatrix} \dot{\tilde{\underline{x}}}_i \\ -\dot{\hat{\underline{x}}}_{zi} \\ \dot{\underline{x}} \end{bmatrix} = \begin{bmatrix} A_i - L_{ix}C_i & F_iZ_i & F_i(H_i - D_{ziy}C) \\ B_{zix} - L_{iv}C_i & A_{zi} & -B_{zix}E_i - B_{ziy}C \\ 0 & 0 & A \end{bmatrix} \begin{bmatrix} \tilde{x}_i \\ -\hat{x}_{zi} \\ \underline{x} \end{bmatrix} + \begin{bmatrix} -F_iD_{ziu} \\ -B_{ziu} \\ B \end{bmatrix} \underline{u} \qquad (12)$$

,

where matrix E_i selects the subsystem's states $\underline{x}_i$ from $\underline{x}$, i.e.

$$\underline{x}_i = E_i\underline{x} \ . \qquad (13)$$

As usual in observer design our main objective for subobserver performance is to keep the error $\tilde{\underline{x}}_i(t)$ as small as possible. This goal can be achieved by appropriate selection of weighting matrices L_{ix} and L_{iv}.

However, before attacking the actual subobserver design we will discuss an apparently different problem, namely a new design method for an optimal output-feedback control law for LTI systems. Lateron we will show that the same design approach can be applied to the design of the subobserver's weighting matrices.

3. OPTIMAL OUTPUT FEEDBACK DESIGN

We consider the control of a general LTI system as given by eqs. (1,2) with a constant output feedback control law

$$\underline{u} = -K\underline{y} \ . \qquad (14)$$

This leads for the closed loop system to

$$\dot{\underline{x}} = (A - BKC)\underline{x} \ . \qquad (15)$$

The elements of matrix K can be chosen such that some cost function J is minimized. Earlier methods for computing an optimal matrix K were reported by Levine, Athans /5/ and Kosut /6/. Both approaches, however, include beside others the input vector $\underline{u}$ in the cost function, which has no counterpart in the observer design problem to be discussed in the next section.

Another method, useful for both the design of output feedback control laws and observer weighting matrices was developed in /7/ and will be outlined next. We introduce as a cost function

$$J = E\{\int_0^\infty \underline{x}^T Q \underline{x} dt\} + tr\{K^T R_B K R_C\} , \tag{16}$$

where Q and R_B, R_C are symmetric, positive semidefinite and positive definite matrices, respectively. In order to avoid a dependency on the initial value $\underline{x}_o = \underline{x}(t=0)$, K in eq. (15) is optimized under the assumption that $\underline{x}_o$ is a random vector with zero mean and a known covariance matrix V, i.e.

$$E\{\underline{x}_o\} = \underline{0} \qquad E\{\underline{x}_o \underline{x}_o^T\} = V . \tag{17}$$

Thus, the expectation of the integral in eq. (16) is taken over all random initial states. The necessary conditions related to this optimization problem /7/ are given by the following system of nonlinear matrix equations:

$$AY^* + Y^*A^T - BR_B^{-1}B^TX^*Y^*C^TR_C^{-1}CY^* - Y^*C^TR_C^{-1}CY^*X^*BR_B^{-1}B^T + V = 0 \tag{18}$$

$$X^*A + A^TX^* - C^TR_C^{-1}CY^*X^*BR_B^{-1}B^TX^* - X^*BR_B^{-1}B^TX^*Y^*C^TR_C^{-1}C + Q = 0 \tag{19}$$

$$K^* = R_B^{-1}B^TX^*Y^*C^TR_C^{-1} . \tag{20}$$

If the unknown matrices X* and Y* are computed from (18,19), the optimal feedback matrix K* follows from (20).

4. DESIGN OF AN OPTIMAL SUBOBSERVER

Based on the results reported in section 3 we will now return to the design of an optimal subobserver taking into account the effect of the total system. As a first step let us consider the case, that there are no inputs into the total system (1,2), i.e. $\underline{u}(t) \equiv \underline{0}$. Then the homogeneous part of eq. (12) can be written as

$$\begin{bmatrix} \dot{\tilde{\underline{x}}}_i \\ -\dot{\hat{\underline{x}}}_{zi} \\ \dot{\underline{x}} \end{bmatrix} = \left[\begin{bmatrix} A_i & F_iZ_i & F_i(H_i - D_{ziy}C) \\ B_{zix} & A_{zi} & -B_{zix}E_i - B_{ziy}C \\ 0 & 0 & A \end{bmatrix} - \begin{bmatrix} I & 0 \\ 0 & I \\ 0 & 0 \end{bmatrix} \begin{bmatrix} L_{ix} \\ L_{iv} \end{bmatrix} [C_i \quad 0 \quad 0] \right] \begin{bmatrix} \tilde{\underline{x}}_i \\ -\hat{\underline{x}}_{zi} \\ \underline{x} \end{bmatrix} . \tag{21}$$

This equation set-up has the same basic structure as eq. (15), the state equation of the closed loop formed by output feedback. Thus, the matrices L_{ix} and L_{iv} can be computed by the same optimization procedure as described in the preceding section for computation of an optimal matrix K*. Matrix Q in the cost function (16) must be chosen such that only the substate $\underline{\tilde{x}}_i$ gets non-zero weight. No other modifications are required to adapt the general output feedback problem to the design of an optimal subobserver.

Next we consider the problem of designing a subobserver for the case when the global input $\underline{u}(t) \neq 0$. We restrict our discussion, however, to a step-type input vector:

$$\underline{u}(t) = \underline{u}_0 \cdot \sigma(t) \ . \tag{22}$$

With this input modelled by

$$\dot{\underline{u}} = \underline{0}, \qquad \underline{u}(t=0) = \underline{u}_0 \ , \tag{23}$$

the equations for the subobserver error (12) can be expanded, leading to the new homogeneous system

$$\begin{bmatrix} \dot{\underline{\tilde{x}}}_i \\ -\dot{\underline{\hat{x}}}_{zi} \\ \dot{\underline{x}} \\ \dot{\underline{u}} \end{bmatrix} = \begin{bmatrix} A_i - L_{ix}C_i & F_iZ_i & F_i(H_i - D_{ziy}C) & -F_iD_{ziu} \\ B_{zix} - L_{iv}C_i & A_{zi} & -B_{zix}E_i - B_{ziy}C & -B_{ziu} \\ 0 & 0 & A & B \\ 0 & 0 & 0 & 0 \end{bmatrix} \begin{bmatrix} \underline{\tilde{x}}_i \\ -\underline{\hat{x}}_{zi} \\ \underline{x} \\ \underline{u} \end{bmatrix} \ . \tag{24}$$

A transformation of this set of equations into a form similar to eq. (21) is not appropriate, since the system described by (24) is not asymptotically stable. Thus design via the output-feedback approach fails here.

In /8/ a solution to this problem is given, which will be described next. Let us assume that the following two conditions are met

(i) the cost function (16) contains the subobserver error $\underline{\tilde{x}}_i$ only
(ii) F_i has full rank and

$$\operatorname{rank}\begin{bmatrix} Z_i \\ A_{zi} \end{bmatrix} = \operatorname{rank}\begin{bmatrix} Z_i & [(D_{ziy}C - H_i)A^{-1}B - D_{ziu}] \\ A_{zi} & [(B_{zix}E_i + B_{ziy}C)A^{-1}B - B_{ziu}] \end{bmatrix} \tag{25}$$

then the subobserver according to eq. (24) can be designed by the same method as the subobserver according to eq. (21). All equations are the same, exept that an $\underline{u}(t) = \underline{u}_o \cdot \sigma(t)$ implies a modification of covariance matrix V in (18).

The above two conditions assure that the zero eigenvalues of the system (24) are not observable in the integral part of the cost function, which means that the subobserver error goes to zero for $t \to \infty$ and so the value of the integral in the cost function will be finite. To meet these conditions the interaction model must contain one integrator of the form

$$\dot{x}_{zik} = 0 \tag{26}$$

for each interaction, which is not modeled exactly in the interaction model for $t \to \infty$.

The algorithm for computing the modified matrix V' which must be used in eq. (18) instead of V was derived in /8 / and is as follows:

(i) compute the solution Ψ_i of the equations

$$Z_i\Psi_i = (D_{ziy}C - H_i)A^{-1}B - D_{ziu} \tag{27}$$

$$A_{zi}\Psi_i = (B_{zix}E_i + B_{ziy}C)A^{-1}B - B_{ziu} \; . \tag{28}$$

Note: The solution exists and is unique because of the observability of the interaction model and eq. (25).

(ii) Define matrix T as

$$T = \begin{bmatrix} I & 0 & 0 & 0 \\ 0 & I & 0 & \Psi_i \\ 0 & 0 & I & A^{-1}B \end{bmatrix} \; . \tag{29}$$

(iii) Matrix V' is given by:

$$V' = T\bar{V}T^T \tag{30}$$

with

$$\bar{V} = E\left\{\begin{bmatrix} \underline{\tilde{x}}_{io} \\ -\underline{\hat{x}}_{zio} \\ \underline{x}_o \\ \underline{u}_o \end{bmatrix} \begin{bmatrix} \underline{\tilde{x}}_{io} \\ -\underline{\hat{x}}_{zio} \\ \underline{x}_o \\ \underline{u}_o \end{bmatrix}^T\right\} \; . \tag{31}$$

By this algorithm it is possible to design an optimal subobserver that takes into account the effect of initial values of the subobserver and the total system as well as special inputs $\underline{u}$ of the total system.

5. EXAMPLE: SURVEILLANCE OF A CHEMICAL PLANT

The design of a decentralized observer will be demonstrated based on the linearized model of a large chemical plant / 8 / as shown in Fig. 1. This plant has 15 states and it consists of three subsystems, a continuous flow stirred tank reactor, a heat exchanger and distillation column, coupled by flows of product. Our goal is to design a subobserver for subsystem 1, the reactor. There are two scalar interactions going into subsystem 1, temperature and concentration of the product coming from the distillation column. Since we are interested in a strictly decentralized subobserver and we know that the interactions change slowly with time, we can use as an interaction model:

$$\dot{x}_{z11} = 0 \qquad z_{11}^{+} = x_{z11}$$
$$\dot{x}_{z12} = 0 \qquad z_{12}^{+} = x_{z12} \quad . \tag{32}$$

As a first step the subobserver is designed based on the assumption that this interaction model is exact and that the total system must not be considered. Minimization of eq. (16) results in subobserver 1. Next the actual behaviour of the interactions is taken into account together with the affect of a step input $\underline{u}$. This leads to subobserver 2.

For purposes of a comparison of the two observer designs we compare their performance by

$$J_o = \int_0^\infty \underline{\tilde{x}}_1^T \underline{\tilde{x}}_1 dt \tag{33}$$

for

$$\underline{\tilde{x}}_{1o} = \underline{0} \qquad \underline{\hat{x}}_{z1o} = \underline{0} \qquad \underline{x}_o = \underline{0}, \qquad \underline{u}_o = [- 0,4 \quad 1]^T . \tag{34}$$

The performance values are as follows

subobserver 1: $J_o = 3,16$, subobserver 2: $J_o = 0,46$.

These sample results demonstrate the better performance of subobserver 2, which was designed by considering the effects of the total system. Further results and simulations are given in / 8/ and will be reported in the oral presentation.

6. CONCLUSION

This paper presents a general discussion of decentralized observation and considers in particular the case of a subobserver including an interaction model. The main result of this paper is a design approach for the weighting matrices of the subobserver. It is shown that the subobserver design problem can be handled as a special case of a new design method for optimal linear output-feedback control. The new design procedure is demonstrated in connection with the development of a decentralized observer of a chemical plant.

LITERATURE

/1/ Kuhn, U.: Verfahren zur dezentralen Zustandsbeobachtung linearer Systeme mit komplexer Struktur. Regelungstechnik 31(1983), 44-50.

/2/ Kuhn, U.; Schmidt, G.: Decentralized observation: A unifying presentation of five basic schemes. Large Scale Systems 5(1984).

/3/ Viswanadham, N.; Ramakrishna, A.: Decentralized estimation and control for interconnected systems. Large Scale Systems 3(1982), 255-266.

/4/ Litz, L.: Dezentrale Regelung. R. Oldenbourg-Verlag, München, Wien, 1983

/5/ Levine, W.A.; Athans, M.: On the determination of the optimal constant output feedback gains for linear multivariable systems. IEEE Trans. Autom. Control, vol. AC-15(1970), 44-48.

/6/ Kosut, R.L.: Suboptimal control of linear time-invariant systems subject to control structure constraints. IEEE Trans. Autom. Control,vol. AC-15(1970), 557-563.

/7/ Kuhn, U.: Ein neuer Weg zur Bestimmung einer optimalen Ausgangsrückführung für die Regelung linearer Systeme. Regelungstechnik 32(1984).

/8/ Kuhn, U.: Bestimmung optimaler Parameter für einen dezentralen Beobachter mit Koppelgrößenmodell. Regelungstechnik 32(1984).

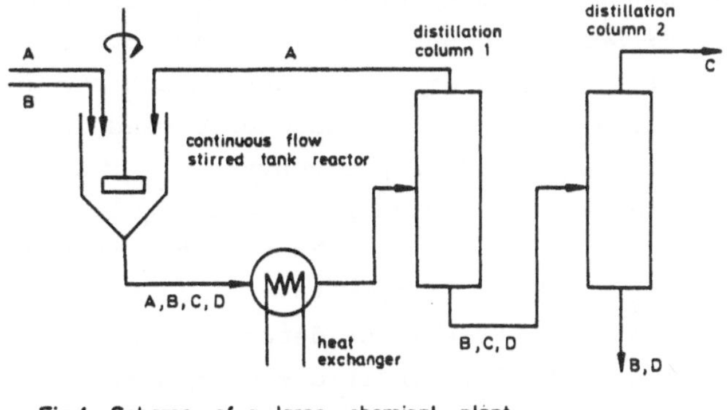

Fig.1: Scheme of a large chemical plant

DECENTRALIZED STABILIZATION BY HIGH GAIN FEEDBACK

Arno Linnemann
Forschungsschwerpunkt Dynamische Systeme
University of Bremen
28 Bremen, West Germany

Abstract. Sufficient conditions are derived for a linear time-invariant interconnected system to be stabilizable by decentralized state feedback. The conditions generalize known results in decentralized control theory and are obtained by combining a state space version of the small gain theorem with results of J.C.Willems on almost invariant subspaces. The stabilizing feedback law is in general of high gain type and the closed loop system has nice robustness properties.

1. Introduction

Consider the interconnected system described by

$$\dot{x}_i = A_i x_i + \sum_{\substack{j=1 \\ j \neq i}}^{l} A_{ij} x_j + B_i u_i \ , \qquad i=1,2,\dots,l, \tag{1.1}$$

where $x_i \in \mathbb{R}^{n_i}$ is the state and $u_i \in \mathbb{R}^{m_i}$ is the input of the i-th subsystem, and (A_i, B_i) is a controllable pair. The problem to stabilize this system by decentralized state feedback of the form

$$u_i = F_i x_i \qquad , \qquad i=1,2,\dots,l \tag{1.2}$$

will be considered.

A well known approach to this problem is to design the local feedback laws F_i such that the real parts of the spectra of $A_i+B_iF_i$ approach minus infinity (Siljak, 1978; Sezer and Hüseyin, 198o; Ikeda and Siljak 198o; Sezer and Siljak, 1981). This is done in order to achieve (block diagonal dominance and stability of the interconnected closed loop system (1.1), (1.2). Since large closed loop eigenvalues normally require large amplification factors, the resulting feedback law is in general of high gain type. However, as is well known, this method does not always work and systems of the form (1.1) with controllable subsystems do exist, which are not stabilizable by decentralized state feedback of the form (1.2) (Wang, 1978). Conditions on the system (1.1) which ensure decentralized state feedback stabilizability, are presented in (Sezer and Hüseyin,1978,198o; Sezer and Siljak,1979,1981;

Ikeda and Siljak, 198o; Nowak, 1982; Bachmann, 1982).

In this paper, a refinement of the above method (to shift the poles of the subsystems towards $-\infty$) will be presented and a class of systems will be described for which the proposed method is feasible. The main ideas for this will now be shortly explained for the case of two subsystems. For l=2, the closed loop system has the structure shown in Figure 1.

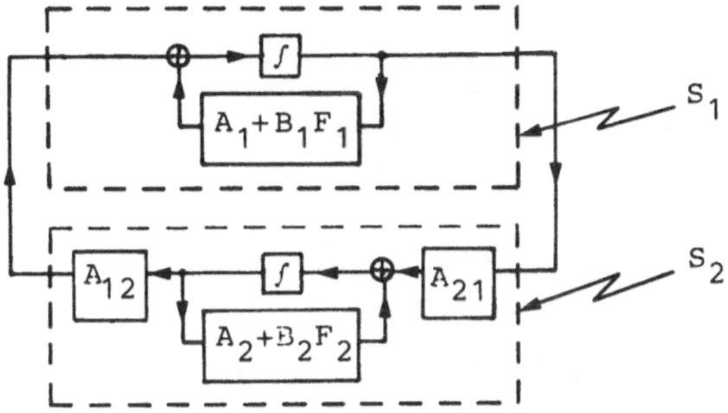

Figure 1

Hence, the closed loop system may be interpreted as a feedback system where S_1 represents the feedforward path and S_2 represents the feedback path. A state space version of the small gain theorem (Desoer and Vidyasagar, 1975) will now show, that this feedback system is stable, provided S_1 and S_2 are stable and the impulse response of S_2 is small enough. Hence closed loop stability can be achieved by stabilizing the subsystems while keeping impulse responses of certain parts of the system small. Since small impulse responses in general can only be achieved by shifting the real parts of the eigenvalues to minus infinity, this method is very similar to the method mentioned at the beginning of this introduction.

The ideas for l=2 directly carry over to the case of more than two subsystems. Hence, in order to show feasibility of the above method, the following subproblems have to be analyzed.

(i) A state space version of the small gain theorem has to be derived. This is done in Section 2.

(ii) A method to reduce the impulse response has to be provided. A nice and rather complete theory on this topic is available (Willems, 1981) which uses the concept of almost invariant subspaces. The results, which are relevant for the present application, are reviewed in Section 3.

Using the preliminaries of Sections 2 and 3, a procedure to stabilize the system (1.1) by decentralized state feedback (1.2) will be presented in Section 4, and a fairly broad class of systems, for which this procedure works, will be described. Moreover, the relation to some known results from literature and robustness properties will be discussed.

2. A state space version of the small gain theorem

In this section, a technical lemma will be derived, which might also be of independent interest.

First, some notation is introduced. Let $|\cdot|$ be a fixed norm on $\mathbb{R}^r (r\geq 1)$, and denote by the same symbol, $|\cdot|$, the induced matrix norm on $\mathbb{R}^{r\times r}$. For $t_1\in(o,\infty]$, $p\in[1,\infty]$ and $X\in\{\mathbb{R}^r,\mathbb{R}^{r\times r}\}$ define $L_p([o,t_1),X)$ to be the set of all mappings $f:[o,t_1)\to X$ satisfying $\|f\|_p<\infty$, where

$$\|f\|_p:=\|f(\cdot)\|_p:=\begin{cases}(\int_o^{t_1}|f(t)|^p dt)^{1/p} & \text{if } p\in[1,\infty)\\ \text{ess sup}\{|f(t)|\,|\,t\in[o,t_1)\} & \text{if } p=\infty.\end{cases}$$

Moreover, for $f:[o,t_1)\to X$ and $T\in[o,t_1)$, define $f_T:[o,t_1)\to X$ by

$$f_T(t)=\begin{cases}f(t) & \text{if } t\leq T\\ o & \text{if } t>T.\end{cases}$$

The following lemma may be viewed (c.f. Section 1) as a state space version of the small gain theorem. Its derivation is also very similar to the proof of the small gain theorem in (Desoer and Vidyasagar, 1975).

Lemma 1. Consider the matrix

$$A=\begin{bmatrix}A_{11} & A_{12}\\ A_{21} & A_{22}\end{bmatrix},$$

where $A_{ij}\in\mathbb{R}^{r_i\times r_j}$, $r_1\geq 1$, $r_2\geq 1$. Then A is asymtotically stable, provided A_{11} and A_{22} are asymtotically stable and

$$\|A_{12}e^{A_{22}(\cdot)}A_{21}\|_1\cdot\|e^{A_{11}(\cdot)}\|_1<1. \tag{2.1}$$

Proof. Asymtotic stability of A_{11} and A_{22} implies

$$\|e^{A_{ii}(\cdot)}\|_1<\infty \quad , \quad i=1,2. \tag{2.2}$$

Hence for i=1,2, the convolution operators

$$G_i : \begin{cases} L_\infty([0,\infty),\mathbb{R}^{r_i}) \rightarrow L_\infty([0,\infty),\mathbb{R}^{r_i}) \\ e_i(\cdot) \quad \rightarrow \int_0^{\cdot} e^{A_{ii}(\cdot-\tau)} e_i(\tau)d\tau \end{cases}$$

are well-defined and satisfy

$$\|G_i e_i\|_\infty \leq \|e^{A_{ii}(\cdot)}\|_1 \cdot \|e_i\|_\infty . \tag{2.3}$$

To conclude asymtotic stability of the matrix A, it suffices to show that the controllable and observable system

$$A, \; B = \begin{bmatrix} I_{r_1} & 0 \\ 0 & I_{r_2} \end{bmatrix}, \quad C = \begin{bmatrix} I_{r_1} & 0 \\ 0 & I_{r_2} \end{bmatrix} \tag{2.4}$$

is L_∞-stable. The system (2.4) may also be interpreted to be a feedback system described by (c.f. Figure 2)

$$e_1 = u_1 + A_{12}y_2 \; , \; y_2 = G_2e_2 , \tag{2.5}$$

$$e_2 = u_2 + A_{21}y_1 \; , \; y_1 = G_1e_1 \; . \tag{2.6}$$

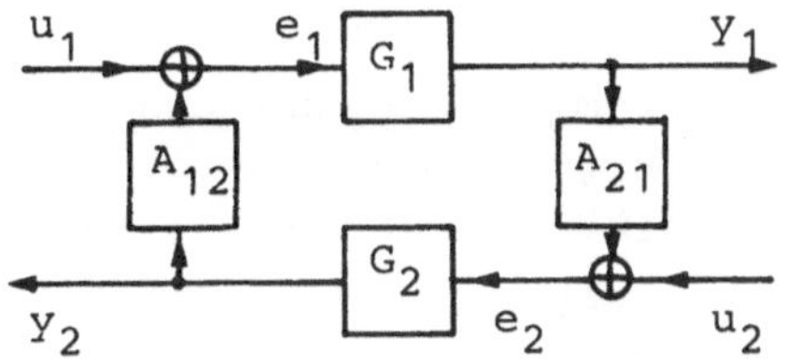

Figure 2

For every $u_i \in L_\infty([0,\infty),\mathbb{R}^{r_i})$, i=1,2, there exist mappings y_1, y_2, e_1, e_2 such that (2.5) and (2.6) are satisfied. Moreover, u_{iT}, y_{iT} and e_{iT} are contained in $L_\infty([0,\infty),\mathbb{R}^{r_i})$ for i=1,2 and all $T \in [0,\infty)$. Hence,

$$\begin{aligned} \|e_{1T}\|_\infty &= \|(u_1+A_{12}G_2(u_2+A_{21}G_1e_1)_T\|_\infty \\ &\leq \|u_{1T}\|_\infty + |A_{12}| \cdot \|e^{A_{22}(\cdot)}{}_T\|_1 \cdot \|u_{2T}\|_\infty \\ &\quad + \|(A_{12}e^{A_{22}(\cdot)}A_{21})_T\|_1 \cdot \|e^{A_{11}(\cdot)}{}_T\|_1 \cdot \|e_{1T}\|_\infty \\ &\leq \|u_1\|_\infty + |A_{12}| \cdot \|e^{A_{22}(\cdot)}\|_1 \cdot \|u_2\|_\infty \\ &\quad + \|A_{12}e^{A_{22}(\cdot)}A_{21}\|_1 \cdot \|e^{A_{11}(\cdot)}\|_1 \cdot \|e_{1T}\|_\infty . \end{aligned}$$

Consequently

$$\|e_{1T}\|_\infty \leq \frac{\|u_1\|_\infty + |A_{12}| \cdot \|e^{A_{22}(\cdot)}\|_1 \cdot \|u_2\|_\infty}{(1-\|A_{12}e^{A_{22}(\cdot)}A_{21}\|_1 \cdot \|e^{A_{11}(\cdot)}\|_1)} \; . \tag{2.7}$$

Since the right-hand side of the inequality (2.7) is finite and does not depend on T, the relation $e_1 \in L_\infty([o,\infty),\mathbb{R}^{r_1})$ follows. Moreover, by (2.2), (2.3), (2.5) and (2.6), $y_i \in L_\infty([o,\infty),\mathbb{R}^{r_i})$ for i=1,2.

□

3. Almost invariant subspaces

This section is concerned with the problem to make impulse responses arbitrary small by employing state feedback. Some results from (Willems, 1981) are reviewed and a corresponding numerically stable algorithm for single input systems is presented.

Theorem 1 (Willems, 1981). Let $A\in\mathbb{R}^{n\times n}$, $B\in\mathbb{R}^{n\times m}$, $H\in\mathbb{R}^{p\ n}$ and $G\in\mathbb{R}^{n\times q}$ be given. Assume that (A,B) is controllable and that

$$\text{im } G \subset R^*_{1,\ker H}(A,B). \tag{3.1}$$

Then for all $\varepsilon > o$ there exists a matrix $F\in\mathbb{R}^{m\times n}$ such that

$$\| He^{(A+BF)(\cdot)} G \|_1 \leq \varepsilon$$

and A+BF is asymtotically stable.

Here, $R^*_{1,\ker H}(A,B)$ denotes the supremal L_1-almost controllability subspace contained in ker H. For a definition and further properties of this linear subspace of the state space $\mathbb{R}^n$, refer to (Willems, 1981). For the present application it is only necessary to know that it can be calculated in terms of the data A,B and H as follows.

Theorem 2 (Willems, 1981). Consider the subspace algorithm

$$R^o = \{o\}$$

$$R^k = \ker H \cap (AR^{k-1} + \text{im} B),\ k=1,2,3,\ldots\ .$$

Then $R^o \subset R^1 \subset R^2 \subset \ldots$ and

$$R^{\dim \ker H} = R^\infty := \lim_{k\to\infty} R^k,$$

$$R^k = R^{k-1} \Rightarrow R^k = R^\infty .$$

Moreover,

$$R^*_{1,\ker H}(A,B) = AR^\infty + \text{im} B.$$

If m=1, the condition of Theorem 1 can easily and reliably be checked by transforming A,B into system Hessenberg form (Laub and Linnemann, 1984), as will now be shown. To each controllable single input system there exists a nonsingular matrix $T\in\mathbb{R}^{n\times n}$ such that $(T^{-1}AT, T^{-1}B)$ has system Hessenberg form, i.e.

$$T^{-1}B=\begin{bmatrix} a_{1o} \\ o \\ \cdot \\ \cdot \\ \cdot \\ \cdot \\ o \end{bmatrix}, \quad T^{-1}AT=\begin{bmatrix} a_{11} & \cdots & \cdots & a_{1n} \\ a_{21} & \ddots & & \vdots \\ o & \ddots & \ddots & \vdots \\ \vdots & \ddots & \ddots & \vdots \\ o \cdots o & & a_{n,n-1} & a_{nn} \end{bmatrix} \qquad (3.2)$$

with $a_{1o}\cdot a_{21}\cdot\ldots\cdot a_{n,n-1}\neq o$. Moreover, T is uniquely determined apart from right multiplication by a nonsingular upper triangular matrix and can even be chosen as orthogonal (van Dooren,1981). Now, take T as above, consider $[h_1,h_2,\ldots,h_n]:=HT$, and define

$$s:=\min\{i=1,\ldots,n \mid h_i\neq o\}. \qquad (3.3)$$

The integer s is independent of the particular transformation matrix T and Theorem 2 gives the following.

Corollary 1. $R^{*}_{1,\ker H}(T^{-1}AT,TB)= \operatorname{im}\begin{bmatrix} I_s \\ o \end{bmatrix}$.

Hence, if T is chosen to transform (A,B) into system Hessenberg form (3.2) and s is defined as in (3.3), then (3.1) means, that TG has the structure

$$TG = \begin{bmatrix} G \\ \\ o \end{bmatrix} \begin{matrix} \}s \\ \\ \}n-s \end{matrix} \quad . \qquad (3.4)$$

Since T can be chosen orthogonal (or as a product of stabilized elementary transformations), an efficient and numerically stable (Stewart, 1973) algorithm to verify (3.4) can be immediately designed. Thus, for the case m=1, the assumptions of Theorem 1 are easily verified. A similar algorithm for more than one input is also possible, but much more complicated. It will be presented elsewhere.

4. Decentralized stabilizability

The following sufficient condition for decentralized stabilizability is the main result of this paper.

Theorem 3. Define the compound matrices G_i and H_i by

$$G_i := \begin{bmatrix} A_{i1} & A_{i2} & \cdots & A_{i,i-1} \end{bmatrix},$$

$$H_i := \begin{bmatrix} A_{1i} \\ A_{2i} \\ \vdots \\ A_{i-1,i} \end{bmatrix},$$

$i=2,3,\ldots,l$. Then the interconnected system (1.1) can be stabilized by decentralized state feedback of the form (1.2), provided

$$\operatorname{im} G_i \subset R^*_{1,\ker H_i}(A_i,B_i), i=2,3,\ldots,l \tag{4.1}$$

and (A_i,B_i) is controllable for $i=1,2,\ldots,l$.

Proof. The proof is by induction on l. For $l=1$, the assertion follows trivially. Now assume that the assertion is true for $l-1$ subsystems. Then there exist feedback matrices $F_1,F_2,\ldots,F_{l-1}$, such that the matrix

$$\tilde{A} = \begin{bmatrix} A_1+B_1F_1 & A_{12} & \cdots & A_{1,l-1} \\ A_{21} & A_2+B_2F_2 & \ddots & \vdots \\ \vdots & \ddots & \ddots & A_{l-2,l-1} \\ A_{l-1,1} & \cdots & A_{l-1,l-2} & A_{l-1}+B_{l-1}F_{l-1} \end{bmatrix}$$

is asymtotically stable. But then, since the closed loop system matrix is given by

$$\begin{bmatrix} \tilde{A} & H_l \\ G_l & A_l+B_lF_l \end{bmatrix},$$

the result follows from Lemma 1 and Theorem 1. □

Since the proofs of Lemma 1 and Theorem 1 are constructive, the proof of Theorem 3 is constructive as well. Hence (at least conceptual algorithms can be given to compute the stabilizing feedback laws.

Note, that the condition (4.1) depends on the ordering of the subsystems. If it is not satisfied, one can relabel the subsystems and try (4.1) again.

According to the discussion at the end of Section 3, a "matrix version" of Theorem 3 can also be given:

Corollary 2. Let the systems (A_i,B_i), $i=1,2,\dots,l$, in (1.1) be controllable and single input. For $i=2,3,\dots,l$, choose a nonsingular matrix $T_i \in \mathbb{R}^{n_i \times n_i}$ such that $(T_i^{-1}A_iT_i, T_i^{-1}B_i)$ has system Hessenberg form and define

$$s_i := \min\{j=1,2,\dots,n_i \mid h_{ij} \neq 0\},$$

where

$$[h_{i1},h_{i2},\dots,h_{in_i}] := H_iT_i .$$

Then the interconnected system (1.1) can be stabilized by decentralized state feedback of the form (1.2), provided $T_2G_2, T_3G_3, \dots, T_lG_l$ have the structure

$$T_iG_i = \begin{bmatrix} G_i \\ \\ 0 \end{bmatrix} \begin{matrix} \} s_i \\ \\ \} n_i - s_i \end{matrix} .$$

This fairly general result is also obtained in (Sezer and Siljak, 1979) and Theorem 3 may be viewed as a generalization of that result to the multivariable case.

For multivariable subsystems, the condition (4.1) in Theorem 3 might be fairly difficult to verify. This is not the case with the sufficient condition of the following corollary.

Corollary 3. The interconnected system (1.1) can be stabilized by decentralized state feedback of the form (1.2), if

$$\mathrm{im}G_i \subset \mathrm{im}B_i, \quad i=2,3,\dots,l,$$

and (A_i,B_i) is controllable for $i=1,2,\dots,l$.

Proof. By Theorem 2,

$$\mathrm{im}\ B_i \subset R^*_{1,\ker H_i}(A_i,B_i),$$

and hence the proof follows from Theorem 3.

□

Corollary 3 shows, that Theorem 3 is a generalization of the well-known result which says, that interconnected systems are decentrally stabilizable, provided the subsystems are controllable and are interconnected only via their inputs and outputs (Sezer and Hüseyin, 1978; Saeks, 1979; Linnemann, 1983).

In the proof of Theorem 3 (c.f. Lemma 1), it is required that certain internal impulse responses of the closed loop system be sufficiently small. In (Nowak, 1982) conditions are derived which ensure that these impulse responses are identically zero, and Theorem 3 may also be viewed as a generalization of that result. By this generalization, the applicability is remarkably broadened, since the condition (4.1) is by far not as stringent as the corresponding condition in (Nowak, 1982). The price to be paid for this is, that the feedback matrix is in general of high gain type and more difficult to compute.

The proposed stabilization procedure has nice robustness properties:

Theorem 4. Assume that the conditions of Theorem 3 are satisfied and that the stabilizing feedback laws $F_1, F_2, \ldots, F_l$ are obtained as in the proof of Theorem 3. Then for all $e_{ij} \in [-1,1]$, $i,j=1,2,\ldots,l$, $i \neq j$ the system

$$\dot{x}_i(t) = (A_i + B_i F_i)x_i(t) + \sum_{\substack{j=1 \\ j \neq i}}^{l} e_{ij} A_{ij} x_j(t), \quad i=1,2,\ldots,l$$

is asymtotically stable.

Proof. Let $U = \mathrm{diag}(u_1, u_2, \ldots, u_{r_1})$, $V = \mathrm{diag}(v_1, v_2, \ldots, v_{r_1})$ be matrices with $u_i \in [-1,1]$, $v_i \in [-1,1]$ and let the norm $|\cdot|$ satisfy $|U| \leq 1, |V| \leq 1$ for all such matrices.
It is then easy to see that the assertions of Lemma 1 even imply asymtotic stability of

$$\left[\begin{array}{c|c} A_{11} & U A_{12} \\ \hline A_{21} V & A_{22} \end{array}\right]$$

for all U,V. Theorem 3 is obtained by repeated application of Lemma 1, and hence the result follows.

□

A closed loop system with the stability properties of Theorem 4 is called connectively stable in (Siljak, 1978). A connectively stable system remains stable even under structural perturbations in the interactions.

5. References

Bachmann,W. (1982). Stabilität dezentral geregelter und beobachteter Systeme. Regelungstechnik 3o, 411.

Desoer,C.A., and M.Vidyasagar (1975). Feedback Systems: Input-Output Properties. Academic Press, New York.

Ikeda,M., and D.D. Siljak (198o). On decentrally stabilizable large-scale systems. Automatica 16, 331.

Laub,A.J., and A.Linnemann (1984). Hessenberg forms in linear systems theory. in preparation.

Linnemann,A. (1983). Decentralized control of dynamically interconnected systems. University of Bremen, Forschungsschwerpunkt Dynamische Systeme, Report Nr.111.

Nowak,H.(1982). Dezentrale Regelung durch Eigenwertentkopplung. Regelungstechnik 3o, 22.

Saeks,R.(1979). On the decentralized control of interconnected dynamical systems. IEEE Trans.Aut.Control AC-24, 269.

Sezer,M.E., and Ö.Hüseyin (1978). Stabilization of linear time-invariant interconnected systems using local state feedback. IEEE Trans.Syst., Man. Cybern. SMC-8, 751.

Sezer,M.E., and Ö.Hüseyin (198o). On decentralized stabilization of interconnected systems. Automatica 16, 2o5.

Sezer,M.E., and D.D.Siljak (1979). Decentralized stabilization and structure of linear large-scale systems. Proc.13th Asilomar Conf., Pacific Grove, California, 176.

Sezer,M.E., and D.D. Siljak (1981). On decentralized stabilization and structure of linear large scale systems. Automatica 17, 641.

Siljak, D.D.(1978). Large-Scale Dynamic Systems: Stability and Structure. North-Holland, New York.

Steward, G.W. (1973). Introduction to matrix computations. Academic Press, New York.

van Dooren,P.M. (1981). The generalized eigenstructure problem in linear system theory. IEEE Trans.Aut.Control AC-26, 111.

Wang,S.-H. (1978). An example in decentralized control systems. IEEE Trans.Aut.Control AC-23, 938.

Willems, J.C. (1981). Almost invariant subspaces: An approach to high gain feedback design - Part 1: Almost controlled invariant subspaces. IEEE Trans.Aut.Control AC-26, 235.

FIXED MODES IN DECENTRALIZED CONTROL SYSTEMS

M.Tarokh
Department of Electrical Engineering
Tehran University of Technology
P.O.Box 3406, Tehran, Iran

INTRODUCTION

In the study of decentralized control of large-scale systems, the concept of fixed modes plays a fundamental role in problems like stabilization and pole placement. Fixed modes are the invariant poles of a system under the decentralized control law and are structural property of the given system.

The problem of characterizing the fixed modes has been the subject of considerable research. Wang and Davison(1973) give a computational test for determining the fixed modes. This test, however, does not give any information about the nature and the reason for the occurance of fixed modes. Anderson and Clements (1981) have presented algebraic tests for the existence of fixed modes. A frequency - domain characterization of fixed modes is reported by Seraji(1982) for systems using scalar local controllers.

In this paper, we consider the general problem of characterizing fixed modes of a multivariable system with respect to decentralized controllers. The characterization given here provides a simple explanation for the existence of fixed modes in terms of the familiar concepts of system poles and zeros.

2. DEVELOPMENT

In this section, we derive an expression relating the closed loop characteristic (pole) polynomial to the controller parameters and suitably-defined zero polynomials. It will be seen that zero polynomials play an important role in the characterization of fixed modes.

Consider the mulivariable system

$$\dot{x} = Ax + Bu$$
$$y = Cx \qquad (1\text{-a})$$

where x is the nxl state vector, u is the mxl input vector and y is the ℓ x 1 output vector. The transfer-function matrix of the system is

$$G(s) = C(sI-A)^{-1} B = \frac{N(s)}{p(s)} \tag{1-b}$$

where $N(s)=C\ \mathrm{adj}(sI-A)\ B$ is the ℓxm numerator transfer-function matrix and $p(s)=|sI-A|$ is the characteristic or pole polynomial of the open-loop system.

Now consider the set of i-input i-output subsystems of (C, A,B), $i=1,2,\ldots,\min(\ell,m)$, referred to as i-dimensional subsystems and denoted by $(C^i_\gamma, A, B^i_\beta)$ where C^i_γ, $\gamma=1,2,\ldots,\nu_c$ are the set of i x n submatrices formed from i rows of the matrix C and where B^i_β, $\beta=1,2,\ldots,\nu_b$ are the set of n x i submatrices formed from i columns of the matrix B. The number ixn submatrices of C and nxi submatrices of B are $\nu_c = \frac{\ell!}{(\ell-i)!i!}$ and $\nu_b = \frac{m!}{(m-i)!m!}$ respectively. The transfer-function matrices of the i-dimensional subsystems are $G^i_{\gamma,\beta}(s) = C^i_\gamma(sI-A)^{-1}B^i_\beta \triangleq G^i_j(s)$, where for notational convenience we have replaced the subscripts γ,β by j. Note that $G^i_j(s)$, $j=1,2,\ldots,\nu_i$ are the set of ixi submatrices of G(s) and $\nu_i = \nu_b\ \nu_c$. From the definition of the zero polynomial of a system (Kwakernaak and Sivan 1972), we obtain the following expression for the set of zero polynomials of the i-dimensional subsystems

$$z^i_j(s) = \begin{vmatrix} sI-A & B^i_\beta \\ C^i_\gamma & 0 \end{vmatrix} = |sI-A|\ |C^i_\gamma(sI-A)^{-1} B^i_\beta| = p(s)|G^i_j(s)| \tag{2}$$

where the roots of $z^i_j(s)=0$ are the set of transmission zeros of the i-dimensional subsystems. Note that the zero polynomials of one-dimensional subsystems are the entries of the numerator transfer-function N(s).

The control law to be applied to the system(1) is

$$u=v-Ky \tag{3}$$

where v is the mxl command vector and the constant mxℓmatrix K can take any desired structure.

Substituting (3) in (1), we obtain the closed-loop characteristic (pole) polynomial as

$$q(s) = |sI-A+BKC| = p(s)\ |I+G(s)K| \tag{4}$$

The determinant in (4) may be expanded to give

$$q(s) = p(s)\left(1+\Sigma_{i=1}^{r} \quad h^{i}(s)\right) \tag{5}$$

where $r=\min(\ell,m)$ and where $h^{i}(s)$ is the sum of principal i-th order minors of $G(s)K$. These minors can be written in terms of determinants of submatrices of $G(s)$ and $\bar{K}$ as*

$$h^{i}(s)= \sum_{j=1}^{\nu i} |\bar{K}_{j}^{i}| \; |G_{j}^{i}(s)|= \sum_{j=1}^{\nu i} |\bar{K}_{j}^{i}||C_{\gamma}^{i}(sI-A)^{-1} \; B_{\beta}^{i}| \tag{6}$$

where $G_{j}^{i}(s)$ and $\bar{K}_{j}^{i}$ are i x i submatrices of $G(s)$ and $\bar{K}$ obtained from the same selection of rows and columns of $G(s)$ and $\bar{K}$ respectively. The proof of (5) and (6) is by elementary matrix manipulations and is omitted. Now suppose that i rows and i columns of $\bar{K}$ have been selected to form $\bar{K}_{j}^{i}$. Let B_{β}^{i}, C_{γ}^{i} denote the submatrices of B and C obtained by selecting the same i columns of B and the same i rows of C as in $\bar{K}_{j}^{i}$. Then we call $(C_{\gamma}^{i},A, B_{\beta}^{i})$ or equivalently $G_{j}^{i}(s)$ the set of subsystems of (C,A,B) corresponding to the submatrices $\bar{K}_{j}^{i}$ of the controller matrix $\bar{K}$. Substituting (2) and (6) in (5), we obtain.

$$q(s)=p(s)+\sum_{i=1}^{r} \; \sum_{j=1}^{\nu i} \; | \; \bar{K}_{j}^{i} \; | \; z_{j}^{i}(s) \tag{7}$$

where $z_{j}^{i}(s)$ are the zero polynomials of the subsystems of (C,A,B) corresponding to the controller submatrices $\bar{K}_{j}^{i}$. Expressing (7) forms the basis for the characterization of fixed modes. Note that $|\bar{K}_{j}^{1}|$, $j=1,2,\ldots,m\ell$ are the entries of $\bar{K}$. We also observe that (7) is a generalization of the result obtained for the single-input, single-output system $g(s)=\frac{z(s)}{p(s)}$ and the scalar controller k where we have $q(s)=p(s)+kz(s)$.

3. CHARACTERIZATION OF FIXED MODES

In this section, we consider the problem of finding conditions for the existence of fixed modes of a multivariable system with respect to the decentralized controller $K=\text{block diag}(K_1,K_2, \ldots K_N)$, where $K_1,K_2,\ldots K_N$ are the local controller matrices of appropriate dimensions.

* Throughout this paper , the transpose of a matrix M is denoted by $\bar{M}$.

The following theorem gives both time and frequency-domain characterization of fixed modes.

Theorem 1

The necessary and sufficient condition for an eigenvalue at λ_o of the system matrix A to be a fixed mode of the system(C, A,B) with respect to the controller K is that:

$$\text{Rank}\begin{pmatrix}\lambda_o\, I-A & B^i_\beta \\ C^i_\gamma & 0\end{pmatrix} < n+i\ ,\ i=1,2,\ldots,\ \min(\ell,m) \qquad (8)$$

where $(C^i_\gamma,\ A, B^i_\beta)$ are the subsystems of (C,A,B) corresponding to non-singular submatrices of $\bar{K}$.

The above theorem maybe restated as follows: The necessary and sufficient condition for the system(1) to have a fixed mode with respect to the controller K is that all subsystems of G(s) corresponding to non-singular submatrices of $\bar{K}$ have a common transmission zero coinciding with a system pole.

Proof The proof follows directly from (7). To show the sufficiency, suppose that $|\bar{K}^i_j| \neq 0$ for a set of i and j and all the corresponding subsystems $G^i_j(s)$ have a common transmission zero at $s=\lambda_o$ which coincides with an open-loop pole at $s=\lambda_o$. Then (7) becomes:

$$q(s) = (s-\lambda_o)\left(\hat{p}(s) + \sum_{i=1}^{r}\sum_{j=1}^{\nu_i} |\bar{K}^i_j|\ \hat{z}^i_j(s)\right)$$

where $\hat{p}(s) = \dfrac{p(s)}{s-\lambda_o}$ and $\hat{z}^i_j(s) = \dfrac{z^i_j(s)}{s-\lambda_o}$. Hence $s=\lambda_o$ is a fixed mode and cannot be moved by the controller K.

To prove the necessity, suppose a particular subsystem, $G^2_1(s)$ say corresponding to $|\bar{K}^2_1| \neq 0$, has no transmission zero coinciding with a system pole at $s=\lambda_o$, but the remaining subsystems have a transmission zero at $s=\lambda_o$, where λ_o is a root of p(s). In this case (7) becomes:

$$q(s) = (s-\lambda_o)\hat{p}(s) + |\bar{K}^2_1|\ z^2_1(s) + (s-\lambda_o)\sum_{\substack{i=i\\ i\neq 2}}^{r}\ \sum_{j=2}^{\nu_i} |\bar{K}^i_j|\ \hat{z}^i_j(s)$$

where $z^2_1(s)$ is the zero polynomial of the subsystem $G^2_1(s)$. It is seen that all closed-loop poles are affected by K, hence the system has no fixed mode.

The following corollaries are deduced from Theorem 1.

Corollary 1 : The fixed polynomial whose roots are the fixed modes, is the greatest common divisor of the pole polynomial and the zero polynomials of all subsystems of G(s) corresponding to non-singular submatrices of $\bar{K}$.

Corollary 2 : Conditions (i) and (ii) below are necessary for a system to have a fixed mode at $s=\lambda_o$:

i) The entries of G(s) corresponding to non-zero elements of $\bar{K}$ must have pole-zero cancellations at $s=\lambda_o$

ii) for the case $m=\ell$; $s=\lambda_o$ must be a transmission zero of the entire system.

Corollary 3 : If any of the subsystems of (C,A,B) corresponding to non-singular submatrices of $\bar{K}$ is minimum-phase, the system has no fixed modes located in the closed right-half complex plane. Hence by Wang and Davison's theorem(1973), the system (C,A,B) is stabilizable by the decentralized (possibly dynamic) controller K.

The above corollary is based on the fact that in minimum-phase systems all the transmission zeros lie in the open left-half complex plane.

Example

Consider the system

$$\dot{x}=\begin{pmatrix}-1 & 1 & -1 & 0 & 0\\ 0 & 0 & 1 & 0 & 0\\ 0 & 0 & 0 & 0 & 0\\ -1 & 0 & -1 & 0 & 0\\ 0 & 0 & 0 & 0 & 0\end{pmatrix}x+\begin{pmatrix}1 & 1 & 0\\ 0 & 0 & 0\\ 0 & 0 & 1\\ 1 & 0 & 0\\ 0 & 1 & 0\end{pmatrix}u$$

$$y=\begin{pmatrix}1 & 0 & 1 & 0 & 0\\ 0 & 0 & 1 & 1 & 0\\ 0 & 0 & 0 & 1 & 1\end{pmatrix}x$$

Determine if the system has fixed modes with respect to the decentralized controller

$$K= \text{block diag}(k_{11}, K_2);\ K_2 = \begin{pmatrix} k_1 \\ k_2 \end{pmatrix} (k_3\ k_4) \qquad (9)$$

where $k_{11}, k_1, \ldots k_4$ are arbitrary scalars.

The transfer-function matrix of the system G(s) and the transpose of K are:

$$G(s) = \frac{1}{s^4\ (s+1)} \begin{pmatrix} s^4 & s^4 & s^2(s^2+1) \\ s^4 & -s^3 & s(s^3-1) \\ s^4 & s^4 & -s(s^2+1) \end{pmatrix};\ \bar{K} = \begin{pmatrix} k_{11} & 0 & 0 \\ 0 & k_1k_3 & k_2k_3 \\ 0 & k_1k_4 & k_2k_4 \end{pmatrix}$$

The set of zero polynomials of one-dimensional subsystems of G(s) corresponding to non-zero elements of $\bar{K}$ are $z_1^1(s)=s^4$, $z_2^1(s)=-s^3$, $z_3^1(s)=s(s^3-1)$, $z_4^1(s)=s^4$ and $z_5^1(s)=-s(s^2+1)$. The set of zero polynomials of appropriate two-dimensional subsystems are:

$$z_1^2(s)=p(s) \begin{vmatrix} g_{11}(s) & g_{12}(s) \\ g_{21}(s) & g_{22}(s) \end{vmatrix} = -s^3;\ z_2^2(s)=p(s) \begin{vmatrix} g_{11}(s) & g_{13}(s) \\ g_{21}(s) & g_{23}(s) \end{vmatrix} = -s$$

$$z_3^2(s)=p(s) \begin{vmatrix} g_{11}(s) & g_{12}(s) \\ g_{31}(s) & g_{32}(s) \end{vmatrix} = 0;\ z_4^2(s)=p(s) \begin{vmatrix} g_{11}(s) & g_{13}(s) \\ g_{31}(s) & g_{33}(s) \end{vmatrix} = -s(s^2+1)$$

Note that the submatrix $\begin{pmatrix} k_1k_3 & k_2k_3 \\ k_1k_4 & k_2k_4 \end{pmatrix}$ and the overall matrix $\bar{K}$ are both singular and hence we need not consider their corresponding zero polynomials. Since all the appropriate subsystems have a transmission zero at s=0 which coincides with a system pole, we conclude that the system has a fixed mode at s=0 with respect to the controller (9).

It is to be noted that if the unity-rank constraint is removed from the structure of K_2, we have $|\bar{K}| \neq 0$ and the corresponding zero polynomial $z_1^3(s)=p(s)|G(s)|=s^4+1$ has no common root with p(s). Hence the system has no fixed mode with respect to K=block diag(k_{11} ,K_2), where K_2 is an unconstrained 2x2 matrix.

4. CONCLUSIONS

Time and frequency-domain characterizations of fixed modes of a multivariable system with respect to decentralized controllers are given in this paper. The characterizations given here provide an insight into the reason for existence and mechanism of formation of fixed modes. The occurance of fixed modes is shown to be the result of pole-zero cancellations in certain subsystems. The number of these subsystems greatly increases with the number of singular controller submatrices. Thus only in special cases where the information flow from the outputs to the inputs is severely limited, a system is likely to have a fixed mode.

REFERENCES

1. Anderson B.O.D. and D.J. Clements (1981). Alegebraic characterization of fixed modes in decentralized control, Automatica, 17,703.

2. Kwakernaak, H. and R. Sivan (1972). Linear optimal control systems, Wiley-Interscience, New-York.

3. Seraji, H.(1982). On fixed-modes in decentralized control systems. Int. J.Control 35,775.

4. Wang, S.H. and E.J. Davison (1973). On stabilization of decentralized control systems. IEEE Trans. Aut. Control AC - 18,473.

ROBUST CONTROL OF LARGE SCALE SYSTEMS VIA COOPERATIVE DECENTRALISED CONTROL METHOD

Z. Binder, J.F. Coudurier* and A.N. Hagras+
Laboratoire d'Automatique de Grenoble
E.N.S.I.E.G. - I.N.P.G.
38402 - Saint-Martin-D'Hères, France

ABSTRACT

An organisation of a multilevel control system is proposed. The redundancy of information exchange between neighbouring elements is used to compensate for partial communication breakdown. On the base of the CODECO method with cooperation, the approach is illustrated on a simulated distillation plant.

INTRODUCTION

An application of the distributed computer network in the control of large scale systems raises several new problems. A control system organisation, decentralised control method distributed data base and information exchanges are the more important aspects in the design and realization of the distributed control system. Several works treat these aspects, based on the particular methodology and description. There is no general methodological approach to solve the global problem of design of the distributed control system and all its elements from algorithm, hardware and software points of view. One of the difficulties in this area is the absence of a common description means for the different specialists participating in this design and for the users of the proposed elements.

Another problem faced by the distributed control system is the reliability and robustness. Usually, the reliability aspect is treated independently only from one point of view ; algorithm, software or hardware. The solutions based on the redundancy solves one part of the problem and

* present address CETIM , 42029 - Saint-Etienne-Cedex
+ died on January 6th, 1984

seldom the proposition takes into account hardware software and algorithm in the same time. The cost paid to satisfy the reliability and robustness may be reduced by a global approach to solve above problem.

One direction on this way is a common study for hardware redundancy and algorithmic information coherence. This way exploits hardware in a better manner to improve the control performance. Our works in this area use a multi-microprocessor network as a basic structure for the real time control algorithms. The structure of these algorithms tolerates a breakdown of some network elements, microcomputers or communication lines. A representation of the physical system by several models has a central role in our study. An association between models and microcomputors in a real time parallel structure gives a basic tool to design a control system. A multi-model parallel control structure [2] is proposed for multivariable control of nonlinear or time variant systems [3] [4] . A simplified representation in the hierarchical control leads to a proposition of the coordinated decentralised control (CODECO) structure [8] [5] and the development of corresponding tracking algorithms. The main characteristics of this approach and its practical advantages from reliability point of view are presented in this paper. The cooperation [6] between interconnected systems plays here an important role.

MULTILEVEL CONTROL STRUCTURE

To realise a progressive implementation of the hierarchical control system we propose a multilevel network with nets formed by "Control and Decision Centers". The functional relation between the neighbouring centers on the same or on different levels and redundancy of corresponding information exchange contributes to the robustness of the proposed structure (fig.1). [7]

The basic idea of the functioning lies on simulating the large scale system by different levels. Models situated on the lower level represent the systems, on the intermediate levels they represent groups of systems and on the top of the hierarchy there is a model of the whole process. Going up in the hierarchy, the simplification of models is completed by the hierarchy of control strategies. The control criterion of the lower levels takes into account physical factors of the process more than the higher level criterion which is concerned with the satisfaction of the economical objectives.

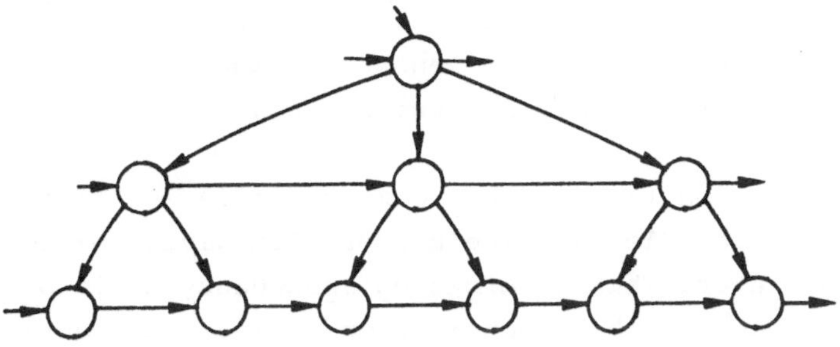

Figure 1. Multilevel Control and Decision Center Network.

The control strategy of each control and decision center is elaborated using a model and the corresponding criterion, which expresses the relations with center environment. The strategy consists of a realization of the goals issued by the center situated on the higher level of the control hierarchy. The neighbouring centers on the same level realize the goals independently of each others and in general their interaction constraints are not satisfied. Then the cooperation between these centers are proposed. A modification of the original goals carry out the modification of interaction variables to respect them.

In figure 2, a part of the control and decision centers network is presented. It controls two interconnected systems (S_1) and (S_2). The

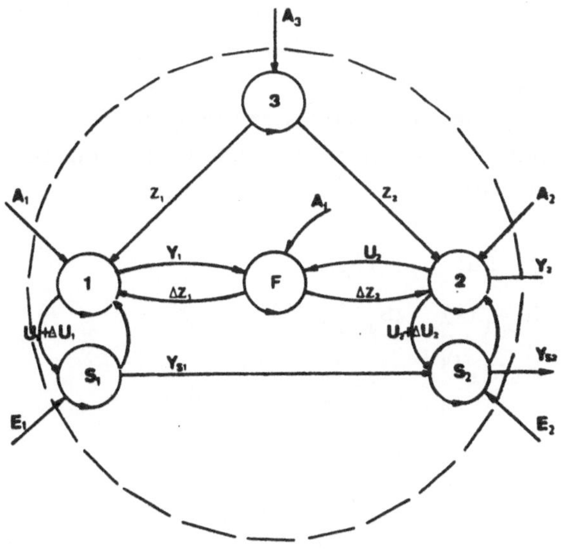

Figure 2. Coordination and cooperation in the control system.

cooperation (F) between the centers (1) and (2) modifies goals $(z_1+\Delta z_1)$ and $(z_2+\Delta z_2)$ and consequently the control variables $(u_1+\Delta u_1)$ and $(u_2+\Delta u_2)$ are applied to the interconnected systems.

Remark the information transfer between the different centers on the same or on the different levels. The cooperation gives complementary information to each center. This information redundancy could be used in the case of disturbances on the communication between the centers.

TRACKING CONTROL ALGORITHM

The relation between models may be represented by the tracking criterion based on the deviation of model variables. Tracking the higher level model by those of a lower level assure the functioning of the system. The input and output variables of each model track the reference variables given by a higher level model.

For linear models and quadratic criterion, the solution of tracking depends on the reference variables evolution during an optimisation horizon. The use of the upper level model state variables in the feed forward part of the algorithm [8] is a consequence of this requirement. To avoid this dimensional problem an horizon reduction could be proposed. One step horizon gives an acceptable results in our case [5] due to the simultaneous input/output tracking (Fig. 3) of the coherent reference variable issued at the same origin, the upper level model. In this case a real time control algorithm is independent of the position of the center in the hierarchy and the information transfer between the centers is limited to the actual reference variables. The basic algorithmic solution is a classical one.

The linear model (M_i) has the form :

$$x_i(k+1) = A_i x_i(k) + B_i u_i(k)$$
$$y_i(k) = C_i \, x_i(k)$$

where

$$u_i = \begin{bmatrix} w_i \\ u_i^{\ell} \end{bmatrix} \qquad y_i = \begin{bmatrix} v_i \\ y_i^{\ell} \end{bmatrix}$$

w and v represent the interaction variables of dimension r_w and m_v.

The input (u) and the output (y) track the reference variables (zu) and (zy) issued by the higher level model

$$zu_i = \begin{bmatrix} zw_i \\ zu_i^{\ell} \end{bmatrix} \qquad zy_i = \begin{bmatrix} zv_i \\ zy_i^{\ell} \end{bmatrix}$$

The tracking cost function is

$$J_i = \sum_{k=o}^{N-1} [zu_i(k) - u_i(k)]^T R[zu_i(k) - u_i(k)] + [zy_i(k) - y_i(k)]^T Q[zy_i(k) - y_i(k)]$$

Taking a one step optimization horizon, the model input is given by [5]

$$u_i(k) = - L_i x_i(k) + M_{1i}\ zu(k) - M_{2i}\ zy(k)$$

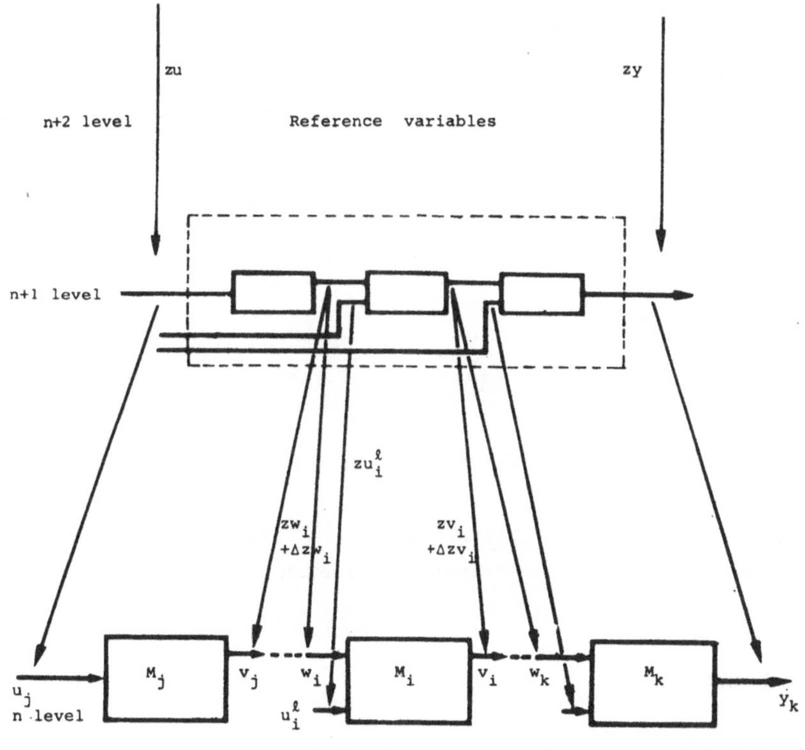

Figure 3. Coordinated Decentralised Control with Cooperation (CODECO)

with

$$L_i = S_i B_i^T K_i A_i$$

$$M_{1i} = S_i B_i^T (G_i^T - I)^{-1} L_i^T R_i + S_i R_i$$

$$M_{2i} = S_i B_i^T (G_i^T - I)^{-1} C_i^T Q_i$$

$$S_i = (R_i + B_i^T K_i B_i)^{-1}$$

$$G_i = A_i - B_i L_i$$

K - Steady state solution of Riccati equation.

The input variable (u) including the interactions (w) is a function of the reference variables (zu) and (zy). The modification of interactions (w) and also (v) could be realized by a correction of corresponding reference variables, there are (Δzw) and (Δzv). In figure 3 the three models (M_j, M_i, M_k) connected in series track a higher level model which satisfies the interaction constraints

$$v_j \neq w_i \quad \text{and} \quad v_i \neq w_k$$

To achieve the equality of interactions, the correction of reference variables could be calculated as a function of the cooperation matrix (F) which expresses a relative effort of each element to satisfy these interactions [6]

$$\begin{bmatrix} \Delta zw_i(k) \\ \Delta zv_i(k) \end{bmatrix} = M_i^{-1} \begin{bmatrix} F_{ij} & 0 \\ 0 & -(I-F_{ik}) \end{bmatrix} \begin{bmatrix} v_j(k) - w_i(k) \\ v_i(k+1) - w_k(k+1) \end{bmatrix}$$

where

$$M_i = \begin{bmatrix} (M_{1i})_w & -(M_{2i})_v \\ (C_i B_i M_{1i})_v & -(C_i B_i M_{2i})_v \end{bmatrix}$$

with $(\quad)_w$, $(\quad)_v$ submatricies of dimensions of the interactions between (i) and the neighbouring elements.

EXPERIMENTAL STUDIES

The proposed "Coordinated Decentralized Control method with cooperation" (CODECO) is tested on a pilot plant with two binary distillation columns connected in series (fig. 4). [1]

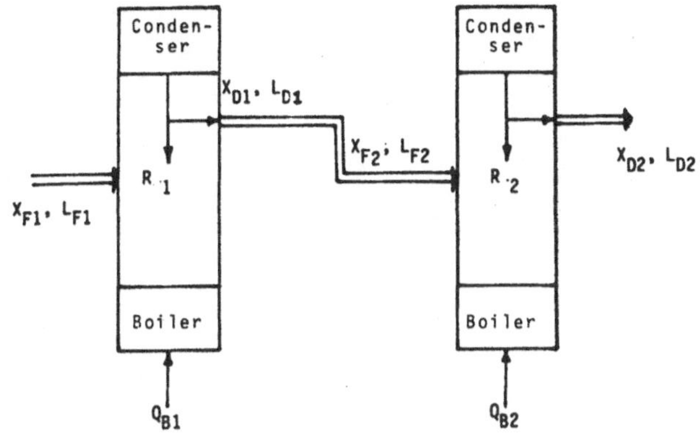

Figure 4. Distillation pilot plant.

An identification of these columns around an operating point gave a 4 states linear model for each column and 2 states simplified model for higher level (fig. 5).

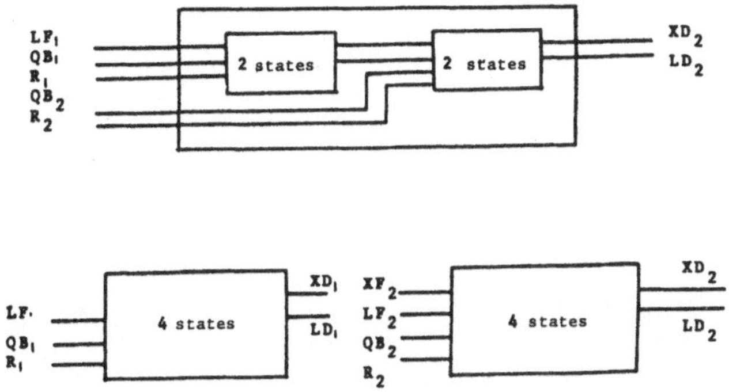

Figure 5. State variable models of distillation columns

Figure 6 presents the evolution of the input and output variables of these two distillation columns when changing the set point. [5]. A local tracking of the reference variables supplied by the coordinating higher level model do not satisfy generally the interaction constraint between these columns. The difference between the output of the first column (concentration XD and output flow LD) and the input of the second

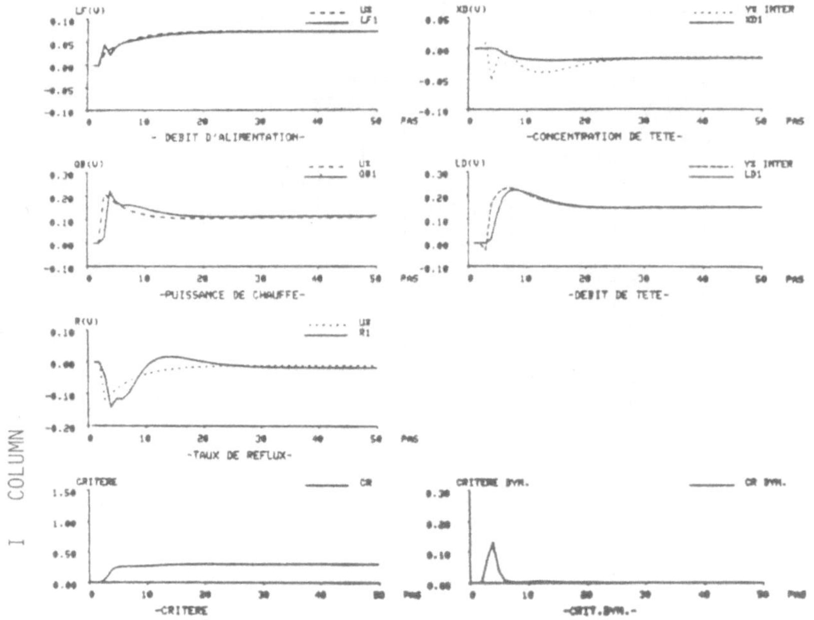

F = { 1/2 , 1/2 }

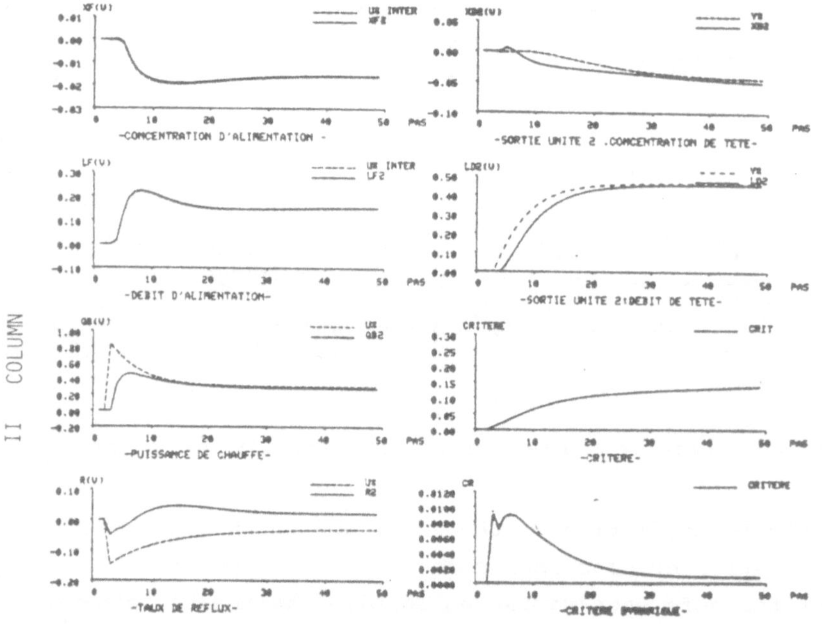

<u>Figure 6</u> : Simulation results of CODECO method with cooperation

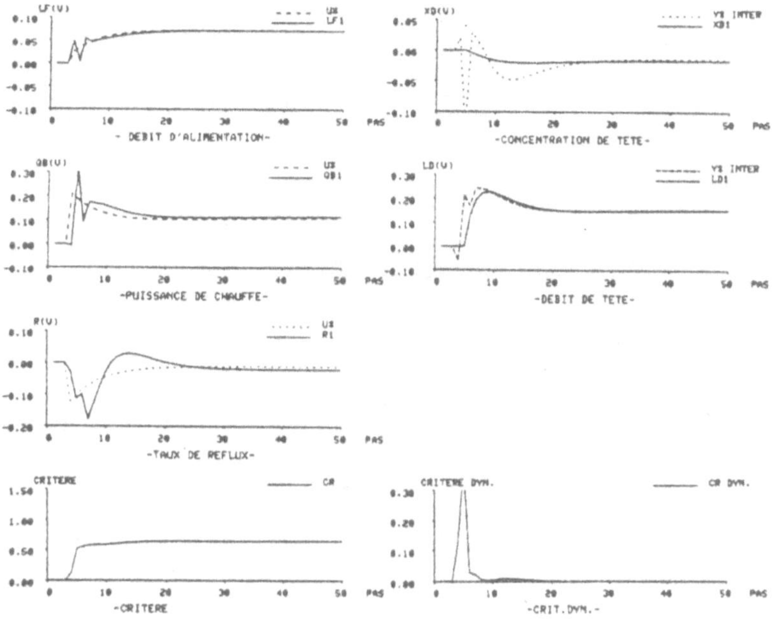

$ZV_1 = 0$ $F = \{0, 0\}$

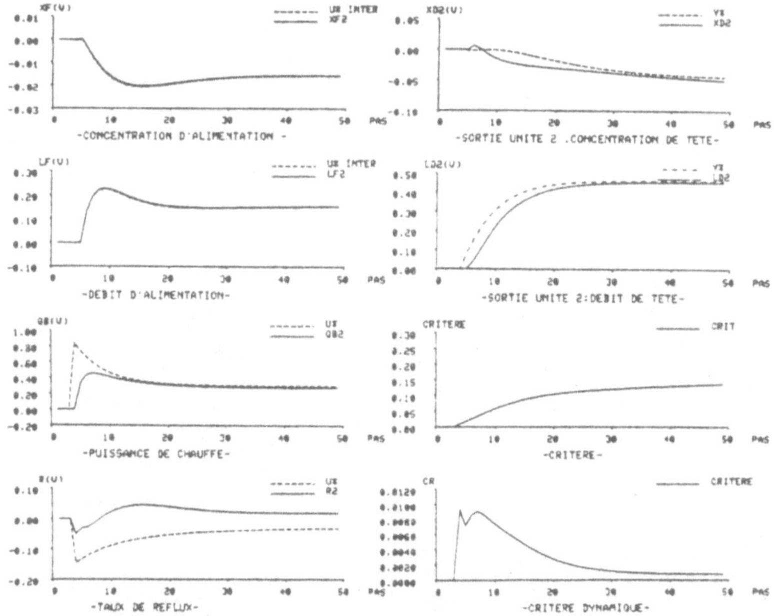

Figure 7 : Cooperative compensation of communication breakdown in the 1st column

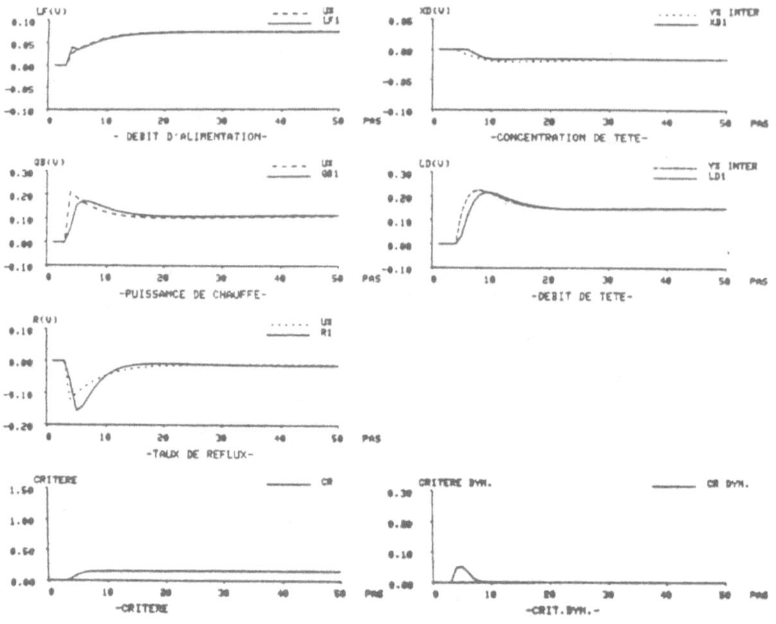

$ZW_2 = 0$ $\qquad\qquad$ $F = \{ 1, 1 \}$

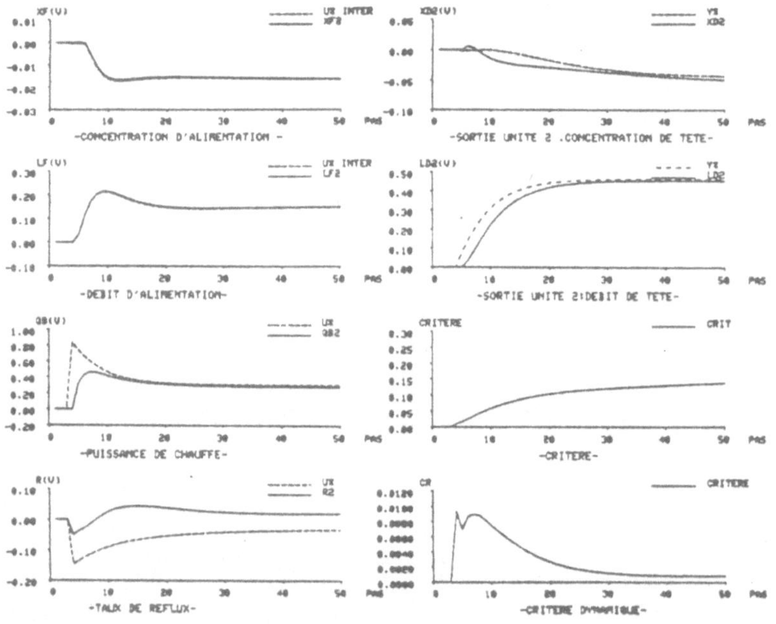

<u>Figure 8</u> : Cooperative compensation of communication breakdown in the 2nd column

(concentration XF and feed flow LF) could be reduced by the cooperation technique. Choosing the cooperation matrix elements F = {1/2 , 1/2} puts the interactions concentration XD = XF and flow LD = LF in the mid-way between those proposed by each column. The evolution of the criterion for the first and second columns mesures the tracking deviation between the coordinating model and the local model including the effect of cooperation.

Figure 7 represents a similar experiment in the case when the interaction reference variables are not known on the first column control centers. They take zero values. There communication errors between the coordinator and first column centers are compensated due to the cooperation effects. The preceding value of cooperation matrix is replaced by F = {0,0} to increase this effect. In this case, which is shown in this figure, the first column gives the interaction value desired by the second column.

Figure 8 shows a cooperative compensation of the communication breakdown between the coordinator and the second column centers. The interaction input reference variables (XF, LF) are not known. The cooperation matrix elements are chosen as F = {1, 1}, the second column accepts the proposition of the first for this interaction.

These three examples show why the cooperation in the decision center may be used to make up for the absence of certain data. We can remark a very small modification of the criterion value and a good reconstruction of the interaction value very close to the value in figure 6 with correct data transfer.

CONCLUSION

Using a redundancy in the control algorithms presents very interesting aspect in the design of the control system. Often the hardware redundancy could be joined by an algorithmic study to solve the problem of the robustness and reliability in the control of industrial processes.

This paper shows why algorithmic and information redundancy could be used to solve same communication problems. This is by making use of a coherence between different data used in the control algorithm. Two functions are complementary, the tracking of higher level model and the cooperation on the same level. Using this, the control with limited information transfert may be possible. The absence of one part of the data could be compensated by an appropriate processing. The cooperation effects assure the control if the reference variables from the higher level are not received. The proposed approach is more interesting for a complex network. The information redundancy allows a design of an information disturbance insensitive control system with detection and dynamic compensation of these disturbances.

REFERENCES

[1] ACQUADRO J.P., BINDER Z., FRANCO A., REY D. :
"Coordinated Decentralized Control (CODECO) : Method, implementation on Microcomputers and experimentation on a physical process"
In Components and Instruments for Distributed Control Systems, Proceeding edided by BINDER and PERRET - Pergamon Press, 1983, pp. 37 - 43.

[2] BINDER Z., BAUDOIS D., FONTAINE H., FERREIRÃ-MAGALHAES M. :
"About a multimodel control methodology, algorithms, microprocessors implementation and application"
In Control Science and Technology for the Progress of Society, Proceeding of I.F.A.C. VII Congress, KYOTO 1981.

[3] BINDER Z., FERREIRA-MAGALHAES M., REY D. :
"Multivariable control of distillation columns"
Preprints of I.F.A.C. VIII Congress, BUDAPEST, 1984.

[4] BINDER Z. :
"Commande Multimodèle et Commande Décentralisée Coordonnée"
Revue RAIRO 2/1984, A.F.C.E.T. - PARIS.

[5] COUDURIER J.F. :
"Robustesse de la commande décentralisée coordonnée (CODECO)
Thèse de Docteur 3° Cycle, I.N.P. GRENOBLE , 1982.

[6] HAGRAS A.N., BINDER Z. :
"A decentralized cooperative control method"
Large Scale Systems 4 (1983) pp. 263-277.

[7] PERRET R. :
"Sur une méthode de commande hiérarchisée"
Bulletin ASSPA/SGA - ZURICH, 1983-3, pp. 3-10.

[8] REY D. :
"Sur la commande décentralisée coordonnée - Application à un procédé pilote de distillation"
Thèse de Docteur-Ingénieur, I.N.P. GRENOBLE, 1978.

MODELLING, CONTROL AND TRANSIENT BEHAVIOUR OF MULTI-TIME-SCALES SYSTEMS

P. BORNE, G. DAUPHIN-TANGUY

Laboratoire d'Automatique et d'Informatique Industrielle
Institut Industriel du Nord (I. D. N.)
B.P. 48
59651 VILLENEUVE D'ASCQ CEDEX - FRANCE

ABSTRACT

A method of modelling of fast transient behaviour for multi-time scales systems, called (SP + R) method is proposed. It uses singular perturbation technique and reciprocal transformation. It leads to a fast decoupled subsystem, which is a very good estimation of the fast initial variables in a time interval corresponding to the boundary layer domain. The determination of fast (SP + R) optimal trajectory is presented. A comparison with usual results is made on an example.

I - INTRODUCTION

The determination of optimal control for large scale systems needs important calculation numerical tools. Particularly in real time problems (aeronautic domain, for example), it is often necessary to simplify the model. In case of multi-time scales systems (1) (2), the singular perturbation (SP) method allows to reduce order by separation of different dynamics, and leads to a quasi-optimal composite control, by determination of an optimal law of control on each reduced part.

But when the study concerns the fast part (parasistics, fast transients, high frequency behaviour), this technique of reduction gives no sufficient accuracy, because the slow dynamic evolution is not taken into account. We propose here a method of modelling of the fast decoupled part, applying the SP technique and the reciprocal transformation, called (SP + R) method, which allows to determine a more precise fast optimal control. In aim of simplicity, the two-time scales case is presented, but it can be extended easily to multi-time scales case.

An example compares optimal trajectory, (SP) and (SP + R) quasi-optimal trajectories.

II - PRESENTATION OF (SP + R) METHOD

II.1 - Singular perturbation method

We want just here to remember the principle of this method, which is now well known (3).

Let us consider the linear singularly perturbed system :

$$\begin{cases} \begin{bmatrix} \overset{\circ}{x} \\ \mu\, \overset{\circ}{z} \end{bmatrix} = \begin{bmatrix} A_{11} & A_{12} \\ A_{21} & A_{22} \end{bmatrix} \begin{bmatrix} x \\ z \end{bmatrix} + \begin{bmatrix} B_1 \\ B_2 \end{bmatrix} u \\ y = \begin{bmatrix} C_1 & C_2 \end{bmatrix} \begin{bmatrix} x \\ z \end{bmatrix} + D\,u \\ x(0),\ z(0) \quad \text{initial values} \end{cases} \tag{1}$$

where $x \in R^{n_1}$ is the slow state vector and $z \in R^{n_2}$ is the fast state vector.

$$\mu \in \left]0\ ,\ 1\right]$$

By setting μ equal to zero, we obtain the slow reduced part as :

$$\begin{cases} \overset{\circ}{x}_s = A_s\, x_s + B_s\, u \\ y_s = C_s\, x_s + D_s\, u \\ z_s = -A_{22}^{-1}\,(A_{21}\, x_s + B_2\, u) \\ x_s(0) = x(0) \\ \text{with} \begin{cases} A_s = A_{11} - A_{12}\, A_{22}^{-1}\, A_{21} \\ B_s = B_1 - A_{12}\, A_{22}^{-1}\, B_2 \\ C_s = C_1 - C_2\, A_{22}^{-1}\, A_{21} \\ D_s = D - C_2\, A_{22}^{-1}\, B_2 \end{cases} \end{cases} \tag{2}$$

The fast variables are decomposed into a slow part (z_s), obtained by an algebraic relation, deduced from the dynamic behaviour of slow variables x_s, and in a fast part $z_f = z - z_s$, which verifies a state equation, called "boundary layer equation" as :

$$\begin{cases} \mu\, \overset{\circ}{z}_{f_{SP}} = A_{22}\, z_{f_{SP}} + B_2\, u \\ y_{f_{SP}} = C_2\, z_{f_{SP}} \\ z_{f_{SP}}(0) = z(0) + A_{22}^{-1}\, A_{21}\, x(0) \end{cases} \tag{3}$$

II.2 - Reciprocal transformation

Many definitions have been presented in different papers (4) (5) (6) (7) for various models like differential equations, transfert function or bond-graph.

Briefly, we define from an initial matrical state equation,

$$\begin{cases} \mathring{x} = A\,x + B\,u \\ y = C\,x + D\,u \\ x(0) \end{cases} \tag{4}$$

the reciprocal system by the reciprocal state equations

$$\begin{cases} \mathring{\tilde{x}} = \tilde{A}\,\tilde{x} + \tilde{B}\,u \\ \tilde{y} = \tilde{C}\,\tilde{x} + \int_{t_o}^{t} D\,u\,dt \\ \tilde{x}(0) \\ \text{with} \begin{cases} \tilde{A} = A^{-1} & \text{(condition : A non singular)} \\ \tilde{B} = B \\ \tilde{C} = -C\,A^{-1} \\ \tilde{x}(0) = x(0) \end{cases} \end{cases} \tag{5}$$

which gives, by introducing a new state variable corresponding to a mode equal to zero :

$$v = \int_{t_o}^{t} u\,dt \tag{6}$$

a canonical form for (5)

$$\begin{cases} \mathring{\tilde{x}} = \tilde{A}\,\tilde{x} + \tilde{B}\,u \\ \mathring{v} = u \\ \tilde{y} = \tilde{C}\,\tilde{x} + D\,v \end{cases} \tag{7}$$

II.3 - (SP + R) method

The reciprocal transformation inverses dynamic and frequential behaviour. By applying

SP technique on the reciprocal model, we obtain a disconnected system, with a good precision on the slow reciprocal decoupled subsystem, which will become, by a new application of reciprocal transformation, the fast reduced part.

The state equations of $z_{f(SP+R)}$ are :

$$\begin{cases} \mu \, \mathring{z}_{f(SP+R)} = A_{22}\, z_{f(SP+R)} + (B_2 + \mu A_{22}^{-1} A_{21} B_1)\, u \\ y_{f(SP+R)} = C_2\, z_{f(SP+R)} + \int_{t_o}^{t} (C_1 - C_2 A_{22}^{-1} A_{21})\, B_1\, u\, dt \\ z_{f(SP+R)}(0) = z(0) \end{cases} \tag{8}$$

We can remark that the slow influence appears in the control term and in the expression of output. The initial values are not modified, which allows to say that the system (8) gives for the fast variables in boundary-layer domain a very suitable model, which takes into account the slow and fast dynamics z_s and z_f.

III - TRAJECTORY OPTIMIZATION

Let us consider here the problem of determination of optimal trajectory for the two time scales system (1) when the cost function to be minimized is :

$$J = \frac{1}{2} \int_0^{+\infty} (y^T y + u^T R u)\, dt \qquad R > 0 \tag{9}$$

III.1 - Optimal trajectory

If Ψ_x and Ψ_z are the adjoint vectors associated to x and z, the hamiltonien H is defined by :

$$H = \frac{1}{2} (y^T y + u^T R u) + \Psi_x^T (A_{11} x + A_{12} z + B_1 u) + \frac{\Psi_z^T}{\mu} (A_{21} x + A_{22} z + B_2 u) \tag{10}$$

The maximum principle gives a well-known result for the optimal control as :

$$u^* = - R^{-1} (B_1^T \Psi_x + \frac{B_2^T}{\mu} \Psi_z) \tag{11}$$

which leads to global optimal trajectory $\begin{bmatrix} x^*(t) \\ z^*(t) \end{bmatrix}$.

III.2 - Quasi-optimal control by separation of dynamics (8) (9)

The global optimal problem is then decomposed into two reduced problems. The slow optimal control u_s^* minimizing the criterion :

$$J_s = \frac{1}{2} \int_0^{+\infty} \left[x_s^T C_s^T C_s x_s + 2 x_s^T C_s^T D_s u_s + u^T (R + D_s^T D_s) u \right] dt \tag{12}$$

associated to equations (2) allows to calculate the optimal trajectory x_s^* and then to deduce the optimal trajectory of the slow part z_s^* of the fast variables.

The boundary-layer equations (3) using a dilated time $\tau = t/\mu$, with fast criterion :

$$J_{f_{SP}} = \frac{1}{2} \int_0^{+\infty} (y_{f_{SP}}^T y_{f_{SP}} + u^T R u) dt \tag{13}$$

leads to a quasi-optimal solution $z_{f_{SP}}^*(\tau)$.

The fast quasi-optimal trajectory z_{SP}^* is then :

$$z_{SP}^*(t) = z_s^*(t) + z_{f_{SP}}^*(\tau) \tag{14}$$

III.3 - Quasi-optimal control by (SP + R) method (5)

The slow reduced system is not modified by this method. So we consider only the fast decoupled equations (8).

By introducing a new fast variable,

$$v = \frac{1}{\mu} \int_{t_o}^{t} u \, dt \qquad u \in R^r \tag{15}$$

and by the basis change :

$$Z_{f_{(SP+R)}} = z_{f_{(SP+R)}} - (B_2 + \mu A_{22}^{-1} A_{21} B_1) v \tag{16}$$

we obtain a new expression for the fast decoupled system :

$$\begin{cases} \mu \, \mathring{z}_{f(SP+R)} = A_{22} \, z_{f(SP+R)} + A_{22} \, (B_2 + \mu \, A_{22}^{-1} A_{21} \, B_1) \, v \\ y_{f(SP+R)} = C_2 \, z_{f(SP+R)} + (C_2 \, B_2 + \mu \, C_1 \, B_1) \, v \\ v(0) = 0 \\ z_{f(SP+R)}(0) = z_{f(SP+R)}(0) \end{cases} \tag{17}$$

with criterion :

$$J_{f(SP+R)} = \frac{1}{2} \int_0^{+\infty} \Big[z_{f(SP+R)}^T \, C_2^T \, C_2 \, z_{f(SP+R)} + 2 \, z_{f(SP+R)}^T \, C_2^T \, (C_2 \, B_2 + \mu \, C_1 \, B_1) \, v + v^T (C_2 \, B_2 + \mu \, C_1 \, B_1)^T \, (C_2 \, B_2 + \mu \, C_1 \, B_1) \, v + u^T R \, u \Big] \, dt \tag{18}$$

This is a linear quadratic problem with infinite horizon of dimension $(n_2 + r)$ $(v \in R^r)$. The fast quasi-optimal trajectory is then :

$$z^*_{SP+R}(t) = z^*_s(t) + z^*_{f(SP+R)}(\tau) \tag{19}$$

III.4 - Comparison on an example

Let us consider the global problem :

$$\begin{cases} \mathring{x} = -\,0.1 \, x + 0.03 \, y + 0.5 \, u \\ \mu \, \mathring{z} = 0.1 \, x - 0.15 \, y + 0.1 \, u \\ y = x + z \\ x(0), \; z(0) \\ \mu = 0.1 \\ \text{criterion to be minimized} \quad J = \frac{1}{2} \int_0^{+\infty} (x^2 + z^2 + u^2) \, dt \end{cases} \tag{20}$$

We present here only the results related to the fast optimal trajectories, determination of optimal control laws being achieved on usual scheme.

- optimal trajectory :

$$z^*(t) = \big[0.1033 \, x(0) - 0.0106 \, z(0) \big] \, e^{-0.5483 \, t} + \big[-\,0.1033 \, x(0) + 1.0106 \, z(0) \big] \, e^{-1.8081 \, t} \tag{21}$$

- <u>(SP) quasi-optimal trajectory</u> :

$$z^*_{SP}(t) = \underbrace{0.0642\ e^{-0.5499\,t}\,x(0)}_{z^*_s(t)} + \underbrace{(z(0) - 0.0642\ x(0))\ e^{-1.8028\,t}}_{z^*_{f_{SP}}(t)} \tag{22}$$

- <u>(SP+R) quasi-optimal trajectory</u> :

$$\left.\begin{aligned} z^*_{(SP+R)}(t) = {} & 0.0642\ e^{-0.5499\,t}\,x(0) + \\ & + (-\,0.0099\,x(0) + 0.1538\,z(0))\ e^{-06158\,t} \\ & + (-\,0.0543\,x(0) + 0.8462\,z(0))\ e^{-2.030\,t} \end{aligned}\right\} \begin{aligned} &\to z^*_s(t) \\ &\to z^*_{f_{(SP+R)}}(t) \end{aligned} \tag{23}$$

which gives the curves presented Figure 1.

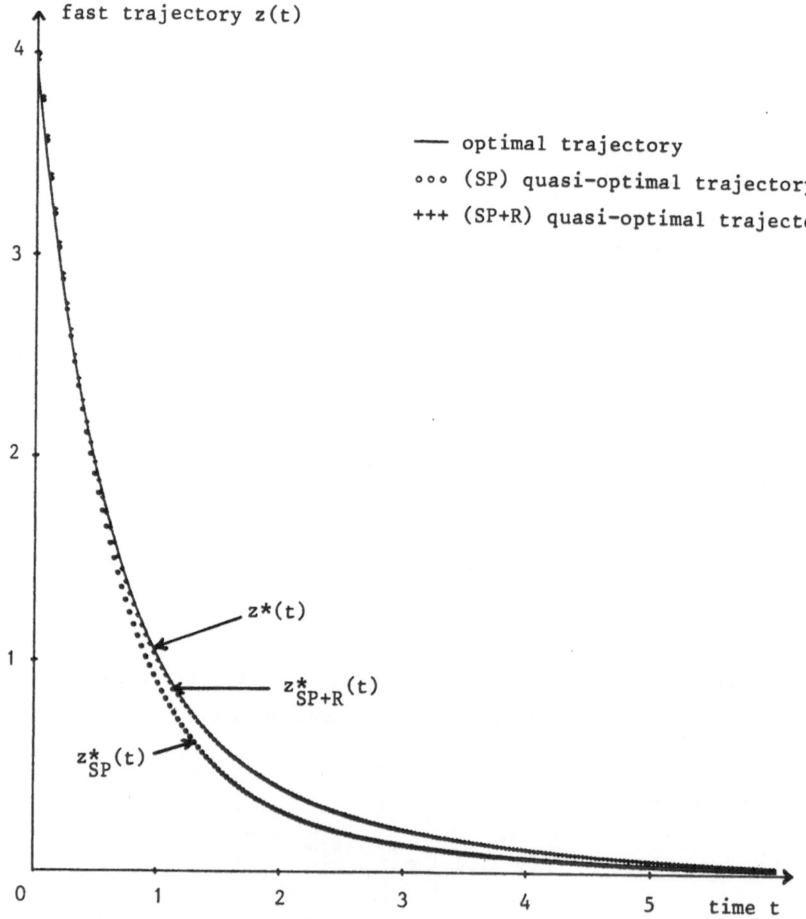

FIGURE 1 - Fast global and decoupled trajectories

Interpretation of results :

The optimal trajectory (21) is composed by two exponential terms with coefficients :

$$\lambda_1 = -0.5483 \quad \text{and} \quad \lambda_2 = -1.8081$$

associated respectively to slow and fast dynamics.

The (SP) quasi-optimal trajectory (22) takes only into account, for the fast part $z^*_{f(SP)}(t)$, the fast value $\lambda_{2_{SP}} = -1.8028$.

But, in the (SP+R) quasi-optimal trajectory (23) appears a term associated to $\lambda_{1(SP+R)} = -0.6158$ which corresponds to the slow evolution of the fast variables.

These remarks show that this approach (SP + R) gives, on a time interval associated to boundary-layer domain, a more precise value of the fast part. For this example, as we can see on Figure 1, the two curves, optimal $z^*(t)$ and quasi-optimal $z^*_{(SP+R)}(t)$ cannot be distinguished.

IV - CONCLUSION

The presented (SP + R) method applied to systems, which have different dynamics, is very interesting in several study domains :

- for modelling because it gives a precise estimation of the transient behaviour, with its slow and fast components, the simulation results are very significant (5) (6),
- for optimal control as shown previously (and in (6) for singular optimal problem), the results are very perceptible when the dynamics are not well separated. They can be used in non linear cases, when the studied system is modellized by a non-constant matrical triplet (A(.), B(.), C(.)).

BIBLIOGRAPHY

(1) KOKOTOVIC P.V., O'MALLEY R.E., SANNUTI P.
"Singular perturbations and order reduction on control theory. An overview"
Automatica 1976, Vol. 12, pp. 123-132.

(2) SAKSENA V.R., O'REILLY J., KOKOTOVIC P.V.
"Singular perturbations in control theory : Survey 1976 - 1982"
IFAC Workshop on Singular Perturbations, Juillet 1982, Ohrid, Yougoslavie.

(3) FOSSARD A.J., MAGNI J.F.
"Modélisation, commande et applications des systèmes à échelles de temps multiples"
RAIRO Automatique/Syst. Anal. and Cont., Vol. 16, n° 1, 1982, pp. 5-23.

(4) DAUPHIN-TANGUY G., BORNE P., MEIZEL D.
"On Order Reduction of multi-time scale systems by singular perturbation and frequency-like methods"
IEEE LSS Symposium, Virginia Beach (U. S. A.), 1982, pp. 190-196.

(5) DAUPHIN-TANGUY G.
"Sur la représentation multi-modèle des systèmes singulièrement perturbés"
Thèse de Doctorat ès Sciences Physiques, Lille (France), 1983.

(6) DAUPHIN-TANGUY G., MOREIGNE O., BORNE P.
"Singular perturbation method and reciprocal transformation on two-time scale systems"
Contribution in "Multivariable control - New concepts and tools"
D. Reidel Publishing Company, to be published.

(7) DAUPHIN-TANGUY G., LEBRUN M., BORNE P.
"Order reduction of multi-time scale systems using bond-graphs, reciprocal system and singular perturbation method"
Journal of Franklin Institute, to be published.

(8) KOKOTOVIC P.V., YACKEL R.A.
"Singular perturbation of linear regulators : basic theorems"
IEEE Trans. on Aut. Control, Vol. AC-17, 1972, n° 1, pp. 29-37.

(9) MAGNI J.F., FOSSARD A.J.
"Commande en deux étapes des systèmes linéaires à deux dynamiques"
RAIRO Automatique / Syst. Anal. and Control, Vol. 16, n° 1, 1982, pp. 25-38.

AN INFORMATION APPROACH TO LARGE SYSTEM DECOMPOSITION

G.R. Sotirov, D.V. Shivkov
Department of Automation of Production
Higher Institute of Mechanical and Electrical Engineering
Sofia, Bulgaria
D.P. Filev, K.B. Konstantinov
Research Laboratory of Instrumentation and Biological Experiment
Automation, Sofia, Bulgaria

Abstract: Information interactions between elements of a complex system are examined and the general information characteristics are determined. Probability estimations of information relations are treated by entropy of information channels. A criterion for information linkage between different elements is given as a base for a procedure of a large system decomposition. An application software is developed for the proposed method. An example of a real large system decomposition in biotechnological industry is presented.

1. Introduction

Synthesizing of effective control systems involves the utilization of adequate methods for decomposition of the complex systems, whereupon the set of interrelated variables $\{X_1,\ldots,X_n\}$ constituting the system X, is broken down into groups $Y_1,\ldots,Y_m$, $\bigcup_j Y_j = X_1,\ldots,X_n$, provided that the condition $Y_i \cap Y_j = \emptyset$, if $i \neq j, \forall i,j=\overline{1,m}$. The interrelation between the X system variables can be of various natures (technological, functional, etc.), and, in the process of the system functioning, it is realized, in the common case, as a stochastic process. Description of the structural properties of X obviously presupposes the utilization of indiceswhich would reflect the mutual influence between the states of system variables in the process of its functioning. That is why the decomposition of X can be realized on the basis of a quantitative measure of the interrelation between the variables X_i, $i=\overline{1,n}$, assigned in a certain manner. Various methods for solving the formulated problem are offered (Emden, 1970). In this paper, we offer a method for complex system decomposition based on an information measure of interaction between the system variables.

2. Information Method Fundamentals

A complex stochastic system X with a specified aim of functioning and algorithm for achieving it is considered. The system variables X_i, $i=\overline{1,n}$ assigned as a function of the discrete time t (discrete systems are considered), that is, $X_i(t)$, $i=\overline{1,n}$, $t=0, 1, 2, \ldots$, are also known. On the basis of the possible conditions of X and the variables $X_i(t)$, $\forall i$, the probability densities occurrence rates are assigned. Then, the behaviour of the X system will be described by means of the model which adequately determines the states of X at each moment of the time t and by means of the probability distributions for each individual state.

For determination of the reciprocal effect between X_i, $i=\overline{1,n}$ in the process of functioning of the X system, the square matrix $D(X) = $ $= d_{ij}$, $i,j=\overline{1,n}$, can be used, its elements d_{ij} being the quantitative factor of influence of the $X_i(t)$ variable upon the $X_j(t+1)$ variable. When $X_j(t+1)$ does not depend on $X_i(t)$, then, obviously, d_{ij} should assume a minimal value and in the opposite case - a maximum value. These conditions are met by Shannon's quantity of reciprocal information - $I(X_iX_j) = d_{ij}$. This factor characterizes the average level of statistical interdependence between system variables at different moments. The measure offered is characterized by its generality and integrity because it can be used in the study of complex systems of a linear and non-linear, discrete and continuous, determined and stochastic character of the interdependences. In all these cases the system state distribution is assigned by means of an established real number.

If the set of possible states of all variables are discrete and final, the elements of the matrix $D(X)$ are determined by the following equation:

$$d_{ij}=I[X_i(t)X_j(t+1)] = H[X_i(t)] + H[X_j(t+1)] - H[X_i(t)X_j(t+1)], \quad (1)$$

where: $H[X_i(t)]$ is entropy of the $X_i(t)$ variable;
$H[X_i(t)X_j(t+1)]$ - entropy of the integrated variables $X_i(t)$ and $X_j(t+1)$. The respective entropies in equation (1) are determined in the following manner:

$$H[X_i(t)] = -\sum_{k\in S_i} P[X_i(t)=k]\log P[X_i(t)=k] \;;$$

$$H[X_j(t+1)] = -\sum_{l\in S_j} P[X_j(t+1)=l]\log P[X_j(t+1)=l] \;;$$

$$H[X_i(t)X_j(t+1)] = -\sum_{k\in S_i}\sum_{l\in S_j} P[X_i(t)=k, X_j(t+1)=l] \times$$

$$\times [\log P[X_i(t)=k, X_j(t+1)=l]$$

where: S_i, S_j are the state sets of variables $X_i(t)$ and $X_j(t)$ respectively;

$P[X_i(t)=k]$ - the probability for the variable X_i to be in k-th state at the moment t , $k\in S_i$;

$P[X_i(t)=k, X_j(t+1)=l]$ - the probability for the variables $X_i(t)$ and $X_j(t+1)$ to be respectively in k-th and l -th states at the same time, $k\in S_i$, $l\in S_j$.

3. Boundary Estimations of the Information Decomposition Level

Under the assumptions made the quantity of average reciprocal information between the states of all variables within the X system is determined by the equation:

$$I(X_1,\ldots,X_n) = \sum_{\{X\}} P(X_1,\ldots,X_n) I^* ,$$

where: $I^* = \log\{ \prod P(X_i X_j)][\prod P(X_i,X_j,X_k,X_l)] \;\ldots\; / \quad P(X_i) \times$

$\times[\prod P(X_i,X_j,X_k)] \;\ldots\}$, $i,j,k,l,\ldots = \overline{1,n}$, if $i<j<k<l \;\ldots < n$, and the products $\prod$ are calculated for all possible index combinations.

In an analogical way we put down the equation for determining the quantity of average conditional reciprocal information between the states of a certain aggregate of system variables with known states of all the rest:

$$I(X_i,\ldots,X_l/X_q,\ldots,X_r) = \sum_{\{X\}} P(X_1,\ldots,X_n) I(X_i,\ldots,X_l/X_q,\ldots,X_r),$$

$$I(X_i,\ldots,X_l/X_q,\ldots,X_r) = \log \{[\prod P(X_i,X_j/X_q,\ldots,X_r) \quad \times$$

$$\times\left[\prod P(X_i,X_j,X_k,X_l/X_q,\dots,X_r)\right]\dots/\left[\prod P(X_i/X_q,\dots,X_r)\right]\times$$
$$\times\left[\prod P(X_i,X_j,X_k/X_q,\dots,X_r)\right]\dots\Big\}\ ,\quad i,j,k,l,\dots=\overline{1,n},$$
$$i<j<k<l\ \dots<n.$$

The value of the system entropy depends on the extent to which the interdependences between the variables states are accounted in the process of decomposition. If the system is considered as an integrity, that is, taking into consideration all interdependences, then the entropy $H(\mathbf{X})$ is determined as below:

$$H(\mathbf{X}) = \sum I(X_i) - \sum I(X_i,X_j) + \sum I(X_i,X_j,X_k) + \dots + (-1)^{n-1} I(X_1,\dots,X_n)\ . \tag{3}$$

summing up being effected for all possible combinations of indices $i<j<k<l\ \dots<n$.

In the case when the interdependences in the $\mathbf{X}$ system are disregarded, that is, $\mathbf{X}$ is considered as an aggregate of independent variables,

$$H(\mathbf{X}) = \sum_i I(X_i) = \sum_i H(X_i)\ . \tag{4}$$

Then the level of informational decomposition of the system $\mathbf{X}$ can be determoned as follows:

$$R(\mathbf{X}) = \left[H(\mathbf{X}) - H(X_1,\dots,X_n)\right]\left[\sum_i H(X_i) - H(X_1,\dots,X_n)\right]^{-1}\ .$$

Since equations (3) and (4) determine, respectively, the lower and upper boundary values of entropy $H(\mathbf{X})$, then, obviously,

$$R(\mathbf{X}) = \begin{cases} 0, & \text{when } H(\mathbf{X}) = H(X_1,\dots,X_n), \\ 1, & \text{when } H(\mathbf{X}) = \sum_i H(X_i) \end{cases}$$

Therefore, every decomposition of the system results in increasing its entropy by $\Delta H = H(X) - H(X_1,\dots,X_n)$, ΔH being a monotonous function of $R(\mathbf{X})$. With respect to information, the entropy variation ΔH can be treated as loss of information upon breaking the interdependent variables into formally independent subsystems.

4. Example

The model of a biotechnological process (penicillin production) is considered. The model is a linear regressive approximation of the well-known model of Fishman-Birjukov where X_1, X_2, X_3, X_4 designate state variables and U designates the system input. The linearized model has the following form:

$$X_1 = -X_1 + 7X_2; \quad X_2 = -X_2 + U;$$
$$X_3 = X_1 - 2X_3; \quad X_4 = 8X_1 - 2X_4.$$

To determine the system structure, the model is simulated as a discrete system. The input signal is generated as an uniformly distributed random value within the range (0; 1). The initial conditions are respectively: $X_1(0) = X_2(0) = X_3(0) = X_4(0) = 1.$

The ranges of variation of X_i, $i = 1, \ldots 4$ and U are respectively spaced at the following intervals:

$$X_1 \in \{(-\infty;\ 3,2),\ [3,2;\ 3,7],\ (3;7;\ +\infty)\};$$

$$X_2 \in \{(-\infty;\ 0,5),\ [0,5;\ 0,7],\ (0,7;\ +\infty)\};$$

$$X_3 \in \{(-\infty;\ 1;5),\ [1,5;\ 1,8],\ (1,8;\ +\infty)\};$$

$$X_4 \in \{(-\infty;\ 10),\ [10;\ 13],\ (13;\ +\infty)\};$$

$$U \in \{(0;\ 0,333),\ [0,333;\ 0,666],\ (0,666;\ 1]\}.$$

Upon 5000 randon implementations of the equation, the values of the reciprocal information quantities in the distributed system are shown in the following table: (see next page)

	u(t+1)	$X_1(t+1)$	$X_2(t+1)$	$X_3(t+1)$	$X_4(t+1)$
u(t)	$1,3.10^{-4}$	0.175	0.097	0.18	0.15
$X_1(t)$	$1,3.10^{-4}$	0.34	0.086	0.59	0.48
$X_2(t)$	$5,3.10^{-4}$	0.32	0.14	0.37	0.27
$X_3(t)$	$8,8.10^{-5}$	0.25	0.06	0.53	0.46
$X_4(t)$	$3,3.10^{-4}$	0.26	0.054	0.46	0.44

Table 3

From the results obtained it can be seen that the variable $X_2(t)$ is least related to the remaining variables U, X_1, X_3, X_4 . Therefore, the system can be decomposed into two subsystems:

$$Y_1 = \{U, X_1, X_3, X_4\} ;$$

$$Y_2 = \{X_2\} .$$

5. Conclusions

The offered approach is marked for its universality and is applicable to complex system decompositions as well.

Its development based upon fuzzy estimations and possibility distributions of the variables will be the subject of further studies.

REFERENCES:

1. Emden M. H., Hierarchical Decomposition of Complexity, Machine Intelligence, No5,1970, Amer.Elsevier,N.Y.
2. 1st IFAC Worcshop - Modelling and Control of Biotechnical Process, Abstr., Helsinki, Aug. 17-19,1982.

MODELING AND CONTROL OF LARGE FLEXIBLE STRUCTURES

R.E. Skelton and A. Hu
Purdue University
School of Aeronautics & Astronautics
West Lafayette, Indiana 47907
U.S.A.

Abstract

It is commonplace in structural dynamics to judge the fidelity of finite element models by the accuracy with which they predict the modal data of the first few modes. However, if the model is being developed for control design the important criteria for judging the finite element models might *not* be the modal data of the first few modes. This paper shows that the modal cost method of judging the importance of modes provides a useful criteria, and that the convergence properties of the modal costs is a reasonable means to determine finite element number and grid distribution.

1.0 Introduction

This paper concerns the finite element modeling of flexible structures when the model is to be used in control design or estimation problems. We wish to show that in this case the finite element decisions (choice of shape functions, and number of finite elements) are quite different than those the structural analysist might make using the standard modal convergence criterion.

We shall illustrate concepts using the simplest possible example, and show that even the simply supported beam lacks a complete theory of finite element modeling for control design. Much has been written about the control of flexible structures, and the beam problem has been used to illustrate the stability problems present with feedback control [1-3]. However, there is yet to emerge a suggestion *how* to improve the original finite element model to improve these stability properties. Instead, attention has been focused on the model reduction problem beginning with a *given* finite element model. [4 - 8]

The motivation for this study therefore, is to begin the development of guidelines for finite element modeling for control design purposes. The development of finite element modeling guidelines for the purposes of structural analysis is rather mature and sophisticated [9 - 11]. Among the principal conclusions are these (as applied, where appropriate, to the simply-supported Euler-Bernoulli beam):

(a) a uniform grid of finite elements is the workhorse in structural dynamics.

(b) uniform grid of finite elements using cubic polynomial shape functions gives no greater than 11% error in the model frequency contained in the model [12].

(c) The convergence of the modal frequencies improves monotonically with increasing number of finite elements. That is, for a given choice of uniform finite elements a 10 finite element model is always better than a 5 finite element model [12].

We will show that the modes which are most critical to the control problem can be measured by their "modal cost," and these critical modes:

(i) are not ordered by frequency

(ii) have modal costs which do not converge monotonically with increasing number of finite elements (uniform)

(iii) have more rapid convergence with non-uniform finite elements.

(iv) The modal cost errors can be greater than 11%.

In other words, *none* of the standard finite element guidelines (a)-(c) hold when the modeling and control problems are combined. The two disciplines must be integrated at a more fundamental level, since the premises of *many* of the theorems developed by each discipline (structural dynamics and control) fail to hold when the two disciplines are integrated.

Modal Cost Analysis (MCA) was introduced in [13-14] to determine the critical modes in a flexible structure by decomposing the control objective function, called the "cost function" in terms of the contribution of each mode, $V = \sum_{i=1}^{N} V_i$ and V_i is called the "modal cost." The inequalities $V_i \geq V_j \geq V_k \geq \ldots$ indicate that V_i is the most critical mode with respect to the control cost function V. The nature of the control problem dictates the choice of the quadratic cost function V. For example, V might represent the line-of-sight errors in the optical path of a flexible space telescope, or it might represent the mean-squared deflections over the surface of an elastic antenna dish. It is clear from the cost decomposition $V = \sum_{i=1}^{N} V_i$ that the importance of a particular mode depends upon the particular cost function of interest, V, and that the modal cost analysis utilizes this dependence. Indeed, almost all of the model reduction literature prior to 1979 *ignored* the control objectives in the model reduction decisions. Only recently has this trend been reversed and a systematic approach sought for the *inseparable* modeling and control problems.

The Ritz procedure

$$(1.1) \qquad \mu(r, t) = \Psi(r)q(t) \qquad \begin{aligned} &\Psi(r) \in R^{3xN} \\ &q(t) \in R^{N} \end{aligned}$$

tolerates a *wide* choice of basis functions $\psi_i(r)$ when N is very large, and in fact when N is infinite the only requirement on the $\psi_i(r)$, i = 1, ... is that they be linearly independent and satisfy the constraint boundary conditions. On the other hand, one may ask "how to choose the $\psi_i(r)$ so that the critical modal costs ψ_i

converge as rapidly as possible. In other words if the fidelity of a given model is measured by the cost error index

$$e_V = \frac{V(\infty)-\hat{V}(N)}{V(\infty)} \tag{1.2}$$

where

V = value of exact cost function (infinite modes)

$\hat{V}$ = value of cost function computed from a N dimensional model

then how should $\psi_i(r)$ be chosen so that for a given value of e_V, the number N is the smallest? Of course, convergence of the total V may not be as important as convergence of each of the critical modal costs V_i. The errors in V_i are defined by

$$e_{V_i} = \frac{V_i(\infty)-\hat{V}_i(N)}{V_i(\infty)} \tag{1.3}$$

where

$V_i(\infty)$ = exact modal cost

$\hat{V}_i(N)$ = modal cost computed from a N-mode model.

The convergence properties of modal costs is the focus of this paper.

The paper is organized as follows: Section 2 briefly describes the beam example used to illustrate the concepts. Section 3 computes the exact value of the cost function V and the exact model costs V_i for the beam. Sections 4 and 5 discuss the convergence properties of the modal costs, and Section 6 suggests an iterative design procedure to obtain faster convergence of the finite element method for control design models.

2.0 A Beam Example

Fig. 1 denotes a torque applied at $r = r_c$ and an output $y(t) = \mu(r_o,t)$ which is the displacement at $r = r_o$. The cost function is

$$V = E_\infty y^T Q y\ , \quad E_\infty \triangleq \lim_{t\to\infty} E \tag{2.1}$$

where E is the expectation operator, needed whenever the inputs at $r = r_c$ have a "noisy" (random) component. Since the inputs on spacecraft are usually from electromechanical actuators, some noise is always present, hence we treat this case. The equations of motion are

$$\rho(r)\ddot{\mu}(r, t) + \mathcal{K}\mu(r,t) = f(t)\delta(r - r_c) \tag{2.2}$$

$$y \triangleq \mu(r_o, t) \qquad 0 \le r \le L$$

where $\rho(r)$ is the mass density along the length r, and f(t) is the magnitude of the force applied at $r = r_c$.

Applying the Ritz method,

(2.3) $$\mu(r,\ t) = \Psi(r)q(t) \quad , \quad \Psi(r) = [\psi_1(r),\ \psi_2(r),\ \ldots.]$$

for any complete set of basis functions $\psi_i(r)$, $i = 1, \ldots, \infty$ it is well known that the partial differential eq. (2.2) is equivalent to the infinite set of ordinary differential equations

(2.4a) $$\mathcal{M}\ddot{q} + \mathcal{K}q = \mathcal{D}w(t) \qquad q \in R^{\infty}$$

$$y = \mathcal{P}q$$

(2.4b) $$\mathcal{M} = \int \Psi^T(r)\rho(r)U(r)dr$$

$$\mathcal{P} \triangleq \psi(r_o)$$

(2.4c) $$\mathcal{K} = \int \psi^T(r)\ K\,\Psi(r)dr$$

(2.4d) $$\mathcal{D} = \int \frac{\partial}{\partial r}\Psi^T(r)\delta(r-r_c)dr = [\frac{\partial}{\partial r}\Psi^T(r)]_{r=r_c}$$

where w(t) is the (noisy) torque applied at $r = r_c$. In this case $Ew(t) = 0$, $Ew(t)w(\tau) = W\delta(t-\tau)$. K is the so called stiffness operator and $\mathcal{K}$ is the stiffness matrix. For the Euler-Bernoulli beam $K \triangleq EI\ \partial^4/\partial r^4$.

Now the finite element point of view of the Ritz theory places certain restrictions on the choice of the basis functions $\psi_i(r)$. In the finite element context these are called "shape functions" since the $\psi_i(r)$ are picked to represent some geometric shape that the ith finite element is assumed to take on. The $\psi_i(r)$ chosen by the finite element method are not usually orthogonal but "nearly" orthogonal since only finite elements which are adjacent to the ith one contribute to nonzero values of $\psi_i(r)$. For these reasons, $\mathcal{M}$ and $\mathcal{K}$ in (2.4) are usually "banded" matrices; zero everywhere except "near" the diagonal. Using cubic shape functions for the beam with constant EI, ρ and length L, and N uniform finite elements,

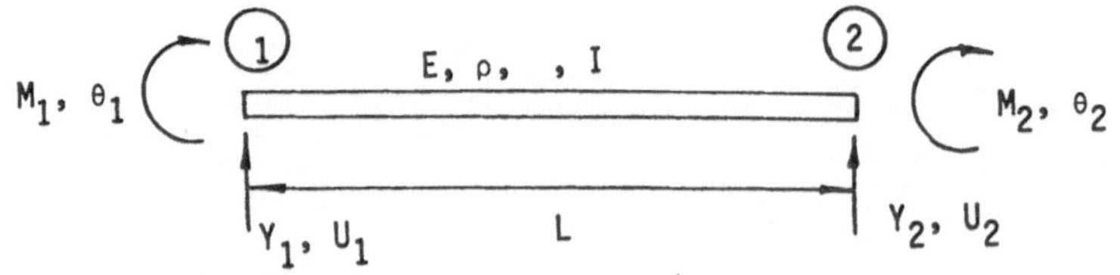

yields for the ith element

(2.5) $$\mu(r, t) = \sum_{i=1}^{N} (\psi_{1i}(r)u_{1i}(t) + \psi_{2i}(r)\theta_{1i}(t) + \psi_{3i}(r)u_{2i}(t)-\psi_{4i}\theta_{2i}(t)) ,$$

(2.6a) $$\psi_{1i}(r) = 1 + 2\,\xi_i^3 - 3\xi_i^2 , \qquad \xi_i \overset{\Delta}{=} r_i/L_i$$

(2.6b) $$\psi_{2i}(r) = -r_i(1 - \xi_i)^2 \qquad r_i = r - (i - 1)L_i , \quad (i - 1)\,L_i \le r \le iL_i$$

(2.6c) $$\psi_{3i}(r) = 3\,\xi_i^2 - 2\,\xi_i^3 \qquad L_i = L/N$$

(2.6d) $$\psi_{4i}(r) = -\,\xi_i + \xi_i^2 \qquad i = 1 , \ldots, N$$

Hence from (2.4)

(2.7) $$M_i \overset{\Delta}{=} \frac{\rho L_i}{420} = \begin{bmatrix} 156 & -22L_i & 54 & 13L_i \\ -22L_i & 4L_i^2 & -13L_i & -3L_i^2 \\ 54 & -13L_i & 156 & 22L_i \\ 13L_i & -3L_i^2 & 22L_i & 4L_i^2 \end{bmatrix}$$

$$K_i = \frac{EI}{L_i} = \begin{bmatrix} 12/L_i^2 & -6/L_i & -12/L_i^2 & -6/L_i \\ -6/L_i & 4 & 6/L_i & 2 \\ -12/L_i^2 & 6/L_i & 12/L_i^2 & 6/L_i \\ -6/L_i & 2 & 6/L_i & 4 \end{bmatrix} \begin{pmatrix} u_1 \\ \theta_1 \\ u_2 \\ \theta_2 \end{pmatrix}$$

The overall mass and stiffness matrices in (2.4) are easily constructed from (2.5)-(2.7). The output matrix, for example with 6 finite elements and r_o = .35L ($r = r_o$ lies within the third finite element), becomes

(2.8) $$P = \Psi(r_o) = [0, 0, 0, \psi_{13}(r_o), \psi_{23}(r_o), \psi_{33}(r_o), \psi_{34}(r_o), 0, 0, 0, 0, 0]$$

Now apply the modal transformation

(2.9) $$q = E\eta \quad , \qquad E = [e_1, e_2, \ldots]$$

where E is chosen so that

(2.10) $$E^TME = I \quad , \qquad E^TKE = \Omega^2$$

where I is the identity matrix and Ω^2 is diagonal with $\Omega_{ii} = \omega_i$. Then (2.4) premultiplied by E^T becomes

$$\ddot{\eta}_i(t) + 2\zeta_i\omega_i\dot{\eta}_i + \omega_i^2\, \eta_i(t) = d_i^T w(t) \qquad i = 1, \ldots, N$$

(2.11)
$$y = \sum_{i=1}^{N} p_i\eta_i(t) \qquad\qquad p_i = \Psi(r_o)e_i = Pe_i$$

$$d_i^T = e_i^T \{\frac{d}{dr}\Psi^T(r)\} = e_i^T \Phi_i^T(r)$$

where modal damping ζ_i has arbitrarily been added, and $\zeta_i = .005$ in all of our subsequent calculations.

The modal cost formula

(2.12)
$$V_i = \frac{(p_i^T Q p_i)(d_i^T W d_i)}{4\zeta_i\omega_i^3}, \qquad i = 1, \ldots, N$$

derived in [14] holds under any *one* of the following three conditions $\{d_i^T W d_j = 0, p_i^T Q p_j = 0, \zeta_i << 1\}$. We claim here only the small damping property $\zeta_i << 1$.

3.0 Value of the Cost Function and the Modal Cost

It is not necessary to perform the finite element analysis of the previous Section if the exact solution of (2.2) can be obtained. In this Section we wish to compute the exact value of the cost function (2.1) subject to (2.2). This will allow the computation of cost errors (1.2)-(1.3), when compared to finite element models. Of course for realistic, complex structures the actual value of the cost may not be computable except possibly by numerical determination of convergence using increasing numbers of finite elements. Nonetheless, it is important to start with an example for which the exact solution is known so that the convergence properties of the modal costs using the finite element method can be confidently established. If we failed to take this step we could not be optimistic about applying the modal cost analysis to more complex structures.

Assume a white noise with intensity W eminating from the torque actuator located at $r = r_c$ on the beam of Fig. 1, with constant EI, ρ. Under these assumptions, V in (2.1) has the following value.

Theorem 1.

For the Euler Bernoulli beam of Fig. 1, *with a displacement* $\mu(r_o, t)$ *at point* $r = r_o$ *and a torque* $w(t)$ *applied at point* $r = r_c$, *the value of* (2.1) *subject to* (2.2) *is*

$$(3.1a)\qquad V=\frac{QW}{4\zeta\sqrt{\rho(EI)^3}}\Big[\frac{2}{3}\frac{r_c^4}{L^4}+\frac{2r_c^2r_o^2}{L^4}\Big]-\frac{1}{3L^3}\Big[(r_c+r_c)^3+|r_0-r_c|^3+2(r_0^3-r_c^3)\Big]+\frac{2}{3}\frac{r^2}{L^2}\}$$

when $r_0+r_c\leq L$

$$(3.1b)\qquad =\frac{QW}{4\zeta\sqrt{\rho(EI)^3}}\{\Big[\frac{2}{3}(1-\frac{r_c}{L})^4+2(1-\frac{r_c}{L})^2(\lambda-\frac{r_0}{L})^2$$

$$-\frac{\lambda}{3L^3}(\big[2L-r_o-r_c\big]^3+|r_o-r_c|^3+2\big[(L-r_0)^3-(L-r_c)^3\big])$$

$$+\frac{2}{3}(1-\frac{r_c}{L})^2\Big]\}$$

when $r_o+r_c\geq L$

The proof of Theorem 1 is provided in the Appendix. Note that the cost V is an explicit function of input location r_c, output location r_o, and material properties ρ, EI, ζ. V is *not* dependent upon mode shapes and mode frequencies, even though V is the infinite sum of all modal costs V_i which *are* dependent upon the modal data.

The exact expression for the modal cost V_i for a lightly damped flexible structure with displacement outputs y is given by (2.12), which is written for the beam of Section 2.0 in the form (A6). A plot of these modal costs versus the modal frequencies is given in Fig. 2. By comparision with the exact expressions (A6) labeled $\odot$ on Fig. 2, is shown the modal costs computed from a 10-finite element model (points lebeled Δ). Note that the importance of the modes as measured by modal costs is *not* ordered by frequency. A three-mode model constructed from modal cost analysis would contain modes 1, 2, and 4. Whereas, a three-mode model constructed from the lowest frequencies would contain modes 1, 2, 3. For displacement type of outputs y, it was proved in [4] that modal costs tend to zero, $V_i \to 0$, as the mode number tends to infinity, $i \to \infty$, but Fig. 2 shows that this convergence is not monotonic, and that the most accurate prediction of the cost function V would *not* be obtained by truncating modes by a frequency criterion.

Theorem 1 establishes convergence of the cost function for torque inputs and linear displacement outputs. It may similarly be shown that cost convergence is established for Euler-Bernoulli beams with either angular or linear displacement outputs and either torque or force inputs. Cost convergence is also established for linear rate outputs and force inputs, but cases for which costs do not converge are linear rate outputs and torque inputs, or angular rate outputs and either torque or force inputs.

4.0 Model Error Indices

To evaluate the finite dimensional beam model we use the cost error index (1.2), where $V(\infty)$ is given by (3.1) and $\hat{V}(N)$ is given by the sum of (2.12), or equivalently (A6), up to i = 1, ..., N. A plot of e_V versus number of modes is given in Fig. 3 for a variety of output locations r_o on the beam, with $r_c = 0$. However, the modes on the abcissa of Fig. 3 are *not* ordered by frequency. Instead, they are ordered by the modal cost criterion

$$(4.1) \qquad V_1 \geq V_2 \geq V_3 \geq \dots \text{ etc.}$$

Of course the modal cost ordering is only based upon the set of modes computed. Fig. 4 illustrates that the cost error e_V tends to converge to a smaller value as number of modes in the computed set increases. Note especially that the last 14 modes out of 100 (denoted by Δ) and the last 7 modes out of 50 (denoted by +) do not have accurate modal costs. The actual cost error converges to zero as the number of modes converges to infinity, but with 20 modes the cost error converges to approximately $2\times10^{-3}\%$, with 50 modes to approximately $2\times10^{-4}\%$, with 100 modes to approximately $2\times10^{-5}\%$, and with 150 modes to approximately $10^{-5}\%$. The modal cost strategy (retain the modes with the largest modal costs) guarantees a *minimum* of e_V for any specified number of modes, since e_V is proportional to the sum of the modal costs deleted, and this sum is the smallest possible by choice of the sequence (4.1). TABLE 1 numbers the modes by frequency (column 2) and also lists the errors in frequency (e_{ω_i}) where

$$(4.2) \qquad e_{\omega_i} \triangleq \frac{(\omega_i - \hat{\omega}_i)}{\omega_i}$$

and the errors in modal costs (e_V) as the number of uniform finite elements increases from 1 to 6. Note from column 6 where the exact modal costs computed from (A6) that the modal cost magnitudes (4.1) are not necessarily ordered by frequency, as also illustrated in Fig. 2. The exact frequency ω_i is given by (A1), whereas the frequency computed from the finite element model is derived from (2.4)-(2.11). Fig. 5 and TABLE 1 show that the frequency errors monotonically decrease as the number of finite elements increase, and that the frequency errors are no greater than 11%. This recurring number for the last mode is due to the well-known fact that the last mode looks like the first mode for *a simple* finite element. Thus, the magnitude of the first mode error (11%) keeps recurring in Fig. 5. Of course, when the length of the shortest finite element ceases to be large compared to the thickness of the beam, the Euler-Bernoulli assumptions are not valid.

In contrast to Fig. 5, Fig. 6 shows that modal cost errors can *exceed* 11% (e.g. 54% for the last mode of a 3 finite element model), *and* that the modal cost errors do not monotonically decrease with an increase in the number of finite elements. From Fig. 6 note that the 5-element model predicts the modal cost of mode 2 ($4\times10^{-3}\%$) much more accurately than a 10-element model ($3\times10^{-2}\%$). One might therefore conjecture that there exists a *non-uniform* finite element distribution for which the accuracy of the modal costs improve monotonically with the number of finite elements (that is make Fig. 6 look more like Fig. 5). This is a useful research objective under study, but no solution is yet known.

5.0 Relationships Between Classical Frequency Errors and Modal Cost Errors

Let $\{V_i, \omega_i, \zeta_i, \psi_i(r), \phi_i(r_c)\}$ be the exact modal cost, frequency, damping, mode shape and mode slope for mode i and let $\{\hat{V}_i, \hat{\omega}_i, \hat{\zeta}_i, \hat{\psi}_i(r_o), \hat{\phi}_i(r_c)\}$ represent the approximate quantities from finite element methods. Define

$$(5.1) \qquad R_{\zeta i} \triangleq \zeta_i/\hat{\zeta}_i \, , \quad R_{\omega i} \triangleq (\omega_i/\hat{\omega}_i)^3 \, ,$$

$$(5.2) \qquad R_{\phi i} \triangleq \|\phi_i(r_c)\|^2/\|\hat{\phi}_i(r_c)\|^2 \, , \quad R_{\psi i} = \|\psi_i(r_o)\|^2/\|\hat{\psi}_i(r_o)\|^2$$

Then the modal cost error defined by (1.3) is related to (5.1), (5.2) as follows.

$$(5.3) \qquad e_{V_i} = \frac{R_{\zeta i}R_{\omega i}}{R_{\phi i}R_{\psi i}} - 1$$

This result follows immediately by substitution of (5.1), (5.2) into (1.3). Eq. (5.3) makes it clear why e_{V_i} is not proportional to e_{ω_i} (as Figs. 5 and 6 illustrated), since the mode shape and mode slope errors $R_{\phi i}$ and $R_{\psi i}$ can dominate the effects of frequency errors. Yedavalli [16] showed also for plates that the cost function V is most sensitive to errors in ζ_i, and V_i is next most sensitive to errors in ϕ_i or ψ_i and V_i is *least* sensitive to errors in frequency. Murphy's law is at work here: in experiments the frequency is the *easiest* parameter to identify but can be the *least* important parameter. For general cost sensitivity with respect to modal data, see [16].

6.0 Non-Uniform Finite Element Grids for Cost-Equivalent Models

The expression (3.1) for the cost function $V(\rho, E, I, L, r_o, r_c)$ makes it clear that the cost is a function of material properties (ρ,E), material geometry (I,L), and sensor and actuator locations (r_o, r_c). Thus, there is a variety of combinations of structure design parameters $(\rho, E, I, L, r_o, r_c)$ which yield the same value of V. These will be called "cost-equivalent" structures. Recalling that the cost is simply the mean-squared value of the deflection at the output location, we conclude that there is a variety of structure designs which will lead to the same performance as measured by V, and (3.1) serves to guide these designs to accomplish a specified performance value V. Furthermore these results are independent of the modal data.

Now if the structure is modeled with finite elements, then for this model the cost calculation $\hat{V}$ becomes a function of the number of finite elements, N. We have observed in Fig. 6 that the modal cost errors do not monotonically decrease with an increase in the number of (uniform) finite elements. To improve the accuracy of the modal cost calculations one may select nonuniform finite elements. The scheme is illustrated in Fig. 7, where the inner iteration loop changes the grid (all elements are uniform except the last one of length ℓ_2, in the beam example) until the total cost converges. Next the outer iteration loop increases the number N of finite elements until each of the critical modal costs converge. The critical modal costs are the largest modal costs. Of course in the beam example V and V_i may be derived *a priori*. In the absence of exact analytical solutions for V and V_i, the convergence is established by comparing $\hat{V}$ and $\hat{V}_i$ with the last iteration.

TABLES 2 and 3 illustrate the convergence properties of the algorithm in Fig. 7 for the simplest grid distribution rule: increase the length of the last finite element (from left to right in Fig. 1) while all other elements have equal length (recall the input is at $r_c = 0$ and $r_o = .35L$). Suppose we wish to construct a 2-mode model (with accurate modal costs for the first two most critical modes). Modes 1 and 2 have the largest modal costs in this case. TABLE 2 shows after 3 iterations that a model without cost error ($\approx 3\text{x}10^{-5}\%$) is obtained with $\ell_2 = .3547L$, and for this 3-finite element configuration the modal cost calculations for modes 1 and 2 are more accurate than the uniform grid case (although the total cost error reduced by 3 orders of magnitude, the individual modes 1 and 2 have only slightly more accurate modal costs). Note also that the 3rd mode, which is truncated in this problem has a larger modal cost error. Thus, an important feature of the modal cost convergence goals is to worry about convergence *only* in the critical modes (as determined by modal costs and not by frequency). TABLE 3 shows a five-finite element case where the total cost error was reduced after 3 iterations to ($5.7\text{x}10^{-6}\%$), but the modal costs for the critical modes 1 and 2 have not converged after 3 iterations. More iterations are required in this case.

7.0 Conclusions

Modal costs rank the modes of a structure in order of their relative importance in the control problem. Herein we have studied the convergence properties of modal costs, after first developing an analytical expression for the infinite sum of all modal costs. This expression reveals the particular combination of material properties, (ρ, E), material geometry (I, L), and sensor and actuator locations (r_o, r_c) which yield "cost-equivalent" beams. Cost convergence is established for Euler-Bernoulli beams with either angular or linear displacement outputs and either torque or force inputs. Cost convergence is also established for linear rate outputs and force inputs, but cases for which costs do not converge are: linear rate outputs and torque inputs, or angular rate outputs and either torque or force inputs.

With uniform finite element grids modal costs do not converge monotonically with an increase in finite elements, although such monotonic convergence may be obtained with nonuniform grid distributions. This topic is currently under study.

References

[1] P. Likins, "The Application of Multivariable Control Theory to Spacecraft Attitude Control," Proceedings IFAC, 4th Symposium on Multivariable Technological Systems, New Brunswick, Fredericton, N.B. Canada, 4-7 July 1977.

[2] M. Balas, "Active Control of Flexible Systems," AIAA Symposium on Dynamics and Control of Large Flexible Spacecraft," Blacksburg, Va., 1977.

[3] R. Skelton, "Control Design of Flexible Spacecraft," Chapter 8, Theory and Applications of Optimal Control in Aerospace Systems, ed. P. Kant, AGARD publication no. 251, KBN 92-835-1391-6, July 1981. Distributed in the United States, by NASA, Langley Field, Va. 23365, attn: Report Distribution and Storage Unit.

[4] R. Skelton, P. Hughes, H. Hablani, "Order Reduction for Models of Space Structures Using Modal Cost Analysis," J. Guidance and Control, Vol. 5, No. 4, July 1982.

[5] P. Hughes, "Modal Identities for Elastic Bodies with Application to Vehicle Dynamics and Control," J. Applied Mechanics, Vol. 47, No. 1, Mar. 1980, pp. 177-184.

[6] J. Sesak, P. Likins, T. Coradetti, "Flexible Spacecraft Control by Model Error Sensitivity Suppression," Proceedings, 2nd VPI&SU/AIAA Symposium on Dynamics and Control of Large Flexible Spacecraft, Blacksburg, Va., 1979, pp. 349-368.

[7] L. Meirovitch and H. Baruh "Optimal Control of Damped Flexible Gyroscopic Systems," J. Guidance and Control, Vol. 3, No. 2, Mar. 1980, pp. 140-150.

[8] L. Pinson, A. Amos, and V. Venkayya, "Modeling, Analysis and Optimization Issues for Large Space Structures," NASA Conference Publication 2258, Proceedings of Workshop, Williamsburg, Va., May 13-14, 1982.

[9] J. Oden and J. Reddy, An Introduction to the Mathematical Theory of Finite Elements, Wiley, N.Y. 1976.

[10] O. Zienkiewicz, The Finite Element Method in Engineering Science, McGraw-Hill, N.Y., 1971.

[11] J. Przemienicki, Theory of Matrix Structural Analysis, McGraw-Hill, N.Y. 1968.

[12] J. Archer, "Consistent Matrix Formulation for Structural Analysis Using Finite Element Techniques," AIAA Journal, Vol. 3, No. 10, Oct. 1965, pp. 1910-1918.

[13] R. Skelton, "Cost Decomposition of Linear Systems with Application to Model Reduction," Int. Journal Control, 1980, Vol. 32, No. 6, 1031-1055.

[14] R. Skelton and P.C. Hughes, "Modal Cost Analysis for Linear Matrix-Second-Order Systems," J. Dyn. Sys. Meas. Control, Vol. 102, Sept. 1980.

[15] M.R. Spiegel, Advanced Mathematics, McGraw-Hill, Inc. 1971.

[16] R. Yedavalli and R. Skelton, "Determination of Critical Parameters in Large Flexible Space Structures with Uncertain Modal Data," J. Dyn. Sys. Meas. Control, Vol. 105, Dec. 1983.

Appendix: Proof of Theorem 1:

The exact mode shapes, mode slopes, and mode frequencies are known for the beam:

$$\omega_i = \sqrt{\frac{EI}{\rho}}\left(\frac{i\pi}{L}\right)^2 \qquad i = 1, \ldots, \infty \tag{A1}$$

$$\psi_i(r) = \sqrt{\frac{2}{\rho L}} \sin\left(\frac{i\pi}{L} r\right), \quad i = 1, \ldots, \infty \tag{A2}$$

$$\phi_i(r) = \frac{\partial}{\partial r} \psi_i(r) = \frac{i\pi}{L}\sqrt{\frac{2}{\rho L}} \cos\left(\frac{i\pi}{L} r\right) \tag{A3}$$

Hence, from (2.11) (noting that the modal matrix E = I in (2.9), (2.10) since we choose the shape functions $\psi_i(r)$ to be the *mode* shapes),

$$p_i = \psi_i(r_o) = \sqrt{\frac{2}{\rho L}} \sin\left(\frac{i\pi}{L} r_o\right) \tag{A4}$$

$$d_i = \phi_i(r_c) = \frac{i\pi}{L}\sqrt{\frac{2}{\rho L}} \cos\left(\frac{i\pi}{L} r_c\right) \tag{A5}$$

and (2.12) becomes

$$V_i = \frac{QWL^2}{i^4 \zeta_i \pi^4 \sqrt{\rho(EI)^3}} \sin^2\left(\frac{i\pi}{L} r_o\right) \cos^2\left(\frac{i\pi}{L} r_c\right) \tag{A6}$$

and the cost decomposition property [14] of the modal costs $V = \sum_{i=1}^{\infty} V_i$ yields

$$V = k \sum_{i=1}^{\infty} i^{-4} \sin^2(ix_o) \cos^2(ix_c), \quad x_o \triangleq \pi r_o/L \tag{A7}$$

$$x_c \triangleq \pi r_c/L$$

$$k \triangleq \frac{QWL^2}{\zeta\pi^4\sqrt{\rho(EI)^3}}$$

The following identities from [15] are useful

$$\sum_{i=1}^{\infty} i^{-4} = \pi^4/90 \qquad \text{(from [15] p. 194)} \tag{A8}$$

$$\pi|x| - x^2 = \pi^2/6 - \sum_{i=1}^{\infty} i^{-2} \cos 2ix, \quad \text{for } 0 \le |x| \le \pi \text{ (from [15] p. 198)} \tag{A9}$$

Twice, integrate both sides of (A9) from 0 to x and obtain

$$-\frac{x^4}{12} + \frac{\pi}{6}|x|^3 = \frac{\pi^2}{12}x^2 + \frac{1}{4}\sum_{i=1}^{\infty} i^{-4}\cos 2ix - \frac{1}{4}\sum_{i=1}^{\infty} i^{-4}, \quad 0 \le |x| \le \pi \tag{A10}$$

Rearrange this expression and use (A8) to obtain

$$\sum_{i=1}^{\infty} i^{-4} \cos 2ix = -\frac{x^4}{3} + \frac{2}{3}\pi|x|^3 - \frac{\pi^2}{3}x^2 + \frac{\pi^4}{90}, \quad 0 \le |x| \le \pi \tag{A11}$$

From well known trigonometric identities

$$\sin\alpha \sin\beta = \frac{1}{2}[\cos(\alpha-\beta) - \cos(\alpha+\beta)],$$

$$\sin\alpha \cos\beta = \frac{1}{2}[\sin(\alpha+\beta) + \sin(\alpha-\beta)], \quad \sin^2\alpha = \frac{1}{2}[1 - \cos 2\alpha],$$

the V in (A7) becomes

$$V = \frac{k}{4}\sum_{i=1}^{\infty} i^{-4}[\sin i(x_o+x_c) + \sin i(x_o-x_c)]^2$$

$$= \frac{k}{4}\sum_{i=1}^{\infty} i^{-4}[\sin^2 i(x_o+x_c) + \sin^2 i(x_o-x_c) + 2\sin i(x_o+x_c)\sin i(x_o-x_c)]$$

$$= \frac{k}{4}\sum_{i=1}^{\infty} i^{-4}[\frac{1}{2}(1 - \cos 2i(x_o+x_c)) + \frac{1}{2}(1 - \cos 2i(x_o-x_c))$$

$$+ 2\frac{1}{2}(\cos i2x_c - \cos i2x_o)]$$

Using identities (A8) and (A11) (using when appropriate $x = x_o+x_c$, $x = x_o-x_c$, $x = x_o$, $x = x_c$) V becomes

$$V = \frac{k}{4}[\pi^4/90 - \frac{1}{2}\{-\frac{(x_o+x_c)^4}{3} + \frac{2}{3}\pi|x_o+x_c|^3 - \frac{\pi^2}{3}(x_o+x_c)^2 + \pi^4/90\}$$

$$- \frac{1}{2}\{-\frac{(x_o-x_c)^4}{3} + \frac{2}{3}\pi|x_o-x_c|^3 - \frac{\pi^2}{3}(x_o-x_c)^2 + \pi^4/90\}$$

$$+ \{-\frac{x_c^4}{3} + \frac{2}{3}\pi|x_c|^3 - \frac{\pi^2}{3}x_c^2 + \pi^4/90\} - \{-\frac{x_o^4}{3} + \frac{2}{3}\pi|x_o|^3 - \frac{\pi^2}{3}x_o^2 + \frac{\pi^4}{90}\}]$$

Hence

$$V = \frac{k}{4}[\frac{(x_o+x_c)^4}{6} + \frac{(x_o-x_c)^4}{6} + \frac{x_o^4}{3} - \frac{x_c^4}{3}$$

$$- \frac{\pi}{3}\{|x_o+x_c|^3 + |x_o-x_c|^3\} + \frac{2}{3}\pi(|x_o|^3 - |x_o|^3)$$

$$+ \frac{\pi^2}{6}\{(x_o+x_c)^2 + (x_o-x_c)^2\} - \frac{\pi^2}{3}x_c^2 + \frac{\pi^2}{3}x_o^2] \tag{A12}$$

where the limits imposed by $0 \le |x| \le \pi$ in (A9) imply that $0 \le |x_o+x_c| \le \pi$, $0 \le |x_o-x_c| \le \pi$, $0 \le |x_o| \le \pi$, $0 \le |x_c| \le \pi$. All these inequalities are automatically satisfied by the definitions of x_o, x_c in (A7) if $|x_o+x_c| \le \pi$. With these constraints and by rearranging terms for the 4-th and second power of r's (A11) agrees with (3.1a). The proof of (3.1b) is similar and is obtained by defining r from the right end of the beam instead of the left end.

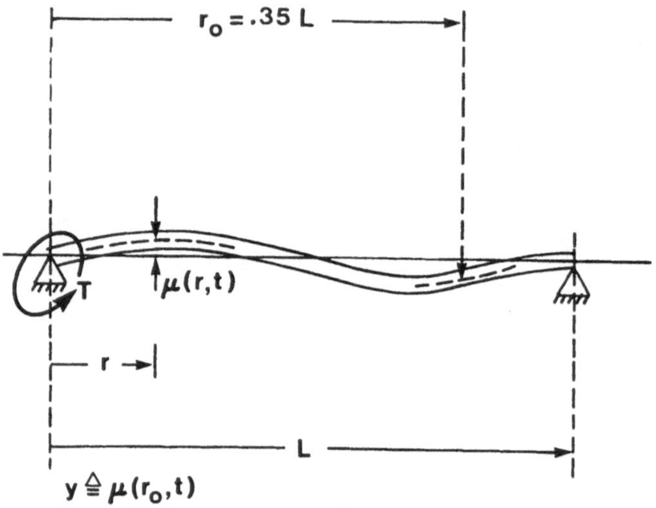

Fig. 1 Simply Supported Beam

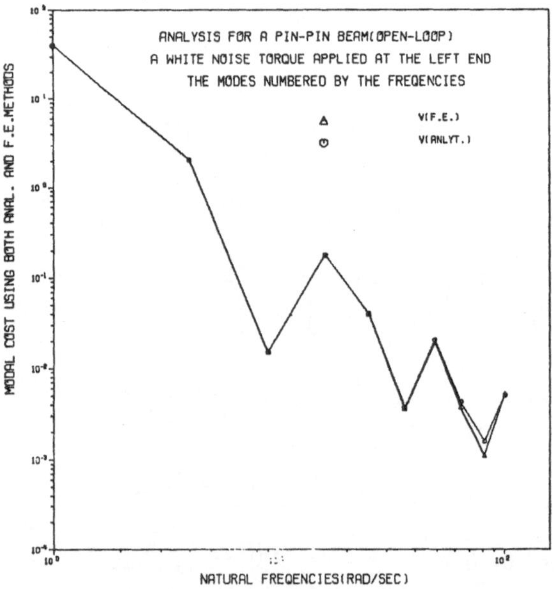

FIGURE 2 MODAL COST VS. NATURAL FREQENCIES

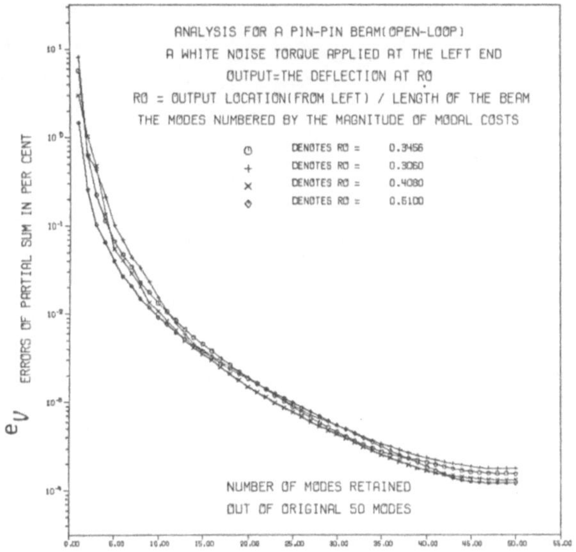

FIGURE 3

CONVERG. FOR OPEN LOOP COST FUNCN. AS THE MODES INCREASES

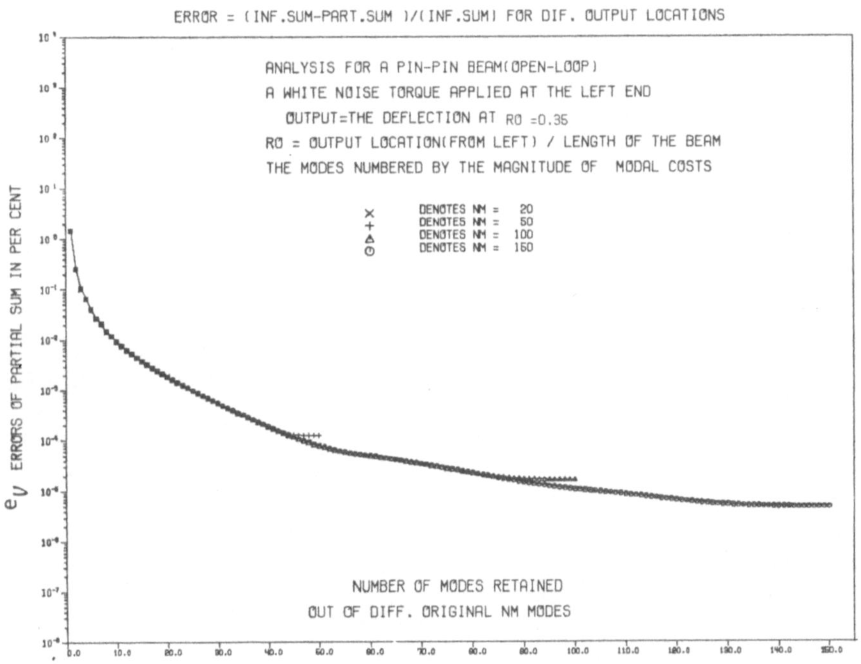

FIGURE 4

CONVERG. FOR OPEN LOOP COST FUNCN. AS THE MODES INCREASES

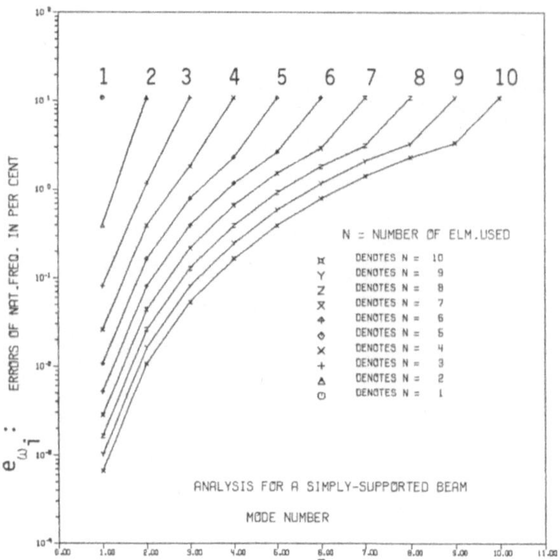

FIGURE 5

ERROR IN FREQENCY VS. MODE NO. FOR DIFF. NO. OF ELMTS

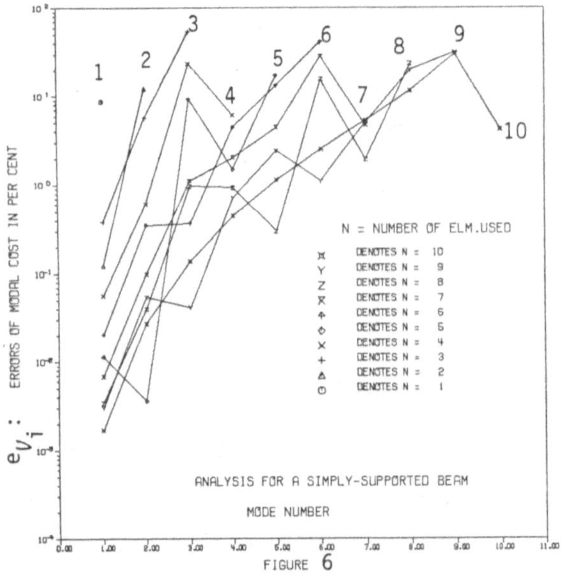

FIGURE 6

ERROR IN MODAL COST VS. MODE NO. FOR DIF. ELMNS NO.

TABLE 1

Comparison of freq. and modal costs using both exact and finite element methods (pinned pinned beam, $EI = \rho$, $\rho = \frac{2}{L}$, $L = \pi$)

N #of elem.	Modes	ω_i rad/sec	$\hat{\omega}_i$ rad/sec	$\lvert e_{\omega_i} \rvert$ (%)	V_i (ANAL.)	$\hat{V}_i$ (F.E.)	$\lvert e_{V_i} \rvert$ (%)
1	1	1	1.10992	10.99	39.69463	43.1462	8.690
2	1	1	1.00395	0.395	39.69563	39.7435	0.1233
	2	4	4.4397	10.99	2.04534	2.2977	12.300
3	1	1	1.0008	0.081	39.69463	39.8453	0.3796
	2	4	4.047	1.182	2.04534	2.1623	5.721
	3	9	9.986	10.99	0.01511	0.0232	53.720
4	1	1	1.00026	0.026	39.69463	39.6721	0.0565
	2	4	4.0158	0.395	2.04534	2.0329	0.6069
	3	9	9.16445	1.830	0.01511	0.01154	23.580
	4	16	17.75869	10.99	0.17666	0.1875	6.170
5	1	1	1.00011	0.011	39.69463	39.69916	0.011
	2	4	4.00663	0.166	2.04534	2.04541	0.004
	3	9	9.07147	0.794	0.01511	0.01651	9.282
	4	16	16.3686	2.304	0.17666	0.17929	1.490
	5	25	27.7479	10.99	0.04000	0.04689	17.231
6	1	1	1.00005	0.005	39.69463	39.70276	0.020
	2	4	4.00324	0.082	2.04534	2.05246	0.348
	3	9	9.03552	0.390	0.01511	0.01516	0.365
	4	16	16.18912	1.18	0.17666	0.18452	4.448
	5	25	25.66363	2.65	0.04000	0.04532	13.296
	6	36	39.95705	10.99	0.00368	0.00521	41.420

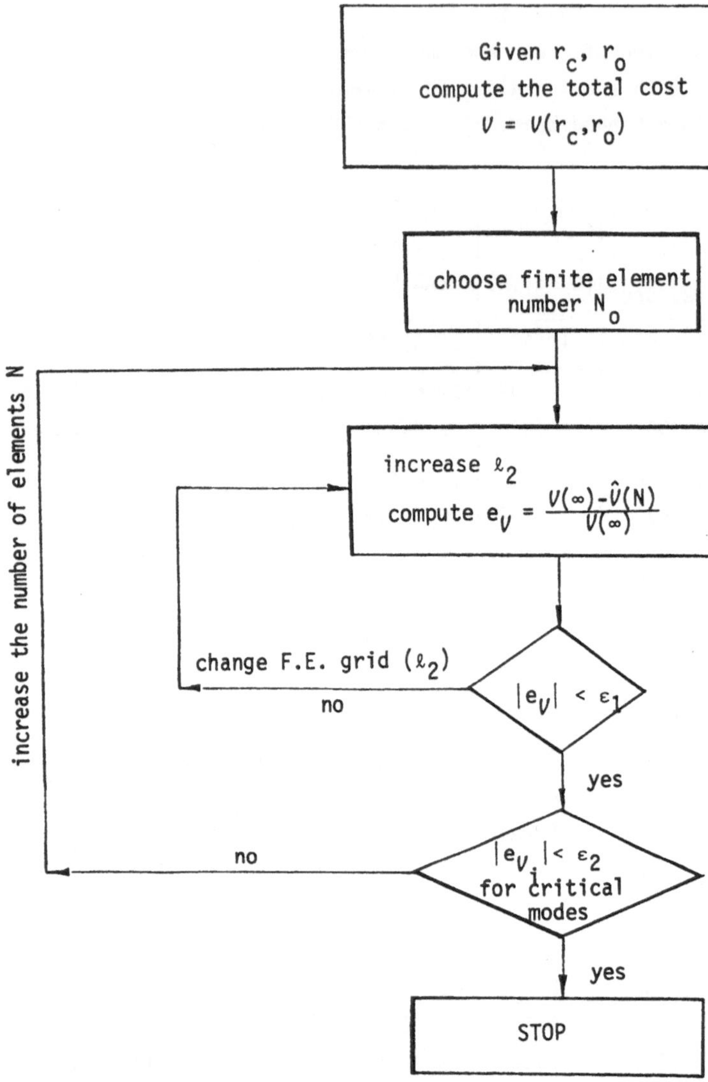

Fig. 7 Finite Element Grid Design

TABLE 2 Non-uniform Finite Elements: Cost Convergence (three finite elements)

		uniform grid	Non-Uniform Grid NUMBER OF ITERATION		
			1	2	3
ℓ_2/L		0.3333	0.3658	0.3536	0.3547
	e_V	0.043	-0.019	0.0021	0.00003
	e_{V_1}	0.380	0.318	0.337	0.335
%	e_{V_2}	5.722	5.480	5.60	5.59
	e_{V_3}	53.721	74.587	68.3	68.97

(Modes are ranked according to modal cost)

TABLE 3 Non-uniform Finite Elements: Cost Convergence (five finite elements)

		uniform grid	Non-uniform Grid NUMBER OF ITERATION		
			1	2	3
ℓ_2/L		0.2000	0.2350	0.23420	0.23419
	e_V	-0.0607	0.00142	0.000062	0.0000057
	e_{V_1}	0.011	0.029	0.029	0.029
	e_{V_2}	0.004	0.394	0.385	0.385
%	e_{V_4}	1.490	11.847	11.640	11.636
	e_{V_5}	17.23	- 3.323	- 2.929	- 2.913
	e_{V_3}	9.282	15.029	14.837	14.829

(Modes are listed in order of modal costs)

Acknowledgments

This work was sponsored in part by AFOSR Grant 83-0104. The authors also wish to thank H.T. Yang for many helpful discussions on finite element methods.

METHOD OF AVERAGING AND OPTIMAL STABILIZATION
OF MOTION OF LARGE-SCALE SYSTEMS

A.A.Martynyuk
Institute of Mechanics, Academy of
Sciences of the Ukr.SSR, Kiev, the Ukr.SSR

1. Introduction.

The present paper contains an algorithm of solutions of the large-scale system [11] with visco-elastic elements [3] stabilization. The base of this algorithm constitute ideas of the method of averaging of nonlinear mechanics [1] together with the method of Lyapunov functions [7]. We think that one of the positive aspects of the given approach is that which allows us to consider problems of stabilization in cases when the non-controlled nonlinear approach of the system has only a neutrally stable solution or it becomes such as a result of the operation of local decentralized controls, i. e. actually, we consider an optimization of solutions in the critical case. The effective application of this approach suggests the solution of two problems: alongside with the Lyapunov function's construction for a contracted system it is necessary to construct its general solution.

2. Notation.

R is a set of real values; R^n is an n -measured real vector space; $R \times R^n$ is a Cartesian product of R and R^n, $x=(x_1,\dots,x_n)^T$ is an n -measured vector-colomn; (^{T}) is a sign of a vector or matrix transposition; $d(x,M)=\inf_{y\in M}\|x-y\|$ is a distance from the point $x\in R^n$ up to the set $M\subset R^n$, $T=[0,+\infty)$, $T_0=[t_0,+\infty)$, $\nabla \upsilon$ is a gradient of the Lyapunov function υ ; (x,y) is a scalar product of vectors x and y ;

$\mathcal{U}$ is a set of continuous functions $u(t,x)$ in the domain $\Omega \subset T \times R^n$.

3. Stating the problem.

We consider a controlled visco-elastic system, the perturbed state of which is described by the equations

$$S: \frac{dx}{dt} = f(t,x,u,\mu,X(t,x,\int_0^t R(t,s,x(s))ds), \qquad (1)$$

where $x \in R^n$, $u \in R^m, t \in T_o, f: T_o \times R^n \times R^m \times \Lambda \times R^n \to R^n$, $\mu \in \Lambda$ is a small parameter. Suppose that system (1) can be decomposed on m - subsystems

$$S_s: \frac{dx_s}{dt} = f_s(t,x_s)+m_s(t,x_s)u_s^{\ell} + \mu X_s(t,x, \int_0^t R(t,s,x,(s))ds), \qquad (2)$$

$$s=1,2,\ldots,m$$

Here $x_s \in R^{n_s}$ is a state of an s-subsystem, so as

$R^n = R^{n_1} \times R^{n_2} \times \ldots \times R^{n_m}$; $f_s: T_o \times R^{n_s} \to R^{n_s}$; $m_s(t,x_s)$ is a matrix, $u_s^{\ell} \in R^{r_s}$ is a vector of the decentralized local control of the s-subsystem; $X_s: T_o \times R^n \times R^n \to R^{n_s}$. Functions f_s, m_s, R are defined and continuous in the correspondent domains $\Delta_1 \subset T_o \times R^{n_s}$, $\Delta_2 \subset T_o \times R^{n_s} \times R^{r_s}$, $\Delta_3 \subset T_o \times R^n \times R^d$, $\Delta_4 \subset T_o \times T \times R^n$ and they vanish for $x_s = x = 0$ $\forall s \in [1, m]$.

The quality of the process of transition for any can be characterised by the functional

$$J_s = \int_0^{\infty} \{ \psi_s(t,x[t], \mu) + (B_s u_s^{\ell}, u_s^{\ell}) \} \, dt . \qquad (3)$$

In the expression (3) the non-negative functions ψ_s must be defined, $(B_s u_s^{\ell}, u_s^{\ell})$ are definite-positive quadratic forms with the symmetrial, given before matrices B_s, $s \in [1,m]$.

The algorithm construction of solutions stabilization follows from the fact that we need to determine the controls $u \in U$ in such a way:

A) the zero solution of system (1) will be uniform μ -stable;

B) the functional [11]

$$J = \sum_{s=1}^{m} J_s\left[t_o, x_{so}, u_s^{\ell}, \mu\right]$$

on the trajectories of system (1), starting from the point $(t_o, x_o) \in \mathcal{X}_1^o \subset T_o \times R^n$, will take the minimal meaning for $\mu < \mu_o$, $\mu_o \in \Lambda$. .

4. Auxiliary transformations.

Suppose that for system (2) for $\mu=0, u_s^{\ell}\ 0\ \forall s \in [1,m]$ we find the Lyapunov functions $U_s(t,x_s)$. of positive terms, which are continuous and differentiable for $(t,x_s) \in \mathcal{X}_1$ and for which

$$\frac{\partial U_s}{\partial t} + (\nabla U_s, f_s(t,x_s)) \equiv \omega_s(t,x_s) \leq \quad \forall s \in [1,m]. \tag{4}$$

From condition of minimum of expressions

$$\mathcal{B}_s\left[U_s; t, x_s, u_s^{\ell}\right] = \omega_s(t,x_s) + (\nabla U_s, m_s(t,x_s(t,x_s)u_s^{\ell}) + (\nabla U_s, X_s(t,x,y)) +$$
$$+ \psi_s(t,x_s,\mu) + (B_s u_s^{\ell}, u_s^{\ell})\, s \in [1,m]$$

we determine

$$(\nabla U_s^{\ell}(t,x_s) = -\frac{1}{2} B_s^{-1} m_s^T (t,x_s) \nabla U_s, \quad s \in [1,m]. \tag{6}$$

Taking into account the fact that [10]

$$(\nabla U_s, m_s(t,x_s)u_s^{\ell}) + (B_s u_s^{\ell}, u_s^{\ell}) = (B_s(u_s^{\ell} - \bar{u}_s^{\ell}), u_s^{\ell} - \bar{u}_s^{\ell}) +$$
$$+ (B_s \bar{u}_s^{\ell}, \bar{u}_s^{\ell}), \quad s \in [1,m].$$

Functions ψ_s in the criterion of the quality (3) can be determined by expressions

$$\psi_s(t,x,\mu) = -\omega_s(t,x_s)-\mu(\nabla U_s,X(t,x,y))+(B_s\bar{u}_s^{\ell},\bar{u}_s^{\ell}), \qquad (7)$$

$$s=1,2,\ldots,m$$

The criterion of the quality (3) (taking into account (7)) takes the concrete contents. By means of the expressions (6) we transform system (2) to the form

$$S:\ \frac{dx}{dt} = g(t,x)+\mu X(t,x,\int_0^t R(t,s,x(s))ds), \qquad (8)$$

where $g(t,x)= (g_1(t,x_1),\ldots,g_m(t,x_m))^T$,

$$g_s(t,x_s)= f_s(t,x_s)- \frac{1}{2} m_s B_s^{-1} m_s^T \nabla U_s, \quad s \in [1,m] .$$

Further on with system

$$\tilde{S}:\qquad \frac{dx}{dt}= g(t,x),\ x(t_o) = x_o,\ t_o \geq 0 \qquad (9)$$

which is obtained from (8) for $\mu=0$, we shall associate the Lyapunov function

$$U= \sum_{s=1}^{m} u_s(t,x_s) \qquad (10)$$

the respective suppositions on which are made in the theorem.

We suppose the solution $x(t)=x(t;t_o,x_o)$ of system (8) to be unknown, but assume that we know the solution $\bar{x}(t)=\bar{x}(t;t_o,x_o)$ of system (9), and also $\bar{x}(t_o)=x(t_o)=x_o$, $x(t)=\tilde{\bar{x}}(t)$ for $0\leq t \leq t_o$; $\tilde{\bar{x}}(t)=\bar{x}(t)$ for $t\in T_o$.

Significance in the further algorithm construction has the calculated on solutions of system (9) mean [4]

$$\varphi_o(t_o,x_o)= \lim_{T\to\infty} \frac{1}{T} \int_{t_o}^{t_o+T} \varphi(r,\bar{x}(r), \int_0^r R(r,s,\tilde{\bar{x}}(s))ds)dr,\ t_o\geq 0 \qquad (11)$$

of the function

$$\varphi(t,x,y)=(\nabla U,X(t,x,\int_0^t R(t,s,x(s))ds).$$

Let $U^*(x)$ be the non-positive scalar function, determined and continuous on $\mathcal{D}_0 \subset R^n$ and $\mathcal{E}(U^*=0)$ in $\mathcal{D}_0$ be a set of points x, for which $U^*(x)$ 0.

4.1. Definition [9]. In the set $\mathcal{E}(U^*=0)$ the mean $\varphi_0(t_0,x_0)$ is definitely smaller than zero if for any values α and $\varepsilon(0<\alpha<\varepsilon<H)$ we can find values $r(\alpha,\varepsilon)$ and $\delta(\alpha,\varepsilon)$ $(r>0,\delta>0)$ such that $\varphi_0(t_0,x_0)<$ $<-\delta(\alpha,\varepsilon)$ for $a<\|x_0\|<\varepsilon, \rho(x_0, \mathcal{E}(U^*=0))<r(d,\varepsilon)$ for all $t_0 \geq 0$.

5. The main result.

According to 8 we state the following definitions:

5.1. Definition. The zero solution of system (8) is:

a) μ-stable with respect to T_i, if for all $t_0 \in T_i$ and any $\varepsilon>0$ there exist $\delta(t_0,\varepsilon)>0$ and $\mu^*(\varepsilon)<\bar{\mu}$ such that $\|x(t;t_0,x_0)\|<$ $<\varepsilon \ \forall t \in T_0$ follows from $\|x_0\|<\delta$ and $\mu<\mu^*$;

b) uniform μ-stable with respect to T_i, if conditions of the Definition 5.1 a) are fulfilled, and the maximal δ_M, satisfying the Definition 5.1 a) and being such that

$$\inf(\delta_M(t_0,\varepsilon):\ t_0, \in T_i)>0$$

corresponds to any $\varepsilon>0$.

In Definitions 5.1 a), b) we can omit the expression "with respect to T_i" if and only if $T_i=R$ [2].

5.2. Theorem. Let the following conditions be fulfilled:

1) there exists the definite-positive differential function $U(t,x)$ admitting the infinitely small highest bound;

2) there exists the function $U^*(x)$ such that the full derivative of the function $U(t,x)$ in virtue of system (9) satisfies the condition

$\frac{\partial U}{\partial t} + (\nabla U, g(t,x)) \leq U^{*}(x) \leq 0$ in the domain $\mathcal{D}_1^{\circ}$;

3) partial derivatives $\partial U/\partial x_i, i \in [1,n]$ are continuous along $(t,x) \in \mathcal{D}_1^{\circ}$ and bounded;

4) there exist constants $M_1, M_2, L_1 \dots L_4$ such that

a) $g(t,x) \in Lip_x(L_1)$, $g(t_0)=0$ $\forall t \in T_0$;

b) $\|X(t,x,y)\| \leq M_1$ $\forall (t,x,y) \in \mathcal{D}_3$, $X(t,0,0)=0 \forall t \in T_0$;

c) $\varphi(t,x,y) \in Lip_x(L_2)$;

d) $\varphi(t,x,y) \in Lip_y(L_3)$;

e) $R(t,s,x) \in Lip_x(L_4)$;

f) $\varphi(t,x,y) \leq M_2$ $\forall x \mathcal{D}_0$ $(U^{*}=0)$ $\forall t \in T_0$;

5) the mean (11) exists uniformly with respect to $(t_0, x_0) \in \mathcal{D}_1^{\circ}$;

6) $\varphi(t_0, x_0) < 0$ is determined in the set $\mathcal{E}$ $(U^{*}=0)$.

Then the zero solution of system (8) is uniform μ -stable. If we add one more condition to the enumerated ones:

7) $\min_{u \in \mathcal{U}} \sum_{s=1}^{m} B_s[U_s; t, x_s, u] = 0$

then the functional

$$J = \sum_{s=1}^{m} \int_0^{r_3} \{\psi_s(t, x[t], \mu) + (B_s u_s^{\ell})\} dt, \quad 0 < r_s < +\infty$$

takes the minimal meaning on the solutions of system (1).

Proof. In order to prove the uniform μ -stability of the zero solution of system (8) we should set $0 < \varepsilon < H$ and construct $\delta(\varepsilon)$ and $\mu^{*}(\varepsilon)$, for which conditions of the Definition 5.1 b) are fulfilled. For the function $U(t,x)$ satisfying the Condition 1 of the Theorem 5.2, we can show the constants $C(\varepsilon_1) > 0$ $(0 < \varepsilon_1 < \varepsilon)$ and $0 < \delta(\varepsilon_1) < \varepsilon_1$ such that the mobile surfice $U(t,x)=C$ will be located in the domain $\|x\| < \varepsilon_1$ as far as the domain $\|x\| < \delta$ will be contained inside the surfice $U(t,x)= = C \; \forall t \in T$. Let the solution $x(t;t_o,x_o)$ of system (8) with the initial conditions $t_o \geq 0, \|x_o\| < \delta$ leave the domain $\|x\| < \delta$ and at the moment $t = t_o'$ intersect the surface $U(t,x)=C$ at the point x_o'. We consider the function's $U(t,x)$ behaviour along the solution $x(t;t_o,x_o)$ of system (8). In virtue of Condition 2 of the Theorem 4.2 we obtain an estimation

$$U(t,x(t)) \leq U(t_o',x_o') + \mu\int_{t_o'}^{t} U^*(x(r))dr + \mu\int_{t_o'}^{t} \varphi(r,x(r),y(r),y(r))dr. \qquad (12)$$

It is necessary to make an estimation of the third term in the inequality (12) in two cases [5]:

1. The meaning $\rho(x_o', \mathcal{E}(U^*=0)) < r(\delta,\varepsilon)$.

2. The meaning $\rho(x_o', \mathcal{E}(U^*=0) \geq r(\delta,\varepsilon)$

In the first case we illustrate the method of choice of $\mu^*(\varepsilon)$ and the values $\ell > 0$ such that the solution $x(t;t_o,x_o')$, having left the surface $U(t,x) = C$ for $t \in [t_o', t_o'+2\ell]$, will remain in the domain $\|x\| < \varepsilon$ and at some moment $t_1 \in (t_o', t_o+2\ell]$ will return back inside the surface $U)t,x = c$. With this aim in view we shall represent the integral

$$I(t) = \int_{t_o'}^{t} \varphi(r,x(r),y(r))dr \qquad (13)$$

in the form [4]

$$I(t) = \int_{t_o'}^{t} \varphi(r,\bar{x}(r), \int_{o}^{t} R(r,s,\tilde{\bar{x}}(s))ds)dr +$$

$$+\int_{t_o'}^{t}\Big[\varphi(r,x(r),\int_0^r R(r,s,x(s))ds)-\varphi(r,\bar{x},(r),\int_0^r R(t,s,x(s)ds\Big]dr$$

$$+\int_{t_o'}^{t}\Big[\varphi(r,\bar{x}(r),\int_0^r R(r,s,x(s))ds)-\varphi(r,\bar{x},\int_0^r R(r,s,\bar{\bar{x}}(s))ds\Big]dr$$

$$=I_1(t)+I_2(t)+I_3(t).$$

We define from the Condition c) of the Theorem 5.2 that

$$I_1(t)\leq -(t-t_o')\,\frac{3\delta}{4} \qquad \text{for} \qquad t>t_o'+\ell \tag{14}$$

where $\delta=\delta(\varepsilon_1),\ell=\ell(\delta)$. Let us note that in virtue of Conditions 4 a), b) of the Theorem 5.2 for a norm of the deviation $\|x(t)-\bar{x}(t)\|$ an estimation

$$\|x(t)-\bar{x}(t)\|\leq \mu K \qquad \text{holds for} \qquad t_o'\leq t\leq t_o'+2\ell$$

where $K=M_1 2\ell\exp[2L_1\ell]$. In virtue of the Condition 4 c), having chosen $\mu_o=\delta/4L_2K$ we obtain

$$I_2(t)\leq\int_{t_o'}^{t}\|\varphi(t,x,y)-\varphi(t,\bar{x},y)\|dt\leq(t-t_o')\,\frac{\delta}{4}\,,$$

$$t_o'\leq t\leq t_o'+2\ell$$

From the Condition 4 d), having chosen $\mu_o'=\delta/4L_3L_4K$ we obtain

$$I_3(t)\leq\int_{t_o'}^{t}\|\varphi(t,x,y)-\varphi(t,x,\bar{y})\|dt\leq(t-t_o')\,\frac{\delta}{4}\,, \tag{16}$$

$$t_o'\leq t\leq t_o'+2\ell.$$

Having collected the estimations (14)-(16), we find out that

$$U(t,x(t))\leq U(t_o',x_o')-\frac{\delta}{4}(t-t_o'),\quad t_o'\leq t\leq t_o'+2\ell\,. \tag{17}$$

In the second of all possible cases at any interval $[t_1,t_2]\subset T_o$, at wich for a solution $x(t;t_o,x_o)$ the condition $\|x\| > \delta, \rho(x, \mathcal{E}(U^* = 0))$ $>r_1 (0<r_1<r(\delta,\varepsilon))$ is fulfilled, we can choose $\mu_o'(\varepsilon)$ so, that the solution $x(t;t_o',x_o')$ will not leave the domain $\|x\| < \varepsilon$. Really, according to the Condition 2) of the Theorem 5.2, there exists the value $\sigma > 0$ such that $\sup (U^*(x))$ for $x \in Q) \leq -\sigma$, where $Q=\{x:x=\{R^n$, $\rho(x, \mathcal{E}(U^*=0)) \geq r_1, \delta \leq \|x\| < \varepsilon\}$. We determine

$$U(t,x(t)) \leq U(t_o',x_o') - \frac{\sigma}{2}(t-t_o'). \tag{18}$$

from the estimation (12), according to the Condition 5 f) of the Theorem 5.2 for $\mu < \mu_o''$, where $\mu_o'' = \sigma/2M_2$.

It follows from estimations (17), (18) that $U(t,x(t))$ does not increase along the solution of system (8) in every of two possible cases. Hence, the solution $x(t;t_o,x_o)$ will neither leave the level $U(t,x)=c$ nor the domain $\|x\|<\varepsilon_1$. Having chosen $\mu^*(\varepsilon)=\min(\mu_o,\mu_o',\mu_o'')$, we define that for $\mu<\mu^*(\varepsilon)$ at every subsequent interval of time the Case 1 or 2 will hold and the solution $x(t;t_o,x_o)$ will not leave the domain $\|x\|<\varepsilon$. All estimations are uniform along t_o' and x_o' of the value δ, and μ^* are chosen independently of t_o. By this very fact we prove the uniform μ-stability in the sense of the Definition 5.1 b).

The proof of optimality is similar to that given in the works [6], [10].

REFERENCES

1. N.N.Bogolyubov, Yu.A.Mitropolsky, Asymptotic methods in the theory of nonlinear oscillations, (Nauka, Moscow, 1974) (In Russian).
2. Lj.T.Grujić, Novel development of Lyapunov stability of motion, Int.J.Contr., 22, N 4 (1975) 525-549.
3. A.A.Ilyushin, B.Ye.Pobedrya, Foundations of the mathematical thermo-visco-elasticity theory, (Nauka, Moscow, 1970) (In Russian).
4. M.M.Khapayev, A.I.Falin, On stability analysis of systems of integro-differential equations, Dokl. AN SSSR, 250, 2 (1980), 295-299 (In Russian).
5. V.I.Kosolapov, To stability of motion in the neutral case , Dokl. AN Ukr.SSR, Ser.A, N 1 (1979) 27-31 (In Russian).
6. N.N.Krasovsky, Supplement IY - In : Theory of Stability of Motion by I.G.Malkin (Nauka, Moscow, 1968) 475-514.
7. A.M.Lyapunov, General problem on stability of motion (Izd-vo AN SSSR, Moscow, 1956) 7-264 (In Russian).
8. A.A.Martynyuk, On stability of motion in nonlinear mechanics, Dokl. AN SSSR, 264, N 5 (1982) 1073-1077 (In Russian).
9. A.A.Martynyuk, V.I.Kosolapov, Principle of comparison and method of averaging in the problem on stability of non-asymptotically stable motions under the constantly effecting perturbations, AN Ukr.SSR, Preprint 78.33 (1978) 24 (In Russian).
10. S.M.Moldabayev, On optimal stabilization of nonlinear systems. In : Dynamics of the Controlled Systems (Nauka, Novosibirsk, 1979) 192-198 (In Russian).
11. D.D.Šiljak, Large-scale dynamic systems : stability and structure (North-Holland, New York, 1978).

THE MATHEMATICAL-HEURISTIC MODELLING AND ITS APPLICATIONS IN REAL TIME CONTROL OF LARGE-SCALE SYSTEMS

Florin Stănciulescu, Ph. D.,

Central Institute for Informatics,
Bd. Miciurin 8-10, Bucharest (Romania)

The modelling of large-scale systems is a difficult problem, because of its dimension, complexity nonlinearity and uncertainty (due to interaction with other systems). This problem becomes more difficult in Real-Time Large-Scale Systems (RTLSS), which require a short response-time. The classical methods of operational research (linear programming, nonlinear programming, dynamic programming etc.) are not well adapted to this kind of problems, because of the following reason: the classical techniques of modelling suppose that we dispose of a complete, sure and reliable information, and in this sense <u>few of these are adapted to uncertainties</u> [3,4]. Our approach is based on a philosophy which corresponds better to reality, if we take into account that in all systems, there are some behavioural aspects which can be mathematically described (e.g. the steady-state regime), but also there are aspects which can be described only using another language than that of mathematical analysis, for example the natural language (e.g. heuristic decision rules). This philosophy generates the so-called "mathematical-heuristic modelling method".

MATHEMATICAL-HEURISTIC MODEL

The mathematical-heuristic model is composed of two (compatible) parts, namely: (a) a simulation model of the process and (b) a heuristic (or logical-linguistic) model of the strategical decision.

<u>The standard simulation model</u> is composed of a set of differential equations [11] :

$$\dot{x}_i(t) = A_i x_i(t) + B_i u_i(t) + v_i(x) + f_i(x_i, \alpha_i), \quad (1)$$

$$x_i(0) = x_{io} \quad (i = 1,2,\dots, n) , \quad (2)$$

$$v_i(x) = \sum_{j=1, j\neq i}^{n} A_{ij}\, x_j , \quad (3)$$

with constraints on u, of the form :

$$u_{i\ell} \leq u_i \leq u_{iu}, \quad (i = 1,2,\ldots, m) , \qquad (3')$$

where α_i represents the specific parametra of the process.

The equations of the type (1) represent the state equations, while $x_i(0)$ and $v_i(x)$ represent the initial conditions and the interactions; A_i and B_i are matrices of suitable dimensions, $f(.,.)$ describes the nonlinearities, α_i being the process parameters.

Remark. The simulation model (1)-(3) can be written under the form of a discrete-time system :

$$x_i(k+1) = A_i x_i(k) + B_i u_i(k) + v_i(x) + f_i(x, \alpha_i), \qquad (1')$$
$$(i = 0,1,\ldots, n; \quad k = 0,1,2,\ldots, N).$$

The admisible variation interval, for the state variables x_i $(i = 1,2,\ldots, n)$ (called also limits or tolerances) is of the form:

$$\bar{x}_i - \Delta\bar{x}_i \leqslant \bar{x}_i \leqslant \bar{x}_i + \Delta\bar{x}_i \ , \qquad (4)$$

where $\bar{x}_i$ are the values of steady-state regime, derived from the solution of (1) or (1'), for t(or k) :

$$\bar{x}_i(u_i, \alpha_i) = \lim_{t\to\infty} x_i(u_i, \alpha_i, t) \qquad (5)$$

The values for the steady-state regime $\bar{x}_i(u_i, \alpha_i)$ can also be computed by means of a theorem from the operational calculus :

$$\bar{x}_i(u_i, \alpha_i) = \lim_{s \to 0} s\, \varphi_i(s, u_i, \alpha_i). \qquad (5')$$

The determination of the dependences

$$x_i = g_i(u_i, \alpha_i), \quad (i = 1,2,\ldots, n), \qquad (5'')$$

where $g_i : R^{m_i} \times R^{\ell_i} \to R^{n_i}$ are known functions, can also be made by identification.

We consider also the global simulation model, obtained from (1) by aggregation. Let be C_i an aggregation matrix so that the dimension of the aggregated vector :

$$z_i = C_i \, x_i$$

is ℓ = dim z < n (dim x) By linearization of (1) and by multiplication of (1) by C_i it results :

$$\dot{z}_i(t) = C_iA_iC_i^{\#} z_i(t) + C_iB_iu_i(t) + \sum_{j\neq i}^{n} C_iA_{ij}C_j^{\#} z_j(t) + C_if_i(t), \qquad (1'')$$

$$z_i(0) = z_{io} \quad (i = 1,2,\dots, n). \qquad (2'')$$

Or in other form :

$$\dot{z}_i(t) = \overline{A}_iz_i(t) + \overline{B}_iu_i(t) + \sum_{j\neq i}^{n} \overline{A}_jz_j(t) + \overline{f}_i(t), \qquad (1''')$$

$$z_i(0) = z_{io}, \quad (i = 1,2,\dots, n),$$

with constraints of the form (3). If we add an objective function, e.g.

$$\min_{u_i\in U_i} J(z_i, u_i) , \qquad (6)$$

an aggregated optimization model of the large-scale system is obtained, under the form (1"), (2"), (6).

The standard heuristic model, is composed of a set of heuristic (or logical - linguistic) rules, derived from the process :

- behavioural (input/output) axioms (B) :

$$< \varphi_j^{B}(x_i,u_i), \ \psi_j^{B}(x_i,u_i) > \qquad (7)$$

$$(i = 1,2,\dots, n; \quad j = 1,2,\dots, m) ;$$

- control axions (C) :

$$<\varphi_\ell^{C}(x_i,u_i) \wedge u_i(k+1) = u_i(k) + \Delta u_i(k) \rightarrow \psi_\ell^{C}(x_i,u_i) > \qquad (8)$$

$$(i = 1,2,\dots, n; \quad \ell = 1,2,\dots, p).$$

The last kind of rules have the following structure :

"Condition (C) $\wedge$ Action (A) $\rightarrow$ Result (R)". (9)

From the heuristic model (7), (8) we can derive the set of heuristic control rules for strategical decisions, under the form e.g. :

"If C_i, then let increase u_i, to increase x_i and x_j",

"If C_j, then let decrease u_k, to mentain x_j constant",

The heuristical (or logical-linguistic) model of control consists in the following :

(1) A finite set of pairs of formulas specific for the language $\mathcal{L}$, called logical-linguistic axioms of control. These formulas constitute an abstract representation of the approximate knowledges;

(2) A finite set of pairs of formulas of $\mathcal{L}$ called logical-linguistic axioms, concerning the behaviour (stimulus-response) of the process. These formulas constitute an abstract representation of the approximate knowledges ;

(3) A formula called objective, which has the significance that the state and the output variables should belong to certain prescribed intervals: $x_i(t) \in \mathbb{I}_i$, $y_j(t) \in \mathbb{I}_j$.

Remark. The uncertainty in the RTLSS rises from the large-dimension, complexity, nonlinearity and interaction with other systems or the exosystem, and is subject to the principle of uncertainty [11].

THE HYBRID ALGORITHMS FOR SOLVING PROBLEMS OF LSCS

Using the theory of mathematical-heuristic modelling, algorithms for solving simulation and control problems of RTLSS can be developed.

a) The algorithm for simulation and control of dynamic regime

Step 1: Solve an aggregated optimization problem, described by means of the model (1"), (2"), (6), and let be z_i^* the optimal aggregated values of the state vector ($z_i = C_i x_i$).

Step 2: Solve a simulation problem, described by means of a nonlinear differential model of the form (1) - (3), and let be x_i and u_i the values obtained by simulation. The stop-test of computation is :

$$\Delta z_i = \min !$$

where $\Delta z_i = \| z_i^* - C_i x_i \|$.

Step 3: Verify if the error $\Delta z_i = \min!$, or if $\Delta z_i < \varepsilon_i$ (ε_i given):

- If YES, then GO TO Step 6 !
- If NO , then GO TO Step 4 !

Step 4: Using the heuristic rules (7), (8), search a new value for u_i ($i = 1,2,\dots, m$), let be $\tilde{u}_i = u_i + \Delta u_i$ - which ensures that x_i verify the condition from step 3 - by a logical searching procedure.

Step 5: Return to Step 3!

Step 6: STOP!

This algorithm can be used in the pursuit of the system's trajectory, in order to bring back this trajectory into the given interval (4).

b) The algorithm for control of steady-state regime :

Step 1: Compute the steady-state values of x_i, noted with $\overline{x}_i$, from the relation (5) or (5').

Step 2: Compute the intervals of variation (tolerances, limits) of the steady-state values $\overline{x}_i$, from the formula (4).

Step 3: Verify if the actual steady-state values belong to the intervals $I_i = [\overline{x}_i - \Delta\overline{x}_i, \overline{x}_i + \Delta\overline{x}_i]$ computed at step 2 :

. if YES, then GO TO STEP 6 !
. if NO, then GO TO STEP 4 !

Step 4: Using the heuristic rules (7), (8), search a new value for u_i $(i = 0\ 1,2,\ldots, m)$, let be $\widetilde{u}_i = u_i + \Delta u_i$ - which ensure that x_i verify (4) - by a logical searching procedure.

Step 5: RETURN to step 3 !

Step 6: STOP !

This algorithm can be used in real-time control of large-scale and complex systems, because of its short response time.

APPLICATION

In APENDIX a mathematical-heuristic model for the control of a power distribution system is presented. This model is composed of a simulation model (formulas for steady-state regime for tensions, active and reactive powers, the phase angles a.s.o.), and a heuristic model (composed of 8 heuristic control rules, derived from the physical and technical rules which govern the controlled process).

CONCLUSIONS

In this paper we develop a new method of modelling RTLSS, called by us "the mathematical-heuristic modelling method", implying a simultaneous use of a mathematical simulation model and several heuristical control rules for strategical decision, derived from the process.

This kind of model can be used both for simulation and for control of a large-scale system.

The concern for mathematical-heuristic modelling is justified, in the frame of Large-Scale and Complex system, because the Real-Time control of these systems can not be achieved by means of deterministic models - which require prohibitive computer-time - but rather by means of heuristic models, based on logical rules of strategical decision.

The simulation model can be a set of differential or difference equations, with constraints on the control variables and given initial states; the heuristic model is composed of a set of logical equations. The mathematical-heuristic model must be seen as a tool of a Decision Support System (DSS).

In this paper an example concerning the mathematical-heuristic model for the control of a large-scale power distribution system is given. Possible applications of the new modelling method are in the field of power systems (production and distribution), chemical plants (petrochemical installations, e.g.), socio-economical systems, ecological systems a.s.o.

REFERENCES

1. G.Fromm, W.L.Hamilton, D.E.Hamilton, Federally Supported Mathematical Models: Survey and Analysis, Report for RANN, June 1974.
2. S.Stone, J.Naughton, Techniques for analysing system dynamics models: A critical survey. In: Proceedings International AMSE Conference "Modelling & Simulation, Paris-Sud, July 1-3, 1983.
3. A.P.Sage, Systems Engineering: Fundamental Limits and Future Prospects. In: Proc. of IEEE, vol. 69, No. 2, February 1981, pp. 158-166.
4. F.J.Charlwood, M.Noton, Uncertainty in World Modelling. In: New Trends in Mathematical Modelling. Polish Academy of Science, Warsaw, 1978.
5. F.Stănciulescu, M.Sularia and others, Research Report nr. 23(C)/83. ICI, Bucharest, December 1983.
6. F.Stănciulescu, Simulation of Large-Scale Systems in Conditions of Uncertainty using a Mathematical-Heuristic Model. Proceedings of International Conference "Modelling and Simulation" (Editor: G. Mesnard), Nice (France), 1983.
7. J.Duda, The heuristic simulation algorithm for a certain complex Scheduling problem. In: Preprints AI 83 IASTED Symposium, Lille (France), March 1983.
8. M.G. Singh, A. Titli, Systems: Decomposition and Control, Pergamon Press., London, 1978.

9. R.W. Blaning, Model Structure and User Interface in Decision Support Systems. DSS-81 Transactions, Institute of Management Science, USA, 1981.

10. A. Sydow, Mathematical and Simulation Models in Systems Analysis, Proceedings of the International Symposium, Berlin (GDR), September 1-5, 1980.

11. F.Stănciulescu, The Large-Scale Systems Dynamics. The Publishing House of Romanian Academy, Bucharest, 1982 (in Romanian).

12. S.A.Arafeh, Hierarchical Control of Power Distribution Systems. In: IEE Trans. on. Automatic Control vol. AC-23, no.2, April 1978, pp. 333-343.

APPENDIX

AN EXPERIMENTAL MATHEMATICAL-HEURISTIC MODEL OF A POWER-DISTRIBUTION SYSTEM

The simulation model

The general diagram of a power-distribution system is presented in figure 1, while figure 2 presents a distribution station. The differential model for simulation (derived from the Kirchhoff's laws) has been presented in [6].

The primary state variables are the following: medium electrical tensions (u_k), phase angles (φ_k), frequency (f), the secondary state variables being: active powers (P_k), reactive powers (Q_k), the electrical tension at the consumer k($\tilde{u}_k$), the power factor at the consumer k ($\cos\varphi_k$) (k = 1,2,..., n). The control variables are : switching a capacitor (ΔC_k)in/off-service, which improves the power factor (if $\varphi > 0/\varphi < 0$); switching a inductor (ΔL_k) in/off-service, which improves the power factor (if $\varphi < 0/\varphi > 0$); positioning of the tapes (Δn_k) of the power transformers, which allow to decrease or increase electricale tensions, and active and reactive powers.

a) The steady-state regime :

The computational relations for the steady-state regime variables are known :

$$P_k = u_k\, i_k \cos\varphi_k \,, \tag{7}$$

$$Q_k = u_k\, i_k \sin\varphi_k \,, \tag{8}$$

$$\tilde{u}_k = u_k - Z_k\, i_k \,, \tag{9}$$

$$\cos\varphi_k = \frac{1}{\sqrt{1 + tg^2\varphi_k}} \,, \tag{10}$$

where : $i_k = u_k / \sqrt{R_k^2 + (L_k\omega - \frac{1}{C_k\omega})^2}$, $tg\,\varphi_k = \dfrac{L_k\omega - \frac{1}{C_k\omega}}{R_k}$,

$\omega = 2\pi f$, z_k, R_k are the impedance and the resistence of the consumer, C_k and L_k being the inductivity and the capacity of substation k (including also the consumer and the control variables C_k^1 and L_k^1). Let be C_k^1 and L_k^1 the capacity of the condensator, and respectively of the inductor battery, and u_k the electrical tension in the secondary of the transformers, corresponding to the reference position of the tapes. Then, following relations hold :

$$C_k^1 = \Delta C_k \,.\, \overline{C}_k^1 \,,\quad \Delta C_k \in \{0,1,\ \ldots,\ \overline{NC}_k\} \,, \tag{11}$$

$$L_k^1 = \Delta L_k \,.\, \overline{L}_k^1 \,,\quad \Delta L_k \in \{0,1,\ \ldots,\ \overline{NL}_k\} \,, \tag{12}$$

$$u_k = a_k \Delta n_k + b_k \ , \ \Delta n_k \in \{ - j_k, \ldots, -1, 0, +1, \ldots j_k \} \qquad (13)$$

where : $\overline{C}_k^1$, $\overline{L}_k^1$ represent the capacity, respectively the inductivity of a single condensator/reactor; $\overline{NC}_k$, $\overline{NL}_k$ the maximum available number of condensators, respectively inductors; j_k, the maximum number of possible positions for the transformers tapes; a_k, a specific constant, and b_k, the tension in the substation line, depending directly upon u_k^o.

By introducing the expressions (11) - (13) into (7) - (10) we obtain the relationships between the state and control variables for the steady - state regime. Then, the goal of the distribution system is to meet the following inequalities :

$$P_k^o \leqslant \frac{4\pi^2 R_k (C_k^o + \Delta C_k \overline{C}_k^1)(a_k \Delta n_k + b_k)^2 f^2}{4\pi^2 R_k (C_k^o + \Delta C_k \overline{C}_k^1)^2 f^2 + \{ 4\pi^2 (L_k^o + \Delta L_k \overline{L}_k^1)(C_k^o + \Delta C_k \overline{C}_k^1) f^2 - 1\}^2} \leqslant P_k^1 \ , \qquad (14)$$

$$Q_k^o \leqslant \frac{2\pi (C_k^o + \Delta C_k \overline{C}_k^1)(a_k \Delta n_k + b_k)^2 \{4\pi^2 (L_k^o + \Delta L_k \overline{L}_k^1)(C_k^o + \Delta C_k \overline{C}_k^1) f^2 - 1\}}{4\pi^2 R_k^2 (C_k^o + \Delta C_k \overline{C}_k^1) f^2 + \{ 4\pi^2 (L_k^o + \Delta L_k \overline{L}_k^1)(C_k^o + \Delta C_k \overline{C}_k^1) f^2 - 1\}^2} \leqslant Q_k^1 \qquad (15)$$

$$u_k^o \leqslant \left(1 - \frac{2 Z_k^o (C_k^o + \Delta C_k \overline{C}_k^1) f}{\sqrt{4\pi^2 R_k^2 (C_k^o + \Delta C_k \overline{C}_k^1) f^2 + \{4\pi^2 (L_k^o + \Delta L_k \overline{L}_k^1)(C_k^o + \Delta C_k \overline{C}_k^1) f^2 - 1\}^2}}\right) . \qquad (16)$$

$$. (a_k \Delta n_k + b_k) \leqslant u_k^1$$

$$\cos \varphi_k^o \leqslant \frac{2\pi R_k (C_k^o + \Delta C_k \overline{C}_k^1) f}{\sqrt{4\pi^2 R_k^2 (C_k^o + \Delta C_k \overline{C}_k^1)^2 f^2 + \{4\pi^2 (L_k^o + \Delta L_k \overline{L}_k^1)(C_k^o + \Delta C_k \overline{C}_k^1) f^2 - 1\}^2}} \leqslant \cos \varphi_k^1$$

$$- \frac{\pi}{2} \leqslant \varphi_k \leqslant \frac{\pi}{2} \ , \ \Delta n_k \in \{ - j_k, \ldots, -1, 0, 1, \ldots, j_k \}, \qquad (17)$$

$$\Delta C_k \in \{ 0, 1, \ldots, \overline{NC}_k \} , \ \Delta L_k \in \{ 0, 1, \ldots, \overline{NL}_k \} ,$$

$$(k = 1, 2, \ldots, n)$$

where : P_k^o, P_k^1, Q_k^o, Q_k^1, $\tilde{u}_k^o$, $\tilde{u}_k^1$, $\cos \varphi_k^o$, $\cos \varphi_k^1$, are given.

b) The mathematical - heuristic model

<u>The mathematical model</u>

The state parameters can be controlled in the following manner :

- switching of a capacitor (C_k) in or off-service, which influences the bus power factor and the voltage :

$$\Delta C_k = \Delta \overline{C}_k \, . \, \overline{C}_k \ , \ \overline{C}_k = \text{constant}, \ \Delta \overline{C}_k \in \{ 0, 1, \ldots \ \} \ ;$$

. switching of a reactor L_k in or off-service, which influence the same parameters :

$$L_k = \Delta \bar{L}_k \cdot \bar{\bar{L}}_k,\ \bar{L}_k = \text{constant},\ \Delta \bar{L}_k \in \{0, 1, 2, \ldots\} ;$$

. increasing/decreasing the tape position (n_k), which influences the bus voltage, and the active and reactive power :

$$\Delta n_k \in \{-j_k, \ldots -1, 0, +1, \ldots, +j_k\}.$$

The dependence relations, in steady-state regime state parameters, out of control variables and between other parameters (as load parameters R_k^o, L_k^o, C_k^o and frequency f), are of the form [5] :

$$u_k = f_1(\Delta n_k, \Delta C_k, \Delta L_k, R_k^o, L_k^o, C_k^o, f) ,$$

$$\cos \varphi_k = f_2(\Delta n_k, \Delta C_k \Delta L_k, R_k^o, L_k^o, C_k^o, f) ,$$

$$P_k = f_3(\Delta n_k, \Delta C_k, \Delta L_k, R_k^o, L_k^o, C_k^o, f) ,$$

$$Q_k = f_4(\Delta n_k, \Delta C_k, \Delta L_k, R_k^o, L_k^o, C_k^o, f),$$

$$-\frac{\pi}{2} \leq \varphi_j \leq +\frac{\pi}{2} ,\ \Delta n_k, \Delta C_k, \Delta L_k \text{ given previously.}$$

The goal of a power distribution system is to ensure a continuous supply of electrical power, with sufficiently small variations of the state parameters:

$$u_k^o \leq u_k \leq u_k^1 ,$$

$$\cos\varphi_k \geq \cos\varphi_k^o ,$$

$$P_k^o \leq P_k \leq P_k^1 ,$$

$$Q_k^o \leq Q_k \leq Q_k^1 ,$$

where the symbol "O" indicates the lower limit, and the symbol "1" the upper limit.

The heuristic model

For the power distribution system, the heuristic rules derived from the physical properties of the process (combined with the experience of the human operator-dispatcher) are the following (but not exhaustive) [12] :

Rule 1: For a lagging load phase angle (φ_k), switching of a capacitor (ΔC_k) in-service improves the substation bus power factor ($\cos\cdot\varphi_k$) and increases the voltage (u_k).

Rule 2: For a leading load phase angle, switching of a capacitor off-service, improves the substation bus power factor but decreases the voltage.

Rule 3: For a lagging load phase angle switching of a reactor (ΔL_k) off-service, improves the substation bus power factor and increases the voltage.

Rule 4: For a leading load phase angle, switching of a reactor in-service, improves the substation bus power factor and decreases the voltage.

Rule 5: Increasing the tap movements (Δn_k) of all parallel transformers increases the bus voltage.

Rule 6: Decreasing the tap movements (Δn_k) of all parallel transformers decreases the bus voltage.

Rule 7: S_{jk} loading of a transformer is increased/decreased by increasing/decreasing the tape position.

Rule 8: Active power (P_k) of a transformer is decreased/increased by switching off/in-service of a part of the load.

The formalized image, of the type (9), of rules (R_1) - (R_8) constitutes the logical-linguistic model :

(R_1) "(C): If $(\varphi_k > 0) \wedge (\Delta C_k < \Delta \bar{C}_k) \wedge (A) : \Delta C_k \leftarrow \Delta C_k + 1$

(R): $\cos \varphi_k(t+1) > \cos \varphi_k(t),\ u_k(t+1) > u_k(t)$" ;

(R_2) "(C): If $(\varphi_k < 0) \wedge (\Delta C_k > 0) \wedge (A) : \Delta C_k \leftarrow \Delta C_k - 1$

(R): $\cos \varphi_k(t+1) > \cos \varphi_k(t),\ u_k(t+1) < u_k(t)$" ;

(R_3) "(C): If $(\varphi_k > 0) \wedge (\Delta L_k > \Delta \bar{L}_k) \wedge (A) : \Delta L_k \leftarrow \Delta L_k - 1$

(R): $\cos \varphi_k(t+1) > \cos \varphi_k(t),\ u_k(t+1) > u_k(t)$" ;

(R_4) "(C): If $(\varphi_k < 0) \wedge (\Delta L_k < \Delta \bar{L}_k) \wedge (A) : \Delta L_k \leftarrow \Delta L_k + 1$

(R): $\cos \varphi_k(t+1) > \cos \varphi_k(t),\ u_k(t+1) < u_k(t)$"

(R_5) "(C): If $(\Delta n_k < j_k) \wedge (A) : \Delta n_k \leftarrow \Delta n_k + 1$

(R): $u_k(t+1) > u_k(t),\ P_k(t+1) > P_k(t),\ Q_k(t+1) > Q_k(t)$"

(R_6) "(C): If $(\Delta n_k > -j_k) \wedge (A) : \Delta n_k \leftarrow \Delta n_k - 1$

(R): $u_k(t+1) < u_k(t),\ P_k(t+1) < P_k(t),\ Q_k(t+1) < Q_k(t)$"

(R_7) "(C): If $(\Delta n_k > -j_k) \wedge (A) : \Delta n_k \leftarrow \Delta n_k - 1$

(R): $S_k(t+1) < S_k(t)$"

(R_8) "(C): If $(P_k > P_k^1) \wedge (A) : S_k(t+1) < S_k(t)$

(R) : $P_k(t) > P_k(t+1) < P_k^1$ "

These rules allow to compute the new values, $u_i + \Delta u_i$ of the control variables. With that end in view we can use the logical computation rules, given in this paper, or otherwise, to find the values of Δu_i using a searching procedure, and verifying these values by means of the formula (5").

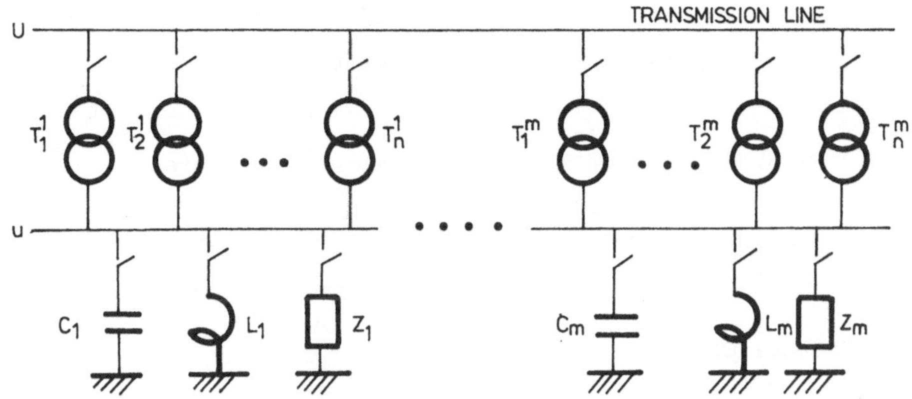

Figure 1 General diagram of a power distribution system, composed of m distribution stations

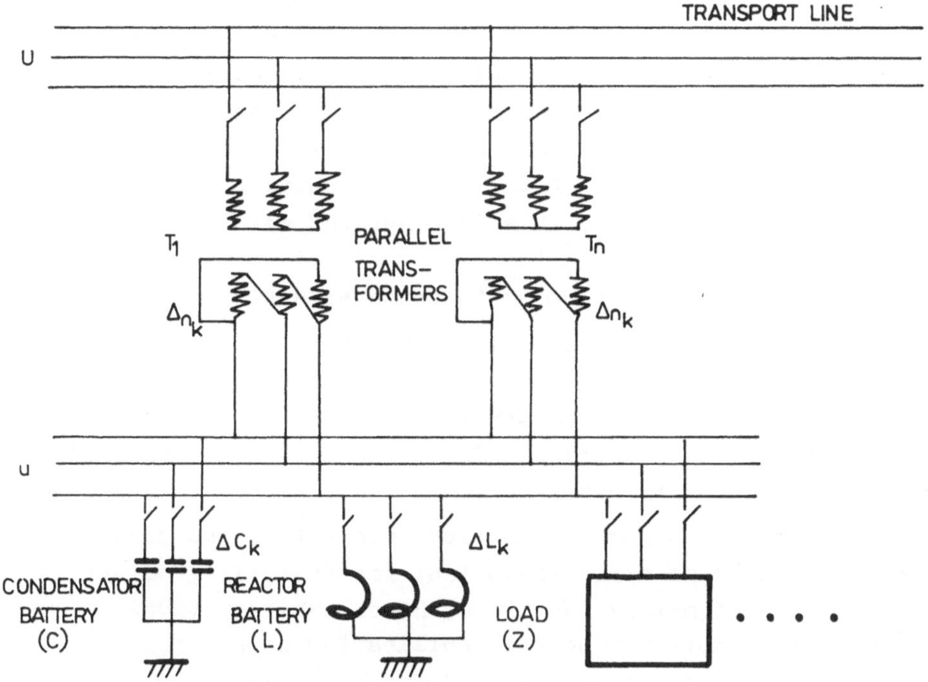

Figure 2 General diagram of a power distribution station

AN APPROACH FOR PARAMETER OPTIMIZATION OF LARGE SCALE SYSTEMS BY CONSTRAINTS

D.V. Shivkov
Higher Institute for Mechanical and Electrical Engineering, Department for Automation of Continuous Processes, 1156 Sofia, Bulgaria

INTRODUCTION

Predetermined constraints on the control functions or on some of the plant state coordinates are often to take into account in the design of control systems. The methods used for such a purpose require important calculations and have solutions which are difficult to realize. Because of that an approach based on an approximation of the obtained control algorithm with nearly optimal solutions seems inevitable. Another solution is the formulation of the synthesing problem so that the realization of the control function do not surpass a predetermined complexity. This is connected wit the necessity of choosing the constraints on the system structure and solving the parameter optimization problem by these constraints.

Gradient methods are very effective with respect to a fast convergence for the purposes of the parameter optimization in case of existence of analytic formulas for gradient calculations. So the difficulties in the numeric approximation,arising of the step length determination are avoided by analytic gradients.

Höfler (1980,1981) found a new mathematical concept of the gradient chain operator which is an appropriate tool for analytic gradient calculation of composed matrix functions. He and Shivkov (1982) showed also how this operator concept apllies for separation of computation algorithm and software into two independent segments: one of them depending only on the controller structure but not on the optimization criterion and the other one depending only on the criterion but no more on the structure.

In this paper a parameter optimization approach based on this concept is applied for the case of large scale systems consisting of connected with themselves subsystems. The constraints on the controller parameters of the subsystems such as the performance indices are complicatedly composed matrix functions of these parameters. It is shown how for this case a separation procedure based on the corresponding rules applies for a segmentation of the algorithm and the software.

PARAMETER OPTIMIZATION BY CONSTRAINTS

For the mathematical description of the parameter optimization problem for each subsystem, composed of controller and a plant, the following equations are used:

For the plant of the i-th subsystem (i=1,...,k)

(1a) $$\underline{\dot{x}}_i(t) = A_i\underline{x}_i(t) + B_i\underline{u}_i(t),$$

(1b) $$\underline{y}_{i_r}(t) = C_{i_r}\underline{x}_i(t) + D_{i_r}\underline{u}_i(t),$$

(1c) $$\underline{y}_{i_m}(t) = C_{i_m}\underline{x}_i(t);$$

For the controller of the i-th subsystem (i=1,...,k; j=1,...,k; j≠i)

(2a) $$\underline{\dot{v}}_i(t) = F_i\underline{v}_i(t) + G_i\underline{y}_{i_m}(t) + M_i\underline{r}_i(t) + N_j\underline{y}_j(t),$$

(2b) $$\underline{u}_i(t) = H_i\underline{v}_i(t) + K_i\underline{y}_{i_m}(t) + L_i\underline{r}_i(t) + P_j\underline{y}_j(t),$$

where the input and output vectors are denoted by
$\underline{x}_i$: state space vector of the plant;
$\underline{v}_i$: state space vector of the controller;
$\underline{y}_{i_r}$: controlled variables;
$\underline{y}_{i_m}$: measurement variables;
$\underline{u}_i$: actuating signals;
$\underline{r}_i$: reference signals;
$\underline{y}_j$: actuating signals from the j-th subsystem.

The parameter optimization goal is the choice of elements from the matrices $F_i, G_i, H_i, K_i, L_i, M_i$ (i=1,...,k) and N_j, P_j (j=1,...,k; j≠i) such that the predetermined performance indices take minimal values.
For convenience these matrices might be combined into the following controller parameter matrix R_i

(3) $$R_i := \left[\begin{array}{ccc|c} F_i & G_i & M_i & N_j \\ \hline H_i & K_i & L_i & P_j \end{array}\right], \quad \text{or} \quad R_i := \left[\begin{array}{c|c|c} R_{ii} & R_{ij} & \cdots \end{array}\right]$$

for the more general case of many connections between the subsystems $(i=1,\dots,k; j=1,\dots,k; j\neq i)$.
Now if the overall system state is defined as

$$(4)\qquad \hat{\underline{x}}_i(t) = \left[\underline{x}_i(t) \;\vdots\; \underline{v}_i(t)\right]^T$$

and $\hat{A}_i, \hat{B}_i, \hat{C}_i$ and $\hat{D}_i$ are matrices of the system state equations

$$(5a)\qquad \dot{\hat{\underline{x}}}_i(t) = \hat{A}_i\hat{\underline{x}}_i(t) + \hat{B}_i\underline{r}_i(t) + \hat{B}_i^j\underline{y}_j(t),$$

$$(5b)\qquad \underline{y}_{i_r}(t) = \hat{C}_i\hat{\underline{x}}_i(t) + \hat{D}_i\underline{r}_i(t) + \hat{D}_i^j\underline{y}_j(t),$$

then the overall system matrix is

$$(6)\qquad T_i(R_i) := \left[\begin{array}{c|c|c} \hat{A}_i & \hat{B}_i & \hat{B}_i^j \\ \hline \hat{C}_i & \hat{D}_i & \hat{D}_i^j \end{array}\right] = \bar{A}_i + \bar{B}_i R_i \bar{C}_i .$$

The matrices $\bar{A}_i, \bar{B}_i$ and $\bar{C}_i$ are functions of the plant parameters only

$$(7)\qquad \bar{A}_i = \left[\begin{array}{c|c|c|c} A_i & 0 & 0 & 0 \\ \hline 0 & 0 & 0 & 0 \\ \hline C_{r_i} & 0 & 0 & 0 \end{array}\right], \bar{B}_i = \left[\begin{array}{c|c} 0 & B_i \\ \hline I & 0 \\ \hline 0 & D_{r_i} \end{array}\right], \bar{C}_i = \left[\begin{array}{c|c|c|c} 0 & I & 0 & 0 \\ \hline C_{m_i} & 0 & 0 & 0 \\ \hline 0 & 0 & I & 0 \\ \hline 0 & 0 & 0 & I \end{array}\right].$$

In the case of structure constraints it is necessary to solve the optimization problem for the i-th controller which is normally characterized by a set of free parameters $\underline{s}_i$ $(i=1,\dots,k)$. This is due to the fact that the structural constraints can be expressed as an equation system of these free parameters (Shivkov and Höfler, 1982). The matter is only that the elements of R_i are no more free, but depend on the low dimensioned set $\underline{s}_i$. With respect to the structure of the matrix R_i in (3) the vector $\underline{s}_i$ is also composed from the vectors of the free parameters, connected with the matrices R_{ii}, R_{ij},

$$(8)\qquad \underline{s}_i := \left[\underline{s}_{ii} \;\vdots\; \underline{s}_{ij} \;\vdots\; \dots\right]^T, \quad R_i = \left[R_{ii}(\underline{s}_{ii}) \;\vdots\; R_{ij}(\underline{s}_{ij}) \;\vdots\; \dots\right].$$

The performance indices η_i $(i=1,\dots,k)$ are preliminary defined as functions of the matrices R_i respective of the sets $\underline{s}_i$. The total performance index might be defined as follows

$$(9)\qquad z := \underline{\eta}^T W \underline{\eta}, \quad \underline{\eta} := (\eta_1(T_1(R_1(\underline{s}_1))) \dots \eta_k(T_k(R_k(\underline{s}_k))))^T$$

where W is a weighting matrix.
The constraints on the controller parameter or on connected with them variables such as the control actions $\underline{u}_i$, whose values must be limitted for given moments ($\hat{\underline{x}}_i(t) = \hat{\underline{x}}_i^o$, i=1,...,k), might be defined as functions of the same matrices

(10) $$g_i^l := g_i^l(T_i(R_i(\underline{s}_i))) \ , \ i=1,\dots,k \ ; \ l=1,\dots,p_i \ .$$

The gradients of the performance indices and of the constraints might be obtained by the following chain operations (Höfler,1980)

(11a) $$\nabla_{\underline{s}_i} z = \nabla_{\underline{s}_i} R_i \supset \nabla_{R_i} T_i \supset \nabla_{T_i} z \ ,$$

(11b) $$\nabla_{\underline{s}_i} g_i^l = \nabla_{\underline{s}_i} R_i \supset \nabla_{R_i} T_i \supset \nabla_{T_i} g_i^l, \ i=1,\dots,k; \ l=1,\dots,p_i$$

where

(11c) $$\nabla_{T_i} z = \nabla_{T_i} \eta_i \supset \nabla_{\eta_i} \underline{\eta}^T W \underline{\eta} \ = \ \nabla_{T_i} \eta_i \supset 2W^T.$$

On the base of the gradient chain rules (Höfler,1980) and with respect to (6) it follows that

(12a) $$\nabla_{R_i} T_i \supset \nabla_{T_i} \eta_i \ = \nabla_{R_i} \eta_i \ = \bar{B}^T \left[\nabla_{T_i} \eta_i \right] \bar{C}^T \ ,$$

(12b) $$\nabla_{R_i} T_i \supset \nabla_{T_i} g_i^l \ = \nabla_{R_i} g_i^l \ = \bar{B}^T \left[\nabla_{T_i} g_i^l \ \right] \bar{C}^T \ .$$

These gradients are composed from the gradients of the functions η_i g_i^l with respect to the structure of R_i in (8)

(13a) $$\nabla_{\underline{s}_i} \eta_i = \begin{bmatrix} \nabla_{\underline{s}_{ii}} \eta_i \\ \hline \nabla_{\underline{s}_{ij}} \eta_i \\ \hline \vdots \end{bmatrix} = \begin{bmatrix} \nabla_{\underline{s}_{ii}} R_{ii} \supset \nabla_{R_{ii}} R_i \\ \hline \nabla_{\underline{s}_{ij}} R_{ij} \supset \nabla_{R_{ij}} R_i \\ \hline \vdots \end{bmatrix} \supset B^T \left[\nabla_{T_i} \eta_i \right] C^T$$

(13b) $$\nabla_{\underline{s}_i} g_i^l = \begin{bmatrix} \nabla_{\underline{s}_{ii}} g_i^l \\ \hline \nabla_{\underline{s}_{ij}} g_i^l \\ \hline \vdots \end{bmatrix} = \begin{bmatrix} \nabla_{\underline{s}_{ii}} R_{ii} \supset \nabla_{R_{ii}} R_i \\ \hline \nabla_{\underline{s}_{ij}} R_{ij} \supset \nabla_{R_{ij}} R_i \\ \hline \vdots \end{bmatrix} \supset B^T \left[\nabla_{T_i} g_i^l \right] C^T.$$

It follows from (13) that the gradients $\nabla_{T_i} \eta_i$ and $\nabla_{T_i} g_i^l$ depend on the functions $\eta_i(T_i)$ and $g_i^l(T_i)$ only, but no more on the controller structure. By analogy the determination of the gradients

$\nabla_{\underline{s}_i} R_i$ depends only on the functions $R_i(\underline{s}_i)$. In this case the calculation of the gradients is acomplished in separate segments: a functional one for determination of the criterion and the constraints gradients with respect to T_i and a structure segment for their calculation with respect to $\underline{s}_i$. The last one allowes the separate calculation of the gradients for the controllers structures.

OPTIMIZATION PROCEDURE

The above calculated gradients might be applied for parameter optimization problems with constraints. For this purpose an algorithm, proposed by Shivkov and Höfler (1983) may be used. The unknown parameters $\underline{s}_i$ are then solutions of differential equations,which contain the gradients of the performance indices and the constraints,and with $t \to \infty$ they take their optimal values.

Based on this approach for parameter optimization a procedure for real time piecewise continuous control with constraints on the actuating variables may be realized. This problem exists for example in the case of programme control. The constraints may be considered as in the following form

(14) $$\left| u_i^l(t) \right| \leq u_{i_{max}}^l \quad , \; i=1,\dots,k; \; l=1,\dots,p_i .$$

The synthesing problem is now to shift the vector $\hat{\underline{x}}_i(t)$ from the initial $\hat{\underline{x}}_i(t_o)$ to the steady state $\hat{\underline{x}}_i(t_\infty)$ in such a way that the performance index

(15) $$\eta_i := \int_{t_o}^{\infty} f_i(\tilde{\underline{x}}_i, \underline{u}_i)\, dt$$

take a minimal value.

In (15) $\tilde{\underline{x}}_i(t) = \hat{\underline{x}}_{i\infty} - \hat{\underline{x}}_i(t)$ and it satisfies the same equation (5) as $\hat{\underline{x}}_i(t)$. For the purposes of the next considerations it is nessesary that the functions $\underline{u}_i(t)$ are admissible with respect to $\tilde{\underline{x}}_i(t_o)$,i.e. they have the following qualities (Kiendl,1972): The equation system (5) has a unique defined trajectory $\tilde{\underline{x}}_i(t)$ with $\tilde{\underline{x}}_i(t_o) = \tilde{\underline{x}}_i^o$ and $\tilde{\underline{x}}_i(\infty) = 0$; the elements of the control functions are limited with respect to (14); the values of the state coordinates or their combinations are also limited. Then the criterion (15) exists and has a finite value.

The synthesing problem is based now on the optimality principle of Bellman. In correspondence with this principle it is nessesary for the particular moments $t_o < t_1 < t_2 < \ldots$ of switching over to define admissible functions $\underline{u}_i(t)$

$$\underline{u}_i(t) = \begin{cases} \underline{u}_i^o(t) & \text{for } t_o \leq t < t_1 \\ \underline{u}_i^1(t) & \text{for } t_1 \leq t < t_2 \\ \underline{u}_i^2(t) & \text{for } t_2 \leq t < t_3 \\ \vdots & \end{cases} \tag{16}$$

which minimize the criterion (15) by the constraints (14) so that

$$\eta_i(\tilde{\underline{x}}_i, \underline{u}_i^{r-1}(\underline{s}_i^{r-1})) = \eta_i(\tilde{\underline{x}}_i, \underline{u}_i^r(\underline{s}_i^r)) + \varepsilon_r \;, \; \varepsilon_r \geq 0 \;, \; r=0,1,2, \tag{17}$$

It follows from (17) that the admissible functions $\underline{u}_i(t)$ must be determined so that the rest of the criterion value for $\underline{u}_i^r(t_r)$ to be less or equal, but no greater than the value for $\underline{u}_i^{r-1}(t_r)$. Hence it is possible to obtain a nearly optimal solution of the control problem on the base of parameter optimization during the control interval.

REFERENCES

Höfler, A.B. (1980)
Gradientenkettenoperatoren und ihre Anwendung bei der Reglerparameteroptimierung, Dissertation, Lehrstuhl für Mess- und Regelungstechnik, Ruhr-Universität Bochum, FRG.

Höfler, A.B. (1981)
A Software Segmentation Technique with High Control Structure Flexibility for Optimization by Gradients, IFAC-Congress, Kyoto, Japan.

Kiendl, H. (1972)
Suboptimale Regler mit abschnittweise linearer Struktur, Springer Verl.

Shivkov, D.V., Höfler, A.B. (1982)
Parameteroptimierung spezieller Reglerstrukturen unter zusätzlichen Beschränkungen, Regelungstechnik (FRG), N-9, pp.319-324.

Shivkov, D.V., Höfler, A.B. (1983)
Ein Algorithmus zur Parameteroptimierung von Regelungssystemen mit Beschränkungen, Regelungstechnik (FRG), N-2, pp.62-68.

SYNTHESIS OF FUZZY AND PROBABILISTIC FUZZY CONTROLLERS BY MEANS OF DECOMPOSITION OF THE CONTROL RULES DERIVED FROM A HUMAN OPERATOR'S ACTIONS

E. Czogala and L. Walichiewicz
RWTH Aachen, Templergraben 64, 5100 Aachen, FRG
Silesian Technical University, Pstrowskiego 16, 44-100 Gliwice, Poland

Abstract

In many real control problems we face the lack of precise and detailed knowledge about the process and this is the usual reason that these problems can not be satisfactorily solved by the use of standard control theory.
The paper deals with the several classes of heuristic algorithms of real time control called fuzzy controllers and probabilistic fuzzy controllers. Such algorithms seem to be convenient in the design of control systems for complex, ill-defined processes.
The synthesis technique called decomposition of control rules, presented in this paper, provides the unified expressions for both single-input, single-output and multi-dimensional controllers and it improves the computational efficiency of the control system. The original method of inference, based on this decomposition, is also presented.

Introduction

The concept of a fuzzy controller was proposed for the first time by Mamdani [13]. This controller is a decision-making algorithm using linguistic rules to describe the control policy. This concept is based on the theory of the fuzzy sets proposed by Zadeh [19] and make use of two essential concepts i.e. fuzzy implication and compositional rule of inference.

Czogala and Pedrycz [4] proposed a generalization of a fuzzy controller called probabilistic fuzzy controller which allows to express a control strategy in terms of distribution functions [2] of probabilistic sets according to Hirota [9] . This concept of the probabilsitic fuzzy controller can be helpful to aggregate the various control strategies and seems to be convenient concerning the uncertainty implied by the subjectivity of human observers ambiguity and variety of the process being controlled etc.

On leave from Silesian Technical University, Gliwice, Poland.
The paper was written while the first author was granted a research fellowship by the Alexander-von-Humboldt-Foundation.

The results of a fuzzy and probabilistic fuzzy approach to control problems have been presented in many papers [5] - [8], [10] - [18].
It should be noted here that computer implementation of the above mentioned algorithms demands the processing of a great amount of information. So it is necessary to look for reasonable methods of simplification of the mentioned control algorithms. In this paper one of the possible ways for obtaining the simplified expressions for fuzzy and probabilistic fuzzy controller, so called decomposition of control rules, is presented.

Synthesis of a fuzzy controller

First let us assume that in a multi-dimensional case a control statement has the form:

"IF"

X_{1i} and....and X_{ni}

"THEN"

U_{1i} and....and U_{pi}

where:

X_{ki} is the fuzzy value of the process output or state variable defined on the fixed universe of discourse $\mathbb{X}^k$ k=1,2,...,n

U_{li} is the fuzzy value of the control variable defined on the fixed universe of discourse $\mathbb{U}^l$ l=1,2,...,N.

Each such statement may also be written as the cartesian product:

$$R_i = X_{1i} \times \dots \times X_{ni} \times U_{1i} \times \dots \times U_{pi} \tag{1}$$

The total rule i.e. fuzzy relation is the following

$$R = \bigvee_{i=1}^{M} R_i \tag{2}$$

Applying the compositional rule of inference to the following inputs $X_1^a,\dots,X_n^a$ (index "a" means actual) we get the composed fuzzy $\bar{U}^a$ in the p-dimensional universe $\mathbb{U}^1 \times \dots \times \mathbb{U}^p$.

$$\bar{U}^a = X_1^a \circ \dots \circ X_n^a \circ R \tag{3}$$

$$\bar{U}^a = \bigvee_{\mathbb{X}^1} X_1^a \wedge (\dots \wedge (\bigvee_{\mathbb{X}^n} X_n^a \wedge R)\dots)$$

Taking into account equation (1) and basing it on latticeal properties of fuzzy sets we may develop the last equation as follows:

$$
\begin{aligned}
\bar{U}_a &= \bigvee_{x_1} X_1^a \wedge (\ldots\wedge(\bigvee_{x_n} X_n^a \wedge R)\ldots) = \\
&= \bigvee_{x_1} X_1^a \wedge (\ldots\wedge(\bigvee_{x_n} X_n^a \wedge \bigvee_i (X_{1i}\wedge\ldots\wedge X_{ni}\wedge U_{1i}\wedge\ldots\wedge U_{pi}))\ldots) \\
&= \bigvee_i (\bigvee_{x_1} X_1^a \wedge (\ldots\wedge(\bigvee_{x_n} X_n^a \wedge (X_{1i}\wedge\ldots\wedge X_{ni}\wedge U_{1i}\wedge\ldots\wedge U_{pi}))\ldots)) \\
&= \bigvee_i (\bigvee_{x_1}(X_1^a \wedge X_{1i}))\wedge\ldots\wedge(\bigvee_{x_n}(X_n^a \wedge X_{ni}))\wedge(U_{1i}\wedge\ldots\wedge U_{pi})) \\
&= (\bigvee_i(\bigvee_{x_1}(X_1^a \wedge X_{1i})\wedge U_{1i})\wedge\ldots\wedge\bigvee_i(\bigvee_{x_n}(X_n^a \wedge X_{ni})\wedge U_{1i})) \qquad (4) \\
&\quad \wedge \\
&\quad \vdots \\
&\quad \wedge \\
&\quad (\bigvee_i(\bigvee_{x_1}(X_1^a \wedge X_{1i})\wedge U_{pi})\wedge\ldots\wedge\bigvee_i(\bigvee_{x_n}(X_n^a \wedge X_{ni})\wedge U_{pi}))
\end{aligned}
$$

Let us denote:

$$G_j = \bigvee_i(\bigvee_{x_1}(X_1^a\wedge X_{1i})\wedge U_{ji})\wedge\ldots\ \bigvee_i(\bigvee_{x_n}(X_n^a\wedge X_{ni})\wedge U_{ji}) \qquad (5)$$

Now we can express equation (4) in the form

$$\bar{U}^a = G_1\wedge\ldots\wedge G_p$$

It is obvious that G_j is a function of u^j only, so G_j is a fuzzy set (membership function) defined in the universe U^j. Let us stress this fact by the following denotation:

$$\bar{U}^a(u^1,\ldots,u^p) = G_1(u^1)\wedge\ldots\wedge G_p(u^p) \qquad (6)$$

The function of a fuzzy controller is to infer from the actual values of its inputs $X_k^a (k=1,2,\ldots,n)$ the actual values of controls. Basing on this from the last equation we deduce that:

$$U_j^a = G_j \qquad (7)$$

Now, let us denote

$$\alpha_{ki} = \bigvee_{X^k} (X_k^a \wedge X_{ki}) \qquad k = 1,2,\dots,n \quad i = 1,2,\dots, \tag{8}$$

As we know from Zadeh's paper [19] α_{ki} is essentially the same as the degree of separation between fuzzy sets X_k^a and X_{ki} and the same as the intersection coefficient introduced by Cheng and others in [1]. Using this notion we can simplify the expression for the fuzzy controller as follows:

$$U_j^a = \bigwedge_k \bigvee_i \alpha_{ki} \wedge U_{ji} \qquad j = 1,2,\dots,p \quad k = 1,2,\dots,n \quad i = 1,2,\dots,M \tag{9}$$

Basing on this expression we can study the influence of each actual input for each actual output as well as the influence of each value used in control rules for the actual output.

The received results may also be extended for a probabilistic fuzzy controller.

Synthesis of a probabilistic fuzzy controller

Let us assume that $X_{li},\dots,X_{ni}$, $U_{li},\dots,U_{pi}$ are probabilistic sets defined in the respective universe of discourse and represented by distribution functions $F_{X_{li}}(z),\dots,F_{X_{ni}}(z)$, $F_{U_{li}}(z),\dots,F_{U_{pi}}(z)$, respectively.

Now, let us recall for details cf [2], [3] the concept of random intersection coefficient, which is the extension of α_{ki} for probabilistic sets, presented by Czogala and Zimmermann in [6].

If X_{ki} and X_k^a are probabilistic sets given by their defining functions $X_{ki}(x^k,\omega)$ and $X_k^a(x^k,\omega)$, respectively, the random intersection coefficient $\alpha_{ki}(\omega)$ between these probabilistic sets takes the form

$$\alpha_{ki}(\omega) = \sup_{x^k \in X^k} (X_k^a(x^k{}_{gw}) \wedge X_{ki}(x^k,\omega)) \tag{10}$$

where $\alpha_{ki}(\omega) \in [0,1]$ is a random variable (ω denotes a simple event belonging to the same space Ω).

Probabilistic sets may be represented by their distribution functions, under some additional assumptions, i.e. a universe $\mathcal{X}^k$ is finite and the random variables $X_k^a(x^k,\omega)$ and $X_{ki}(x^k,\omega)$ are independent. Then we can find the distribution function representation of the random intersection coefficient in the form

$$F_{\alpha_{ki}}(z) = \left(F_{X_k^a}(z) + F_{X_{ki}}(z) - F_{X_k^a}(z) \cdot F_{X_{ki}}(z)\right) \tag{11}$$

For specific calculations we can carry out the moment analysis. As an example, the mean value of the random intersection coefficient may be found as

$$E[\alpha_{ki}] = \int_0^1 z \, dF_{ki}(z) \tag{12}$$

Similarly the higher moments can be found.
If the distribution functions of X_k^a and X_{ki} are the unit step functions we get the deterministic coefficient α_{ki} i.e.

$$\alpha_{ki} = \sup_{x^k \in \mathcal{X}^k}(X_k^a(x^k) \wedge X_{ki}(x^k)) = \bigvee_{\mathcal{X}^k}(X_k^a \wedge X_{ki}) \tag{13}$$

Thus, we shall point out that the concept of fuzzy intersection coefficients is embedded in the random intersection coefficient presented above.
Basing on the presented above equations and assuming the independence of control rules we may get the following expression for the probabilistic controller

$$F_{U_j^a(u^j)}(z) = 1 - \prod_{k=1}^{n} (1- \prod_{i=1}^{M} (F_{\alpha_{ki}}(z) + F_{U_{ji}}(z) - F_{\alpha_{ki}}(z)F_{U_{ji}}(z))) \tag{14}$$

Such expression for the probabilistic fuzzy controller should be helpful in both analysis and design of control systems and as a considerable simplification operating with the distribution functions representing the probabilistic sets.

Concluding remarks

The idea of decomposition of control rules presented in this paper allowes the simplification of both the fuzzy and probabilistic fuzzy algorithms of control, which seems to be useful in systems with uncertainty of both kinds i.e. uncertainty of fuzzy nature and uncertainty of random nature. Moreover, it improves the computational efficiency of the control systems as an important task in the design of real time control systems of complex, ill-defined processes.

References

1. Cheng, W.M., Reh, S.-J., Wu, C.F., Tsuei, T.H.; An expression for fuzzy controller, Fuzzy Information and Decision Processes, Gupta, M.M. and Sanchez, E. (eds.), North Holland Publishing Company, 1982, pp. 411-413.

2. Czogała, E.; On distribution function description of probabilistic sets and its application in decision making, Fuzzy Sets and Systems 10, 1983, pp. 21-29.

3. Czogała, E.; A generalized concept of a fuzzy probabilistic controller, Fuzzy Sets and Systems (to appear).

4. Czogała, E.; Pedrycz, W.; On the concept of fuzzy probabilistic controllers, Fuzzy Sets and Systems 10, 1983, pp. 109-121.

5. Czogała, E., Pedrycz, W.; Walichiewicz, L.; Control of complex processes by using fuzzy probabilistic controller, IFAC Symposium: Fuzzy information, Knowledge presentation and decision analysis, 18-21 July 1983, Marseille (France).

6. Czogała, E., Zimmermann, H.-J.; Some aspects of the synthesis of a probabilistic fuzzy controller, Fuzzy Sets and Systems (to appear).

7. Deng, J., Li, C.; Liguistic phase plane - fuzzy control for nonlinear systems, Proc. of IFAC Symposium on Fuzzy Information, Knowledge Representation and Decision Analysis 1983, pp. 67-72.

8. Gupta, M.M.; Feedback control applicationas of fuzzy set theory: a survey, Proc. of 8th Triennal World Congress off IFAC 1981, Vol. V, pp. 1-6.

9. Hirota, K.; Concept of probabilistic sets, Fuzzy Sets and Systems 5, 1981, pp. 31-46.

10. Kickert, W.J.M., Mamdani, E.H.; Analysis of fuzzy logic controller, Fuzzy Sets and Systems 1, 1978, pp. 29-44.

11. Larsen, P.M.; Industrial applications of fuzzy logic control, in B.R. Gaines and E. H. Mamdani, Eds., Fuzzy reasoning and its applications (Academic Press, London, 1981).

12. Lemke van Nautah, H.R., Kickert, W.J.M.; Application of fuzzy controller in a warm water plant, Automatica 12, 1976, pp. 301-308.

13. Mamdani, E.H.; Application of fuzzy algorithms for control of simple dynamic plant, Proc. IEEE 1979, 121,2,1979, pp. 1585-1588.

14. Murakami, S.; Application of fuzzy controller to automobile speed control system, Proc. of IFAC Symposium on Fuzzy Information, Knowledge Representation and Decision Analysis 1983, pp. 43-48.

15. Negoita, C.V.; Ralescu, D.A.; Application of fuzzy sets to system analysis, 1975, Birkhauer Verlag, Basel und Stuttgart.

16. Tong, R.M.; A control engineering review of fuzzy systems, Automatica 8, 1977, pp. 559-569.

17. Tong, R.M.; Analysis and control of fuzzy systems using finite discrete relation, Int. J. Man-Machine Studies 27, 1978, pp. 431-440.

18. Yonekura, M.; The application of fuzzy set theory to the temperature control of box annealing furnaces using simulational techniques, Proc. 8th Triennal World Congress of IFAC 1981 Vol V, pp. 13-17.

19. Zadeh, L.A.; Fuzzy Sets, Infromation and Control, 8, 1965, pp. 338-353.

DEAD-BEAT SERVO PROBLEM FOR 2-D LARGE SCALE SYSTEMS

T. KACZOREK

Technical University of Warsaw
ul.Koszykowa 75,00-662 Warszawa
Poland

Sufficient conditions are given for the existence of a solution to the dead-beat servo problem for two-dimensional (2-D) linear large scale systems. An algorithm for finding local linear controllers such that the tracking errors are zero after a finite "time" for any boundary conditions of the system, reference genetors and controllers.

1. Introduction

The dead-beat control and the dead-beat servo problem have been considered in many papers [4,9,13,15,17]. Kučera in [11] has formulated and solved the dead-beat servo problem in a novel way using the polynomial equation approach. Kučera's method has been extended for multivariable linear systems in [17,4,12] and for two-dimensional (2-D) systems in [9]. Large scale system theory and 2-D and 3-D system theory have been receiving extensive attention in the last few years [1-3,5-8,14,18-21].
The purpose of this paper is to formulate and partly solve the dead-beat servo problem for 2-D linear large scale systems.

2. Preliminaries

Let $R^{m\times n}[d_1,d_2]$ be the set of mxn polynomial matrices in d_1 and d_2 with real coefficients. Consider polynomial matrices $\bar{A}\in R^{l\times q}[d_1,d_2]$, $\bar{B}\in R^{l\times m}[d_1,d_2]$ with $t=m+q\geq l\geq 1$ and

$$\bar{D}=[\bar{A}\ \bar{B}]\in R^{l\times t}[d_1,d_2]$$

Definition 1 [23]

The matrices $\bar{A},\bar{B}$ are said to be zero left-coprime (ZLC) if there exists no a pair (d_1,d_2) which is a zero of all the lxl minors of the matrix $\bar{D}$. The matrices $\bar{A},\bar{B}$ are zero right-coprime (ZRC) if the transposed matrices $\bar{A}^T,\bar{B}^T$ are ZLC.

Theorem 1 [23]

The matrices $\bar{A},\bar{B}$ are ZLC if and only if there exist two polynomial matrices $\bar{X}\in R^{q\times l}[d_1,d_2]$, $\bar{Y}\in R^{m\times l}[d_1,d_2]$ such that

$$\bar{A}\bar{X}+\bar{B}\bar{Y}=I_l$$

where I_l is the lxl identity matrix.

Theorem 2 [23]

There exist four polynomial matrices $\bar{X}\in R^{q\times l}[d_1,d_2]$, $\bar{Y}\in R^{m\times l}[d_1,d_2]$, $C_2\in R^{(t-l)\times t}[d_1,d_2]$, $D_2\in R^{t\times(t-l)}[d_1,d_2]$ such that

$$\begin{bmatrix} \bar{A} & \bar{B} \\ \hline & \bar{C}_2 \end{bmatrix} \begin{bmatrix} \bar{X} & \\ \bar{Y} & D_2 \end{bmatrix} = I_t$$

if and only if the matrices $\bar{A},\bar{B}$ are ZLC.

Definition 2

The matrices $\hat{A} \in R^{lxq}[d_1,d_2]$, $\hat{B} \in R^{pxl}[d_1,d_2]$ $(p+q > l)$ will be called externally skew prime (ESP) if and only if there exists a pair of polynomial matrices $\hat{X} \in R^{qxl}[d_1,d_2]$, $\hat{Y} \in R^{lxp}[d_1,d_2]$ such that

$$\hat{A}\hat{X} + \hat{Y}\hat{B} = I_l$$

It is an extension for 2-D case of the definition given for 1-D case in [22].

3. Problem formulation

Consider a 2-D large scale system S which consists of two subsystems S_1,S_2 connected by A_{12} and A_{21} (Fig.1). The subsystems are described by

S_1: $x_1' = A_1'x_1 + B_1'u_1$ (1a)

$y_1 = C_1'x_1$ (1b)

S_2: $x_2' = A_2'x_2 + B_2'u_2$ (2a)

$y_2 = C_2'x_2$ (2b)

where $x_k' = \begin{bmatrix} x_k^h(i+1,j) \\ x_k^v(i,j+1) \end{bmatrix}$,

$x_k = \begin{bmatrix} x_k^h(i,j) \\ x_k^v(i,j) \end{bmatrix}$

(k=1,2)

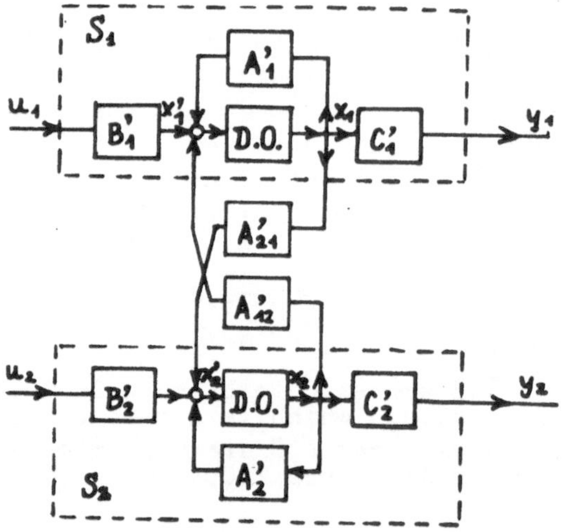

Fig.1. D.O. - unit delay 2-D operator

$x_k^h(i,j) \in R^{n_{1k}}$ is the horizontal state vector of S_k

$x_k^v(i,j) \in R^{n_{2k}}$ is the vertical state vector of S_k

$u_k = u_k(i,j) \in R^{m_k}$ is the input vector of S_k

$y_k=y_k(i,j)\in R^{l_k}$ is the output vector of S_k

$A'_1,A'_2,B'_1,B'_2,C'_1,C'_2$ are real matrices of appropriate dimensions.
The boundary conditions for S_k are given by $x^h_k(0,j)$, $x^v_k(i,0)$ for $i,j=0,1,2,...$
The system S is described by

$$S: \quad \begin{aligned} x' &= A'x+Bu \quad &(3a)\\ y &= C'x \quad &(3b)\end{aligned}$$

where $x=\begin{bmatrix}x_1\\x_2\end{bmatrix}$, $u=\begin{bmatrix}u_1\\u_2\end{bmatrix}$, $y=\begin{bmatrix}y_1\\y_2\end{bmatrix}$

$$A'=\begin{bmatrix}A'_1 & A'_{12}\\A'_{21} & A'_2\end{bmatrix}, \quad B'=\begin{bmatrix}B'_1 & 0\\0 & B'_2\end{bmatrix}, \quad C'=\begin{bmatrix}C'_1 & C'_2\end{bmatrix}$$

Let us assume that the k-th reference generator has n^g_k states and is described by

$$w'_k=A''_k w_k, \quad r_k=C''_k w_k \quad (k=1,2) \quad (4)$$

with non-zero boundary conditions $w^h_k(0,j)$, $w^v_k(i,0)$ for $i,j=0,1,2,...$
where

$$w'_k=\begin{bmatrix}w^h_k(i+1,j)\\w^v_k(i,j+1)\end{bmatrix}, \quad w_k=\begin{bmatrix}w^h_k(i,j)\\w^v_k(i,j)\end{bmatrix}$$

$r_k=r_k(i,j)\in R^{l_k}$ is the k-th reference input vector, A''_k,C''_k are real matrices of appropriate dimensions.
Consider two 2-D linear controllers with n^c_k states described by

$$\begin{aligned} v'_k &= H_k v_k+J'_k y_k+J''_k r_k \quad &(5a)\\ u_k &= K_k v_k+L'_k y_k+L''_k r_k \quad &(5b)\end{aligned} \qquad (k=1,2)$$

with non-zero boundary conditions $v^h_k(0,j),v^v_k(i,0)$ for $i,j=0,1,2,...$

where $v'_k=\begin{bmatrix}v^h_k(i+1,j)\\v^v_k(i,j+1)\end{bmatrix}$, $v_k=\begin{bmatrix}v^h_k(i,j)\\v^v_k(i,j)\end{bmatrix}$

$H_k,J'_k,J''_k,K_k,L'_k,L''_k$ are real matrices of appropriate dimensions.

The problem is to design the controllers so that

$$y_k(i,j)=r_k(i,j) \text{ for } i \geqslant p_1,\ j \geqslant p_2 \quad (k=1,2) \tag{6}$$

p_1, p_2 are positive integers and for any boundary conditions of the system, reference generators and controllers.
Using 2-D Z transformation and introducing the unit delay operators $d_1=z_1^{-1}$, $d_2=z_2^{-1}$ we can write (3) in the form

$$Ay = Bu+C \tag{7}$$

where $A \in R^{l\times l}[d_1,d_2]$ $(l=l_1+l_2)$, $B \in R^{l\times m}[d_1,d_2]$ $(m=m_1+m_2)$ and $C \in R^{l\times 1}[d_1,d_2]$ are defined by

$$C'\left(\begin{bmatrix} I_{n_1}z_1 & 0 \\ 0 & I_{n_2}z_2 \end{bmatrix}-A'\right)^{-1}B'=A^{-1}B\ ,\quad C'\left(\begin{bmatrix} I_{n_1}z_1 & 0 \\ 0 & I_{n_2}z_2 \end{bmatrix}-A'\right)^{-1}\begin{bmatrix} D_1 \\ D_2 \end{bmatrix}=A^{-1}C$$

$$D_k=\begin{bmatrix} z_1x_k^h(0,z_2) \\ z_2x_k^v(z_1,0) \end{bmatrix}\quad ,\quad n_k=n_{1k}+n_{2k} \quad (k=1,2)$$

and $y=y(d_1,d_2)$, $u=u(d_1,d_2)$ are 2-D Z transforms of $y(i,j)$ and $u(i,j)$, respectively.

In a similar way from (4) and (5) we obtain

$$F_kr_k=G_k \quad (k=1,2) \tag{8}$$

and

$$P_ku_k=-Q_ky_k+R_kr_k+T_k \tag{9}$$

where

$F_k \in R^{l_k\times l_k}[d_1,d_2]$, $G_k \in R^{l_k\times 1}[d_1,d_2]$, $P_k \in R^{m_k\times m_k}[d_1,d_2]$,

$Q_k \in R^{m_k\times l_k}[d_1,d_2]$, $R_k \in R^{m_k\times l_k}[d_1,d_2]$ and $T_k \in R^{m_k\times 1}[d_1,d_2]$

are defined by

$$C_k''\left(\begin{bmatrix} I_{n_{1k}^g}z_1 & 0 \\ 0 & I_{n_{2k}^g}z_2 \end{bmatrix}-A''\right)^{-1}\begin{bmatrix} z_1w_k^h(0,z_2) \\ z_2w_k^v(z_1,0) \end{bmatrix}=F_k^{-1}G_k$$

$$K_k\left(\begin{bmatrix} I_{n_{1k}^c}z_1 & 0 \\ 0 & I_{n_{2k}^c}z_2 \end{bmatrix}-H_k\right)^{-1}J_k'+L_k'=-P_k^{-1}Q_k$$

$$K_k\left(\begin{bmatrix} I_{n^c_{1k}} z_1 & 0 \\ 0 & I_{n^c_{2k}} z_2 \end{bmatrix} - H_k\right)^{-1} J_k'' + L_k'' = P_k^{-1} R_k$$

$$K_k\left(\begin{bmatrix} I_{n^c_{1k}} z_1 & 0 \\ 0 & I_{n^c_{2k}} z_2 \end{bmatrix} - H_k\right)^{-1} \begin{bmatrix} z_1 v_k^h(0,z_2) \\ z_2 v_k^v(z_1,0) \end{bmatrix} = P_k^{-1} T_k$$

$$n^g_{1k}+n^g_{2k}=n^g_k \quad , \quad n^c_{1k}+n^c_{2k}=n^c_k$$

and $r_k=r_k(d_1,d_2)$, $u_k=u_k(d_1,d_2)$, $y_k=y_k(d_1,d_2)$ are 2-D Z transforms of $r_k(i,j)$, $u_k(i,j)$ and $y_k(i,j)$, respectively.

Now the problem can be restated as follows.
Given the matrices A,B and $F=\begin{bmatrix} F_1 & 0 \\ 0 & F_2 \end{bmatrix}$, find the matrices $P=\begin{bmatrix} P_1 & 0 \\ 0 & P_2 \end{bmatrix}$, $Q=\begin{bmatrix} Q_1 & 0 \\ 0 & Q_2 \end{bmatrix}$ and $R=\begin{bmatrix} R_1 & 0 \\ 0 & R_2 \end{bmatrix}$ so that the tracking error $e=r-y=\begin{bmatrix} r_1 \\ r_2 \end{bmatrix}-\begin{bmatrix} y_1 \\ y_2 \end{bmatrix}$ is a polynomial vector in d_1 and d_2 independently of C, $G=\begin{bmatrix} G_1 \\ G_2 \end{bmatrix}$ and $T=\begin{bmatrix} T_1 \\ T_2 \end{bmatrix}$ (Fig. 2.).

4. Problem solution

From the theorem 2 it follows that if A,B are ZLC then there exist $A_2 \in R^{mxm}[d_1,d_2]$, $B_2 \in R^{lxm}[d_1,d_2]$, $P \in R^{mxm}[d_1,d_2]$, $Q \in R^{mxl}[d_1,d_2]$, $\overline{P}_2 \in R^{lxl}[d_1,d_2]$ and $\overline{Q}_2 \in R^{mxl}[d_1,d_2]$ such that

$$\begin{bmatrix} A & B \\ Q & -P \end{bmatrix}\begin{bmatrix} \overline{P}_2 & B_2 \\ \overline{Q}_2 & -A_2 \end{bmatrix} = \begin{bmatrix} I_l & 0 \\ 0 & I_m \end{bmatrix} \tag{10}$$

Taking into account that $AB_2=BA_2$ and substituting

$$Fr = G$$

and

$$y=B_2A_2^{-1}u+A^{-1}C$$

into

$$Pu = -Qy+Rr+T$$

we obtain

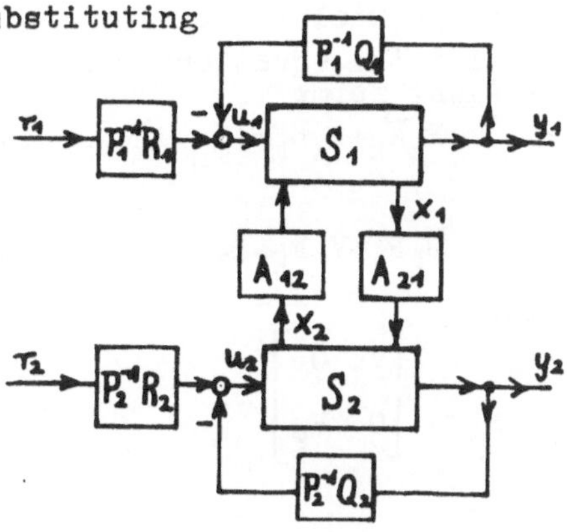

Fig.2.

$$e=r-y=\left[I_1-B_2(PA_2+QB_2)^{-1}R\right]F^{-1}G-\left[I_1-B_2(PA_2+QB_2)^{-1}Q\right]A^{-1}C-B_2(PA_2+QB_2)^{-1}T \tag{11}$$

From (10) we have

$$PA_2+QB_2=I_m \quad (12a) \quad \text{and} \quad \overline{P}_2A+B_2Q=I_1 \quad (12b)$$

After substitution of (12) into (11) we obtain

$$e=\left[I_1-B_2R\right]F^{-1}G-\overline{P}_2C-B_2T \tag{13}$$

Let R and V be a solution of the equation

$$B_2R+VF=I_1 \tag{14}$$

Substituting (14) into (13) we obtain

$$e=VG-\overline{P}_2C-B_2T \tag{15}$$

i.e. a polynomial vector in d_1 and d_2 for any C,G and T

Theorem 3

The problem has a solution if

1) there exist polynomial matrices

$$A_2=\begin{bmatrix}\overline{A}_1 & 0\\ 0 & \overline{A}_2\end{bmatrix},\quad B_2=\begin{bmatrix}\overline{B}_1 & 0\\ 0 & \overline{B}_2\end{bmatrix},\ \overline{A}_k\in R^{m_k\times m_k}[d_1,d_2],\ \overline{B}_k\in R^{l_k\times m_k}[d_1,d_2]$$
$$(k=1,2)$$

such that

$$\left[AB\right]\begin{bmatrix}B_2\\ A_2\end{bmatrix}=0 \tag{16}$$

2)
a) A,B are ZLC and $\overline{A}_k$, $\overline{B}_k$ are ZRC for k=1,2

b) F_k, $\overline{B}_k$ are ESP for k=1,2.

Proof

It has been shown that if A,B are ZLC and F,B_2 are ESP then e is a polynomial vector for any C,G and T. If the condition 1) is satisfied the equations (12a) and (14) can be written in the equivalent forms

$$P_k\overline{A}_k+Q_k\overline{B}_k=I_{m_k} \quad (k=1,2) \tag{17}$$

and

$$\overline{B}_kR_k+V_kF_k=I_{l_k} \quad (k=1,2) \tag{18}$$

where

$$V=\begin{bmatrix}V_1 & 0\\ 0 & V_2\end{bmatrix}$$

The equations (17) and (18) have solutions if the conditions 2) are satisfied. Solving (17) and (18) we can find the desired matrices P,Q and R. □

Let us assume

$$A=\begin{bmatrix}A_{11} & A_{12}\\ A_{21} & A_{22}\end{bmatrix},\quad B=\begin{bmatrix}B_{11} & 0\\ 0 & B_{22}\end{bmatrix}\quad \begin{matrix}A_{kk}\in R^{l_k\times l_k}[d_1,d_2]\\ B_{kk}\in R^{l_k\times m_k}[d_1,d_2]\ (k=1,2)\end{matrix} \tag{19}$$

Lemma

If $(A_{11},B_{11}),(A_{22},B_{22})$ are ZLC and $A_{21}=0$ or $A_{12}=0$, then the matrices (19) are ZLC.

Proof

If (A_{11},B_{11}) and (A_{22},B_{22}) are ZLC then exist four polynomial matrices X_1,Y_1 and X_2,Y_2 such that

$$\begin{bmatrix}A_{11}B_{11}\end{bmatrix}\begin{bmatrix}X_1\\ Y_1\end{bmatrix}=I_{l_1}\quad \text{and}\quad \begin{bmatrix}A_{22}B_{22}\end{bmatrix}\begin{bmatrix}X_2\\ Y_2\end{bmatrix}=I_{l_2}$$

Choosing

$$X=\begin{bmatrix}X_1 & -X_1A_{12}X_2\\ 0 & X_2\end{bmatrix},\quad Y=\begin{bmatrix}Y_1 & -Y_1A_{12}X_2\\ 0 & Y_2\end{bmatrix}$$

we obtain for $A_{21}=0$

$$\begin{bmatrix}AB\end{bmatrix}\begin{bmatrix}X\\ Y\end{bmatrix}=I_l$$

For $A_{12}=0$ the proof is similar. □

Algorithm

If the conditions of the theorem 3 are satisfied the matrices P,Q and R can be found by the use of the following algorithm

Step 1 For given A,B find A_2,B_2 satisfying (16).

Step 2 Solving (17) for k=1,2 find P_k,Q_k,P and Q.

Step 3 Solving (18) for k=1,2 find R_k,V_k and R.

The above considerations with slight modifications can be extended for 2-D large scale system which consists of n subsystems.

References

1. Aoki M., On feedback stabilizability of decentralized dynamic systems. "Automatica",vol.8, 1972, 163-173.

2. Bose N.K., Multidimensional Systems: Theory and Applications, IEEE Press 1979.

3. Chan W.S. and Desoer C.A., Eigenvalue assignment and stabilization of interconnected systems using local feedbacks. IEEE Trans.

Autom.Contr., vol.24,1979, 312-317.

4. Eichstaedt B., Multivariable closed-loop dead-beat control: a polynomial-matrix approach. Automatica, vol.18, 1982,589-593.
5. Eising R., Realization and stabilization of 2-D systems. IEEE Trans.Autom.Contr. AC-23, 1978,793-799.
6. Fessas P.S., Decentralized control of linear dynamical systems via polynomial matrix methods I: Two interconnected scalar systems. Int.J.Control, vol.30,1979,259-276.
7. Fessas P.S., Matrix fraction description approach to decentralized control. IEE Proc.D. vol.129,1982,206-210.
8. Fornasini M. and Marchesini G., State space realization theory of two-dimensional filters. IEEE Trans.Autom.Contr.,AC-21,1976, 484-492.
9. Kaczorek T., Dead-beat servo problem for 2-dimensional linear systems. Int.J.Control, vol.37,1983,1349-1353.
10. Kaczorek T., Deadbeat control in multivariable non-linear time-varying systems with constraints of control inputs. Int.J.Control vol.38,1983,449-458.
11. Kučera V., A dead-beat servo problem. Int.J.Control,vol.32,1980, 107-113.
12. Kučera V. and Šebek M., On deadbeat controllers. IEEE Trans. Autom.Contr.,AC-28 (in print).
13. Leben B., Multivariable deadbeat control. Automatica,vol.13, 1977,185-188.
14. Paraskevopoulos P.N. and Mertzios B.G., Transfer function factorization of 2-D systems using state feedback. Int.J.Systems Sci.,vol.12,1981,1135-1147.
15. Porter B. and Bradshaw A., Design of dead-beat controllers and full-order observers for linear multivariable discrete-time plants. Int.J.Control,vol.22,1975,149-155.
16. Ramakrishna A. and Viswanadham N., Decentralized control of interconnected dynymical systems. IEEE Trans.Autom.Contr., vol.27,1982,159-164.
17. Šebek M., Multivariable dead-beat servo problem, Kibernetika, vol,16,1980,442-453.
18. Sezer M.E. and Šiljak D.D., On decentralized stabilization and structure of linear large scale systems. Automatica,vol.17, 1981,641-644.
19. Tzafestas S.G. and Pimenides T.G., Exact model-matching control of three-dimensional systems using state and output feedback. Int.J.Systems Sci.,vol.13,1982,1181-1187.
20. Tzafestas S.G. and Pimenides T.G., Transfer function computation and factorization of 3-dimensional systems in state-space IEE Proc. vol.130 Pt.D.,1983,231-242.
21. Wang S.H. and Davison E.J., On the stabilization of decentralized control systems. IEEE Trans.Autom.Contr.,vol.18,1973,473-478.
22. Wolowich W.A., Skew prime polynomial matrices. IEEE Trans.Autom. Contr.,vol.AC-23,1978,880-887.
23. Youla D. and Gnavi G., Notes on n-dimensional system theory. IEEE Trans.Cir.Syst. CAS-26,No 2,1979,105-111.

DISTRIBUTED SYSTEM FOR A TENTACLE MANIPULATOR

Mircea Ivănescu
Automatic and Computer Department
University of Craiova,1100 Craiova
ROMANIA

Abstract

A class of manipulator arms with an infinite number of degrees-of-freedom (a tentacle) is studied. A distributed parameter model is determined using the generalised form of Lagrange's equation. This model is characterised by an integral-differential equation which generates the dynamic of the manipulator.

A discret model with adequated feedback control is proposed. An optimal algorithm for a problem of final point and minimum energy is developed and the optimal feedback control is determined. A strategy of real time control, based on this algorithm,is them discussed.

Introduction

During recent years the problem of manipulator action control has been attracting considerable attention. There are a great number of publications and achievments in the field of manipulator control. Out of a great number of them we notice papers in which the control problem of motion is a problem of controlling multivariable systems possessing redundant degrees-of-freedom in a finite number.

This paper presents a new manipulator with ideal flexible elements and distributed mass and torques,a tentacle model. This model is equivalent to a mechanism with infinite number of degrees-of-freedom. It is shown that the dynamic of the system is defined by a set of integral-differential equations obtained by using the generalised Lagrange's principle.

Such systems offer a great possibility of control in some points in the space,a good flexibility for a large number of applications etc.The control of this system is an important problem. It requires to find the control law of distributed torques so that the general trajectory in space should become the imposed trajectory.Several difficulties arise associated with the complex form of dynamic equations and with the highly non-linear,interactive nature of the system.

In this paper it is shown that by applying the power of dedicated micro-computers it is possible simultaneously to linearise and decouple the overall system.The procedure is based on the discret - spatial model which approximates the original system and upon using some adequate feedback controllers. The control law is determined for an optimal problem: final-point and minimum energy. A dynamic programming procedure is used to compute the feedback controllers.

The analysis of the control methods for a planar tentacle is described but this system possesses all the important features so that the application of the methods to more complex systems in a 3-dimensional space may be readily inferred.

Dynamic model

We shall consider the tentacle arm of Fig.1. The system can operate in a 2-dimensional vertical space,XOY. We assume the arm as perfectly flexible,without any viscous friction. Its great flexibility determines an analysis similar to that used in infinite-dimensional systems,the distributed parameter systems.

Technologically,such systems can be obtained by using a cellular structure for every element of the arm. The control can be produced using an electro-hydraulic or pneumatic action which determines the contraction or dilatation of the peripheral cells. In this paper we shall discuss only the dynamic of the control without dealing with these technological problems.

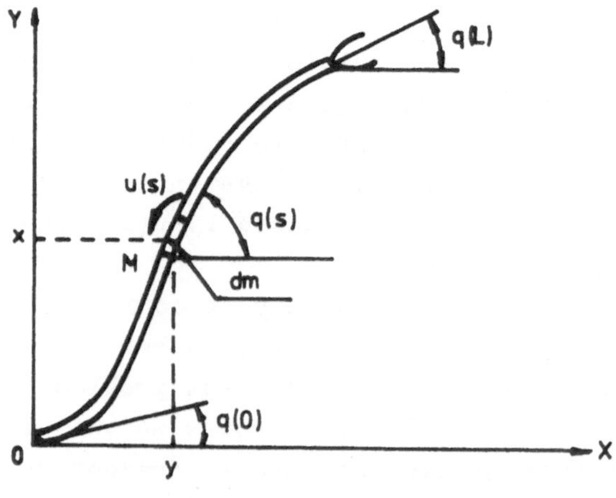

Fig.1. The tentacle arm

Let us take the arm in Fig.1. We assume a uniform distributed mass with a linear density ϱ[Kg/m] .Also,we shall neglect the effects of the section in dynamic motion of the arm. We denote by s the spatial variable on the length of the arm, $s \in [0,L]$.

The position of some point M of the arm can be defined in Cartesian coordinates (x,y) or in Lagrange generalised coordinates q,s, where q represents the absolute angle.

$$x = \int_0^s \cos q' \, ds' \tag{1}$$

$$y = \int_0^s \sin q' \, ds' \tag{2}$$

where q' represents

$$q' = q(s') \quad , \quad s' \in [0,s] \tag{3}$$

In a dynamic analysis,the generalised coordinate q will be a function of time t,

$$q = q(s,t) \tag{4}$$

The dynamic behaviour of the tentacle arm is determined by the effects of a distributed torque on overall length of the arm u(s,t) and by the potential energy of the system.

Let v_x, v_y be the velocity components with respect to X and Y axes. From (1) and (2) results,

$$v_x = \dot{x} = - \int_0^s \sin q' . \frac{\partial q'}{\partial t} \, ds' \tag{5}$$

$$v_y = \dot{y} = \int_0^s \cos q' . \frac{\partial q'}{\partial t} \, ds' \tag{6}$$

For an element dm, the kinetic energy will be,

$$dT_c = \frac{1}{2} \, dm \, . \, v^2 \tag{7}$$

Using (5),(6),we have

$$T_c = \frac{1}{2} \varrho \int_0^L \left\{ \left(\int_0^s \frac{\partial q'}{\partial t} \sin q' ds' \right)^2 + \left(\int_0^s \frac{\partial q'}{\partial t} \cos q' ds' \right)^2 \right\} ds \tag{8}$$

where T_c represents the kinetic energy for the overall arm.Obviously, this energy is a functional by the form

$$T_c = T_c \left(q, \frac{\partial q}{\partial t}, s\right) \tag{9}$$

Similarly we obtain the potential energy,

$$dT_p = dm \, . \, g \, . y \tag{10}$$

and from (2) results

$$T_p = \int_0^L \rho g \int_0^s \sin q' ds' ds \tag{11}$$

where

$$T_p = T_p(q,s) \tag{12}$$

The dynamic behaviour of the arm will be obtained using Lagrange's principle developed for infinite-dimensional systems [3],

$$\frac{\partial}{\partial t}\frac{\delta T_c}{\delta(\frac{\partial q}{\partial t})} - \frac{\delta T_c}{\delta q} - \frac{\delta T_c}{\delta s} - \frac{\delta T_p}{\delta q} - \frac{\delta T_p}{\delta s} = Q \tag{13}$$

where $\delta T/\delta q$ denotes a functional partial (variational) derivative [1], which is defined as the variation of the functional T with respect to the function q at a point $s \in [0,L]$, and Q is the generalised input of the system,

$$Q = \int_0^L u(s,t)\, ds \tag{14}$$

Substituting the relations (8) and (12) into (13) and expanding them we obtain the general form for the system motion,

$$\int_0^L \left\{ \int_0^s \left[\frac{\partial^2 q'}{\partial t^2}\cos(q'-q) - \left(\frac{\partial q'}{\partial t}\right)^2 \sin(q'-q) - \frac{\partial q}{\partial t}.\frac{\partial q'}{\partial t}\cos(q'-q)\right] ds' - g(\sin q + \cos q) \right\} ds = \frac{1}{\rho}\int_0^L u\, ds \tag{15}$$

where

$$q' = q(s',t) \quad , \quad s' \in [0,s]$$

$$q = q(s,t) \quad , \quad s \in [0,L]$$

The initial conditions are derived from the initial state of the manipulator,

$$q(0,s) = q^o(s)$$
$$\frac{\partial q}{\partial t}(0,s) = q_1^o(s) \tag{16}$$

where $q^o(s)$ and $q_1^o(s)$ represents initial position angle and initial velocity.

The equation (15) allows to compute the absolute angle q and other parameters of the motion as velocity and acceleration for every point of the arm, at any time, for an adequated distributed control applied on overall domain. Clearly, there arises the difficulty of the problem determined by the complexity of the integral-differential equation and by its nonlinearity associated with the strong iteractive relation between the points of the system.

Control problem formulation

The control problem represents a classical problem for the manipulator arms.This problem asks for determining the torque-inputs so that the trajectory of the system be specified trajectory. We shall assume that there are not the constraints in space of the state q and inputs u.Consequently we shall define a performance index by the functional (17)

$$J = \int_0^L \int_0^L q(t_f,s)Fq(t_f,s')dsds' + \frac{1}{2}\int_0^{t_f}\int_0^L\int_0^L u(t,s)Wu(t,s')ds'dsdt$$

where $F = F(s,s')$ and $W = W(s,s')$ denote the weight functions.The first term of this equation represents the end point constraints of the arm so that a final position be targeted and the second denotes some restrictions on the system energy. The control problem is to find a distributed torque to satisfy (15),(16) and to minimize the functional (17).

Approximate system

We shall use an approximate model based on a spatial discretization which has the advantage to be strongly connected to the physical structure of the system. Let $s_o,\dots,s_N$ be the spatial discrete variable and Δ the size of the space increment.We denote by

$$q_i(t) = q(t,s_i) \quad , \quad i = 0,1,\dots,(N-1) \tag{18}$$

We select N and Δ so that

$$\left| q_i(t) - q_{i-1}(t) \right| < \varepsilon \qquad \begin{array}{l} i = 1,\dots,(N-1) \\ t \in [0,t_f] \end{array} \tag{19}$$

where ε is a positive constant sufficiently small.
By considering different values of i,from (15),(16) we obtain the following set of differential equations

$$\sum_{i=o}^{m}\left\{\frac{d^2q_i}{d\,t^2}\cos(q_m-q_i) - \frac{dq_m}{dt}.\frac{dq_i}{dt}\cos(q_m-q_i) - (\frac{dq_i}{dt})^2\sin(q_m-q_i)\right.$$

$$- g(\sin q_m + \cos q_m) = \frac{1}{\rho\Delta}u_m \quad , \quad m = 0,1,\dots,(N-1) \tag{20}$$

with initial conditions

$$q_i(0) = q_i^o \tag{21}$$

$$\frac{dq_i}{dt}(0) = q_{1i}^o \tag{22}$$

where

$u_m(t) = u(t,s_m)$.

A new approximate model of the system can be obtained by considering the specifical form of (20) and using the inequality (19). This is

$$\frac{d^2q_o}{dt^2} - (\frac{dq_o}{dt})^2 = \frac{1}{\rho\Delta}$$

$$\frac{d^2q_i}{dt^2} - (\frac{dq_i}{dt})^2 = \frac{1}{\rho\Delta}(u_i - u_{i-1}) \quad , \quad i = 1,2,\ldots,N-1 \tag{23}$$

where u_i, u_{i-1} denotes two consecutive sequences of the control. We define

$$u_i = u_i^* + u_i^{**} \tag{24}$$

where u_i^* is the optimal component of the control and u_i^{**} is a local feedback control selected as

$$u_i^{**} = -\rho\Delta\ ((\frac{dq_i}{dt})^2 + a\frac{dq_i}{dt} + bq_i) + u_{i-1}^{**} \tag{25}$$

where the first term is a compensator of the nonlinearity and the next terms assure the stability conditions [2] . If we denote

$$q_i(t) = q_{1i}(t) \tag{26}$$

$$\frac{dq_i(t)}{dt} = q_{2i}(t) \tag{27}$$

the relations (23),(25) become

$$\dot{q}_{1i} = q_{2i} \tag{28}$$

$$\dot{q}_{2i} = -aq_{2i} - bq_{1i} + \frac{1}{\rho\Delta}(u_i^* - u_{i-1}^*) \tag{29}$$

$$u_i^{**} = -\rho\Delta(q_{2i}^2 + aq_{2i} + bq_{1i}) + u_{i-1}^{**} \tag{30}$$

Equations (28),(29) define a linear model which be made as stable as necessary by suitable choice of the a and b.

Strategy of control

Let the performance index (17) be modified according to the spatial discretization

$$J = q^T(t_f)\ Fq(t_f) + \frac{1}{2} \int_o^{t_f} U^T\ W\ U\ dt \tag{31}$$

where

$$q = \mathrm{col}(q_{1o},\ldots,q_{1N-1},q_{2o},\ldots,q_{2N-1}) \tag{32}$$

$$U = \mathrm{col}(u_o^*,u_1^*,\ldots,u_{N-1}^*)$$

$$F = \begin{bmatrix} F^1 & F^{12} \\ F^{12} & F^2 \end{bmatrix} \tag{33}$$

F^1,F^2,F^{12} are (N x N) constant positive definite matrices and W is the identity matrix.

The dynamic programming equation associated to this problem will be

$$-\frac{\partial J}{\partial t} = \min_U\ (U^T U + J_q^T \cdot \dot{q}) \tag{34}$$

From (30),(31),(34) we obtain the control u_i^* which minimizes the right side of (34),

$$u_i^* = -\frac{1}{\rho\Delta}\,(J_{q_{2i}} - J_{q_{2,i+1}}) \tag{35}$$

If we assume that the solution to equation (34) has the form

$$J = \frac{1}{2}\ q^T K\ q \tag{36}$$

$$K = \begin{bmatrix} K^1 & K^{12} \\ K^{12} & K^2 \end{bmatrix} \tag{37}$$

where K^1,K^{12},K^2 are undetermined (N x N) symmetric matrices,then from (34),(35) results the optimal control u_i^*,

$$u_i^* = -\frac{1}{\rho\Delta} \sum_{j=o}^{N-1} (k_{ij}^{12}\ q_{1j} + k_{ij}^2 q_{2j} - k_{i+1,j}^{12} - k_{i+1,j}^2 q_{2j}) \tag{38}$$

In this relation k_{ij}^{12}, k_{ij}^2 are the elements of the matrices K^{12} , K^2 and verify the following equations,

$$-\frac{dk_{ij}^1}{dt} = \frac{1}{(\rho\Delta)^2}(k_{ij}^{12}-k_{i+1,j}^{12})^2-2bk_{ij}^{12}+\frac{2}{(\rho\Delta)^2}\sum_{r=o}^{N-1} k_{ri}^{12}(k_{r+1,j}^{12}-2k_{i,j}^{12}+k_{r-1,j}^{12}) \tag{39}$$

$$- \frac{dk_{ij}^{12}}{dt} = \frac{1}{(\rho\Delta)^2}(k_{ij}^{12}-k_{i+1,j}^{12})(k_{ij}^2-k_{i,j+1}^2)+k_{ij}^1-ak_{ij}^{12}-bk_{ij}^2+ \tag{40}$$

$$+ \frac{1}{(\rho\Delta)^2}\sum_{r=o}^{N-1} k_{ri}^{12}(k_{r+1,j}^2-2k_{r,j}^2+k_{r-1,j}^2)+k_{ri}^2(k_{r+1,j}^{12}-2k_{rj}^{12}+k_{r-1,j}^{12})$$

$$- \frac{dk_{ij}^2}{dt} = \frac{1}{(\rho\Delta)^2}(k_{ij}^2-k_{i,j+1}^2)^2+2(k_{ij}^{12}-ak_{ij}^2) + \tag{41}$$

$$+ \frac{2}{(\rho\Delta)^2}\sum_{r=o}^{N-1} k_{ri}^2(k_{r+1,j}^2 -2k_{r,j}^2 + k_{r-1,j}^2) \ , \ i,j=o,1\ldots(N-1)$$

These equations are integrated backwards in time from t_f to 0 for final conditions

$$k_{ij}^1(t_f) = F_{ij}^1$$

$$k_{ij}^2(t_f) = F_{ij}^2 \tag{42}$$

$$k_{ij}^{12}(t_f) = F_{ij}^{12}$$

The relation (38) associated with (39)-(42) offers a very good control law,which can be implemented in a computer control. This type of implementation is attractive for a control in a real-time because the calculations of k_{ij} can be done off-line and only operation defined by the relations (38) is used in real-time.The computational algorithm can be synthesized as:

1. For an imposed trajectory,the finalconditions (42) are computed.
2. Integrate the equations (39)-(42) backwards in time.The values of k_{ij} are stored.

The steps 1 and 2 are computed off-line.

3. Implemente the control law (38) in real-time.Computing requires simple operations which can be done with a specialized computer.

The system configuration is given in Fig.2. This system requires a memory for k_{ij} and an arithmetic block which can implemente the local and optimal feedback controller. The values of a and b computed for this model,assure the global stability of the system.

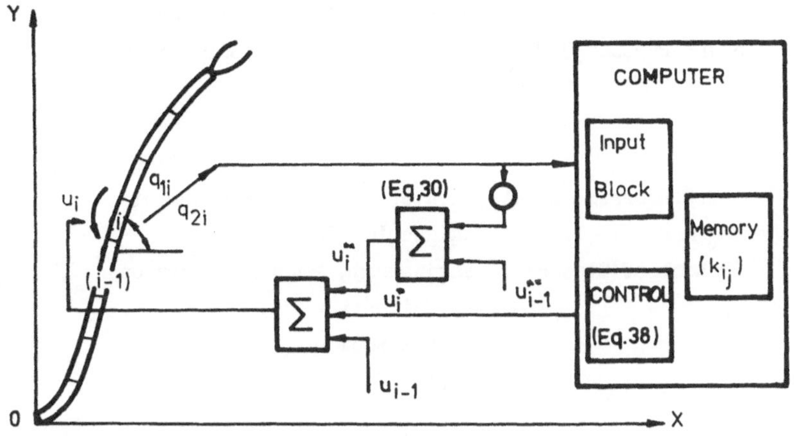

Fig.2. The control configuration

Conclusions

The paper presents the dynamic of a new manipulator, a tentacle. A parameter distributed system associated to this manipulator is derived and a feedback control system is proposed.

This control law offers some facilities so that a real-time control for the system is adequated.

References

1. Wang, P.K.C., Optimum Control of Distributed Parameter Systems, IEEE Trans.Automatic Control, vol.AC-9, pp 13-22, 1964.
2. Hewit, J.R., Padovan, J., Decoupled Feedback Control of Robot and Manipulator Arms, 3rd Symp. on Theory of Robots and Manipulators Udine, Sept.1978, pp 251-266.
3. Ivănescu, M., Extension of Lagrange's Equation to Continuous Systems, The 2nd Symp. Theory of Control, Craiova, pp 62-67, 1982.

ON-LINE DISTRIBUTED HIERARCHICAL CONTROL AND OPTIMISATION OF LARGE SCALE PROCESSES USING A MICRO-COMPUTER BASED SYSTEM

P.D.Roberts, C.W.Li, I.A.Stevenson, D.S.Wadhwani
Control Engineering Centre
School of Electrical Engineering & Applied Physics
The City University
Northampton Square
London EC1

1) INTRODUCTION

Because of economic reasons and intensive competition within the market sector, increasing pressure is stressed upon industry to improve efficiency and productivity of industrial plants. A method for the practical implementation of an on-line optimisation and control scheme for an industrial process considers the overall design as a two layer hierarchical system. Modern computer technology which has resulted in low cost computer power may then be employed with advantage in order to employ micro or mini-computers, implemented within a distributed hierarchical computer structure, to coordinate, optimise and regulate individual decision tasks and sub-processes.

Problems associated with on-line hierarchical optimisation and control of industrial processes such as disturbances, model reality differences and timing are considered and ways to counteract these problems are suggested in this paper.

Recently, several hierarchical control schemes have been suggested for on-line control of steady state systems (Findeisen 1978), (Roberts 1983), (Shao 1983). This paper implements some of the proposed schemes through a distributed control system where the aim is to study closed loop hierarchical control of simulated industrial processes.

2) DISTRIBUTED HIERARCHICAL COMPUTER SYSTEM

Within the Computer Control Laboratory of The City University there is a distributed two-level hierarchical computer system. The infimal level of the hierarchy contains four I-MIC micro-computers and a LSI-11/02 mini-computer which are used as local decision units for direct control of pilot plants and a small analogue computer system. A DEC LSI-11/23 mini-computer is employed at the supremal level which acts as a host machine to coordinate and supervise computers at the lower level.

At present the I- MICs control a pilot scale freon vaporiser, a mixing process and simulated interconnected dynamic processes on the analogue computer. The LSI-11/02 controls a pilot-scale eight zone electrically heated travelling load furnace.

The LSI-11/23 has a full complement of RAM (256Kb) and runs the TSX-PLUS time shared operating system. Programs and data are stored on two 20Mb Winchester disks and a 1.2Mb floppy disk drive. The peripherals which are available to the LSI-11/23 include a Tektronix storage tube graphics terminal, a hard copy unit and an Intecolor colour graphics terminal. The latter is used to display the current measurements from any process on a mimic diagram. Although several high-level languages are available for programming on the LSI-11/23, Fortran is used since, at present, this is the only one which can provide the real-time support required (timer functions, interrupt handling etc.).

The I-MIC is an Intel 8085 based micro-computer which has 16Kb of RAM. It can be programmed using the interpretive language - Control Basic or using assembly language. Each I-MIC is interfaced to its respective plant by means of plug-in memory mapped I/O boards. In the case of the analogue computer simulation rig, the I/O boards used provide a 16 multiplexed channel 12 bit A/D converter and a dual 12 bit D/A converter.

The computer used to simulate interconnected industrial processes is an EAL Pace TR48 general purpose analogue computer which is composed of solid state computing components. It is of modular design with eight different computing modules so that the computer configuration can be varied to suit a particular problem.

Communication between the I-MICs and the host computer takes place over 20 mA current loop serial lines at 1200bd. The data transfer is initiated by the local processor and takes place according to a protocol developed in the laboratory. In the laboratory environment we have not experienced any problem with transmission errors caused by electrical noise, therefore no error correction algorithms have been needed.

Interactive programs have been developed which enable several users to control their respective pilot-scale processes and the hybrid computer system simultaneously, thus maximising the usage of the computer facilities. The schematic representation of the distributed hierarchical computer system is shown in fig.1.

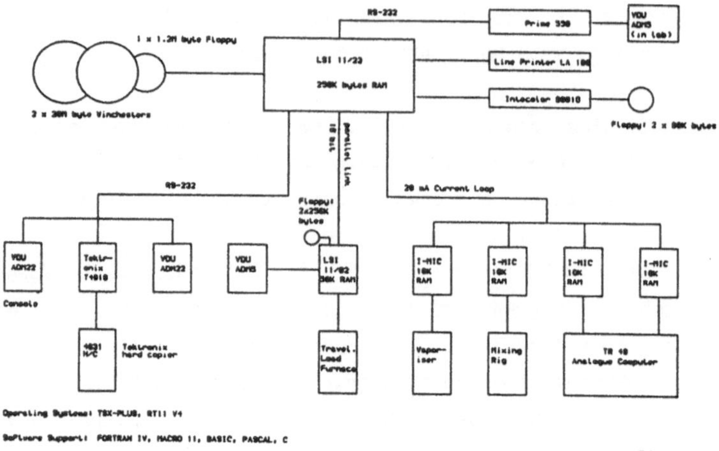

DISTRIBUTED COMPUTER NETWORK fig.1

3) FORMATION OF THE CONTROL PROBLEM

A steady state mathematical model is required in order to optimise the real system under consideration. It is assumed that external disturbances affecting the system are constant over the considered time interval and are therefore omitted from the system equations. Consider the ith sub-system, the output vector, $\underline{y}_i$ is described by,

$$\underline{y}_i = \underline{F}_i(\underline{c}_i, \underline{u}_i) \qquad (1)$$

where $\underline{c}_i$, $\underline{u}_i$ are the controls and interaction inputs to the sub-system respectively. $\underline{F}_i$ is the ith sub-system input/ output mapping.

It is assumed for each overall control vector, $\underline{c}$, applied to the system, there exists an unique value of overall output vector, $\underline{y}$. Hence a mapping, $\underline{K}$, between these two vectors can be established as,

$$\underline{y} = \underline{K}(\underline{c}) \qquad (1.a)$$

The interconnections between sub-systems are defined by,

$$\underline{u}_i = \sum_{j=1}^{N} H_{ij}\underline{y}_j \quad , \quad i \in \overline{1,N} \qquad (2)$$

where N is the number of sub-systems and H_{ij} is the interconnection matrix matrix composed of zeros and ones.

The local inequality constraints, $\underline{g}_i$ in sub-system i are,

$$\underline{g}_i(\underline{c}_i, \underline{u}_i, \underline{y}_i) \leq 0 \qquad (3)$$

The performance index, P_i of sub-system i can be expressed as,

$$P_i = \varphi_i(\underline{c}_i, \underline{u}_i, \underline{y}_i) \qquad (4)$$

The ith sub-system has the optimisation task of determining the control vector, $\underline{c}_i$ to optimise the performance index, P_i.

It is assumed that the global system performance index P is the sum of the sub-system performance indices, i.e.

$$P(\underline{c}, \underline{u}, \underline{y}) = \sum_{i=1}^{N} P_i(\underline{c}_i, \underline{u}_i, \underline{y}_i) \qquad (5)$$

The task of the global control problem is to optimise the overall system performance index of the model subject to constraints.

$$\left.\begin{aligned} &\min_{\underline{c}} \sum_{i=1}^{N} P_i(\underline{c}_i, \underline{u}_i, \underline{y}_i) \\ &\text{subject to} \quad \underline{y}_i = \underline{F}_i(\underline{c}_i, \underline{u}_i, \underline{y}_i) \\ &\qquad\qquad\quad \underline{g}_i(\underline{c}_i, \underline{u}_i, \underline{y}_i) \leq 0 \\ &\qquad\qquad\quad \underline{u}_i = \sum_{j=1}^{N} H_{ij}\underline{y}_j \end{aligned}\right\} \qquad (6)$$

3.1) Mathematical Model

A steady state system model consisting of two interconnected sub-processes, simulated by the analogue computer, is used to investigate different coordination methods for closed loop hierarchical control. The computer system structure is shown in fig.2.

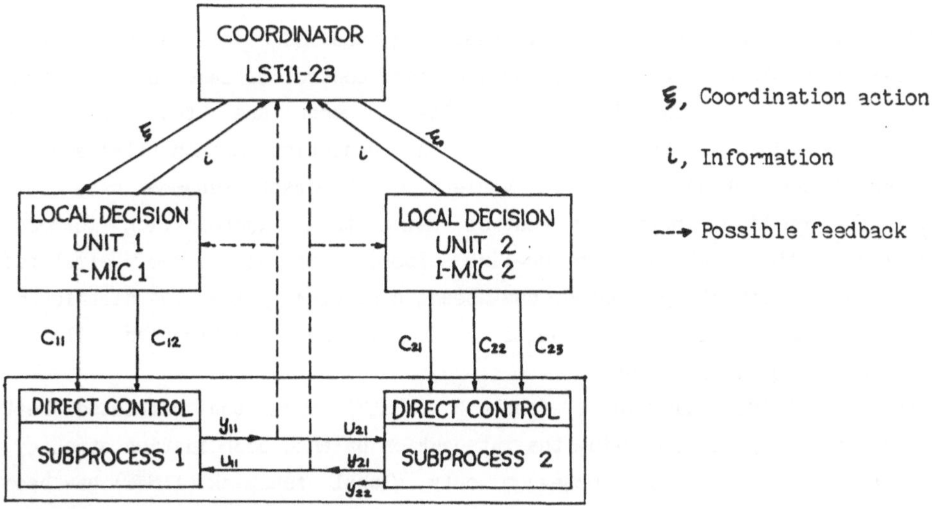

First order time constants are introduced to the interaction inputs and the controls. The mathematical model and reality equation and performance indices of the sub-processes are:

$$\begin{bmatrix} y_{11} \\ y_{21} \\ y_{22} \end{bmatrix} = \begin{bmatrix} f_{11}(\underline{c}_1,\underline{u}_1) \\ f_{21}(\underline{c}_2,\underline{u}_2) \\ f_{22}(\underline{c}_2,\underline{u}_2) \end{bmatrix} = \begin{bmatrix} c_{11} - c_{12} - 2u_{11} \\ c_{21} - c_{22} + u_{21} \\ 2c_{22} - c_{23} - u_{21} \end{bmatrix}$$

$$P_1(\underline{c}_1,\underline{u}_1,\underline{y}_1) = (y_{11}^2 - 1) + c_{11}^2 + c_{12}^2$$

$$\begin{bmatrix} y_{11}^* \\ y_{21}^* \\ y_{22}^* \end{bmatrix} = \begin{bmatrix} f_{11}^*(\underline{c}_1,\underline{u}_1) \\ f_{21}^*(\underline{c}_2,\underline{u}_2) \\ f_{22}^*(\underline{c}_2,\underline{u}_2) \end{bmatrix} = \begin{bmatrix} 1.4c_{11} - 0.6c_{12} + 1.8u_{11}^* \\ 1.3c_{21} - 1.1c_{22} + 1.1u_{21}^* \\ 2.3c_{22} - 0.7c_{23} - 1.1u_{21}^* \end{bmatrix}$$

$$P_2(\underline{c}_2,\underline{u}_2,\underline{y}_2) = 2(y_{21}-2)^2 + (y_{22}-3)^2 + c_{21}^2 + c_{22}^2 + c_{23}^2$$

The system constraints are:

$$\{\underline{g}_1(\underline{c}_1,\underline{u}_1,\underline{y}_1) \in R^4 : |c_{11}| \leq 1,\ |c_{12}| \leq 1,\ y_{11} \geq 0,\ (0.8 - c_{12} - 0.6u_{11}) \geq 0\}$$
$$\{\underline{g}_2(\underline{c}_2,\underline{u}_2,\underline{y}_2) \in R^5 : |c_{21}| \leq 1,\ |c_{22}| \leq 1,\ |c_{23}| \leq 1,\ y_{21} \geq 0,\ y_{22} \geq 0\}$$

The coupling equation is:

$$\begin{bmatrix} u_{11} \\ u_{21} \end{bmatrix} = \begin{bmatrix} 0 & 1 & 0 \\ 0 & 1 & 0 \end{bmatrix} \begin{bmatrix} y_{11} \\ y_{21} \\ y_{22} \end{bmatrix}$$

4) ON-LINE COORDINATION METHODS

Different coordination methods have been suggested and developed in the last two decades for steady state optimisation and control of large scale systems (Mesarovic,1970; Findeisen,1978). Basically, there are two principle coordination methods, namely, the Interaction Prediction Method (IPM) and the Interaction Balance Method (IBM). A variety of feedback schemes have been suggested for on-line optimising control using these coordination methods. Investigation of these methods for on-line closed loop control of simulated interconnected industrial processes has been performed using the distributed computer system and difficulties encountered during the investigation will be discussed in subsequent sections.

Both global (IBMGF) and local feedback (IBMLF) from the simulated real process have been implemented using the interaction balance coordination strategy. At the present stage of the research only global feedback (IPMGF) has been investigated using the interaction prediction coordination method. Details concerning the formulation of the optimisation problem for the coordinator and the local decision units, existence and convergency conditions for all these coordination methods can be obtained from the literature (Findeisen,1980).

4.1) Synchronisation and Interprocesses Communication

In the real-time implementation of the closed loop coordination methods using the distributed computer system (fig.2), two local decision units are used in the infimal level. Parallel computation can be performed at the local decision level once coordination parameters have been received from the supremal level. Because first order transfer lags have been introduced in the controls and the interaction inputs within the simulated real process the decision units need to wait sufficient time for the system transients to die down in order to take steady state measurements. The determination of a minimum waiting time for different transfer lag time constants has been carried out off-line using the simulation software package 'ISIS' available on a Prime 550 mini-computer (3).

Using the distributed hierarchical optimising control approach, interprocess communication problems arise at the decision unit level because each decision unit optimisation iteration finishes at a different time interval. Two methods have been considered for synchronising the decision units so that controls can be sent to the real sub-processes simultanaeously,

1) Elapse Time:
Estimate the time required for each decision unit to finish one optimisation iteration. Each decision unit then waits until the slowest unit completes its task before sending the controls to the real sub-processes and taking measurements at steady state.

2) Semaphore:
After each optimisation computation, each decision unit sends a completion flag to the coordinator and waits. A task of the coordinator is to check that all the completion flags from the decision units have been received before transmitting a start flag to each decision unit, all of which then send controls simultaneously to the real process and take measurements at steady state. The start flag is then reset for the next data set.

The advantage of the first proposed synchronisation scheme is that total decentralisation at the local decision unit level can be obtained. However, practically it is very difficult to determine the waiting time for each decision unit because the computation time required at each optimisation iteration will often vary. This may affect stability and convergence at the infimal level. Therefore only the second method has been adopted for synchronising the decision units although this requires extra information exchange between the coordinator and the local decision units. This synchronisation scheme has been shown to work efficiently during real time control of the simulated real process under different coordination methods with feedback.

Using the local feedback scheme (IBMLF), a further synchronisation problem arises because each decision unit converges to its final solution over a different time horizon. Because theory suggests that the coordinator should adjust its parameters only when all the decision units have converged to their solution, the sempahore method is again used to synchronise the coordinator and the local decision units.

An investigation has been carried out where the decision units were not synchronised before applying the controls to the real process and it was observed that, as a result, the decision units became unstable. Although decision units could be stabilized by reducing their iterative loop gain parameters, this decreased the convergence rate of the local optimisation problem. In this particular simulation study, the synchronised scheme required far fewer iterations in the local decision level, and thus the total iterations required for the global optimisation problem was reduced.

4.2) Simulation Results and Discussion

Using the open loop solution set as the first estimate for the coordinator parameters, the following results have been obtained:

Coordination method	Performance index	Suboptimality (%)	Iterations required	Computation time(m.,s.)
IBMGF	5.989	1.057	7	3,16
IBMLF	5.962	0.613	33	81,50
IPMGF	5.933	0.110	32	15,50

Considering coordination by the interaction balance method, steepest descent and simplex algorithms have been used to solve the optimisation problem of the coordinator for global feedback and local feedback respectively. The fact that two different optimisation algorithms have been used should be taken into account when comparing the convergence properties of these two methods. Using local feedback, the infimal level requires an average of eight iterations before converging to its solution. This implies that, on average, eight times more on-line computation time is required for the overall optimisation problem than required using global feedback. However, in this particular example, local feedback has resulted in a better performance. On the other hand, local feedback is not as robust as global feedback because of stability problems occuring in the local decision problems (Roberts,1983).

The interaction prediction method with global feedback has given the best performance (5.933) among the results obtain in this particular example. However, in general this coordination method requires solvability and feasibility analysis for constrained optimisation problems which may be very difficult in many practical cases.

5) CONCLUSIONS

This paper has demonstrated that the suggested coordination methods can be used for distributed on-line optimising control within a real time environment. Problems relating to system transient and local decision unit asychronisation, and their effects on system stability and convergence, have been investigated and methods for dealing with these problems are suggested. The convergency, accuracy and robustness properties of each coordination method have been discussed. Engineering judgement is needed to choose a particular coordination method with feedback to best suit the class of problem concerned. Further research should look into other coordination methods with feedback and to develop stablising control techniques for the decision units under asynchronous operation, so as to reduce the waiting time in the infimal level and the information exchange between the two levels.

ACKNOWLEDGEMENT

The authors are grateful to the Science and Engineering Research Council which supports the work reported in this paper.

REFERENCES

1) Findeisen, W., F. N. Bailey, M. Brdys, K. Malinowski, P. Tatjewski, and A. Wozniak (1980). Control and Coordination in Hierarchical Systems. Wiley, London.
2) Findeisen, W., M. Brdys, K. Malinowski, P. Tatjewski and A. Wozniak (1978). On-Line Hierarchical Control for Steady State Systems. IEEE Trans. on Auto Control, AC-23, 189-209.
3) ISIS Users Manual (1980).
4) Madnick, S. E. and J. J. Donovan (1974). Operating Systems. Mcgraw Hill.
5) Mesarovic, M. D., D. Macho and Y. Takahara (1970). Theory of Hierarchical, Multilevel Systems. Academic Press.
6) Roberts, P. D., J. E. Ellis, C. W. Li, F. Q. Shao, P. Zheng and B. W. Wan (1983). Algorithms for Hierarchical Steady State Optimisation of Industrial Processes. IFAC/IFORS LSSTA, 425-431.
7) Shao, F. Q. and P. D. Roberts (1983). A Price Correction Mechanism with Global Feedback for Hierarchical Control of Steady State Systems. Large Scale Systems, 4, 67-80.

Appendix A : Local Decision Feasibility Set under Constraints

Due to the lack of programming memory, mathematical functions and subroutines available in the micro-computers (local decision units), standard library numerical optimisation algorithms cannot be used to solve the local decision problems. Hence, each local optimisation problem has to be solved analytically using the theory of extrema.

Starting with the unconstrained solution, the performance index of each local decision problem is differentiated with respect to the manipulated inputs (controls) to obtain the optimal controls in terms of interaction variables. Then these control values are substituted in the constraint sets of the decision unit to search for the active constraints which bound the unconstrained solution. Then the active constraints are plotted with the interaction variables to indicate the bounded unconstrained solution. Once the unconstrained solution region has been found, the active constraints are subsequently used as new constraint boundaries for the generation of new solution regions next to the unconstrained solution region. Defining a new solution region by one of the active constraint boundaries previously obtained, the procedure is repeated and a new bounded solution region is found. The complete procedure is then repeated until no further new solution regions are formed.

This analytical approach is used to solve the local decision optimisation problems for each coordination strategy. Decision unit solution regions using the interaction balance method with global feedback are shown in fig.3.

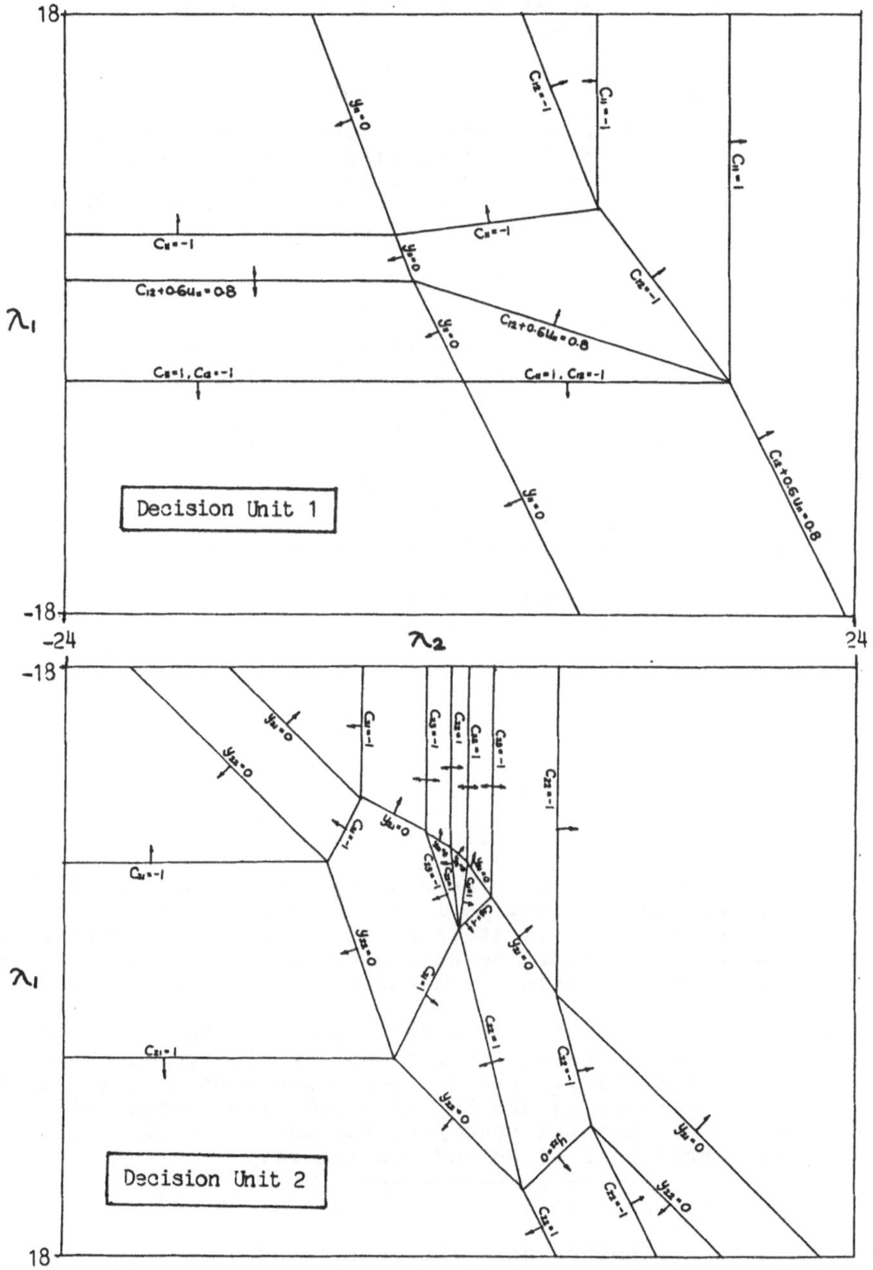

The drawback of this analytical method for solving the local decision optimisation problem is restricted to simple low dimensional problems because it is impossible to visualise greater than two dimensional solution regions.

ADA+ FOR SPECIFYING SYSTEMS

S.J. Goldsack.

Department of Computing,
Imperial College of Science and Technology,
London, England.

This paper presents a short summary of part of the work of a study group, supported by the Commission of the European Communities, into the use of Ada for system specification. The work was carried out by the team of members from industry and the universities, listed below*.

1. Introduction

It is widely accepted that specification of the functional requirements for a large information processing system is an essential precursor of its implementation. There are, however, different views of the role of the specification. Some writers consider that the purpose is to ensure that the customer, the eventual user of the system, is able to define his needs, while others consider that the specification is required to ensure that the system designer starts from an unambiguous definition of his objectives.

Consider the project development model in the figure below.

```
      written               proven consistency         program
      specification     ------------------------->     system
              |                                           |
              |                                           |
    modelling |                                           | algorithm
              |                                           |
              |                                           |
      actual                                           running
      requirements      ------------------------->     system
                              validation by test
```

It illustrates the view taken here of a specification as a model of the world, rigorously defined, so that all the subsequent steps of design and implementation are, at least in principle, capable of formal verification. It must be expressed in a language whose semantics are unambiguously defined, to constitute a formal linguistic system.

For many reasons, it is considered essential that the model should be a faithful image of the required real world system, both in structure and in functionality. For a clear discussion of this topic, see Jackson's book, describing his System Development methodology(1). The language used to express the specification should be able to model the objects in the world in a clear and concise way.

\+ Ada is a registered trade mark of the US Government, AJPO.

* The authors and contributers were the following:
B. Cherrie, N. Davis, B. Dacker, S.J. Goldsack, V. Haase, H. Halling, J. Jarray, J. Jones, G. Koch, J. Kramer, J. Ludewig, A. McGettrick, K. Mander, J.Page, I.Pyle, S. Savoysky, G. Winterstein.

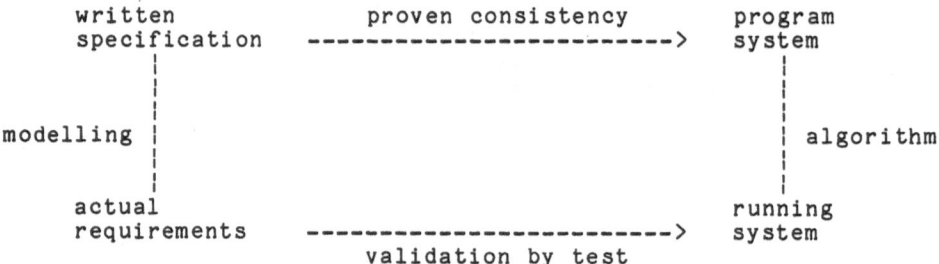

The work described here has sought to answer the question of whether Ada itself can serve the purpose, and if not, then what is needed to supplement it.

2. The rationale for the use of Ada

There are a number of arguments for using the language in which the programs are to be written as early as possible in the development process. These are listed in full in the report delivered to the CEC. For the present summary, the most important are the following:

- The process or creating the programs to implement a required system, consists typically of a stepwise progress through a series of transformations, each step being regarded as provably consistent with the previous. It is much easier to conduct this process if there are no abrupt changes in the representation at each stage.

- The need to hold several langauges in mind at the same time adds to the difficulties of the designer/programmer.

- Programs and specifications can benefit from the same set of tools in the programming support environment.

- Ada has, or will have, a formal definition, so that it can serve as a suitable formal linguistic system.

Programs are models, and Ada was designed to permit the description of system structures in an elegant and well defined way. Later sections examine the requirements of a specification language, and how well Ada meets the needs.

3. The rationale for being different

Before proceeding to analyse the use of Ada for specifying, consider first some counter-arguments, suggesting that a programming language should **not** be used in this way. The main points are these:

- The processes of programming and specifying are different, and should not be confused in the mind of the specifier. There should be no danger that the designer thinks as a programmer while specifying the system, leading to premature design decisions. The safest way to avoid that is to use a distinct language.

- A programming language intended for writing programs to execute on traditional Von Neuman machines, is deliberately intended for expressing how the prescribed effects can be obtained on such machines. Specification is concerned with what is required, not with how it is to be achieved. Thus, the objectives of the languages for specification and programming are quite different.

The results of the work show that both arguments have some validity, but that it is not impossible to reconcile the two views.

4. What must a specification language provide?

An object forming part of a system is described by its name, together with a set of variables, whose values define the instantaneous state of the object (the state variables) and a set of laws determining how those values are related, and how they change in time as a result, both of internal dynamical laws and of external influences from other objects in the system.

A language for specification must therefore provide the following features.

Abstract data definitions
It must have data definition features sufficiently powerful to express the model of the system state in all problem areas within its intended scope.

Procedural abstraction
It must have facilities to define the behaviour of the elements of the model in terms of the laws governing the transformations of the state variables.

Abstract objects
It must have structuring features, to define the boundaries of the components, and the interfaces through which they interact with others.

Concurrency
It must model the 'co-existence' of the separate elements, each obeying its own laws concurrently with others in the system.

Modelling of interactions between objects
It must provide means of describing the interactions between elements, and their contribution to the system laws.

Exceptions
At least for real-time applications, it must provide facilities to define the consequences or error, malfunction, or unexpected data conditions.

Low level details
Extending the language to include low level specifications, it may be necessary to define detailed sequences of actions, and to specify hardware structures.

Modularity
The designer must be able to concern himself with the system a part at a time, and construct his fully specified system by assembling piecewise specifications.

Generality
Specification of units capable of filling the needs of different systems, or different places in the present system, by paramaterisation, are desirable.

Modelling of interaction with environment
It must provide means of describing the interfaces and interactions between an entity and its environment. This of course, includes the human interfaces.

Performance and reliability
It must provide means for specifying response requirements across specified interfaces, size estimates, failure rates of specified components.

are addressed to this problem.

Packaging

Ada packages provide units for modelling objects. By packaging together the descriptions of the state variables and of the operations which determine the laws of the sub-system,units are formed which model objects in the sense discussed above.

Concurrency

A system typically contains many sub-systems, each of which can be modelled as an object. However, each has an instantaneous state of its own, which develops in time according to its own laws. The objects in the problem domain co-exist; their models exist concurrently within the overall system model. Ada provides tasks to represent such objects.

A collection of similar concurrently active objects may be derived from a common task type definition, and arrays of tasks can model collections of similar active objects.

Object interaction

Interactions between objects in the world can be recognised in two forms.

1. Message Passing.
2. Direct Interactions.

Message passing may have several forms, summarized in the following, together with the Ada features which can be used to model it.

- A message may be passed in a handshake between willing partners.	(rendezvous)
- A message may be transmitted at the initiative of the sender who may wait to ensure its receipt, or to receive a reply.	(entry call)
- A message may be inserted into the data space of the reciever by the sender.	(procedure call)
- A message may be obtained by the receiver by inspecting the data in the space of the sender.	(function or procedure call)
- A message may be entrusted by a sender to an agent who will ensure its delivery.	(messenger task)
- A message may be placed in a shared message box, to be taken by the receiver when he wishes. The box may have capacity for one or more messages.	(buffer)
- A message may be posted on a notice board for all interested parties to inspect.	(global data)
- A message may be broadcast to a selected class of recievers.	(?)

If objects interact in a more symmetrical mode, such as the force between a pair of objects, then Ada fails to provide primitive mechanisms to define the interaction. However, a generic unit can be provided implementing the desired effects.

Exceptions

Ada's exception mechanism enables the specification to recognise the existence of potential error conditions, but does not provide means of specifying the circumstances in which the

Constraints
It must provide means for expressing operational and development restrictions, time scales, and cost restrictions.

As a language defining logical behaviour and structure, Ada provides no features for the last two purposes. It provides, however, features for modelling which satisfy many of the other requirements. In the following, these subjects are reviewed briefly, to discover to what extent Ada is inadequate. In a few cases it may be found that shortcomings exist at the primitive level, but that Ada has obvious and simple ways of extending its own power by the use of its programming features to 'plug the gap'.

5. How well can Ada serve as a specification language?

In this section the facilities provided by Ada are reviewed, in relation to the above list of requirements. It will be found that many of the needs are satisfied, but that others cannot be met by primitve constructs of Ada. In this case, it may be possible to use Ada's own ability to define new concepts which are not primitive in the language, but which may be defined by writing passages of Ada program to define their operation.

Modelling the system state with data
Ada has a comprehensive system of data typing facilities, which are sufficiently flexible to model any set of data capable of computer representation. The use of enumeration types and user-defined numeric types make the definitions of data machine independent. Structuring facilities enable data models of complex objects to be defined.
In spite of Ada's reasonably complete facilities for type definitions, there are a few shortcomings which are discussed in the full report. The most important are the following:

- Iterated data may be represented by arrays. These have implications for the method of storage, so are insufficiently abstract.
- Unions of types are awkward to define.
- Sets are commonly needed; they must be implemented using non primitive Ada features.
- Relationships between objects may be handled by Ada access types, but these are also too much concerned with methods of implementation.

Abstract operations
By defining a procedure, a function, a task or a block, composite operations may be treated as a single abstract operation. The name and external interations of the object define how it may be used, but what it actually does can only be determined by inspecting the code:

```
procedure SORT (L: LIST);
```

The semantics of this procedure is not given by its definition. This is the main and crucial weakness of Ada for specification. It provides a syntactic frame, modelling system structure; it supports the definition of new concepts by definition of new named objects, but only defining the semantics by the procedural code. Most of the existing work on languages for annotating Ada exception is raised, nor the actions to be taken on the raising of the exception, except by giving sample implementations. This is similar to the problem of specifying procedures in an abstract way.

Low level details

Ada provides an effective sub-language for specifying representation of data, and other low level details.

The value of this feature has been illustrated by a number of examples from different domains; perhaps the most interesting is the re-working the definition of the Ethernet protocol, previously defined using Pascal.

There are three conceptual layers, the client layer generates messages, the data link layer maps them onto their physical representation, and the physical layer provides the interface to the network hardware.

The "layer" idea is naturally captured in Ada with **with** clauses, controlling the visibility of interfaces between separately compiled objects.

Modularity and generality

The facilities for separate compilation and for the definition of generic units go a long way to satisfy these requirements.

6. Conclusions

The complete report covers additional aspects, such as the role of specifications in software engineering, the tools required in the APSE for the support of specification, and methods of annotating Ada programs to extend the power of specification. This summary has addressed topics in the central area of the use of Ada itself for specifying.

The main conclusions are following:

Except for the actual behavioural details of objects, Ada provides primitive, (and so formally defined) features for modelling objects and their interactions; a few shortcomings can be handled by defining higher level concepts in Ada itself, using the generic features.

A number of systems of annotation are available for extending the power of the language to include the behavioural semantics.

Possibly the definition of language extensions to provide the non primitive features as primitives of an Ada-related specification language, whose implementation in Ada can be easily created, may be the right way forward. 'Dad', an Ada related language developed by S.Savoysky as part of the project, is an example of such a language.

When it is the details of machine structure and procedural behaviour of devices that are to be specified, Ada provides a powerful means of expression.

References:

1) M Jackson, System Development. Prentice Hall International, 1983.

ABSTRACT DATA TYPES IN INDUSTRIAL REAL TIME BASIC

Dr. G. M. Bull,
Reader in Computer Science,
School of Information Sciences,
The Hatfield Polytechnic,
Hatfield, Herts. England.

Abstract

The importance of abstract types is stressed. A number of languages which provide abstract data type facilities are identified and Modula and Ada are examined in more detail. Industrial Real Time Basic is described together with two radically different approaches which could be used to implement abstract data types within the language.

Keywords

Abstract Data Types, Encapsulation, Information Hiding, Concurrency.

Introduction

The importance of abstract data types (also called data encapsulation and information hiding) has been widely recognised (1). It is an important concept both as a program design tool (2) and as a program structuring tool.

An abstract data type is a data type which is defined only in terms of the operations which may be performed on objects of that type, without any indication of how such objects are represented. The important point is that the way the data structure is realised is hidden. What is presented to the user is typically a number of procedures and functions which provide the operations. As an example, a stack may be defined as an abstract data type with the operations push, pop, isempty, isfull, and clear provided for the user of the stack with appropriate syntax and semantics for these operations as follows:

clear - the state of the stack is set to 'empty'.

push (item) - if the state of the stack is 'full' then no action takes place else item is placed on the 'top' of the stack. If there was an item on the 'top' of the stack prior to the push operation, that item shall be the predecessor of the current item on the 'top' of the stack. If as a result of the push operation there are 100 items on the stack, the state of the stack is set to 'full'.

pop (item) - if the state of the stack is 'empty' then the value of item is undefined and the stack remains 'empty' else item is assigned the current value on the 'top' of the stack and that item is removed from the stack. If as a result of the pop operation there is at least one item on the stack, then the 'top' of the stack becomes the predecessor of the item just removed.

isfull - if the state of the stack is 'full' this returns the value 'true' else 'false' isempty - if the state of the stack is 'empty' this returns the value 'true' else 'false'.

Language Features

A number of languages include facilities for creating abstract data types. Languages such as Ada (4), Basic 7 (5), CLU (6), Modula (7), and Path Pascal (8) all provide such facilities. Those languages which are purely sequential in nature provide the facility by a grouping of procedures and variables accessible only to those procedures.

For languages which provide for concurrency, there are basically two different approaches which can be taken - Monitors (9) (or some variant of them) or dynamic structures based on independent processes.

The first of these approaches follows the method used by sequential languages and groups together a set of procedures with variables accessible only to these procedures. Since any number of processes may be executing concurrently and calling procedures within the monitor, it is necessary to incorporate some mechanism for ensuring that only a single process is active within the monitor at any one time in order to ensure that the shared data is accessed in a controlled way. Monitors also allow a process to relinquish control of the monitor whilst the process waits for a resource to become free. Since this resource is made available by another process entering the monitor to release it, a rule must be made about which process has control of the monitor at the point where the resource is made available.

An example of a language which takes this approach is Modula. An interface module is an encapsulation mechanism which allows constants, types and procedures to be defined (in a DEFINE list) and used by processes in other modules. The stack abstract data type can be realised in Modula in the following way.

```
INTERFACE MODULE stack;
   USE element;
   DEFINE push, pop, clear, isfree, isempty;
   CONST
        full  = 100;
        empty = 0;
   TYPE
        st = RECORD
                s   : ARRAY [1:full] OF element;
               ptr  :        0..full
              END;
   VAR
        stk :  st ;
   PROCEDURE clear
   BEGIN
        stk.ptr:=empty
   END clear;
   PROCEDURE push (CONST item : element)
   BEGIN
        WITH
            stk
        DO
            IF
               ptr < full
            THEN
               inc(ptr)
               s[ptr]:=item
            END
        END
   END push
```

```
    PROCEDURE pop (VAR item : element);
    BEGIN
         WITH
             stk
         DO
             IF
                 ptr=empty
             THEN
                 item::=0
             ELSE
                 item::=s[ptr];
                 dec(ptr)
             END
         END
    END pop;
    PROCEDURE isfull : boolean
    BEGIN
         isfull:=(stk.ptr=full)
    END isfull;
    PROCEDURE isempty : boolean;
    BEGIN
         isempty:=(stk.ptr=empty)
    END isempty;
 END stack;
```

The second approach is to have an active element to realise the abstract data type. Rather than have a set of procedures which are called from any number of processes, the abstract data type is provided by a separate process with which other processes communicate in order to invoke the operations. Such an approach is only possible with a language which provides concurrency and is therefore not appropriate for a sequential language. One language which provides an active process which can be used to realise an abstract data type is Ada. Using the rendezvous mechanism we could implement the stack abstract data type as follows:

```
task stack is
     entry push (item : in element);
     entry pop (item : out element);
     entry clear;
     entry isfull  (full : out boolean);
     entry isempty (empty : out boolean);
end;
task body stack is
     full : constant :=100;
     empty : constant : = Ø;
     type st is
          record
              s : array (1..full) of element;
              ptr : integer range empty .. full
          end record;
     stk : st;
begin
  loop
     select
           accept clear do
               stk.ptr :=empty
           end clear
     or
           accept push (item : in element) do
                 if
                     stk.ptr < full
                 then
                     inc (stk.ptr);
                     stk.s(ptr) := item;
```

```
                end if;
            end push;-
        or
            accept pop(item:out element) do
                if
                  stk.ptr = empty
                then
                  item:=Ø;
                else
                  item:= stk.s(s.ptr);
                  dec(s.ptr);
                end if;
            end pop;
        or
            accept isfull  (full:out boolean) do
                full:(stk.ptr = full);
            end isfull ;
        or
            accept isempty (empty:out boolean) do
                empty:=(stk.ptr=empty);
            end isempty;
        end select;
      end loop;
    end stack;
```

It is also possible to create a stack in Ada using a package and procedures which are called much like the **Modula example.**

Industrial Real Time Basic (IRTB) (10)

IRTB is intended for real-time applications that can be described in terms of a number of concurrent activities which are largely independent and asynchronous, but which can communicate and synchronise. The program for such an application does not have an overall thread of control. The program must be capable of running indefinitely - it is not a problem-solving program that starts, operates on some data to produce some output, and is then finished.

An IRTB program is divided into a number of concurrent single-thread activities which co-operate to achieve the overall objective of the application.

Statements are provided to start concurrent activities, and to enable them to respond to internally or externally generated events. Once started, concurrent activities execute in parallel (at least conceptually).

Each activity is a program module that communicates with its environment through three types of 'ports':

(a) process I/O ports that communicate with plant interface hardware,

(b) message ports for synchronisation and communication between concurrent activities, and

(c) shared-data ports for access to data structures outside the individual activities, for example data in a real-time database system.

The executable code for an activity is written in an extended form of Basic. Activities have the usual facilities to access system resources such as files, the computer console and subprograms.

In order to define concurrent activities a new language structure for Basic, the 'parallel section', has been introduced. A parallel section is a program unit in which all variables, internal functions, channel numbers, data-statements, etc., are local to the section. Execution of a parallel section constitutes a concurrent activity.

The concept of a data structure has been introduced to define the interface presented by the three types of ports. A data structure is similar to a record in Pascal for example, in that it is an ordered list of the data types numeric or string, scalar or array. A data structure is an abstract structure in the sense that it does not define data storage and is not associated with particular variables or shared data sections - it is a 'template' that defines the structure of data transferred through a port.

The use of data structures allows a language processor to check the consistency of statements transmitting data through message, shared data and process-I/O ports. It also allows the checking of compatability between interfaces of communicating activities, particularly when they are in separately compiled program units. For large systems, and especially in the distributed case, the declarations for shared data, message paths and process-I/O paths will be in a separate global section that becomes the 'system definition'. The concept of a data structure will facilitate consistency checking by the language processor between the global section and the code for the individual activities.

In statements and out statements are used to perform I/O to plant interface equipment. Process I/O is distinguished from conventional I/O since they are semantically and functionally different.

Get-statements and put-statements are used to access data that exists independently of the executing activities. The view of the shared data from the point of view of an individual activity is declared in data-port declarations. A data-port declaration defines the name of a data port and the structure of the data accessible through it.

A message mechanism is provided for synchronising concurrent activities, and for passing data at the point of synchronisation. Message communication is a subset of the Ada 'rendezvous' mechanism.

Normally two activities participate in a message transfer, the message path being the logical connection between a 'send' port in one activity and a 'receive' port in the other. When both activities reach corresponding send-statements and receive-statements, data is moved from the sending activity to simple variables and/or arrays in the receiving activity. The transmission of the data is an indivisible operation.

A single receive port in one activity can be connected to many send ports in other activities. Because of the synchronising constraints and the indivisibility of message data transfer, this configuration can be used to implement a Monitor for resource management. The sending activities will be forced to queue, the data being accepted from each in turn, allowing that queued activity to proceed. An example is a logging printer activity that accepts data-log information from a number of other activities, with the requirement that the printing of the data from each activity must be completed without interruption before the next set of data is accepted.

Broadcasting of messages from one send port to many receive ports is not permitted. Such a configuration would lead to non-deterministic behaviour of the program since it could not be known how many receive ports were supposed to receive data. If the message were received only by those activities that had reached receive statements when the send statement is executed, timing variations could cause some activities to miss the information.

Abstract Data Types in IRTB

In order to illustrate how abstract data types may be realised in IRTB, as before, a stack is used as the example.

Each abstract data type is mapped to a parallel activity which responds to messages sent to it requesting the operations required. The coding of the stack example might take the following form:

```
PARACT stack
    !Use an array for storage, max of 100 items
    DIM S(1 TO 100)

    !Give names to the stack ranges
    LET empty=0
    LET full=100

    !Loop forever, processing requests for operations
    DO
      SELECT
        CASE PORT clear
          RECEIVE FROM clear
          LET ptr=empty
        CASE PORT push
          RECEIVE FROM push TO item
          !Make sure stack is not full, otherwise discard item
          IF PTR<full THEN
            LET ptr = ptr + 1
            LET s(ptr) = item
          END IF
        CASE PORT pop
            !Make sure stack is not empty
            IF ptr = empty THEN
              LET item = Ø
            ELSE
              LET item = s(ptr)
              LET ptr = ptr - 1
            END IF
            SEND TO pop FROM item
        CASE PORT isempty
          IF ptr = empty THEN
            LET reply$ = "true"
          ELSE
            LET reply$ = "false"
          END IF
          SEND TO isempty FROM reply$
        CASE PORT isfull
          IF ptr = full THEN
            LET reply$ = "true"
          ELSE
            LET reply$ = "false"
          END IF
          SEND TO isfull FROM reply$
        END SELECT
      LOOP
    END PARACT
```

Since at most one message port is responded to on one pass through the loop, synchronisation is provided implicitly.

A process wishing to push an item would execute

```
SEND TO push FROM item.
```

A process wishing to pop an item would execute

```
RECEIVE FROM pop TO item.
```

Similary calls on the other operations would be of the form

```
SEND TO clear
RECEIVE FROM isfull TO full$
RECEIVE FROM isempty TO empty$.
```

The above solution shows how abstract data types may be realised in IRTB. However, it may be that Basic itself could benefit from the introductions of such a facility. Basic as proposed in the draft ANSI standard already has subprograms and functions with parameters; what is required is an encapsulation mechanism which makes certain data global to a number of subprograms and functions and which makes certain subprograms and functions visible to the rest of the program.

A new construct, called a Monitor, is introduced to provide the functionality required. As before the stack example is used as an illustration.

```
MONITOR stack
   EXPORT push (item), pop(item), init, isfull$, isempty$
   LOCAL ptr, stack (1 TO 100)
   !Local suprograms.
   SUB inc (x)
     LET x = x + 1
   END SUB
   SUB dec(x)
     LET x = x - 1
   END SUB
   SUB push(item)
     !Make sure stack is not full
     IF ptr < 100 THEN
        CALL inc(ptr)
        LET stack (ptr) = item
     END IF
   END SUB
   SUB pop(item)
     !Make sure stack is not empty
     IF ptr = 0 THEN
       LET item = Ø
     ELSE
       LET item = stack(ptr)
       CALL dec (ptr)
     END IF
   END SUB
   SUB init
     LET ptr = 0
   END SUB
   FUNCTION isempty$
     IF ptr = 0 THEN
       LET isempty$ = "true"
     ELSE
       LET isempty$ = "false"
     END IF
   END FUNCTION
   FUNCTION isfull$
     IF ptr = 100 THEN
```

```
         LET isfull$ = "true"
      ELSE
         LET isfull$ = "false"
      END IF
   END FUNCTION
END MONITOR
```

Such a structure is already present in one implementaton of Basic (5).

If such a structure were introduced into Basic it would automatically be available in IRTB since everything that is in Basic is in IRTB. In view of this, it would be necessary for the semantics of a Monitor to be such that only one parallel activity could be executing code in a given Monitor at a time.

Conclusions

Abstract data types are important both as a structuring tool and as a design tool. As such it is an important feature of any language. The method used by most languages has been the monitor, which is suitable for both sequential and concurrent languages. An extension of Basic (and IRTB) is suggested which provides a monitor facility. It is shown that without changes to IRTB, the features provided allow the creation of abstract data types through the offices of an active process. This facility enables a design based on abstract data types to be directly coded in IRTB with ease.

References

(1) Guttag J V, Horowitz E, and Musser D R
Abstract data types and software validation
CACM Vol 21 No 12 December 1978

(2) Mitchell R J
Program design - a practical approach
In E Knuth and E J Neuhold (Eds)
Specification and Design of Software Systems
Lecture notes in Computer Science, No 152
Springer Verlag 1983

(3) Welsh J and Mckeag M
Structured system programming
Prentice Hall (324 pages) 1980

(4) Formal Definition of the Ada programming language
U S Department of Defense 1982

(5) Bull G M and Garland S J
Specification for Dartmouth Basic Version 7
TM112 Dartmouth College 1980

(6) Liskov B H, Snyder A, Atkinson R and Schaffert C
Abstraction Mechanisms in CLU
CACM Vol 20 No 8 August 1977.

(7) Wirth N
Modula: a language for modular multiprogramming
Software Practice and Experience Vol 7 P3-35, 1977

(8) Kolstad R B and Campbell R H
Path Pascal user manual
Sigplan Notices Vol 15 No 9 September 1980

(9) Hoare C A R
Monitors - an operating system structuring concept
CACM Vol 17 No 10 October 1974

(10) Bull G M and Lewis A
Industrial Real Time Basic
Software Practice and Experience Vol 13, p 1075 - 1092 , 1983

(11) American National Standards Institute
Draft Standard for Basic X3J2/82-17, October 1982.

TWO CAD SYSTEMS OF LARGE SCALE CONTROL STRUCTURES

L.Orăşanu, R.Gaşpar, and F.G.Filip
Central Institute for Management and Informatics
Bd.Miciurin nr.8-10, 71316, Bucharest, Romania

ABSTRACT

The paper presents some results in CAD of multilevel control laws for large scale interconnected systems by using hierarchical optimization algorithms (MESINT - software package) as well as for optimizing the physical computer - based control structures in order to minimize the"processing costs" (OPTCON - system).

1. INTRODUCTION

In recent years, hierarchical approach have been extensively used to simplify the design and implementation of decentralized-coordinated control schemes for interconnected subsystems. The multilevel optimization methods convert a large problem to one requiring little storage but generally implying longer computation time [11] . Some extensions of the existing hierarchical computation methods of control laws in the case of interconnected (sub)systems have been presented in [7], [8]. A brief presentation of their implementation in a software package (MESINT) and some numerical results are given in section 2.

The design of the physical control structure, including many control units disposed on different levels, is a complex problem and generally consists in estimating the number, the models, and the locations respectively of the devices (equipments) the control system is composed of. The hardware configuration has to be designed so that to minimize the overall cost of the system, that is to make the system more performant by quicker response time and higher reliability. The OPTCON software system [9] can automatically take over such phases in the design process over hierarchical control structure as : the terminals subsystem design, the best location of the central computer, the design of the local control/concentrator units, the multidrop lines lay-out. In section 3 the use of some OPTCON components is discussed in connection with optimal lay-out of processing equipment.

2. MESINT - THE SYNTHESIS METHODS OF CONTROL LAWS FOR INTERCONNECTED SYSTEMS

Consider a large scale system, composed of N linear interconnected subsystems, described by the equations :

$$\dot{\underline{x}}^i(t) = A^i\underline{x}^i(t) + B^i\underline{m}^i(t) + G^i\underline{u}^i(t) + B^i_p\underline{u}^i_p \qquad i = \overline{1,N} \qquad (1)$$

$$\underline{x}^i(t_0) = \underline{x}^i_0$$

$$\underline{y}^i(t) = C^i\underline{x}^i(t) \qquad (2)$$

$$\underline{z}^i(t) = D^i\underline{x}^i(t) + E^i\underline{m}^i(t) \qquad (3)$$

$$\underline{u}^i(t) = \sum_{j=1}^{N} L^{ij}\underline{z}^j(t) = \sum_{j=1}^{N}\left[H^{ij}_x\underline{x}^j(t) + H^{ij}_m\underline{m}^j(t)\right] \qquad (4)$$

where: $\underline{x}^i$, $\underline{m}^i$, $\underline{y}^i$, $\underline{u}^i_p$, $\underline{u}^i$, $\underline{z}^i$ are the state, control, output, known disturbance, input and output interconnection vectors (with dimensions: n^i_x, n^i_m, n^i_y, n^i_x, n^i_p, n^i_u and n^i_z respectively) of the i-th subsystem. A^i, B^i, C^i, H^{ij}, D^i and E^i are matrices of appropriate dimensions (time variable in the general case). Note that $\underline{z}^i$ depends on $\underline{x}^i$ as well as on $\underline{m}^i$.

When $\underline{y}_r$ is the reference vector, the optimal tracking problem is to find the control history $\overset{*}{\underline{m}}(t)$, $t \in [\, t_0\ t_f)$, which minimizes the following cost functional :

$$J = \sum_{i=1}^{N} J^i = \frac{1}{2}\sum_{i=1}^{N}\left\{\|\underline{e}^i(t_f)\|^2_{Q^i_f} + \int_{t_0}^{t_f}\left[\|\underline{e}^i(t)\|_{Q^i} + \|\underline{m}^i(t)\|^2_{R^i}\right]dt\right\} \qquad (5)$$

where: $\underline{e}^i(t) = \underline{y}_r(t) - \underline{y}(t)$; Q^i, $Q^i_f \geqslant 0$ and $R^i > 0$.

In order to solve the dynamic optimization problem defined by (1) ÷ (5) the following Lagrangean is defined :

$$L = \sum_{i=1}^{N}\Big\{J^i + \int_{t_0}^{t_f}\Big[\langle \underline{p}^i, A^i\underline{x}^i + B^i\underline{m}^i + G^i\underline{u}^i + B_p\underline{u}^i_p - \dot{\underline{x}}^i \rangle$$

$$+ <\underline{\varrho}^i,\underline{u}^i> - \sum_{j\neq i} <\underline{\varrho}^j, H^{ji}\underline{v}^i> \Big] dt , H = \begin{bmatrix} H_x & 0 \\ 0 & H_m \end{bmatrix} \underline{v} = \begin{bmatrix} \underline{x} \\ \underline{m} \end{bmatrix}$$

As an extension of existing results [11], [2], an hierarchical optimization algorithm is presented in [7]. When the convergence of the iterative algorithm takes place, the control of the i-th subsystem is :

$$\underline{m}^i(t) = - F^i(t)\underline{x}^i(t) + \underline{g}^i(\underline{x}_o,t) \qquad (6)$$

$$F^i = (R^i)^{-1} B^{iT} P^i$$

The open loop part of the control is :

$$\underline{g}^i = (R^i)^{-1} \Big[- B^{iT} \overset{*}{\underline{s}}{}^i + \sum_{j\neq i} (H^{ji})^T \underline{\varrho}^j \Big]$$

where the symmetric matrix $\overset{*}{P}(t)$ and the vector $\overset{*}{\underline{s}}{}^i$ are the solutions to the equations :

$$\dot{P}^i = - P^iA^i - A^{iT}P^i + P^iB^i(R^i)^{-1}B^{iT}P^i - C^{iT}Q^iC^i$$

$$P^i(t_f) = \left[C^i(t_f)\right]^T Q^i_f C^i(t_f)$$

$$\dot{\underline{s}}^i = - \left[A^i - B^i(R^i)^{-1}B^{iT}P^i\right]^T\underline{s}^i + C^{iT}Q^i(\underline{y}^i_r - B^i_p\underline{u}^i_p)$$

$$- P^iB^i(R^i)^{-1} \sum_{j=i}^{N} \left[H^{ji}_x + H^{ji}_m\right]^T \underline{\varrho}^j + P^iG^i\underline{u}^i + P^iB^i_p\underline{u}^i_p$$

The control history (6) which is well suited for finite time problems in a batch environment, where $\underline{y}_r$ is a known program, may be implemented in a decentralized microprocessor-based structure (Fig.1).

In the case of known, constant references and disturbances $\overline{\underline{y}}_r$ and $\overline{\underline{u}}_p$ respectively, the optimal control law is :

$$\underline{m}(t) = - \overline{F}\underline{x}(t) + \overline{F}_r\underline{y}_r - \overline{F}_p\underline{u}_p \qquad (7)$$

where $\overline{F}$, $\overline{F}_r$ and $\overline{F}_p$ are time invariant matrices.

The procedure of identifying the columns of $\overline{F}$ matrix by using the values $\underline{m}(t = t_o)$ obtained for a set of n_x orthogonal initial conditions in a long but finite time problem [11], may be extended to calculate the matrices $\overline{F}_r$ and $\overline{F}_p$, which are not dependent on the initial state or on the values of $\underline{y}_r$ or $\underline{u}_p$ [3]. Thus, the hierarchical optimization algorithm is applied with the "augmented initial condition"; $\underline{x}^{[k]}(0) = \underline{1}_k = [0...1^k...0]^T$; $\overline{\underline{y}}_r{}^{[k]} = \underline{0}$; $\overline{\underline{u}}_p{}^{[k]} = \underline{0}$. The resulting values $\underline{m}^{[k]}(0)$ can be used to calculate the k-th column of $\overline{F}$ matrix (F_k) :

$$\underline{m}^{[k]}(0) = -F.\underline{1}_k = -F_k \; ; \quad k = \overline{1,n_x} \; ; \quad n_x = \sum_{i=1}^{N} n_x^i \tag{8}$$

Similarly, by using the initial conditions: $\underline{x}(0) = \underline{0}$, $\underline{y}_r = \underline{1}_k$, $\underline{u}_p = \underline{0}$ (where $k = \overline{1,n_y}$), and $\underline{x}(0) = \underline{0}$, $\underline{y}_r = \underline{0}$, $\underline{u}_p = \underline{1}_k$ (where $k = \overline{1,n_x}$), we "identify" column by column the matrices $\overline{F}_r$ and $\overline{F}_p$ respectively.

In, the case of a decentralized implementation (Fig.1), the control law (7) can be written as :

$$\underline{m}^i(t) = -\overline{F}_L^i\, \underline{x}^i(t) - \overline{F}_C^i\, \underline{x}(t) + \overline{F}_r^{ii}\, \overline{\underline{y}}_r^i - \overline{F}_p^{ii}\, \overline{\underline{u}}_p^i$$

$$+ \sum_{j \neq i}^{N} \overline{F}_r^{ij}\, \overline{\underline{y}}_r^i - \overline{F}_p^{ij}\, \overline{\underline{u}}_p^j \qquad i = \overline{1,N} \tag{9}$$

The matrix $\overline{F}_C = \mathrm{col}\left[\overline{F}_C^i\right]$ is obtained from $\overline{F}$ by substracting bloc diag $\left[\overline{F}_L^i\right]$, where $\overline{F}_L^i = (R^i)^{-1} B^{iT} P^i$.

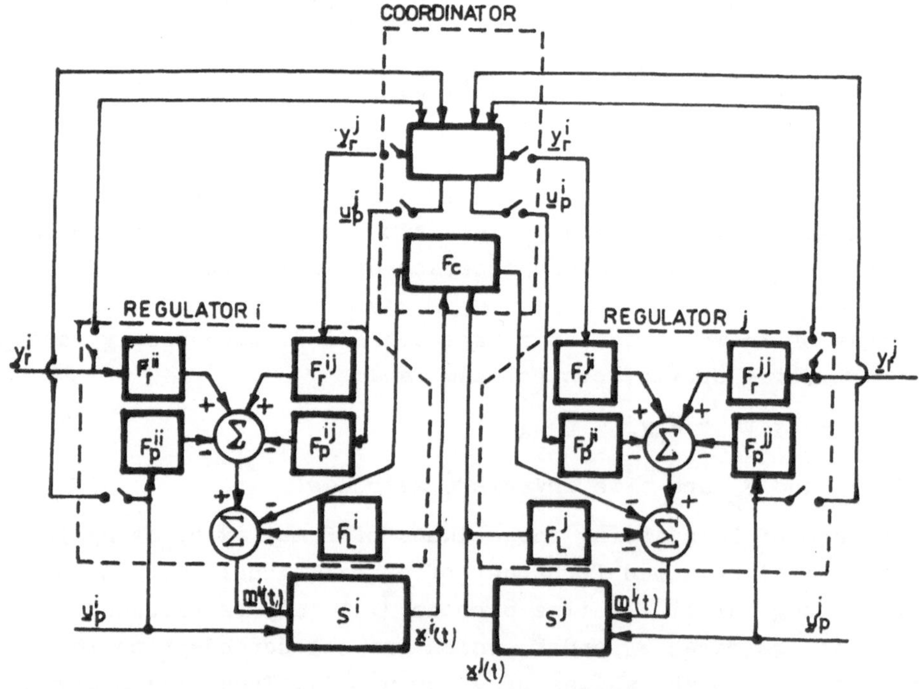

Fig.1. Two level control structure

A conversational software package, MESINT, was produced based on the above described methods. There are other four cases implemented in MESINT such as: regulator problems, tracking problems with persistent nonconstant disturbances and references by using the augmented state approach etc [8]. The data base of the system (Fig.2) contains the following files: FIMET (method data), FIMOD (model data), FIPRED (predicted trajectories), FISINT (control matrices). There are 3 main programs corresponding to different specific functions: CREM - data entry/updating; SINT - the synthesis of control law; SIM - the simulation of the system behaviour by applying the synthetised control law.

The hardware support consists in a PDP 11 machine or some similar machine (such as the Romanian minicomputer I-100 or CORAL 4011/4030), with 32KBytes main memory, a cartridge disk, a printer, a display or a teletype terminal. The software support consists in the RSX-11M operating system or similar, with FORTRAN IV processor.

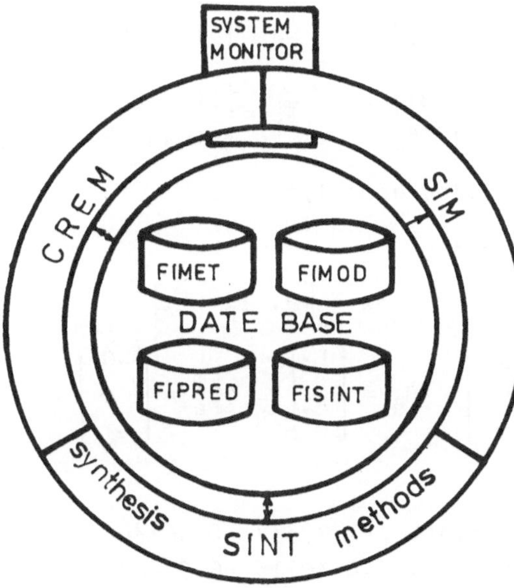

Fig.2. MESINT - structure

Numerical example. The following example illustrates the algorithms performances [11] :

Subsystem 1 :

$$\dot{x}_1^1(t) = -x_1^1(t) + m_1^1(t) + u_1^1(t) + u_{p_1}^1$$

$$u_1^1(t) = x_1^2(t)$$

$$y_1^1(t) = x_1^1(t)$$

$$J^1 = \frac{1}{2}\int_{t_o}^{t_f}\left[Q_1^1(y_{r_1}^1 - x_1^1(t))^2 + R_1^1(m_1^1(t))^2\right]dt$$

Subsystem 2 :

$$\dot{x}_1^2(t) = -x_1^1(t) + m_1^2(t) + u_{p_1}^2$$

$$y_1^2(t) = x_1^2(t)$$

$$y^2 = \frac{1}{2}\int_{t_o}^{t_f}\left[Q_1^1(y_{r_1}^2 - x_1^2(t)^2 + R_1^2(m_1^2(t))^2\right]dt$$

For constant disturbance and reference and $Q_1^1 = Q_1^2 = 50$, $R_1^1 = R_1^2 = 1$, the following results can be reached :

$$\bar{F} = \begin{bmatrix} 6.98131 & 0.26142 \\ 0.26431 & 6.97285 \end{bmatrix} \qquad \bar{F}_C = \begin{bmatrix} -0.01797 & 0.26142 \\ 0.26431 & 0.06356 \end{bmatrix}$$

$$\bar{F}_r = \begin{bmatrix} 7.59669 & -0.57187 \\ 0.45836 & 7.54746 \end{bmatrix} \qquad \bar{F}_p = \begin{bmatrix} 0.83663 & 0.01144 \\ 0.14178 & 0.84905 \end{bmatrix}$$

For a equivalent extended description :

$$\dot{\underline{x}}_a = A_a\underline{x}_a + B_a\underline{m} + G^i\underline{u}^i$$

$$\underline{y}_a = C_a\underline{x}_a \qquad \underline{x}_a = \left[\underline{x}, \underline{x}_p, \underline{x}_r\right]^T$$

$$A_a = \begin{bmatrix} A & B_p & 0 \\ 0 & 0 & 0 \\ 0 & 0 & 0 \end{bmatrix} \quad B_a = \begin{bmatrix} B \\ 0 \\ 0 \end{bmatrix} \quad C_a^T = \begin{bmatrix} -C \\ 0 \\ +I \end{bmatrix}$$

We have obtained :

$$\bar{F} = \begin{bmatrix} 6.98131 & 0.83406 & -7.72537 & 0.26146 & 0.01128 & 0.56390 \\ 0.26431 & 0.14281 & -0.40713 & 6.97284 & 0.84645 & -7.67751 \end{bmatrix}$$

$$\bar{F}_C = \begin{bmatrix} -0.01797 & -0.01086 & 0.02883 & 0.26146 & 0.01128 & 0.56390 \\ 0.26431 & 0.14281 & -0.40713 & 0.06355 & 0.00154 & 0.07669 \end{bmatrix}$$

3. THE OPTIMAL DESIGN OF PHYSICAL CONTROL STRUCTURE

If the microprocessor - based structure of Fig.1 is to implement the hierarchical control law (9), in order to minimize wire, equipment and installation costs, the OPTCON system, which is first of all concerned with optimizing large hierarchical computer control systems, can help. For each subsystem a set of terminal (sensor/actuator) locations and a set of possible concentrator (local control units - LCU) locations are given. The difficulties lies in finding the best location of LCU and the optimal (minimal cost) lay-out of the lines connecting the terminals to LCU. Then, all LCU-s are to be connected to one of the possible locations of the central computer/coordinator (CO).

In order to prevent structural disturbances a robust control law can be designed by using for example the procedure proposed in [6]. An alternative way to increase the interlevel communication reliability is to create redundant alternative links between CO and the LCU-s. The OPTCON package also evaluates interconection alternatives. The overall reliability is defined [12] :

$$r_o = \sum_{i=1}^{N} (1 - P^i)n^i / \sum_{i=1}^{N} n^i \tag{10}$$

where : N - the number of LCU-s

n^i - the number of terminals connected to the i-th LCU ;

$$P^i = \sum_{j=1}^{N^i} p_{ji}$$

p_{ji} - failure probability of the j-th prime branch cutset with respect to pair of nodes LCU i - central computer/coordinator ;

N^i - number of prime branch cutsets with respect to a pair of nodes LCU i and central computer.

System reliability r_o is understood as the expected percentage of terminals "connected" to CO. An example is given in Fig.3 and Table 1 [4].

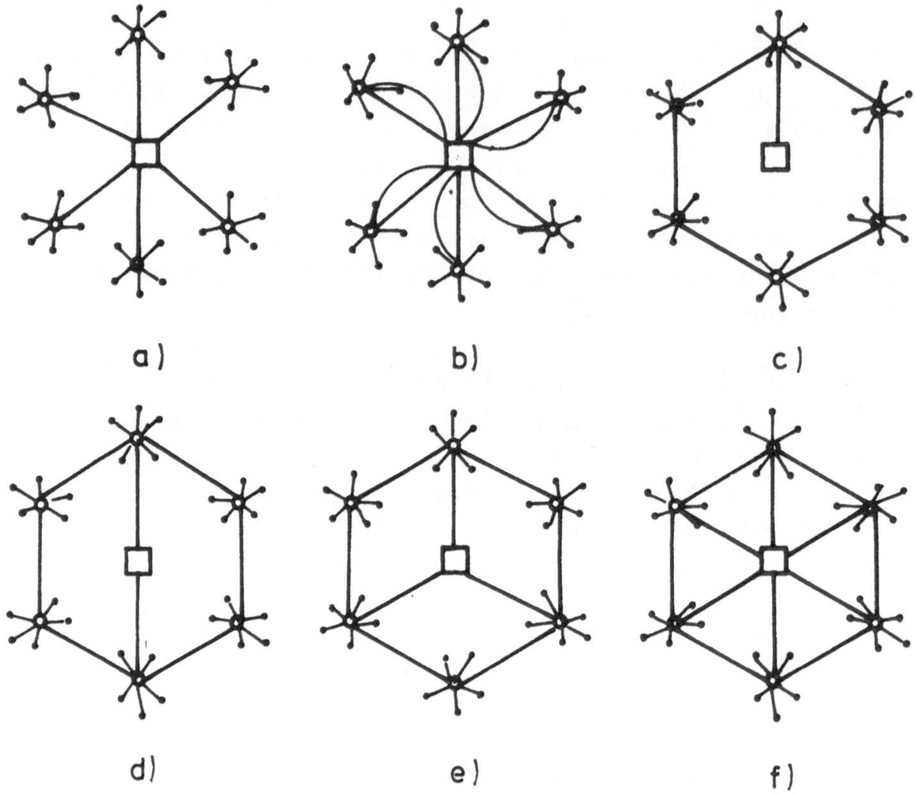

Fig.3. Various connections of the LCU to the central computer/coordinator (. terminal ; o local control unit; □ central computer)

Table 1

Reliability and costs of the configurations in Fig.3

case	r_o	cost
a	0.4014	1.00
b	0.8090	2.00
c	0.7339	1.21
d	0.7691	1.44
e	0.8300	1.55
f	0.9908	2.05

4. CONCLUSIONS

It is felt that the results in hierarchical systems and the advances in computer science and technology are complementary. Having presented some extensions of existing results, this paper was an attempt to bring to light some common issues which may be useful in the design of practical control systems.

REFERENCES

1. Athans, M., and P.L.Falb (1966). Optimal Control An Introduction to the Theory and its Applications. Mc. Graw Hill, New York.
2. Filip, F.G., A.D.Donciulescu, M.Muratcea, L.Orăşanu, D.N.Predoiu (1980). Hierarchical control of Large Scale Chemical and Mettalurgical Systems. In A.Titli and M.G.Singh (Eds.), Large Scale Systems Theory and Applications. Pergamon Press, p.563-575.
3. Filip, F.G., D.Donciulescu, M.Muratcea, L.Orăşanu, D.Predoiu (1981). Hierarchical Control in Large Scale Systems. Technical Report. RT & HS-01, ICI.
4. Guran, M., F.Filip, A.Donciulescu, L.Orăşanu, C.Petrescu (1978). Computer Aided Design of Data Transmission Network for A Large Real Time Computer System. Problems in Automatic Control, 10, 221-236.
5. Kwakernaak, H. and R.Sivan (1972). Linear Optimal Control Systems. Willey Interscience, New York.
6. Hassan, M.F., M.G.Singh (1979). Stability, Stabilisation and Performance of Multi-Level Controllers Subject to Structural Perturbations. In W.Findeisen (Ed.), 2-nd Workshop on Hierarchical Control, Warszawa.
7. Orăşanu, L., R.Gaşpar and F.G.Filip (1982). The Synthesis of Optimal Control Laws for Interconnected (Sub) Systems by Hierarchical Computation. Technical Report. RT & HS 05, ICI.
8. Orăşanu, L., R.Gaşpar and F.G.Filip (1983a) MESINT-A CAD System of Optimal Control Laws Based on Hierarchical Approach. Preprints of 5-th Int. Conf. on Control Syst. and Comp. Sci., Bucharest.
9. Orăşanu, L., R.Gaşpar and F.G.Filip (1983b) OPTCON - User's Manuals, ICI.
10. Singh, M.G., M.F.Hassan and A.Titli (1976). Multilevel Feedback Control for Interconnected Dynamical Systems Using the Prediction Principle. IEEE Transaction on Systems, Man and Cybernetics. SM-6(4), 233-239.
11. Singh, M.G., and A.Titli (1978). Systems: Decomposition, Optimisation and Control. Pergamon Press, Oxford.
12. Wilkov, R.S. (1971). Reliability Considerations in Computer Network Design. Proceedings of IFIP Congress, Ljubljana.

MICROCOMPUTER CONTROL OF MOTOR DRIVE SYSTEMS

George A. Perdikaris
Engineering Science Division
University of Wisconsin-Parkside
Kenosha, Wisconsin 53141

Algorithms for controlling the speed or position of motor drive systems by a microcomputer are developed and implemented. Performance characteristics concerning the ability of the control system to track various types of inputs and reject external disturbances are verified experimentally.

1. INTRODUCTION

To implement digital computer control for a servosystem, the controlled variable, that is, velocity or position, is periodically sampled and compared to a desired reference value. A control action (or manipulation) is next calculated according to a control law (or algorithm) and subsequently used to command the motor drive system. The control algorithm is designed to meet the problem requirements and it is stored inside the computer.

The objective of this paper is to design a control algorithm that is practical and versatile. The concepts developed are applied to control the velocity of a DC motor drive plant. The same concepts can be adopted, however, to control the position of such a plant, too. The ability of the control system to track various types of inputs and/or reject external disturbances, is analyzed and evaluated.

2. ANALOG CONTROLLERS

The control scheme emphasized in this paper is shown in Figure 1. The algorithm itself is referred to as PDFF because it is obtained from the standard PDF (see Reference [1]) by adding the Feed-forward term K_F. The PDFF algorithm differs from the more conventional PID algorithm which places all of its control actions in the forward path of the control system.

When designing a velocity controller for a DC motor drive system, the simplified model for the plant is often represented by the transfer function

$$G_p(s) = K_\omega/s, \tag{1}$$

where K_ω denotes the velocity-plant gain.

Then, from Figure 1 and with $K_D = 0$, the closed-loop velocity-to-input transfer function becomes

$$\frac{\Omega(s)}{R(s)} = \frac{K_\omega(K_F s+K_I)}{s^2+K_\omega K_P s+K_\omega K_I} . \tag{2}$$

One can show that the comparable transfer function for the PID control loop is given by

$$\frac{\Omega(s)}{R(s)} = \frac{K_\omega(K_P s+K_I)}{s^2+K_\omega K_P s+K_\omega K_I} , \tag{3}$$

and that both the PDFF as well as the PID velocity control loops have identical velocity-to-disturbance transfer functions given by

$$\frac{\Omega(s)}{D(s)} = \frac{K_\omega s}{s^2+K_\omega K_P s+K_\omega K_I} . \tag{4}$$

When designing a position loop controller, the simplified model for the plant is given by

$$G_P(s) = K_\theta/s^2, \tag{5}$$

where K_θ represents the position plant gain.

Again, from Figure 1, the position-to-input transfer function becomes

$$\frac{\theta(s)}{R(s)} = \frac{K_\theta(K_F s+K_I)}{s^3+K_\theta K_D s^2+K_\theta K_P s+K_\theta K_I} , \tag{6}$$

which can be compared to that obtained from the corresponding PID control system, that is,

$$\frac{\theta(s)}{R(s)} = \frac{K_\theta(K_D s^2+K_P s+K_I)}{s^3+K_\theta K_D s^2+K_\theta K_P s+K_\theta K_I} . \tag{7}$$

The position-to-disturbance transfer function for both the PDFF and PID control loops is given by

$$\frac{\Theta(s)}{D(s)} = \frac{K_\theta s}{s^3+K_\theta K_D s^2+K_\theta K_P s+K_\theta K_I} \ . \tag{8}$$

3. DIGITAL CONTROLLERS

The digital implementation of the PDFF control scheme is shown in Figure 2. The difference equations describing the control algorithm are

$$m_k = m_{k-1}+K_i e_k$$

and

$$m_n = m_k-K_p c_n-K_d(c_n-c_{n-1})+K_f r_n \ , \tag{9}$$

where m_n represents the manipulation variable. Note that the digital contoller gains are denoted by lower-case subscripts, compared to the analog controller gains that have been denoted by upper-case subscripts.

It can be shown that the difference equations describing the PID controller are given by

$$m_k = m_{k-1}+K_i e_k$$

and

$$m_n = m_k+K_p e_n+K_d(e_n-e_{n-1}) \ . \tag{10}$$

4. TUNING OF CONTROLLERS

Since control systems are inherently time-domain systems, their transient and steady-state performance characteristics can be adjusted according to standard filter forms, such as Binomial, Butterworth, or ITAE [1]. Thus, tuning of a velocity PDFF controller, for instance, implies matching the denominator coefficients of the transfer function of Equation (2) to the coefficients of the appropriate second-order filter transfer function. That is, for a Binomial filter type of response, the digital controller gains are

$$K_p = 2\omega_n/K_\omega \ ,$$

and

$$K_i = \omega_n^2 T/K_\omega \ , \tag{11}$$

where T represents the control system sampling time.

For tuning a positional PDFF controller, the third-order denominator of Equation (6) is matched with the appropriate filter denominator. Note that the digital controller gains are obtained from the analog controller gains using the relationships

$$K_p = K_P \ , \tag{12}$$

$$K_i = TK_I \ , \text{ and} \tag{13}$$

$$K_d = K_D/T \tag{14}$$

5. EXAMPLE

The velocity-loop PDFF algorithm was implemented and tested using a DC motor with a three-phase, half-wave, SCR power supply. The motor shaft velocity was monitored with a DC tachometer generator which returns velocity in volts. The tachometer output was next sampled every T sec using an Analog-to-Digital Converter (ADC) and the manipulation, calculated from the control algorithm, was sent directly to the drive plant via a Digital-to-Analog Converter (DAC).

The value of the velocity plant gain was estimated experimentally as $K_\omega = 20$. This was accomplished by applying a step input into the motor drive plant and measuring the slope of the velocity response. Next, with a sampling time of T = 5 msec, and assuming a velocity-loop natural frequency of $\omega_n = 50$ rad/sec, Equations (11) resulted in the digital controller gains

$$K_p = 5 \text{ and } K_i = 0.625. \tag{15}$$

It should be noted that the open-loop plant response exhibited a small time delay. To correct for it, and to fine-tune the control loop even further, a non-zero feedback derivative was added, as shown in Figure 3.

The transfer functions of the PDFF control system demonstrate that the control scheme is practical and versatile. If $K_f = 0$, for instance, the transfer functions are the same as those of the PDF control scheme. On the other hand, if $K_f = K_p$, the PDFF control scheme performs like the conventional PID control scheme, as shown in Figure 4. Finally, if $0<K_f<K_p$ the PDFF controller performance characteristics can be set between those of the PDF and PID controllers.

Whether in velocity or position loops, the PDFF controller can follow or track step inputs with zero steady-state error and remove disturbances that may enter the control loop in stepwise fashion. In addition, the algorithm can be tuned to track ramp inputs with zero steady-state error, and parabolic inputs with finite steady-state error, just like the PID controller.

For additional information about the steady-state performance of the PDFF controller, and for its application to control the position of a DC motor drive plant using a digital tachometer generator, see Reference [2].

6. CONCLUSIONS

The experimental results were obtained using integer arithmetic on a 16-bit microcomputer.

The sampling time of T = 5 msec was small enough for the computer to handle, and for the performance of the digital control system to resemble its analog counterpart.

The disturbance-rejection characteristics of the velocity loop were judged excellent.

REFERENCES

[1] Perdikaris, G.A. and Van Patten, K.W., "Computer Schemes for Modeling, Tuning, and Control of DC Motor Drive Systems", Proc. of PCI/ MOTORCON '82, pp. 83-96.

[2] Perdikaris, G.A., "Computer Methods for Modeling, Tuning, and Control of Motor Drive Systems", Proc. of International Symposium on Measurement and Control (MECO '83).

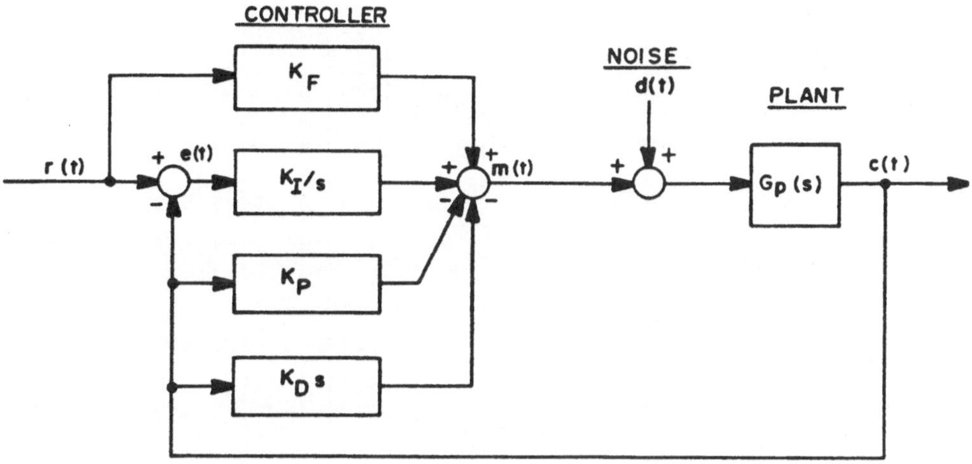

Figure 1. Analog PDFF control system.

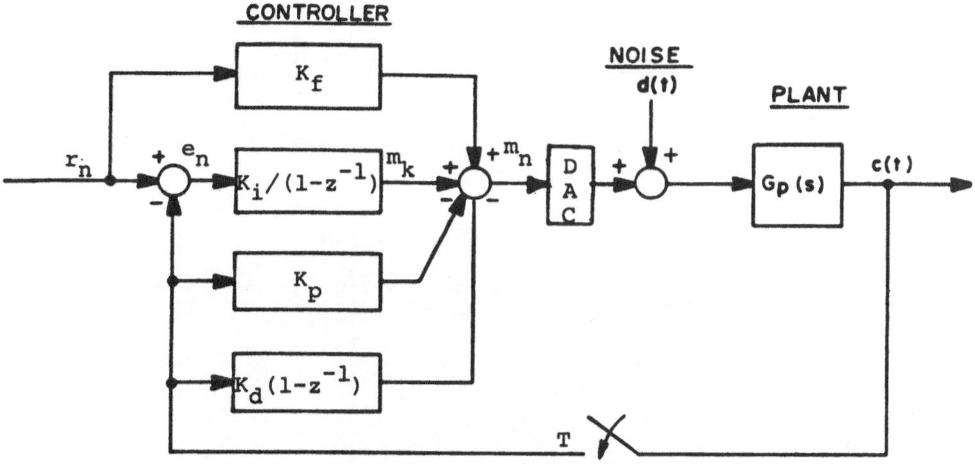

Figure 2. Digital PDFF control system.

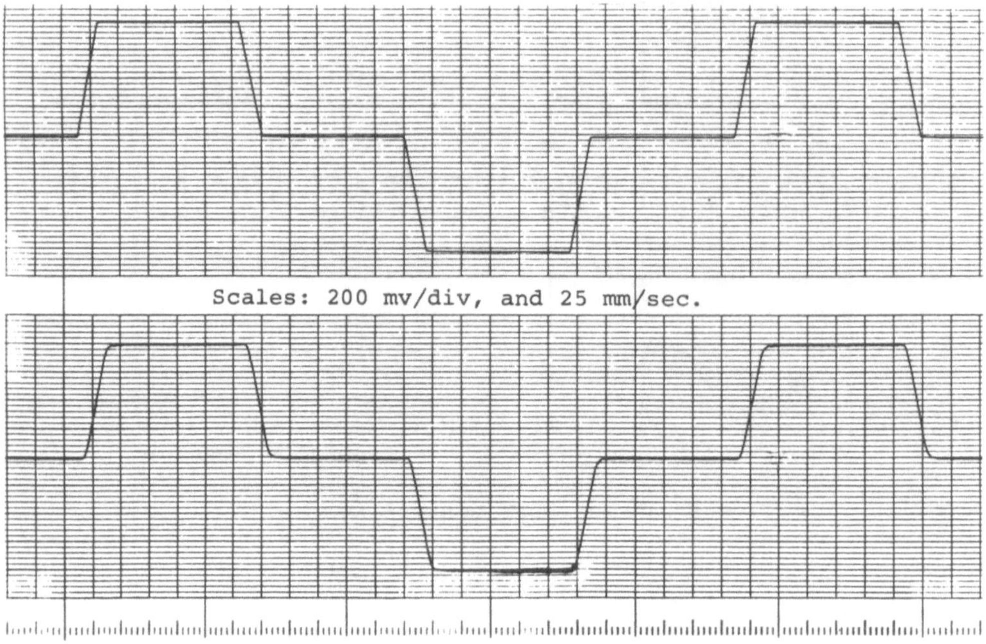

Figure 3. Velocity loop: $K_p = 5$, $K_i = 0.6$, $K_d = 10$, $K_f = 0$.
Velocity reference (top), and velocity feedback (bottom).

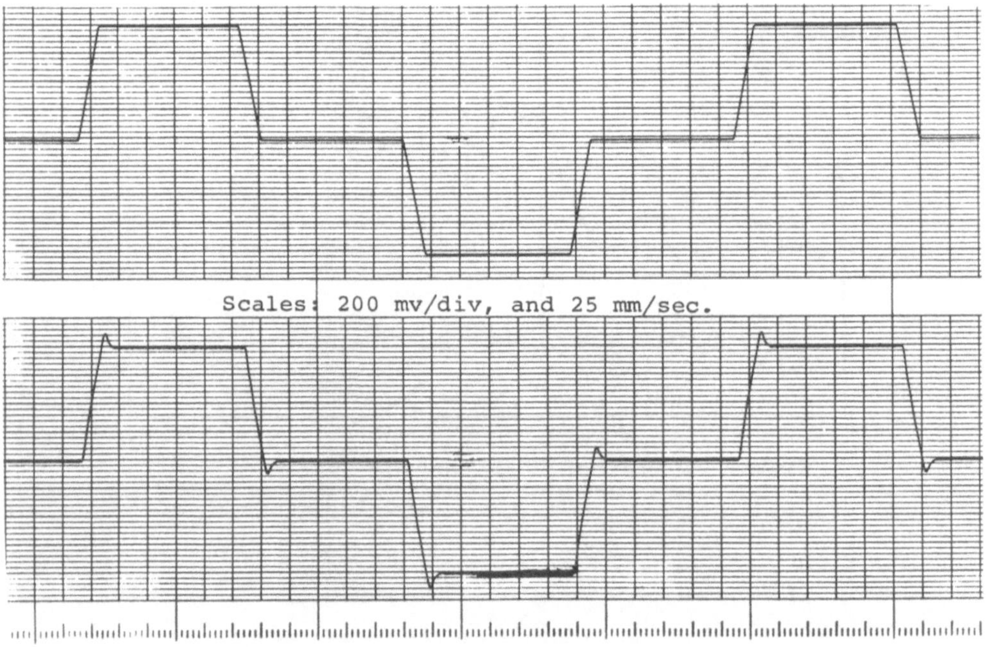

Figure 4. Velocity loop: $K_p = 5$, $K_i = 0.6$, $K_d = 10$, $K_f = 5$.
Velocity reference (top), and velocity feedback (bottom).

DISTRIBUTED COMPUTATION FOR REAL TIME CONTROL OF ELECTRIC POWER SYSTEMS

Prof. M.J.H. Sterling and Dr. M.R. Irving
University of Durham
Department of Engineering
South Road
Durham, DH1 3LE, U.K.

1. INTRODUCTION

The increasing complexity of modern power systems has led to a greater dependence on automatic control at all levels of operation. The higher levels of control have traditionally relied heavily on human judgement especially in respect of economic factors. However, recent research experience has shown that centralised computer control is feasible for small and medium scale networks but that costs would rise alarmingly for the implementation of schemes on a national basis. The geographical distribution of the plant presents enormous data gathering and control problems, which for the large scale systems preclude any direct approach to centralised control. The real time monitoring and control of large scale electrical power generation, transmission and distribution networks does however lead very naturally to the concept of a distributed computer solution. The system is geographically dispersed, often into a number of well defined areas, and also separates into a hierarchy of networks at different voltage levels. It is anticipated that the application of distributed algorithms will therefore result in lower telemetry costs, reduced computer loadings, and a more reliable and easily managed control system.

The present paper describes the configuration of a network of four minicomputers which are being used for research into the application of distributed and hierarchical solution techniques to many of the problems occurring in the real time control of electrical power systems. A particular aspect of the overall control problem: the constrained economic dispatch of active power generation, is developed in detail, and the application of a sparse matrix implementation of Dantzig-Wolfe decomposition is considered. New computational results are presented which enable the performance of the algorithm to be estimated both for tightly-coupled and loosely-coupled multiprocessor implementations. It is concluded that the distributed solution of the economic dispatch problem is feasible and advantageous even within the limitations imposed by the restricted bandwidth of present communication channels.

2. DISTRIBUTED COMPUTER SYSTEM

A distributed computing facility which has been established at the University of Durham for research into the distributed control of electrical power systems is shown in Figure 1. The system consists of four 32 bit minicomputers (1 x Perkin Elmer 3230 and 3 x Perkin Elmer 3220) interconnected via serial communication links. The majority of applications in power system analysis and control impose a large burden of floating point operations on the processor and each computer has therefore been equiped with floating point hardware. The numerical results given in this paper have been obtained using FORTRAN 77 with floating point data stored in single precision (32 bit) format except for a small amount of information held in double precision (64 bit) format. The inter-processor communication channels normally operate at a maximum of 19.2 k baud, but the data transmission speed can be deliberately reduced to simulate the limitations imposed by present day telecontrol communication networks.

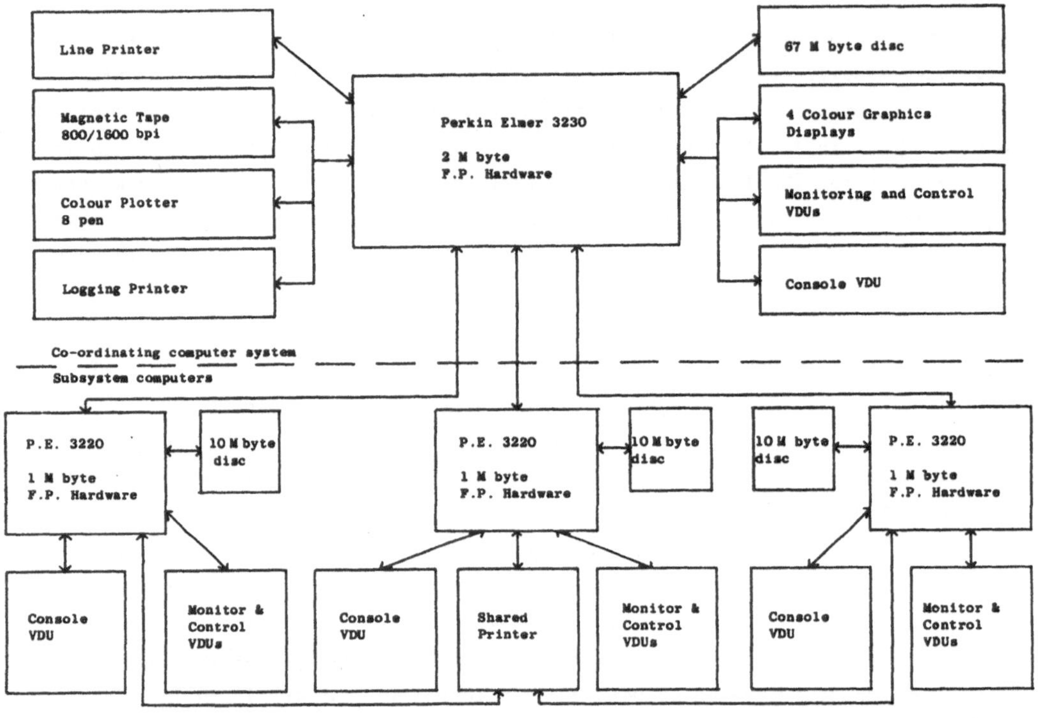

Figure 1: Hardware configuration

3. ECONOMIC DISPATCH USING SPARSE DANTZIG-WOLFE DECOMPOSITION

The economic dispatch function is concerned with the allocation of target output powers for generators, in order to satisfy the predicted load demand at minimum cost, while remaining within operational constraints. Computation of the economic dispatch of active power is often achieved using linear programming techniques[1]. The production cost which must be minimised, may be approximated by a linear function of the generator outputs.

$$\text{Min } J = c^T P \tag{1}$$

where c represents a vector of linear cost coefficients associated with each generator and P represents a vector of the generator active power outputs.

The minimisation is subject to a large number of operational constraints, which are often linearised and considered in groups depending on the scope of the optimisation being attempted. For dispatch of active power only the following set of constraints are normally applied.

$$P_{min} < P < P_{max} \qquad \text{(generator limits)} \tag{2}$$

$$\sum_{i=1}^{N} P_i = P_o \qquad \text{(total generation requirement)} \tag{3}$$

$$\sum_{\text{some } i} P_i > G_{min} \qquad \text{(group import constraints)} \tag{4}$$

$$\sum_{\text{some } i} P_i < G_{max} \qquad \text{(group export constraints)} \tag{5}$$

$$\nabla P_{min} < \nabla P < \nabla P_{max} \qquad \text{(rate limits)} \tag{6}$$

The geographically distributed nature of the problem is reflected in the division of these constraints into a set of overall constraints and a number of sets of local or subsystem constraints. The block structure of the dispatch problem may be exploited by a decomposition technique in which a master problem supplies co-ordinating information to a set of subproblems. A suitable algorithm is provided by the Dantzig-Wolfe decomposition principle[2] which can be stated:

If there is a linear program which can be expressed as:

$$\text{Min.} \sum_{i=1}^{n} c_i^T x_i \tag{7}$$

$$\text{S.t.} \sum_{i=1}^{n} A_i x_i = b_o \tag{8}$$

$$B_i x_i = b_i \qquad \text{(all i)} \tag{9}$$

$$x_i > 0 \qquad \text{(all i)} \tag{10}$$

Then an equivalent form is:

$$\text{Min.} \sum_{i=1}^{n} \sum_{k=1}^{k_i} c_{ik} Z_{ik} \tag{11}$$

$$\text{S.t.} \sum_{i=1}^{n} \sum_{k=1}^{k_i} P_{ik} Z_{ik} = b_o \tag{12}$$

$$\sum_{k=1}^{k_i} Z_{ik} = 1 \qquad \text{(all i)} \tag{13}$$

$$Z_{ik} > 0 \qquad \text{(all i,k)} \tag{14}$$

This is the required master problem which is a linear program in the convex weights Z_{ik} based on the corner points X_{ik} of the subproblem constraints, defined by:

$$\text{Min.} \quad (c_i^T - \pi A_i)\, x_i \tag{15}$$

$$\text{S.t.}\quad B_i\, x_i = b_i \tag{16}$$

$$x_i > 0 \tag{17}$$

The vector π is the price vector, which is readily available in the computer solution of the master problem using the revised simplex method. An overall solution is arrived at iteratively in major iterations (for the master problem), and minor iterations (for the subproblems). The distributed structure of the method is shown in Figure 2. It can be shown that the algorithm converges in a finite number of steps, with computation proceeding as follows:

Major Iteration - The master problem is solved using the current set of generation proposals from the subproblems, producing a new price vector which is transmitted to the subproblems.

Minor Iterations - The subproblems are solved, possibly in parallel, using the latest price vector and new generation proposals are returned to the master problem.

Full details of the algorithm are available in reference 2 and applications in power system control are considered in references 1, 3, 4 and 5.

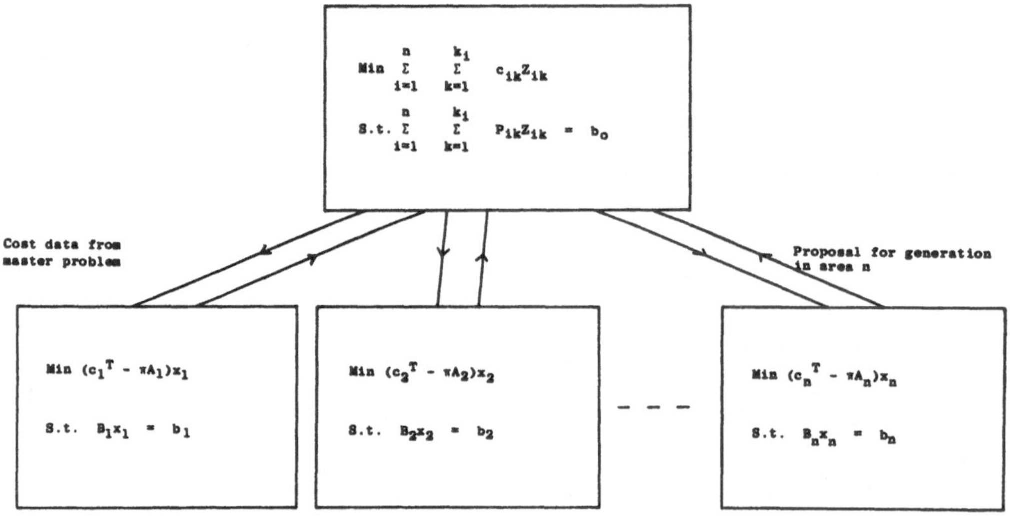

Figure 2: Dantzig-Wolfe Decomposition

The subproblems created using the decomposition principle retain the sparsity pattern of the original linear program formulation and can therefore be solved efficiently using a sparse linear programming technique. However, it is interesting to note that the master problem is also sparse, and may be solved economically using the primal revised simplex method with the basis matrix held as sparse triangular factors. This is particularly appropriate if the maximum number of convexity constraints (13) are applied[2], and if the linking constraints (8) are sparse. The inclusion of the maximum number of convexity constraints (i.e. as many as there are subproblems) also allows a number of master problem basis exchanges (or cycles) in each major iteration. This property considerably reduces the quantity of data transferred between master problem and subproblems, which is of importance in a distributed computing environment.

4. COMPUTATIONAL RESULTS

The application of Dantzig-Wolfe decomposition to the economic dispatch problem using modern linear programming techniques, including sparse matrix factorisation, has produced favourable results. Table 1 shows the typical convergence of a Fortran 77 implementation on a dispatch problem of 100 generators and 24 subproblems (100 variables and 269 constraints including 21 linking constraints). The value of allowing many cycles in the master problem is clearly illustrated with an average of 6 useful basis exchanges occurring in each major iteration between data communication steps. By contrast, if the master problem is limited to a single basis exchange, a total of 49 major iterations are required for this example. The Dantzig-Wolfe principle also enables the calculation of a lower bound on the final cost at any iteration[2]. This may be used to terminate the iterative process at an early stage if some degree of sub-optimality is acceptable. Table 1 illustrates the increasing accuracy of the lower bound estimate of the final optimal cost as the solution proceeds (ratio 1) compared with the true position as represented by the ratio of the present cost to the actual final optimal cost (ratio 2). The actual optimal cost is, of course, unknown during the solution procedure and consequently it can be seen that the lower bound cost estimate represents a useful indicator of optimisation progress.

Table 1: Convergence of the Dantzig-Wolfe Algorithm

Iteration	No. of Cycles	Simplex Phase	Cost (units)	Lower Bound Cost(units)	Optimality Ratio 1*	Optimality Ratio 2+
0	0	1	6090.0	-2390.0		
1	24	1	1452.6	-647.3		
2	7	1	194.9	-103.1		
3	1	1	0.0	0.0		
4	0	2	95202.8	79969.1	1.19049	1.01258
5	7	2	94386.1	93802.4	1.00622	1.00389
6	8	2	94050.0	93802.4	1.00264	1.00032
7	2	2	94045.0	94007.5	1.00040	1.00027
8	1	2	94020.0	94020.0	1.00000	1.00000

* Optimality ratio 1 = present cost/lower bound cost

\+ Optimality ratio 2 = present cost/minimum cost

Note: Costs in simplex phase 1 are artificial.

Convergence and processor time requirements (for a single Perkin Elmer 3230) are shown in Table 2, for various numbers of linking constraints in the dispatch problem. The total solution time increases only moderately with the number of linking constraints.

Table 2: Variation of Solution Requirements With The Number of Linking Constraints

No. of Linking Constraints	No. of Master Problem Iterations	No. of Master Problem Cycles	Total Solution Time (sec.) (Single Processor)
1	5	40	6.0
3	5	40	6.1
5	7	42	8.0
7	7	42	9.4
9	8	41	11.2
11	7	44	9.8
13	5	41	8.9
15	9	51	12.0
17	7	47	10.5
19	7	48	10.9
21	8	50	11.9

Tables 3(a) and 3(b) give total solution times applicable to tightly-coupled and loosely-coupled multi-processor systems. Master problem and subproblem solution times have been accurately measured whereas the data communication time has been estimated based on a net communication speed of 4800 bits per second. If 24 subproblem processors and one master processor are available overall solution times of 3.5 seconds for a tightly-coupled system (negligible communication delay) and 4.5 seconds for a loosely-coupled system are possible. Alternatively, grouping the subproblems to utilise 3 subproblem processors and one master processor leads to solution times of 6.1 seconds (tightly-coupled) and 10.2 seconds (loosely-coupled). These figures must be compared to an overall solution time of 11.9 seconds on a single processor with no communications overhead. The distributed approach therefore offers a faster real time solution even with a limited number of processors and relatively slow data communications. It is expected that higher speed communications equipment which may become available in the future will lead to considerable savings in overall solution time using the distributed computing approach.

Table 3(a): Solution Time For 25 Processors

Iteration	Read Data (sec)	Master Problem (sec)	Longest Subproblem (sec)	Estimated Communication Time (sec.) (tightly-coupled)	(loosely-coupled)	Output Results (sec.)
0	0.817	0.069	0.152	0.0	0.11	
1		0.550	0.136	0.0	0.11	
2		0.189	0.150	0.0	0.11	
3		0.059	0.014	0.0	0.11	
4		0.034	0.051	0.0	0.11	
5		0.219	0.156	0.0	0.11	
6		0.264	0.105	0.0	0.11	
7		0.087	0.156	0.0	0.11	
8		0.054	0.138	0.0	0.11	0.104

Total solution time (tightly-coupled) ≈ 3.5 sec

" " " (loosely-coupled) ≈ 4.5 sec

Table 3(b): Solution Time for 4 Processors

Iteration	Read Data (sec)	Master Problem (sec)	Longest Subproblem (sec)	Estimated Communication Time (sec.) (tightly coupled)	(loosley coupled)	Output Results (sec.)
0	0.817	0.069	0.562	0.0	0.46	
1		0.550	0.498	0.0	0.46	
2		0.189	0.497	0.0	0.46	
3		0.059	0.100	0.0	0.46	
4		0.034	0.141	0.0	0.46	
5		0.219	0.505	0.0	0.46	
6		0.264	0.335	0.0	0.46	
7		0.087	0.492	0.0	0.46	
8		0.054	0.475	0.0	0.46	0.104

Total solution time (tightly-coupled) ≈ 6.1 sec

" " " (loosely-coupled) ≈ 10.2 sec

5. CONCLUSIONS

This paper has described a distributed computer system which has been designed for research into the distributed control of electrical power generation and transmission networks. A particular aspect of the control problem, the economic dispatch of generator outputs, has been discussed, and the application of a decomposed solution algorithm has been considered. The computational results which have been obtained to date are favourable and indicate that overall savings in solution times may be achieved even with a communication system of relatively low bandwidth.

6. REFERENCES

1. M.J.H. Sterling and M.R. Irving: 'Constrained Dispatch of Active Power by Linear Decomposition', Proc. IEE, 124, 3, 1977, pp 247-251.

2. L.S. Lasdon: 'Optimization Theory for Large Systems', MacMillan, 1970.

3. R. Romano, V.H. Quintana, R. Lopez and V. Valades: 'Constrained Economic Dispatch of Multi-Area Systems Using the Dantzig-Wolfe Decomposition Principle', IEEE trans. PAS, 100, 4, 1981, pp 2127-2137.

4. J.G. Waight, A. Bose and G.B. Sheble: 'Generation Dispatch with Reserve Margin Constraints Using Linear Programming', IEEE trans. PAS, 100, 1, 1981, pp 252-258.

5. B.F. Wollenburg and W.O. Stadlin: 'A Real Time Optimizer for Security Dispatch', IEEE trans. PAS, 93, 5, 1974, pp 1640-1649.

TRANSIENT SECURITY ASSESSMENT IN POWER SYSTEMS

J.M.G.Sá da Costa
CAUTL - IST
Technical University of Lisbon
1096 Lisbon Codex - Portugal

Abstract

Transient security assessment is an important part of power systems operation. In particular, on-line assessment is a present need in large power systems. This paper presents a feasible method based on pattern recognition techniques where all the on-line calculations for transient security assessment are replaced by the only evaluation of the sign of a security function dependent on measurements directly obtained from the power system. This security function has learning capacity to adapt to different network configurations and to new load and generation schedules, becoming, after sometime, independent of the power system model and of the methodology adopted in the off-line design.

1. INTRODUCTION

The modern power system is a complex one, consisting of a large number of units of energy production, interconnected by a transmission and distribution network. Over ninety percent of the time, we find the system is operating in its normal state, i.e., all loads are met, frequency is constant, voltage magnitudes are within prescribed limits and no component is overloaded. However, inevitable disturbances may affect production as well as transmission systems, deteriorating the electric power quality delivered to the customers. Thus, there exists a pressing need to develop an on-line security monitoring system, which helps the operating enginner to detect conditions which may lead to possible failures or deteriorations of the quality of the power supply, before they actually occur. System security involves a vast spectrum of subproblems. However, the most important factor affecting the question of on-line security assessment is that of system fault dynamics and in particular the transient security of the system.

To assess the transient security of a power system, each disturbance of a prespecified contingency set is applied to the model of the system, in the state where it will be in a close future and a transient stability analysis made. If any of the disturbances of the prespecified contingency set leads to system instability, meaning instability the loss of synchronism of one or more generators, the system is insecure and will enter into "alert state". In such a case, preventive measures should be looked for and applied to the system.

Transient security evaluation using simulation of current system status has long been in use. However, this usually involves rather large amount of computing time, and the volume of data generated needs a significant and complex human analysis to assess the security of the system. This turns the simulation method unpracticable for large power systems. An alternative would be to use any of the available direct methods for transient stability studies, even though these methods have sometimes proved unrealistic for large power systems. However, all classical direct methods of transient stability analysis need some kind of simulation of the system status, being for this reason unsuitable for a fast transient security assessment.

If we look carefully into the problem of power systems security, it is obvious that in their operation it is desirable that the future operating state be both steady state secure and transient secure. Judgement on stability depends on the two classes to which an operating state may belong. These two classes comprise the secure and the insecure states, respectively. If an equation of a separating surface between these two classes is evaluated as a security function, it may be used to assess the stability of the power system. This kind of problem is a typical problem of pattern recognition.

Generally speaking a pattern recognition system classifies a pattern (operating state, in our case), belonging to a certain environment, into one of the possible classes to which each pattern may belong. In our case, the pattern recognition system should discriminate an operating state as belonging to the secure or the insecure classes.

2. <u>PATTERN RECOGNITION</u> (1)

There are two aspects to pattern recognition problems; developing a decision rule and using it. The "recognition" occurs in the use of the rule. The formulation of the pattern recognition problem begins with class definition and the formation of a set of labelled samples of these classes in some workable representation. The problem is solved when a decision rule has been derived, which assigns a unique label to new patterns.

In power systems security assessment, the final objective is to build a security function which classifies the operating states of the power system as transient secure or insecure. This security function will be a function of a small set of variables which we call "features".

Two phases exist to obtain the features. In the first one, "measurement selection", the measurement space on which the system is totally described is reduced to a much smaller subspace. In the second phase, "feature extraction", the reduced measurement space is transformed into a feature space in order to decorrelate the space and increase the discriminant power between the two possible classes.

2.1 Measurement Selection

As with any other physical system, a power system can be fully described by a set of variables which we shall call measurements. In on-line operation, these measurement will be provided to the security system by the so-called "State Estimator". The number of measurements describing a power system for transient security studies will depend on the detail with which we want to represent the system's behaviour, as well as the size of the power system. Nevertheless, a great number of measurements is always needed, which implies a high dimensional space to represent the power system state.

One of the first steps in the design of the security function is to reduce the measurement space to a lower order space by selecting the most significant variables representing the system. This can be done by means of a measure of separability between classes, of each measurement, like the one proposed by Becker (2)

$$F_k = |u_k^{(1)} - u_k^{(2)}|/(\sigma_k^{(1)} + \sigma_k^{(2)})$$

with

$u_k^{(i)}$ = estimated mean value of variable k for class i, (i=1,2).

$\sigma_k^{(i)}$ = estimated standard deviation of variable k for class i, (i=1,2).

This measure of separability gives information about the discriminant power of the measurement considered to represent the state of the power system. The measurements are ranked upon decreasing values of their discriminant power. We take as selected measurement the first one in the ranking. However, we know from statistics that two highly correlated variables give little more information than one. To guarantee that each extra measurement that we choose adds the maximum amount of information, after selecting one measurement the remaining are modified in order to remove any redundant information. This modification of the remaining measurements changes the measure of separability, which needs to be recalculated, as well as the new ranking. After, a new measurement may be selected. This process, called "repetitive ranking", is carried out until we reach the desired number of measurements to represent the operating state in a lower order measurement space.

2.2 Feature Extraction

Having reduced the number of measurements needed to described the power system by discarding those with the least discriminant power, it is desirable to transform this reduced space to a different coordinate system in order to decorrelate the space and to allow a better discrimination between states belonging to different classes.

The most popular feature extraction methods are based on the Karhunen--Loeve expansion. In all these methods the resulting transformation is linear and optimal, in the sense that minimizes the loss of information contained in the training set. The difference between the different methods using Karhunen-Loeve expansion is the type of information they use. For a more detail description of these methods see reference (3). Here we will adopt the method which does the optimal compression of the discriminant information contained in the class mean.

The resulting space, called "feature space", has its components uncorrelated. This components, called "features", will be the variables of the security function, to be developed.

2.3 Security Function

Having chosen the features, the problem now is how to build a security function which reliably discriminates the secure from the insecure states. Also, it is desirable that this security function has the learning capacity to adapt to different system conditions, like different network configurations, new load and generation schedules, etc..

Using the Bayes decision rule (3) which minimizes the total expected loss, i.e., it assigns each operating state to the class with smallest conditional loss, the decision rule will be given by

$$d(x) = f_1(x) - K\, f_2(x)$$

with

$$d(x) = \text{class 1 if } d(x) > 0$$
$$\qquad \text{class 2 if } d(x) < 0.$$

x = feature vector

f_i (.) = estimated probability density function of class i, (i=1,2).

k = bias factor.

Estimating the density function of each class by means of superposition of potencial functions of type 1, (4), it is possible to write the security function in a polynomial form, given by

$$d(x) = d_{0\ldots 0} + d_{10\ldots 0}\, x_1 + d_{010\ldots 0}\, x_2 + \ldots + d_{0\ldots 01}\, x_p + d_{20\ldots 0}\, x_1^2 +$$

$$+ d_{10\ldots 0}\, x_1 x_2 + \ldots + d_{z_1 z_2 \ldots z_p}\, x_1^{z_1} x_2^{z_2} \ldots x_p^{z_p} + \ldots$$

where

x_i = features, (i=i,...,p)

z_i = power of feature x_i

$d_{z_1 \ldots z_p}$ = weights.

As explained in reference (5) this polynomial security function has the advantage of the weigths being computed recursively. This allows an adaptation of the security function to new system conditions by adding the effect of each new operating point to the weights already computed, without the need of a training set (1).

2.4 Training Set

In statistical pattern recognition theory we assume that the measurements describing the system are random variables, which have an underlying probability density function. In order to determine the statistical characteristics of these measurements, a set of characteristic operating states of the power system is needed. Each of these operating states must be associated with one of the possible two classes. This set of operating states, which help to design, in off-line mode, the security function is called "training set".

Based on statistical considerations (6), it is required that the number of operating states available in the training set, for each class, be greater than three times the number of features used in the security function. These operating points can be obtained from real data of the system under study or by simulation of the power system (1), taking into account system characteristics. A conventional technique should be used to analyse the transient security of each operating state of the training set.

3. RESULTS FOR THE 225 KV - CIGRE SYSTEM

Here, the usefulness and application of the security function, briefly described in section 2 is illustrated to assess the transient stability of the 225 KV - CIGRE system. This system is a moderate size power system with ten burbars, thirteen lines and up to thirty three generators. The total active load of the system varies from 675 to 2030 MW. The one line diagram of the 225 KV - CIGRE system is shown in figure 1. Five load levels have been chosen to represent typical working conditions of the system. For each of these load levels a contingency set of five three - phase faults, cleared by tripping a line, were considered. The table of figure 2 gives information related to these faults. All together 180 operating states in a 196- - dimensional space were considered. Each operating state was represented by a set of variables usually considered in steady state and transient analysis of power systems.

Fault No.	At Bus	Line Cleared	Clerance Time (sec.)
1	1	1-3	0.18
2	2	2-3	0.27
3	3	3-4	0.16
4	4	4-9	0.14
5	6	6-4	0.37

Figure 2. Contingency set (Faults and clearence times).

The design of the pattern recognition security system is made much more efficient if a security function is built up for each fault in the contingency set. Thus, the overal security function will here be made up of five partial security functions, corresponding to each of the five faults in the contingency set. If any of the partial security functions shows the system as insecure then the value of the total security function is insecure.

The first fifteen measurements selected by the repetitive ranking method are related with the dynamic variables (acelerating power, inertia constant, generating power, rotor angle, etc.) at the faulty busbar and/or with the electrically closest generating busbar.

Fault	No. of Features	Secure/Insecure
1	6	94.55/100
2	5	83.89/100
3	7	94.82/100
4	4	85.55/100
5	5	82.77/100

Figure 3. Percentage of correct classification for partial security functions.

The table of figure 3 presents the percentage of good classification of the partial security functions for the five faults. The results presented are for a second order polinomial security function. The bias factor was selected so that all the insecure states are identified correctly, i.e., no false dismissals were allowed. This means that the classifiers are biased in order to avoid false dismissals, leading to an increase of false alarme (classification of secure states as insecure states). Assuming equal probability of occurence for every fault in the contingency set we have that the performance of the total security function will be 88.2% of correct classification of secure states and 100% of correct classification of insecure states.

4. CONCLUSIONS

The results obtained, by using a pattern recognition approach to design a transient security system for the 225 KV - Cigre system show high accurancy for the chosen security function. The main disadvantage of this approach is related with the generation of the training set, even when the simulation approach is used. The central problem in designing a security function is related with the selection of characteristic measurements to be used to extract the features. The main advantages of this approach are: all the one-line calculations for transient security assessment are replaced by the only evaluation of the sign of a security function dependent on measurements directly obtained from the power system; and, the security function has learning capacity to adapt to different network configurations and to new load and generation schedules without the need redesigning in an off-line mode.

REFERENCES

(1) J.M.G.SÁ DA COSTA, "Application of Pattern Recognition to Transient Security Assessment in Power Systems", PhD Thesis, University of Manchester Institute of Science and Technology, 1982.

(2) P. BECKER, "Recognition of Patterns", Polytenisk Forlag, Copenhagen 1968.

(3) P.A. DEVIJVER and J.KITTLER, "Pattern Recognition. A Statistical Approach", Prentice - Hall 1982.

(4) W.S. MEISEL, "Computer - Oriented Approaches to Pattern Recognition", Mathematics in Science and Engineering, Academic Press 1972.

(5) D.F. SPECHT, "Generation of Polynomial Discriminant Functions for Pattern Recognition", IEEE Trans., Vol. EC - 16, 1967, pp. 308-319.

(6) D.H. FOLEY, "Considerations of Sample and Feature Size", IEEE Trans., Vol. IT - 18, 1972, pp. 618-626.

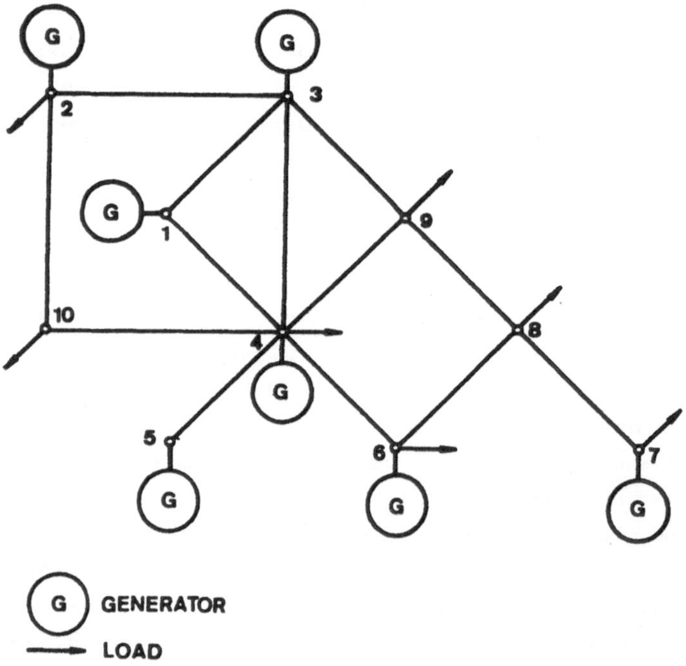

Figure 1. 225 KV - CIGRE system.

DECENTRALIZED CONTROL SCHEME FOR TURBOGENERATORS IN ELECTRIC POWER SYSTEMS

S. M. Osheba
Menoufia University
Faculty of Engineering and Technology
Shebin El-Kom , EGYPT .

ABSTRACT

The paper describes the design and implementation of a decenteralized controller for a turbogenerator, operating in a multi-machine environment . The controller use only control signals measured at the machine terminals and there is no need for remote measurements . Controllers have been designed for a multi-machine system and the results are presented which illustrate substantial improvements in performance by using decenteralized controllers .

INTRODUCTION

The size of interconnections in electric power systems is now so large that the largest machine is only a fraction of the total capacity in the system . The number of generators is large and interaction exist among machines which are geographically separated . The design of a controller for a synchronous generator requires linear model of the plant which describes the system dynamical behaviour about a nominal operating point . Traditionally, this model is obtained by considering the well-known power system model (i.e. single machine infinite bus arrangement) and substantial contribution has been made (1-6) . This, however, does not take account of interconnections with other machines in the system .

The outcome of research in this area has shown that in order to devise a suitable controller for a synchronous machine the system practicalities and physical constraints must be considered . This includes continous changes in system configuration and operating conditions, geographical distances between machines , interaction among generators and continual growth in the grid . Physical realization and practical constraints have been considered (5, 6) and problems relating to multi-machine environments have had less attention .

A centeralized controller was designed for a multi-machine system , which use measurements from the machine to be controlled and all other generators in the system (7) . Also, techniques have been suggested to overcome mathematical difficulties arising from high dimentionality in large scale systems (8) . Although the numerical difficulties envolved could be overcomed , the centeralized controllers are not a feasible practical arrangement . Local controllers which employ measurements at the machine terminals is more physically realizable structure and such controllers could perform satisfactorily in a multi-machine environment (9, 10) .

This paper illustrates the design of a decenteralized physically realizable controller for a synchronous machine operate in a multi-machine environment . The controller is designed without mathematical difficulties , employs measurements of local variables and the results shows substantial improvement in performance .

PROBLEM FORMULATION

The linear model of a multi-machine system with n-interconnected machines can be obtained in the following discrete-time form :

$$\begin{bmatrix} \underset{\sim}{x}^{1}_{k+1} \\ \cdot \\ \underset{\sim}{x}^{i}_{k+1} \\ \cdot \\ \underset{\sim}{x}^{n}_{k+1} \end{bmatrix} = \begin{bmatrix} \phi_{11} & \cdot & \phi_{1i} & \cdot & \phi_{1n} \\ \cdot & \cdot & \cdot & \cdot & \cdot \\ \phi_{i1} & \cdot & \phi_{ii} & \cdot & \phi_{in} \\ \cdot & \cdot & \cdot & \cdot & \cdot \\ \phi_{n1} & \cdot & \phi_{ni} & \cdot & \phi_{nn} \end{bmatrix} \begin{bmatrix} \underset{\sim}{x}^{1}_{k} \\ \\ \underset{\sim}{x}^{i}_{k} \\ \\ \underset{\sim}{x}^{n}_{k} \end{bmatrix} + \begin{bmatrix} \Delta_{11} & \cdot & \Delta_{1n} \\ \cdot & \cdot & \cdot \\ \Delta_{i1} & \cdot & \Delta_{in} \\ \cdot & \cdot & \cdot \\ \Delta_{n1} & \cdot & \Delta_{nn} \end{bmatrix} \begin{bmatrix} \underset{\sim}{u}^{1}_{k} \\ \cdot \\ \underset{\sim}{u}^{i}_{k} \\ \cdot \\ \underset{\sim}{u}^{n}_{k} \end{bmatrix} \qquad (1)$$

A controller can be designed by using minimization technique to minimize the cost function :

$$J = \sum_{k=o}^{N} \underset{\sim}{x}_{k+1}^{t} \; Q \; \underset{\sim}{x}_{k+1} + \underset{\sim}{u}_{k}^{t} \; R \; \underset{\sim}{u}_{k} \qquad (2)$$

With a large value of N the minimization algorithm leads to the following control law for the ith generator :

$$\underset{\sim}{u}_{k}^{i} = F_{ii} \; \underset{\sim}{x}_{k}^{i} + \sum_{\substack{j=1 \\ j\neq i}}^{n} F_{ij} \; \underset{\sim}{x}_{k}^{j} \qquad (3)$$

Where F_{ii} and F_{ij} are constant feedback matrices for the ith generator and between i and j machines . $\underset{\sim}{x}_{k}$ and $\underset{\sim}{u}_{k}$ are the deviations in control-and-state vectors from reference vectors $\underset{\sim}{X}_{o}$ and $\underset{\sim}{U}_{o}$.

The control law , Eqn. 3 , shows that the control vector for the ith generator requires measurements of control signals from the machine to be controlled and all other units in the system which is not a feasible control arrangement , because of the practical constraints involved . Generally , problems relating to control a generating unit operates in a multi-machine environment can be classified into :

(i) Difficulties in implementation and design of centeralized controllers which appear in handling matrices with high dimensions, choice of suitable weighting factors and measurements of control signals from remote machines, some of them may not be in service

(ii) Physical realization of practical constraints in the system :

1- The generator operates over a wide range of operating conditions and it is essential that the controller function satisfactorily while the generator is shifting from one operating point to another .

2- Permenant changes in system configuration, such as switch out of one transmission line, is another problem and the controller must cope with .

3- The generators number is large, geographically separated and some of them may be out of duties . Controllers which employ measurements of control signals at the machine terminals are more physically realizable than centeralized controllers .

DESIGN OF DECENTERALIZED CONTROLLERS

Equivalent Model

Consider a power system with n-nodes and generators are connected to m of these nodes . The network equation could be written as follows :

$$\begin{bmatrix} I_1 \\ \cdot \\ I_m \\ \hline 0 \end{bmatrix} = \left[\begin{array}{c|c} Y_{11} & Y_{12} \\ \hline Y_{21} & Y_{22} \end{array}\right] \begin{bmatrix} V_1 \\ \cdot \\ V_m \\ \hline V_n \end{bmatrix} \qquad (4)$$

Eqn. 4 can be reduced to :

$$\underset{\sim}{V} = ZR \ \underset{\sim}{I} \qquad (5)$$

Where

$$\underset{\sim}{V} = \begin{bmatrix} V_1 \\ \cdot \\ V_m \end{bmatrix} \quad \text{and} \quad \underset{\sim}{I} = \begin{bmatrix} I_1 \\ \cdot \\ I_m \end{bmatrix} \qquad (6)$$

$$ZR = \left[Y_{11} - Y_{12} \ Y_{22}^{-1} \ Y_{21} \right]^{-1} \qquad (7)$$

= Reduced impedance matrix

From Eqn. 5 , the complex voltage at the i<u>th</u> generator terminals could be written as :

$$\underset{\sim}{V}^i = Z_{ii} \ \underset{\sim}{I}^i + \sum_{\substack{j=1 \\ j\neq i}}^{m} Z_{ij} \ \underset{\sim}{I}^j \qquad (8)$$

Perturbing Eqn. 8 , about an operating point and neglecting variation of voltage at remote nodes due to small changes in the inputs to the i<u>th</u> generator yieds :

$$\Delta \underset{\sim}{V}^i = Z_{ii} \cdot \Delta \underset{\sim}{I}^i \qquad (9)$$

Components of network voltage and currents are usually expressed in common reference DQ which is different from the machine d-q frame (7) as shown in Fig. 1 . Eqn. 9 , can be transformed from the network to the machine reference using the following transformation :

$$\begin{bmatrix} V_D \\ V_Q \end{bmatrix}^i = \begin{bmatrix} \cos\delta & -\sin\delta \\ \sin\delta & \cos\delta \end{bmatrix}^i \begin{bmatrix} V_d \\ V_q \end{bmatrix}^i \qquad (10)$$

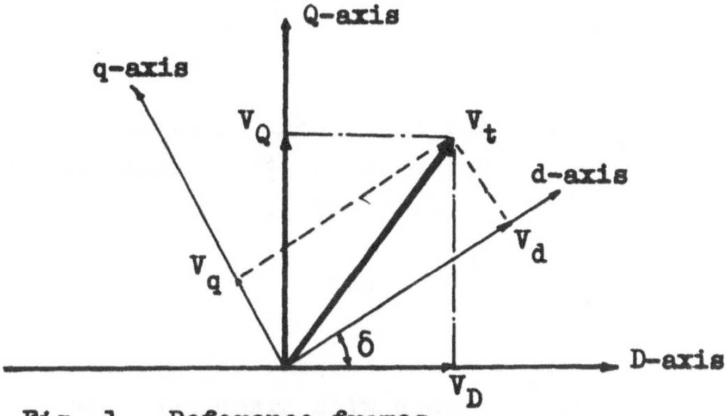

Fig. 1 Reference frames
DQ Network reference
dq Machine reference

Substituting from Eqn. 10 into Eqn. 9 yields :

$$\Delta \underset{\sim}{V}^i_{DQ} = Z_\varepsilon \quad \Delta \underset{\sim}{I}^i_{DQ} \qquad (11)$$

and

$$\Delta \underset{\sim}{V}^i_{dq} = \xi \quad \Delta \underset{\sim}{I}^i_{dq} + \zeta \ \Delta\delta \qquad (12)$$

Where

$$\Delta V^i_{DQ} = \Delta \begin{bmatrix} V_D \\ V_Q \end{bmatrix}^i , \quad \Delta I^i_{DQ} = \Delta \begin{bmatrix} I_D \\ I_Q \end{bmatrix}^i , \quad Z_\varepsilon = \begin{bmatrix} r_\varepsilon & -x_\varepsilon \\ x_\varepsilon & r_\varepsilon \end{bmatrix}$$

Eqn. 12 represents the dynomic behaviour of network axes voltage at the <u>ith</u> node including the effects of interconnections to other generators in the system . Similar equation could be obtained from the machine side , Appendex , leading to :

$$\Delta \underset{\sim}{V}^i_{dq} = \tilde{\xi} \quad \Delta \underset{\sim}{I}^i_{dq} + \tilde{\zeta} \ \Delta\psi_f \qquad (13)$$

From Eqns. 12 and 13 :

$$\underset{\sim}{I}^{i}_{dq} = \begin{bmatrix} K_1 & K_2 \\ K_3 & K_4 \end{bmatrix} \cdot \Delta \begin{bmatrix} \delta \\ \psi_f \end{bmatrix}^i \tag{14}$$

$$\underset{\sim}{V}^{i}_{dq} = \begin{bmatrix} K_5 & K_6 \\ K_7 & K_8 \end{bmatrix} \cdot \Delta \begin{bmatrix} \delta \\ \psi_f \end{bmatrix}^i \tag{15}$$

Perturbing the ith generator non-linear equations, Appendix, and from Eqns. 14 and 15 the following linear time-invariant state-space model for the ith generator is obtained :

$$\Delta \dot{\underset{\sim}{x}}^{i} = A_{\varepsilon} \, \Delta \underset{\sim}{x}^{i} + B_{\varepsilon} \, \Delta \underset{\sim}{u}^{i} \tag{16}$$

Eqn. 16 is an equivalent model for the ith generator which include the effects of interconnections with other machines . Matrices A_{ε} and B_{ε} are the equivalent matrices in which the effects of interconnection with the other machines are included .

Control Algorithm

A physically realizable controller for a turbogenerator has been developed and extensively tested (6, 9) . In this , the state vector is augmented by a vector Y which contains the variables by which the desirable operating conditions can be defined . The technique is described fully in Reference 6 , where it is applied to a single machine . With a multi-machine Eqn. 16 may be developed and following the procedures given in (6) the following control law is obtained for the ith generator :

$$\underset{\sim}{U}^{i}_{k} = \begin{vmatrix} K_1 & K_2 \end{vmatrix} \cdot \begin{bmatrix} \underset{\sim}{x}^{i}_{k} \\ \underset{\sim}{y}^{i}_{k} \end{bmatrix} \tag{17}$$

$$\underset{\sim}{y}^{i}_{k} = \underset{\sim}{Y}^{i}_{k} - \underset{\sim}{Y}_{D}$$

$\underset{\sim}{Y}_{D}$: Contains information about the desired operating point in terms of rotor angle and terminal voltage .

NUMERICAL EXAMPLE

The interconnected power system considered in this study is shown in Fig. 2 . It consists of three generating units , each generator is deriven by a three stage steam turbine with reheat . The generator is represented by a 7th order non-linear model based on Parks equations . Full description of models representing each generating unit is shown in Reference 6 . Simulation of this non-linear system is complicated as it requires the solution of the whole non-linear models of the system , together with the network equations and control loops . The network equations are described in common reference which is different from the d-q axes in which the equations of machines should be solved . Axes transformation between these references are employed . The simulation is described elsewhere (11) .

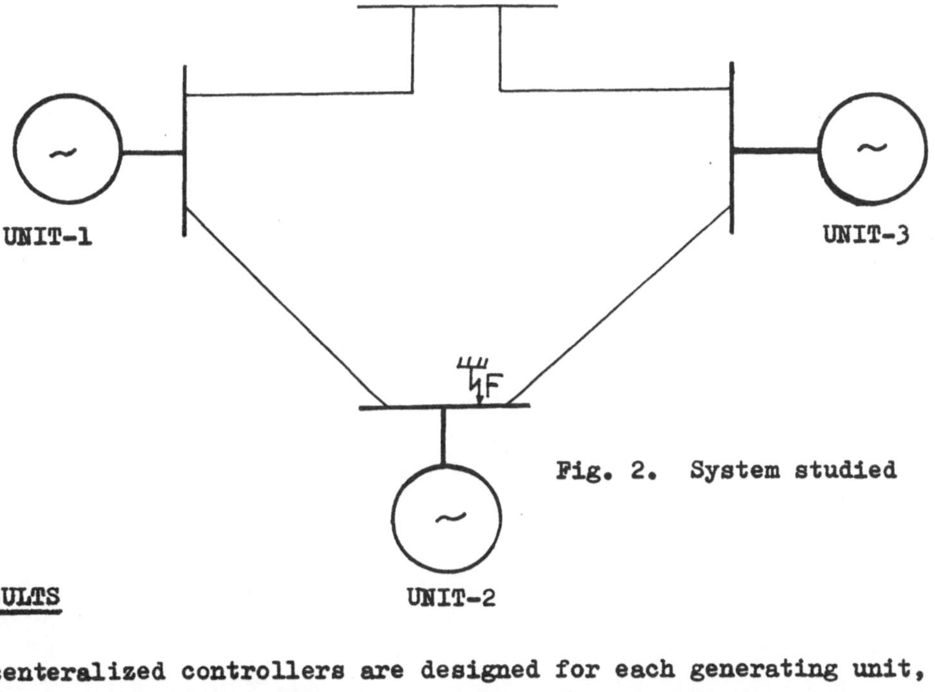

Fig. 2. System studied

RESULTS

Decenteralized controllers are designed for each generating unit, using the method described, by minimizing the performance index using a dynamic programming procedures to compute the control matrices . The performance is shown in Fig. 3 , which illustrates the response to a three-phase short circuit at F (Fig. 2) , cleared after 200 ms. The effect of controllers is shown to provide substantial damping to oscillation with significant reduction in rotor first swing, thus

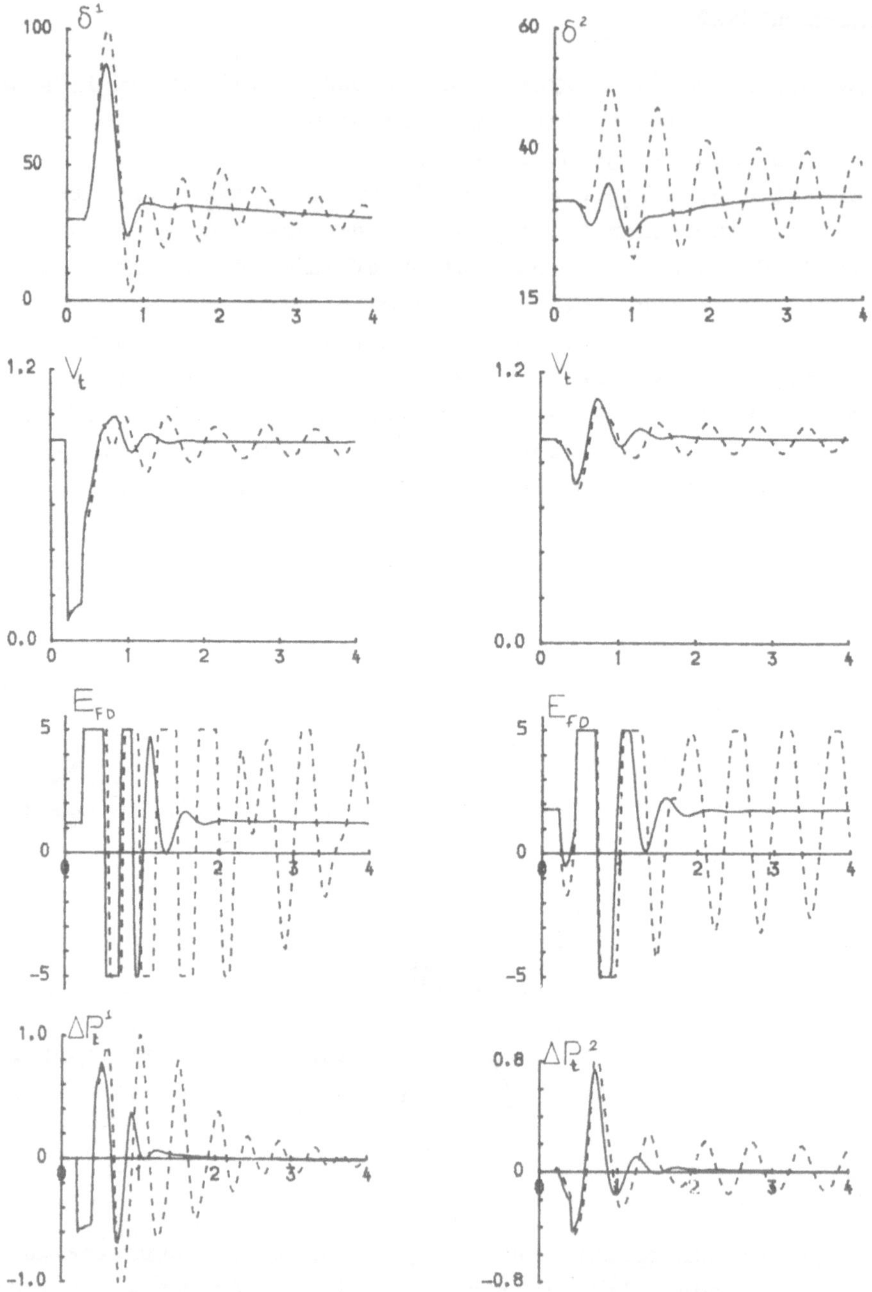

Fig. 3. System response to a three-phase short circuit
------ Conventional controller
——— Decenteralized controller

illustrating increasing in stability limits . The rotor angle δ of machine 1 and 2 were measured w.r.t. machine 3 as there is no infinite bus in the system . The performance of controllers is compared with similar results obtained with conventional controllers which are normally operate in practice . This forms a useful guide to assess novel controllers .

CONCLUSION

The paper describe a method for designing decenteralized controllers for a generating unit operating in a multi-machine environment . The controllers accept changes in system configuration , operating conditions and use measurable control signals at the terminals of machines only . The control law does not refere to remote measurements . The results illustrate the effectiveness of the technique described and significant improvement in performance could be achieved by using decenteralized controllers to replace the conventional controllers which are in common use .

APPENDEX

Non-linear model for the ith machine :

State-Vector : $\underline{X}^t = (\delta , p\delta , \psi_{fd} , G_v , T_M)$

Input-Vector : $\underset{\sim}{U}^t = (u_e , u_g)$

$$\dot{X}_1 = X_2$$
$$\dot{X}_2 = (w_o/2H)(X_5-C_1.X_3.\text{Sin } X_1-C_2 \text{ Sin } X_1 \text{ Cos } X_1-K_d. X_2)$$
$$\dot{X}_3 = C_3 . \text{Cos } X_1 - C_4 . X_3 + C_5 . U_e$$
$$\dot{X}_4 = C_6 . X_4 + C_7 U_g$$
$$\dot{X}_5 = C_8 X_4 + C_9 X_5$$

Where

$$C_1 = \frac{V . X_{ad}}{X_{fd}(X_T+X_d - X_{ed}^2/X_{fd})} , \quad C_2 = \frac{V^2(X_d-X_q-(X_{ad}^2/X_{fd}))}{(X_T+X_d- X_{ad}^2/X_{fd})(X_T+X_q)}$$

$$C_3 = \frac{w_o . V . X_{ad} . r_{fd}}{X_{fd}(X_T+X_d-X_{ad}^2/X_{fd})} , \quad C_4 = \frac{w_o . r_{fd}(X_d+X_T)}{X_{fd}(X_T+X_d-X_{ad}^2/X_{fd})}$$

$$C_5 = w_o . r_{fd}/X_{ad} , \quad C_6 = - 1/T_G$$

$$C_7 = - C_6 , \quad C_8 = 1 / T_h$$

$$C_9 = - C_8$$

REFERENCES

1- Habibullah, B., and Yu, Y., "Physically realizable wide power range optimal controllers for power systems" IEEE Trans. PAS-93, 1974, PP. 1498-1506 .

2- Anderson, J.H. and Raina, V.M., "Power system excitation and governor design using optimal control theory" Int. J. Control, 19, 1974, PP. 289-308 .

3- Elmetwally, M.M., Rao, N.D. and Malik, O.P., "Experimental results on the implementation of an optimal control for synchronous machines" IEEE Trans. PAS-94, 1975, PP. 1192-1200 .

4- Newton, M.E. and Hogg, B.W., "Optimal control of a micro-alternator" IEEE Trans. PAS-95, PP. 1821-1833 .

5- Pullman, R.T. and Hogg, B.W., "Discrete State-space controller for a turbogenerator" Proc. IEE., 126, 1979, PP. 87-92 .

6- Osheba, S.M. and Hogg, B.W., "Multi-variable controller for a turbogenerator" IEEE Winter Power Meeting, 1980, New York, A80-106-5 .

7- Moussa, H.A.M. and Yau-nan Yu, "Optimal stabilization of a multi-machine system" IEEE Trans. PAS-91, 1972, PP. 1174-1182 .

8- Declacour, J.D. and Darwish, M., "Control strategies for large scale systems" Int. J. Control , 27, 1978, PP. 753-767 .

9- Osheba, S.M. and Hogg, B.W., "Performance of state-space controllers for turbogenerators in multi-machine power systems" IEEE Trans. PAS-101, 1982, PP. 3276-3287 .

10- Osheba, S.M. and Hogg, B.W., "Equivalent turbogenerator model for design of multi-variable controllers in multi-machine power systems" Int. J. Electric Power and Energy Systems (To be Published) .

11- Osheba, S.M., "State-space controllers for A.C. Turbogenerators in multi-machine electric power systems" Ph.D Thesis, Liverpool University, May 1981 .

OPTIMAL STABILISATION OF A MULTI-MACHINE POWER SYSTEM THROUGH THE MATRIX SIGN FUNCTION THEORY

A.V. Machias E.N. Dialynas

National Technical University
Electric Energy Systems Laboratory
42 Patission Street-GR 106 82
Athens - Greece

ABSTRACT - The objective of the paper is the optimal linear regulator design of a multi-machine power system through a computational algorithm based on the matrix sign function theory, which can give solutions without the evaluation of the system eigenvalues. For this purpose, a computer program has been written to incorporate the developed computational techniques which are based on the matrix sign function theory and can obtain the optimal controllers and the dynamic responses of the power system. The system under study consists of one thermo plant, two hydro plants and an infinite bus. The state equations of the multi-machine system are obtained using the linearised equations of each machine together with the equations for the exciter and voltage regulator system and the equations for the governor - hydraulic system.

1. INTRODUCTION

The importance of small signal dynamic stability of multi-machine power systems starts having practical interest as the control requirements of the power plants become more sophisticated and demanding. The application of the optimal control theory to stabilise power systems using exciter-regulator and governor systems is relatively new[6]. The multimachine system is described by equations of the state variable form

$$\dot{Y} = AY + Bu \tag{1}$$

where Y is the n-state variable vector, u is the control vector with r signals, A(nxn) is the system coefficient matrix and B(nxr) is the coefficient matrix of controls.

The paper considers the low order machine model for the controller design which is obtained by applying a matrix elimination technique on the system algebraic equations and by neglecting the flux linkage variations except the field flux linkage variable (ψ_F).

The real problem is the development of a computational technique to determine the stabilisation variables. It has been shown[1] that the optimal control problem requires the solution of a Riccati matrix equation. This solution can be determined by using the matrix sign function theory which has been proved to be very effective in the stabilisation study of a single machine power system [2]. The same technique is used in this paper for the multi-machine optimal controllers design.

2. PRINCIPLE OF THE DEVELOPED ALGORITHM

It is well known that by means of a suitable linear feedback of the states, the controllable linear dynamic system can be stabilised to have an optimal performance. This is succeeded by minimising the cost function of the form [1], [4].

$$J = \frac{1}{2} \int_0^{\infty} (Y^T QY + u^T Ru)\, dt \qquad (2)$$

where Q and R are nxn and rxr diagonal matrices of constant coefficients.
The optimal control is found by applying Pontyagin's maximum principle. Engineering experience is useful in choosing Q and R matrices while computational methods have also been developed to evaluate the appropriate values of Q and R. This paper considers that all the coefficients of Q and R matrices are units with the exception of these belonging to Q and corresponding to the torque angle and speed variables which are considered to be ten.
The purpose of minimising the cost function [6] is to minimise the control effort and error response of the system after a disturbance.
The optimal control is given by,

$$u = -R^{-1} B^T K Y \qquad (3)$$

where K is the Riccati matrix, that is computed by solving the non-linear algebraic matrix Riccati equation

$$-KA - A^T K + KBR^{-1} B^T K - Q = 0 \qquad (4)$$

Matrix K is symmetric because K^T satisfies also the same equation (4).
The closed loop system equations with optimal control become:

$$\dot{Y} = A_1 Y \qquad (5)$$

where

$$A1 = A - BR^{-1} B^T K \qquad (6)$$

In reference [5] a procedure is reported to compute the K matrix. This procedure is based on shifting all the eigenvalues of the initial system to the left complex plane through an eigensystem sensitivity method [6]. However, the evaluation of system eigenvalues requires a considerable amount of computational effort which increases rapidly as the system size increases. To overcome this serious problem, a computational algorithm has been developed, which obtains the solution of equation (4) (K matrix), without the calculation of system eigenvalues. This algorithm is based on the matrix sign function theory [7] and is summarised in the flowchart shown in Figure 1.
The sign function matrix, sign (H) of a matrix H (2nx2n) is evaluated applying the

following iterative Newton-Raphson technique:

$$S(i+1) = \frac{1}{2}\left[S(i) + S^{-1}(i) \right]$$

where $S(0) = H$ and $S(i)$ converges to sign (H).

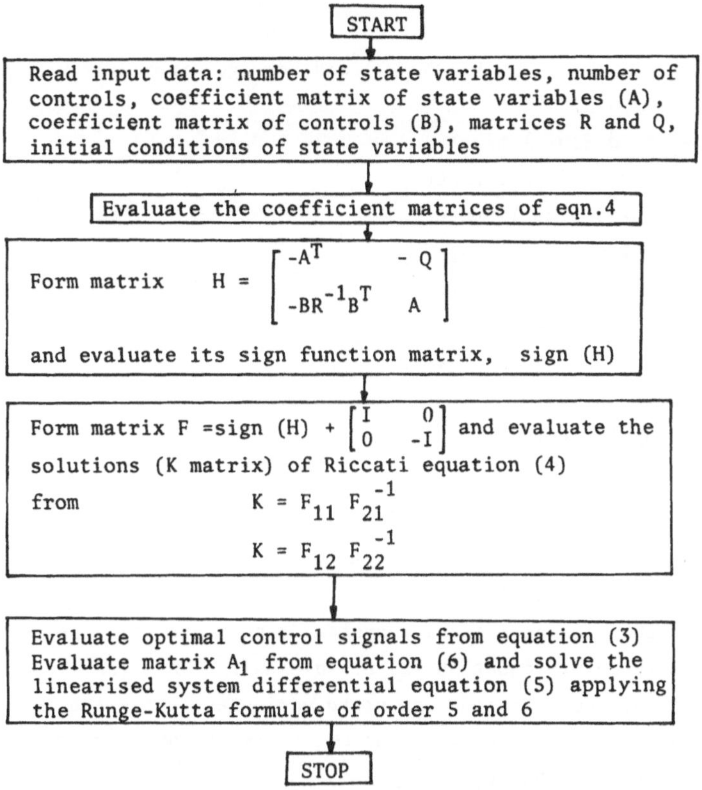

Figure 1. Flowchart showing computational algorithm

In order to improve the convergence properties of this iterative technique, several speeding modifications have been tested and the following one has been proved more useful and has been applied:

$$S(i+1) = \alpha\, S(i) + \beta\, S^{-1}(i)$$

where:

$$\alpha = ||S^{-1}(i)|| / [\,||S(i)+S^{-1}(i)||\,]$$

$$\beta = ||S(i)|| / [\,||S(i)+S^{-1}(i)||\,]$$

and

$$||S(i)|| = \sum_{j=1}^{2n} \sum_{m=1}^{2n} S_{jm}^{2}(i)$$

3. MULTI-MACHINE STATE EQUATIONS

The state equations can be easily assembled into a matrix equation, which has the form of equation (1) and represents [6].

a) Machine windings

b) Exciter and voltage regulator system

c) Torque of the complete system

d) Governor system

There are altogether thirteen sets of equations and the corresponding nine machine variables (ψ_F, ψ_d, ψ_q, ψ_D, ψ_Q , V_F, V_R, δ, ω) and four governor system variables (a, a_f, g, h) are:

ψ_F = field flux linkage

ψ_d = d-axis flux linkage

ψ_q = q-axis flux linkage

ψ_D = D-axis flux linkage

ψ_Q = Q-axis flux linkage

V_F = field voltage

V_R = exciter-regulator voltage

δ = torque angle

ω = angular velocity

a = gate actuator signal

a_f = dashpot feedback signal

g = gate movement

h = hydraulic head

It must be emphasised that for an n-machine system, each state variable is represented by an n-vector and therefore equation (1) has dimension 13n x 13n. If the machine equations are linearised, the corresponding state variables are of the linearised form $\Delta\psi_F$, $\Delta\psi_d$, $\Delta\psi_q$, $\Delta\psi_D$, $\Delta\psi_Q$, ΔV_F, ΔV_R, $\Delta\delta$ and $\Delta\omega$ and represent a deviation from the initial state under investigation [6].

For the study of a system with n-machines and an infinite bus, the system coefficient matrix has also the same form because the existence of an infinite bus affects only the network node impedance matrix.

4. NUMERICAL EXAMPLE

Consider the three-plants system example [6], [8] consisting of one thermo plant (≠1), two hydro plants (≠2,≠3) and an infinite bus. Each plant is modelled as a fourth order system (ψ_F, V_F, δ,ω), a third order synchronous machine plus a first order exciter-regulator system. For the linearised system the state variables are $\Delta\psi_F$, ΔV_F, $\Delta\delta$,$\Delta\omega$ and the dimension of system coefficient matrix is 12x12.

The numerical values of matrix A of equation (1) are given in reference [6], while Q and R matrices are chosen as follows:

$$Q = \text{diag.}\,[1\;\;1\;\;10\;\;10\;\;1\;\;1\;\;10\;\;10\;\;1\;\;1\;\;10\;\;10]$$

$$R = \text{diag}\,[\,1\;\;1\;\;1]$$

Matrix Q is chosen so that, more emphasis is given to the torgue angle ($\Delta\delta$) and angular velocity ($\Delta\omega$) state variables of each machine.
The B matrix of equation (1) is [6].

$$B = \begin{bmatrix} 0 & 36.1 & 0 & 0 & 0 & 0 & 0 & 0 & 0 & 0 & 0 \\ 0 & 0 & 0 & 0 & 0 & 78.9 & 0 & 0 & 0 & 0 & 0 \\ 0 & 0 & 0 & 0 & 0 & 0 & 0 & 0 & 0 & 100 & 0 & 0 \end{bmatrix}^T$$

A three phase fault near plant ≠3 is considered and the system becomes unstable (oscillatory) without stabilising signals. This can be verified by obtaining the system variable responses from an integration of equation (1).
The optimal system responses are obtained by applying the computational algorithm shown in Fig. 1. The algorithm evaluates two solutions of the Riccati eqn.(4) (K matrix). For the first solution the system remains unstable while the second one makes it stable. The system variable ($\Delta\delta$, $\Delta\psi_F$, ΔV_F, $\Delta\omega$) responses corresponding to the second solution are shown in Figures 2, 3, 4 and 5 respectively.
The optimal system responses of control signals are shown in Fig. 6.
As mentioned above, the multi-machine system becomes unstable. Therefore, a one machine optimal excitation control can be designed for plant ≠3 in order to stabilise the multi-machine system. In the design, of course, all system dynamics are included. The numerical values of matrices A and Q remain the same while matrices B and R are as follows:

$$B = \begin{bmatrix} 0 & 0 & 0 & 0 & 0 & 0 & 0 & 0 & 0 & 100 & 0 & 0 \end{bmatrix}^T$$

$$R = 1$$

The optimal system responses with this type of control signals are obtained using the same procedure as above. The responses of torque angle ($\Delta\delta$) for all three plants are shown in Fig. 7. From Fig. 7 it can be seen that plant ≠3 becomes stable very quickly while the other two plants oscillate for considerable time before becoming stable. This shows that multi-machine systems are better stabilised using multi-optimal controllers.

5. CONCLUSIONS

The paper presents a computational algorithm which is based on the matrix sign function theory and evaluates very efficiently the optimal controllers required in the optimal linear regulator design of a multi-machine power system. This algorithm has been implemented into a computer program which has the following advantages. (a) Analyses power systems with any number of machines (b) Models each machine with any number of differential equations.(c) Permits the existence of any number of control signals to each system machine (d) the storage requirements are smaller compared with the corresponding program of Puri's and Gruver's method (e) it is very economical in computing time because it is not required to evaluate the system eigen-

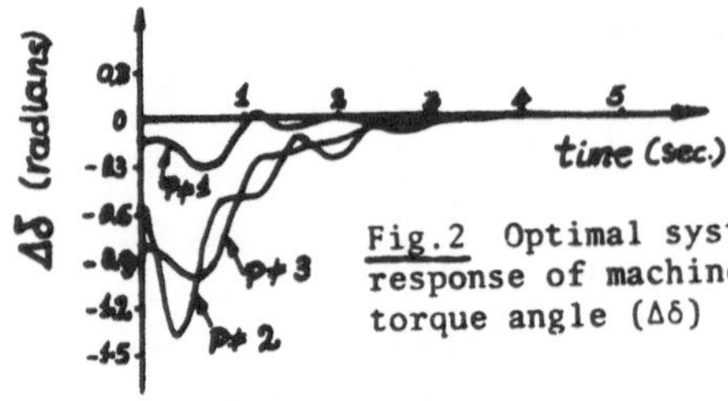

Fig.2 Optimal system response of machines' torque angle (Δδ)

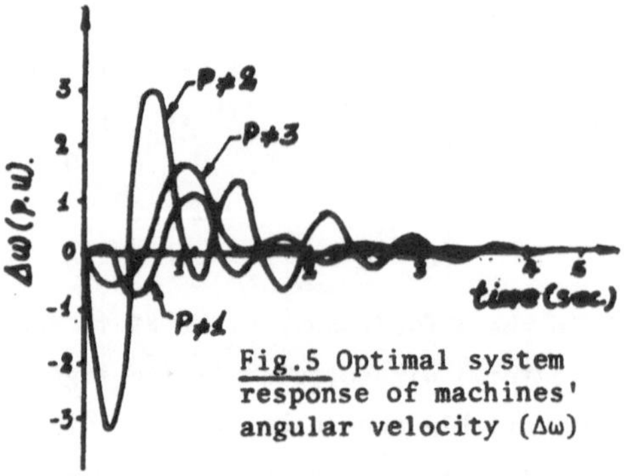

Fig.5 Optimal system response of machines' angular velocity (Δω)

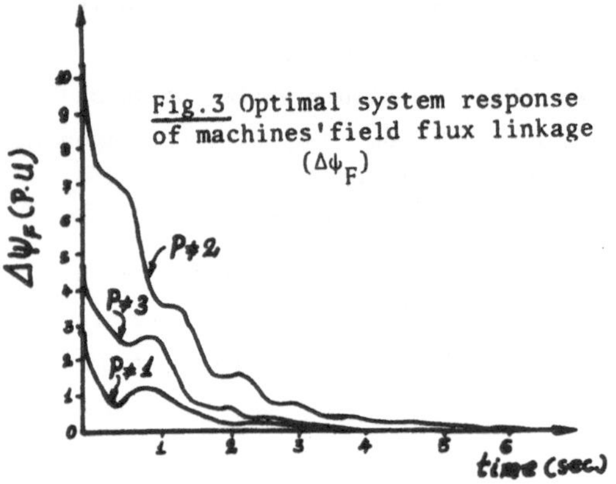

Fig.3 Optimal system response of machines' field flux linkage ($\Delta\psi_F$)

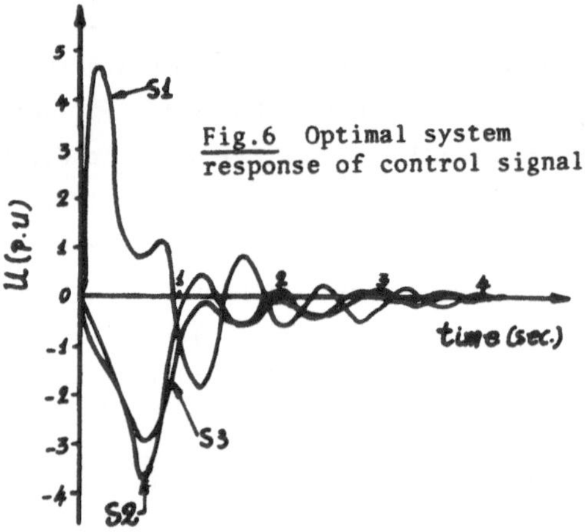

Fig.6 Optimal system response of control signals

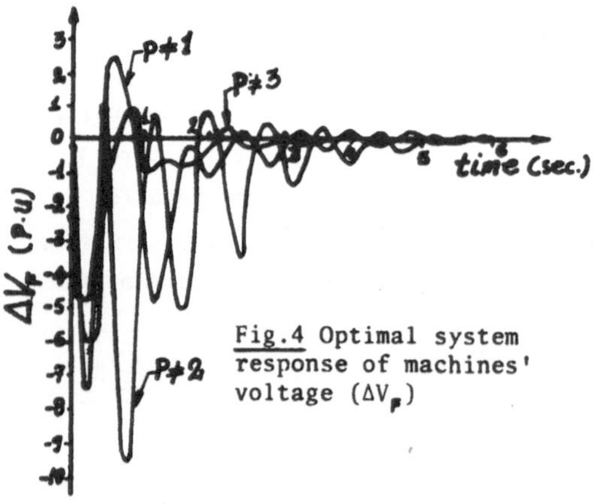

Fig.4 Optimal system response of machines' voltage (ΔV_F)

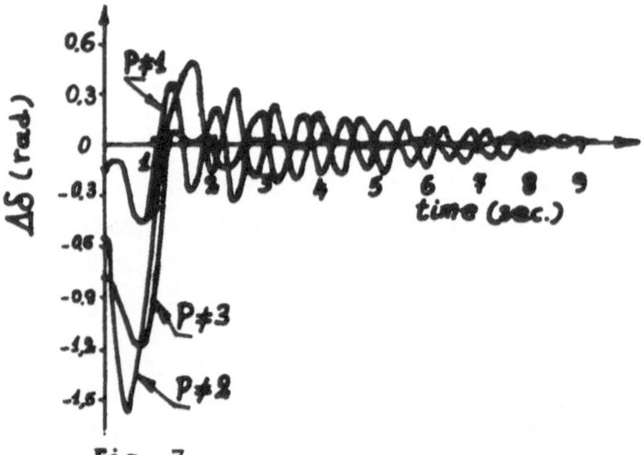

Fig. 7
Optimal System response of machines' torque angle (Δδ) with optimal control only on plant ≠ 3.

where: P≠1 = plant ≠ 1, P≠2 = plant ≠ 2, P≠3 = plant ≠ 3,
S1 = signal 1, S2 = signal 2, S3 = signal 3

values.

Therefore, the system designer can use the presented computational algorithm and program to analyse any size of power systems and study a great number of alternative control designs in order to obtain the optimal one.

6. REFERENCES

[1] YAO-NAN YU, KHIEN VONGSURIYA and L.N.WEDMAN: "Application of an optimal control theory to a power system" (IEEE Trans. on power Apparatus and Systems, Vol. PAS-89, No. 1, 1970, pp. 55-62).

[2] E.N. DIALYNAS, A.V.MACHIAS and B.C.PAPADIAS "Optimal power system stabilization using the matrix sign function theory", (MELECON'83 conference, Proc. Vol. II, D5.05 - Athens, May 1983).

[3] N.N. PURI and W.A. GRUVER "Optimal Control design via seccessive approximations" (JACC 1967 Philadelphia, Pa., pp. 335-344).

[4] M.ATHANS and P.L. FALD, "Optimal Control"(N.Y. Mc Graw-Hill, 1966).

[5] H.A. MOUSSA and YAO-NAN YU, "Optimal power system stabilization through excitation and/or governor control"(IEEE Trans. on Power App. and Systems, Vol. PAS-197, 1971, pp. 1166-1173).

[6] YAO-NAN YU and H.A.M. MOUSSA, "Optimal Stabilization of a Multi-Machine System" (IEEE Summer Meeting, July 18-23, 1971, pp. 1174-1182)

[7] E.D. DENMAN, "The matrix sign function and computations in Systems" (Applied mathematics and Computation 2, 63-94, 1976).

[8] Y.N. YU and C. SIGGERS, "Stabilization and Optimal Control signals for a power System" (IEEE Trans. Vol. PAS-90, July/Aug. 1971, pp. 1469-1481).

DECENTRALIZED SUBOPTIMAL CONTROL FOR THE COMPLEX POWER SYSTEM BY USING MODIFIED BELLMAN-LYAPUNOFF EQUATION

V.Veselý, K.M.Soliman and J.Murgaš
EF, SVŠT, Katedra ASR-TP, 81219 Brat., ČSSR

ABSTRACT:

This paper is concerned with the Decentralized Control of the dynamic behaviour of the Complex Power System(CPS). It is suggested here two different approaches for designing Suboptimal Decentralized Controllers by using the Modified Bellman-Lyapunoff Equation(MBLE) via the optimal techniques. In the first approach,Suboptimal Control Laws for turbine powers are deduced by the aid of scalar Lyapunoff Function(LF) for the whole system. In the second one,Suboptimal Control Laws are deduced with the aid of using vector LF which can be constructed by decomposing the power system to some interconnected subsystems.

INTRODUCTION:

Many theoretical studies have been carried out concerning with the design of dynamic behaviour Controllers for large scale power systems [1,3,5-12],Some classes of these works are interesting with the design of Decentralized Controllers through the construction of BLEs[3,7,9] or MBLEs [3,8,10] as functions of scalar or vector LFs. The trend of using MBLEs as a class of applying the Modern Control Theory for Controllers design fulfils the main advantage of achieving some amount of optimization for the chosen system LF. That is by constructing the performance index under optimization in the form of BLE according to Malkin Sense [4] ,and then optimizing this equation with respect to some suggested factors(a_i) which are multiplied by the LF-components. Nevertheless, the validity of this technique when it is applied on power systems is until now under test, and it was handled in literatures by little amount. So, the objective of this paper is to apply, extend and advance the utilization of this technique in power systems, that is in purpose of modifying and advancing the system dynamic operation through the transient periods produced from sudden or sustained large accidents.

BASIC MATHEMATICAL CONSTRUCTIONS:

Neglecting the internal resistances of the machines in the CPS which contains n-Synchronous Generating Machines(SGMs),and assuming constant back e.m.f. behind the transient reactances,two well known differential equations[5,9,12] can be used for representing the dynamic transient state of each machine. As a function of the state variables chosen around the steady state equilibrium point(δ_{io}, ω_{io} and P_{tio}), these equations which will be used in the first approach can be stated as:

$$dx_{i1} / dt = x_{i2} \qquad ,i=1,2,..,n \quad (1)$$

$$dx_{i2} / dt = x_{i3}/T_i - \lambda_i x_{i2} - \sum_{j=1, j\neq i}^{n} A_{ij}\left[\sin(x_{i1}-x_{j1}+x_{ijo}) - \sin(x_{ijo})\right] \qquad ,i=1,2,..,n \quad (2)$$

where: $x_{i1} = \Delta\delta_i = \delta_i - \delta_{io}$, $x_{i2} = \Delta\omega_i = \omega_i - \omega_{io}$

$x_{ijo} = \delta_{io} - \delta_{jo}$, $A_{ij} = E_i E_j y_{ij}/T_i$, $\lambda_i = D_i/T_i$

As a mathematical model for representing each controlled turbine, the following first order state model equation will be chosen[6]:

$$dx_{i3} / dt = -x_{i3}/T_{pi} - x_{i2}/\sigma_i T_{pi} + u_i/T_{pi} \qquad ,i=1,2,..,n \quad (3)$$

where: $x_{i3} = \Delta P_{ti} = P_{ti} - P_{tio}$

P_{tio} is the steady state input power to the SGM number i.

$u_i = K_{ti}P_{gi} - P_{tio}$ is the feed back control signal acting on the turbine which is vanished in cases of uncontrolled turbine.

K_{ti}, P_{gi} and σ_i are the mechanical turbine efficiency, input power to the turbine, and output static regulation coefficient resp.

In the second approach, the n-th generating plant will be chosen as an infinite Bus-Bar[1,3], and the overall system will be decomposed into (n-1) interconnected subsystems each of them contains one SGM. Hence the decomposed model will be reformulated to the following forms:

$$dx_{i1} = x_{i2} \qquad ,i=1,2,..,n-1 \quad (4)$$

$$dx_{i2}/dt = x_{i3}/T_i - \lambda_i x_{i2} - A_{in} f_i(x_{i1}) + h_i(\bar{X}) \qquad ,i=1,2,..,n-1 \quad (5)$$

$$dx_{i3}/dt = -x_{i3}/T_{pi} - x_{i2}/\sigma_i T_{pi} + u_i/T_{pi} \qquad ,i=1,2,..,n-1 \quad (6)$$

where the self and interconnection functions between subsystems will be:

$$f_i(x_{i1}) = \sin(x_{i1}+x_{ino}) - \sin(x_{ino}) \quad (7)$$

$$h_i(\bar{X}) = -\sum_{j=1, j\neq i}^{n-1} A_{ij}\left[\sin(x_{i1}-x_{j1}+x_{ijo}) - \sin(x_{ijo})\right] \quad (8)$$

FIRST APPROACH: SUBOPTIMAL CONTROLLERS DESIGN FOR THE WHOLE SYSTEM:

For the system state equations(1) to (3), the LF necessary for confirming the transient stability conditions can be suggested as follows:

$$V = \sum_{i=1}^{n} (V_{i1} + V_{i2}) \quad (9)$$

$$V_{i1}= .5T_i x_{i2}^2 - x_{i1}P_{tio} - \sum_{j=i+1}^{n} F_i T_i A_{ij}\left[\cos(x_{i1}-x_{j1}+x_{ijo})-\cos(x_{ijo})\right] \quad (10)$$

$$V_{i2}= (T_{pi}/2D_i)\ x_{i3}^2 + x_{i3}x_{i2}/\lambda_i + .5T_i\sigma_i' x_{i2}^2 \quad (11)$$

where: $F_i = \begin{pmatrix} 1 & \text{for} & i\neq n \\ 0 & \text{for} & i=n \end{pmatrix}$, $\sigma_i' = (\sigma_i T_i + T_{pi}) \;/\; (\sigma_i D_i T_{pi})$

The time derivative of this LF can be deduced in the form:

$$\dot{V} = -\sum_{i=1}^{n} \left(D_i' x_{i2}^2 + S_i(\bar{X})\right) \quad (12)$$

$$S_i(\bar{X})=(\sigma_i' x_{i2}+x_{i3}/D_i) \sum_{j=1,\, j\neq i}^{n} T_i A_{ij}\left[\sin(x_{i1}-x_{j1}+x_{ijo})-\sin(x_{ijo})\right] \quad (13)$$

where: $D_i' = D_i(1 + \sigma_i') + 1 \;/\; (\lambda_i \sigma_i T_{pi})$

The prove of the positive definite condition of this LF,and the negative semidefinite condition of its time derivative,can be found in details in References [2,3]. Adding the suggested positive modification factors(a_i), and applying on the system by the control signal($u_i \neq 0$), the modified LF and its time derivative will be:

$$V_m = \sum_{i=1}^{n} a_i(V_{i1} + V_{i2}) \qquad \text{with } a_i > 0 \qquad ,i=1,2,..,n \quad (14)$$

$$\dot{V}_m\big|_c = -\sum_{i=1}^{n}\left[a_i D_i' x_{i2}^2 + a_i S_i(\bar{X}) - (a_i u_i/D_i)(x_{i3}+T_i x_{i2}/T_{pi})\right] \quad (15)$$

and the Optimal Control Law(OCL) as a function of the MBLE(B_m) will be:

$$\underset{\bar{A}}{\mathrm{Max}}\left[\underset{\bar{U}}{\mathrm{Min}}\left[B_m(\bar{X},\bar{U},\bar{A},t) = \dot{V}_m\big|_c + \sum_{i=1}^{n}(\bar{X}_i^T[Q_i]\bar{X}_i + \bar{U}_i^T[R_i]\bar{U}_i)\right]\right] \doteq -0 \quad (16)$$

where: $[Q_i]$ and $[R_i]$ are the positive definite energy-or power-matrices which construct the different components of the performance optimized function corresponding to state and control vectors resp.

Applying the sufficient conditions of minimizing B_m with respect to u_i (zero first derivative and positive second derivative), the following local minimum expression for the control variable will be obtained:

$$u_i^* = -(a_i/2r_iD_i)(x_{i3}+T_i x_{i2}/T_{pi}) \qquad \text{with } r_i > 0 \quad ,i=1,2,..,n \quad (17)$$

Back substitution into(16) leads to the following form of the OCL:

$$\underset{\bar{A}}{\mathrm{Max}}\left[B_m^*(\bar{X},\bar{A},t) = -\sum_{i=1}^{n}\left[\bar{X}_i^T(a_i^2[N_i] + a_i[M_i] + [-Q_i])\bar{X}_i + a_i S_i(\bar{X})\right]\right] \doteq -0 \quad (18)$$

$$\text{where: } [N_i] = \begin{bmatrix} 0 & 0 & 0 \\ 0 & 1/(4r_i\lambda_i^2T_{pi}^2) & 1/(4r_iD_i\lambda_iT_{pi}) \\ 0 & 1/(4r_iD_i\lambda_iT_{pi}) & 1/(4r_iD_i^2) \end{bmatrix},$$

$$[M_i] = \begin{bmatrix} 0 & 0 & 0 \\ 0 & D_i' & 0 \\ 0 & 0 & 0 \end{bmatrix} \quad \text{and} \quad [Q_i] = \begin{bmatrix} 0 & 0 & 0 \\ 0 & q_{22} & q_{23} \\ 0 & q_{23} & q_{33} \end{bmatrix}$$

Due to the term $S_i(\bar{X})$, the design of optimal values for a_i will be accompanied with very much amount of mathematical complications. In our belief, the negligence of this term in the following steps may be accepted for the following reasons:

1-This negligence will not affect the system stability because it is already taken into account through the fulfilment of negative semi-definite condition of $\dot{V}$.

2-This term had not any effect on the deduction of u_i^*.

3-Since the summation part in this term expresses the transient change in the system electric power, so its value may be considered small, and consequently with little effect on the designed factors a_i^*.

Hence, chosing the matrix $[Q_i]$ as mentioned before, and using the property of the eigen value extreme, suboptimal values a_i^* can be obtained by pushing B_m^* to be near to zero as possible, then the OCL can be converted to the following scalar norm equation:

$$\underset{\bar{A}}{\mathrm{Max}}\left[B_m^*(\bar{X},\bar{A},t)\right] = -\sum_{i=1}^{n}\left(a_i^{*2}\lambda_{min}[N_i]+a_i^*\lambda_{min}[M_i]+\lambda_{min}[-Q_i]\right)\|\bar{X}_i\|^2=0 \quad (19)$$

where λ_{min} is the minimum eigen value of the corresponding matrix.

Since the norm of the state vector cannot be considered zero, and avoiding the choice of the zero eigen values which produce meaningless trivial solution, a real unique solution of equation(19) will be:

$$a_i^* = \sqrt{\left[(2r_iD_i'D_i^2T_{pi}^2)/(T_i^2+T_{pi}^2)\right]^2+(4r_iD_i^2T_{pi}\lambda_{max}[Q_i])/(T_i^2+T_{pi}^2)} - (2r_iD_i'D_i^2T_{pi}^2)/(T_i^2+T_{pi}^2) \quad ,i=1,2,..,n \quad (20)$$

SECOND APPROACH: SUBOPTIMAL CONTROLLERS DESIGN BY DECOMPOSITION:

Till the preparation of this paper, the concept of designing these Controllers through the construction of VLF for the whole state model given by equations(4) to(6) did not lead to good results. So we will follow here a design procedure through two main steps [10]. In the first step, the turbine power(ΔP_{ti} or x_{i3}) will be considered as an Intermediate

Control Variable(u_i') applying on the swing equations(4) and(5) which represent the decomposed system to the following state form:

$$\dot{\bar{X}}_i=\begin{bmatrix}0 & 1\\ 0 & -\lambda_i\end{bmatrix}\bar{X}_i+\begin{bmatrix}0\\ -A_{in}\end{bmatrix}f_i(x_{i1})+\begin{bmatrix}0\\ 1\end{bmatrix}h_i(\bar{X})+\begin{bmatrix}0\\ 1/T_i\end{bmatrix}u_i' \quad ,i=1,2,..,n-1 \quad (21)$$

where: $\bar{X}_i^T=[x_{i1} \quad x_{i2}]=[\Delta\delta_i \quad \Delta\omega_i]$, $\bar{X}^T=[\bar{X}_1 \quad \bar{X}_2 \quad .. \quad \bar{X}_{n-1}]$

Since $f_i(x_{i1})$ fulfils Popov Sector[3], a chosen modified VLE with its time derivative for this controlled system can be expressed as:

$$\bar{V}_m=\sum_{i=1}^{n-1} a_iV_i \qquad \text{with} \qquad \dot{\bar{V}}_m\Big|_c=\sum_{i=1}^{n-1} a_i\dot{V}_{mi} \quad (22)$$

$$V_i=\bar{X}_i^T[H_i]\bar{X}_i+g_i\int_0^{x_{i1}} A_{in}f_i(y_i)\,dy_i \quad (23)$$

$$\dot{V}_{mi}=-\bar{X}_i^T[G_i]\bar{X}_i-A_{in}x_{i1}f_i(x_{i1})+(x_{i1}+g_ix_{i2})h_i(\bar{X})+(x_{i1}+g_ix_{i2})u_i'/T_i \quad (24)$$

where: $[H_i]=.5\begin{bmatrix}\lambda_i & 1\\ 1 & g_i\end{bmatrix}$, $[G_i]=\begin{bmatrix}0 & 0\\ 0 & \lambda_ig_i-1\end{bmatrix}$,

g_i is voluntary constant,which must be greater than $1/\lambda_i$ to confirm positive definite$[H_i]$ and positive semidefinite$[G_i]$.

Similar OCL as given by equation(16) leads to the following expression:

$$u_i'^{*}=-(a_i/2r_iT_i)(x_{i1}+g_ix_{i2}) \quad \text{with } r_i>0 \qquad ,i=1,2,..,n-1 \quad (25)$$

Back substitution leads to corresponding formula to(18) as follows:

$$\underset{\bar{A}}{\text{Max}}\left[B_m^{*}(\bar{X},\bar{A},t)=\sum_{i=1}^{n-1} B_i\right]\doteq -0 \qquad \text{for} \quad \bar{A}\doteq\bar{A}^{*} \quad (26)$$

$$B_i=-a_ix_{i2}^2(\lambda_ig_i-1)-a_i^2(x_{i1}+g_ix_{i2})^2/(4r_iT_i^2)-a_iA_{in}x_{i1}f_i(x_{i1})$$
$$+a_i(x_{i1}+g_ix_{i2})\,h_i(\bar{X})+\bar{X}_i^T[Q_i]\bar{X}_i \quad (27)$$

Since it is required to push the negative function B_m^{*}(or B_i) to a value near to the zero as possible,so smaller values for the negative terms and greater values for the positive terms may be accepted as approximations which lead to suboptimal solution instead of an optimal one.Hence in the above equation(27), we can substitute the corresponding right hand side terms of the following inequalities instead of the L.H.Ss:

1- $x_{i1}f_i(x_{i1}) \geqslant x_{i1}^2\cos(x_{ino})$ (by using Taylor expansion)

2- $\bar{X}_i^T[Q_i]\bar{X}_i \leqslant \lambda_{max}[Q_i]||\bar{X}_i||^2$ (trajectory extreme)

3- $\bar{X}_i^T[M_i]\bar{X}_i \geqslant \lambda_{min}[M_i]\|\bar{X}_i\|^2$ (trajectory extreme)

4- $h_i(\bar{X}) \leqslant A''_{ij}\|x_{i1}\| + \sum_{j=1,j\neq i}^{n-1} A'_{ij}\|x_{j1}\|$ (Taylor expansion and trajectory extreme)

5- $x_{i1}+g_i x_{i2} \leqslant \sqrt{1+g_i^2}\,\|\bar{X}_i\|$

where: $A'_{ij} = A_{ij}\cos(x_{ijo})$, $A''_{ij} = \sum_{j=1,j\neq i}^{n-1} A'_{ij}$

$$[M_i] = \begin{bmatrix} A'_{in}+a_i/(4r_iT_i^2) & g_i a_i/(4r_iT_i^2) \\ g_i a_i/(4r_iT_i^2) & \lambda_i g_i - 1 + a_i g_i^2/(4r_iT_i^2) \end{bmatrix}$$

By using these approximations, the OCL will take the form:

$$\underset{\bar{A}}{\mathrm{Max}}\left[B_m^*(\bar{X},\bar{A},t) = -\left[\|\bar{X}\|\right]^T[N]\left[\|\bar{X}\|\right]\right] \triangleq -0 \quad \text{for } \bar{A} \triangleq \bar{A}^* \qquad (28)$$

where $[N]$ is symmetralized matrix of the following elements:

$$N_{ii} = a_i\lambda_{min}[M_i] - a_i A''_{ij}\sqrt{1+g_i^2} - \lambda_{max}[Q_i] > 0 \quad ,i=1,2,..,n-1 \qquad (29)$$

$$N_{ij} = -.5\left(a_i A'_{ij}\sqrt{1+g_i^2} + a_j A'_{ji}\sqrt{1+g_j^2}\right) \quad ,i,j=1,2,..,n-1,i\neq j \qquad (30)$$

equation (28) leads to the following final condition of optimality:

$$B_m^*(\bar{X},\bar{A},t) = -\lambda_{min}[N]\,\left\|\left[\|\bar{X}\|\right]\right\|^2 = 0 \quad \text{i.e. } \lambda_{min}[N] = 0 \qquad (31)$$

Since equation(31) is a function of (n-1) values of a_i^*, hence a trial and error procedure can be used for chosing (n-1) values of g_i and (n-2) values of a_i^*, and then calculating the residual one in condition of fulfilling the positive condition of equation(29).

In the second step of suboptimal Controllers design, the obtained formula of $u_i'^*$ by equation(25) will be substituted into equation(6) in the place of x_{i3} through the Linear Laplace Transform to get a suggested formula for the overall system Suboptimal Control Signal as follows:

$$u_i^* = x_{i2}/\sigma_i - (a_i^*/2r_iT_i)\left[x_{i1} + (g_i+T_{pi})x_{i2} + T_{pi}g_i\, dx_{i2}/dt\right] \qquad (32)$$

NUMERICAL RESULTS AND COMMENTS:

For testing the validity of the proposed approaches, meanwhile making comparisin between them, it is used an application example of two SGMs connected together with an infinite Bus-Bar. Numerical data and complete results can be found at the authors [2]. Figures(1) to (4) show some of the transient variations for both of $(\Delta\delta_1)$ and $(\Delta\delta_2)$. From all the obtained results, the following comments can be summarized:

1- In general, better transient characteristics are achieved by using both of the proposed control systems.

2- In the first approach, suitable values for [R]-elements can be chosen in the range of .01 to 1, whilst in the second approach their suitable range is 10 to 100. It is important to notice that very small values for these elements represent the cases of violent controlling effect on the turbine power which may be practically difficult.

3- In the second approach, much amount of mathematical trials-which may be sometimes unsuccessful- must be carried out for calculating the suboptimal values of a_1, whilst in the first approach only direct and simple calculations are enough.

FINAL CONCLUSION:

The two proposed approaches for designing Decentralized Suboptimal Controllers for the CPS have the advantage of ensuring the system stability conditions in normal cases, and also when the system is imposed to strong accident disturbances. Better Controllers effects are obtained when the Controllers design is based upon the construction of VLF with system decomposition, but also this idea is accompanied with more mathematical complications.

REFERENCES:

1-Jocic Lj.B. and Šiliak D.D.,(1978),"Decomposition and Stability of Multimachine Power System",Congress IFAC,Helsinki,Finland.
2-Kamel M.S.,(1984),"Decentralized Control and Stability Studies of the Complex Power System",CSc.Thesis,Bratislava,Czech.
3-Koščova M.,(1980),"Decentralized Control for transient processes in Electric Systems",CSc.Thesis,Bratislava,Czech.
4-Malkin J.G.,(1966),"Stability Theory of Movement Systems",Moscow.
5-Mansour M.,(1972),"Stability Analysis and Control of Power Systems", Proc.of the Symposiom of Real-Time Control of Electric Power Systems, Edited by Edmund Hondschin.
6-Sterninon L.D.,(1975),"Transient Process under Control of Frequency and Power in Power Systems",Energia,Moscow.
7-Veselý V.,(1979),"Dynamic Behaviour Control of Power Systems",Electroenergetickej system III,Int.Conf.,Glivice,Poland.
8-Veselý V.,Murgaš J. and Bízik J.,(1981),"Decentralized Control of Dynamic Linear Systems",IFAC Congress VIII,42-4,Kyoto,Japan.
9-Veselý V. and Soliman K.M.,(1981),"Design of an Optimal Controller for the Complex Power System",Proc.of First Comemop Conf.,Czech.
10-Veselý V. and Soliman K.M.,(1983),"Design of Suboptimal Controllers for the Complex Power System by using Modified Bellman-Lyapunoff Equation",Proc.of Second Comemop Conf.,Stupava,Czech.
11-Wang S. and Davison E.J.,(1973),"On the Stability of Decentralized Control Systems",IEEE Trans.,Autom.Control,18,473-478.
12-Willems J.L.,(1974),"A Partial Stability Approach to the problem of Transient Power System Stability",Int.J.Control,vol.19,No.1.

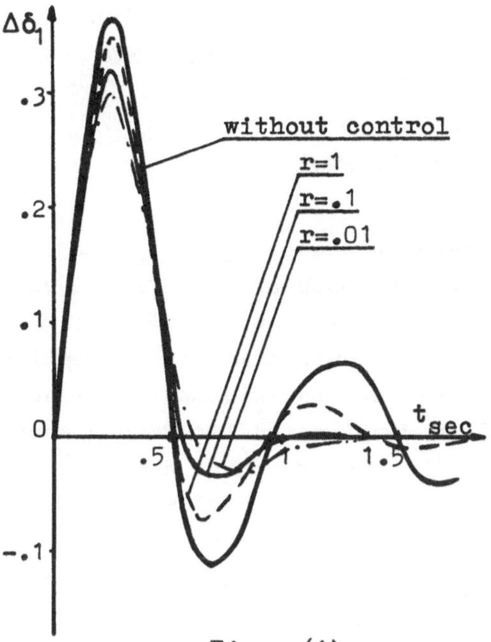

Figure(1)
(First approach, δ_{1o}=0.8rad)

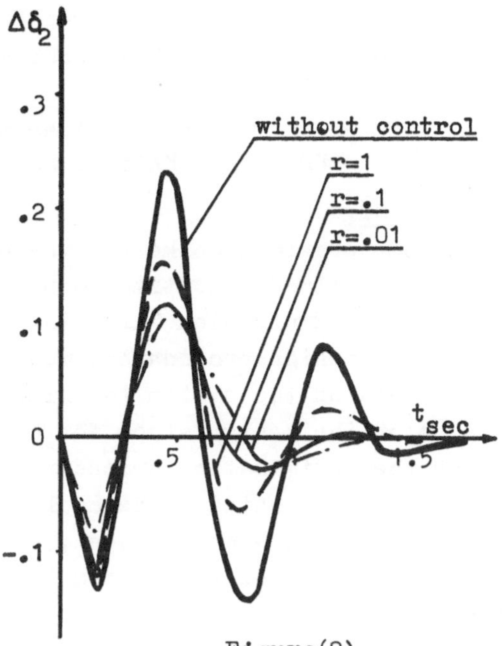

Figure(2)
(First approach, δ_{2o}=0.6rad)

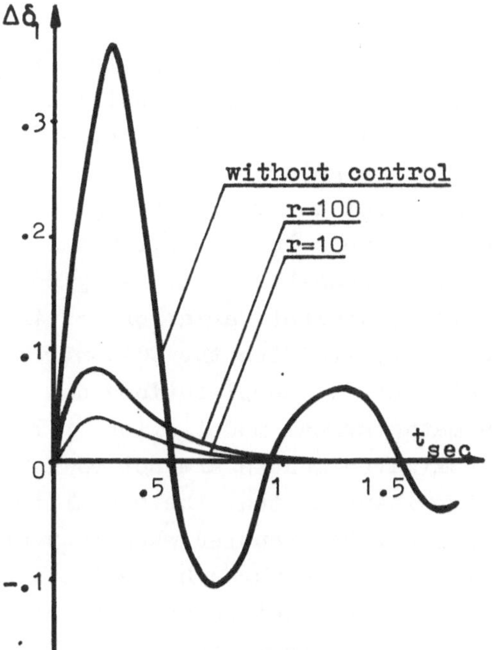

Figure(3)
(Second approach, δ_{1o}=0.8rad)

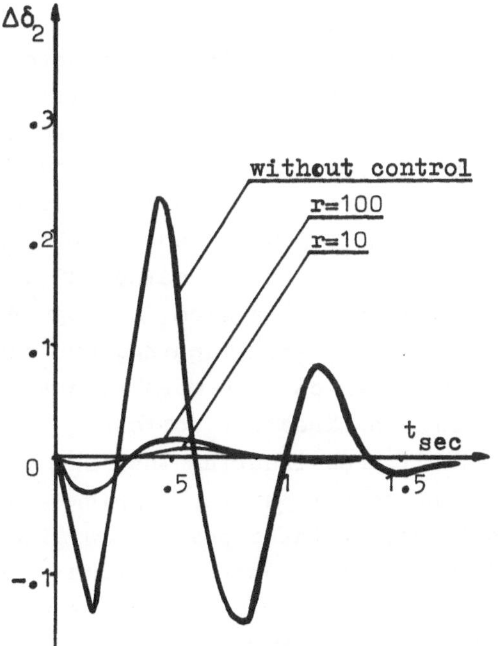

Figure(4)
(Second approach, δ_{2o}=0.6rad)

PRACTICAL REALIZATION FOR DESIGNED SUBOPTIMAL PRIME-MOVER AND EXCITATION CONTROLLERS FOR THE COMPLEX POWER SYSTEM

K.M.Soliman and V.Veselý
EF, SVŠT, Katedra ASR-TP, 81219 Brat., ČSSR

ABSTRACT:

In some previous works for the authors,it was designed by different ways Decentralized Suboptimal Controllers for both of the Prime Movers (PMs) and the Excitation Circuits(ECs) in Complex Power Systems (CPSs). Through the design procedures, much amount of approximation assumptions were imposed on the system for simplifying the CPS-mathematical model, and also on the design steps for avoiding complicated treatments appear through the optimization procedures. In this paper, these designed Controllers are tested to be applied on the original system which is represented by more exact model, with proving that these approximations have small effect on the system transient characteristics, and have no effect on the global system stability. Moreover, a practical EC-Control system is added to apply on each Synchronous Generating Machine(SGM), that is to present a realized system which is able to model some physical systems.

INTRODUCTION:

In the field of designing Decentralized Suboptimal Controllers for the PMs in the CPS by using Bellman-Lyapunoff Equations or Modified Bellman-Lyapunoff Equations, References [2,5,6] have been presented different approaches for carrying out this design as continuation, modification and advancement of other presented works[1,3,4]. All of these works follow the acceptable assumption in a bulk of literatures of simplifying the CPS-mathematical model as possible, that is by considering loss-less system, and ignoring the transient variations of the back e.m.f. of each machine, and consequently avoiding the EC-Controllers design which need much amount of mathematical complications due to the high nonlinearity contained in the mathematical modellation of the EC itself. For insuring that these approximated models are not so far from the actual physical systems, and for insuring that the stability conditions of these physical systems will not be troubled when they are affected by the Control Systems which are designed through hypothetical approximated models, this paper presents a more exact 7n-th order state model for modelling the CPS includes both of the last designed PM-Decentralized Suboptimal Controllers,and suggested EC-practical Controllers.

MATHEMATICAL MODEL OF SGMs WITH CONTROLLED TURBINES:

Considering only the assumption of zero direct axis back e.m.f. (E'_{di}), the i-th SGM in the CPS which contains n-machines can be represented as functions of the state variables($\Delta\delta_i$, $\Delta\omega_i$, $\Delta E'_{qi}$, ΔP_{ti} and ΔU_{bi}) chosen around the steady state equilibrium stable point(δ_{io}, ω_{io}, E'_{qio}, P_{tio} and U_{bio}) by the following third order state model [1,2]:

$$d\,\Delta\delta_i/dt = \Delta\omega_i \tag{1}$$

$$T_i\, d\,\Delta\omega_i/dt = P_{tio}+\Delta P_{ti}-D_i\,\Delta\omega_i-(\Delta E'_{qi}+E'_{qio})\; i_{qi} \tag{2}$$

$$T_{bi}\, d\Delta E'_{qi}/dt = K_{gio}(\Delta U_{bi}+U_{bio})-(\Delta E'_{qi}+E'_{qio})+(x_{di}-x'_{di})\; i_{di} \tag{3}$$

with the following terminal machine current and voltage expressions:

$$i_{qi}=(\Delta E'_{qi}+E'_{qio})\,y_{ii}\cos\varphi_{ii}-\sum_{k=1,k\neq i}^{n} y_{ik}(\Delta E'_{qk}+E'_{qko})\cos(\Delta\delta_i-\Delta\delta_k+\Theta_{ik}) \tag{4}$$

$$i_{di}=-(\Delta E'_{qi}+E'_{qio})\,y_{ii}\sin\varphi_{ii}+\sum_{k=1,k\neq i}^{n} y_{ik}(\Delta E'_{qk}+E'_{qko})\sin(\Delta\delta_i-\Delta\delta_k+\Theta_{ik}) \tag{5}$$

$$U_{qi}+jU_{di}=\left[x'_{di}i_{di}+(\Delta E'_{qi}+E'_{qio})-i_{qi}R_i\right]-j\left[x_{qi}i_{qi}+i_{di}R_i\right] \tag{6}$$

$$U_{gi}=\sqrt{U_{di}^2+U_{qi}^2} \tag{7}$$

where: Y_{ii} is the driving point admittance $= y_{ii}\, e^{-j\varphi_{ii}}$

Y_{ik} is the transfer admittance $= y_{ik}\, e^{-j\varphi_{ik}}$

R_i, K_{gio} are the total stator resistance and EC-overall gain resp.

$\Theta_{ik} = \delta_{io}-\delta_{ko}+\varphi_{ik}$

For testing the effect of the previous designed PM-Suboptimal Controllers[2,5,6], it will be considered here the same first order dynamic equation for representing each controlled turbine as follows:

$$T_{pi}\, d\Delta P_{ti}/dt = -\Delta P_{ti}-\Delta\omega_i/\sigma_i+\Delta P^{*}_{tic} \tag{8}$$

where: ΔP^{*}_{tic} is the input control signal to the turbine, it is substituted by some previous designed formulae. For example, for applying the designed formula by using scalar Lyapunoff Function for the whole system it will be expressed as[2]:

$$\Delta P^{*}_{tic} = -(a_i^{*}/2r_iD_i)(\Delta P_{ti}+T_i\,\Delta\omega_i/T_{pi}) \qquad \text{with } r_i > 0 \tag{9}$$

$$a_i^{*}=\sqrt{\left[2r_iD'_iD_i^2T_{pi}^2/(T_i^2+T_{pi}^2)\right]^2+\left[4r_iD_i^2T_{pi}^2/(T_i^2+T_{pi}^2)\right]\lambda_{max}[Q_i]\, -}$$

$$(2r_i D_i' D_i^2 T_{pi}^2)/(T_i^2 + T_{pi}^2) \qquad (10)$$

Since negative turbine power do not represent practical condition,and also for limiting the turbine power not to exceed some practical positive limits, the following constraint equation will be imposed on the problem for confirming practical logic for the turbine operation:

$$0 \leqslant P_{tio} + \Delta P_{ti} \leqslant P_{timax} \qquad (11)$$

when the maximum limit can be chosen around the value of $P_{timax}=2P_{tio}$.

SUGGESTED EC-PRACTICAL CONTROLLERS:

In purpose of fulfilling practical simulation to some real systems, the proposed EC-Control System shown in figure(1) is added to each SGM. The main elements construct these Controllers can be exposed as follows:

1-The filtering feed back circuit:which is used for adjusting the output generator voltage(U_{gi}) to be suitable as a negative feed back control signal (practically, U_{gi} is about one thousand times U_{Di}).

2-The simulation circuit:which is used for realizing the considered mathematical model to be equivalent to some physical systems.

3-Amplifier:for amplifying the input excitation error signal.

4-The adjustment switch:which can be used in cases of strong accidents. If the EC-Controllers effects are slow, and not enough to govern the dynamic characteristics,this switch is used for passing to the machine strong signal which can help in damping and improving the transient variations in the first few overshoots. For synchronizing this signal with the changeable sign of the machine angle,it can be expressed as:

$$U_{bimax} = U_{mi} \operatorname{sign} (d\Delta\delta_i/dt) \qquad (12)$$

where the maximum magnitude(U_{mi}) may be chosen 1.5-2 times (U_{bio}).

Using the classical block diagram reduction methods, and considering the auxiliary intermediate state variables(U_{1i},U_{2i}),the proposed Controller can be represented by the following third order state model:

$$d\Delta U_{bi}/dt = U_{1i} \qquad (13)$$

$$dU_{1i}/dt = U_{2i} \qquad (14)$$

$$T_{3i}dU_{2i}/dt=-T_{2i}U_{2i}-T_{1i}U_{1i}-U_{bio}-\Delta U_{bi}+K_{2i}U_{Di}-K_{3i}U_{gi}-K_{4i}dU_{gi}/dt \qquad (15)$$

where: $K_{2i}= K_{1i}K_{Bi}$, $K_{3i}= K_{2i}K_{fi}$, $K_{4i}= K_{3i}T_{spi}$,

$T_{1i}= T_{Bi}+T_{fi}+T_{spi}+ K_{Bi}K_{spi}$, $T_{3i}= T_{fi}T_{Bi}T_{spi}$,

$T_{2i}= T_{fi}T_{spi}+T_{fi}T_{Bi}+T_{spi}T_{Bi}+ K_{Bi}K_{spi}T_{fi}$

Exact expression for the term (dU_{gi}/dt) can be deduced from the time derivative of equation (7) as a function of the time derivatives of equations(4) to(6). For simplicity, this term can be approximated to:

$$(dU_{gi}/dt)^{j+1} = \left[(U_{gi})^{j+1} - (U_{gi})^{j}\right] / \ \Delta t \qquad (16)$$

where: Δt is the time interval of the transient calculations.
$j+1$ represents the transient point under calculations.
j represents the last already calculated point.

SUMMARY OF THE MATHEMATICAL MODEL:

For each SGM, the mathematical model which will be used for carrying out the transient calculations can be summarized as **follows:**

- The 7-th order state dynamic model given by equations(1)-(3),(8) and (13)-(15), with fulfilling the practical constraint equation(11).
- adding to equation (12) when using the switch adjustment.
- with the auxiliary mathematical expressions given by equations (4)-(7) and(16) (or exact expression instead of(16) as mentioned before).
- and the input turbine control signal is substituted from equations (9) and(10) or any suggested formula according to the tested approach.

RESULTS AND COMMENTS:

The validity of the proposed Control System is tested by carrying out the transient calculations on an example of 2-SGMs connected with an infinite Bus-Bar using Runge-Kutta Method. Figures(2)-(7) show some of the obtained transient characteristics. From all the obtained results, the following comments can be summarized:

1-As an effect of the suggested EC-Controllers, all the amplitudes of the transient characteristics are damped by about 20-30%.

2-The maximum transient change in the back e.m.f.(E'_{qi}) do not exceed about 1% from the steady state value, and its transient variations are approximately unobservable. This proves that the assumption of constant back e.m.f. considered through the PM-Controllers design is practically very acceptable, and dont represent any reckless assumption.

3-Some observable change in the back e.m.f. transient variations can be achieved by using the switching adjustment procedure. So, the very complicated mathematical steps necessary for designing theoretical EC-Controllers can be saved by using some practical Controllers.

FINAL CONCLUSION:

The more exact mathematical model used for representing the CPS, with both of suggested EC-Practical Controllers and previous designed PM-Suboptimal Controllers help by great amount in improving and advancing the CPS-transient characteristics. Much amount of mathematical complications are **saved** by ignoring the theoretical design of EC-Controllers depending on the accepted assumption of constant back e.m.f., and replacing that by suggested EC-Practical Controllers.

REFERENCES:

1-Koščova M.,(1980),"Decentralized Control for Transient Processes in Electric Systems",CSc.Thesis,Bratislava,Czech.

2-Soliman K.M.,(1984),"Decentralized Control and Stability Studies of the Complex Power System",CSc.Thesis,Bratislava,Czech.

3-Veselý V.,(1979),"Dynamic Behaviour Control of Power Systems",Electroenergetickej System III,Int.Conf.,Glivice,Poland.

4-Veselý V.,(1981),"Suboptimal Stabilization for Dynamic Systems",Electrotechnicky Časopis,No.4,Czech.

5-Veselý V. and Soliman K.M.,(1981),"Design of an Optimal Controller for the Complex Power System",Proc.of first Comemop Conf.,Czech.

6-Veselý V. and Soliman K.M.,(1983),"Design of Suboptimal Controllers for the Complex Power System by using Modified Bellman-Lyapunoff Equation",Second Comemop Conf.,Czech.

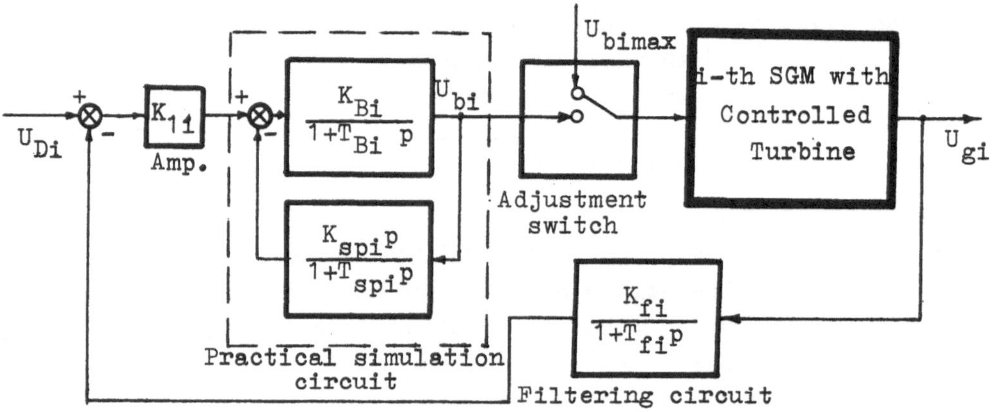

Figure(1)
SGM with practical EC-Controller

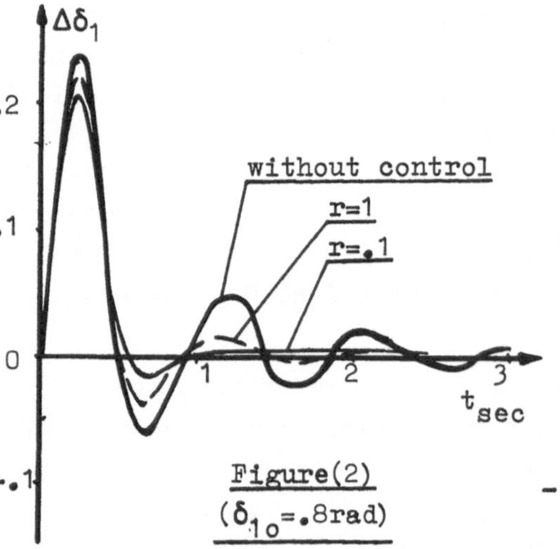

Figure(2)
(δ_{1o}=.8rad)

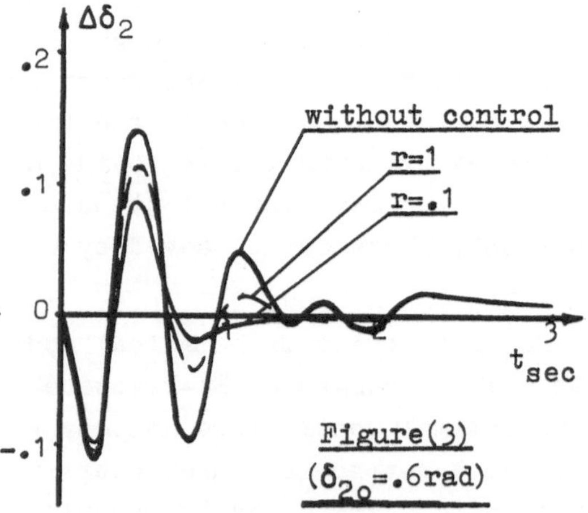

Figure(3)
(δ_{2o}=.6rad)

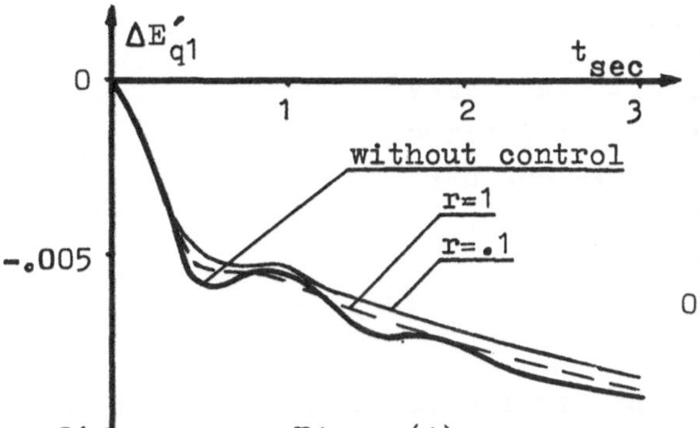

Figure(4)

(E'_{q1o}= 1.055 p.u.)

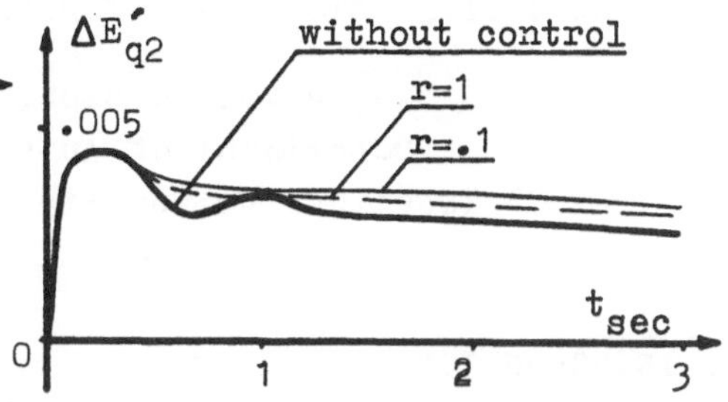

Figure(5)

(E'_{q2o}= 1.3767p.u.)

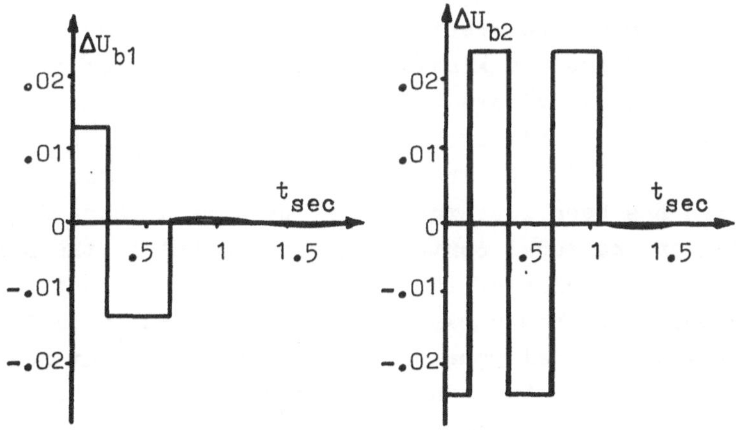

Figure(6)

(Switching applied excitation signal)

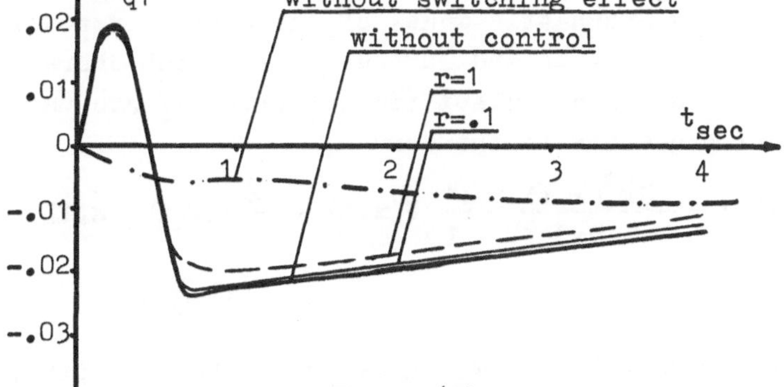

Figure(7)

(E'_{q1o}= 1.055p.u.)

SUBOPTIMAL STOCHASTIC CONTROL OF A NUCLEAR POWER REACTOR

Józef Korbicz, M.Z. Zgurovsky*
Department of Automatic and Metrology
Technical University of Zielona Góra, Poland

1. INTRODUCTION

Today a lot of works in which the problem of controlling the power distribution of a nuclear reactor is considered [1,2,3], have a theoretical character and are based on the single group diffusion model of reactor without stochastical disturbances. The fundamental difficulty in the synthesis of the reactor control system is nonlinearity diffusion equation of neutron kinetics. The modal control algorithms for nonlinear reactor model are studied in [4,5]. But one of the often used methods for solving this problem is a linearization of the model around a stationary state [2,6]. The stochastic control problem of the power distribution has been researched only by some authors [6,7,8]. In [7] the problem of the modal control was solved but in the survey work [8] only some main features of the stochastic reactor control are presented.

The purpose of this paper is to study the suboptimal stochastic control system of axial power distribution in nuclear reactors with the large sized core. The control system is synthesised for linearizated reactor model and it consists of the suboptimal state estimation algorithm and the feedback controller.

2. STOCHASTIC REACTOR MODEL

The diffusion stochastic model of the kinetic reactor [7] with one group of delayed neutrons and lumped-models which characterize the changes of the fuel and coolant temperatures [4] are used for describing the dynamical processes in the core

$$\frac{\partial F(x,t)}{\partial t} = a\,\frac{\partial F^2(x,t)}{\partial x^2} - \frac{\beta}{l}\,F(x,t) + \frac{\beta}{l}\rho(x,t)\,F(x,t) + \lambda\cdot C(x,t) + d(x)\,\delta\rho(x,t) \qquad /1/$$

---

*Department of Control Systems, Kiev Polytechnic Institute, USSR.

$$\frac{\partial C(x,t)}{\partial t} = \frac{\beta}{l} F(x,t) - \lambda \cdot C(x,t) \qquad /2/$$

$$\frac{\partial Q_B(x,t)}{\partial t} = (1/T_{BQ})\left[Q_K(x,t) - Q_B(x,t)\right] + w\, F(x,t) \qquad /3/$$

$$\frac{\partial Q_K(x,t)}{\partial x} = b\left[Q_B(x,t) - Q_K(x,t)\right] \qquad /4/$$

$$\rho(x,t) = k_{TK}\, Q_K(x,t) + k_{TB}\, Q_B(x,t) + \rho_R(x,t) \qquad /5/$$

The boundary and initial conditions have the form

$$\begin{aligned} &F(x,t) = 0 \quad \text{for } x=0 \text{ and } x=H \\ &Q_K(x,t) = \varphi_1(t) \quad \text{for } x=0 \\ &F(x,0) = F_0(x), \quad C(x,0) = C_0(x) = (\beta/l\cdot\lambda)F_0(x) \\ &Q_B(x,0) = Q_{B0}(x) \end{aligned} \qquad /6/$$

In equations /1/-/6/ the following notations are accepted: t - the time; x - the axial coordinate; H - the highness of the reactor core; $F(x,t)$ - the neutron density; $C(x,t)$ - the concentration of delayed neutrons; $Q_B(x,t)$ - the fuel temperature; $Q_K(x,t)$ - the coolant temperature; $a=D\cdot\vartheta$, D - the diffusion coefficient; ϑ - the thermal velocity of the neutrons; l - the average life time; β - the proportional part of the carries of delayed neutrons during fission; $1/\lambda$ - the average life time of delayed neutrons; T_{BQ} - the thermal time constant; w characterizes the effect of $F(x,t)$ on $Q_B(x,t)$; b - characterizing of the heat transfer from the fuel to the coolant; $\delta\rho(x,t)$ - the stochastical component of the reactivity; $d(x)=(\beta/l)F_0(x)$ characterizes the effect of $\delta\rho(x,t)$ on $F(x,t)$; $\rho(x,t)$ - the total reactivity; k_{TK}, k_{TB} - the reactivity coefficients determined by the temperature feedback of the coolant and fuel temperature, respectively; $\rho_R(x,t)$ - the reactivity determined by the effect of the control rods.

The neutron density is measured at distinct axial points x^j, $j=\overline{1,N}$ by detectors. The measured variable is corrupted by noises introduced by the detectors and therefore the output signal $Z(t)$ may be described by

$$Z(t) = H(t)F_N(t) + V(t) \qquad /7/$$

where $Z(t)=\left[Z(x^1,t), Z(x^2,t),\dots, Z(x^N,t)\right]^T$ is an N-dimensional output signal vector, $F_N(t)$ is an N-dimensional vector of the neutron density at the measurement points, $H(t)$ is an $(N\cdot N)$ known diagonal matrix and $V(t)$ is an N-dimensional measurement noise vector.

It is assumed that the random process $\delta\rho(x,t)$ and the measurement noise V(t) may be approximated by white-Gaussian processes in time and space. Statistic characteristics of those noises are given by

$$E[\delta\rho(x,t)] = 0, \quad E[V(t)] = 0, \quad E[\delta\rho(x,t)V(t)] = 0,$$
$$E[\delta\rho(x,t)\delta\rho(y,\tau)] = \sigma^2(x,t)\delta(x-y)\delta(t-\tau), \tag{/8/}$$
$$E[V(t)V^T(\tau)] = Q(t)\delta(t-\tau)$$

where $E[\cdot]$ denotes an expectation operator and $\delta(\cdot)$ is the Dirac delta function. It is assumed that $\sigma^2(x,t)$ is the dispersion of noise and $Q(t)$ is symmetric and positive definite matrix.

For simplification of the non-linear system /1/-/6/ linearizated reactor equations with respect to small deviation from the stationary parameters value will be considered. If the input and output variables are presented by sum of their stationary values and deviations $y_1(x,t)$ - $y_4(x,t)$, then from /1/-/7/ a perturbation linear reactor model may be described as follows

$$\frac{\partial y_1(x,t)}{\partial t} = a\frac{\partial^2 y_1(x,t)}{\partial x^2} + a_1 y_1(x,t) + a_5 y_2(x,t)\ b_1(x)u_1(x,t) + d_1(x)w_1(x,t) \tag{/9/}$$

$$\frac{\partial y_2(x,t)}{\partial t} = \lambda[y_1(x,t) - y_2(x,t)] \tag{/10/}$$

$$\frac{\partial y_3(x,t)}{\partial t} = (1/T_{BQ})[a_2 y_4(x,t) - y_3(x,t)] + a_3 y_1(x,t) \tag{/11/}$$

$$\frac{\partial y_4(x,t)}{\partial x} = b[a_4 y_3(x,t) - y_4(x,t)] \tag{/12/}$$

$$Z_1(t) = H(t)\,Y_{1N}(t) + V_1(t) \tag{/13/}$$

with respectively boundary and initial conditions. The parameters a_1 - a_5 and functions $b_1(x)$, $d_1(x)$ depend on parameters of model /1/-/6/ and $Y_{1N}(t)$ is an N-dimensional vector of neutron density deviation.

For synthesising computer algorithms of the optimal control of reactor the equations /9/-/12/ are transformed into discrete form by applying the Krank-Nicolson's difference-finite formulae. After executing this simple operation we finally obtain

$$Y_1(k+1) = A_1 Y_1(k) + B_1 U_1(k) + C_1 S_1(k) \tag{/14/}$$

$$Y_2(k+1) = q_2 Y_2(k) + C_2 S_2(k) \tag{/15/}$$

$$Y_3(k+1) = q_3 Y_3(k) + C_3 S_3(k) \tag{/16/}$$

$$Y_4(k+1) = C_4 S_4(k+1) \tag{/17/}$$

with initial conditions

$$Y_1(0) = Y_1^0, \quad Y_2(0) = Y_2^0, \quad Y_3(0) = Y_3^0$$

where k is the discrete time ($k=\overline{0,N_k}$, $t_k=k\,\Delta t$), n is the discrete space coordinate ($n=\overline{0,N_n}$, $x^n=n\,\Delta x$), $Y_1(k)-Y_4(k)$ are N_n-dimensional vectors of state deviation, $S_1(k)-S_4(k)$ are N_s-dimensional disturbance vectors and $U_1(k)$ is an N_m-dimensional control vector. A_1, B_1 and C_1-C_4 are $(N_n\times N_n)$, $(N_n\times N_m)$, $(N_n\times N_s)$ known matrices respectively.

3. STATEMENT OF PROBLEM

Based on the reactor model /9/-/13/ the Linear-Quadratic-Gaussian (LQG) problem can be stated as follows: Given a linear reactor model define the control signal $u_1(x,t)$ to transform reactor from the initial stationary state $F_0(x)$ to new desire stationary state $F_k(x)$ and to minimize a quadratic cost function of the type

$$J(y_1,u_1) = E\left[\int_{t_0}^{t_f}\int_0^H\int_0^H \left[y_1(x,t)\,S(x,x_1,t)\;y_1(x_1,t) + u_1(x,t)\cdot R(x,x_1,t)\,u_1(x_1,t)\right]dx\,dx_1\,dt\right] \qquad /18/$$

where $S(\cdot)$ and $R(\cdot)$ are symmetrical and positive definite matrices. In this case the separation theorem is satisfied and therefore the problem may be transformed for solving two separation problems: synthesis of the state estimation system and synthesis of the optimal regulator.

4. SUBOPTIMAL STATE ESTIMATION ALGORITHM

The realization of optimal filter for the distributed parameter system requires solving the nonlinear Riccati equation [9]. It is a difficult computable problem and therefore we shall apply the suboptimal algorithm [10], which is given by set equations for the reactor model /9/-/13/.

<u>Equation for estimate</u> $\hat{y}_1(x,t)$

$$\frac{\partial \hat{y}_1(x,t)}{\partial t} = a\,\frac{\partial^2 \hat{y}_1(x,t)}{\partial x^2} + a_1\hat{y}_1(x,t) + a_5\hat{y}_2(x,t) + b_1(x)\times u_1(x,t) + \hat{P}(x,t)H^T(t)Q^{-1}(t)\left[Z_1(t) - H(t)\hat{y}_{1N}(t)\right] \qquad /19/$$

State estimates $\hat{y}_2(x,t)$, $\hat{y}_3(x,t)$ and $\hat{y}_4(x,t)$ are defined from equations /10/-/12/ taking into consideration $\hat{y}_1(x,t)$.

<u>Equation for the estimation error</u> $\Delta y_1(x,t) = y_1(x,t) - \hat{y}_1(x,t)$

$$\frac{\partial \Delta y_1(x,t)}{\partial t} = a\frac{\partial^2 \Delta y_1(x,t)}{\partial x^2} + a_1 \Delta y_1(x,t) + a_5 \Delta y_2(x,t) - \hat{P}(x,t)H^T(t)Q^{-1}(t)\left[Z_1(t) - H(t)\hat{y}_{1N}(t)\right] + d_1(x)w_1(x,t) \qquad /20/$$

The structure of equations for determining the estimation errors $\Delta y_2(x,t)$, $\Delta y_3(x,t)$ and $\Delta y_4(x,t)$ is adequate to the equations structure /10/-/12/.

<u>Equation for the dispersion estimate</u> $\hat{P}(x,t)$

$$T\frac{\partial \hat{P}(x,t)}{\partial t} + \hat{P}(x,t) = \frac{1}{X}\int_{x-X/2}^{x+X/2} \Delta y_1(x_1,t)\cdot\Delta y_1(x_1,t)\, dx_1 \qquad /21/$$

where T and X are initial known constants. The boundary and initial conditions can be defined as in [10].

For solving those set equations /19/-/21/ the difference method /the Krank-Nicolson's formulae/ has been used.

5. OPTIMAL DISCRETE REGULATOR

For discrete model /14/-/17/ the problem of definition such strategy $U_1(k)$ that transforms the reactor from stationary state $Y_1(0)$ to the new stationary state $Y_1(N_k)$ and minimize cost function /18/ in discrete form is solved. Additionally, this optimization problem is solved with constraints on the control and state vectors. It results in the state estimates feedback control law given by the set recurrent equations [11]

$$U_1^{opt}(k-1) = K(k-1)Y_1(k-1) + L(k-1)S_1(k-1) \qquad /22/$$

$$K(k-1) = -B_1^T SB_1 + R(k-1)^{-1}B_1^T SA_1,\quad L(k-1) = - B_1^T SB_1 + R(k-1)^{-1} \qquad /23/$$

$$Y_1(k) = (A_1 + B_1K)Y_1(k-1) + (B_1L + C_1)S_1(k-1) \qquad /24/$$

It follows from the above formulaes that the multi-stage optimization problem was solved by one transition without storaging a lot of information. Besides the procedure of some spectrum calculation r(k) of the penelty matrix on control R(k) /discrete form matrix of cost function /18// is proposed. In this procedure constraints on the control and state vectors on each step of optimization are taken into account. For spectrum calculation of the penelty matrix the following relations can be used

$$r = \begin{cases} r_1, & |r_1| < |r_u| \\ r_u, & |r_1| \geq |r_u| \end{cases} \qquad /25/$$

where

$$r_l = X_l^{-1}(k)\, d_l(k)\ , \quad r_u = X_u^{-1}(k)\, d_u(k),$$

$$X_l = \mathrm{diag}\left[b_{1l}\ b_{2l}\ \dots\ b_{N_n l}\right],\quad X_u = \mathrm{diag}\left[b_{1u}\ b_{2u}\ \dots\ b_{N_n u}\right]$$

$$b_l = |U_{1u} - U_{1l}|E_l + U_{1l},\qquad b_u = |U_{1u} - U_{1l}|E_u - U_{1u} \qquad /26/$$

$$d_l = B_1^T(S+P)\left[B_1 b_l + A_1 Y_{1u} + C_1 S_1\right],\ d_u = B_1^T(S+P)\left[-B_1 b_u + A_1 Y_{1l} + C_1 S_1\right]$$

$$P = (A_1 - B_1 K)^T (S+P)(A_1 - B_1 K) + K^T R K, \qquad P(0)=0$$

$$E_u = (U_{1u} - U_1)/|U_{1u} - U_{1l}|,\qquad E_l = (U_1 - U_{1l})/|U_{1u} - U_1|$$

$$U_{1l} \leqslant U_1 \leqslant U_{1u},\qquad Y_{1l} \leqslant Y_1 \leqslant Y_{1u},$$

and Y_{1l}, U_{1l}, Y_{1u}, U_{1u} are the lower and upper boundaries on the state and control vectors.

6. SIMULATION RESULTS

Based on the above mentioned algorithms the simulation system was developed. This system may be used for the simulation of the following processes:

- the simulation of the dynamical processes in the reactor core /1/-/5/ and the measurement system /7/;
- the simulation of the dynamical processes reactor state estimation /1/-/7/ and /19/-/21/;
- the simulation of the feedback control system without state estimation /1/-/7/ and /22/-/26/;
- the simulation of the feedback control system with the suboptimal filter /1/-/7/, /19/-/21/ and /22/-/26/.

While simulating the following values of parameters were assumed: $H=2.75m$, $D=0.32\times10^{-2}$, $\nu=3.04\times10^{-3}m/s$, $\beta=0.0064$, $l=10^{-4}$, $\lambda=0.08s^{-1}$, $w=0.226\times10^{-11}m^3\ {}^oC/neutron$, $T_{BQ}=4.48s$, $b=0.0327m^{-1}$, $k_{TK}=10^{-4}\ {}^oC^{-1}$, $k_{TB}=-1.5\times10^{-5}\ {}^oC^{-1}$. And characteristcs of noises were also taken into account: $\sigma^2=5\times10^{-4}$, $R(t)=\mathrm{diag}\left[\sigma_V^2\right]=\mathrm{diag}\left[5\times10^{23}\right]$. Some simulation results in fig.1 are shown /$\Delta x=0.111m$, $\Delta t=0.01s$/. It follows from fig.1 that the error estimation of the neutron density for sufficiently level of noises and for 10 discrete measurement points is not exceeded 0.3%. If number of the detectors is smaller than 10 the estimation error increases very slovly. In the stationary state the neutron density is given by the expression: $F_0(x)=A_0\sin(\pi x/H)$. The feedback control system with the suboptimal filter for two cases was investigated i.e.

the transformation reactor from the initial stationary state A_o=3.94× 10^{14}neutron/m^3 in new stationary states A_k=3.84 10^{14} and A_k=4.04×10^{14} neutron/m^3. It was obtained high regulation quality and the steady state error is less than 0.5%.

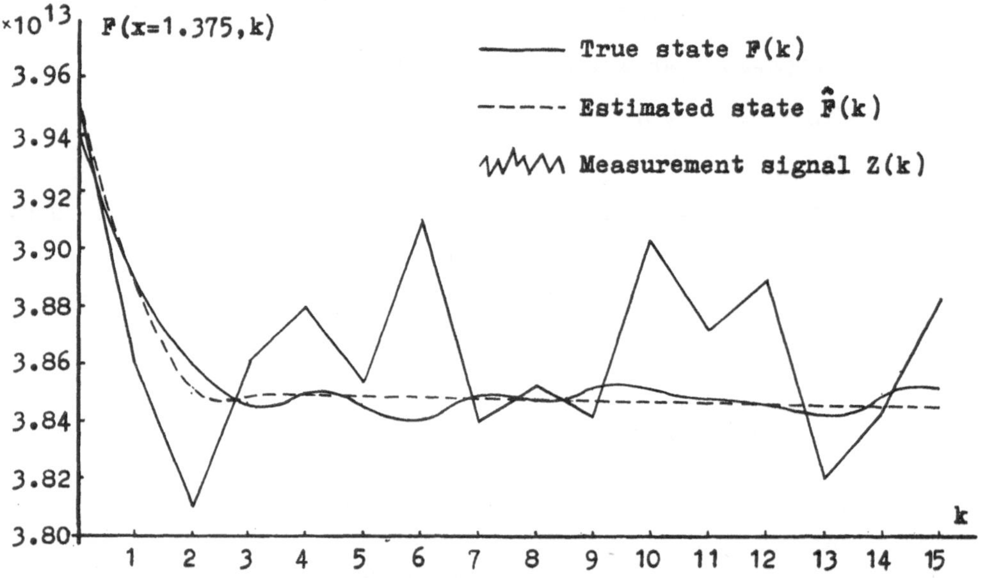

Fig.1 Control and estimated the neutron density at x=1.375

7. CONCLUSION

The presented method of the synthesis of the feedback control system allows to take into account the fundamental properties of the real system, i.e. the stochastic nature of the processes occuring in the reactor, the measurement inaccuracies and its discretization. The reality of the accepted simplification at the synthesis of the filter and regulator /only one variable state - the neutron density was taken into cosideration/ has been corroborated. Received results will be used in the simulator of a nuclear power station for learing operators.

REFERENCES

1. Jegorov A.I. Optimal Control of Thermal and Diffusion Processes, Moscow, Nauka, 1978 /in Russian/.

2. Filipchuk J.V., Potapienko P.T., Postnikov V.V. Control of Nuclear Reactor Neutron Field, Moscow, Energoizdat, 1981 /in Russian/.
3. Kuroda Y., Makino A. Some Problem Arising in Distributed Parameter Reactor Systems, "Lect. Notes in Contr. and Inf. Sci.", Springer-Verlag, 1977, No.2, pp.93-101.
4. Stark K. Modal Control of a Nuclear Power Reactor, "Automatica", 1976, v.12, pp.613-618.
5. Yorke G.L., Cherchas D.B. An Algorithm for Non-linear Space-time Nuclear Reactor Control, "Automatica", 1981, v.17, No.3, pp.471-482.
6. Karppinen J., Blomsnes B. An Application of Optimization Methods to Spatial Control of Nuclear Reactor Cores, "Lect. Notes in Contr. and Inf. Sci.", Springer-Verlag, 1977, No.2, pp.478-498.
7. Korbicz J. Optimal Control of a Nuclear Power Reactor, "Systems Science", 1983 /in press/.
8. Tzafestas S.G. Distributed-parameter Nuclear Reactor Optimal Control "Lect. Notes in Contr. and Inf. Sci.", 1977, No.2, pp.19-48.
9. Sawaragi Y., Soeda T., Omatu S. Modeling, Estimation, and their Application for Distributed Parameter Systems, "Lect. Notes in Contr. and Inf. Sci.", Springer-Verlag, 1978, No.11.
10. Kraskevitch V.E., Korbicz J. Suboptimal Kalman-filter for the Distributed Parameter Systems, "Systems Science", 1980, v.6, No.3, pp.225-234.
11. Zgurovsky M.Z. Control Systems Designed Automation of Dynamic Processes with Distributed Parameters, "Adaptive Automatic Control Systems", 1983, v.11, pp.113-124 /in Russian/.

HIERARCHICAL CONTROL IN TELEPHONE NETWORKS

P. GAUTHIER, P. CHEMOUIL
Centre National d'Etudes
des Télécommunications
92131-ISSY-LES-MOULINEAUX (FRANCE)

I. INTRODUCTION

The fast development of new digital switching systems and the design of a new signalling system able to quickly carry control informations make possible the implementation of adaptive call routing algorithms in telecommmunication networks [1],[2],[3].

The basic principle of adaptive traffic routing is to optimize the allocation of network capacity to calls in order to meet grade of service requirements under abnormal conditions. In recent years, a lot of work has been dedicated to adaptive traffic routing and a wide range of methods have been proposed in the literature [1],[2],[3],[4]. Most of them are suited for an all-digital network in a long term horizon (5 to 10 years). In this paper a hierachical method is proposed to be used in the near future when electromechanical switching machines with stored program controllers are still in place. The aim of the method is to react to topology changes in the network (trunk group or switching centre failure) and to persistent traffic variations. Then the paper is focused on the coordinator level in order to get the new reference traffic routing rather than on the local control level which updates the routing according to real-time local measurements.

In this paper we shall first describe the model of the network and the problem to be solved. Next we detail the control algorithm at the coordinator level. Last we present a study of the behaviour of this algorithm on a testbed network designed and dimensioned for an adaptive real-time routing control.

II. PROBLEM FORMULATION

The telecommunication network which is considered is a one level transit network consisting in source (and destination) nodes inter-connected through tandem switches by trunk groups (see figure 1). Note that direct trunk groups between two nodes may exist according to efficiency constraints.

The routing policy that is used to route calls throughout the network is the celebrated alternative policy which consists in :

- at each switching centre, an outgoing call is directed to a "first choice" route defined according to the call's destination.
- if this route is congested, the call is overflowed to a "second choice" route.
- if the latter route is blocked too, the call is rejected.

The following notations are used : X and A are respectively the measured carried traffic and the offered traffic, N stands for the trunk group capacity and P is the blocking probability. These variables are subsripted with indices I, K or J according to respectively origin, transit or destination considerations.

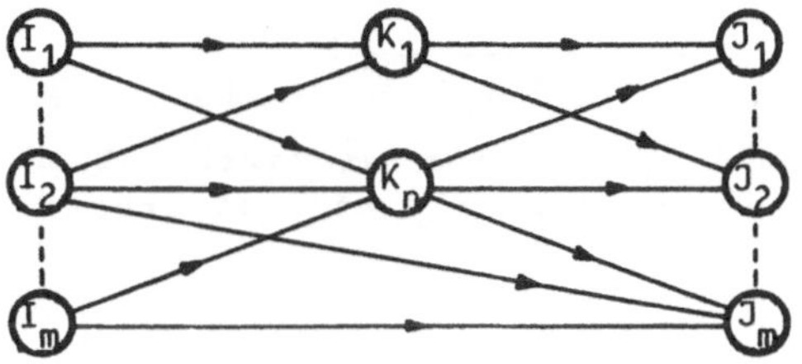

FIGURE 1 - One level transit network

II.1 Models of the Network

In order to model and to evaluate the performance of the routing scheme, classical traffic assumptions are made : Poisson arrivals and exponentially distributed service times.

Then using Chapman-Kolmogorov equations, a dynamic model could be written to describe the behaviour of the network. Such a model is convenient for a myopic analysis but inasmuch as the complete behaviour of the network is studied, a dynamic model is unrealistic since the size of this model would be enormous. Two main static models are therefore considered for the analysis of the system :

- the first one is based on the famous Equivalent Random Theory [5]. This approximation method permits the calculation of the carried traffic, the overflowed traffic and the blocking probability over alternative trunk groups by considering the mean and the variance of the traffic flows.

As an example we consider the case of m direct trunk groups of capacity N_i with offered traffic A_i overflowing over a trunk group of capacity N_o and offered traffic A_o (see figure 2). The offered traffic is Poisson traffic and the carried traffic over the direct trunk groups is obtained using Erlang's loss formula :

$$X_i = A_i(1-E(N_i, A_i)) \qquad i = 1,m$$

The mean and the variance of each overflowed traffic is:

$$A'_i = A_i E(N_i, A_i) \qquad i = 1,m$$

$$V'_i = A'_i\left[1 - A'_i + \frac{A_i}{N_i + 1 + A'_i - A_i}\right] \qquad i = 1,m$$

The mean and the variance of the whole traffic offered to the alternative trunk group are equal to :

$$A = A_o + \sum_i A'_i \qquad i = 1,m$$

$$V = A_o + \sum_i V'_i \qquad i = 1,m$$

This traffic is considered as the one overflowing from an equivalent trunk group of capacity N^* with an offered traffic A^* (see figure 3). Then :

$$A = A^* E(N^*, A^*)$$

$$V = A\left[1 - A + \frac{A^*}{N + 1 + A - A^*}\right]$$

This method is well suited for hierarchical networks with an alternative routing scheme and is widely used at present.

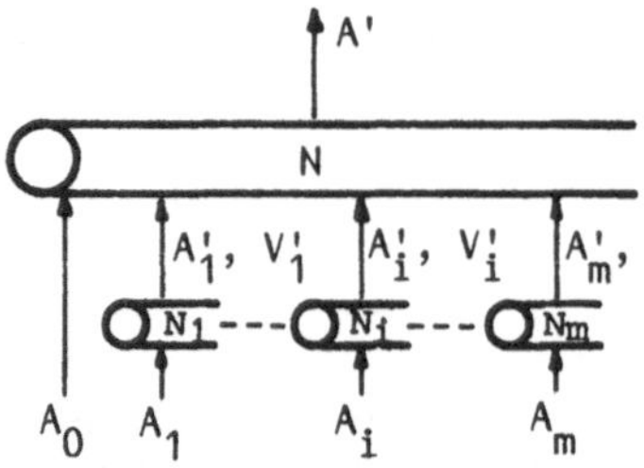

FIGURE 2 - Example of alternative routing

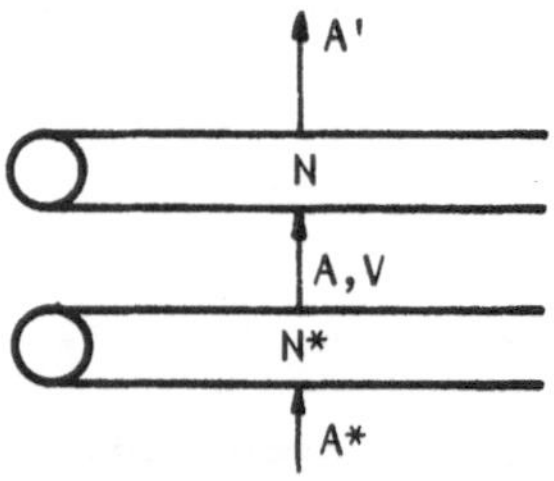

FIGURE 3 - Equivalent trunk-group

- a new method was developed by Le Gall [6] to model the telephone traffic. This approach is a dynamic one moment model whose steady state is obtained by using relaxation techniques. Fictitious offered traffic is introduced in the model in order to take into account the peaky characteristics of overflowed traffic. This simple model may be used for networks with general routing policies and permits the determination of end-to-end blocking probabilities in a network. For the same example as above the one moment model is written as :

$$P_i = E(N_i, A_i) \qquad i = 1,m$$

$$X_i = A_i(1 - P_i) \qquad i = 1,m$$

$$X_o = A_o(1 - P_o) + \sum_1^m A_i(P_i - P_{oi})$$

$$A_{oo} = X_o/(1 - P_o)$$

$$P_o = E(N_o, A_{oo})$$

$$A_{oi} = (X_o + X_i)/(1 - P_{oi}) \qquad i = 1,m$$

$$P_{oi} = E(N_o + N_i, A_{oi}) \qquad i = 1,m$$

where A_{oi} ($i = 1,m$) represent fictitious offered traffic to fictitious trunk groups of capacity $N_o + N_i$. The solution is obtained with relaxation techniques until a convergence test is satisfied.

Use of these two models is made in this paper. The first one is considered for the evaluation of the network grade of service and the one moment model is used in order to improve the control.

II.2 Control Problem

The problem that is to be solved is to best allocate the network capacity to traffic flows when only second choice changes are considered.

The problem can be formulated as :
Given the finite capacity of the network and considering the carried traffic measurements over all the trunk groups, what is, for each traffic flow, the best alternative route in order to obtain an optimal steady-state grade of service ?

III. DESCRIPTION OF THE ALGORITHM

III.1 Principle of the Method

The algorithm works in two steps:

- first find the optimal allocation of traffic to the transit nodes-destination nodes trunk groups [7].
- second determine the new alternative routing table in order to reach the optimal traffic allocation found at the first step.

III.2 Optimal Allocation of Traffic

Whatever a perturbation happens in the network (traffic overload or failure), its action is always affecting the carried traffic over the transit node-destination node trunk group. Then a new allocation of traffic to capacity is necessary.

With the assumption that losses are negligible in the switches and over the origin node-transit node subnetwork, the multicommodity flow problem to be solved is the maximization of the function :

$$S_X = \sum_{KJ}\sum X_{KJ}$$

under the constraints :

$$X_{KJ} = (1 - P_{KJ})A_{KJ}$$

$$P_{KJ} = E(N_{KJ}, A_{KJ})$$

$$A_{KJ} = \sum_{I} X_{IK}(J)$$

where S_X is the carried traffic over the whole network and $X_{IK}(J)$ stands for the carried traffic over the trunk group IK towards node J. Hennet [7] solved the problem by using the approximation that the traffic destined to each node J is constant ($\sum_{K} A_{KJ} = B_J$ = constant).

For each destination node J, the sub-problem is formulated as follows :

$$\min_{A_{KJ}} \sum_{K} E(N_{KJ}, A_{KJ})$$

under the constraint $X_{KJ} = (1 - P_{KJ})A_{KJ}$

Each sub-problem is solved by a feasible gradient method leading to the near-optimal traffic A^*_{KJ}. A complete description of this part of the algorithm may be found in [7].

III.3 Alternative Routing Updating

The updating of the alternative routing table is made by considering the KJ linkset where the gap between the optimal offered traffic A^*_{KJ} (determined above) and the actual offered traffic A_{KJ} (estimated from the measured carried traffic by inverting Erlang's loss formula) is the larger.

$$D_{KJ} = A_{KJ} - A^*_{KJ}$$

$$SD_{KJ} = \max_{K,J} |D_{KJ}|$$

If D_{KJ} is negative the problem consists in finding an originating node I where there exists a second choice traffic that can be transfered from a route IK'J to the route IKJ in order to add traffic on the KJ trunk group (and the contrary if D_{KJ} is positive).See figure 4.

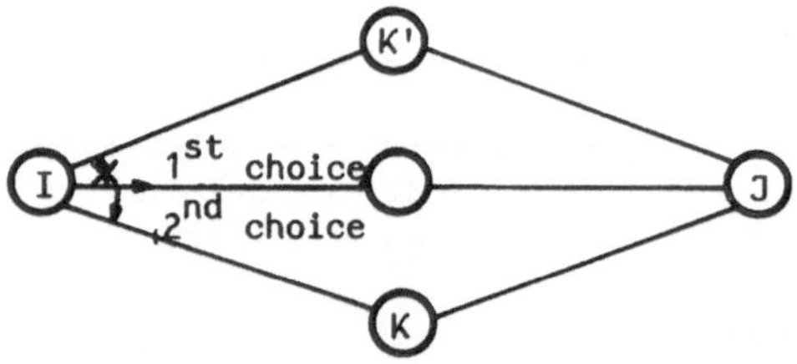

FIGURE 4 - Alternative routing updating

To check the validity of the updating, the coordinator evaluates the effect of a routing change before sending the control to the concerned switch. This evaluation is made in the following way using the one moment method [6] :

- the offered traffic A_{IK} over each link IK is calculated using Le Gall's model.
- the coordinator chooses an origin node I whose second choice route towards node J does transit through node K'.

* losses on the routes IKJ and IK'J are calculated (over trunk groups IK and IK' the traffic loss equals the offered traffic minus the carried traffic; over trunk groups KJ and K'J the traffic loss is obtained by inverting Erlang's loss function).
* the coordinator fictitiously changes the alternative route IK'J to the route IKJ
* the new carried traffic and new losses over links IK and IK' are determined using the one moment method, giving therefore the new traffics offered to links KJ and K'J.
* the new losses over these links are calculated.
* the former losses are compared to the new ones over the two routes involved by the control. If the difference is positive and better than a predetermined margin (0.1 Erlang in the test application), the change of alternative route is confirmed and is recommended to the concerned switch. If not the change is infirmed and the coordinator looks for another adequate node I and starts the procedure again.

IV. APPLICATION OF THE CONTROL ALGORITHM

IV.1 Testbed Network

The node network is a subset of 8 electronic switches (S) and 3 electronic tandem switches (T) from the Paris network and was chosen in order to have a model network close to a real possible trial network that could be used in a few years. The traffic matrix was taken out from the projected matrix for 1985; the total traffic is 407.5 Erlangs (Table 1).

	S1	S2	S3	S4	S5	S6	S7	S8
S1	0.0	1.9	2.9	2.0	2.7	5.2	25.5	6.5
S2	2.0	0.0	5.4	7.4	24.7	6.9	2.3	4.5
S3	2.3	6.9	0.0	5.7	5.2	11.7	3.1	6.4
S4	2.9	9.5	5.0	0.0	13.2	8.8	4.0	7.5
S5	2.1	21.5	4.4	9.9	0.0	6.6	2.0	4.6
S6	3.4	5.3	9.9	5.8	5.4	0.0	4.0	19.2
S7	29.3	2.2	3.8	3.9	2.3	6.1	0.0	4.7
S8	4.1	5.0	5.9	6.8	6.0	27.1	4.3	0.0

Table 1. Offered traffic matrix

We first sized the links with the usual rules of hierarchical routing and obtained 866 trunks. This network was in fact useless for the adaptive routing policy since each origin node was connected to the only hierarchical transit node. Then we sized the trunk groups according to a real-time routing scheme in order that all the nodes be connected to all tandem nodes, leading to 848 trunks (i.e. 2% less than in the hierarchical network). See Table 2 for the trunk group size. The direct trunk groups are shown in Figure 5.

	S1	S2	S3	S4	S5	S6	S7	S8	T1	T2	T3
S1	—	—	—	—	—	8	31	9	15	4	4
S2	—	—	7	9	30	9	—	—	5	15	5
S3	—	9	—	9	7	18	—	10	4	10	4
S4	—	12	—	—	16	12	—	11	5	4	19
S5	—	26	—	12	—	9	—	—	4	20	5
S6	—	7	15	9	7	—	—	22	5	6	13
S7	36	—	—	—	—	9	—	—	23	4	4
S8	—	7	10	10	8	30	—	—	6	5	16
T1	5	10	13	12	12	6	5	11	—	—	—
T2	12	5	9	5	6	6	13	16	—	—	—
T3	17	5	10	5	6	6	18	6	—	—	—

Table 2. Trunk group capacities

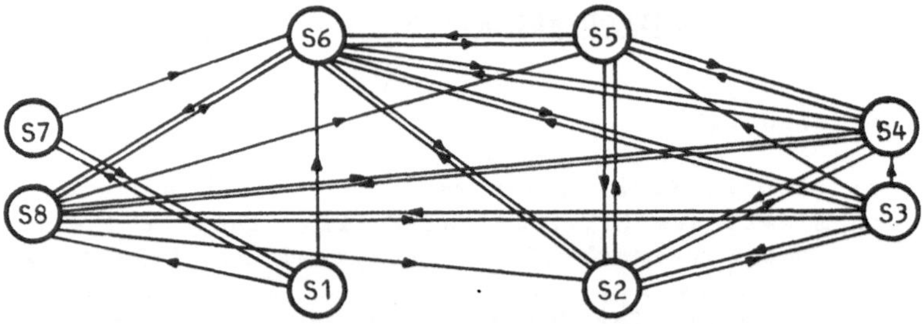

FIGURE 5 - Direct links between nodes

IV.2 Experiments

As the subnetwork we study could not be isolated from the real network, we used an existing computer tool developed at the CNET [8] to simulate the behaviour of the network. Regarding the topology of the network (trunk groups, nodes, routing tables),this tool named MALEPESTE gives the carried traffic over each link and the total losses in the network, making use of the Equivalent Random method [5] with Rapp's approximations [9]. The outputs of MALEPESTE (i.e. carried traffics) are used as measurements for our algorithm (see Figure 6).

We examined 8 realistic cases of disturbances :

- the first one concerns only a disturbance in the routing table;
- the second case consists in reducing the size of the two links S8-S6 (15 trunks instead of 30) and S8-T3 (8 trunks instead of 16).
- in the third case the capacity reduction concerns the links S4-T3 (9 trunks instead of 19) and T1-S5 (6 trunks instead of 12).
- the 3 following cases consider capacity changes respectively over the links T3-S7 (3 trunks instead of 6), T2-S7 (8 trunks instead of 16) and T1-S7 (5 trunks instead of 11). Note that the first two linksets have no alternative traffic towards S7 as the third link has no first choice traffic to S7.
- in the case 7 all trunks groups going out of tandem node T2 are set to 0.
- the last case concerns a disturbed traffic matrix.

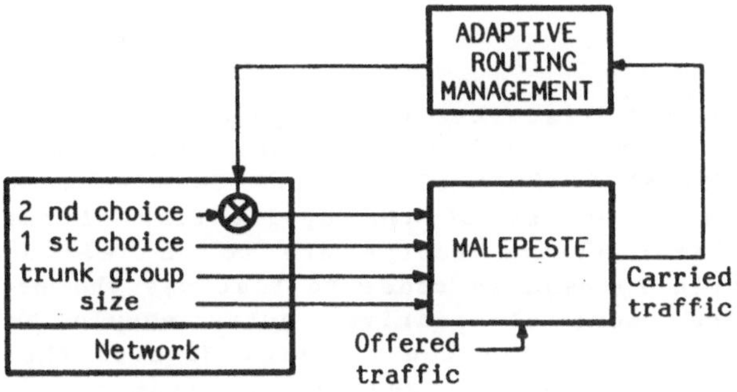

FIGURE 6 - Experiment configuration

IV.3 Results

The results of the algorithm are given in Table 3. The general conclusions are :

* the network is rather loaded and traffic is well shared upon the different links except in the first case where the algorithm is very efficient. Loss reduction is then difficult to obtain.
* when a link outgoing from a tandem node is affected by a failure, gain is obtained if the link carries alternative traffic (case 6). if not the improvement is very small.
* most of the loss cannot be prevented since first choice traffics are not affected by the algorithm.
* a final remark is that mutual overflow is not allowed in MALEPESTE (designed for hierarchical networks and ordering the trunk groups is necessary). Some possible reroutings have been then eliminated.

disturbance case	initial loss Erlang	final loss Erlang	gain %	iteration number
1	11.33	5.4	52	23
2	15.30	15.26	0.3	5
3	8.97	7.99	10.9	6
4	9.25	9.14	1.2	4
5	7.67	7.56	1.4	3
6	7.14	5.99	16.1	6
7	38.44	32.90	14.4	18
8	7.21	6.72	6.8	18

Table 3. Algorithm performance

V. CONCLUSION

In this paper the first step to design a hierarchical control of traffic routing has been described. The coordinator level is used to determine a near-optimal traffic allocation. Some enhancements are to be made regarding the results of the application. First in case of topology failure, first choice routes are to be reallocated. Second mutual overflow should be used in order to best use the network. In case of important failure, some alternative routes should be temporarily avoided rejecting thus some overflow traffic. However the major improvements of the coordinator control should come from the local controllers that can update the routing tables according to real-time local measurements. For example, one important measurement is the total occupancy of outgoing trunk groups and such information should permit to take into account real-time traffic variations.

All these improvements are under study at the CNET and will be evaluated using an event by event simulation.

REFERENCES

[1] HENNION B. : "Le Partage de Charge dans la Téléphonie Adaptative". L'Echo des Recherches April 1978

[2] SRIKANTAKUMAR P.R., NARENDRA K.S. : "A Learning Model for Routing in Telephone Networks". SIAM J. of Cont. and Opt. vol. 20 n°1 Jan 1982

[3] CAMERON W.H., REGNIER J., GALLOY P., SANDIE A.M. : "Dynamic Routing for Intercity Telephone Networks". Proc. of International Teletraffic Congress Montréal June 1983

[4] BEL G., CHEMOUIL P., BERNUSSOU J., GARCIA J.M., LE GALL F. : "Adaptive Traffic Routing in Telephone Networks". submitted to Large Scale Systems

[5] WILKINSON R.E. : "Theories for Toll Traffic Engineering in the U.S.A.". Bell Syst. Tech. Journal vol; 35 1956

[6] LE GALL F. : "One Moment Model for Telephone Traffic". Applied Math Modelling vol. 6 Dec. 1982

[7] HENNET J.C. : "Etudes de Politiques d'Acheminements d'Appels dans un Réseau Téléphonique Interurbain à 2 Niveaux Hiérarchiques". Dr Ing. Thesis Toulouse 1978

[8] CAMOIN B., MASSON A. : "Etude de la Robustesse des Structures de Réseaux". Tech. Report CNET NT/RCC/RTR/415

[9] RAPP Y. : "Planning of Junction Network in a Multi-Exchange Area". Ericsson Technics n°2 1965

AN ALGORITHM FOR OPTIMIZATION OF PACKET FLOW IN TELECOMMUNICATION NETWORK

Krzysztof Bochenek
Institute of Control,Systems Engineering
and Telecommunications
University of Mining and Metallurgy
Kraków, Poland

Abstract

The flow of packets in the network is a discrete process,because each packet is treated separately by all the elements of network.Therefore it seems that future development of methods of the packet routing in the telecommunication networks is connected with the analyse of such a flow that gives us information about every packet in the network. In this paper we present model of the telecommunication network for the problem formulated above and then we propose an algorithm of optimal control of packet flow in the network.

Introduction

The optimization problem of dynamic flow in telecommunication network is very important from the practical point of view. Despite many techniques of packet routing in telecommunication network are used there is still a lack of models and methods of analysing these techniques allowing for the efficient packet control in the network. According to kind of model / static,dynamic,diffusion or stochastic/ we can get better or worse approximation of reality.

The flow of packets in the network is a discrete process,because each packet is treated separately by all the elements of network. Therefore it seems that future development of of methods of the packet routing in the telecommunication networks is connected with the analyse of such a flow that gives us information about every packet in the nerwork.

Solving this problem we can use dynamic programming method [2] or branch-and-bound method [3],[4]. However this methods are not useful for real time control of telecommunication network because their complexity.

In this paper we present algorithm which is based on branch-and-bound method. Taking into account the phisycal aspect of packet transmission process we reduce expenditure of computation in radical way.

The model of the telecommunication network

The flow of each packet in the network will be represent by a chain which consists of arcs representing: transmission of the packet by channel, transmission of the packet by node, waiting for release of node or channel.

Let introduce the following notation:

$L=\{(i,j),(j,k),\ldots$ is the set of the channels of the network,

$k=1,2,..K$ - the numbers which are assign to the packets,

$OUT(i)=\{j:(i,j)\in L\}$ is the set of channels going out of the node i,

i_t denotes the appearing of the packet in the node i at the time t,

i'_t denotes the finishing of the packet processing in the node i at the time t,

i^j_t denotes the beginning of the packet processing in the node i at the time t,

b_i the packet processing time in the node i,

b_{ij} the packet transmission time through the channel (i,j),

$arc(i_t,i'_{t+b_i})_k$ represents the transmission of the packet k through the node i,

$arc(i^j_t,j_{t+b_{ij}})_k$ represents the transmission of the packet k through the channel (i,j),

$arc(i'_t,i^j_t)_k$ means that the transmission of the packet k was finished in the node i at time t and the transmission of the packet through the channel (i,j) was started,

$arc(i_t,i_{t+\tau})$ denotes that the packet k was waiting for time τ for the release of the node i,

q_k - the destination node for the packet k.

Because the arcs represent execution times of some operations in the network we assign them the following lengths:

$$d(i_t,i'_{t+b_i})=b_i \tag{1}$$

$$d(i^j_t,j_{t+b_{ij}})=b_{ij} \tag{2}$$

$$d(i_t,i_{t+\tau})=\tau \tag{3}$$

$$d(i_t,i^j_t)=0 \tag{4}$$

For transmission of k-th packet we create chain $\pounds^{k}_{iq_k}$ which is the sequence of arcs representing operations made on that packet on its way from the source node i to the destination node q_k:

$$\pounds^{k}_{iq_k} = \langle i_t, (q_k)'_T \rangle_k \qquad (5)$$

We denote by $\pounds$ the set of chains representing one possible way of sending of all packets through the network.

$$\pounds = \{ \langle i_{t_{ok}}, (q_k)'_{T_k} \rangle_k \; , \; k=1,2,\ldots K \} \qquad (6)$$

where: k- number of packet,

t_{ok}- appearing time of k-th packet in node i,

T_k - time at which k-th packet reaches destination.

Now we describe the properties of set $\pounds$.

Property 1

$$(i_{t_a}, i'_{t_a+b_i})_{k_1} \cap (i_{t_b}, i'_{t_b+b_i})_{k_2} = 0 \quad \text{for each } k_1=1,2,..K,\ k_2=1,2,..K,\ t_a>0,\ t_b>0 \qquad (7)$$

Property 2

$$(i^j_{t_a}, j_{t_a+b_{ij}})_{k_1} \cap (i^j_{t_b}, j_{t_b+b_{ij}})_{k_2} = 0 \quad \text{for each } k_1=1,2,..K,\ k_2=1,2..K,\ t_a>0,\ t_b>0 \qquad (8)$$

The meaning of the property (1) is following: if in node i takes place processing of some packet the processing another packet cannot be begun. The meaning of the property (2) is following: if in channel (i,j) takes place transmission of some packet the transmission of another packet cannot be begun.

Let us take the set of all permissible sets $\pounds$ which satisfy property (1) and property (2).

$$\mathcal{F} = \{ \pounds : \text{property (1) and property (2) are fulfilled} \} \qquad (9)$$

Optimisation problem consists in finding, in set $\mathcal{F}$, the set $\pounds$ which maximizes the expresion (max T_k), where k=1,2,...K.
This problem can be briefly written as:

$$OP_{min} : \left. \pounds \; ; \; (\max_{k=1,2,..K} T_k) \right]_{\pounds \in \mathcal{F}} \qquad (10)$$

Solution of the problem

In order to find optimal flow of packets, we use branch-and-bound method. We take into account physical aspect of transmission process. It helps us find optimal solution at the first steps of searching of tree of feasible solution.

At the first we define the length of path joining nodes i and q_k

$$d[\mathcal{J}(i,q_k)] = \sum_{(m,n)\in\mathcal{J}(i,q_k)} b_{mn} \tag{11}$$

For each pair of nodes $\{i,q_k\}$ we introduce set $M(i,q_k)$ channels which start the shortest paths /in sence of definition (11)/ from node i to node q_k. Selecting channel for packet transmission from node i to node q_k we have to use additional conditions. We describe them below.

The set of chains representing flow of packets through the network will be created gradually and simultaneously for all K packets. Let $\pounds^k_t$ denote partial chain representing the way passed by the packet k untill time t. We take set $\pounds_{cz}$ which contains K partial chains $\pounds^k_{t_k}$, k=1,2,...K.

$$\pounds_{cz} = \{ \pounds^1_{t_1}, \pounds^2_{t_2}, \ldots, \pounds^k_{t_k} \} \tag{12}$$

We will add new arc to partial chain for which the following is satisfied

$$\min_k t_k = t_l \tag{13}$$

then $\pounds_{cz} = \{\pounds^1_{t_1}, \pounds^2_{t_2}, \ldots, \pounds^1_{t_l} + (A,B)_l, \ldots, \pounds^k_{t_k}\}$

If the last node of chain $\pounds_{t_l}$ is:

- node i_t then $(A,B)_l = (i_{t_l}, i'_{t_l+b_i})_l$ 14

  or $(A,B)_l = (i_{t_l}, i_{t_l+\tau})_l$ 15

- node i'_t then $(A,B)_l = (i'_{t_l}, i^j_{t_l})_l + (i^j_{t_l}, j_{t_l+b_{ij}})_l$ 16

  or $(A,B)_l = (i'_{t_l}, i^j_{t_l})_l + (i^j_{t_l}, i^j_{t_l+\tau})_l + (i^j_{t_l+\tau}, j_{t_l+\tau+b_{ij}})_l$ 17

From above we can see that during creation of the set $\pounds$ of chains representing optimal flow of K packets in the network we must take one of the following decisions:

- to start the processing of the packet or to wait for the moment when the node will be free,
- to sent the packet into one of the channels from set $M(i,r_1)$ or to wait for the miment when the selected channel will be free.

We assume that expression (12) is fulfilled for k-th packet at the time t. Let node i_t be the last node of the chain f_i^k. Then:

if exist $(i_{t-a}, i'_{t-a+b_i})_k$ such that $0 < a < b_i$, $l=1,2,..K$, $l \neq k$

$$\text{then } f^k_{t-a+b_i} = f^k_t + (i_t, i_{t-a+b_i})_k \tag{18}$$

$$\text{otherwise } f^k_{t+b_i} = f^k_t + (i_t, i'_{t+b_i})_k \tag{19}$$

Let us cosider situation when two packets l,m appear at the time t in node i.

$$\text{If } d\left[\mathcal{T}(i,q_l)\right] > d\left[\mathcal{T}(i,r_m)\right] \tag{20}$$

$$\text{then } f^l_{t+b_i} = f^l_t + (i_t, i'_{t+b_i})_l \tag{21}$$

$$f^m_{t+b_i} = f^m_t + (i_t, i_{t+b_i})_m \tag{22}$$

$$\text{otherwise,if } d\left[\mathcal{T}(i,q_l)\right] < d\left[\mathcal{T}(i,r_m)\right] \tag{23}$$

$$\text{then } f^l_{t+b_i} = f^l_t + (i_t, i_{t+b_i})_l \tag{24}$$

$$f^m_{t+b_i} = f^m_t + (i_t, i'_{t+b_i})_m \tag{25}$$

Now we describe action which should be taken in situation when expresion (12) is fulfilled for k-th packet and node i_t is the last node of the chain f^k_t. Then appears the following problem: we have to add to the chain f^k_t one of the arcs (i'_t, i^m_t), where $m \in OUT(i)$ and channel $(i,m) \in M(i,q_k)$.

If one of the channels $(i,n) \in M(i,q_k)$ is busy at the time t, then we do not take this channel into consideration, choosing node m to which packet k should be send i.e

If exist $(i^n_{t_1}, n_{t_1+a_{in}}) \wedge t_1 < t < t_1 + a_{in}$ then

$$M'(i,q_k) = M(i,q_k) - (i,n) \tag{26}$$

If all channels are busy then k-th packet waits for the moment when one of them will be free

$$f^k_{t+\tau} = f^k_t + (i'_t, i'_{t+\tau})_k \quad \text{where: } t+\tau \text{ -time when channel } (i,n) \in M(i,q_k) \text{ becomes free} \tag{27}$$

$$f^k_{t+\tau+b_{in}} = f^k_{t+\tau} + (i^n_{t+\tau}, n_{t+\tau+b_{in}})_k \tag{28}$$

In order to find the channel to which packet k should be send we use information about paths taken by other pakets. This information could be obtained from the chains $\underset{t_l}{l}$, $l=1,2..,K$, $l \neq k$.

At first we consider the situation, when processing of l-th packet in node j at time $t_l \geqslant t$ will be finished.

If exist n such that:

$$(i^n_t, n_{t+b_{in}})_k \in M(i,q_k) \quad \text{and} \quad (j^n_{t_l}, n_{t_l+a_{jn}}) \in M(j,q_l)$$

$$\text{then} \quad M'(i,q_k) = M'(i,q_k) - (i,n) \tag{29}$$

It means that we reject, from set OUT(i), the node to which other packet will complete for access at the time $t_l \geqslant t$.

If in node j the end of packet transmission in channel (m,j) takes place at the time t, then

$$\text{if exist } j: \quad (i^j_t, j_{t+b_{ij}}) \in M(i,r_k) \quad \text{then}$$

$$M'(i,r_k) = M'(i,r_k) - (i,j) \tag{30}$$

If $M'(i,r_k) = 0$ we send packet k to the first channel from the set $M'(i,r_k)$. The channels in the set $M'(i,r_k)$ are arranged according to the length of the paths joining node i and destination node r_k.

If $M'(i,r_k) = 0$ then we send packet k to the channel which was last removed from the set $M'(i,r_k)$.

Testing operation

For paths which are started with channel from the set $M(i,r_k) - (i,j)$ we must designate lower estimation of transmission time of packet k.

Let $(i,m) \in \{M(i,r_k) - (i,j)\}$

If we send packet k throught channel (i,m) then lower estimation of transmission time of packet k is:

$$D_t^{im} = \tau_{im} + \sum_{(r,s)\in \mathcal{T}(i,r_k)} (b_{rs} + b_s) \quad (31)$$

where: τ_{im}- waiting time for vacation of the channel (i,m)

After finding packet flow we test if exist else better flow. To this end we look for k which fulfils condition

$$\max_{i=1,2,..,K} T_i = T_k \quad (32)$$

In chain $\mathcal{L}_{T_k}^{k}$ we look for such nodes for which

$$D_t^{im} < T$$

if such nodes exist we try to find better flow. In order to attain new packet flow we choose node m as next node to which we send packet k from node i at time t, i.e:

$$\mathcal{L}_{t+b_{im}}^{k} = \mathcal{L}_t^k + (i_t^{\prime}, i_t^m)_k + (i_t^m, m_{t+b_{im}}) \quad (33)$$

We can prove that using above algorithm we find optimal packet flow at the first steps of algorithm.

Conclusion

In this paper we present an algorithm of optimal control of packet flow in the network. This algorithm is based on branch-and-bound method. Taking into account the phisycal aspect of packet transmission process we reduce expenditure of computation in radical way.

The flow packet is described by giving transmission and processing time for each packet. It seems that future development of methods of the packet routing in the telecommunication networks is connected with the analyse of such a flow that gives us information about every packet in the network.

References

1. K. Bochenek, 1983. Design of optimal control systems for large telecommunication networks. IMA/SERC Symposium on control theory, University

of Warwick,Coventry 83.
2.Schoute,1978.Decentralized control in packet switched satellite communication,IEEE Trans.on Automatic Control,Vol.AC-23,No.2,1978,pp 362-371.
3.T.Batycki,A.Kasprzak,1980.The modelling of a dynamic multicommodity flow in packet-switched computer communication network,Systems Science,Vol.6,No.2,1980 pp.193-199.
4.L.R.Ford,D.R.Fukelson,1969.Przepływy w sieciach.PWN,Warszawa,1969.

THE EVOLUTION OF AUTOMATIC MONITORING IN THE OFFICE OF THE FUTURE

Chris J. Georgopoulos
Dept. of Electrical Engineering
University of Thrace
Xanthi, Greece

SUMMARY

This paper deals with techniques of automatic monitoring and related principles as they may be applied to the office systems. Due to the variety of equipment that will be used in the office of the future, the system designer has to meet the objectives of a good network-control system in order to keep availability up.

INTRODUCTION

The last decade has produced many unexpected changes in the areas of information processing and information transfer. As computer and communications equipment become faster, more efficient, and more functional, the price/performance ratios improve. However, there is another price to pay as an interactive network delivers increasing benefits through growth and that is figured in downtime. The office operations manager is responsible for keeping all equipment and terminals up and running uninterrupted as long as possible, that is, minimizing the workstation and terminal hours lost to downtime [1],[2].

In an automated office where voice, data, and video networks may be anticipated, the more routine functions such as monitoring, fault isolation, and equipment restoration must be accomplished without the need for operator attention. The ideal network control system must provide functions like the following:

Pre-operational Self-Test Capability: The ability to check the system prior the putting it to normal operation.

Monitoring : Gives the office manager an indication of a problem condition in the network while in operation.

Functional Problems-Failure Isolation: Isolation of a problem or a failure to the specific element within the network.

Restoration : Ability to restore the network operation.

Performance : Provision of the necessary information to insure the optimization of the transfer of information and operation of the network.

The above functions must be provided automatically, independently of the type of equipment used and with flexibility to be able to accommodate changes and growth within the network.

PRE-OPERATIONAL SELF-TEST CAPABILITY.

In order to determine the transmission quality of digital communication systems in the modern electronic office, the error performance should be considered, because it is one of the most important parameters. In practice, a distinction is made between pre-operational or out-of-service error measurements and in-service error measurements. In the case of pre-operational testing, the live traffic is replaced by a known test pattern. In order to simulate the random structure of message signals as closely as possible, pseudo-random sequences are often used when performing error measurements. The most common patterns are shown in Table 1 [3].

In addition to pseudo-random patterns, periodic sequences can serve as

TABLE 1

PSEUDO-RANDOM SEQUENCES RECOMMENDED BY THE CCIT FOR THE MEASUREMENT OF DIGITAL ERRORS

APPLICABLE BIT RATES	PATTERN LENGTH	CCITT RECOMMENDATION
up to 20 kbps	2^9-1 bits	V.52
20-72 Kbps	2^{10}-1 bits	V.57
64 Kbps	2^{11}-1 bits	0.xb
1500-8500 kbps	2^{15}-1 bits	0.151
8500-139,000 kbps	2^{23}-1 bits	0.151

test signals. Because the input signal applied to the object under test is known in both cases, the output from the test object can be compared with a reference signal which is identical with the original input signal. This permits the detection of any error in the data stream. Therefore, this way of assessing digital errors is often called a true error measurement. During a pre-operational self-check, true error measurements are not possible if the test object still carries traffic, because the content of the messages is normally not known. However, in most cases digital signals are structured in a specific, known way and obey certain rules such as coding or framing. Code violations or, for example, errors in frame-alignment signals can be measured in-service and provide information about the error performance. The smaller the error ratio is, the smaller the difference between true error measurements.

MONITORING

The interface to a digital link in the office is via DTE or Data Circuit -terminating Equipment (DCE), usually interpreted as a modem or a front-end processor. Monitoring the performance of these interfaces must be the system designer's first consideration.

Basic Monitoring

The earliest and most basic approach to managing a network starts with equipment that resides at the central site. The network manager must first provide access to both the digital and analog interfaces of all modems in the network. Modems are in the unique position of bridging the analog and digital domains at every network site. In addition to the normal main data channel supported by the modems, they create a special test channel in both domains.

Automatic Monitoring and Alarming with Centralized Control.

An important feature of the network-control system here is the test module that resides at every terminal interface and communicates over the test channel. This test module must recognize its own address when instructed, then accept commands and perform required functions.

Figure 1 shows an office network with two local groups of data lines, which report their network interface unit (NIU) circuit monitoring and alarming conditions back to a control network in the workstation. If a data monitor is used that has auto-monitoring capabilities, no hands-on re-adjusting of the data monitor is required, regardless of what type of traffic protocol is monitored, even synchronous. Automonitors can automatically detect the protocol, speed, bits per character and traffic on the port.

Sophisticated data systems allow for the retrieval of various network statistics. Link or port problems are frequenly intermittent in nature and very hard to detect. Statistics provided by a statistical multiplexer, however, will usually help locate such problems. These systems also have network control options which allow for complete system operation,diagnostics and statistics monitoring from a single location via a CRT, printer or dial-up modem [4].

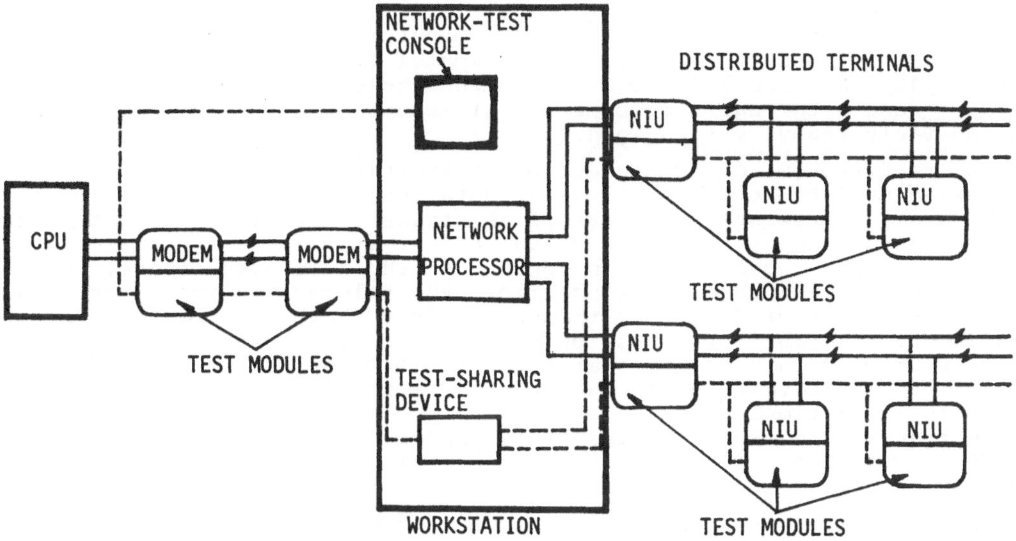

Figure 1. Simple office network with individual test modules. Each test module recognizes its own address when instructed, then accepts commands to perform its assigned functions.

FUNCTIONAL PROBLEMS AND FAILURE ISOLATION VIA DIAGNOSTICS.

To recognize that a monitoring system is required for an automated office is one thing. In addition two key questions about the control system for the networks in such environment must be answered : a) what are the features of such a system? and b) what tests and remedial controls can a user expect in a state-of-the-art network control system?

To select a good network control system, the user should consider the following three design elements : (1) the test channel, (2) the addressable test modules in the terminals and peripherals, (3) the central controller in the workstation, and (4) the µP-based equipment.

Network diagnostics Using µP Technology.

The growing use of µP technology is dominating the movement towards incorporating extensive diagnostic abilities in data network products. Despite increasing product complexity, µPs have provided the methodology for developing improved test capabilities. Automatic and user-friendly testing procedures have lessened the need for highly technical personnel, a big help for solving problems in the office environment staffed only with non-technical personnel.

Diagnostics fall into two basic categories: internal (or self-testing) and network. In µP- based systems such as statistical multiplexers, self-testing is very common. Self-tests usually only occur during the period known as initialization, that is, after either a power-on of the unit or a re-start or re-set command has been given. During initialization a separate software routine, stored in what is known as shadow PROM , is activated. Shadow PROM is responsible for performing the test of the hardware and software and, once completed, is not part of the normal operating system.

Fault Tolerant Approach.

The primary goals of a fault tolerant system are to maintain data integrity and maximize system uptime. To meet these goals, a fault tolerant computer must detect and compensate for or correct faults. The methods of meeting these goals and needs are as varied as the online transaction processing, realtime process [5]. Many computer companies see fault tolerance as a worthwhile feature for a computer to have in any environment. A CPU-32/16, for example, could be designed to correct both hardware and software faults through a combination of redundant architecture, monitoring and fault recovery techniques.

The tolerant systems design takes as its starting point the assertion that the permanent data in the system (the data resident in files, tables,etc., on the disk) represents the state of an office's business (or part of the business) and that it is this data which must at all times be, from an external viewpoint, consistent and correct in the automated office. To guarantee fault tolerance of a transaction, the tolerant system automatically performs a number of functions transparently to both the terminal and the programmer. Tolerant's transaction executive (an integral part of the operating system) maintains the state of execution environment by recording process state information. In the event of a failure, the system performs the following actions [6].

a) The transaction executive identifies which transactions and which processors have been affected. The communications system and the file/database management system are informed of the identified transactions.

b) The transaction executive recreates failed processes on a different system building block. The re-instituted processes are reconnected to their logical I/0 channels.

c) The file/database management system rolls back the affected transactions to their start points by use of the before images.

d) The communications system initiates system re-execution of the affected transactions by reading the transaction input log. Output generated by the re-execution is discarded up until the previous point of failure.

RESTORATION

With only a few lines, it is difficult to justify a network monitoring system. There are, however, many attractive and inexpensive alternatives. For example, it may be possible to combine monitoring access with a restoration capability. Here, a simple application of standard A/B switching (or fallback switching) will enable the system designer to provide a spare for the front-end processor.

Switches are available which also have monitoring capability for, say, transmit data, receive data, and Carrier Detect (CD). It is even possible to get A/B switching with an adjustable audible alarm on RS-232C leads, such as the CD. Thus, if the CD goes down (indicating that either a telco line is lost or a modem has problems), the operator would be immediately notified.

In general,an ideal system in the office would automatically scan and monitor for present alarm conditions, run auto-diagnostic tests to isolate a problem and, once the source of the problem is located, perform re-routing or equipment substitution via intelligent switching to restore communications within the network. The system would perform these tasks automatically with as little human intervision as possible. Of course,the operator must have the ability to override system commands that may interrupt communications should an incorrect switching command be issued or the primary control method ever fail. Remedial functions are extremely important in helping to directly control downtime by reducing the repair time for several kinds of failures.

PERFORMANCE

In an automated office system, where various alternative communication schemes may be used [7], as indicated in Fig.2, the primary network performance measurement is throughput, S, given by the following relationship which represents a theoretical model:

$$S = \frac{\lambda e^{-\lambda T}}{\lambda(1+2T) + e^{-\lambda T}} ,$$

where λ= packets per slot time
T= critical period time for possible collisions. (Beyond period T, the channel is sensed as busy and no collissions can occur).

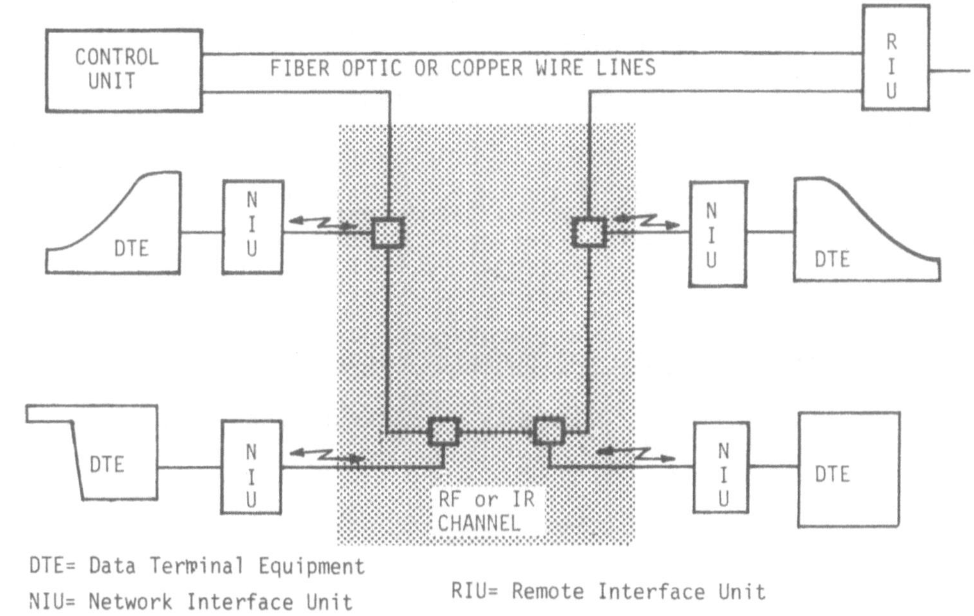

Figure 2. Alternative Communication schemes in an automated Office.

It is very desirable for a local network to use 100% of the available channel bandwidth. In practice, however, a carier-sense multiple access (CSMA)-type LAN fails to achieve this goal by about 20% because of data -packet collisions [8].But collisions rarely happen unless the channel becomes heavily loaded. The main problem arises with a busy channel which blocks the entry of packet arrivals.

In an ideal network, as fast as the data-traffic load imposes on the transmission channel (at λ packets/slot), the network transmits the traffic to its destination. However, practical system factors, such as busy and idle periods, collisions and propagation delays, diminish the theoretical model's throughput (S) and cause a deviation from the line S=λ.

Recent developments in electronic switching systems have made it possible for designers and office managers to have a more positive attitude toward tech control and specifically toward automatic monitoring. They do not see it as merely preventive maintenance and correction procedure; rather, it is considered as being inseparable from office network management that increases efficiency and data communicating productivity.

CONCLUSIONS

Self-testing, network testing and statistics reporting will be essential requirements for data communication systems in the office of the future. Automatic and user-friendly testing procedures can lessen the need for highly technical personnel, a big help for solving problems in the office environment with diversified equipment and staffed only with non-technical personnel.

REFERENCES

1. S.L. Teger, "Factors Impacting the Evolution of Office Automation," Proceedings of the IEEE , Vol.71, No.4, pp. 503-511, April 1983.
2. F.H. Lochovsky, "Improving Office Productivity: A Technology Perspective," Proceedings of the IEEE, Vol.71, No.4, pp. 512-518, April 1983.
3. F. Coenning, "Digital Error Performance," Telecommunications ,p.69, July 1983.
4. J. Veni, "Self-Testing Components Pinpoint Network Faults," Digital Design, p.84, August 1983.
5. G. Brown, "Recovery of Data in a Fault-Tolerant Network," in Proc. INTELEXPO'81, Los Angeles, CA., September 14-17, 1981.
6. D. Eidsmore, "Fault Tolerant Architectures," Digital Design, p.70, August 1983.
7. C.J. Georgopoulos, "Alternative Communications Techniques in the Office of the Future," in Proc. HETELCON'83, Athens, Greece, August 24-26, 1983.
8. F.N. Rounds, "Use Modeling Techniques to Estimate Local-Net Success," EDN, p.143, April 14, 1983.

HIERARCHICAL OPTIMIZATION OF NaOH DISSOLUTION

D. Matko, R. Karba, B. Zupančič, P. Omersel*
Faculty of electrical engineering, Ljubljana, Tržaška 25
61000 Ljubljana,
*Razvojni center Celje, 63000 Celje, Ul. XIV. div. 14
Yugoslavia

ABSTRACT

The NaOH dissolution is a typical endpoint problem. In the paper first the mathematical model of the exotherm reaction in the batch reactor including some nonlinearities and mathematical model of the cooling system are given. Using this model the optimal temperature and concentration responses are obtained by means of an optimization package and approximation of input signal with Tschebyshev polynomials. These responses are used for corresponding reference values and a PID controller is designed using the same package. In the conclusion some on line implementation possibilities are discussed.

1. INTRODUCTION

Simulation can be successfully used for the design of technology, planning of equipment and design of automatic control procedures for supervised control of chemical reactions. In the paper problems with the hierarchical optimization of NaOH dissolution process are described. First the mathematical model of the exotherm reaction in the batch reactor including some nonlinearities and mathematical model of the cooling system are given. Then the optimization criteria and optimization procedure are described. The results of this procedure are the optimal responses of the temperature and concentration. As the parameters of the reactor may differ from the ones used in the simulation, the PID controller is designed for tracking the optimal responses. In the conclusion some on line implementation possibilities are discussed.

2. MATHEMATICAL MODEL

The sodium hydroxide solution NaOH, used in processing technology of ironoxide pigment production, however, in exactly defined concentration is prepared in iron dissolving basin. A suitable mathematical model of reactor should be prepared in order to simulate the exothermic chemical reaction of dissolving. Due to the fact that only the top questions of chemistry are taken into account, the individual data or values are to be estimated and latter on corrected by the real data, obtained by the experimental pilote line measurements.

At a given example, a final concentration of solution 350 to 400 g NaOH/l or 7 mol/kg is needed.

NaOH can be used in a crystalline state, at the concentration of 95% NaOH, approximately or in a liquid state at the concentration of 60% NaOH, approximately. While dissolving in a basin, a reaction heath is evolved. It is in proportion to: a specific heath of the reaction - hr, a substance quantity - Mr (occuring during the reaction), a concentration Ca and to a specific velocity of reaction k.

Due to the theory of kinetics of simple reactions it can be resumed that the dissolving of NaOH in water represents a first order reaction with the specific velocity of reaction k in proportion to a temperature and presented according to Arrhenius equation

$$k = k_o \exp(-Ea/RT)$$

where Ea is energy needed to activate the reaction, R is a general gas constant, T is absolute temperature and k_o is factor of frequency or effectiveness of molecule collision. It is evident from the equation that velocity constant grows exponentially with temperature growth. Due to this equation the molecules are to reach critical energy Ea before they could react.

The activating energy represents a threshold of potential energy, the molecules have to reach, in order to get into the activated state. Heath effects of the reaction equal the difference of activating energies of the initial and return reaction. The Arrhenius equation is mostly used to define the activating energy on the basis of a diagram with the linear relation between the logarithm of specific reaction velocity and with the reciprocal temperature value. Here we generally specify or measure the specific reaction velocity in lab conditions or in pilote production at constant temperature.

At the given example, where 1600 kg crystalline NaOH are being dissolved in

4000 kg H_2O, some data of the kinetic reaction of NaOH dissolving were not known, however, we foresee them on the basis of practical observations. The energy of reaction activation (Ea = 20930 J/mol) is estimated. The specific reaction velocity is not known. Taking into account the demanded or given technological data - inside temperature of reactor without cooling would reach 75°C after 32 hours and considering the solution of Arrhenius equation we defined the constant - k_o and the course of specific reaction velocity k subjected to reactor temperature.

During the following technological procedure a solution at room temperature 25°C is used. Therefore the reactor is usually equiped with a cooling system as it is evident from the mathematical model shown in Fig.1

Equns 6,7,8 and 10 represent the reactor, equns 1,2 and 5 the cooling system and eq. 9 the heat exchange with the environment. Also the PID controller, the dynamics of the temperature sensor and the valve characteristic are included in the scheme.

3. OPTIMIZATION

The NaOH dissolution is a typical endpoint problem. The objective of the control procedure is that the temperature in the reactor after 40 hours is 25°C and that the quantity of undissolved NaOH is minimal. As the high temperature accelerates the dissolution according to the Arrhenius equation it is not recommended to keep the temperature of the reactor on 25°C during the initial phase of the dissolution. On the other hand the capacity of the cooling system is limited not only through the maximal flow (0.67 l/s) but mainly through the limited temperature (12°C) of the mixed cooling water returning to the cooler.

According to these requirements we choose the following optimization criterion:

$$I = \int_{40h}^{50h} ((T_r - 25)^2 + C_a{}^2)\,dt$$

where T_r is the reactor temperature and C_a is the quantity of the undissolved NaOH. Also the constraint

$$T_{hii} < 12^\circ C$$

is taken into account, where T_{hii} is the temperature of the mixed water returning to the cooler. The optimization was done by an interactive pac-

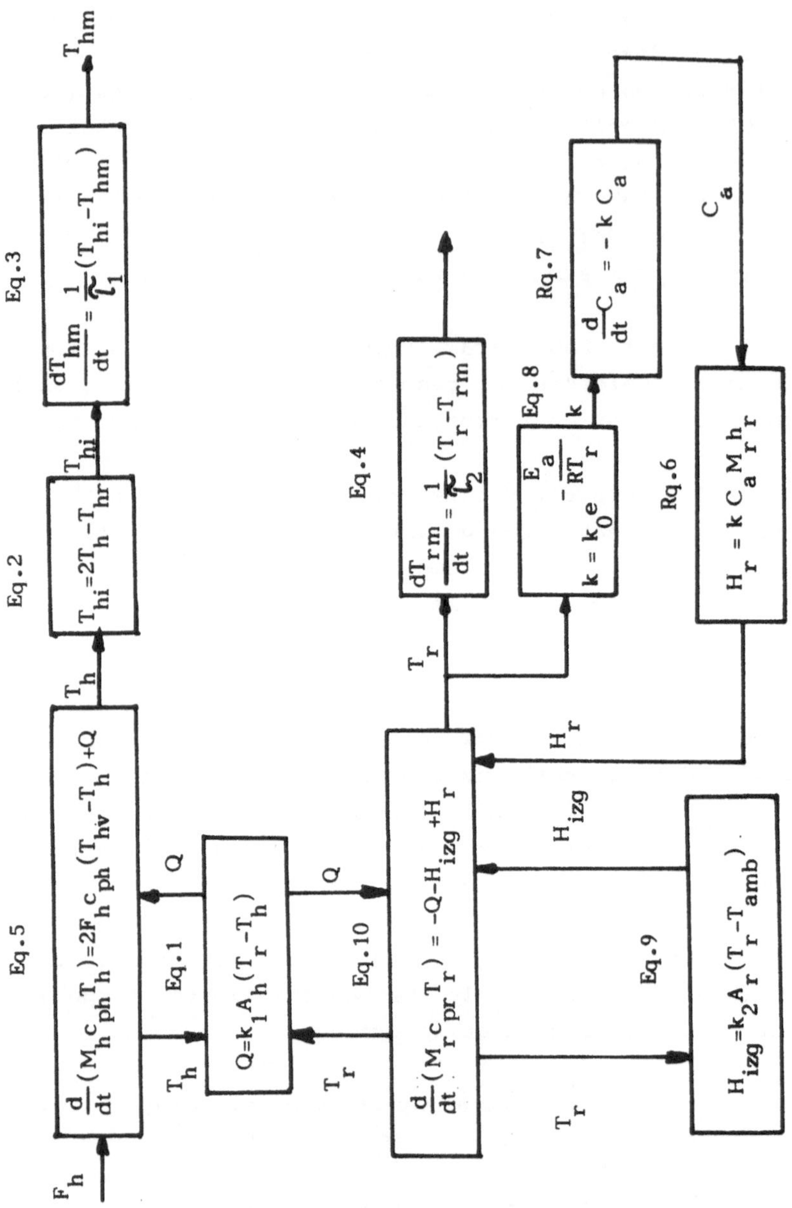

Fig.1. The mathematical model of a chemical reactor with exothermic reaction

kage designed for CAD purposes. The input was chosen in the form of third order Tschebyshev polynomial sum. The optimal open loop responses are shown in Fig. 2 where T_r^*, F_h^* and T_{hii} mean temperature of reactor, three way valve position and temperature of water returning to cooler respectively.

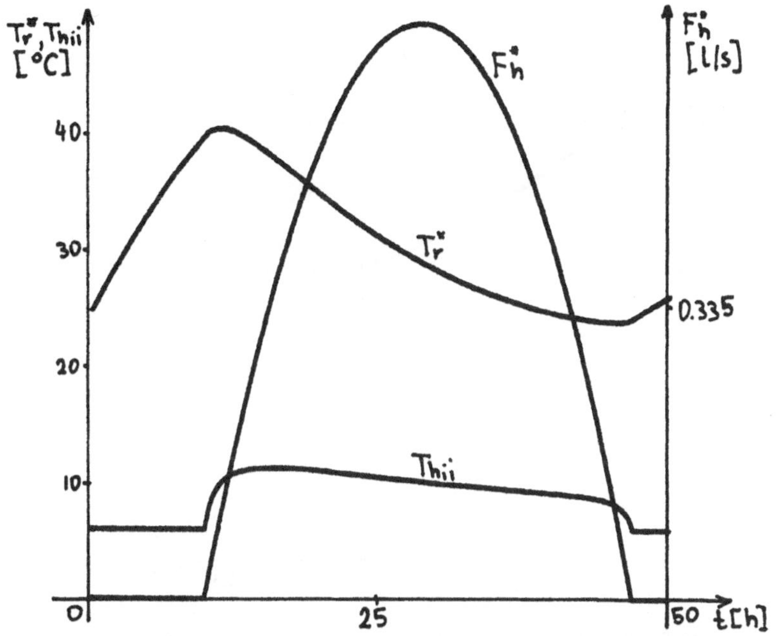

Fig.2. Optimal open loop responses of the reactor

This is naturally a suboptimal solution, since only 3 terms were used as the input function. So we decided to modify the optimal reactor temperature response as shown in Fig.3, i.e. we force the temperature after 40h to be 25°C.

As the real reactor parameters differ from the ones used in the simulation, a PID controller was implemented as shown in Fig.4.

Fig.5 shows the open loop response (no PID) of the reactor temperature (T_r^-) at the +10% change in thermal conductivity of the cooling system. After the PID controller was applicated into the control loop the response (T_r^+) became very similar to the optimal one. Also the three way valve position (F_h) is shown in the same figure. In Fig.6 the optimal response and value position at the -10% change in thermal conductivity are shown.

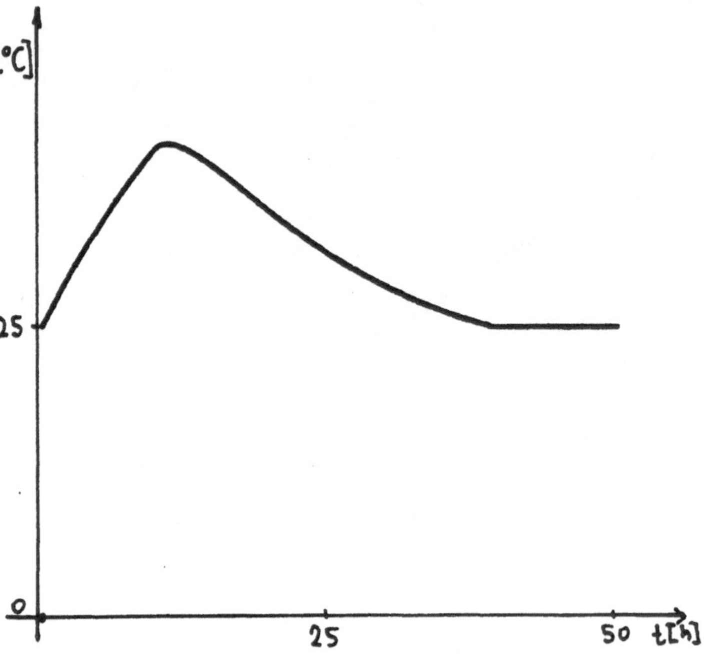

Fig.3. Modified optimal reactor temperature response

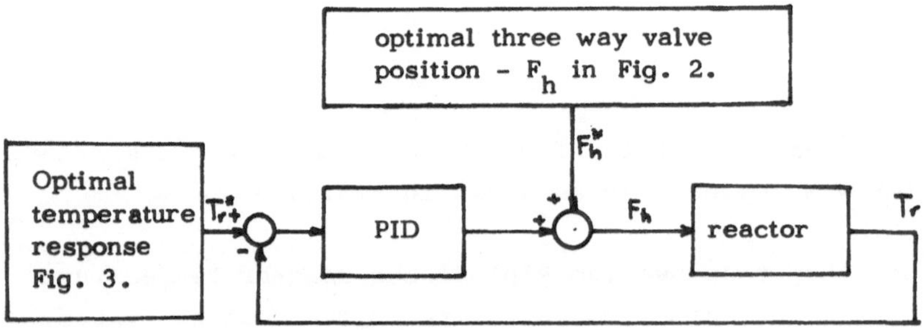

Fig.4. Implementation of the PID controller

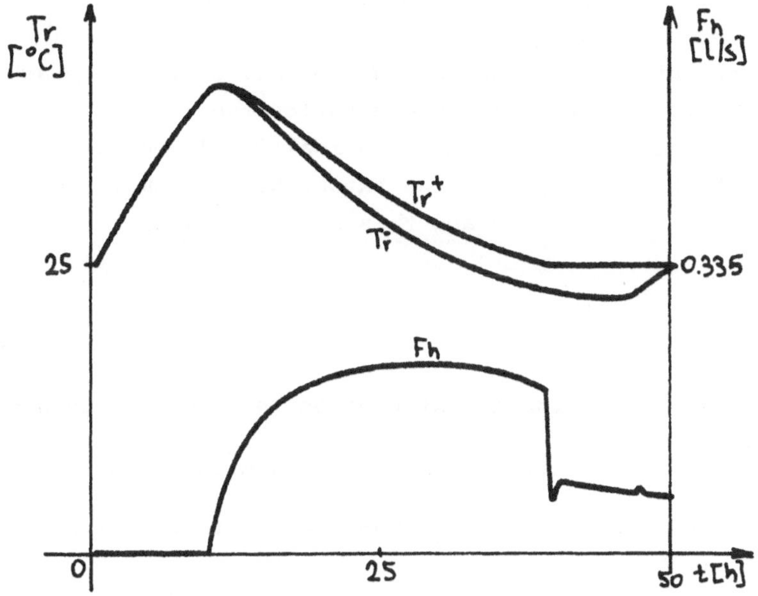

Fig.5. Responses at the +10% change in thermal conductivity

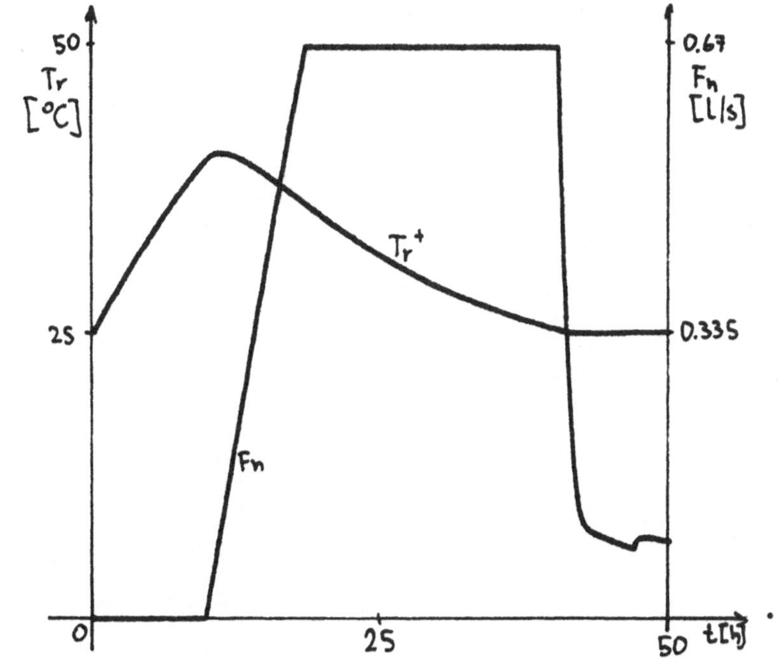

Fig.6. Responses at the -10% change in thermal conductivity

4. CONCLUSION

By means of an interactive CAD package, the optimal control of NaOH dissolution was designed. For on line implementation the reference value is computed according to the chosen mathematical model and corresponding parameters, which may vary with the batch. So a parameter estimation procedure would be preferable and this is the goal of our next investigations.

5. REFERENCES

Franks, R.G.E. (1972): Modelling and simulation in chemical engineering, Wiley Interscience, New York.

Weber T.W. (1973): An introduction to process dynamics and control, Willey Interscience, New York.

ANA - interactive CAD package - manual, Faculty of electrical engineering, Ljubljana.

OPTIMAL PLANNING FOR OPERATING AN OXYGEN PLANT

Nobuo Sannomiya
Yoshikazu Nishikawa
Department of Electrical Engineering
Kyoto University, Sakyo-ku
Kyoto 606, Japan

Hirochika Akagi
KAWASAKI STEEL CORPORATION
1, Kawasaki-cho, Chiba-shi
Chiba 260, Japan

Yoshio Takeyama
Takashi Tsuda
FUJI FACOM CORPORATION
1, Fuji-machi, Hino-shi
Tokyo 191, Japan

Takao Yashima
FUJI ELECTRIC CO., LTD.
1, Fuji-machi, Hino-shi
Tokyo 191, Japan

1. Introduction

In the last few years, computer control systems have been developed in various kinds of production plants with good results from the viewpoint of energy saving. For decreasing further energy loss and operating cost, we need to make a planning for operating each equipment in a plant with due regard to optimizing the combination of various equipments with different performances. This idea is important especially for the case where there are redundant equipments under lowering of demand.

A number of combinatorial optimization problems are formulated as mixed-integer programs. However, solving them exactly becomes difficult with increase in the problem size. Therefore, formulating the problem as an integer program has been avoided so far in the case of executing process control and planning with use of a process computer.

This paper deals with an optimal planning for operating an oxygen plant which produces several gases used in a process of iron works. The plant consists of air turbo compressors and separators, and generates both gas and liquid states of oxygen, nitrogen and argon from compressed air. There are several types of equipments which can be substituted with one another. But, they have differences in the performance and the operating cost. Therefore, we need a scheduling for operating each equipment so as to minimize the total operating cost. Further, we have additional constraints associated with the performance of the equipments. They are usually given as bounded conditions for the decision variables. In this case, since the minimum load of each equipment is not zero but takes a positive value, the feasible region of the decision variables becomes disconnected. Consequently, the problem needs to be formulated as a mixed-integer program.

For real time control and planning of the plant, we develop a method for solving the problem of mixed-integer type quickly. A decomposition procedure[1] is applied to the problem with due regard to an angular form of the problem structure. The master problem is a linear program and the subproblems are mixed-integer programs with a single binary variable. Therefore, we do not need any specific algorithm for solving mixed-integer programs. A sufficient condition for optimality is obtained. The termination of the procedure is checked in two stages. That is to say, if the optimality test is satisfied, the procedure terminates and the optimal solution is obtained. If not, the search for improving the solution is continued within a restricted extent, and the best solution obtained so far is provided as a suboptimal solution. In either case, the procedure always gives a feasible solution. Therefore, the method is efficient for scheduling and control of the plant.

2. Problem Statement

We consider an optimal planning for operating an oxygen plant as shown in Fig. 1. The plant consists of n air turbo compressors and m separators. The separator produces both gas and liquid states of oxygen, nitrogen and argon from compressed air generated by the air compressor. The gas product is used in a process of iron works. In the case of shortage of the gas product, the liquid product is substituted for it. On the other hand, if the gas product is superfluous, the liquid product is sold at high price for other purposes.

In order to satisfy the present demand condition, we do not need to use all the existing equipments. Therefore, we make a planning for operating each equipment in such a way that the following requirements should be satisfied:

1) to meet the demand of each product,
2) to minimize the total operating cost,
3) to maximize the profit obtained by selling the liquid product.

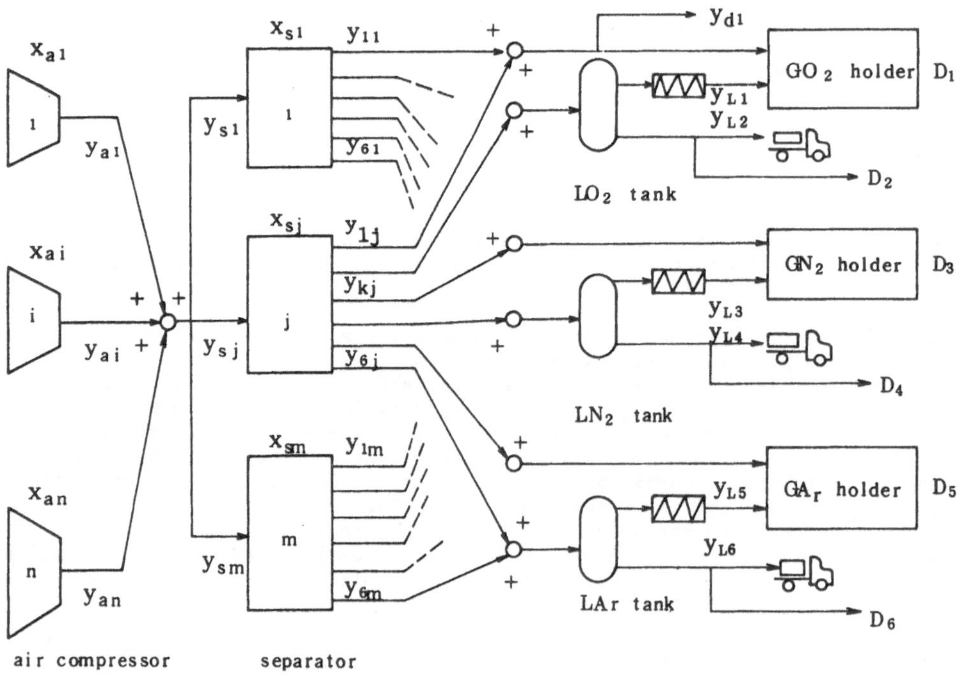

Fig. 1 The schematic diagram of an oxygen plant

Thus, the problem is formulated mathematically by treating the item 1) as the constraint and the items 2) and 3) as the objective. Further, we have additional constraints associated with the performances of the equipments.

Constraint for air turbo compressor: Let y_{ai} be the quantity of the compressed air generated by the i-th compressor, and x_{ai} be the binary variable associated with the i-th compressor (i.e. $x_{ai} = 1$ when the i-th compressor is used, and $x_{ai} = 0$ otherwise). Then we have

$$x_{ai}\, y_{ai}^{-} \leqq y_{ai} \leqq x_{ai}\, y_{ai}^{+} \qquad (y_{ai}^{-} > 0) \qquad i = 1, 2, \ldots, n \tag{1}$$

$$x_{ai} = 0 \text{ or } 1 \qquad i = 1, 2, \ldots, n \tag{2}$$

where the superscripts - and + mean the values associated with the minimum and the maximum load of the corresponding equipment, respectively.

Constraint for separator: Let y_{sj} and x_{sj} be the quantity of the compressed air used at the j-th separator and the associated binary variable, respectively. Further, let y_{kj} be the quantity of the k-th product generated by the j-th separator. Then the following constraints are obtained:

$$x_{sj}\, y_{sj}^{-} \leqq y_{sj} \leqq x_{sj}\, y_{sj}^{+} \qquad (y_{sj}^{-} > 0) \qquad j = 1, 2, \ldots, m \tag{3}$$

$$x_{sj} = 0 \text{ or } 1 \qquad j = 1, 2, \ldots, m \tag{4}$$

$$y_{sj} = \sum_{k=1}^{6} \alpha_{kj}\, y_{kj} + \alpha_{sj}\, x_{sj} \qquad j = 1, 2, \ldots, m \tag{5}$$

$$x_{sj}\, y_{kj}^{-} \leqq y_{kj} \leqq x_{sj}\, y_{kj}^{+} \qquad k = 1, 2, \ldots, 6; j = 1, 2, \ldots, m \tag{6}$$

$$y_{2j} + y_{4j} + y_{6j} \leqq H_j \qquad j = 1, 2, \ldots, m \tag{7}$$

where α_{kj} is the required air rate corresponding to y_{kj}, and α_{sj} is the required base air when the j-th separator is used. H_j is the upper bound for the total quantity of the liquid products generated by the j-th separator. As for the subscript k, see Table 1.

Table 1 The products generated by the separator

k	Product	
1	gas oxygen	(GO_2)
2	liquid oxygen	(LO_2)
3	gas nitrogen	(GN_2)
4	liquid nitrogen	(LN_2)
5	gas argon	(GAr)
6	liquid argon	(LAr)

Relationship between supply and demand for each product: Let $y_{L,2k-1}$ be the quantity of the (2k-1)-th gas product supplied by evaporating the 2k-th liquid product, $y_{L,2k}$ be the quantity of the 2k-th product sold, and y_{d1} be the quantity of GO_2 diffused. Then the following constraints are obtained:

$$0 \leqq y_{Lk} \leqq y_{Lk}^{+} \qquad k = 1, 2, \ldots, 6 \tag{8}$$

$$\left.\begin{array}{ll} \sum_{j=1}^{m} y_{1j} + y_{L1} - y_{d1} = D_1 & \\ \sum_{j=1}^{m} y_{kj} + y_{Lk} \geq D_k & k = 3, 5 \\ \sum_{j=1}^{m} y_{kj} - y_{L,k-1} - y_{Lk} = D_k & k = 2, 4, 6 \end{array}\right\} \quad (9)$$

where D_k is the demand quantity of the k-th product. The interconnection between the air compressors and the separators is given by

$$\sum_{i=1}^{n} y_{ai} = \sum_{j=1}^{m} y_{sj} \quad (10)$$

Thus we have the following problem:

$$\min z = \sum_{i=1}^{n} a_i\, y_{ai} + \sum_{j=1}^{m} s_j\, y_{sj} + d\, y_{d1} + \sum_{k=1}^{6} r_k\, y_{Lk} \quad (11)$$

subject to (1) - (10)

where

- a_i : operation cost coefficient for the i-th air compressor,
- s_j : operation cost coefficient for the j-th separator,
- d : loss coefficient due to diffusion of GO_2,
- r_{2k-1}: loss coefficient due to evaporation of the 2k-th product,
- r_{2k} : profit coefficient obtained by selling the 2k-th product.

Note that r_2, r_4, and r_6 are negative. The other coefficients are positive.

3. Decomposition Procedure

The problem described in Sec. 2 has an angular structure. In fact, it is rewritten as

$$\text{(P1)} \quad \min z = \sum_{t=1}^{T} c(t)'y(t)$$

$$\text{subject to} \quad \{y(t), x(t)\} \in F_t \qquad t = 1, 2, \ldots, T \quad (12)$$

$$\sum_{t=1}^{T} A(t)y(t) = b \quad (13)$$

In (P1), the decision variables are partitioned into $T(\triangleq n + m + 1)$ subsets called blocks. For the t-th block, $y(t)$ is a continuous vector of appropriate dimension, and $x(t)$ is a binary scalar variable. The constraint (12) is called a block constraint, which corresponds to (1) to (8). On the other hand, the relation (13), called a coupling constraint, corresponds to (9) and (10).

We apply a decomposition algorithm[1] to the problem (P1). The basic idea of the algorithm is to decompose the problem in the way similar to the Dantzig-Wolfe decomposition technique in linear programs[2, 3], and to solve a restricted master program and subproblems iteratively. The procedure is as follows:

First, we find an initial feasible integer solution $\hat{x}(t)$ for each t. By substituting $x(t) = \hat{x}(t)$ into (12), the block constraint for $y(t)$ is obtained as

$$y(t) \in \hat{F}_t \qquad t = 1, 2, \ldots, T \tag{14}$$

Thus, we have the following linear programming problem:

$$\text{(P2)} \quad \min z = \sum_{t=1}^{T} c(t)'y(t) \qquad \text{subject to (13) and (14)}$$

This is called a restricted master program for (P1). By using the Dantzig-Wolfe decomposition technique, we solve (P2) to obtain the minimum objective value z^* as a current solution. We also obtain the simplex multipliers π_0 and π_t $(t = 1, 2, \ldots, T)$ associated with the coupling constraint and the block constraints, respectively. As for the detailed procedure, refer to [1].

Secondly, the optimality test for the current solution is checked. For this purpose, we solve the T subproblems given by

$$\text{(P3)} \quad \min z_t(\pi_0) = [c(t)' - \pi_0'A(t)]y(t) \qquad \text{subject to (12)}$$
$$t = 1, 2, \ldots, T$$

It is noted that the problem (P3) is a mixed-integer program with only one binary variable $x(t)$. Accordingly, we do not need to use the branch and bound method in order to solve (P3). The subproblem solution $\{y^*(t), x^*(t)\}$ with the minimum objective value $z_t^*(\pi_0)$ is determined merely by comparing two results obtained for two values of $x(t)$.

It is known from [1] that the current solution is optimal for (P1) if

$$f(t) \triangleq z_t^*(\pi_0) - \pi_t \geqq 0 \qquad \text{for all } t \tag{15}$$

If the condition (15) does not hold, there is a possibility of improving the current solution. Therefore, we try to improve the solution of (P2) by replacing the value $\hat{x}(t)$. The procedure is as follows:

The block numbers with $f(t) < 0$ are listed as

$$H \triangleq \{t_i \mid t_i \in \{1, 2, \ldots, T\},\ f(t_1) \leqq f(t_2) \leqq \ldots \leqq f(t_{i_f}) < 0\} \tag{16}$$

Then, the replacement of the integer value is made for a block t^* H, and a new restricted master program (P2) is constructed by fixing

$$x(t) = \begin{cases} x^*(t^*) & \text{for } t = t^* \\ \hat{x}(t) & \text{for } t \neq t^* \end{cases} \tag{17}$$

The block number t^* is chosen according to the order listed in H. If the solution of (P2) thus constructed improves the current solution, update the current solution and proceed to the optimal test. If not, the procedure of solving (P2) is continued by replacing the integer value for another block listed in H. Consequently, if the procedure does not succeed even for the last block t_{if}, return to the first block in H and solve (P3) again, excluding the value of $x(t)$ obtained previously by solving (P3). As a result, if $f(t) < 0$, proceed to solving (P2). On the other hand, if $f(t) \geqq 0$, remove the block number t from the list H.

The procedure mentioned above is terminated when the list H becomes empty. In this case, we have examined all possibilities for improving the current solution by replacing the integer value for one block at a time. However, a simultaneous replacement of the integer values for multiple blocks is not made in this procedure. Therefore, if the procedure terminates with $H = \phi$, the current solution obtained so far is adopted as a suboptimal solution. In this case, the difference between the optimal and the suboptimal objective value is less than ε, where ε is given by[1]

$$\varepsilon = -\sum_{t=1}^{T} f(t) \tag{18}$$

Figure 2 shows the entire procedure of the algorithm.

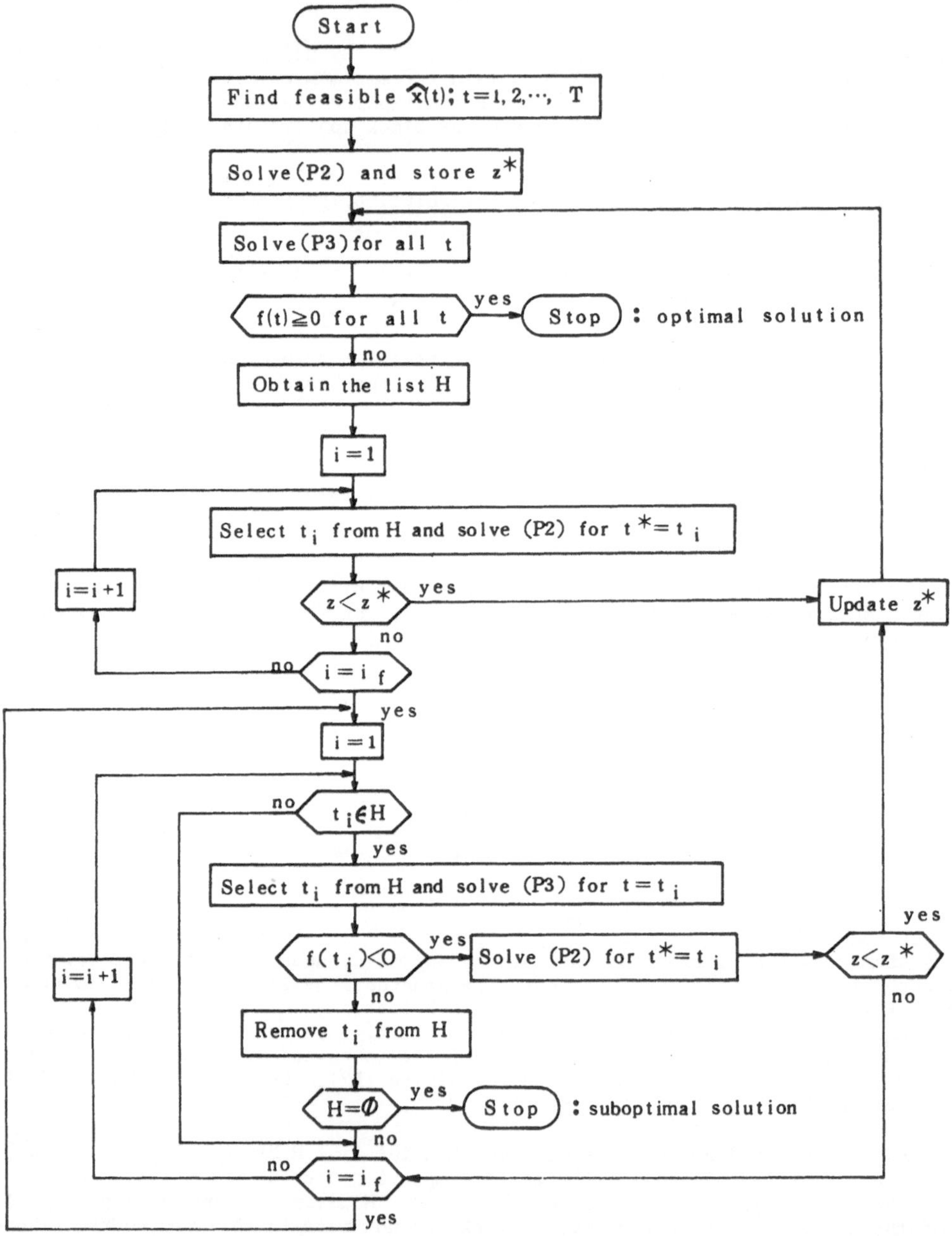

Fig. 2 Flowchart of the algorithm.

4. Numerical Result

Numerical results are shown for an application of the proposed method to the problem which is formulated for a real oxigen plant. The plant consists of eight air turbo compressors and five separators(i.e. $n = 8$ and $m = 5$). Then, we have thirteen binary variables and fifty continuous variables as the decision variables of (P1). We also have seven coupling constraints. On the other hand, the total number of the block constraints is twenty with the exception of the bounded condition for the continuous variables. The real system has weak nonlinearities in (5) and (11), but the linear approximation is permissible for them.

The plant is now working under operation control whose algorithm is based on a heuristic search method for finding almost all possible efficient combinations of equipments. In the existing algorithm, we have treated the problem without decomposition but with considering the nonlinearity.

Let us compare the results obtained for the linearized problem between the proposed algorithm(method A) and the existing algorithm(method B). For this purpose, eight problems are prepared by varying the demand quantities D_k ($k = 1, 2, \ldots, 6$). Table 2 shows a comparison of the computational results obtained for these problems. The detailed description for the demand condition is omitted in the table, though only the value of D_1 is shown for each problem. As shown in Table 2, the method A tends to have less computation time than the method B. The numerical computation has been made by a large-scale computer(FACOM M-200) whose computation speed is approximately fifteen times higher than that of an actually used process computer, such as FACOM U-1500 (sixteen-bit machine with main memory of 128 KB). Further, it is observed from a comparison of the objective value that the method A and the method B are comparable with respect to accuracy.

In this plant, the schedule for operating the equipments needs to be determined at intervals of an hour under fluctuations in the demand of each product. For this purpose, the values of D_k should be estimated ten minutes before the operation mode is changed, and the optimal solution of operation mode for the new demand condition is desired to be obtained in a minute. The result obtained suggests that the proposed algorithm is applicable to this situation by using a computer with speed twice higher than U-1500.

Table 2 Comparison of the computational results

		method A				method B			
Problem No.	D_1	(i)	(ii)	(iii)	(iv)	(i)	(ii)	(iii)	(iv)
1	20	00000001	01100	143.98	0.99	00000001	01100	143.98	4.98
2	30	00000001	01001	153.78	3.11	00001001	00101	181.95	20.84
3	40	00001001	01101	253.96	6.76	00001101	01101	255.61	54.19
4	50	00010001	01110	332.68	3.36	00010001	01011	328.34	80.12
5	60	00010001	01110	332.68	5.24	00011001	00111	377.35	90.80
6	70	00011101	10011	490.26	8.06	00011101	01111	454.66	60.78
7	80	10011101	10111	579.35	8.27	00011111	11011	572.50	19.96
8	90	00011111	11111	658.65	14.97	10011111	11111	660.99	4.49

Notes: The items (i) to (iv) are as follows.
(i) The values of x_{ai} ($i = 1, 2, \ldots, 8$)
(ii) The values of x_{sj} ($j = 1, 2, \ldots, 5$)
(iii) The value of the objective function.
(iv) Computation time in seconds.

5. Conclusion

A decomposition algorithm for solving mixed-integer linear programming problems with an angular structure has been applied to the problem of the optimal operation of an oxygen plant. As a result of decomposing the problem, the subproblem of mixed-integer type is easy to solve because it has one binary variable. Therefore, we do not need to use any subroutine of the branch and bound method. Only a subroutine of the revised simplex method is needed for solving linear programs of small size.

The plant considered here is now working under a computer control system whose function is to determine the optimal operation mode of equipments. The existing algorithm is based on solving nonlinear programming problems repeatedly for almost all possible efficient combinations of equipments. Therefore, it requires much computation time. This paper aims at showing an improvement of the existing algorithm. According to the numerical results, the proposed algorithm is expected to be efficient from the standpoint of not only computation time but also accuracy.

References

[1] N. Sannomiya and M. Tsukabe: A method of decomposing mixed-integer linear programming problems with angular structure, Int. J. Systems Science, Vol. 12, pp. 1031-1043(1981).

[2] G. B. Dantzig and P. Wolfe: Decomposition principle for linear programming, Ops. Res., Vol. 8, pp. 101-111(1960).

[3] L. S. Lasdon: Optimization Theory for Large Systems, Macmillan, New York, 1970.

DECENTRALISED CONTROL OF PETROLEUM REFINERY

T.Dyrga, B.Kazimierczak, N.Kostyk, F.Milkiewicz, M.Szymański
Technical University of Gdańsk, Gdańsk, Poland

Abstract. In the paper, a production control system elaborated for a Polish petroleum refinery is described. Some of the technological installations in this refinery, especially, the installations producing components of lubricating oils, should be working alternately in several regimes. Since the succession and frequency of switching over of the regimes are influencing on the production effects of the whole refinery, the schedule of switching over of regimes must be calculated in the refinery production planning problem. The calculations of the production plans and schedules should be done frequently enough because of different disturbances influencing on the petroleum refinery, especially, the disturbances of products shipping. From this reason the planning and scheduling system must be directly connected with a real time system of refinery production state supervising.
The functional structure of the whole system is discussed. The system realization is described, too.

1. INTRODUCTION

The problem of a decomposed operative production control of a petroleum refinery is considered. The considerations are carried on the example of a standard fuels-lubricating oils refinery. The flow-sheet of the refinery is shown on Fig.1. It is assumed that the refinery consists of twelve production sections $ZP^{(k)}$, $k \in \overline{1,M}$, M = 12. Each production setion $ZP^{(1)}$, $ZP^{(2)}$, $ZP^{(3)}$, $ZP^{(5)}$ can be operated only in one regime. The production sections $ZP^{(4)}$, $ZP^{(6)}$ - $ZP^{(8)}$, $ZP^{(9)}$, $ZP^{(10)}$, $ZP^{(11)}$, $ZP^{(12)}$ can be operated in several regimes, for example $ZP^{(6)}$ - $ZP^{(8)}$ can be operated in 5 regimes /5 kinds of base lubricating oils are produced/, $ZP^{(9)}$ - in 2 regimes /production of regular gasoline, or premium gasoline/ and so on, but only one regime can be operated simultanously in each production section. It is as-

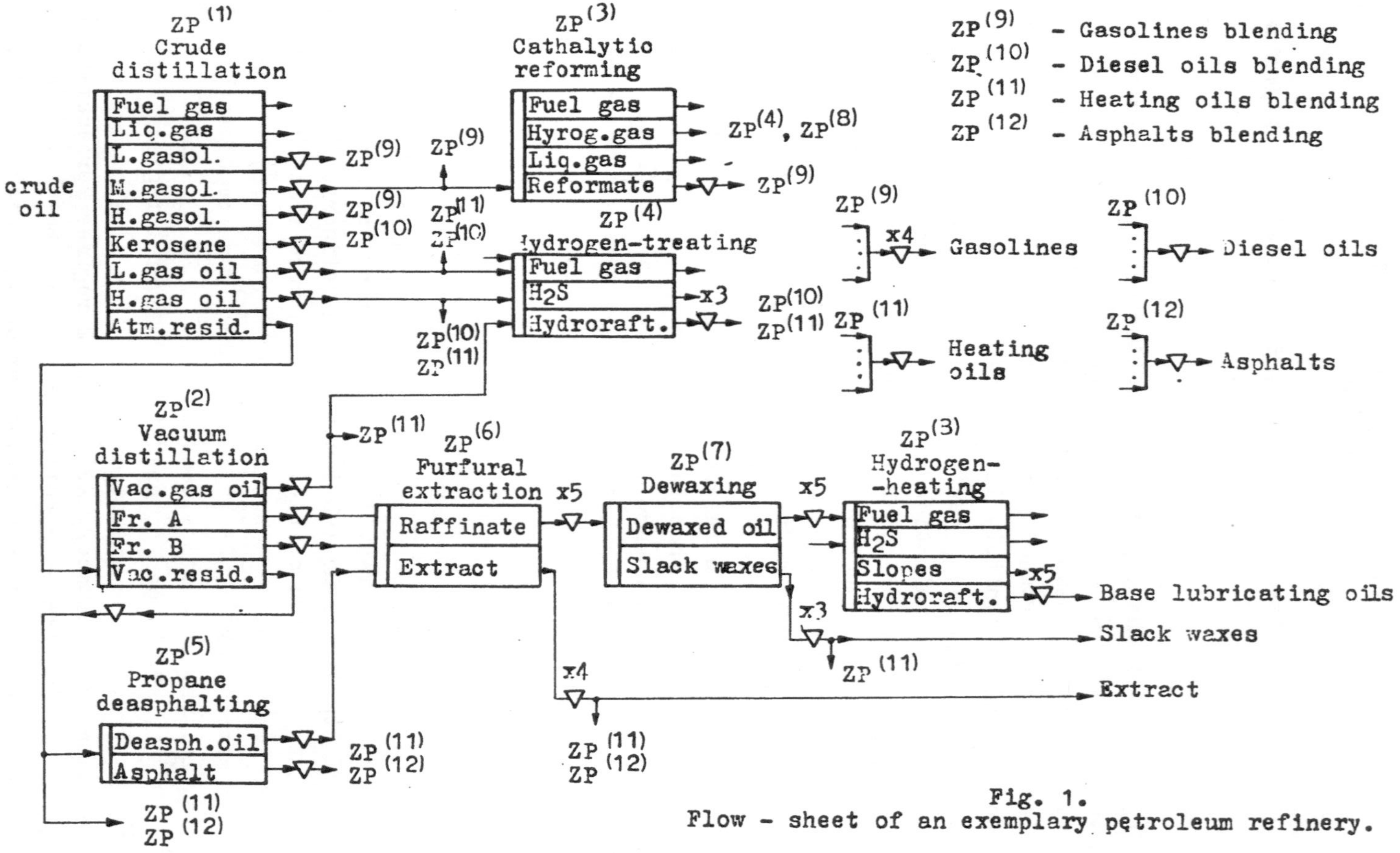

Fig. 1.
Flow - sheet of an exemplary petroleum refinery.

sumed that each blending section is connected with its own shipping section.

The aim of the operative production control of the refinery is to take the following decisions concerning the nearest twentyfour hours: a/ choosing instants of switching on and off of each regime of each production and shipping sections, b/ choosing loads of each regime of each production section, c/ choosing amounts of each product which are to be shipped to each customer.

The above mentioned decisions must assure the realization of the selling tasks in time interval $[t_o, t_o + T]$ with the simultaneous achievement of the maximum value of the profit in this time interval. The operative production control system must respond to each disturbance inside the refinery and to each disturbance in product selling. The manner of taking the above mentioned decisions as well as the designed operative production system realization for a Polish refinery are described below.

2. THE MAIN ELEMENTS OF OPERATIVE PRODUCTION CONTROL DECISION TAKING PROBLEM

The operative production decisions are taken on the basis of the multihorizon production control method /Milkiewicz, 1980/ and on the basis of the decomposed production scheduling method /Milkiewicz, 1983/. In agreement with these methods, the whole problem of the operative production control decision taking of a petroleum refinery is decomposed in the way shown in Fig.2. This decomposition manner of the problem is correct when the prices of diesel fuels, heating oils and asphalts are considerably lewer than the prices of lubricating base-oils.

A. Calculating of the whole refinery production plans for time interval $[t_o, t_o + T]$

The input data for calculating the production plans are: a/ Amounts of products which must be sold and amounts of products which can be sold in time subintervals $[t_v, t_{v+1}] \subset [t_o, t_o + T]$, $v \in \overline{0, N-1}$, $t_N = t_o + T$. For fixing attention it is assumed that T = 1 quarter, $t_{v+1} - t_v$ = 1 month, N = 3. b/ Available amounts of materials in $[t_v, t_{v+1}]$. c/ Initial states/at t_o /of the stocks of raw materials, semiproducts and products. d/ The forecasted time intervals of even-

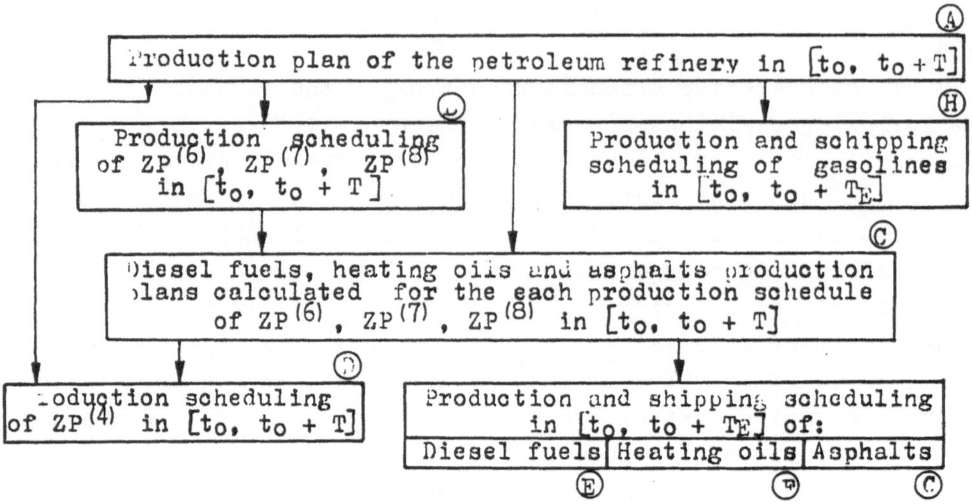

Fig. 2.
Production planning and production scheduling.

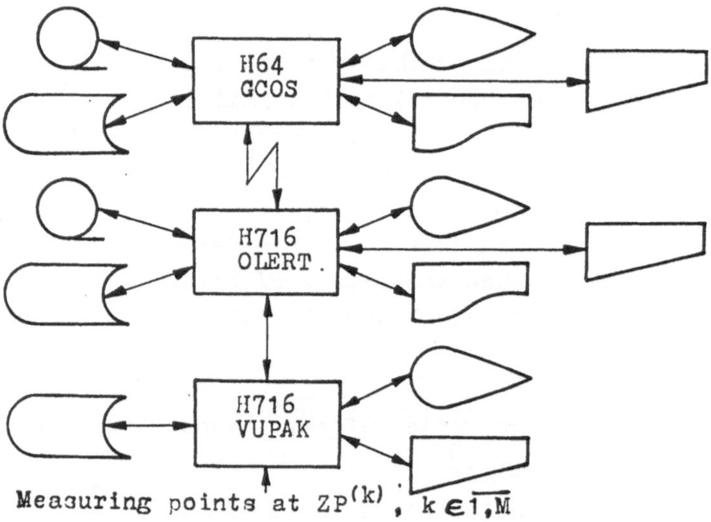

Fig. 3.
The configuration of the computers.

tual standstills of $ZP^{(k)}$, $k \in \overline{1,M}$. e/ The forecasted product shipping capacities in $[t_o, t_o + T]$.
The linear methematical model of the refinery is created on the basis of the above mentioned data and the flow-sheet of the refinery taken into account. This model is defined for time interval $[t_o, t_o + T]$. Time interval $[t_o, t_o + T]$ is divided into subinetrvals $[t_v, t_{v+1}]$, $v \in \overline{0, N-1}$. The optimal production plan of the refinery is calculated by means of the linear programming algorithm. The objective function in the problem is the profit achieved in $[t_o, t_o + T]$.
In the mathematical model of the refinery production planning it is assumed that $ZP^{(k)}$, $k \in \overline{1,M}$ can operate in all the regimes simultaneously, while in reality, $ZP^{(k)}$ can operate simultaneously only in one regime each. Thus the production plans based on such a model can be unrealizable.
The production planning mathematical model has been suitably modified to avoid the unrealizability of the calculated production plans. But these modifications cause that the calculated production plan gives lower value of the profit than it is possible to achieve in reality. The introduction of the step by step production planning as well as of the additional planning of diesel fuels, heating oils and asphalts /see C in Fig.2/ after solving problem B /see Fig.2/. made up for the mentioned above loss in profit.

B. Production scheduling of $ZP^{(6)}$, $ZP^{(7)}$ and $ZP^{(8)}$

The problem of production scheduling of $ZP^{(6)}$, $ZP^{(7)}$, $ZP^{(8)}$ is based on the choice of such instants t_μ, $\mu \in \overline{0, \mathcal{M}-1}$ at which particular regimes of $ZP^{(6)}$, $ZP^{(7)}$, $ZP^{(8)}$ are switched on and off, and on the choice of such their loads at time subintervals $[t_\mu, t_{\mu+1}]$, $\mu \in \overline{0, \mathcal{M}-1}$, $t_{\mathcal{M}} = t_o + T$, at which the following conditions are fulfilled: a/ the amounts of each of the base lubricating oils produced and sold at time intervals $[t_v, t_{v+1}]$ must be equal to the amounts calculated in problem A, b/ the upper and lower limits of loads and other constraints existing in $ZP^{(6)}$, $ZP^{(7)}$, $ZP^{(8)}$ are not to be exceeded, c/ the upper and lower limits of tank capacities are not to be exceeded before and after each $ZP^{(k)}$, k = 6, 7, 8, d/ the number of switchingsover of $ZP^{(k)}$, k = 6, 7, 8 must be minimum.
The production schedule of $ZP^{(6)}$, $ZP^{(7)}$, $ZP^{(8)}$ is calculated by means of the algorithm which is based on the generation of all the branches of the regimes switching over tree for the set number of the switchings over and on the simultaneous successive elimination of the bran-

ches which do not satisfy the constraints mentioned above. The search for the optimum production schedule characterized by the minimum number of switchings over is initiated by the least of all set number of the switchings over. In general, there can be more than one admissible production schedule for the same number of switchings over. As the optimum one is chosen the one, for which the best production plan C /see below/ is achieved.

C. Calculating the production plans of diesel fuels, heating oils and asphalts

There are given: a/ Calculated in B amounts of fractions received from each regime performed at $ZP^{(7)}$ and $ZP^{(8)}$ and consumed by each regime of $ZP^{(6)}$ in $[t_{\mu}, t_{\mu+1}]$, b/ Calculated in A the amounts of fractions received from $ZP^{(1)}$ and $ZP^{(2)}$ at $[t_v, t_{v+1}]$, c/ The amounts of each kind of diesel fuel, heating oil and asphalt which must and which can be sold at $[t_v, t_{v+1}]$, d/ The initial states of stocks /at t_o/ of all the components of diesel fuels, heating oils and asphalts.
On the basis of the above mentioned data the production plan C for diesel fuels, heating oils and asphalts produced at $t_{\mu}, t_{\mu+1}$ is calculated for each production schedule received in B /see Fig.2/ by means of the linear programming algorithm. The optimal production schedule of $ZP^{(6)}$, $ZP^{(7)}$ and $ZP^{(8)}$ is the one for which the calculated plan C has maximum profit. The optimal plan C received in such a way is the base for the decisions concerning acceptance of the customers product orders. This plan gives us the production schedule for $ZP^{(5)}$ too.

D. Calculating the production schedule of the diesel fuels hydrogen--treating section

On the basis of the data received from the solution of the problems A, B and C one must calculate such instants of switching on and off and such loads of each regime of $ZP^{(4)}$ in $[t_o, t_o + T]$ at which the following requirements are fulfilled: 1/ The amounts of hydrofined fractions supplied to $ZP^{(10)}$ and $ZP^{(11)}$ at $[t_{\mu}, t_{\mu+1}]$ must be equal to the amounts calculated in C, 2/ All the constraints existing in $ZP^{(4)}$ and the upper and lower limits of capacities of all the tanks before and after $ZP^{(4)}$ are not to be exceeded, 3/ The number of switchings on and off of $ZP^{(4)}$ in $[t_o, t_o + T]$ must be minimum.
The production schedule of $ZP^{(4)}$ is calculated by means of the algorithm similar to algorithm B.

E, F, G. Calculating of blending and shipping schedules for diesel fuels, heating oils and asphalts

It is assumed that diesel fuels, heating oils and asphalts are served by individual shipping sections. Then the blending and shipping schedules for these products can be calculated independently. In this case the following data are necessary: a/ Desired and contracted time-tables of supplying the products to customers at time interval $[t_o, t_o + T_E]$, where $T_E < T$. It is assumed that $T_E = 1$ month, b/ Forecasted time-tables of the transport means supply to the refinery required for products shipping at $[t_o, t_o + T_E]$, c/ Calculated in C proportions of each component in each product and the amounts of each product at $[t_\mu, t_{\mu+1}]$, d/ The initial states of stocks at t_o of the components and products.
The blending and shipping schedule of the individual group of products is calculated by means of an heuristic algorithm. This algorithm minimizes charges of penalties caused by breaking the contracted time-tables of supplying the customers with products, and charges of a standstill of transport means /Milkiewicz, 1979/.

H. Calculating of blending and shipping schedules for gasolines

These schedules are calculated directly after solving problem A - see Fig.2. The algorithm for solving the problem is the same as algorithm in E, F, G.

3. THE MANNER OF OPERATIVE PRODUCTION CONTROL DECISION TAKING

It is assumed that the refinery operative production control decisions are taken at instants $t_o, t_1 \ldots, t_\Pi, \ldots$ The distance between succeeding instants $t_{\Pi+1} - t_\Pi$ is constant and it is Δt /for example $\Delta t = 24$ hours/. The decisions taken at instant t_Π are valid for the whole time subinterval $[t_\Pi, t_{\Pi+1}]$, $\Pi = 0,1,\ldots$ Taking into account Fig.2 and earlier considerations the operative production control decisions are taken at instants t_Π, $\Pi = 0,1,\ldots$ in the following way: /1/ At instant t_Π one ought to measure: the state of the products sale realization, the stock state of raw materials, semiproducts and products, the technical state of production sections. /2/ One brings up to date at t_Π for $[t_\Pi, t_o + T]$ the forecasts of: the sale of products, supplying the refinery with transport means, the characteristics of production sections. /3/ It is checked, if the differences between the planned values and the values fixed at /1/ and /2/ exceeded the limited levels. If not, then at instant t_Π the operative production control decisions will be taken on the basis of lately calculated production plans A, C and production schedules B, D, E, F, G, H. If yes, then according to that, which limited levels were exceeded at instant t_Π will be calculated: or only one, or a few, or all production schedules D, E, F, G, H, or production plan C and consequently production schedules D, E, F, G, or production schedule B and consequently production plan C and production schedules D, E, F, G, or production plan A and consequently B, C, D, E, F, G, H. /4/ The operative production control decisions for time subinterval $[t_\Pi, t_{\Pi+1}]$ are taken on the basis of valid, at instant t_Π, all the production plans and all the production schedules: /4.1/ The decisions about loads of $ZP^{(1)}$, $ZP^{(2)}$, $ZP^{(3)}$ are taken from plan A, /4.2/ The decisions about loads of $ZP^{(6)}$, $ZP^{(7)}$ and $ZP^{(8)}$ for the regimes which are being contimued at instant t_Π are taken from schedule B. A decision about switching over at instant $t \in [t_\Pi, t_{\Pi+1}]$ of a regime performed at ZP is taken if such a switching over was anticipated by production schedule B. In this case the decision about leading this ZP at time subinterval $[t, t_{\Pi+1}]$ is taken too. /4.3/ The production schedule of loads of $ZP^{(5)}$ at $[t_\Pi, t_{\Pi+1}]$ is determined directly from plan C. /4.4/ The decisions about the load and about the eventual instants of switching over of regimes of $ZP^{(4)}$ are taken in the same way as /4.2/ but from production schedule D. /4.5/ The decisions about instants of switching on t_ϱ and off t'_ϱ of each product blending process and about the amount of these products at time subintervals $[t_\varrho, t'_\varrho]$ are taken on the basis of production schedules E, F, G, H. The decisions concerning the shipping of products are also taken in the similar way.

4. THE REALIZATION OF THE OPERATIVE PRODUCTION CONTROL SYSTEM

Three cooperating HONEYWELL computers: computer H64 and two minicomputers /"plant computer" and "master computer"/ were used for the technical realization of the system described in the chapters 2 and 3. Figure 3 presents the configuration of these computers and their connections. Minicomputers H716 are installed in the refinery while computer H64 is about 700 km away from the refinery. Minicomputers H716 cooperate with each other by the inter-computer communication unit,

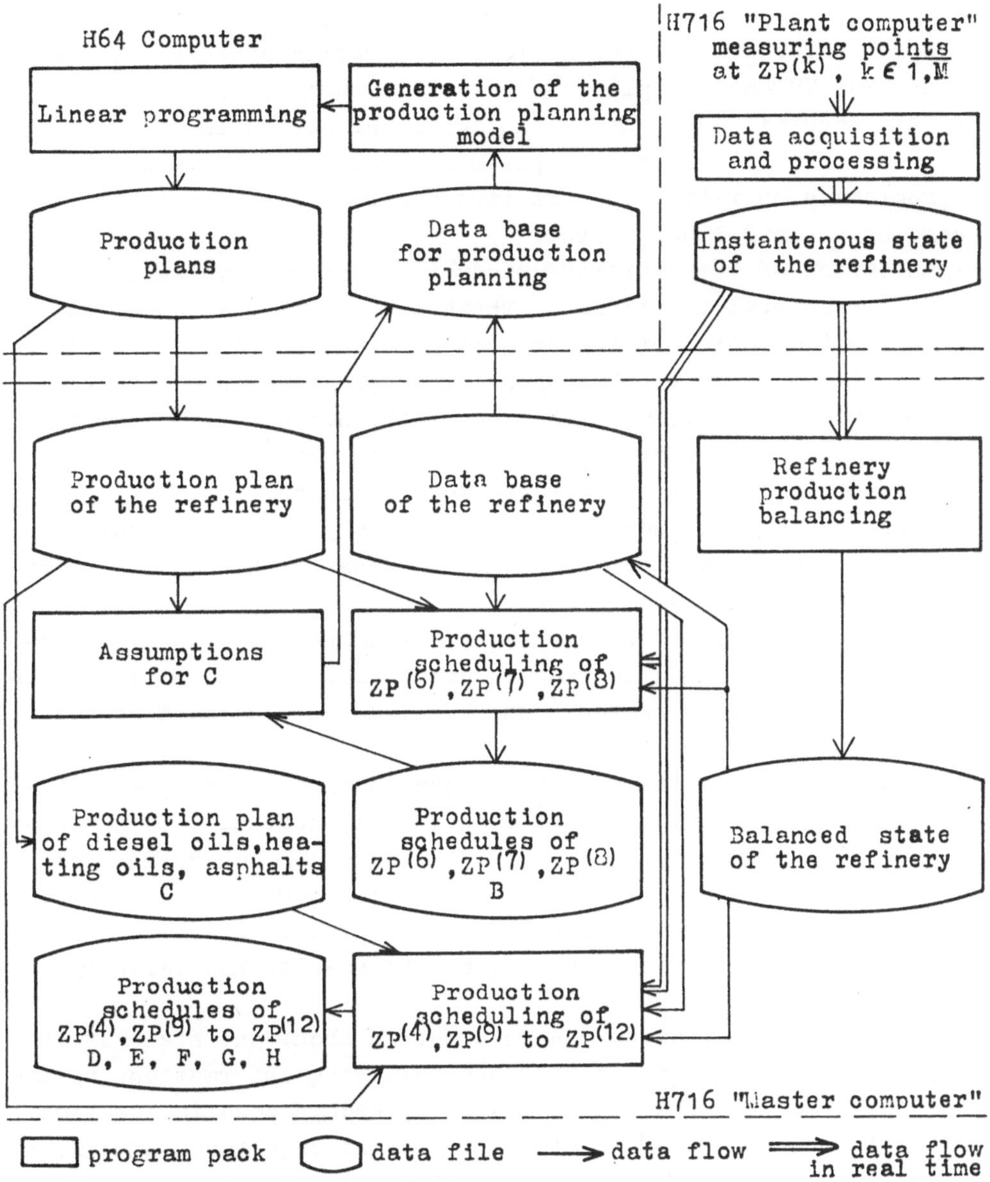

Fig. 4.
The organization of the operative production control system of the refinery.

while computer H64 cooperates with "master computer" by the telecommunication link. There are the following operating systems of the minicomputers: VUPAK /at "plant computer"/ ensuring the realization of the data acquisition about technological processes in $ZP^{(k)}$, $k \in \overline{1,M}$ and OLERT /at "master computer"/ enabling the service of various programmes operating in real time and realizing the tasks in the range of the off-line control.

The organization of the operative production control system of the refinery is presented in figure 4. There are two main items of the system: the optimization of the production plans and the production scheduling of the production sections. The optimization of the production plans is realized with the use of the H64 computer. For these purposes, the H64 computer is equipped with software packages: the linear model generation software package and the linear programming software package standard /Biernacki, 1980/. The optimization of the production plans is carried out by means of the data base for the production planning which is also included in the software of computer H64. Its contents is determined in agreement with the refinery data base stored in H716 "master computer". The production scheduling of the production sections was realized on the basis of H716 "master computer".

This computer comprises the software packages of the data preparations for the production planning of diesel oils, heating oils and asphalts too. The software packages of the production scheduling of the production sections generate the production schedules of the corresponding production sections on the basis of the selective information acquisition out of the connected data base of the refinery /Duzinkiewicz, 1982/.

The up-to-date data set gives the information about the state of the refinery at instant t_π and includes the following information: a/ the state of the production task realization by each production section at instant t_π, b/ the state of stocks in the storage tanks at instant t_π, c/ the state at instant t_π of the realization of the product shipping.

The contents of this data set is formed by the production balance software package of production sections /Faron, 1975/ on the basis of the data set: instantaneous state of the refinery. The production balance of the production sections is carried out every 24 hours, or, if required at any instant. Each of the mentioned software packages realized on H716 "master computer" is used on the user's request in the conversational mode. The solution time of the production section operation scheduling ranges from a few to a dozen or so minutes depending on the degree of the problem complexity. To test the system, on the basis of the real data there were carried out some exemplary calculations of the optimization production plan of the whole refinery, and afterwards the production schedules of $ZP^{(6)}$ and $ZP^{(7)}$ were determined. The obtained production schedule of $ZP^{(6)}$ and $ZP^{(7)}$ fixes the number of subperiods which ought be taken into account in the optimization of the production plan of diesel oils, heating oils and asphalts. The number of the considered subperiods influences fundamentally the size of the production planning problem /in the considered refinery there was 1000-1500 constraints/. In the case of computer H64 the solution of the production planning problem having 1000-1500 constraints requires 2-3 hours. Taking into account the necessity of the data transmission both the assumptions to calculations as well as their results, it is estimated that the total time of one production plan calculation is about 6 hours. In the proposed method, to choose the best production schedule of $ZP^{(6)}$, $ZP^{(7)}$ and $ZP^{(8)}$ it is necessary to calculate for each of these schedules the corresponding production plan of diesel oils, heating oils and asphalts. In practice, it will be necessary to calculate successively a few production plans, thus the total calculation will last about several hours. Accordingly, it will be possible to take the decisions concerning the choice of the production schedule of sections $ZP^{(6)}$, $ZP^{(7)}$ and $ZP^{(8)}$ not more frequently than once in a few days - - and this fully meets the requirements of the proposed method.

REFERENCES

Biernacki W., Kawalec K., Miś J., Nagórska A., Pirowski J., Wiaderek L. /1980/. Operating documentation of optimal load distribution and optimal fuels blending. "Petroinform" Kraków. Techn.Rep., pp. 70 /in polish/.

Duzinkiewicz K., Dyrga T., Fudali J., Kazimierczak B., Kostyk N., Milkiewicz F., Szymański M. /1982/. GZR fuels and lubricating oils production planning and production scheduling computerized system. Techn.Univ. of Gdańsk, Techn.Rep., pp. 380 /in polish/.

Faron J., Kostecki J.M., Miś J., Nizińska-Bardel I., Szefler M., Walczak G., Wiaderek L. /1975/. GZR production information system. "Petroinform Kraków". Techn.Rep., pp. 223 /in polish/.

Milkiewicz F. /1980/. Control of production systems.2nd IFAC Symp. on Large Scale Systems, Toulouse, June 24-26.

Milkiewicz F. /1983/. Decomposition of Production Scheduling in Production Systems with Switched over Continuous Technological Processes. Preprints of the IFAC/IFORS Symp. on Large Scale Systems: Theory and Applications, Warsaw, Poland, 11-15 July, pp. 338-345.

MODELS FOR EVALUATION OF LARGE SCALE SYSTEMS IN CHEMICAL PROCESSES BY RELIABILITIES

ZENG HONG
Department of Chemical Engineering
Chengdu University of Science & Technology
Chengdu, Sichuan,610065
People's Republic of China

ABSTRACT. It may be lacking in suitable model for evaluation of the chemical processes. Here, we derive the model for evaluation of the large scale systems in chemical processes by reliabilities.

INTRODUCTION. The importance of evaluation of the chemical processes by the reliabilities is evident. The chemical processes fail in operation, if the reliabilities of the system or its subsystems would not available.

In accordance with the fundamental conceptions in reliability, we may have

$$R = R_s = \prod_{t=1}^{\lambda} R_t$$

$$R = R_p = 1-\prod_{t=1}^{\lambda} (1-R_t)$$

where, R = the reliability of certain system in chemical process;
R_s = the reliability of a series system in chemical process;
R_p = the reliability of a parallel system in chemical process;
R_t = the reliability of the t-th subsystem.

It is not necessary to describe it in more detail[1].

Under some conditions, the optimal system or subsystem in chemical processes may only be determined by its reliabilities, if the reliabilities are the fators of control. In this case, the calculations may be simple. Sometimes, the effective factors are not only the reliabilities, but also the other objective values. For this case, the optimal system or subsystem may be firstly determined by its reliabilities and then subsequently checked by its objective values.

There are many ways to establish a system or subsystem. We may choose the best one by reliabilities.

DERIVATION. Put R to be the set of reliabilities in chemical processes. From the theoretical point of view, R may be infinite; but in practical cases, the R can be finite. For producing certain chemical product, we may use different processes. Let Q be the set of it. The Q also denotes the corresponding set of systems, in the chemical processes. Therefore, we can assign the $R(Q)$ to be the set of reliabilities corresponding to the set of chemical system , Q . We can distinguish the different systems by using the upper index n , such as Q^n. In the same case, we may define $R^n(Q^n)$, where n indicates the nth system. In fact, there are many subsystems

$$Q^n_{\zeta(n)}$$

belonging to Q^n, where ζ is the subscript . It can be used to distinguish the different subsystem of the n-th system. It is evident that we may define

$$R^n_{\zeta(n)}(Q^n_{\zeta(n)})$$

in the same way.

Basing on the fact of the chemical processes, we have

$$R^n_{\zeta(n)}(Q^n_{\zeta(n)}),\ R^n(Q^n),\ R(Q),\ R \neq \emptyset$$

$$R^n(Q^n) \in R(Q)$$

$$\forall \exists R^n_{\zeta(n)}(Q^n_{\zeta(n)}) \subset R^n(Q^n)$$

$$R^n(Q^n) = \bigcup_{\zeta} R^n_{\zeta(n)}(Q^n_{\zeta(n)})$$

$$R^n_I(Q^n_I) \sqcap R^n_J(Q^n_J) = \emptyset$$

where I and J indicate the different subsystems of the ηth system, and

$$n \in \{ 1,2,\ldots,k \}$$

$$\zeta \in \{ 1,2,\ldots,L \}$$

$$k,L \in \{ \text{Integers} \}$$

$$0 < k,L < \infty \quad .$$

It is very important to confirm the optimal reliability of the system in chemical processes, for determining the best method. We may take $R^*(Q)$ to be the optimal reliability of the chemical system Q . Therefore, We can obtain

$$R^*(Q) = \text{Opt } R(Q) \quad .$$

For the definite processes or systems in chemical industry, it is evident that one can write

$$R^*(Q) = \text{Opt } \{ R^n(Q^n) \mid R^n(Q^n) \in \{R^1(Q^1),\ R^2(Q^2), \ldots, R^k(Q^k) \},$$

$$0 < k < \infty \ , \qquad k \in \{ \text{Integers} \} \} . \qquad (1)$$

For the convenient of consideration, put Q^* to be the optimal method or system in chemical processes; then , we have

$$Q^* = \text{Opt } Q \quad . \qquad (2)$$

It is noted that Q^* must be in accordance with $R^*(Q)$ of the chemical process in questions.
Model (1) may be used only for the case that the chemical system can be evaluated by the reliabilities and that it is not needing other conditions.

For the case requiring the additional conditions , except the reliabilities, we must consider the other objective value of the system Q. We assign it as $\theta(Q)$. With respect to the optimal condition, one may take the notation $\theta^*(Q)$ representing the optimal-objective-value.
Similarly, we can get

$$R^n(Q^n) \geq R_G \qquad (3)$$

where R_G is the given reliability of question and it is a standard one by which the chemical system can be operated in a reasonable way.
Basing the R_G, we can select many chemical processes those may be used but not the optimal one. After we choose them by the correlative chemical systems's reliabilities, those selected systems must be evaluated further by the following model to determine the optimal system,

$$\theta^*(Q) = \text{Opt } \theta(Q) = \text{Opt } \{ \theta^n(Q^n) \mid \theta^n(Q^n) \in \{ \theta^1(Q^1),\ \theta^2(Q^2), \ldots, \theta^k(Q^k) \},$$

$$n \in \{ 1,2,\ldots,k \} , \qquad 0 < k < \infty \ ,$$

$$k \in \{ \text{integers} \} \} . \qquad (4)$$

Model (4) may be used to determine the optimal system in chemical processes by combining with the model (1)or model (3).

We can describe the chemical processes or systems by the following model,

$$Q = \{ Q^n \mid Q^n \in Q^1, Q^2, \ldots, Q^k \}, \quad 0 < k < \infty , \quad k \in \{\text{Integers}\} \} . \tag{5}$$

It is evident that one can write

$$Q^* = \text{Opt}\{ Q^n \mid Q^n \in \{ Q^1, Q^2, \ldots, Q^k \}\}, \ 0 < k < \infty , \ k \in \{\text{Integers}\}\} . \tag{6}$$

It is the model for evaluation of the optimal system or process in chemical industry. As the above description, model (6) may be used with the model (4), (3)or (1). Besides these, we note that model (2) equivalent to the model (6).

EXAMPLE. For the series system in chemical process, we take

$$\theta^n_{\zeta(n)}(Q^n_{\zeta(n)})$$

to be the objective function of the arbitrary subsystem $Q^n_{\zeta(n)}$. The optimal-objective-value of it can be

$$\text{Opt } \theta^n_{\zeta(n)}(Q^n_{(n)}) .$$

According to the construction of the large systems in chemical processes, the overall-optimal-objective-value of the ηth system may be taken from their local-optimal -objective -value about its subsystem.
It is clear that the overall-optimal-objective-value of the ηth system may be

$$\sum_n \sum_\zeta \text{Opt } \theta^n_{\zeta(n)}(Q^n_{\zeta(n)}) .$$

By comparing the overall-optimal-objective-values, we can get the optimal system. the steps of comparison may be represented by the following model,

$$\begin{aligned} \theta^*(Q) = \text{Opt} \{ &\sum_n \sum_\zeta \text{Opt } \theta^n_{\zeta(n)}(Q^n_{\zeta(n)}) \mid \theta^n_{\zeta(n)}(Q^n_{\zeta(n)}) \in \{ \theta^1_{\zeta(n)}(Q^1_{\zeta(n)}), \\ &\theta^2_{\zeta(2)}(Q^2_{\zeta(2)}), \ldots, \theta^k_{\zeta(k)}(Q^k_{\zeta(k)}) \}, \ n \in \{ 1,2,\ldots,k \} , \\ &\zeta \in \{ 1,2,\ldots,L \} , \quad 0 < k,L < \infty , \\ &k,L \in \{ \text{Integers} \} \} . \end{aligned} \tag{7}$$

It is the model used to calculate the overall-optimal-objective-value for series systems in chemical processes , combining the reliabilities of the systems to evaluate them.

DISCUSSIONS. The constraint conditions in calculation may be varied with the individual situations of the systems in chemical processes. The cost of investment, consumption of energy,etc., may be taken as the objective values depending upon the questions.

REFERENCES.

[1] Zeng Hong, The Strategies, J. of the Standard, 5 (1981), 8-11, P.R.China.

HIERARCHICAL OPTIMIZATION OF LARGE-SCALE WATER RESOURCES SYSTEMS

by

M. Jamshidi
Department of Electrical
and Computer Engineering
Albuquerque, NM U.S.A.

C. M. Wang
GTE Electronics, Inc.
1 Camino de Lenkurt NE
Albuquerque, NM U.S.A.

ABSTRACT

In this paper a hierarchical scheme is proposed for large-scale water resources systems. The usually nonlinear discrete-time time-delay models of such systems are linearized around the systems' nominal trajectories for water management operating rules. The system (area) is then decomposed into a number of subsystems (sub-areas) such that the water outflows from one subarea and inflows into the next one constitute the systems interactions. A fast converging algorithm is developed for the hierarchical optimization process. The method is demonstrated for a two-subsystem four-reservoir system.

INTRODUCTION

The optimal operation and control of water resources systems has been of interest to many researchers for many years [1-5]. The models of such systems can be best described by nonlinear differential difference equations. This would mean that the theory of time-delay control systems can be used to find an optimal operating rule. The diversified applications of water such as flood control, recreation, hydro-electric power generation, irrigation, industrial use, etc. makes such systems very complex and often large in order. For this application and other similarly important applications the control of large-scale systems has been of recent interest [6-10]. Most of these initial efforts involve the extension of the decentralized control or the goal coordination of hierarchical control to systems with delay.

In this paper the "interaction prediction" method of large-scale systems is applied to the control of a water resources systems described by a nonlinear time-delay system. A fast-converging iterative algorithm is developed to obtain a suboptimal control for such systems. A Taylor series expansion is used to linearize the original system whose sub-optimal control results from the optimization of a sequence of non-delay linear time-varying subsystems problems. The coordination of these sub-problems is identical to that of the non-delayed interaction prediction. The scheme has been applied to a 4-reservoir water resources system.

PROBLEM STATEMENT

Consider a water resources system model described by a non-linear large-scale system with multiple-delays.

$$\dot{X} = F(X(t), X(t-\tau_1),\ldots,X(t-\tau_p), U(t), U(t-h_1), \ldots, U(t-h_q))$$

$$t \geq t_o,\ . \ \tau_p \geq \cdots \geq \tau_i \geq \cdots \geq \tau_1$$

$$h_q \geq \cdots \geq h_i \geq \cdots \geq h_1 \tag{1a}$$

$$X(t) = X_o(t) \qquad t_o - \tau_p \leq t \leq t_o$$

$$U(t) = U_o(t) \qquad t_o - h_q \leq t \leq t_o \tag{1b}$$

Where $X(t) \epsilon R^n$ and $U(t) \epsilon R^m$ are, respectively, state (reservoir volumes) and control (water, releases and diversions) vectors, τ_i, $i=1,2,\ldots,p$ and h_j, $j-1,2,\ldots,q$ are constant positive scalars representing delays, $F(.)$ is assumed to be a continuously differentiable function of its arguments; t_o is the initial process time; $X_o(t)$ and $U_o(t)$ are specified initial functions. The cost functional, to be minimized, is

$$J = 1/2X^T(t_f)SX(t_f)+1/2\int_{t_o}^{t_f}[X^T(t)Q(t)X(t)+U^T(t)R(t)U(t)]dt. \quad (2)$$

where without any loss of generality, the matrices S and $Q(t)$ are assumed to be diagonal and positive semi-definite; and piecewise continuous, and the matrix $R(t)$ is diagonal positive definite and piecewise continuous. The problem is to find a control $U(t)$, i.e. water release and diversion requirements, $t_o \leq t \leq t_f$ which for fixed final time t_f and free final state $X(t_f)$ satisfies (1) and approximately minimizes the cost functional J in (2)

HIERARCHICAL CONTROL VIA INTERACTION PREDICTION METHOD

APPROXIMATE SYSTEM REPRESENTATION

Given the non-linear system described by (1), a linear approximation may be obtained through a Taylor's series expansion about the nominal trajectories (X_n,U_n). Let (z,u) be the trajectories of derivations between (X,U) and (X_n,U_n), i.e.

$$z = X - X_n, \quad u = U - U_n \quad (3)$$

It is assumed that the pair (X_n,U_n) satisfy (1) i.e.

$$\dot{X}_n = F(X_n,X_n(t-\tau_1),X_n(t-\tau_2), \ldots, X_n(t-\tau_p),U_n,U_n(t-h_1), U_n(t-h_2),\ldots,U_n(t-h_q)) \triangleq F_n \quad (4)$$

The expansion of (1a) about the nominal trajectories will be:

$$\dot{X} = F_n+(\frac{\partial F}{\partial X_n})^T z(t)+ (\frac{\partial F}{\partial X_n(t-\tau_1)})^T z(t-\tau_1)+\ldots+(\frac{\partial F}{\partial X_n(t-\tau_p)})^T z(t-\tau_p) +(\frac{\partial F}{\partial U_n})^T u(t)+(\frac{\partial F}{\partial U_n(t-h_1)})^T u(t-h_1)+\ldots+ (\frac{\partial F}{\partial U_n(t-h_q)})^T u(t-h_q)+H.O.T. \quad (5)$$

Considering the definition of (4) and following (5) a linear approximation of the system can be represented as:

$$\dot{z}(t) = A_o(t)z(t)+\sum_{k=1}^{p}A_k(t)z(t-\tau_k)+ B_o(t)u(t)+\sum_{l=1}^{q}B_l(t)u(t-h_l) \quad (6a)$$

with initial trajectories

$$z(t) = X_o(t)-X_n(t) = z_o(t) \cdots t_o-\tau_p \leq t \leq t_o \quad (6b)$$

$$u(t) = U_o(t)-U_n(t) = u_o(t) \ldots t_o-h_q \leq t \leq t_o \quad (6c)$$

where

$$A_o = (\frac{\partial F}{\partial X_n})^T, \ldots, A_k = (\frac{\partial F}{\partial X_n(t-\tau_k)})^T \quad k = 1,2,\ldots,p \quad (6d)$$

$$B_o = (\frac{\partial F}{\partial U_n})^T, \ldots, B_l = (\frac{\partial F}{\partial X_n(t-h_l)})^T \quad l = 1,2,\ldots,q \quad (6e)$$

and the associated cost functional becomes:

$$\hat{J} = 1/2z^T(t_f)(Sz(t_f)+2SX_n(t_f))+1/2\int_{t}^{t_f}[z^T(t)(Q(t)z(t)+2Q(t)X_n(t))$$

$$+u^T(t)(R(t)u(t)+2R(t)U_n(t))]dt \tag{7}$$

where the effect of the nominal trajectories (X_n,U_n) on the original cost function which does not influence the optimization process with respect to (x,u) has been ignored.

SYSTEM DECOMPOSITION

We start with the decomposition of the system (6) into N subsystems by partitioning the state vector x and the control vector u below:

$$\dot{x}_i(t) = A_{0i}(t)x_i(t)+B_{0i}(t)u_i(t)+z_i(t) \qquad t \geq t_0 \tag{8a}$$

$$x_i(t) = x_{i0}(t) \qquad \tau_p \leq t \leq t_0 \tag{8b}$$

$$u_i(t) = u_{i0}(t) \qquad h_q \leq t \leq t_0 \tag{8c}$$

$$z_i(t) = \hat{A}_{0i}(t)x(t)+\hat{B}_{0i}(t)u(t)+ \sum_{k=1}^{p} A_{ki}(t)x(t-\tau_k)+ \sum_{l=1}^{q} B_{li}(t)u(-h_l)$$

$$= z_{ie}(t) \tag{9}$$

where A_{0i} is the diagonal block of matrix A_0,$\hat{A}_{0i}$ is the i-th block row of matrix A_0 with A_{0i} set to zero, i.e.

$$\hat{A}_{0i} = Block\text{-}diag\ [A_{i1},A_{i2}, \ldots, A_{i,i-1},0,A_{i,i+1},\ldots A_{iN}] \tag{10}$$

The similar notation applies to matrices $\hat{B}_{0i}$, B_{0i}; A_{ki} is the i-th block row of Matrix A_k, and the same as B_{li} to B_l for $i=1,2,\ldots,N$, $k=1,2,\ldots,p$, $l=1,2,\ldots,q$ and $x_i \epsilon R^{n_i}, u_i \epsilon R^{m_i}$, $i=1,2,\ldots,N$ so that

$$\sum_{i=1}^{N} n_i = n, \qquad \sum_{i=1}^{N} m_i = m \tag{11}$$

In order to guarantee a satisfactory performance of the system despite the on-off participation of the subsystem, we also assume the cost functional, (7), to be represented by

$$\min_u \hat{J} = \sum_{i=1}^{N} \min_{u_i} \hat{J}_i \tag{12}$$

where

$$\hat{J}_i = 1/2x_i^T(t_f)(S_i x_i(t_f))+1/2 \int_{t_0}^{t_f} [x_i^T(t)(Q_i(t)x_i(t)$$

$$+2Q_i(t)X_{ni}(t))+u_i^T(t)(R_i(t)u_i(t)+2R_i(t)U_{ni}(t))]dt \tag{13}$$

However, a natural decomposition often exists for any given large-scale system. If this is not the case, we can partition the system such that, on the average, a maximum number of zeros occur in the off diagonal blocks of the matrices A_0 and B_0 or the system would be weakly-coupled within the context of non-delay systems [10].

SOLUTION METHOD

We propose an iterative procedure to determine a suboptimal control for the problem defined by (6) and (7). In each iteration the delay terms and the interaction terms will act as the known perturbation inputs obtained in the previous iteration.

First, let's introduce a set of Lagrange multipliers $\alpha_i(t)$ and costate (adjoint) vectors $p_i(t)$ to augment the "perturbation" equality constraint (9) and subsystem dynamic constraint (8) to the cost function's integrand, therefore the i-th subsystem Hamiltonian in the r-th iteration is defined by

$$H_i^r = 1/2x_i^{Tr}(t)(Q_i x_i^r(t)+2Q_i x_{ni}(t))+1/2u_i^{Tr}(t)(R_i u_i^r(t)$$

$$+2R_i U_{ni}(t))+p_i^{Tr}(t)(A_{0i} x_i^r(t)+B_{0i} u_i^r(t)+z_i^{r-1}(t))$$

$$+\alpha_i^{Tr}(t)(z_i^{r-1}(t)-z_{ie}^{r-1}(t)) \tag{14}$$

The necessary conditions for optimality are

$$\dot{p}_i^r = -\frac{\partial H_i^r}{\partial z_i^r} = -Q_i z_i^r - Q_i X_{ni} - A_{oi}^T p_i^r \qquad t \geq t_o \tag{15a}$$

$$p_i^r(t_f) = S_i z_i^r(t_f) + S_i X_{ni}(t_f) \tag{15b}$$

$$O = \frac{\partial H_i^r}{\partial u_i^r} = R_i u_i^r + R_i U_{ni} + B_{oi}^T p_i^r \tag{15c}$$

Equations (8a) and (15a) can be decoupled by defining the adjoint vectors $g_i^r(t)$, $r=1,2,\ldots$ as follows:

$$p_i^r = K_i z_i^r + g_i^r \qquad t_o \leq t \leq t_f \tag{16}$$

where K_i is the symmetric positive-definite solution to the matrix Riccati differential equation

$$\dot{K}_i = -K_i A_{oi} - A_{oi}^T K_i + K_i B_{oi} R_i^{-1} B_{oi}^T K_i - Q_i \tag{17}$$

Equations (15) and (16) imply that

$$\dot{g}_i^r = -(A_{oi} - B_{oi} R_i^{-1} B_{oi}^T K_i\ )^T g_i^r - K_i z_i^{r-1} + K_i B_{oi} U_{ni} - Q_i X_{ni} \tag{18a}$$

The final conditions for (17) and (18a) can be determined by comparison of (16) for $t=t_f$ with (15b), which yields

$$K_i(t_f) = S_i \tag{18b}$$

$$g_i^r(t_f) = S_i X_{ni}(t_f) \tag{18c}$$

Following this formulation, the optimal control for the r-th optimization problem can be written as

$$u_i^{*r} = -R_i^{-1}(B_{oi}^T p_i^r + R_i U_{ni}) = -R_i^{-1} B_{oi}^T K_i z_i^{*r} - R_i B_{oi}^T g_i^r - U_{ni} \tag{19}$$

Hence, from (8a) and (19), the optimal state trajectory $z^{*r}(t)$ is the solution to

$$\dot{z}_i^{*r} = [A_{oi} - B_{oi} R_i^{-1} B_{oi}^T K_i] z_i^{*r} - B_{oi} R_i^{-1} B_{oi}^T g_i^r - B_{oi} U_{ni} + z_i^{r-1} \qquad t_o \leq t \leq t_f \tag{20a}$$

$$z_i^*(t) = z_{io}(t) \qquad t_o - \tau_p \leq t \leq t_o \tag{20b}$$

$$u_i^*(t) = U_{io}(t) \qquad t_o - h_q \leq t \leq t_o \tag{20c}$$

The chosen optimization strategy, however, cannot achieve the optimal performance index

$$J^* = \sum_{i=1}^{N} J_i^* \tag{21}$$

by using only the local control $u^*(t)$, unless all the perturbations are absent ($z_i = 0$, $i=1,2,\ldots,N$). To solve this problem, let's form the second level problem which is essentially updating the new coordination vector $\alpha_i(t)$ and $z_i(t)$. For this purpose, define the additively separable Lagrangian

$$L = \sum_{i=1}^{N} L_i \tag{22}$$

where

$$L_i = 1/2 z_i^T(t_f)(S_i z_i(t_f) + 2S_i X_{ni}(t_f)) +$$
$$1/2 \int_{t_o}^{t_f} \{[z_i^T(Q_i z_i\ \ 2Q_i X_{ni}) + u_i^T(R_i u_i + 2R_i U_{ni})] +$$
$$\alpha_i^T(z_i - z_{ie}) + p_i^T(-\dot{z}_i + A_{oi} z_i + B_{oi} u_i + z_i)\}\, dt \tag{23}$$

The values of $\alpha_i(t)$ and $z_i(t)$ can be obtained by

$$0 = \frac{\partial L_i}{\partial \alpha_i} = z_i - z_{ie} \tag{24a}$$

$$0 = \frac{\partial L_i}{\partial z_i} = \alpha_i + p_i \tag{24b}$$

which provides at r-th iteration

$$z_i^r = z_{ie}^{r-1} = \hat{A}_{oi} z^{r-1} + \hat{B}_{oi} u^{r-1} + \sum_{k=1}^{p} A_{ki} z^{r-1}(t-\tau_k) + \sum_{l=1}^{q} B_{li} u^{r-1}(t-h_l) \tag{25a}$$

$$\alpha_i^r = -p_i^{r-1} \tag{25b}$$

Once the total system perturbation error, in normalized form,

$$E = \sum_{i=1}^{N} \int_{t_o}^{t_f} [(z_i - z_{ie})^T (z_i - z_{ie})] dt / \Delta t \tag{26}$$

is sufficiently small an optimal solution has been obtained. Here, Δt is the step size of integration.

COMPUTATIONAL PROCEDURE AND ALGORITHM

In order to initiate the iterative process we let

$$z_i^o(t) = \phi_i(t, t_o) z_{io}(0) \quad t \geq t_o \tag{27}$$

where $\phi_i(t,t_o)$ is the state transition matrix corresponding to (20). The first step is to solve the Riccati equation (16) to obtain K_i for $i=1,2,...N$. Then determine $z_i^o(t)$ from (27) and choose an arbitrary initial control $u_i^o(t)$ $i=1,2,...N$. Thus z_i^o can be obtained from (9) for each i respectively. Guess α_i^o for each ith subsystem and use boundary condition (18c). Then (18a) can be solved backward in time for iteration number $r=1$ to obtain $g_i^r(t)$. Furthermore, using the boundary condition (8b), $z_i^{*r}(t)$ can be solved for each i. The corresponding optimal control $u_i^{*r}(t)$ can then be obtained from (19). This procedure is continued for consecutive integer values of r.

The following algorithm summarizes the suboptimal control scheme.

ALGORITHM

Step 1. Choose an arbitrary nominal control $U_n(t)$ which satisfies (1).

Step 2. Solve (1) with the known control vector $U_n(t)$ to obtain the nominal state vector $X_n(t)$.

Step 3. Obtain a linear approximation system of (6a) by substituting the nominal trajectories $(x_n(t), U_n(t))$ in (6d), (6e), and decompose them as in (8) and (9).

Step 4. Solve (17) with final condition (18b) to obtain $K_i(t)$ and store. $i = 1,2,...,N$.

Step 5. Determine $z_{io}(t)$ from (27), and predice α_{io} for $i = 1,2,...,N$.

Step 6. Set $r = 1$, use $z_{io}(t)$ and $u_{io}(t)$ (from (8c)) in (9) to obtain $z_{ie}(t)$, $t \geq t_o$

Step 7. Use the final condition (18c) in (18a) to obtain $g_i^r(t)$, and get $p_i^r(t)$ from (16) thereafter. Store $g_i^r(t)$ and $p_i^r(t)$, $i=1,2,...,N$.

Step 8. Solve the i-th suboptimal control problem to find $z^{*r}(t)$ and $u^{*r}(t), t \geq t_o$ as described by (19) and (20). Store $z^{*r}(t)$, $u^{*r}(t)$, $i=1,2,...,N$.

Step 9. Use the stored $u^{*r}(t)$ and $z^{*r}(t)$, and $p_i(t)$ in (25) to update $\alpha_i(t)$ and $z_i(t)$, 1,2,...,N.

Step 10. Check for the convergence at the second level by evaluating the overall perturbation error from (26).

Step 11. If $E < \epsilon$ for a prespecified small constant ϵ then

$$U^*(t) = u^{*r}(t) + U_n(t) \tag{28a}$$

$$X^*(t) = z^{*r}(t) + X_n(t) \tag{28b}$$

and stop, otherwise set $r = r+1$ and go to Step 7.

A NUMERICAL EXAMPLE

Consider a model of a typical control problem occurring in the water resource system, which is shown in Figure 1. The list of symbols used in the system model is given below:

y = input water flow,
s = reservoir storage,
e = evaporation quantity,
u = water release,
d = irrigation release,
α = fraction of water return to main stream,
τ = delay time,

Table 1 presents a summary of state and control variables used in this example.

Vector	State	Control
Subsystem No.1 Physical	S_1, S_2	u_1, d_1, u_2
Mathematical	X_1, X_2	U_1, U_2, U_3
Subsystem No. 2 Physical	S_3, S_4	d_2, u_3, d_3, u_4
Mathematical	X_3, X_4	U_4, U_5, U_6, U_7

Table 1. State and control variables of the Illustrated Example.

The state equation derived from the system model is a form of a non-linear and time-delay as shown below:

$$\dot X_1 = Y_1(t) - U_1(t) - e_1(t)$$
$$\dot X_2 = U_1(t) - U_3(t) - U_2(t) + \alpha_1 U_2(t-\tau_1) - e_2(t)$$
$$\dot X_3 = Y_3(t) - U_5(t) - e_3(t)$$
$$\dot X_4 = U_3(t) - U_4(t) + \alpha_2 U_4(t-\tau_2) + U_5 - U_6 + \alpha_3 U_6(t-\tau_3) - U_7 - e_4(t) \tag{29}$$

where $e_i(t) = \beta_i X_i^{2/3}(t)$ for $i = 1,2,3,4$. A suitable cost function for this system is

$$\max_{U,X} \quad J = 1/2\, X^T(12) S(12) X(12) + 1/2 \int_1^{12} (X^T(t) Q(t) X(t) + U^T(t) R(t) U(t))\, dt \tag{30}$$

where

$$R_1 = diag\,(-1,0,-2), \qquad R_2 = diag\,(-1,-1,-2,-1)$$
$$S_1 = Q_1 = diag\,(-1,-2) \quad S_2 = Q_2 = -I_2 \tag{31}$$

Here the weighting matrices R_i, Q_i, S_i, i=1,2 are all chosen to be negative definite for the maximization problem. The parameters in the state equation are chosen to be:

$$\alpha_1 = 20\% \qquad \alpha_2 = 30\% \qquad \alpha_3 = 25\%$$
$$\beta_i = 15\%,\ i=1,\dots,4 \qquad \tau_j = 1,\ j=1,2,3 \tag{32}$$

Simulating from the data given in Peters et al [5], the input water flow is chosen to be:

$$Y_1 = 1 + 0.6*\sin(\Pi/6*T) \qquad T = 1,2,...,12$$
$$Y_3 = 0.8 + 0.6*\sin(\Pi/6*T) \tag{33}$$

Using the above data, the system can then be rewritten as:

$$\dot{X}_1 = 1 + 0.6*\sin(\Pi/6*T) - U_1 - 0.15*X_1^{2/3}$$
$$\dot{X}_2 = U_1 - U_2 - U_3 + 0.2*U_2(t-1) - 0.15*X_2^{2/3}$$
$$\dot{X}_3 = 0.8 + 0.6*\sin(\Pi/6*T) - U_5 - 0.15*X_3^{2/3}$$
$$\dot{X}_4 = U_3 - U_4 + 0.3*U_4(T-1) + U_5 - U_6 + 0.25*U_6(T-1) - U_7 - 0.15*X_4^{2/3}$$
$$\text{for } T = 1,2,...,12 \tag{34}$$

The initial conditions used here are:

$$X_1 = 0.6 \qquad U_1 = 1 \qquad U_5 = 0.8$$
$$X_2 = 0.7 \qquad U_2 = 0.4 \qquad U_6 = 0.4$$
$$X_3 = 0.8 \qquad U_3 = 0.5 \qquad U_7 = 0.6$$
$$X_4 = 0.7 \qquad U_4 = 0.3 \qquad (Unit = 10^9 ft^3) \tag{35}$$

First, choose the nominal control trajectories to be:

$$U_{n1} = U_1 = 1 \qquad U_{n5} = U_5 = 0.8$$
$$U_{n2} = U_2 = 0.4 \qquad U_{n6} = U_6 = 0.4$$
$$U_{n3} = U_3 = 0.5 \qquad U_{n7} = U_7 = 0.6$$
$$U_{n4} = U_4 = 0.3 \tag{36}$$

Substitute the nominal control vectors in (34) to find the nominal state trajectories $X_{ni}(t)$ $i=1,2,3,4$. From (6d) and (6e) using the nominal trajectories values, a linear model can thus be obtained. The system is then decomposed into two subsystems due to its physical form. The Algorithm is applied to solve the problem, using an HP-9845 computer. The perturbation error converged to $8.438302*10^{-12}$ in twelve iterations. Figures 2 to 6 show the interaction error and trajectories of the state, control and vectors and input water flow.

DISCUSSIONS AND CONCLUSIONS

In this paper an algorithm for the optimization of water resources systems described by a non-linear time-delay system using a near-optimal design technique is developed. The scheme using the interaction prediction is oriented towards the successive approximation method. The advantage of this method is that at the first level we solve simple linear problems with quadratic cost which are centered around the solutions of accompanying matrix Riccati equations required to be solved once only, and the second level we predict in first time and then just simply substitute to get the new values of the coordinating variables which makes this method so computationally attractive. In addition to this, it also has the advantage of the possibility of using a multiprocessor system.

REFERENCES

[1] R. E. Larson and W. G. Keckler, "Applications of dynamic programming to the control of water resources systems," *IFAC J. Automatica,* Vol. 5, pp. 15-26, 1969.

[2] M. Heidari, V.T. Chow, P.V. Kokotovic' and D.D. Meredith, "Discrete differential dynamic programming approach to water resources optimization", *Water Resources Research,* Vol. 7, p. 273, 1971.

[3] M. Jamshidi, "Optimization of water resources systems with statistical inflows," *Proc. IEE,* Vol. 124, pp. 79-82, 1977.

[4] M. Jamshidi and M. Heidari, "Modelling and optimization of Khuzestan water resources systems," *IFAC J. Automatica,* Vol. 13, pp. 287-293, 1977.

[5] R. J. Peters, K.-C. Chu and M. Jamshidi, "Optimal operation of a water resources system by stochastic programming," *Mathematical Programming,* Study 9, pp. 152-175, 1978.

[6] B.D.O. Anderson, "Time delays in large-scale systems," *Proc. CDC,* Ft. Lauderdale, FL., 1979.

[7] M. Jamshidi and M. Malek-Zavarei, "A hierarchical control for large-scale time-delay systems," *Large-Scale Systems,* Vol. 4, pp. 149-163, 1983, see also *Proc. IEEE CDC,* San Diego, CA., pp. 1302-1304, 1981.

[8] M. Jamshidi and C.-M. Wang, "A computational algorithm for large-scale nonlinear time-delay systems," *IEEE Trans. SMC,* 1984 (to appear)

[9] C.-M. Wang and M. Jamshidi, "Optimal Control of large-scale non-linear systems with time-delay" *Int. J. Control,* 1984 (to appear).

[10] M. Jamshidi, *Large-Scale Systems-Modeling and Control,* Elsevier North-Holland, (New York), 1983.

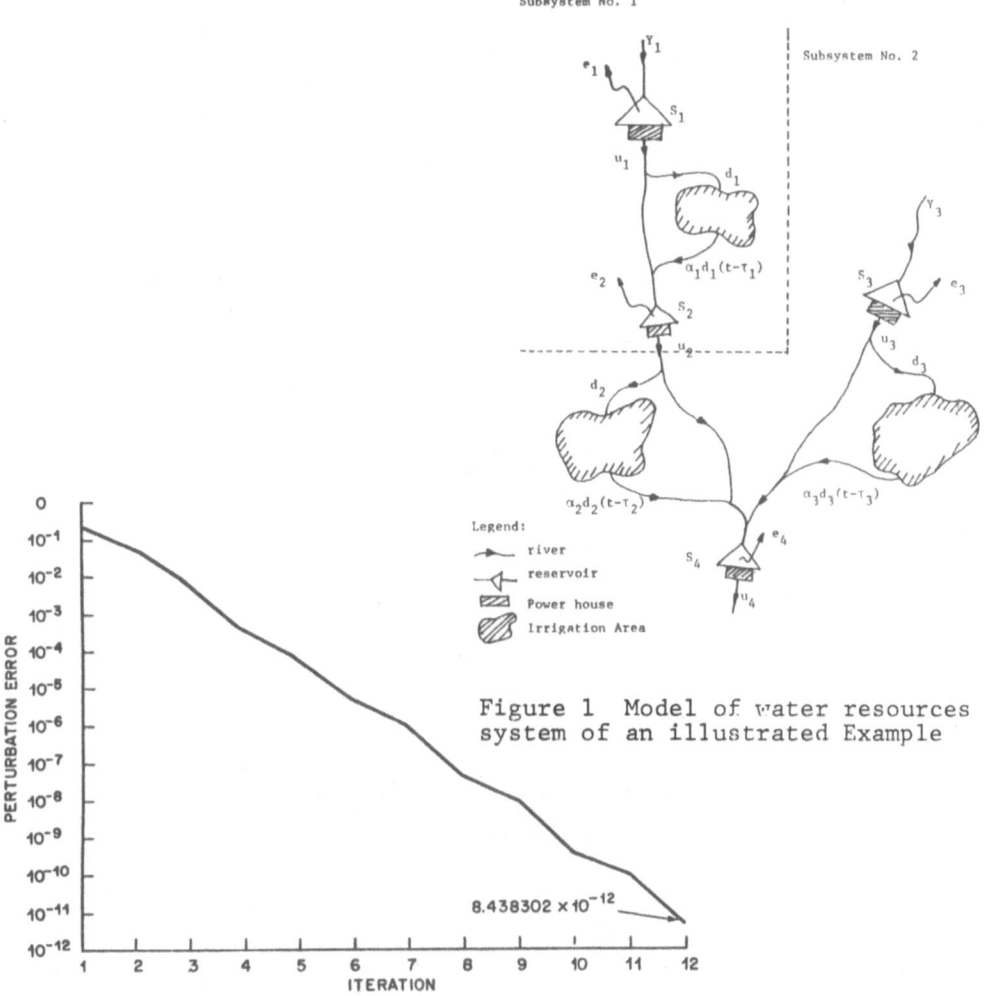

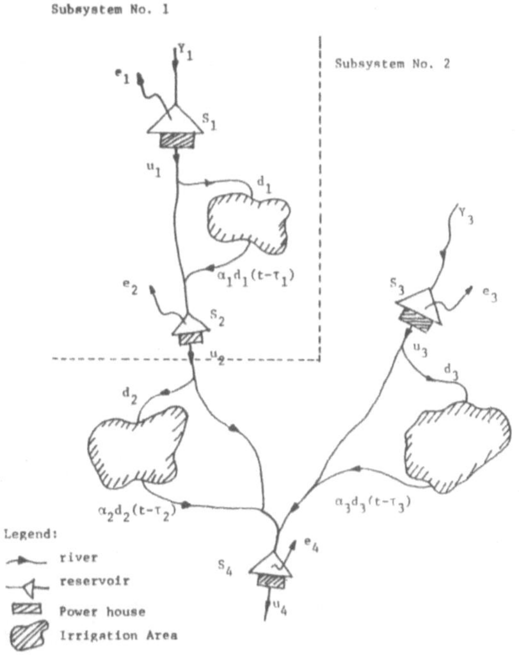

Figure 1 Model of water resources system of an illustrated Example

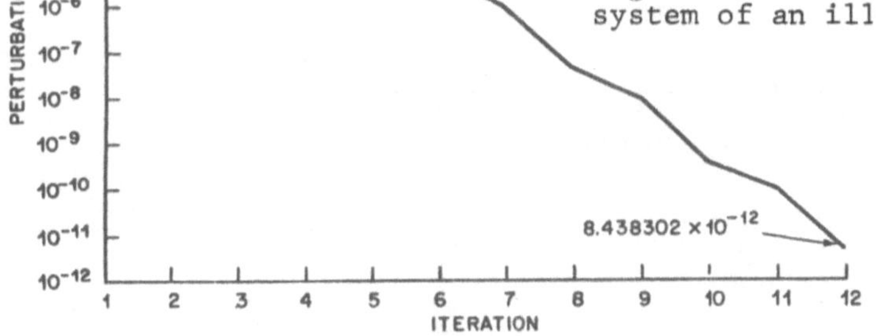

Figure 2 Interaction error vs. iteration for the illustrated example

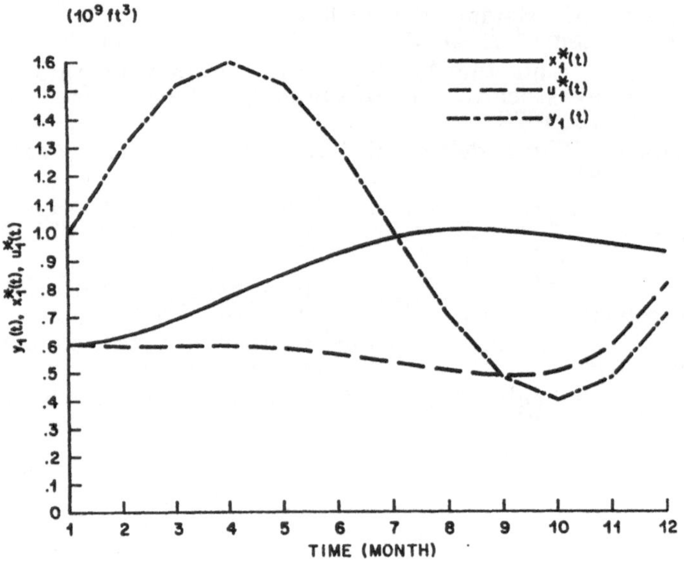

Figure 3 Time responses of $y_1(t)$, $x_1^*(t)$, $u_1^*(t)$

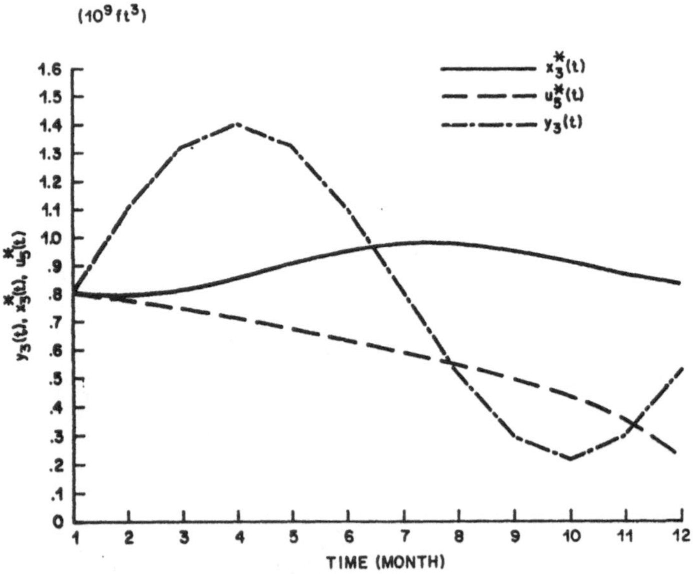

Figure 4 Time responses of $y_3(t)$, $x_3^*(t)$, $u_5^*(t)$

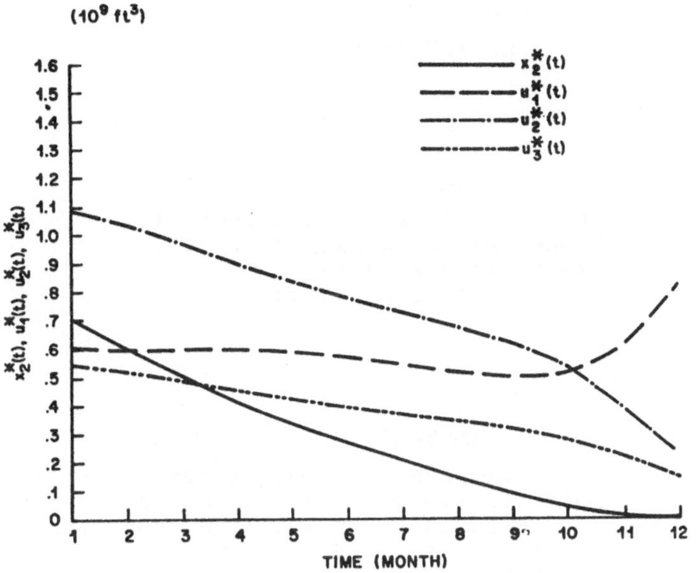

Figure 5 Time responses of $x_2^*(t)$, $u_1^*(t)$, $u_2^*(t)$, $u_3^*(t)$

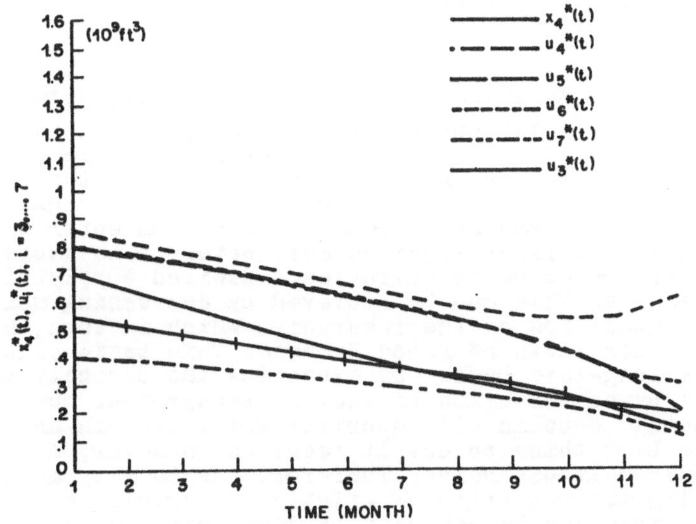

Figure 6 Time responses of $x_4^*(t)$, $u_3^*(t)$, $u_4^*(t)$, $u_5^*(t)$, $u_6^*(t)$, $u_7^*(t)$

HIERARCHICAL STRUCTURE FOR REAL-TIME FLOOD CONTROL IN A MULTIRESERVOIR SYSTEM

K.Malinowski
Institute of Automatic Control
Technical University of Warsaw
ul.Nowowiejska 15/19
00-665 Warszawa Poland

ABSTRACT

The paper presents a hierarchical control structure for real-time flood control in a multireservoir system. Water releases from the individual reservoirs are adjusted by the local operators while the objective of a central authority is to modify local performance functions so as to achieve his overall goals. Due to highly uncertain forecasting of future inflows the decision mechanism of a central dispatcher is based upon fast simulation of the system operation under different flow scenarios. An application of the presented control structure to a case study multireservoir system is shortly described together with some selected simulation results.

Keywords: flood- control, hierarchical control structure, flood wave transformation, flow forecasting, local and central decision mechanisms, coordination

1. Introduction and basic description of the control structure

Reliable and, if possible, optimal real-time flood control in multireservoir systems is of an increasing interest. This is so due to steadily growing population and industrialisation of the close-to--water areas which require therefore more protection against flood damages. On the other hand an increasing demand on water during normal (low inflow) periods requires more water to be stored in retention reservoirs so as to quarantee a desired supply of water within dry seasons. This can be achieved by decreasing mandatory flood reserve capacities of the reservoirs-which in turn, however, provides for higher risks of flood damages. This basic dilemma can be solved on a long-term basis by expanding the capacity of existing reservoirs and by construction of the new reservoirs. Such retention build-up is being done in all important water systems and is definitely the best thing to do. It requires, however, a very long time and is extremely expensive. Therefore one is very much interested in improving the operation of existing reservoirs during the flood periods. This can be achieved by developing reliable control structures for multireservoir systems so as to provide for a proper cooperation of multiple reservoirs during floods. Now, there are two basic difficulties which one immediately encounters. First of all the local objectives (e.g. those of a given reservoir operator) are inconsistent with the overall objectives. In particular a reservoir operator will be usually interested in minimizing the peak of a release from his reservoir during the flood (Real-Time Operation of Hydrosystems, 1981) while the central authority may first of all want to minimize the peak water levels at the most important places located far downstream the reservoirs (IMGW reports 1981, 1982). The second difficulty is due to a very high level of uncertainty (especially at the beginning of a flood) with respect to incoming inflows to reservoirs. In order to provide for a proper cooperation of reservoirs the central authority should possess rather accurate forecasts of the inflows, while, quite unfortunately, such forecasts are not usually available. Apart of the above issues it is also necessary to make it certain that an operation of a control structure during a flood period will not be too much affected by possible disruptions of communication links between the central and the local authorities (reliability issue). Finally, the degree of complication as created by complex high dimensional flood transformation mathematical models (e.g. given by numerous ordinary-differential-equations) makes it very difficult, if not impossible, to compute centrally the optimal reservoir releases so as to minimize the overall goals. For these reasons it seems reasonable (Findeisen 1979) to propose a suitable hierarchical control structure. The theory of hierarchical control is now quite developed (see e.g, Findeisen at al. 1980). Hierarchical management of water resources has already been successfully approached (Haimes 1978; Malinowski 1980, Findeisen and Malinowski 1981, Malinowski at al. 1983).

This paper presents a hierarchical control structure for real-time flood control in a multireservoir system-with the above issues being taken into account. The system consists of several reservoirs located on tributaries to the major river. Water releases from reservoirs are adjusted by local reservoir operators so as to satisfy specified local objectives: protection of dams during flood and minimization of local cost functions expressing losses due to damage created by high water levels. The objective of a central authority (central dispatcher) is to modify local performance functions so as to achieve his overall goal. This includes local goals as well as some primary global goals related to peak water levels at important crosssections. Within proposed hierarchical control structure the central dispatcher is supposed to adjust (every T_G hours - e.g. every 12 hours) the parameters of the local performance indices-basing upon long-term flow forecasting and upon simulation of the operation of local decision rules together with flood wave transformation. Local reservoir operators determine water releases while using current local observation data, local inflow forecasts and the parameters specified by the central dispatcher.

The basic elements of the control structure, as described in the subsequent sections, are:

(1) flood wave transformation simulator (set of about hundred ordinary-differential-equations),
(2) flow forecasting routines used by local operators and by central dispatcher,
(3) decision mechanism as used by the local operators,
and (4) decision mechanism used by the central dispatcher in order to find suitable parameters for the local cost functions. This decision mechanism is based upon fast simulation of the system operation under different flow scenarios.

An application of the presented control structure to a case study multireservoir system in Southern Poland is described shortly in Section 6. Finally in the last section the conclusions are given together with some directions of future improvements.

2. Flood wave transformation

Flood wave transformation in river reaches downstream the reservoirs can be described by partial-differential-equations (Saint-Venant eqs.) while taking into account river bed shape and related properties (e.g. bed slope, friction coefficients etc.) as well as side-inflows (i.e. surface rainfall water inflow and ground water infiltration into a river bed). However, for operational decision making much simpler models are required since the transformation equations are going to be rapidly solved very many times in order to evaluate possible impacts of different decisions under different flow scenarios (see section 5) . Moreover, in view of inaccurate forecasts of inflows to reservoirs and even less accurate forecasts of side-inflows the use of very sophisticated flow-transformation models may not be quite justified. In particular, for a case study system as described in Section 6 the following model has been adopted. Each river reach (between the adjacent watermarks) is represented by a cascade of 10 nonlinear reservoirs; every one of those is described by an ordinary-differential-equation (IMGW reports 1981, 1982):

$$\frac{dS_i}{dt} = Q_{in_i} - Q_{out_i} \tag{1a}$$

$$S_i = k_i Q_{out_i}^{n_i} \tag{1b}$$

where S_i represents the water volume stored in the i-th "reservoir"; Q_{in_i}, Q_{out_i} represent, respec-tively, the inflow and the outflow; k_i, n_i are two parameters relating S_i to Q_{out_i}. Since in a multi-reservoir system there usually are several river reaches to be taken into account, the number of equations (1) may very well exceed a hundred. Q_{in_i} is related to Q_{out_j} as follows

$$Q_{in_i} = Q_{out_{i-1}} + q_i \tag{2}$$

where q_i represents a side-inflow. A junction of any two rivers is described also as a reach with two joining flows treated as additive inputs. It should be noted that a very carefully written simulation program for solving the set of above eqs. (1) and (2) (over, say, a 300 h horizon) on a modern minicomputer still requires the computing time of about 1-2 mins. When a mainframe computer is not available then in order to reduce the computing time it may be worthwhile to use during real-time decision making much faster hybrid (analog-digital) simulator of flood wave transformation. Before using the transformation model one has first of all to find proper values of parameters k_i, n_i, predict side-inflows q_i and finally to estimate current values of state variables S_i. Therefore the "calibration" of the flood wave transformation model has to be done at each stage of a flood control structure operation which requires the use of this model. It is expected that the calibration of the model will be performed at the central dispatcher level with a help provided by a specialisted software package.

3. Flow forecasting

Flow forecasting routines are essential software tools as required by local and central decision mechanisms. As far as a local decision mechanism is concerned it may use long-term forecasts or short-term forecasts only or, even, no forecasts at all. (Pleskov 1975, IMGW reports 1981, 1982, IAPW reports 1981-1983). A particular local decision rule as presented in the next section requires long-term forecasting of individual inflow to a given reservoir. On the other hand, as already mentioned in the introduction, the central decision mechanism must necessarily use long-term forecasts of inflows to reservoirs as well as forecasts of side-inflows to river reaches in order to provide for a proper cooperation of different reservoirs. Within a control structure as considered in this paper the following flow forcasting routines are used:

(i) short-term forecasting developed by specialists in hydrology (IMGW reports 1980, 1981) and consisting of predicting flows (in interesting crossections) during next 24 hours or next 48 hours when basing upon the observed rainfall and using specialised models describing transformation of this rainfall into water outflows to river reaches and, then, using models of transformation of flows,

(ii) long-term flow forecasting (Karbowski 1983 , IAPW reports 1981-1983). This consists of extending the above short-term forecasts by using historical data base together with current observations related to the actual flood situation development. Speaking in broad terms the procedure involves (a) - proper "synchronisation" of historical data base (historical inflows to reservoirs during flood periods), (b) - expressing the currently available present flood waves (together with short-term hydrological forecasts) in terms of the historical data and (c) - using the obtained coefficients so as to produce the long-term flow forecasts-based again upon the historical data.

In the basic version of the control structure only one long-term forecast for each inflow is considered at a given time stage. More advanced versions require multi-forecasting in order to take into account different possible flow scenarios at the central level and, also, at the local level (see Sections 4,7) .

4. Local decision mechanism

The following model of a local decision mechanism of each reservoir operator is proposed:

Every T_L hours, say at time instant k, the i-th reservoir operator determines his decision $\tilde{m}_i(k)$ (the release from the i-th reservoir in m^3/s between times k and k+1) as a solution of the following problem:

LPi :
$$\min_{m_i(k),\dots,m_i(K)} \; \max_{j=k,\dots,K} \; \alpha_{1i}(j)\cdot m_i(j) \qquad (3)$$

subject to

$$w_i(j+1)=w_i(j) + \bar{z}_i(k,j) - m_i(j) \qquad (4)$$

$$w_i(k) = w_j^r(k), \qquad w_i(K+1) = w_{i\,max} \qquad (5)$$

$$w_{i\,min} \leqslant w_i(j) \leqslant w_{i\,max} \qquad (6)$$

$$m_{i\,min} \leqslant m_i(j) \leqslant m_{i\,max}(w_i(j)) \qquad (7)$$

In the above problem $w_i(j)$ denotes volume of water in the i-th reservoir at time j, $w_i^r(k)$ - actual real volume of water in the reservoir at a current time k, $\bar{z}_i(k,j)$ - inflow forecast for time period j as specified at a current time k by a long-term forecasting routine (see section 3) , K - considered time horizon (predicted end time of a flood period). Constraints (6) and (7) express necessity to keep water level between minimum and maximum values, and, respectively, to consider water releases from an admissible range. Weighting function $\alpha_{li}(j)$ as specified by the central dispatcher at time k_l (prior to time k) enables the central authority to influence local decisions. In particular $\alpha_{li}(\cdot)$ is assumed to have the form

$$\alpha_{li}(j) = \begin{cases} 1 & \text{for} \quad k_l \leq j < k^{*}_{li} \\ c_{li} & \text{for} \quad k^{*}_{li} \leq j \leq K \end{cases} \tag{8}$$

Thus $\alpha_{li}(\cdot)$ is completely specified by two parameters: k^{*}_{li} and c_{li} - as decided upon by the central dispatcher at time k_l (see section 5) . When $c_{li} < 1$, then the peak of a release from the i-th reservoir should be delayed and when $c_{li} > 1$ this peak should be accelerated. A typical solution of problem LP_i (for the case of $c_{li} < 1$) is shown in fig.1. The details regarding the presented local decision mechanism are given in (IAPW reports 1981-1983).

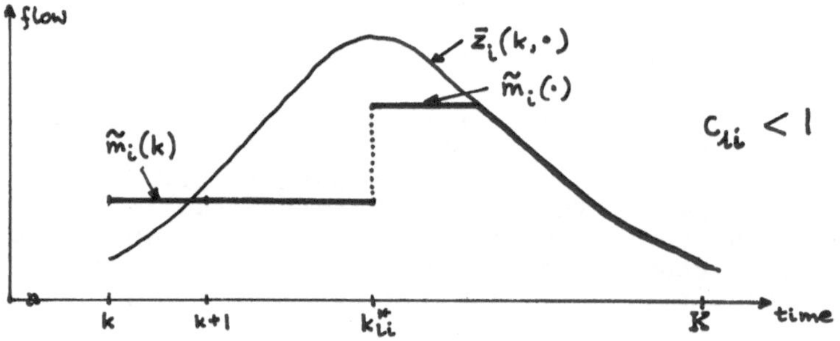

Fig.1. Typical solution of problem LP_i .

After solving problem LP_i the release $\tilde{m}_i(k)$ is administered from the i-th reservoir. Then, at time k+1 the procedure is repeated and so on.

5. Central decision mechanism

The central dispatcher has to find at, say, time k_l (every T_G hours) suitable parameters k^{*}_{li} , c_{li}, $i=1,\dots,\nu$ (where ν is the number of reservoirs) for the local cost functions. In order to achieve this, the central dispatcher has to take into account the fact, that the local decisions will be based upon inaccurate local forecasts $\bar{z}_i(k,\cdot)$ (see section 4) . He has also to consider overall objectives related to peak water levels at important crosssections downstream the resersoirs. The following central level procedure is proposed. At time instant k_l the central dispatcher considers, say, N long-term flood scenarios (obtained from long-term forecasting module-see section 3) and assigns to each of them a probability p^l_n $(n=1,\dots,N)$, $\sum_n p^l_n = 1$. Then (assuming certain values of parameters k^{*}_{li}, c_{li}, $i=1,\dots,\nu$) for every scenario n the central dispatcher performs a simulation of the local level operation until the end of the flood period. This simulation involves repetitive use of local forecasting and decision making routines as well as the use of a flood wave transformation model. After being completed the simulation provides for a value of an overall performance index - this value $J_n^1(k_{11}^*, c_{11}, \ldots, k_{1\nu}^*, c_{1\nu})$ is related to coordination inputs k_{1i}^*, c_{1i} and to the considered flood scenario n. Repeating the above simulation for all scenarios the central dispatcher can compute the value of his performance index

$$J^1(k_{11}^*, c_{11}, \ldots, k_{1\nu}^*, c_{1\nu}) \overset{df}{=} \sum_{n=1}^{N} p_n^1 J_n^1 \tag{9}$$

This can be minimized with respect to parameters k_{1i}^*, c_{1i}, $i=1,\ldots,\nu$. Assuming that the hill-climbing procedure performing this minimization will require S steps the simulation of the local level operation has to be repeated S·N times; in particular a flood wave transformation model has to be computed that many times. After optimal values of the parameters are established, they are passed down to local operators and then, after T_G hours, the whole procedure is repeated.

6. Case-study system; simulation results.

The proposed hierarchical flood-control structure has been applied to a case-study multireservoir system in Southern Poland. It is beyond the very limited scope of this presentation to describe in detail the system itself as well as the simulation results. However, it is worthwhile to describe shortly the main results of a simulation of the control structure operation when applied to a subsystem of two reservoirs Dobczyce and Rożnów - located on two tributaries (Raba and Dunajec) to the Vistula river. From the global point of view it is especially important to minimize the flood damages as represented by peak water levels at important crosssections of Raba, Dunajec and Vistula rivers (see fig. 2) . These preferences were taken into account when formulating the global performance index. The layout of the considered subsystem is shown in fig.2.

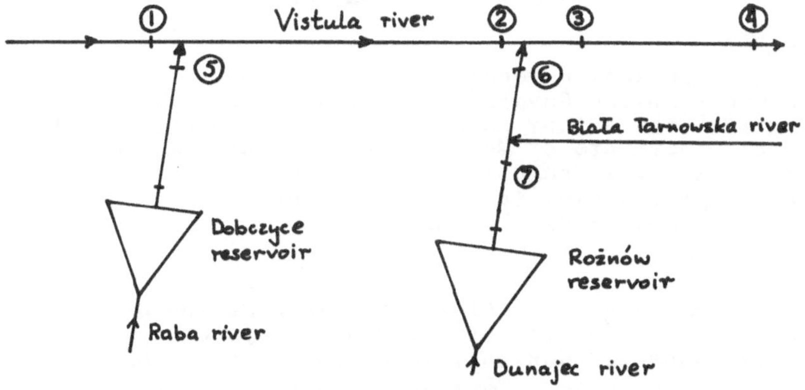

Fig.2. Layout of considered two-reservoir subsystem

The flow at crosssection ① has been obtained by transforming a "natural" flood wave (an example flood scenario has been considered) through the remaining (not shown in fig.2) part of the waye-transformation model. The central dipatcher decision was taken only once-at the beginning of the "flood" - at "0" hour. Also only one set of long-term flow forecasts has been taken into account at the central level (N=1 in eq. (9)). The following results in terms of the performance index values has been obtained

(i) natural flood (without reservoirs) - 2113
(ii) traditional individual decision rules- 1773
(iii) new local decision mechanisms without coordination ($c_{1i}=1$) - 1692
(iv) new control structure with coordination performed at "0" hour - 1726

It can be observed that in case (iii) the performance value is better than in case (iv). This is due to the fact that the central level mechanism was used only once and was based only upon one set of very inaccurate long-term forecasts.

Results of the other more elaborated simulations have shown that the coordination can in fact be effective only if one is sufficiently aware (in terms of considering different flood scenarios) of inevitable uncertainty of long-term forecasts.

7. Conclusions

The following observations can be made-based upon experience which is already available:

(i) - the presented control structure offers potential improvement over traditional individual decision rules as used within flood periods,

(ii) - local operators are left to make their independent decisions and when their communications with the center are disrupted they can still use their local decision mechanisms (with "old" values of upper level inputs k^*_{1i}, c_{1i}) ,

(iii)- it is worthwhile to consider possible improvements of the presented decision mechanisms; in particular it seems promising to replace problems LP_i (eqs. (3-7)) with a decision rule taking into account more than one local flood wave forecast as well as taking into account future interventions of a local operator. The multi-long-term forecasting at the central level is necessary in order to avoid severe misjudgements while specifying the coordinating inputs (see section 6).

Acknowledgements

The research and simulation results presented in this paper have been carried under Polish Governmental Project PR-7 by a research group working with the author in close cooperation with the Kraków Division of the Institute of Meteorology and Water Management headed by Dr H. Słota. I would like to express my gratitude to all colleaques and in particular to Mr A. Karbowski who did most of the programming and computer simulation.

References

1. Findeisen, W. - A view on decision and information structures for operational control in water resources systems, Report, Institute of Automatic Control, Tech. Univ. of Warsaw, 1979

2. Findeisen, W., F.N. Bailey, M.Brdyś, K.Malinowski,P. Tatjewski and A. Woźniak - Control and Coordination in Hierarchical Systems, J. Wiley, 1980

3. Findeisen, W. and K. Malinowski - Hierarchical approach to real-time control for multireservoir systems, Proceedings Internat. Symp. on Real-Time Operation of Hydrosystems, Univ. of Ontario, Canada, 1981

4. Haimes, Y.Y - Hierarchical Analyses of Water Resources System, McGraw-Hill, 1978

5. IA PW reports 1981-1983 - Reports on task PR-7.05.08.11 (by K. Malinowski at al.), Institute of Automatic Control, Techn.Univ. of Warsaw (in Polish)

6. IMGW reports 1981, 1982 - Reports on task PR-7.05.08.03 (by H. Słota at.al), Kraków Division of the Institute of Meteorology and Water Management (in Polish)

7. Karbowski, A. - Hierarchical flood-control in a multireservoir system, M.Sc thesis, Institute of Automatic Control, Techn.Univ. of Warsaw, 1983 (in Polish)

8. Malinowski, K. - Two-time scale control of multireservoir systems, Proceedings of V-th Polish-Italian Conf. on Appl. of Syst. Theory, Toruń, Poland, 1980

9. Malinowski, K., K. Salewicz and T.Terlikowski - Hierarchical dispatching control structure for a multireservoir system, Proceedings of the Third, IFAC/IFORS Symp. on LSSTA, Warsaw, 1983

10. Pleskov, J.F. - Regulirovanie rečnovo stoka, Gidrometeoisdat, Leningrad 1975 (in Russian)

11. Real-Time Operation of Hydrosystems - Proceedings of the International Symposium on Real-Time Operation of Hydrosystems, Univ.of Waterloo, Canada, 1981.

MULTIPLE CRITERIA INTERACTIVE CONTROL
OF THE NATURAL GAS DISTRIBUTION NETWORK

A. Lewandowski
Institute of Automatic
Control, Technical
University of Warsaw
00-665 Warsaw, Nowowiejska 15/19

A. Osiadacz
Institute of Petroleum
and Gas Engineering
01-224 Warsaw, Kasprzaka 25

1. Introduction

Natural gas has become a most attractive source of fuel for both industrial and domestic consumption. Because the sources of natural gas are concentrated in areas which located rather far from the customers, gas must be distributed through the pipeline network. Becouse of energy losses due to friction, pumping stations are necessary to perform this task. These pumping stations play the role of active (control) elements of the system. The action of the pumping station should be performed in such a way, as to ensure the desired quality of transmission process. This control problem is formulated in the paper, as well the methodology for its solution based on the multiple criteria optimization is proposed.

2. Models of network elements

Pipelines

The flow rate in a pipeline depends on 4 main factors, namely the internal diameter of the pipe, the specific gravity of the gas, the friction forces and the pressures on both ends of pipeline. There are number of formulas describing the steady-state gas flow; the Weymouth's equation was applied in our case:

$$Q = K \cdot D^{8/3} \left(\frac{p_1^2 - p_2^2}{LTsZ} \right)^{0.5} \tag{1}$$

where: Q - flow in pipe
D - internal diameter
L - length of pipe
p_1 - absolute pressure at sending end of pipe
p_2 - absolute pressure at receiving end of pipe
T - temperature of gas
S - specific gravity of gas
Z - compressibility factor
K - constant.

Compressors

The compresor raises the inlet pressure p_A, to a higher outlet pressure p_B. This higher pressure then provides a pressure gradient to maintain flow in the next pipeline segment.

Flow through a compressor is given by:

$$Q = \lambda \left(1 - C\left(\left(\frac{p_B}{p_A}\right)^{1/m_2} - 1\right)\right) V_s \cdot n \cdot k \tag{2}$$

where λ - the capacity loss coefficient,
m_2 - the polytropic exponent of expansion
V_s - the volume swept,
n - r.p.m.,
k - number of cylinders,
C - the ratio of clearance volume to swept volume.

Constrains on compressor operation include minimum and maximum values for flow, pressure and compression ratio.

Delivery points

The m-th delivery point in a pipeline network is characterized by constant demand flow D_m, and required pressure p_{dm}.

Sources

The j-th source of gas is modelled in terms of its pressure and flow. It was assumed that the source pressure is constant and that any desired flow can be furnished by the source provided this flow is between limits Q^+ and Q^-.

3. Objective function

The operation of the transmission system must be performed in such a way as to ensure the minimal value of objective function

$$J = \sum_{k=1}^{K} N_k \tag{3}$$

where: k - number of acting compressor stations.
Under the assumption of polytropic compression the energy required to maintain a specific compression ratio for a specified flow Q is given by:

$$N = \frac{m_1 \, p_A \cdot Q}{\eta_m (m_1 - 1)} \left(\left(\frac{p_B}{p_A} \right)^{\frac{m_1 - 1}{m_1}} - 1 \right) \tag{4}$$

where: m_1 - the polytropic exponent of compression
η_m - the mechanical efficiency.

From the other point of wiev, storage of gas in the system pipelines should be as large as possible. High storage will ensure higher reliability of transportation process - in a case of pumping station failure the pipelines act as tanks. These two objectives conflict with each other - improved achievment with one objective can only be accomplished by the expense of another.
Storage in the pipe is computed using the following formula

$$S = \frac{p \cdot V T_n}{Z T \, p_n} \tag{5}$$

where: p - average pressure in the pipe,
V - volume of the pipe
T_n, p_n temperature and pressure of gas under NTP conditions.

The most important component of performance function is connected with the customer losses due to too low gas pressure at the point of delivery. This losses can be calculated using half-quadratic function

$$C(p,p_d) = \begin{cases} 0 & \text{if} \quad p \geqslant p_d \\ (p-p_d)^2 & \text{if} \quad p < p_d \end{cases} \tag{6}$$

where: p_d - desired pressure at the point of delivery

4. Steady-state simulation

Any solution of a steady-state flow must satisfy two physical laws - Kirchoff's laws, and flow equation.

- The algebraic sum of flows in every node is equal to zero, i.e.

$$\underline{A}_1\underline{Q} = \underline{q} \tag{7}$$

where:

$\underline{Q}$ = col $[Q_1, Q_2, \ldots, Q_m]$ - vector of flows in branches
$\underline{q}$ = col $[q_1, q_2, \ldots, q_{n-n_1}]$ - vector of loads in nodes
$\underline{A}_1 = [a_{ij}]_{(n-n_1)xm}$

n - number of nodes
n_1 - number of nodes with known pressures
m - number of branches.

- The algebraic sum of pressure drop in any loop is equal to zero, i.e.

$$\Delta\underline{p} = \underline{A}^T\,\underline{P} \tag{8}$$

where:

$\Delta\underline{p}$ = col $[\Delta p_1, \Delta p_2, \ldots, \Delta p_m]$ - vector of pressure drops
$\Delta p_k = p_i^2 - p_j^2$ (k - number of branche incident with nodes i and j)
$\underline{P}$ = col $[P_1, P_2, \ldots, P_n]$ - vector of node pressures
$P_i = p_i^2$
$\underline{A} = [a_{ij}]_{n \times m}$

- The pipe flow equation (1) can be expressed in matrix form

$$\Delta\underline{p} = f(\underline{Q}) \tag{9}$$

where: f - is a vector of flow functions for each branch.

Solution of the system must satisfy all the above equations (7),(8), (9).
Efficient simulation procedure was developed, it makes possible to simulate sufficiently quickly rather complicated real networks consisting of hundreds of nodes and branches. For details, see Osiadacz, 1982.

5. Control problem

Direct application of the optimal control theory leads usually to unacceptable solution. This is caused by the fact that the performance function formulated above cannot express formally all aspects of system operation. Therefore it is necessary to consider a decision maker (dispatcher of the system) as a part of control structure and apply the interactive and multiple-criteria optimization methods.

Analysing the goals of the system we can state that it is possible to introduce the following, hierarchically ordered groups of goals

- customer losses due to gas deficit
- system reliability (gas accumulation)
- direct operating cost (energy used for transmission).

These goals are ordered according to their importance.

From the other point of view, the first two goals can be treated as vectors

$$CL = \{ \varphi_1(p_1), \varphi_2(p_2), \ldots, \varphi_n(p_n) \} \quad \text{over all customers} \tag{10}$$

$$AC = \{ A_{11}, A_{12}, \ldots, A_{nm} \} \quad \text{over all branches} \tag{11}$$

where: CL is customer losses performance vector
AC is accumulation vector.

Taking into account the hierarchy of goals, we can formulate the following algorithm

- considering only the CL vector apply the interactive multiple criteria procedure to find the compromise distribution of deficits between the gas users. Using the trade-offs coefficients $\hat{\alpha}_i$ (obtained as a byproduct of the optimization algorithm) formulate the equivalent single-criterion performance function

$$CL_a = \sum_{i=1}^{n} \hat{\alpha}_i \varphi_i \tag{12}$$

- repeat the procedure considering only the AC vector and neglecting the rest; the equivalent single-criterion performance function

$$AC_a = \sum_{i,j=1} \hat{B}_{ij} A_{ij} \tag{13}$$

- using the above performance functions (i.e. CL_a and AC_a) and the energy accumulated in the system, find the compromise solution using the same procedure as in above steps.

6. Implementation

The efficiency of the procedure described above depends on the efficiency of the interactive multiple-criteria optimization algorithm. In our case the reference point optimization method was applied. This method (Grauer at all., 1982) utilizes the following idea:

- the decision maker specifies the reference point - desired values of all objective functions ("aspiration level")
- if this reference point is not attainable, the procedure calculates the nearest point from Pareto set,
- if the reference point is attainable, the procedure calculates the point from Pareto set maximizing the distance between this point and reference point,
- if the solution is not satisfactory, the decision maker specifies new reference point and the procedure is repeated.

This procedure, implemented on many computers was applied for solving many practical problems (e.q. Grauer at all., 1982). This procedure seems also to be the ideal tool for solving the gas transmission problem.

From the mathematical point of view this procedure performs a single--criteria optimization

$$\min S(\underline{q}-\hat{\underline{q}}) \tag{14}$$

where: $\underline{q}$ - is a vector of objectives,

$\hat{\underline{q}}$ - is a reference point,

S - suitably choosen <u>scalarizing function</u>.

In the simplest case

$$S(\underline{w}) = -\min_i w_i \tag{15}$$

Other forms of scalarizing functions were proposed by Wierzbicki (Wierzbicki, 1980).

7. Conclusions

The experimental software was developed for simple hypothetical system with 25 nodes, 36 branches, 3 compressor stations and 1 source. The experiments shown full applicability and high efficiency of the presented methodology. On the basis of existing experience the professional version is now being developed.

8. Literature references

M. Grauer, A.Lewandowski (1982) . The reference point optimization approach - methods of efficient implementation. International Institute for Applied Systems Analysis (IIASA), WP-82-26, Laxenburg, Austria.

M. Grauer, A. Lewandowski, L. Schrattenholzer (1982). Use of the reference level approach for the generation of efficient energy supply strategies. IIASA, WP-82-19.

A. Osiadacz (1982). Static and dynamic simulation of an arbitrarily configurated gas transportation system. Warsaw (in Polish).

A. Osiadacz, D.J. Bell (1981) . A simplified algorithm for large--scale gas networks. IFAC, Kyoto.

A. Wierzbicki (1980). A mathematical basis for satisfacing decision making. IIASA, WP-80-90.

HIERARCHICAL PRODUCTION PLANNING AND CONTROL

C. Harrison and Dr. P. J. O'Grady
Department of Production Engineering and Production Management
University of Nottingham
Nottingham
United Kingdom
Tel. (0602) 506101 Telex 37346 UNINOT G

ABSTRACT

The nature of decision making in manufacturing organisations is traditionally hierarchical. This enables decision models at each level to be constructed which can be solved by the use of current techniques, while facilitating overall coordination of the decision making process. The modern manufacturing environment demands control and planning methodologies suited to the flexibility and responsiveness required of organisations. This paper considers some existing techniques and proposes a method for modern manufacturing control.

1. INTRODUCTION

The management of a manufacturing system involves decisions at separate levels of the control structure. The decisions at each level will be concerned with different objectives and planning horizons. As the chain of control passes down through the management structure the objectives become more specific and the planning horizons shorter. The advent of modern manufacturing approaches, such as flexible manufacturing systems, has resulted in reductions in lead times making manufacturing systems more responsive and enabling reductions in the planning horizons at all levels. The need for fast, efficient, informed decision making in such systems naturally leads to the choice of hierarchical computer control. This hierarchical approach consists of planning and control at two distinct levels, the medium term and the short term. The medium term planning process is concerned with producing a viable schedule of work over the planning horizon. This schedule becomes the input to the short term planning process where ordered sequences of jobs are allocated to specific processors. This paper considers the optimisation of the decision model at each level and the interfacing between levels.

2. MEDIUM TERM PLANNING

The medium term planning process is concerned with determining a viable plan of the quantities to be produced over the planning horizon given certain resources and facilities and given a, perhaps very inaccurate, forecast of demand. Several methods have been proposed to aid the medium term planning process.

The approach that has probably gained most attention is that of linear programming; in its usual form it has the advantage of a conceptually simple model and the relatively straightforward inclusion of restraints. The use of readily available mathematical software is also an advantage. However the necessary approximation to linear function and the assumption of deterministic behaviour are major disadvantages. Under stochastic conditions some research [Gunther (1981)] would indicate that L.P. models perform worse than some other models.

A second major approach is that of the HMMS Linear Decision Rule (LDR) developed by Holt et al. (1960). The HMMS model is one of a simple aggregate (or simple product) system with no manufacturing delay and with the production rate and work force level as decision variables. The simple LDR solution and the ability to handle stochastic elements [Hausman and McClain (1971)] are major advantages of the HMMS LDR approach. Furthermore the results of Vergin (1980) suggest that the HMMS LDR performs better than several other approaches. However the HMMS LDR in its original form suffers from two major disadvantages. Firstly it is only applicable to an aggregate system with no manufacturing delay and, secondly, the cost formulation is inflexible.

Recently the attention of some researchers has turned towards heuristic switching methods. Elmaleh and Eilon (1974) set control levels on the inventory whereby if the inventory passes a control level then a change in the production rate is triggered. Although the approach has the advantage of simplicity, its main disadvantage is that it is only applicable to an aggregate system with no manufacturing delay.

Byrne and O'Grady (1982) use a heuristic switching approach to schedule a multi-product system. In this approach each product is given a priority based on the expected stock level over the planning horizon. Products in their usual batch sizes are then scheduled in priority order until the capacity limit is reached. The main advantages of this approach are its low computing requirements, its simplicity of operation and the lack of restriction on the cost function. The main disadvantage is that the approach only gives an approximately optimum solution. This is not thought to be a major disadvantage in many industrial concerns where data is often inaccurate or unavailable and approximations have to be made in any case.

3. SHORT TERM PLANNING

The complex nature of the problem of short term planning (STP) excludes analytical or enumerative techniques. The most successful solution method to date has been to apply a simple sequencing rule to the queues of jobs at each machine. Each waiting job is assigned a priority on some basis and the jobs are selected for processing with respect to this priority. Consequently, these rules are often referred to as priority rules.

A large number of such priority rules have been formulated and compared. Many authors [see for example Rochette and Sadowski (1976)] cite the shortest operation time (SOT) rule as most effective in reducing mean flow, idle and waiting times. This assigns highest priority to the job with the shortest operation time on the processor in question. However, in a system where operation times have a large variance, the SOT rule delays the jobs with longer operation times until all others have been processed. Alternatively, the objective of the STP may be different to those which SOT optimises. Either of these may be unacceptable in practice.

Panwalkar and Iskander (1977) review 36 key works which compare the performance of different sequencing rules. A total of 113 rules are covered by these reviews but no definite 'best' rule is found. This was due to the wide variety of conditions under which they were tested and to the large number of different measures used to assess the rules' performance. Panwalkar and Iskander conclude that it is best to select the review which uses test conditions that most closely match operating conditions and objectives. The results of the selected review can then be used to determine the best rule for the given situation. However, the dynamic nature of the practical STP problem can cause difficulties. Changes in product mix, machine breakdown and a large number of other factors combine to cause the character of the problem to change. From the work of Panwalkar and Iskander it would be reasonable to assume that a sequencing rule which produced optimal results at one time would not necessarily be optimal at a future time, when either the conditions in the shop or the objectives had changed. Thus to ensure that the most suitable rule is being used at all times the state of the system would have to be continually monitored and a suitable rule used. However, this relies on the existence and knowledge of reviews which cover all the likely operating conditions of the shop, which is unlikely in practice.

Hershauer and Ebert (1975) construct a rule which determines priority as a linear combination of any number of decision variables (e.g. imminent operation time, slack, due date etc.). A computer search routine is used to determine the weightings used in the linear sum such that the resulting combinatorial rule gives near optimal values of a predefined objective function. The value of the objective function corresponding to each set of weightings is determined by simulation. This rule has a general applicability because, although Hershauer and Ebert use a linear combination of performance measures as their objective function, any objective function can be defined. The disadvantages of Hershauer and Ebert's approach are twofold. Firstly, for large systems, the computational requirements of the search routine may become prohibitive, indicating the need for some form of decentralised control. Secondly, the decision variables which are to be included in the linear combination have to be selected by some process. This second problem may be overcome by changing the format of the linear combination so that it is a sum of the basic quantities which make up the decision variables. A linear combination of operation times, due dates and costs would be capable of representing combinations of most of the popular priority rules, such as slope, slack, SOT etc., without altering the basic concept expressed in Hershauer and Ebert's work. Furthermore, the approach of Hershauer and Ebert has a major advantage over the single priority rule approach since, by use of the computer search routine, it can be adapted to most situations.

4. HIERARCHICAL PLANNING AND INTERFACING BETWEEN LEVELS

The control and planning process for an entire industrial organisation is a large, complex problem. It requires decisions to be made on a number of different planning horizons and in differing degrees of detail. Traditionally this has been executed by a hierarchy of management. Each level makes decisions within a framework and to meet objectives defined by the level immediately above it. The results of these decisions then become the framework and objectives for the next level down the hierarchy. Top levels of the management structure make decisions which are broadly strategic and operate over a long planning horizon while in the lower levels more detailed decisions are made over shorter horizons within the broader strategic objectives. This process ensures overall coordination of the decision making machinery and produces planning problems at each level whose scope and detail remain within feasible bounds.

The currently available analytical techniques are not capable of solving the problem of complete, detailed control of an industrial organisation in a single level problem. Therefore, it seems reasonable to adopt an approach similar to the traditionally hierarchical management structure. Hax and Meal (1975) and Bitran, Haas and Hax (1981a and 1981b) propose the hierarchical production planning (HPP) methodology. By the use of suitable aggregation and disaggregation of the product items the planning problem at each level can be contained within reasonable bounds. Thus the long term planning (LTP) problem deals with product items in their most aggregated form. The level of aggregation decreases with the planning horizon until short term planning (STP), which is concerned with individual product items.

To ensure that the plans developed at each level of the hierarchy are feasible at the next level down careful problem formulation is required. Suitable aggregation and disaggregation procedures are also necessary. Bitran et al. (1981a) propose the use of effective demand, that which cannot be satisfied from initial inventory, at each stage rather than actual demand. They state that use of the latter is likely to lead to solutions which cannot be disaggregated. They then use the regular knapsack method (RKM) to disaggregate plans for one level into the requirements for the next lower level. Violations of constraints are reconciled by a trial and error reallocation on a penalty cost basis.

The problem at each level of the hierarchy is then formulated such that it can be solved by the use of current techniques. As the degree of detail increase and the planning horizon decreases, problems of different character are formulated to which specific solution techniques are suited. Bitran et al. (1981a and 1981b) and Hax and Meal (1974) use LP to solve the planning problem at higher levels within the hierarchy. In the final disaggregated level a heuristic which equalises the runout times (EROT) for items within a family is used.

While the HPP approaches outlined above are conceptually appealing the techniques applied at each level have certain practical disadvantages. Firstly, the approaches detailed do not solve the problem in sufficient detail for most manufacturing organisations. The final solution only delivers master schedules of product items for each period which do not violate the aggregate capacity constraints. In a truly integrated HPP system the final level would need to be the STP system concerned with job sequencing on production facilities. Also, such solution methods as LP do not necessarily perform well in practical situations where stochastic elements are encountered. In addition, the modelling of constraints as linear functions is often not a sufficiently accurate representation of reality. Finally, the computational requirements of LP can be excessive, especially when a large number of constraints are incorporated. We propose a HPP approach which overcomes the drawbacks to existing HPP approaches by using a switching algorithm at the MTP stage and a search sequencing method for STP.

PROPOSED HIERARCHICAL SYSTEM

We have defined the problem of production planning using a hierarchical approach therefore as involving decision making at the two major levels of the control structure: the medium term and the short term.

The methods that are used to obtain the production plan at these two levels determine the overall control structure. However as outlined above the existing proposed hierarchical approaches have major disadvantages in their ability to handle stochastic elements and in the considerable computation time involved in analysing many manufacturing systems. To overcome these disadvantages we propose in this paper the use of a hierarchical approach based on the Byrne and O'Grady (1983) switching algorithm for the medium term plan and the use of search sequencing [Hershauer and Ebert (1974)] for the short term plan.

The switching heuristic approach of Byrne and O'Grady (1983) has advantages over other medium term planning methods in term of computation time, the ability to handle multi-product systems and in the ease of use. The fact that the solution obtained is only approximately optimal is not considered to be a major problem where in many manufacturing concerns there is the presence of stochastic variables and there is the likelihood that the state of sections of the system is unlikely to be measured.

In a practical hierarchical planning system it is envisaged that the STP level would operate in real time. Currently available data collection systems enable up to date information about the state of the shop to be made available to the STP system. By using a technique similar to Hershauer and Ebert's (1975) a sequencing rule could be determined which gave near optimal results for the current shop state with respect to a predefined objective. Decisions made according to this rule could then be communicated via shop floor terminals. If the data acquisition system detected a change in the system state which invalidated the current sequencing rule the rule determination could be repeated using the new system state. The large computational demands of Hershauer and Ebert's method could be overcome by using such forms of decentralised organisation as group technology (GT), where the machines are organised into autonomous groups and very little, if any, inter group work flow occurs. This would enable each group to be solved independently of the others reducing the problem size and hence the computational requirements.

The hierarchy of control therefore operates by inserting data of the forecast of demand, work in progress and stock levels, batch sizes, safety stock margins, lead times and capacity limitations into the medium term stage. The output from this is a plan of work stretching over the planning horizon giving information about which products are to be produced in each time period. This information is then fed into the short term planning stage where the constraints and objectives are included in the search procedure proposed by Hershauer and Ebert (1975). The jobs are then sequenced on each machine facility to give detailed plans of the factory operation. Feedback from the shop factory can be used to update each plan.

CONCLUSION

We have proposed in this paper the use of a hierarchical approach to production planning and control using the switching heuristic approach of Byrne and O'Grady (1983) at the medium term level and the approach of Hershauer and Ebert (1975) at the short term level. Advantages of the hierarchy of control include low computing requirements, ease of use and the fine degree of final detail that results.

REFERENCES

BITRAN, G.R., HAAS, E.A., and HAX, A.C. (1982 (1)) 'Hierarchical Production Planning. A Single Stage System'. Opns. Res. v29(4) Pp. 717-742

BITRAN, G.R., HAAS, E.A., and HAX, A.C. (1982 (2)) 'Hierarchical Production Planning. A Two Stage System'. Opns. Res. v30(2) Pp. 232-251

BRYNE M.D. and O'GRADY,P. J. (1983) "A New Approach to Stochastic Control of Production Systems" Proceedings Symposium on Production Management, European Institute for Advanced Studies in Management, Brussels 1983.

ELMALEH, J. and EILON, S. (1974) 'A New Approach to Production Smoothing', Int. J. Prod. Res., Vol. 12, No. 6, 1974 Pp 673-681.

GUNTHER, H. O. (1981) 'A Comparison of Two Classes of Aggregate Production Planning Models under Stochastic Demand', Second International Working Seminar on Production Economics, Innsbruck, February 16-20, 1981.

HAUSMAN, W. and McCLAIN, J. L. (1971)'A Note on the Bergstrom-Smith Multi-Item Production Planning Model', Management Science, Vol. 17, 1971, Pp. 783-785.

HAX, A.C. and MEAL, H.C. (1975) 'Hierarchical Integration of Production Planning & Scheduling'. in 'Studies in Management Sciences' vol 1 Logistics Pp. 53-69 ed. M.A. Geisler pub. North-Holland-American, Elsevier.

HOLT, C.C., MODIGLIANI, F., MUTH, J. and SIMON, H.A. (1960) 'Planning Production, Inventories and Work Force', Prentice Hall Int., 1960.

PANWALKAR, S.S and ISKANDER, W. (1977) 'A Survey of Scheduling Rules' Opns. Res. v25(1) Pp. 45-61

ROCHETTE, R. and SADOWSKI, R.P. (1976) 'A Statistical Comparison of the Performance of Simple Dispatching Rules for a Particular Set of Job Shops' I.J.P.R. v14(1) Pp. 63-75

VERGIN R.G. (1980) "A New Look at Production Switching Heuristics for the Aggregate Planning Problem", Management Science, Vol. 26, 1980.

A CONTINUOUS TIME FORMULATION OF SERIAL MULTI-LEVEL PRODUCTION/INVENTORY SYSTEMS

José Luis MENALDI
Department of Mathematics
Wayne State University
Detroit, MI 48202
U S A

and

Edmundo ROFMAN
INRIA
BP 105 78153 LE CHESNAY CEDEX
FRANCE

1° INTRODUCTION :

In broad terms, multi-level production/inventory systems are concerned with production and/or inventory problems involving two or more interrelated activities.

The most common example of a multi-level inventory system is given by a distribution network for a family of products : it involves at the lowest level a number of retail outlets (i.e. stores) in business to satisfy customer demands for goods and which, in turn, act as customers of higher-level wholesale activities (i.e. warehouses). The wholesale activities themselves may be customers of still higher-level wholesale activities (i.e. factories). We can suppose several interconnection schemes for the activities. The simplest one is that in which each node (activity) has at most one predecessor (Fig. 1) ; it gives *an arborescent* system

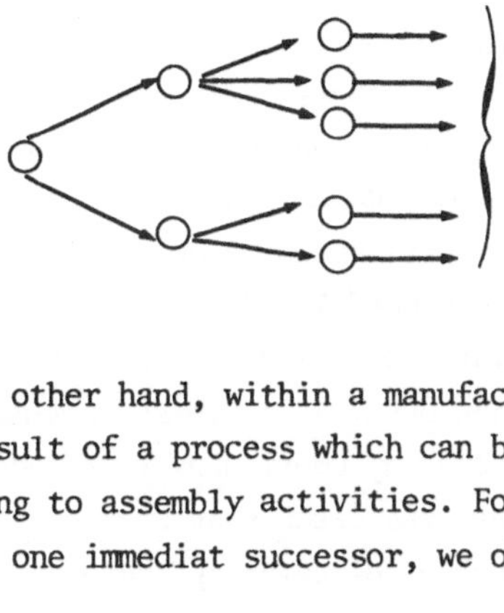

outside customer demand

Fig. 1

On the other hand, within a manufacturing context, a final product is some time the result of a process which can be decomposed into several levels, broadly corresponding to assembly activities. For simplicity, if we consider that each node has at most one immediat successor, we obtain *the assembly* system illustrate in Fig. 2

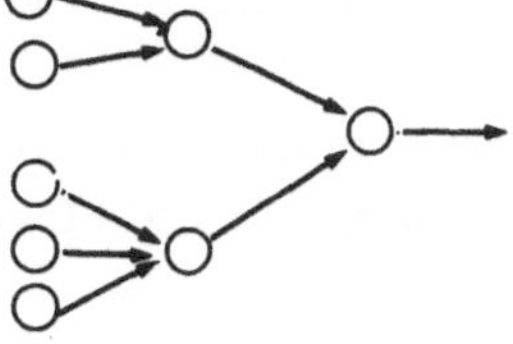

outside customer demand

Fig. 2

In this paper, we will discuss *serial* systems. In a serial system each node has at most one successor and at most one predecessor (Fig. 3). Such a system is both an arborescent and an assembly system.

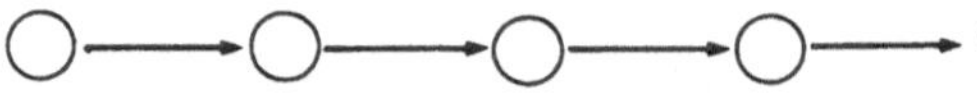

Fig. 3

Several inventory or/and production systems can be modelled by a serial system (Clark and Scarf [11] and Zangwill [21]) But, the main reason to discuss it is to obtain preliminary results allowing us to attack the study of more complex systems.

Let us remark that an extremely large and rich bibliography exists concerning the theory and the applications of multi-level systems.

As in many cases we will need the same approachs used to study single activity models, we will refer here to the surveys of such work given by Scarf [17], Veinnot [20], Aggarwal [1] and the recent book of Bensoussan, Crouhy and Proth [5].

For a general overview of contributions in the theory of multi-level production/inventory systems see Clark [10] (covering published results till 1971) and the book edited in 1981 by Schwarz [18] giving last results and a good picture of what was done in the 1970's.

At this point, let us make two general remarks concerning the whole research work attempted in the field of multi-level systems.

The first is that an extremely large quantity of papers concerns only discrete time models (also presented as *periodic review* models). One, but not the only reason, is the well known difficulty to discuss properly continuous time (*continuous review*) models. In fact, under similar assumptions the two models often require quite different methods to be studied. Consider, for example, the time-delay δ for satisfying an order and remark that the case of "no delivery lag" in periodic review models is not equivalent to the case $\delta = o$ in continuous review ones. As in these last models $\delta = o$ implies instantaneous delivery, in general we study continuous time models with the more real hypothesis $\delta > o$; but doing that the discussion becomes hard. On the other hand, in periodic review models the "no delivery lag" case has a sense and is the easiest one to be studied.

The second remark concerns the dichotomy between "centralized" and "descentralized" solutions.

The first results of the one-item multi-level inventory model involving uncertain demand were introduced in three papers by Clark [9] and Clark-Scarf [11], [12].

With their method the optimality of the system is in some cases achieved by a sequential process of determining optimal polices at each outlet activity (each node). Unfortunately, this nice "descomposition property" of the optimal global policy occurs rarely. Nevertheless, several descentralized procedures can be found in the bibliography. They give *suboptimal* global policies even if they are locally optimal in some suitable sense.

On the other hand, to consider centralized solutions introduces a new difficulty. Because the complexity and dimensionality of some systems, a rigurous recursive computation of such solutions is, in general, completely impractical. In the surveys above mentioned ([10] ; [18]) some efficient algorithms and heuristics are given for particular problems.

The two precedent remarks are close related with the main lines of our papers.

Let us consider for a single product a serial system with d installations (outlets) :

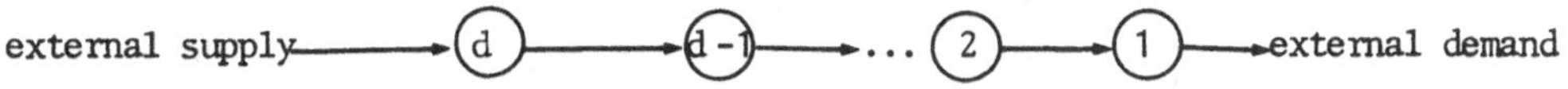

Fig. 4

The demand acts in the system at the lowest level (inst. 1) and at no other point in the system. Arrows show the sense of the supply of stock. Obviously orders follow the opposite sense.

Once the parameters and characteristics of the model have been specified (demand distribution, initial stock at each installation, holding and shortage costs, purchase costs, delay for delivering, capacity of each installation, excess demand is backlogged or not, etc) we consider the problem of determining optimal (centralized or descentralized) purchasing policies.

Discrete time models in a dynamic programming approach have been introduced to solve this problem by Clark [9] and Clark-Scarf [11] [12] . They assume stochastic external demand, convex (holding and shortage) costs, illimited capacity of installations. Shipping times and backlogging are allowed. After introducing the essential notion of "echelon" (at each installation an echelon consists of all stocks in the system at that installation and below, including all on-hand and in-transit amounts) they obtain an optimal descentralized policy if purchase costs are assumed linear while sub-optimal descentralized policies are proposed if fixed reorder costs are allowed.

Centralized solutions are not studied in those papers. The existence of a centralized solution for the discrete time models introduced in [11] [12] is proved in Bensoussan - Crouhy [5]

In Zangwill [21] the theory of concave cost network is applied to optimize a dynamic economic lot size production system under deterministic demand. A multi-echelon structure modelize the manufacture of a final product. No backlogging is allowed. Using a discrete dynamic programming approach an algorithm is proposed for calculating the centralized optimal production schedules.

Some other studies (always periodic review is assumed) have followed these major contributions of Clark-Scarf and Zangwill in serial systems. We will only mention an important paper by Bessler and Veinott [8], in which the period independence notion of Veinott [19] is adapted to the multi-echelon situation. (In their model at the beginning of a period each installation may also order stock from an exogenous source)

As is obvious from the title herein we will consider continuous time models. In the frame of the dynamic programming theory our purpose is to use the quasi-variational technique introduced by A. Bensoussan and J.L. Lions [6] to propose continuous review solutions. Recent numerical and theoretical contributions to the study of quasi-variational inequalities (QVI) will be useful for that. They will be refered later.

In § 2, we present a non stationary system with stochastic continuous external demand, holding and shortage costs (hypotheses on convexity or concavity are not necessary) while purchase costs contain fixed reorder costs.
Backlogging is allowed and fixed shipping times can be considered. The supplies (our controls) are of impulsive type. For this quite general formulation, the existence and properties of an optimal centralized purchase policy are given in Theorems 2.1 and 2.2.

In the simplified model of § 3, continuous and impulsive supplies are considered But we will assume linear purchase costs and no delay for delivering. Even if together these two assumptions give a too much unreal model, the descentralized policy obtained for it (Theorem 3.1) will be useful for understanding the asymptotic behaviour of more complex models.

In § 4 we take advantage of the results of § 2 and § 3 for discussing particular models. Numerical methods are proposed. In particular, for centralized policies, we emphazise the interest of using the general procedure introduced by González and Rofman [13]. The end of § 4 contains features of forthcoming papers.

2° CENTRALIZED SOLUTION

2.1 Notations and assumptions

Suppose d levels of decision in a hierarchic single product system. We denote by $(X_s\ ,\ s \geq o)$ the inventory on hand at the time s, precisely $X_s = (X_s^1, \dots, X_s^d)$ with the following meaning :

X_s^i : sum from level 1 to level i of the inventory on hand and in transit at the time s, i.e., inventory of echelon i at time s.

Under this notation, we must have :

(2.1) $X_s^1 \leq \ldots \leq X_s^d$ for every $s \geq o$,

but each X_s^i needs not to be positive.

The demand arises (see Fig. 4) at the lowest installation (i = 1), each level (i) places order to level (i + 1), i = 1 ; 2 ; ..; d-1 and the exterior supplies the highest installation (i = d). These purchasing decisions are made at any time and they may modify instantaneously the state of the system. However, the demand varies in a continuous fashion and some noises affect the system. Precisely, we assume that between two consecutive orders , each coordinate X^i is a one dimensional diffusion process with the same coefficients i.e.

(2.2) $dX_s^i = -b(s)\,ds + \sigma(s)\,dw(s-t) + d\nu^i(s-t)$, $s \geq t$

where the marginal demand distribution is characterized by the coefficients $(\sigma(s), b(s))$ and the orders (our control) are represented by the process $\nu = (\nu^i, i = 1,\ldots,d)$,

b(s) : mean of the marginal demand,

$\sigma^2(s)$: covariance of the marginal demand ; σ: intensity of the noises,

$(w(s), s \geq o)$: one dimensional standard Wiener process,

$(\nu^i(s), s \geq o)$: cumulative orders from level i.

The process ν involves the impulse control (sequence of random times θ_j and quantities ξ_j^i ordered at these times); precisely :

(2.3) $$\begin{cases} \nu^i(s) = \sum_{j=1}^{\infty} \xi_j^i \, \chi(o \leq \theta_j \leq s) & i = 1,\ldots, d \\ \xi_j^i \geq 0 \quad \theta_j \leq \theta_{j+1} \to \infty ; & j = 1, 2,\ldots, \end{cases}$$

$\nu^i(s)$ is the quantity of the item shipped to the installation i during the period of time [o , s] and χ is the characteristic function.

The processes $(X_s, \nu_s, s \geq o)$ are refered to a fixe probability space satisfying the usual conditions, in particular, they are assumed to be adapted to the Wiener process $(w_s, s \geq o)$, cfr. [16].

Since the process X_s may have discontinuities, we assume :

(2.4) $X_{s-}^{i+1} \geq X_s^i$, $\forall s \geq t \quad \forall i = 1, \ldots d-1$, which is equivalent to

(2.5) $\quad x_{i+1} + \nu^{i+1}(s-) \geq x_i + \nu^{i}(s)$, $\forall s \geq o$, $\forall i = 1,..,d-1$,

where $x = (x_1 \ldots, x_d)$ is the initial state, i.e. $X_t = x$ for the initial time t.

Remark 2.1 In what follows, the parameters x and t will be regarded as variables.

Let us introduce the measurable and non negative functions :

f (x, t) : cost of holding (or shortage) a quantity x at time t ;
k (x,t) : cost of ordering a quantity x at time t.

For any decision $\nu = (\nu^i, i = 1,..,d)$ it is associated a cost

(2.6)
$$J_{xt}(\nu) = E \{ \int_t^T f(X_s, s) \exp (-\int_t^s \alpha (\lambda) d\lambda) + \sum_{j=1}^{\infty} k(\xi_j, \theta_j) \exp (-\int_t^{\theta_j} \alpha (s) ds) : X_t = x\}$$

Where $\xi_j = (\xi_j^1, \ldots, \xi_j^d)$ and T is the horizon. Note that hypotheses as

(2.7)
$$\begin{cases} f(x,t) = \sum_{i=1}^{d} f_i (x_i , t) , & \forall x, t \\ k(\xi,t) = \sum_{i=1}^{d} k_i (\xi^i, t) , & \forall \xi, t \end{cases}$$

are not essential because we are working for centralized solutions.

The optimal cost function is

(2.8) $\quad \hat{u} (x, t) = \inf \{ J_{xt} (\nu) : \nu \text{ satisfying } (2.5)\}$

2.2 Characterization of the optimal cost function

A formal application of the dynamic programming permits us to obtain the following quasi-variational inequality (Q.V.I.)

(2.9)
$$\begin{cases} Au \leq f & \text{in} \quad \bar{\mathcal{O}} \times [o,T], \ u \leq Mu \quad \text{in} \quad \bar{\mathcal{O}} \times [o,T] \\ Au = f & \text{if} \quad u < Mu \end{cases}$$

to be satisfied by the optimal cost $\hat{u}$, where

(2.10) $\quad \bar{\mathcal{O}} = \{ x \in \mathbb{R}^d : x_i \leq x_{i+1} , i = 1, \ldots, d-1\}$

(2.11) $$Au = -\frac{\partial u}{\partial t} - \frac{1}{2}\sigma^2(t)\sum_{i,j=1}^{d}\frac{\partial^2 u}{\partial x_i \partial x_j} + b(t)\sum_{i=1}^{d}\frac{\partial u}{\partial x_i} + \alpha(t)u$$

(2.12) $$Mu(x,t) = \inf\{ k(\xi,t) + u(x+\xi,t) \quad \xi_i \geq o \ , \ \xi \neq o \ , \ x_i + \xi_i \leq x_{i+1}, \ i = 1,..,d-1\}$$

In general, the function $\hat{u}$ is not twice continuously differentiable; however $A\hat{u}$ can be regarded as a distribution in Schwartz' sense, i.e. for any smooth function with compact support ϕ ,

(2.13) $$\begin{cases} \langle A\hat{u}, \phi \rangle = \int_{\mathcal{O}\times]o,T[} \hat{u}\, A^*\phi \, dx\, dt , \\ A^*\phi = \frac{\partial \phi}{\partial t} - \frac{1}{2}\sigma^2(t)\sum_{i,j=1}^{d}\frac{\partial^2\phi}{\partial x_i \partial x_j} - b(t)\sum_{i=1}^{d}\frac{\partial\phi}{\partial x_i} + \alpha(t)\phi \end{cases}$$

Let $C_p(\mathbb{R}^d x[o,T])$ be the space of all continuous functions h satisfying :

(2.14) $$\begin{cases} \text{for every } \varepsilon > o \text{ there exists a constant} \\ C = C(\varepsilon, h, p) \text{ such that for any } x, x' \text{ in } \mathbb{R}^d, t, t' \text{ in } [o,T] \text{ we have} \\ |h(x,t) - h(x',t')| \leq \varepsilon \ (1 + |x|^p) + |t|^p + C\,(|x-x'|^p + |t-t'|^p) \end{cases}$$
if $p > o$.

(2.15) When $p = o$ we obtain the space of all uniformly continuous and bounded functions.

<u>Theorem 2.1</u>

Assume the conditions :

(2.16) $b(t)$, $\sigma(t)$, $\alpha(t)$ are bounded and uniformly continuous in $[o,T]$,

(2.17) either T is finite or $\alpha(t) \geq \alpha_o > o$ for every $t \geq o$,

(2.18) $k(\xi,t)$ is lower semicontinuous and $k(\xi,t) \geq k_o > o$, $\forall t$, $\forall \xi \neq o$,

(2.19) $$\begin{cases} f \in C_p(\mathbb{R}^d x\, [o,T]) \quad \text{and} \\ f(x,t) \geq f_o\,(1 + |x^+|^p) \quad \forall x \in \mathbb{R}^d, \text{ with} \\ x^+ = (x_1^+, \ldots, x_d^+) \ , \ f_o > o \end{cases}$$

hold. Then the optimal cost $\hat{u}$ given by (2.8) is the unique solution of the problem :

$$(2.20)\quad \begin{cases} \text{find } \hat{u} \text{ in } C_p\,(\bar{\mathcal{O}}\times[o,T]) \text{ such that} \\ A\hat{u} \le f \text{ in } \mathbb{D}'(\mathcal{O}\times]o,T[) \quad , \hat{u} \le M\hat{u} \text{ in } \bar{\mathcal{O}}\times[o,T] \\ A\hat{u} = f \text{ in } \mathbb{D}'([\hat{u} < M\hat{u}]) \\ \hat{u}(\cdot,T) = o \quad \text{if } T \text{ is finite }. \end{cases}$$

Moreover, $\hat{u}$ satisfies (2.19) for some other constant $f_o > o$.

The proof of Theorem 2.1 can be founded in [16] .

2.3 Construction of the optimal impulse control policy

We will obtain $\hat{v}$ after introducing the following definitions :

let $\xi(x,t)$ be a Borel measurable function satisfying

$$(2.21)\quad \begin{cases} M\hat{u}\,(x,t) = k(\hat{\xi}(x,t)\ t) + \hat{u}\,(x + \hat{\xi}\,(x,t),t) \\ \hat{\xi}(x,t) \neq o\ ,\ \hat{\xi}_i(x,t) \ge o \quad , \forall\ (x,t) \in \bar{\mathcal{O}}\times[o,T], \end{cases}$$

$$(2.22)\quad \begin{cases} dY_s^{o,i} = -\,b\,(s)\ ds + \sigma(s)\ dw\ (s-t) \quad , \quad s \ge t \\ Y_t^{o,i} = x_i \qquad i = 1,..,d\ , \end{cases}$$

$$(2.23)\quad \theta^o = o \quad , \quad Y_s^o = (Y_s^{o,1},\dots,Y_s^{o,d})\ ,$$

$$(2.24)\quad \begin{cases} \theta^{j+1} = \inf\{s \in [\theta^j,T] : \hat{u}(Y_s^j,\ s) = M\hat{u}(Y_s^j,\ s)\} \\ \theta^{j+i} = T \quad \text{if the above set is empty }, \end{cases}$$

$$(2.25)\quad \begin{cases} dY_s^{j,i} = -b(s)ds + \sigma\,(s)dw\ (s-t) \quad , \quad s \ge \theta^j \\ Y_{\theta^j}^{j,i} = Y_{\theta^j}^{j-1,i} + \hat{\xi}(Y_{\theta^j}^{j-1}\ ,\theta^j) \\ Y_s^{j,i} = Y_s^{j-1,i} \qquad o \le s < \theta^j \\ i = 1,..,d \quad , j = 1,2,..., \end{cases}$$

$$(2.26)\quad \hat{\theta}_j = \begin{cases} \theta^j & \text{if } \quad \theta^j < T \\ +\infty & \text{if } \quad \theta^j = T \end{cases} \qquad j = 1,2,..,$$

$$(2.27)\quad \begin{cases} \hat{\xi}_j = \hat{\xi}\ (Y_{\hat{\theta}j}^{j-1}\ ,\ \hat{\theta}_j) \qquad \text{if } \hat{\theta}_j < \infty \\ \hat{\xi}_j = (\hat{\xi}_j^1,\dots,\ \hat{\xi}_j^d\ ), \qquad j = 1,2,.. \end{cases}$$

and

$$\hat{\nu}^i_s = \sum_{j=1}^{\infty} \hat{\xi}^i_j \; \chi(o \leq \hat{\theta}_j \leq s) \quad , \; i = 1,\dots,d \tag{2.28}$$

and the following theorem holds :

Theorem 2.2

The impulse control $\hat{\nu}$ (constructed using (2.28)) associated with the continuation set of (2.20), is optimal.

Remark 2.2

As a rule, Theorem 2.2 means that we let the process evolves freely inside the region $[\hat{u} < M\hat{u}]$, called continuation set, and when it is going to exit, we take a decision determinated by the quantity $\hat{\xi}$ achiving the minimum of $M\hat{u}$.

Remark 2.3

Theorem 2.1 also applies if a delivery lag time, $\delta > o$, deterministic and unique for each installation, is considered. For the proof of this assertion, see [16]. If we assume a stochastic delay $\delta(\xi,t) > o$ the problem is much more complex. In a first analysis for the existence and characterization of the optimal value function $\hat{u}$, we can roughly consider the impact in (2.6) of this delay as a new ordering cost. So, if now $k(\xi,t)$ also includes this contribution, we will obtain that $\hat{u}$ is solution of (2.20)
Moreover, rules (2.21) —— (2.26) can be useful to define policies in which the delays are considered.

3 - A MODEL WITH OPTIMAL DESCENTRALIZED POLICY

We consider the same serial system of d installations of § 2 (see fig. 4). As before X^i_s denotes the inventory of echelon i at time s.

In this model we will consider controls (orders) involving a continuous part and an impulse part. They will be represented by the process $\eta_s = (\eta^1_s,\dots,\eta^d_s)$
The compatibility relations to be now satisfied are:

If $s' \geq s \geq t$ we must have for $i = 1, 2,\dots d$

$$\begin{cases} X^1_s \leq \;\dots\dots\; \leq X^d_s \\ \eta^i_{s'} \geq \eta^i_s \geq o \end{cases} \tag{3.1}$$

The evolution in time of each coordinate X^i_s is given by the one dimensional ITO's differential (cfr.(2.2)) :

$$(3.2)\quad \begin{cases} dX_s^i = -b(s)ds + \sigma(s)dw(s-t) + d\eta_s^i & s \geq t \text{ (initial time)} \\ X_t^i = x_i + \eta_t^i & x = (x_1,\dots,x_d), \text{ initial state} \\ & \text{with } x_1 \leq x_2 \leq \dots \leq x_d \,. \end{cases}$$

The coupling between succesive installations is defined as follows :

$$(3.3)\quad \begin{cases} X_{s-}^{i+1} \geq X_s^i \\ \eta_s^i - \eta_{s-}^i \leq X_s^{i+1} - X_{s-}^i & \forall s \geq t, \ \forall i = 1,2\dots d-1. \end{cases}$$

We remark that η_s^i and X_s^i are continuous from the right, while for each point they have left hand limit; in particular, $\eta_{t-}^i = o$, $X_{t-}^i = x_i$. Relations (3.3) means that at any time s, installation i cannot place an order exceeding the stock available at installation i+1. Thus, (3.3) is equivalent to :

$$(3.4)\quad \eta_s^i \leq (x_{i+1} - x_i) + \eta_{s-}^{i+1} \qquad \forall s \geq t, \ \forall i = 1,2\dots d-1$$

Now, as $X_{s-}^{i+1} - X_s^i = (x_{i+1} - x_i) + (\eta_{s-}^{i+1} - \eta_s^i)$, $\forall s \geq t$

we obtain that the coupling (3.3) (or its equivalent (3.4)) guarantees the first inequalities of (3.1).

For defining the cost function, we introduce the functions :

$$(3.5)\quad \begin{cases} c_i(s) & \text{cost per unit shipped at time } s \text{ to level } i, \\ f_i(X_s^i,s) & \text{storage and shortage cost at level } i, \\ f(X_s,s) = \sum_{i=1}^{d} f_i(X_s^i,s) & \\ \Phi(t,s) = \exp\left(-\int_t^s \alpha(\lambda)\, d\lambda\right) & \text{discount function ;} \end{cases}$$

then the pay off function is given by

$$(3.6)\quad \begin{aligned} J_{xt}(\eta) = E\Big\{\sum_{i=1}^{d} \Big(&\int_t^T f_i(X_s^i, s)\, \Phi(t,s)\, ds + c_i(t)\, \eta_t^i \\ &+ \int_t^T c_i(s)\Phi(t,s)\, d\eta_s^i\Big\} , \end{aligned}$$

where

E : the conditional expectation in the probability space where the processes are defined,

T : finite horizon ($T = \infty$ could be also considered)

$\int_t^T c_i(s)\, \Phi(t,s)\, d\eta_s^i$ is considered in the sense of Stieltjes.

Note that here we do not consider positive reorder costs. In this model, we could accept such cost at the highest level (level d) ; but for the sake of simplicity we neglect such a possibility.

From (3.6) the optimal cost is :

(3.7) $$\hat{u}(x,t) = \inf \{J_{xt}(\eta) : \eta \text{ admissible}\} .$$

Let us introduce our technical hypotheses :

(3.8) $b(t)$, $\sigma(t)$, $c_i(t)$, $\alpha(t)$ are non negative continuous functions on $[o, T]$; T finite.

(3.9) $f_i : \mathbb{R} \times [o,T] \to \mathbb{R}$ are non negative and convex in x_i $(i = 1,2,..d)$ with polynomial growth i.e, there exist constants $C \geq c \geq o$, $m \geq 1$ such that for every z we have
$$C(z^+)^m - C \leq f_i(z,t) \leq C\,(1 + |z|^m)$$

(3.10) The following estimate $|f_i(z,t) - f_i(z',t)| \leq \tilde{C}\,(1 + |z| + |z'|^{m-1})\,|z - z'|$ hold for every z,z', $i = 1 \ldots d$, some constant $\tilde{C}$ and the same exponent m of (3,9)

It will be useful to rewrite the pay-off function (3.6) as follows :

(3.11) $$\begin{cases} J_{xt}(\eta) = E\,\{\sum_{i=1}^{d} \int_t^T f_i(z_s^i + \eta_s^i, s)\Phi(t,s)ds + c_i(t)\eta_t^i + \int_t^T c_i(s)\Phi(t,s)d\eta_s^i \\ z_s^i = -\int_t^s b(\lambda)d\lambda + \int_t^s \sigma(\lambda)dw(\lambda - t) + x_i \end{cases}$$

with η admissible control

3.1 Characterization of the optimal cost

As in § 2 we will use the QVI technique. Now (as in remark 2.1) x and t will be considered as variables.

Denote by A and M the operators :

(3.12) $$\begin{cases} Av = -\frac{\partial v}{\partial t} - \frac{1}{2}\sigma^2(t)\sum_{i=1}^{d}\frac{\partial^2 v}{\partial x_i^2} + b(t)\sum_{i=1}^{d}\frac{\partial v}{\partial x_i} \\ Mv(x,t) = \inf\{v(x+\xi, t) + \sum_{i=1}^{d} c_i(t)\,\xi_i : \xi > o,\ \xi_i \geq o,\ x_i + \xi_i \leq x_{i+1}, \\ \qquad \forall i = 1,2,\ldots d-1\} . \end{cases}$$

Using a dynamic programming approach we can formally deduce that the optimal cost $\hat{u}(x)$ defined by (3.7) satisfies in $\bar{\mathcal{O}}$

$$(3.13)\quad \begin{cases} Au \le f \\ u \le Mu \\ u(.,T) = o \end{cases}$$

where $\mathcal{O} = \{(x_1, x_2,\dots x_d) \; ; \; x_1 < x_2 < \dots < x_d\}$ and the boundary conditions are determined by the same QVI (3.13) after reducing the dimension. Moreover, similar to a paper of the authors [15] and to Menaldi-Robin [14], we can show that $\hat{u}$ is actually the maximum solution of (3.13) in a suitable sense. For the proof see our paper [16].

3.2 Decomposition of the optimal cost

As it was said in the introduction our purpose is at this point to show the following descompositions of $\hat{u}$:

$$(3.14)\quad \hat{u}(x_1, x_2,\dots,x_d, t) = \hat{u}_1(x_1, t) + \dots + \hat{u}_d(x_d, t)$$

where $(\hat{u}_i, i = 1,2,..d)$ are the solutions of certain one-dimensional state problems that will be introduced immediately. Then an optimal descentralized policy will be deduced from (3.14).

Let us define the function :

$$(3.15)\quad \hat{u}_1(z,t) = \inf \{J^1_{zt}(\eta^1) \; ; \; \eta^1\}$$

with

$$(3.16)\quad \begin{cases} J^1_{zt}(\eta^1) = E^{zt}\{ \int_t^T f_1(Z_s + \eta^1_s, s)\, \Phi(t,s)\, ds + c_1(t)\, \eta^1_t + \int_t^T c_1(s)\, \Phi(t,s) \\ \qquad d\,\eta^1_s\} \\ Z_s = z - \int_t^s b(\lambda)\, d\lambda + \int_t^s \sigma(\lambda)\, dw\,(\lambda - t) \;, \quad (Z \in \mathbb{R}). \end{cases}$$

On the other hand, denote V_m the set of functions $v(z,t)$ satisfying (3.9), (3.10) and consider the problem :

Find $u_1\,(z,t)$ in V_m such that

$$(3.17)\quad \begin{cases} A_z\, u_1 \le f_1 \quad \text{in } \mathcal{D}'(\mathbb{R}) \\ u_1 \le M_1\, u_1 \\ u_1(.,T) = o \end{cases}$$

where $\mathcal{D}'$ is distribution space,

$$A_z = -\frac{\partial}{\partial t} - \frac{1}{2}\sigma^2(t)\frac{\partial^2}{\partial z^2} + b(t)\frac{\partial}{\partial z} \quad \text{and}$$

$$M_1\, w(z) = \inf \{w(Z+\mu) + C_1\mu \;;\; \mu \geq o\} \quad \text{with } w(z) \text{ a non negative function in } \mathbb{R}\,.$$

Proposition 3.1

Under the assumptions (3.8), (3.9), (3.10) the optimal cost $\hat{u}_1(z,t)$ defined by (3.15), (3.16) is the maximum solution of the QVI (3.17). Moreover, the function $\hat{u}_1(z,t)$ is twice continuously differentiable in z and satisfies.

$$(3.18)\quad \begin{cases} A_z\hat{u}_1 = f \quad \text{if } z \geq z^*(t) \\ \dfrac{\partial \hat{u}_1}{\partial z}(z^*(t),t) = -c_1(t)\,, \quad \text{in} \quad o \leq t < T \\ \hat{u}_1(z,t) = \hat{u}_1(z^*(t),t) + c_1(t)\,(z^*(t)-z) \quad \text{if } z < z^*(t) \end{cases}$$

where $z_1^*(t) = \inf\{\, z \in \mathbb{R} : \frac{\partial \hat{u}}{\partial z}(z,t) + c_1(t) > o\}$ □

By induction we define the following functions

$$(3.19)\quad \Lambda_i(z,t) = \begin{cases} f_i(z,t) - A_z\hat{u}_i(z,t) & \text{if } z \leq z_i^*(t) \\ o & \text{o therwise} \end{cases}$$

$$(3.20)\quad \tilde{f}_i(z,t) = f_i(z,t) + \Lambda_1(z,t) + \ldots + \Lambda_{i-1}(z,t)$$

$$(3.21)\quad \hat{u}_i(z,t) = \inf\{\mathcal{J}_z^i(\eta^i) \;:\; \eta^i\}$$

$$(3.22)\quad z_i^*(t) = \inf\{\, z \in \mathbb{R} : \frac{\partial \hat{u}_i}{\partial z}(z,t) + c_i(t) > o\,\}$$

with $\tilde{\mathcal{J}}_z^i(\eta^i)$ given like (3.16) with f_1, c_1, η^1 replaced by $\tilde{f}_i$, c_i, η^i respectively.

As $\Lambda_1(z,t)$ is non negative and convex in z, the function $\tilde{f}_2$ satisfies (3.9), (3.10). Therefore, by means of proposition 3.1, $\hat{u}_2(z,t)$ have similar properties than $\hat{u}_1(z,t)$. Then, in general, we can state :

Proposition 3.2

Assume the conditions (3.8), (3,9), (3.10), and also $Z_i^*(t)$ continuous in [o,T]. Then, the optimal cost $\hat{u}_i$ (z,t) defined by (3.21) is the maximum solution of the following QVI :

$$(3.23) \quad \begin{cases} \text{Find } \hat{u}_i(z,t) \text{ in } V_{\hat{m}} \text{ such that} \\ A_z u_i \le \tilde{f}_i \quad \text{in } \mathbb{D}'(\mathbb{R}) \\ u_i \le M_i\, u_i \quad \text{in } \mathbb{R} \quad . \end{cases}$$

Moreover, the function $\hat{u}_i(z,t)$ is twice continuously differentiable in z and satisfies

$$(3.24) \quad \begin{cases} A_z \hat{u}_i = \tilde{f}_i \qquad \text{in}[z_i^*(t), +\infty) \\ \dfrac{\partial \hat{u}_i}{\partial z}(z_i^*(t),t) = -\, c_i(t) \quad , o \le t < T \\ \hat{u}_i(z,t) = \hat{u}_i(z^*(t),t) + c_i(t)\,(z^*(t)-z) \quad \text{if } z < z^*(t) \quad . \end{cases}$$

Remark 3.1

For every i = 2 ,...d is

$$\tilde{f}_i = f_i + \sum_{j=i}^{i-1} (f_j - A_z \hat{u}_j) \qquad \square$$

Now, we can state one of the main results of our analysis

Theorem 3.1

Let the assumptions of Proposition 3.2 hold. Then under the notation (3.7), (3.15), (3.21), we have the descomposition property

$$(3.26) \quad \hat{u}\,(x_1, x_2, \ldots, x_d, t) = \hat{u}_1(x_1, t) + \ldots + \hat{u}_d(x_d, t)$$

for every $x = (x_1, x_2, \ldots, x_d)$, $x_1 \le x_2 \le \ldots \le x_d$ $\square$

3.3 De composition of the optimal centralized policy

The descomposition of the optimal d- dimensional state value function given by (3.26) is complemented by a result concerning the optimal centralized policy. In fact, for each one of the problems (3.15), (3.21) we are able to construct an optimal control $\hat{\eta}_t^i$ given by :

$$(3.27)\quad \begin{cases} \hat{\eta}_t^i = \max\{(z_i^*(t) - z + \int_t^s b(\lambda)d\lambda - \int_t^s \sigma(\lambda)\, dw\, (\lambda-t))^+ : 0 \le s \le t\} \\ \qquad\qquad \text{if } z \ge z_i^*(t) \\ \hat{\eta}_t^i = (z_i^*(t)-z) + \max\{(\int_t^s b(\lambda)d\lambda - \int_t^s \sigma(\lambda)dw\,(\lambda - t))^+ : 0 \le s \le t\} \\ \qquad\qquad \text{if } z < z_i^*(t) \end{cases}$$

with $z_i^*(t)$ given by (3.22); i.e.

$$\begin{cases} \hat{u}_1(z,t) = J_{zt}^1(\hat{\eta}^1) \\ \hat{u}_i(z,t) = \mathcal{J}_{zt}^i\,(\hat{\eta}_i) \qquad i = 1,\dots,d \end{cases} .$$

Moreover, we can also prove that from (3.27) we can construct an admissible control $\hat{\eta}$ giving the optimal policy for (3.7), i.e. :

$$(3.28)\qquad \hat{u}\,(x,t) = J_{xt}(\hat{\eta}).$$

Roughly, such optimal control $\hat{\eta}$ can be described by the follow feedback law : at time t, for each i = 1, 2,..., d for which $z_i(t) < z^*(t)$ we order up to reach the stock $z_i^*(t)$. If the necessary quantity is not avalaible at level i+1, we order the maximum disposable. Then, at each level, we order in a continuous fashion just to avoid that z_i becomes smaller than z_i^*

In practice, it is not difficult to solve, at each level i, the partial differential equations (3.24). In particular, for stationary problems, we use current numerical algorithms for ordinary differential equations. See, e.g.[15] in which a stationary problem with two installations is completely solved.

3.4 Comment of the proofs

The propositions and theorems above stated have been proved for the stationary case in [15] . To prove them as they are here presented we can in several cases to adapt recent results presented in Menaldi-Robin [14]. See [16] for a more detailled version of the proofs.

4° <u>OTHER MODELS AND NUMERICAL TECHNIQUES</u>

The variational approach used in § 2 for the characterization of the optimal value function can be applied in more complex models. For example, if a purchasing policy includes continuous and impulsive orders, with different unit costs and reorder cost for the impulsive part, we can formulate the QVI relative to these new assumptions.

It is also possible to consider installations with finite maximum capacities. In fact, after introducing stopping time controls the problem can be modelized by means of suitable QVI taking into account this capacity upper bounds.

A final obvious remark is concerning deterministic demand. In this case we will find first order differential operators in the definitions (2.11), (2.13) allowing a simplified discussion of the problem. We will obtain the same essential results of § 2.

Let us give some information about numerical solutions. Centralized solutions for deterministic models can be obtained using the procedure recently presented in Gonzalez-Rofman [13]. The method lean upon the characterization of the optimal value function $\hat{u}$ (x,t) as the maximum solution of a deterministic QVI similar to (2.20).

Numerical examples solved with the state x belonging to spaces of dimension 3 and 5 can be found in Bancora-Imbert et al [2] [3]

An the other hand, we know that descentralized solutions of the problem presented in § 3 are easily computed. If we consider this model giving the asymptotic behaviour of those described in § 2, its solution may give us useful information to obtain approximate solutions of short run models of type (2.1)—(2.6).

In a forthcoming paper, we will discuss short run arborescent models. We intend divide them in groups of activities for which centralized policies may be computed.

REFERENCES

[1] AGGARWAL , S.C. [1974],"A review of current inventory theory and its applications" International Journal of Production Research 12,(4) pp. 443-482

[2] BANCORA-IMBERT, M.C., GONZALEZ R. MIELLOU J.C. and ROFMAN E. [1983],"An algorithm to optimize energy-production systems in real time" in Tzafestas S.G. - Hamza M.H (Eds) : Advances in modelling, planning, decision and control of energy, power and environmental systems, Acta Press. pp. 65-67

[3] BANCORA-IMBERT, M.C., GONZALEZ, R., MIELLOU J.C. and ROFMAN E. [1984]: "Numerical optimization of energy-production systems" Rapport de Recherche INRIA, no 306.

[4] BENSOUSSAN A.- CROUHY M., [1980] "Multi-level control systems for inventory management" in Lainiotis D.G. and Tzannes N.S.(Eds) Applications of information and control systems, pp. 305-315, D. Reidel Pub. Company.

[5] BENSOUSSAN A., CROUHY M. and PROTH J.M, [1983], "Mathematical theory of production planning" North-Holland.

[6] BENSOUSSAN A.- LIONS J.L. [1973] ,"Nouvelles formulations des problèmes de contrôle impulsionnel et applications" C.R. Acad. Sc. Paris A-276 pp. 751-754.

[7] BENSOUSSAN A.- LIONS J.L. [1982]"Contrôle impulsionnel et équations quasi-variationnelles" DUNOD.

[8] BESSLER S.A. - VEINOTT A.F.Jr, [1966] "Optimal policy for a dynamic multi-echelon inventory model" Nav. Res. Log. Quart. 13 pp. 355-389.

[9] CLARK A.J. [1958] "A dynamic, single-item, multi-echelon inventory model" RM-2297, The Rand Corporation, Santa Monica, California.

[10] CLARK A.J., [1972], "An informal survey of multi-echelon inventory theory" Naval Research Logistics Quaterly, 19,(4) pp.621-650

[11] CLARK A.J.-SCARF H.E., [1960], "Optimal policies for a multi-echelon inventory problem, Management Science, 6 (4) pp. 475-490

[12] CLARK A.J.-SCARF H.E. [1962] "Approximate solutions to a simple multi-echelon inventory problem" Ch. 5 in Arrow, K.J., Karlin,S and Scarf, H. (Eds.) : Studies in Applies Probability and Management Science, Stanford Univ. Press.

[13] GONZALEZ R. - ROFMAN E. [1983] "On determistic control problems : an approximation procedure for the optimal cost" Proc. of the 22nd IEEE conference on decision and control, San Antonio, TX (1),pp. 150-154.

[14] MENALDI J.L.- ROBIN M., [1983] "On some cheap control problems for diffusion processes" Trans. of the Amer. Math. Society 278,(2), pp. 771-782

[15] MENALDI J.L. - ROFMAN E. [1982] "A continuous multi-echelon inventory problem" in D.E. Hardt (Ed.) : IFAC Information Control problems in manufacturing technology 1982" Pergamon Press. Oxford

[16] MENALDI J.L. - ROFMAN E. [1984]"Variational approach of multi-echelon inventory systems" Rapport de Recherche INRIA, to appear

[17] SCARF H. [1963] "A survey of analytic techniques in Inventory Theory" chap. 7 in Scarf H., Gilford D. and Shelly M. (Eds) : Multistage inventory models and techniques, Stanford Univ. Press.

[18] SCHWARZ L.B. (Ed) [1981] "Multi-level production/inventory control systems : theory and practice" TIMS Studies in the Management Sciences, North-Holland.

[19] VEINOTT A.F. jr. [1965] "Optimal policy for a multi-product, dynamic non stationary inventory problem" Management Sciences, 12, (3) pp. 206-222.

[20] VEINOTT A.F. [1966] "The status of mathematical inventory theory" Management Sciences, 12, pp. 745-777

[21] ZANGWILL W. [1969] "A backlogging model and a multi-echelon model of a dynamic lot size production system : a network approach" Management Science, 15, (9) pp. 506-527.

KNOWLEDGE REPRESENTATION AND ARTIFICIAL INTELLIGENCE IN THE CONTROL OF THE HIERARCHY PRODUCTION SYSTEMS

Ján Uličný, Oliver Moravčík, Zdenka Králová, Eugen Molnár, Ladislav Dráb
Slovak Technical University, Faculty of Electrical Engineering
Bratislava, ČSSR

Abstract

Complex systems presented today by hierarchy production systems working under conditions of uncertainty with subsystems of relative deciding authority and other control fields have started using artificial intelligence elements. We are pointing out the main utilization possibilities of formal intelligence in the control of complex systems and we will introduce one of the modern unconventional control task solution methods within the information control complex.

Key words: information control complex, solving system, recognition situation system, control system, solving realization system, functional activity mechanism, automatic theorem proving, predicate logic of the first order, internal model of outward environment, intelligence data base.

Introduction

Our complex production systems control concept results from the following assumptions: the production system has a hierarchy structure, the control task is solved from the centrum position depending on the centrum informations of the subsystems functional activity and also depending on the subsystems relative authority of solving its own tasks. The information control complex at the complex system centrum level compiles a control task solution plan for the whole complex system during the planning stage. The system is controlled at the plans realization stage by a tree scheme created upon informations obtained at this stage or, in the case of a extraordinary system situation, unmanageable for the information control complex, by man - the active element - interfering by means of a dialogue system question-answer.

Information Control Complex

The ICC function chart /diagram/ is shown in Fig. 1. Information acquisition and its transformation to input situations is done by the recognition situation system. We have a concept in process, containing the recognition situation system, for text informations by the help of a semantic model created on an analytic base.
We will start from the significant elements analysis of a certain notion field and from relations between these elements. By the conjunction of certain attributes and relations, higher degree notions can be made. In this way the recognition situation system is prepared for the input situation known as a sequenced information set with recognition importance on the information control complex input.
Information processing is performed in the logic of the first order and the solving unit is programmed in the artificial intelligence language LISP. The solving system, called the automatic task solving system, creates a control plan on the basis of the input situation, i.g. an operators tree graph for the control task solution. On the basis of the input information, which for example contains an objective function of g_o centrum and g_i subsystems, and further sources deviding the centrum and sources disposed by the complex systems subsystems, hierarchy activity conditions, method of their application and elements determinating the shape of the objective function, concrete values and parameters etc., are all determinative for the selection of the functional activity mechanism known as the regulation for the solving of the control task and its distribution of results. An example of this kind of regulation is an objective function optimization g_o u,x,z given in the following way:

$$\max_{u \in U} \min_{x \in X} \min_{z \in Z} g_o(u,x,z) \longrightarrow \hat{u} = [\hat{u}_1, \ldots, \hat{u}_n]$$

from the centrum position of the two leveled hierarchy structure, where the centrum solves the control task and the subsystem notes the result:

$$S_o \longrightarrow S_i : \hat{u}_i .$$

The application of the predicate calculus of the first order is closely connected to the elegant automatic theorem solving method, by which the functional activity mechanism application conditions are compared to the given input situation. After the automatic selection of the functional activity mechanism an appropriate selection of conventional methods of static and dynamic optimization follows for the sol-

Fig. 1: Block Scheme of the Information Control Complex

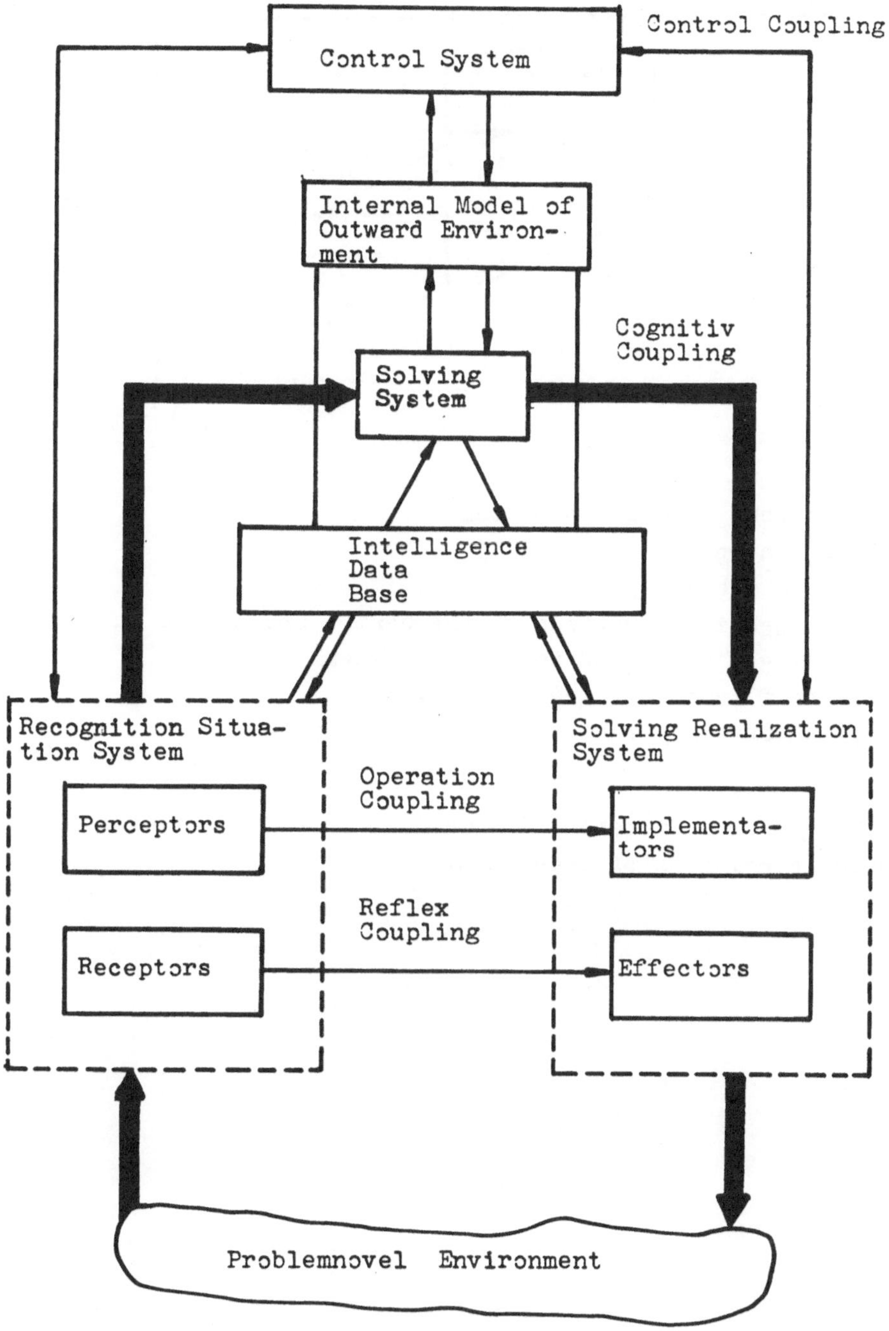

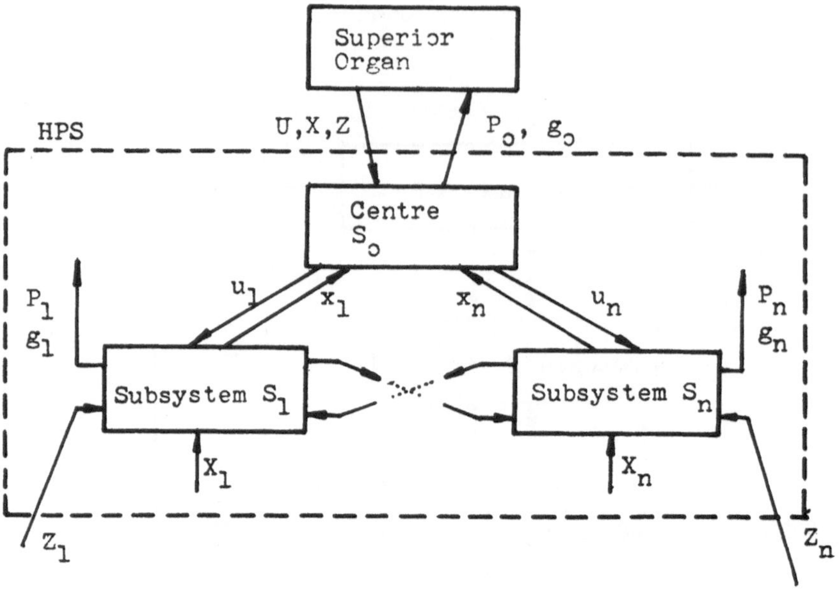

<u>Legend:</u>

HPS - Hierarchy Production System
S_o - HPS centre
$S_1,\dots,S_n$ - HPS subsystems
$g_o,\dots,g_n$ - objective functions of the centre and of the subsystems
$u_i, \{u_i \in U_i\}$ - exogen source of the S_i subsystem
$x_i, \{x_i \in X_i\}$ - self source of the S_i subsystem
$z_i, \{z_i \in Z_i\}$ - uncertainty factor of the S_i subsystem

Fig.2: Scheme of the Hierarchy Production System functioning under conditions of uncertainty

ving of the above given regulation example. That means that a operators sequence presented by the competent optimization conventional method is developed. This selection is also carried out by automatic theorem proving and by competent application conditions of a given optimization method. For the correct solving of the solution task we will need another operators for the control strategy, i.g. for the control tasks future results of time objective layout with regard to priority of each result by their implementation at the plans realization stage. The solved control task with its results and distributions is given to the realization solving system, which secures their application at the plans realization stage.
The control system presents the highest coordinated control system of the information control complex. This system also receives an input situation image at the planning stage, i.g. during the forming period of the deciding proces and its results it receives a so called awaiting result of the production systems functional activity from the solving system. During the realization period of the deciding activity, i.g. at the plans realization stage the control system receives by feedback a real image of the systems functional activity result by means of the situation recognition system. The system removes the disproporcions by itselft according to urgency or it commands the solving system to reevaluate the deciding process of those alterations, causing the disharmony of the systems awaiting result. The control system further secures contact between the coordinator and the complex systems information control complex. The coordinator by the help of the dialogue system DSQA /dialogue system question-answer/ has a chance to enter the process either during the deciding or control phase. The control system further controls the adherence of the deciding and control mode so that they will run in the frame of real time holding conditions. With the help of the DSQA the coordinator is acquainted with the extraordinary deciding and control situations. In unavoidable cases he gives orders, commands, or disposals by a so called substitutional control method used in the case that the system fails to give satisfactory informations about the influence of the environment on it. During feedback the control system verifies each of the information control complex subsystems functional activity from the standpoint of its correct responces on input demands. The solving realization system secures interconnections between the realization system and the hierarchy production system, i.g. it secures the solving of its results which continue to the planned functional activity, specified by a regulation, and therefore they are the solving systems responces to input situations. The solving realization system, by the

help of the two sequences of operators - control operators and operators for distribution of the control results-accomplished by the layout of the results of the solution control task among the elements of the hierarchy production system. This system also secures interconnections from the production system to the control system, from which it receives, during extraordinary situations, reserve commands to avert break-down states.
The utilizing possibility of some formal intelligence elements during the control of the described complex structures /see Fig.2/ is stimulated by the solid built internal model of the outward environment, being always adapted to the concrete controlled complex system. Parameters, informations and constants, to which the individual subsystems of the information control complex have access, are stored in the intelligence data bank. The basis of the intelligence data bank is the knowledge base, storing the regulation informations existing in the objective field. The construction of the knowledge base begins from the knowledge model, where relations between individual facts of the control field and its environment are data like the facts alone.

Conlusion

One of the most important facts of the complex production complexes control is real time, deciding about the complex production systems total economical efficiency.
The information control complex controlling this kind of a production system from the centrum is actually programmed as a program system but its processing development is dependent on the production system /technology and production operations, planning, deciding, limitary conditions/ in the same way as on preparation quality of the program system and on the applied digital computer.
Part of the program system solving the control task by the help of the formal intelligence elements is programmed in the LISP 1.5 artificial intelligence language being the interpretting language, considerably slowing down control activity. We are seeking for a satisfactory solution in this field, which will guarantee the compromise between the keeping of processing at real time conditions and the solving quality of the control task even though the complex production systems work with higher time units, like for example in technology processes.

References:

1. Uličný, J. - Thuan, N.T. - Baláž, J. - Dráb, L: Automatic Solutions Projecting of Controlling Complex in Automated Control Sys-

tems, In: Proceedings of the 3rd IFAC/IFORS Symposium, Large Scale Systems-Theory and Applications, Warsaw 1983

2.Uličný, J. - Moravčík, O. - Králová, Z. - Dráb, L.: The Information Control Complex of the Complex System on the Artificial Intelligence Base, In: Proceedings of the 5th States Conference AKYTEX, Bratislava 1983 /in slovak/

A HIERARCHICAL SYSTEM FOR CONTROL OF CONTINUOUS FERMENTATION PROCESS SYNTHETIZED ON THE BASIS OF THE LINGUISTIC APPROACH

G. Sotirov
Higher Institute for Mechanical and Electrical Engineering
D. Filev, K. Konstantinov
Research Laboratory of Instrumentation
and Biological Experiment Automation

Abstract: The complexity of the continuous fermentation processes makes difficult the synthesis of systems for their automatic control. In this paper, a method for hierarchical control of one of the most important parameters of the continuous fermentation process - the flow rate of the nutrient medium is suggested. The control in the loops for the flow rate of the nutrient medium - dissolved oxygen concentration is carried out at the first levelsupposing constant activity of the microorganisms. A fuzzy controller with nonequidistant discretization in the look-up table is synthetized. The problem of the control strategy in an activity change of the living cells is solved at the second level. Information about this change is obtained by means of check stoppings of the nutrient medium flow rate. Here we call the fermentation process CFP.

1. Introduction

The requirement for high economic efficiency of CFP calls for strict observation of the experimentally established optimal conditions. The total coverage of CFP by an optimal control system is still a problem to be solved. The difficulties here arise from the complexity of the processes of microbial synthesis which take place in the living cell. For this reason, from the point of view of the automatic control theory, CFP can be characterized as non-linear, non-stationary, ill-defined object of control. That is why there is still no satisfactory mathematical model for the purpose of automatic control which could adequately describe all the input-output relationships in CFP.

The up-to-day systems for automatic control of the CFP usually involve only the control over some of the parameters of the medium where the process takes place - pH, temperature and dissolved oxygen concentration (DO).

One of the ways to improvement of CFP automatic control systems is the development of methods for efficient control of the substrate flow rate, since both the productivity and the economical consumption of the nutrient substrate depend on it.

2. Application of the Fuzzy Control

In /1/, an approach for fuzzy controller synthesis on the basis of the linguistic approach is discussed. With membership functions selected by means of expert estimations and after transformation of the control strategy in a set of logical relations, the following look-up table of fuzzy controller was obtained:

↓ →	-2.0	-1.8	-1.6	-1.4	-1.2	-1.0	-0.8	-0.6	-0.4	-0.2	0	0.2	0.4	0.6	0.8	1.0	1.2	1.4	1.6	1.8	2.0
-5.0	0.3	0.3	0.3	0.3	0.3	0.3	0.3	0.3	0.3	0.3	0.3	0.3	0.25	0.25	0.25	0.25	0.25	0.25	0.2	0.15	0.1
-4.5	0.3	0.3	0.3	0.3	0.3	0.3	0.3	0.3	0.3	0.3	0.25	0.25	0.25	0.2	0.2	0.2	0.2	0.2	0.15	0.1	0.05
-4.0	0.3	0.3	0.3	0.3	0.3	0.3	0.3	0.3	0.3	0.25	0.25	0.2	0.2	0.2	0.2	0.2	0.15	0.15	0.1	0.1	0
-3.5	0.3	0.3	0.3	0.3	0.3	0.3	0.3	0.25	0.25	0.25	0.25	0.2	0.2	0.15	0.15	0.1	0.1	0.05	0.05	0	0
-3.0	0.3	0.3	0.3	0.3	0.25	0.25	0.25	0.25	0.25	0.2	0.2	0.2	0.15	0.15	0.15	0.1	0.1	0.05	0.05	0	0
-2.5	0.3	0.3	0.2	0.2	0.2	0.2	0.2	0.2	0.2	0.2	0.2	0.15	0.15	0.1	0.1	0.1	0.05	0.05	0	0	0
-2.0	0.2	0.2	0.2	0.2	0.2	0.2	0.2	0.2	0.2	0.2	0.15	0.15	0.15	0.1	0.1	0.1	0.05	0.05	0	0	0
-1.5	0.2	0.2	0.2	0.2	0.2	0.2	0.15	0.15	0.15	0.15	0.15	0.1	0.1	0.05	0.05	0.05	0.05	0	0	0	0
-1.0	0.2	0.2	0.2	0.15	0.15	0.15	0.15	0.15	0.15	0.1	0.1	0.1	0.05	0.05	0.05	0.05	0	0	0	-0.1	-0.1
-0.5	0.1	0.1	0.1	0.1	0.1	0.1	0.1	0.1	0.05	0.05	0.05	0	0	0	0	-0.1	-0.2	-0.2	-0.2	-0.2	-0.2
0	0.1	0.1	0.1	0.1	0.1	0.05	0.05	0.05	0	0	0	0	-0.1	-0.2	-0.2	-0.2	-0.2	-0.2	-0.2	-0.2	-0.2
0.5	0.05	0.05	0.05	0.05	0.05	0.05	0	0	0	0	-0.01	-0.1	-0.1	-0.2	-0.2	-0.2	-0.2	-0.3	-0.3	-0.4	-0.4
1.0	0	0	0	0	-0.1	-0.1	-0.1	-0.1	-0.2	-0.2	-0.2	-0.3	-0.4	-0.5	-0.6	-0.6	-0.6	-0.6	-0.6	-0.6	-0.6
1.5	-0.1	-0.1	-0.1	-0.1	-0.2	-0.2	-0.2	-0.2	-0.4	-0.5	-0.6	-0.6	-0.6	-0.6	-0.6	-0.6	-0.6	-0.6	-0.6	-0.6	-0.6
2.0	-0.1	-0.1	-0.1	-0.1	-0.2	-0.2	-0.2	-0.2	-0.5	-0.6	-0.7	-0.7	-0.7	-0.7	-0.7	-0.7	-0.7	-0.7	-0.7	-0.7	-0.7

→ discrete values of ΔE
↓ discrete values of E

Table 1

where E is the error for dissolved oxygen measured value DO in relation to the assigned DO_{SP}, ΔE - its variation, and the values in the look-up table - the variations of control (substrate flow rate). The block diagram of the control system is shown in Fig. 1.

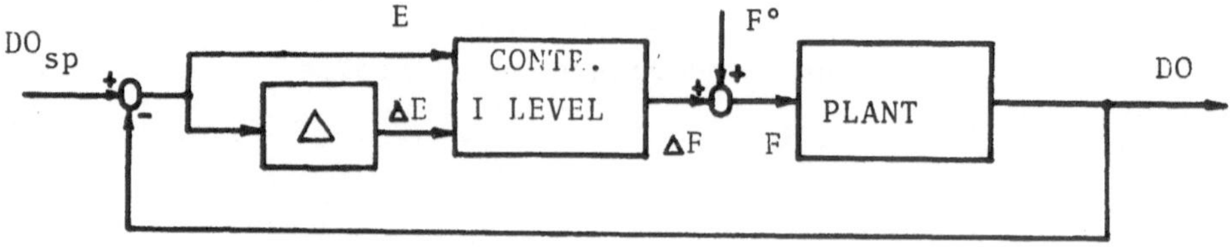

Fig. 1

In the case of synthesis, a constant microorganism activity is assumed. The derived controller compared to an optimal PI controller when the system is simulated, proved the advantages of fuzzy control.

3. The Concept of Hierarchy in a Fuzzy Control System

From technology point of view, the control of CFP is found to require taking into account the activity variations. Since during the biotechnological process it changes according to a law hardly liable to modeling and does not demonstrate a good reproductivity, it is of practical interest to design a strategy for set-point modification DO_{SP} on the basis of the linguistic approach. On these grounds, an algorithm for control on the second level of hierarchy is derived, providing for a supervisor modification of DO_{SP}, in compliance with the block diagram shown in Fig. 2.

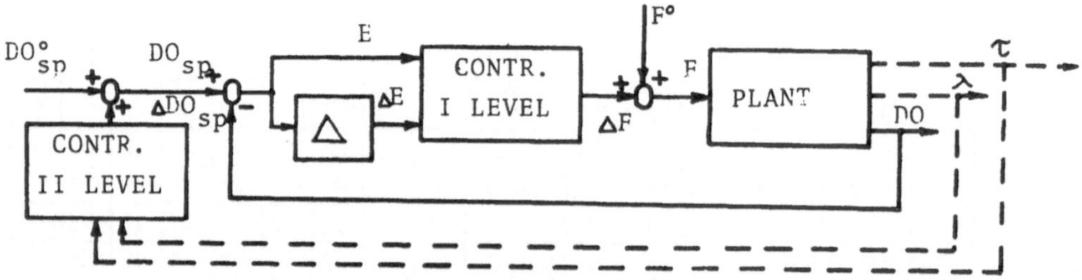

Fig. 2

The algorithm uses the information about DO changes upon stopping the substrate batching which is effected in fixed time intervals. Two typical cases are illustrated in Fig. 3. (see next page)

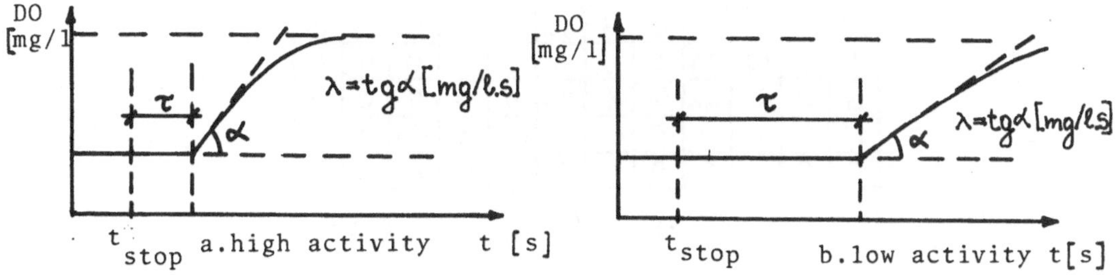

Fig. 3

From the characteristic derived one can estimate the activity of micro-organisms. It is directly proportional to the curve slope λ and inversely proportional to the time delay τ.

The type of the function is complex due to which both parameters are required for the correct determination of activity. According to the result of the experiment the set point of DO_{SP} is corrected in the necessary direction within the interval /2, 7/ mg/1.

The input variables for the fuzzy controller implementing this algorithm are τ [s] and λ [mg/1s], while the output one is ΔDO_{SP}. The expert estimation of the membership levels of the respective discrete values in the selected fuzzy variables gave the results shown in Table 2a, b, c (see next page).

The rules of manual control define defined by the operator have the following linguistic form:

1. If (τ is z and λ is z) or (τ is s and (λ is z or λ is s)) or or (τ is M and (λ is z or λ is M)) or (τ is G and (λ is z or or λ is s or λ is M), than ΔDO_{sp} is G.
2. If (τ is z and λ is s) or (τ is M and λ is M) or (τ is G and and λ is G), than ΔDO_{sp} is S.
3. If (τ is z and λ is M) or (τ is s and λ is M) or (τ is M and and λ is G), than ΔDO_{sp} is z.
4. If (τ is z and λ is G) or (τ is s and λ is G), than ΔDO_{sp} is N.

The procedure of calculation of the look-up table of control leads to the result shown in Table 3.

↱	5	6	7	8	9	10	11	12	13	14	15
Z	1	0.7	0.4	0	0	0	0	0	0	0	0
S	0	0	0.2	0.6	1	0.5	0	0	0	0	0
M	0	0	0	0	0.1	0.3	0.7	1	0.3	0	0
G	0	0	0	0	0	0	0	0.2	0.5	0.7	1

a.

→ discrete values of τ [s]

↓ fuzzy values of τ

↱	0	0.1	0.2	0.3	0.4	0.5	0.6
Z	1	0.8	0.3	0	0	0	0
S	0	0.1	0.4	1	0.2	0	0
M	0	0	0	0.4	1	0.8	0
G	0	0	0	0	0.4	0.6	1

b.

→ discrete values of λ [mg/l.s]

↓ fuzzy values of λ

↱	-1	-0.5	0	0.5	1.0	1.5
N	1	0.6	0.1	0	0	0
Z	0	0.1	1	0.1	0	0
S	0	0	0.2	1	0.1	0
G	0	0	0	0.1	0.7	1

c.

→ discrete values of ΔDO_{sp} [mg/l]

↓ füzzy values of ΔDO_{sp}

Used values:

N - negative, Z - zero, S - small, M - middle, G - great

Table 2

↓→	5	6	7	8	9	10	11	12	13	14	15
0	1.0	1.1	1.1	1.1	1.1	1.15	1.2	1.3	1.3	1.3	1.3
0.1	0.8	0.9	0.9	1.0	1.1	1.1	1.2	1.2	1.2	1.25	1.25
0.2	0.3	0.5	0.8	0.9	1.1	1.1	1.1	1.15	1.15	1.15	1.15
0.3	0	0.1	0.3	0.4	0.6	0.7	0.7	0.75	0.8	0.9	1.0
0.4	-0.2	0	0	0	0.15	0.2	0.3	0.5	0.5	0.6	0.8
0.5	-0.5	-0.3	-0.3	-0.1	0	0	0.2	0.3	0.4	0.4	0.6
0.6	-0.8	-0.75	-0.6	-0.5	-0.4	-0.3	0	0	0.1	0.3	0.4

→ discrete values of τ [s]
↓ discrete values of λ [mg/l.s]

Table 3

4. Conclusion

The method presented based on the already classical linguistic approach to fuzzy control and the principles of supervisor control combines all the advantages of manual and automatuc control. It increases the ability of the system for CFP and can be easily implemented in microprocessor real time control systems.

REFERENCES:

1. Filev D., Konstantinov K., Application of Fuzzy Control Theory to the Fermentation Process, Modelling, Identification and Control, Insbruck, Austria, 1984 (to appear)
2. 1st IFAC Workshop - Modelling and Control of Biotechnical Process, Abstr., Helsinki, Aug. 17-19, 1982.

OPTIMIZATION AND OPTIMUM CONTROL OF ORGANIZATION IN LARGE-SCALE SOCIETAL PLANNING SYSTEMS

J. Jerina
Zavod za družbeno planiranje Ljubljana

Ljubljana, Yugoslavia

Abstract: The purpose of this paper is to develop new methods for solving problems of integral societal planning based on system theory. Integral societal planning system is a large -scale system, constituted of subsystems which are in most cases nonlinear. Such a system must be stable and optimal in various cases of operation. We must ensure such operating of large-scale system. The basic algorithms are derived from automatic control theory and multilevel optimization methods are used.

INTRODUCTION

The scope of this paper is to describe the use of suitably constructed mathematical models in societal planning development systems and their application for describing the behaviaur of the whole planning system. Thereby three basic aspects of socio-economic development shall be taken into account simultaneously:

- social aspect
- economic aspect
- spatial aspect.

Accordingly, it is our task to pass from a classical type development planning, characterized by partial approach to each sphere (social, economic and spatial), to such planning type where all three spheres enter the system simultaneously. This approach is possible through the application of the system theory.

SYSTEM DESCRIPTION

It has been known that the societal planning development system is a large system consisting of several interdependent and interconnected subsystems. At analysing and solving problems of societal planning efforts have been taken to apply the already known methods and procedures from system theory. Nevertheless, the societal planning system cannot always be recorded in a standard form, therefore in such cases the method of system theory cannot be practiced either.

In principle, the societal planning system is a interconnected dynamic non-linear system. Each subsystem is to a certain extent autonomous and shall conform to the requirements for stability; the same requirements have been imposed to the whole system. The societal planning system structure changes in accordance with several factors (economic, monetary, political etc.) as well as with regard to both, the aims to be attained by the system and the restrictions (those related to space, power supply, raw materials etc.). In spite of this, the system shall remain stable and it shall conform to various requirements for attaining of aims. Considering the fact that the societal planning system is very large, the principle of decomposition can be used. This would however, lead to the concept of multilevel hierarchic control.

The development of each subsystem of the integral societal planning system depends on the values of outputs of all the subsystems. Some subsystems may contain time-delay elements. We often encounter the case when complete characteristic of some subsystems are only partially known. These characteristics often involve some difficulties in practical applications.

STABILITY

For studying the stability of large-scale dynamic systems, direct application of the Lyapunov method is out of place. In this situation such stability analysing methods are required which are based on structural conditions of the complex system. With regard to the above statement the vector Lyapunov function is acceptable for each unforced subsystem. Accordingly, the stability of the overall system is defined in the product space of all state spaces of subsystems. For the application of the above principle, it is, however, necessary to know the dynamics of each subsystem and to construct the Lyapunov function for every subsystem. This method is not always applicable because the societal planning system may consist of numerous subsystems for which the above conditions are not always known. In such cases it is recommended to use the concept of stability which is characteristic for large-scale systems. This concept is based on macroscopic stability of interconnected systems.

The societal planning system is a large-scale system consisting of subsystems and their interactions. In the sense of the above approach to the analysis of stability conditions the dynamics of each subsystem shall be described with a simple system, called comparison system.

Let us take a interconnected dynamic system S, consisting of N subsystems S_1, S_2, ..., S_N. Each subsystem is characterized by a control function u_i and output y_i as real scalar and continuous functions. In general, the input function u_i $(i=1, \ldots, N)$ is a linear combination of other subystem outputs:

$$u_i(t) = \sum_{j=1}^{N} \alpha_{ij} y_j(t-\mu_{ij}(t)), \quad \alpha_{ii}=0 \tag{1}$$

$$(i = 1,2, \ldots, N)$$

The time-lag $\mu_{ij}(t)$ between the subsystems S_i and S_j can be defined as a finite value:

$$0 \leq \mu_{ij}(t) < T, \quad T>0, \qquad (i,j = 1,2, \ldots, N). \tag{2}$$

The above approach does not require perfect knowledge of dynamic properties of subsystems. It is based on the definition of comparison system.

The system S_{ic} is called a comparison system of the subsystem S_i, if for each input-output pair $(u_i(t), y_i(t))$, $(t \geq 0)$ of the subsystem S_i a differentiable function $e_i(t)$, $(t \geq 0)$ is available, i.e.:

$$\dot{e}_i = - \alpha_i e_i(t) + \beta_i \left| u_i(t) \right| \tag{3}$$

$$\left| y_i(t) \right| \leq e_i(t), \qquad t \geq 0.$$

The comparison dynamic system S_{ic} is provided with a input - output pair $(v_i(t), e_i(t))$, which shall correspond to the following relation:

$$\dot{e}_i = - \alpha_i e_i(t) + \beta_i v_i(t). \tag{4}$$

The coefficients α_i and β_i are pozitive constants.

Considering the fact that the whole procedure including the proof has been treated in detail in (2) , this paper is designed to concentrate on how the results related to studies of system stability on macro level can be used for solving the societal planning system. As with societal planning system very few information is available about the state and the dynamics of the system, the concept that the state of a interconnected system S is defined with all subsystem output values, is, of course, welcome. Accordingly, the state of the system S in $t = t_1 > 0$ is defined with $y(t_1) \in R^N$. The system S state, defined in this way, is called macroscopic state and corresponding vector of the state $y = (y_1, \ldots, y_N)^T$ is called macroscopic state vector.

If macroscopic state variables $y_i(t)$, $(i = 1,2, \ldots, N)$ converge to zero for each initial state as t tends to infinity, the system S is called to be macroscopically asymptotically stable.

In systems like the societal planning system, this simple and general stability criterion is useful enough.

OPTIMIZATION

Optimization of large interconnected dynamic societal planning systems can be solved in several ways. As a rule, a system can be reduced by decomposition, whereby the problem of optimization is extended to a certain number of optimization procedures. An optimization problem shall be solved for each subsystem apart whereas a global optimum can be atained by coordination of individual solutions trough an iterative procedure. A local optimization is based on standard optimization procedures.

Let us analyze the requirement imposed by the question of optimum control and attaining of the required aim within the prescribed time:

- the system is dynamic and provided with variable coefficients,
- the coefficients are solutions of mathematical programmes whereof the restrictions are parameters which represent external restriction to the system of differential equations,
- in view of defining the aim we have to do with principles that come in questions with finite conditions in time T,
- in view of gradual optimization from one change of conditions to the other we have to deal with similar problems as those arising with dynamic programming.

The societal planning system requires that within the time T, foreseen for attaining the set aim, it is also necessary to answer the current optimum requirements within the time t_1 all with regard to variable coefficients of individual matrices. The solution of optimization problems is based on multilevel optimization, which is reasonable and economical with regard to the fact that the whole system is too extensive.

The problem of dynamic system control can be solved by the use of one of hierarchic optimization methods concerning aim coordination. As the societal planning system is treated as a complex system, consisting of interconnected subsystems, this can lead to organizational hierarchy structure where each subsystem is provided with own purpose or criterial function conforming to which it optimizes its functions.

A superior level operates with information on interactions between subsystems. Conflicts between units of a certain level shall be settled at a higher level. The solution of the coordination is an intervention that can be:

- a modification of the goal,
- a modification of external circumstances,
- a modification of operating conditions at a lower level.

A common aim of the complex societal planning system is attained when individual levels operate with regard to their functions and conforming to the prescribed operational parameters and conditions. These structures include both the vertical and the horisontal decomposition.

One of the aim coordination methods for dynamic systems is the method (1) with a three-level decomposition. This method is based on the treatment of the minimization problem at the first level at a given trajectory as a dual Lagrange problem.

The optimal performance index is as follows:

$$J(x,u) = \sum_{j=1}^{N} \left\{ F_j(x_j(K)) + \sum_{K=0}^{K-1} f_j(x_j(k), u_j(k), k) \right\} \tag{5}$$

and the interconnections between subsystems are:

$$\sum_{j=1}^{N} z_j(x_j(k), u_j(k), k) \leq 0, \qquad k = 0, 1, \ldots, K-1 \tag{6}$$

where N denotes the number of subsystems.

The dual problem is defined in the following way:

$$\max_{P} M(p), \tag{7}$$

where

$$M(p) = \min_{x,u} \left\{ F_j(x_j(K)) + \sum_{k=0}^{K-1} \left[f_j(x_j(k), u_j(k), k) + \right.\right. \tag{8}$$

$$\left. + \lambda^T(k)\, z_j(x_j(k), u_j(k), k) \right] +$$

$$\left. + \sum_{k=0}^{K-1} \left[p^T(k)\, g_j(x_j(k), u_j(k), k) - x_j(k+1), \right] \right\}$$

under the following conditions:

$$h_{jk}(x_j(k), u_j(k)) \leq 0, \qquad k = 0, 1, \ldots, K-1$$

$$h_{jK}(x_j(K)) \leq 0, \tag{9}$$

$$x_j(0) = x_{j0},$$

provided that $j = 1, 2, \ldots, N$.

It is necessary to deal with K + 1 of independent minimizations, yet now for each subsystem apart. The optimization algorithm forms hierarchic structure.

At the first level independent minimizations shall be solved for fixed p and λ, which can be due to their simple character calculated analytically as well. At the second level M (p) shall be maximized at a selected trajectory λ. At the third level the maximization of $\Theta(\lambda)$ is calculated.

A hierarchic scheme for the optimal control of large-scale societal planning systems is proposed, whereby optimization is carried out on both levels: the subsystem and global system level. On the global system level a global controller is created to minimize the effects of time-varying interconnections between the subsystems.

Several extensions and generalizations of the present methods may be considered.

CONCLUSION

As already mentioned above, this paper is concerned with the search for a most general approach to the solution of societal planning systems. The described method is only one of the possible ways practiced for solution of such problems. We have been aware of the fact that societal planning systems have been subject to various modifications concerning their structure, interactions as well as attaining of different aims. Therefore it is very difficult to be efficient in dealing with the whole system. A solution would be the development of various models: the model of economic development, the model of physical development and the model of social development. The solution shall be optimal for each subsystem apart as well as for the whole system. This only enables an optimum societal development.

For specific characteristics of each societal planning system no uniform methodology is possible. Therefore each problem shall be treated separately; only basic definitions try to use the already developed procedures and algorithms, known from the automatic control theory. The above domain is still rather vague; accordingly, it allows for numerous approaches as well as for the application of various methods and researches.

REFERENCES

(1) J.Lenarčič, "Hierarchical control of some dinamical systems" (in Slovenian), University of Ljubljana, Faculty of Electrical Engineering, 1979

(2) H.Tokumaru T.Amemiya, "Macroscopic stability of interconnected systems", Faculty of Engineering, Kyoto University, Kyoto, Japan
Akashi Tecnical College, Akashi, Japan

(3) J.Jerina, "Large-scale systems, stability and dynamics", University of Ljubljana, Faculty of Electrical Engineering, 1977 (in Slovenian)

(4) R.Kulikowski,"Optimum Control of environment development system", Institute of Applied Cybernetics, Polish, Academy of Science, Warszawa, Poland

(5) D.D.Šiljak, "A Multilevel Optimization of Large-Scale Dynamic Systems, IEEE Transactions on automatic control, february 1976

AUTHOR INDEX

Lecture Notes in Control and Information Sciences

Edited by A. V. Balakrishnan and M. Thoma

Vol. 43: Stochastic Differential Systems
Proceedings of the 2nd Bad Honnef Conference
of the SFB 72 of the DFG at the University of Bonn
June 28 – July 2, 1982
Edited by M. Kohlmann and N. Christopeit
XII, 377 pages. 1982.

Vol. 44: Analysis and Optimization of Systems
Proceedings of the Fifth International
Conference on Analysis and Optimization of Systems
Versailles. December 14–17, 1982
Edited by A. Bensoussan and J. L. Lions
XV, 987 pages, 1982

Vol. 45: M. Arató
Linear Stochastic Systems
with Constant Coefficients
A Statistical Approach
IX, 309 pages. 1982

Vol. 46: Time-Scale Modeling of Dynamic Networks
with Applications to Power Systems
Edited by J. H. Chow
X, 218 pages. 1982

Vol. 47: P. A. Ioannou, P. V. Kokotovic
Adaptive Systems with Reduced Models
V, 162 pages. 1983

Vol. 48: Yaakov Yavin
Feedback Strategies for Partially
Observable Stochastic Systems
VI, 233 pages, 1983

Vol. 49: Theory and Application of Random Fields
Proceedings of the IFIP-WG 7/1
Working Conference
held under the joint auspices of the
Indian Statistical Institute
Bangalore, India, January 1982
Edited by G. Kallianpur
VI. 290 pages. 1983

Vol. 50: M. Papageorgiou
Applications of Automatic Control Concepts
to Traffic Flow Modeling and Control
IX, 186 pages. 1983

Vol. 51: Z. Nahorski, H.F. Ravn, R.V.V. Vidal
Optimization of Discrete Time Systems
The Upper Boundary Approach
V, 137 pages 1983

Vol. 52: A. L. Dontchev
Perturbations, Approximations and Sensitivity Analysis
of Optimal Control Systems
IV, 158 pages. 1983

Vol. 53: Liu Chen Hui
General Decoupling Theory of Multivariable
Process Control Systems
XI, 474 pages. 1983

Vol. 54: Control Theory for Distributed
Parameter Systems and Applications
Edited by F. Kappel, K. Kunisch,
W. Schappacher
VII, 245 pages. 1983.

Vol. 55: Ganti Prasada Rao
Piecewise Constant Orthogonal Functions
and Their Application to Systems and Control
VII, 254 pages. 1983.

Vol. 56: Dines Chandra Saha, Ganti Prasada Rao
Identification of Continuous
Dynamical Systems
The Poisson Moment Functional
(PMF) Approach
IX, 158 pages. 1983.

Vol. 57: T. Söderström, P. G. Stoica
Instrumental Variable Methods
for System Identification
VII, 243 pages. 1983.

Vol. 58: Mathematical Theory of
Networks and Systems
Proceedings of the MTNS-83 International
Symposium
Beer Sheva, Israel, June 20–24, 1983
Edited by P. A. Fuhrmann
X, 906 pages. 1984

Vol. 59: System Modelling and Optimization
Proceedings of the 11th IFIP Conference
Copenhagen, Denmark, July 25-29, 1983
Edited by P. Thoft-Christensen
IX, 892 pages. 1984

Vol. 60: Modelling and Performance
Evaluation Methodology
Proceedings of the International Seminar
Paris, France, January 24–26, 1983
Edited by F. Bacelli and G. Fayolle
VII, 655 pages. 1984

Vol. 61: Filtering and Control of Random
Processes
Proceedings of the E.N.S.T.-C.N.E.T. Colloquium
Paris, France, February 23–24, 1983
Edited by H. Korezlioglu, G. Mazziotto, and
J. Szpirglas
V, 325 pages. 1984

Lecture Notes in Control and Information Sciences

Edited by A.V. Balakrishnan and M. Thoma

Vol. 62: Analysis and Optimization
of Systems
Proceedings of the Sixth International
Conference on Analysis and Optimization
of Systems
Nice, June 19–22, 1984
Edited by A. Bensoussan, J. L. Lions
XIX, 591 pages. 1984.

Vol. 63: Analysis and Optimization
of Systems
Proceedings of the Sixth International
Conference on Analysis and Optimization
of Systems
Nice, June 19–22, 1984
Edited by A. Bensoussan, J. L. Lions
XIX, 700 pages. 1984.

Vol. 64: Arunabha Bagchi
Stackelberg Differential Games
in Economic Models
VIII, 203 pages, 1984

Vol. 65: Yaakov Yavin
Numerical Studies
in Nonlinear Filtering
VIII, 273 pages, 1985.

Vol. 66: Systems and Optimization
Proceedings of the Twente Workshop
Enschede, The Netherlands, April 16–18, 1984
Edited by A. Bagchi, H. Th. Jongen
X, 206 pages, 1985.

Vol. 67: Real Time Control of Large Scale Systems
Proceedings of the First European Workshop
University of Patras, Greece, Juli 9–12, 1984
Edited by G. Schmidt, M. Singh, A. Titli,
S. Tzafestas
XI, 650 pages, 1985.